Birds of Nebraska

Birds *of* Nebraska

Their Distribution &
Temporal Occurrence

ROGER S. SHARPE

W. ROSS SILCOCK

& JOEL G. JORGENSEN

University of Nebraska Press, Lincoln and London

© 2001 by the Univer-
sity of Nebraska Press
Manufactured in the
United States of
America ⊗
Photo credits: Plates 8
and 20 are courtesy of
the Nebraska Game
and Parks Commis-
sion. All other photo-
graphs are by Roger S.
Sharpe.
Library of Congress
Cataloging-in-Publi-
cation Data
Sharpe, Roger S.,
1941–   Birds of
Nebraska : their distri-
bution and temporal
occurrence / Roger S.
Sharpe, W. Ross
Silcock, and Joel G.
Jorgensen.   p.   cm.
Includes bibliographi-
cal references (p.   ) and
index. ISBN 0-8032-
4289-1 (cloth: alk. paper)
1. Birds—Nebraska.
I. Silcock, W. Ross.
II. Jorgensen, Joel G.
III. Title.
QL684.N2 S52   2001
598'.09782—dc21
00-064925
"N"

Thank You

The authors and the University of Nebraska Press wish to express their heartfelt thanks to the following organizations for their generous financial support:

Central Nebraska Public Power and Irrigation District, Holdrege, Nebraska

Henry Doorly Zoo® Omaha Zoological Society, Omaha, Nebraska

The Nebraska Chapter of the Nature Conservancy, Omaha, Nebraska Support for this project graciously provided in memory of Ethel R. Mook.

The mission of the Nature Conservancy is to preserve the plants, animals and natural communities that represent the diversity of life on earth by protecting the lands and waters they need to survive.

Nebraska Ornithologists' Union and members who kindly donated to this project

Contents

Maps and Illustrations

Preface

It has always been my belief that a sound general understanding of the biogeography of North American birds would not be complete without a careful analysis of the birds of the central Great Plains and that Nebraska was a key component to that analysis. In the absence of such a work, it has therefore always been a professional goal to produce a comprehensive book on the birds of Nebraska. I completed a version in 1983, at which time 393 species had been identified within the political boundaries. But it was also clear after reviewing more than 600 000 state migration records, contributed by members of the Nebraska Ornithologists' Union over a period of more than 80 years, that most of these came from the eastern, more populated part of the state. Other areas were sadly lacking in records, and despite my periodic efforts to visit these sites, information about the avifauna of those regions was woefully incomplete. One could also see in reviewing those records that most birders who had contributed records were not systematic about their birding efforts. Most were weekend and fair-weather types, understandably, and limited much of their birding to peak migratory periods. Fall records were scanty, and winter records were most often entered as "present" without dates.

At about the same time that my first draft was completed a new generation of active, dedicated birders was developing in the state. New bird records from less populated areas were forthcoming, observers were more careful and critical, records increased in number and quality, and surprising new state records, including those of new species, were being generated. With each new issue of the *Nebraska Bird Review* it was obvious that my manuscript needed serious revision before it was printed. The revision, particularly the passerine section, within a couple of years looked like an entirely new work. Thus I set the project aside.

In about 1991 Ross Silcock approached me about my dormant manuscript. He had already organized a critical review of the state bird list (Bray and others 1986) and was quite aware of the seasonal and geographic deficiencies in the available records. It had also become clear to him that the state needed a comprehensive bird book, and he believed that the records were becoming sufficiently complete to proceed. With his encouragement we dove into an entirely new work, using the record compilations that both he and I had developed independently. I concentrated on the nonpasserines, and Ross tackled the passerines. In 1996 we added Joel Jorgensen to the team because of his knowledge of and considerable experience with shorebirds. The observations from his periodic treks to shorebird hotspots, including almost daily visits to the Rainwater Basin over several years, have significantly changed our understanding of sandpiper migration in the state. Even though we have had our individual responsibilities in the preparation of this work, each of us has also reviewed and critiqued all of the species accounts. My concerns have focused on biological and ecological considerations where appropriate.

Both Ross and Joel are part of that new breed of Nebraska birders—energetic, relentless, and passionate in their pursuit of birds. They think nothing of arising in the middle of the night to drive 300 miles for a sunrise birding exercise at Lake McConaughy. They are also very critical in their review of records; both have served on the Nebraska Ornithologists' Union Records Committee and write Seasonal Field Reports for the *Nebraska Bird Review*.

No state bird work represents the final edition. Birds are very mobile creatures that are subject to individual experiences, individual motivations, and individual losses of memory and thus are likely sources for the unexpected. First and late date records will be broken; rare birds will occur. And we will never be able to fully explain the rare and unusual, however interesting. Nonetheless, it is our hope that this work presents to the reader an accurate summary of the normal and expected about Nebraska's avifauna. It is also our hope that it will not only provide the professional and advanced amateur with an objective summary of the current status of Nebraska birds but also stimulate the beginner to look more closely at the birds in the backyard, foster dreams about interesting habitats, and encourage field forays to go see and explore.

Roger S. Sharpe

Acknowledgments

This work argues strongly that the activities of weekend birders, many now no longer with us, have not gone unnoticed, nor were their efforts wasted. This book is a collection of their efforts, a distillation of thousands of seasonal records that were enthusiastically and often dutifully submitted by individuals over a century's span to the Nebraska Ornithologists' Union for review and publication. Many of these birders were motivated to submit their records simply to share their exciting findings with others. The more experienced also recognized that their contributions would collectively help create a more accurate picture of the geographic distribution and temporal occurrence of the birds that frequent the state. Without the efforts of these individuals (nearly all of them amateur birders), the book would be far from complete. For those birders whose records were scrutinized and questioned, we apologize for our brashness and also thank you for your tolerance. It should also be noted that the present work is not an endpoint but part of a continuum, because bird populations change in number and distribution. Without the continuing contributions of records from Nebraska birders, the next necessary update will be far more difficult.

The development of this manuscript would not have been possible without the help of many people, some who made small but important contributions and others who have given considerably of their time and expertise. We wish to thank Tom Labedz, Division of Zoology, University of Nebraska State Museum, for his help in providing specimens for our review. Various individuals with the Cornell Laboratory of Ornithology provided copies of Nebraska nest card records. Danny Bystrak of the Bird Banding Laboratories, Pautuxent, Maryland, kindly provided computer summaries of Nebraska banding data. Fred Zeillemaker provided important unpublished seasonal reports from the period that he was at Crescent Lake National Wildlife Refuge. Loren and Babs Padleford provided several years of unpublished data on migrating raptors. Angie Fox, scientific illustrator, University of Nebraska State Museum, helped develop and create the maps.

Early drafts of the manuscript were reviewed by several scientists. John Dinan, Division of Non-game Wildlife, Nebraska Game and Parks Commission (NGP), reviewed nearly the entire manuscript and made many useful suggestions. He also provided us the opportunity to review the commission's records, many of which were unpublished and significant. Joe Gabig, formerly waterfowl biologist with NGP, reviewed our introduction and the waterfowl species accounts. Scott Taylor, upland game specialist, NGP, reviewed the species accounts for the gallinaceous birds. Dave Easterla and Steve Dinsmore reviewed the entire manuscript and provided many useful suggestions that greatly improved our work. Lastly, Dan Reinking and Mark Robbins, manuscript referees chosen by the University of Nebraska Press, provided both critical comments and suggestions that helped with the text's overall organization as well as helped to clarify some of the species accounts.

Additionally, each of us has some personal acknowledgments:

From Ross Silcock:

I wish to thank my wife, Lyndabeth, for her patience in letting me spend inordinate amounts of time pursuing my interest in birds. Chris Marsh, while still a high school student, started me on the birding path by introducing me to North American birds about 30 years ago in Raleigh, North Carolina. Loren and Babs Padelford unselfishly provided information and advice about Nebraska birds over a period of many years. Tanya Bray showed me where birds are found in the state and often tolerated my goofy ideas about where to look. Doug Rose provided companionship on many birding expeditions and reinforced my enthusiasm for the sport. Steve Dinsmore, one of the best birders I have been with, shared many birding ventures with me and continually reinforced the importance of accuracy.

From Joel Jorgensen:

I wish to thank my parents for tolerating and encouraging a budding birder and Roger Sharpe and Ross Silcock for allowing me to be part of this project.

From Roger Sharpe:

I wish to thank my wife, Bev, for the tedious hours upon hours she gave in helping to collate bird records and for her patience and encouragement in the long development of the manuscript. Many people have had an impact on my development as a biologist and specifically my interest in birds, including family members, teachers, and countless students. A few individuals stand out. My aunt Emily provided me with my first bird book, *A Golden Guide to North American Birds*, which I turned to rags before I discovered the Peterson guides. My maternal grandmother instilled in me an awe for nature. Paul Johnsgard showed me how to interpret nature systematically and in the larger contexts of space, time, gene pools, and environments. Harvey Gunderson showed me that one's personal relationship with nature need not be buried during the process of doing science. In the birding community, several individuals had an influence on my professional development and ultimately on the development of this work. They include R. G. "Rusty" Cortelyou, Willeta "Willie" Lueschen, Jim Malkowski, and Ruth Green. The flow of students through a professor's career is also a great source of reinforcement, and several of them are often remembered. They include Bill Erwin, John Crawford, Gordon Mueller, Tom Bicak, Greg Hansen, Barb Wilson, John Parker, Jim Ducey, Russ Benedict, Jim Anderson, Keith Geluso, and Joel Jorgensen.

Birds of Nebraska

Introduction

Some 400 bird species regularly occur in the state of Nebraska. Of these, about half are regular or occasional breeders. The remainder include migrants, winter visitors, and predictable summer visitors. This great avifaunal diversity reflects Nebraska's strategic geographic location and its diversity of habitats.

Positioned near the geographic center of the continent and just east of the Rocky Mountains, Nebraska lies in the migratory pathway for a large number of species whose breeding and winter ranges span the entire Western Hemisphere. These include both species of North American cranes; most of the tundra-nesting shorebird species; most of the migratory raptors; a large variety of passerines, including wood warblers and sparrows; and, excluding a handful of saltwater species, nearly all the North American species of waterfowl. Especially impressive to birders of all levels of experience are the spring waterfowl concentrations, which are sometimes so large that they choke the wetlands.

Some migrants that breed to the north find Nebraska's winter climate sufficiently mild and the food supply sufficiently abundant to overwinter. Nomadic longspurs, gathered in flocks of thousands of individuals, may be seen periodically as they range the snow-covered prairies in search of food. Thousands of migrant waterfowl may remain in the state until weather or food supply force them farther south; during mild winters, many remain all season. Some overwintering species, however, are permanent residents. The Red-bellied Woodpecker, for example, may be found during all seasons; it shifts its foraging strategies from season to season to accommodate changing food supplies.

Still other species frequent the state only sporadically. The most frequent transients are large wading birds such as the Little Blue Heron, which drifts into the state during late spring and summer months from breeding areas that are normally well to the south. Still others, such as the Bohemian Waxwing, enter the state during winter months with such nomadic irregularity to defy any predictability.

More than 200 breeding bird species may be found in the state because its wide variety of habitats. Before European immigrants and their agriculture changed much of the state's landscape, the first explorers embarked on their journey to the Rocky Mountains from a point near present-day Omaha and encountered elements of 3 major North American biomes: eastern deciduous forest, grassland, and montane coniferous forest, which are still represented within the state's political boundaries (map 1). As these travelers continued westward up the valley of the Platte, they also encountered a variety of wetland habitats associated with the riverine system; many still persist today and attract a rich variety of breeding and migratory birds.

Today the transition in habitats that the pioneers witnessed, although somewhat obscured by agriculture, is still detectable and roughly definable. Much of the prairie in the eastern third and south-central part of the state and along the broad floodplains of the major rivers has been replaced by cropland (map 2); remnants of these prairies still persist as small, isolated islands surrounded by a sea of agriculture. However, the grasslands of north-central Nebraska, better known as the Sandhills, and those of the Panhandle still remain, but where bison once grazed, cattle now prevail.

The eastern deciduous forest of the southeastern part of the state, dominated by oaks and other hardwoods, has also been reduced. Some of this wooded habitat has been converted to cropland; other woodlands have become pasture. Floodplain forests, however, are much more widely distributed today than in pre–European settlement times, having extended westward along Nebraska's streams and rivers in response to a reduced frequency of grass fires. Those elements of montane coniferous forest, represented in the Pine Ridge of the Panhandle, although having been selectively cut throughout the post–European settlement period, still retain much of their original distribution and character. Thus, despite many changes brought on by European settlement, good examples of all 3 terrestrial biomes still may be found along with their avifaunal constituents.

Although many early explorers and travelers considered the region of Nebraska part of the "Great American Desert," the state's diverse aquatic environments attract a large variety of nesting and migratory bird species. Several major rivers course eastward across the length of the state, emptying into the Missouri River. Hundreds of small lakes and marshes as well as semipermanent and ephemeral wetlands dot the state, the majority being in the Sandhills region. In addition, the numerous impoundments constructed in the 20th century on many rivers, streams, and waterways add additional breeding habitat for summer residents and resting and feeding habitat for migrants.

Clearly many environmental changes have taken place since European settlement that have decidedly altered the biological landscape of the state. Many habitats have been eliminated or now exist as isolated patches. Wetlands have been drained, and rivers have been dammed and tamed. All of these events have had substantial impacts on some avian species. Remarkably, even in light of significant habitat modification and loss, there is no species that we can say with assurance that has been permanently extirpated during the 20th century. On the contrary, as many as a dozen species may have been added to the state list because of the creation of new habitats, namely large reservoirs, and several species extirpated in the 19th century have been successfully reintroduced.

NEBRASKA'S PHYSICAL ENVIRONMENT

Much of Nebraska's habitat diversity reflects its wide range of geomorphic and climatic features. These physical characteristics play an active role in determining the ecology of an area, particularly influencing vegetation, which responds to local soil, topographic, and moisture conditions. In turn, the controlling physical conditions and resulting vegetative assemblage create a variety of niches for an associated grouping of animals. Even though the resulting mixture is not totally explainable, the trained observer may predict the avian component of the community with a fair degree of confidence.

Geomorphology

Nebraska's biological diversity is the product of the extremes in the state's physical environment. The eastern and western boundaries are separated by some 740 km (460 mi) and share few landscape features. Much like an inclined plane, the state tilts from west to east, dropping between 1.5 and 1.6 m per km (8.5 and 9 ft per mi). Elevation above sea level is about 1600 m (5300 ft) near the Colorado and Wyoming borders in Banner and Kimball Counties; in Richardson County, in the extreme southeast, the elevation is only about 250 m (825 ft) (Nebraska Blue Book 1975).

The state's contemporary geomorphic features lie upon an inclined plane of bedrock many millions of years old. Created by a long series of geologic events beginning well into the Paleozoic (more than 300 million years ago) and continuing into the late Tertiary (more than 1 million years ago), this bedrock is exposed only where wind and water erosion has cut deep into the earth's surface. Bedrock exposures may form cliffs adjacent to the floodplain of rivers such as the North Platte in Garden, Morrill, and Scotts Bluff Counties and along the Niobrara River in Cherry and Rock Counties. The Pine Ridge region provides extensive cliff exposure in the northern Panhandle where an ancient, now extinct, river eroded the scarps of the ridge (plate 1). These environments offer ledges and crevices that are ideal nesting sites for birds of prey such as Golden Eagle and Prairie Falcon, which are essentially absent from environments not providing this topographic relief. Limited exposures of bedrock, principally limestone and shale, are also found in southeastern Nebraska; however, few of these exposures provide sufficient relief to support populations of cliff-dwelling birds.

During the Ice Age of the last million years, considerable outwash material was deposited on the western bedrock. Produced in the Rockies and carried eastward during heavy montane glacial melt, outwash material is still in evidence in the Panhandle. Upon the outwash is deposited a thin layer of wind-blown soil that supports communities of short grasses such as buffalo grass (*Buchloe* spp.) and scattered stands of sage (*Artemesia* spp.). In some locations on the high-plains tablelands, water runoff has eroded these thin soils and revealed considerable outwash composed of materials such as heavy gravels, rocks of varying sizes, and soils of mixed textures. The resulting arroyos (plate 2) may accumulate sufficient moisture to support occasional clumps of woody vegetation that serve as small islands of breeding habitat for species such as Swainson's Hawk and Western Kingbird. The banks of the arroyos often form crevices and ledges that may support the nests of species such as Rock Wren and Say's Phoebe. The ancient Rocky Mountain outwash accumulation diminishes in the east, where it is replaced or covered by materials of different origins.

Glaciation has also had its impact in eastern Nebraska but as a direct physical intervener rather than a source of outwash material. At least 2 glaciers extended into the region, bringing with them and leaving in their wake considerable debris, including large amounts of gravels, rocks, and even boulders. This material, known as glacial till, lies exposed in several locations in southeast Nebraska where the boulder-strewn surface precludes cultivation, thus leaving pastures that still sustain stands of original tallgrass prairie. Enough remnant prairie remains in Johnson and northern Pawnee Counties to sustain a resident population of Greater Prairie-Chickens (*Tympanuchus cupido*) (plate 3).

The retreat of these glaciers also exposed vast areas of unprotected soil. During their retreat and during particularly dry interglacial periods, strong prevailing northwest winds carried soil particles into Nebraska, covering much of the east and central portions of the state with soils of varying size. It is upon these windblown soils that grasslands developed. Wind has also contributed to the development of some 49 000 km² (19 000 mi²) of sand dunes across the north-central portion of the state, constituting the largest geomorphic and biogeographic unit in Nebraska, the Sandhills (plate 4).

Increasing evidence argues that the Sandhills may have had their beginnings in very late Pleistocene, perhaps as late as 7000 years ago and likely originated from the sandstones and silt stones upon which they rest (Ahlbrandt and others 1983). It is also likely that the distribution and elevation of the Sandhills have been modified several times as dune activity has alternately increased and subsided in response to drought, fires, and wind. The sand dunes today are relatively inactive; however, extended drought as well as overgrazing may reduce vegetative cover enough to initiate localized sand dune activity, creating what are locally known as blowouts. This sandy soil, coupled with limited precipitation, has proven largely unfavorable for cultivation and thus is still today covered with prairie grasses as it was at the beginning of European settlement. Dotted by numerous lakes, marshes, ponds, and wet meadows, Nebraska's Sandhills are the state's most unusual geomorphic feature and largest biologic reservoir.

Additional accumulations of finer, windborne soil particles, known as loess, are found throughout much of the eastern third of Nebraska. These deposits may be several hundred meters thick but may vary in thickness locally because of later wind and water erosion, forming a landscape of rolling hills. Formerly these hills were covered by midgrass and tallgrass prairies, but because loess has good water-holding properties and is often rich in nutrients, loess now provides the primary resource for Nebraska's agricultural economy.

Dense late-glacial deposits of loess are revealed as bluffs on either side of the Missouri River floodplain in eastern Nebraska, forming the Loess Hills region. Created over the millennia as the river and its tributaries downcut through the loess (Bettis and others 1986), these bluffs support stands of upland hardwoods such as oaks and hickories, depending upon exposure. On more exposed south- and west-facing slopes, grasses may predominate, but on the cooler, moister east- and north-facing slopes, woodlands have developed and may grade into dense forest in the more shaded ravines.

Aquatic Systems

Several major rivers arising in the Rockies meander slowly east and southeastward through the state (map 1). The Platte and Republican flow slowly at rates averaging less than 8 km (5 mi) per hour. Broad and shallow, they carry in their beds great accumulations of outwash. Sand, being one of the lighter materials, is frequently held in suspension by the rivers. Where the river is slow moving, sand may be deposited, creating sandbars. Some sandbars are ephemeral, arising during low flow rates and later vanishing to be created elsewhere during high runoff periods. Other sandbars, because of shifting channels, gradually build up and become vegetated, creating islands of floodplain vegetation (plate 5). Even these islands may be ephemeral in a human lifetime, because as channels slowly shift the island banks are gradually undercut and removed. The ancient meander of the Platte has created a floodplain so broad that an uninformed observer traveling along the interstate highway in the Platte Valley would think that the state was flat.

The Niobrara, arising in eastern Wyoming, flows eastward through the state along the northern border. Not more than a prairie stream of a few meters in width in the Panhandle, the Niobrara gathers sufficient volume from tributaries to cut into the bedrock from central Cherry County eastward, creating a canyon as much as 240 m (150 ft) deep in eastern Cherry County and in Brown, Rock, and Keya Paha Counties. Into the Niobrara canyon country pour numerous freshets and small streams from the adjoining Sandhills, creating, even in the heat of the midsummer, a moist, cool environment amid a semidesert.

The Sandhills are drained primarily by the Loup River system. These prairie streams snake their way southeastward and empty into the Platte. Relatively shallow with sandy beds, the Loup Rivers meander sufficiently to create numerous oxbows and backwater marshes. Entering the Platte near Columbus, the waters of the Loup River system ensure continuous stream flow in the lower Platte, whose bed is often dry above that point during especially arid summers.

The northeastern Sandhills are drained by the Elkhorn River, which cuts through the loess deposits in eastern Nebraska and enters the Platte near Omaha. Smaller river systems that drain part of Nebraska's loess include the Big and Little Blue Rivers in south-central Nebraska, which enter Kansas in Gage County, and the Big and Little Nemaha Rivers in the extreme southeast. These prairie rivers meander considerably but are significantly more turbid than their counterparts elsewhere in the state because of the fine soil mantle that they drain.

Nebraska's largest river, the Missouri, forms the state's eastern boundary with South Dakota, Iowa, and Missouri. Once a large, meandering river that developed a broad floodplain replete with numerous marshes and oxbows, the Missouri today is flood controlled by several dams in the Dakotas and has been channelized except for a short stretch between Yankton, South Dakota, and Sioux City, Iowa. Many of the Missouri floodplain marshes have been drained and converted to agricultural land; except for several managed oxbow lakes, oxbows are likewise being lost to natural aging processes.

Natural lakes and associated wetlands are especially well developed in the Sandhills; 2 Sandhills lakes regions are evident and include a concentration in east-central Cherry County and another in northern Garden and southern Sheridan Counties. Numerous shallow lakes—along with their associated marshes, smaller playas, and ephemeral wetlands—dot these regions (plate 6).

Some lakes may be several hundred hectares in size and may sustain a viable fishery (plate 7). All are alkaline in nature; smaller, shallower lakes are especially alkaline. While these may not sustain a fishery, they may support large concentrations of aquatic insects such as black fly larvae and emerging adults (*Simulium*) and other alkaline-tolerant invertebrates, important foods for many aquatic bird species.

Coincident with those periods of late- and post-glacial loess deposition was the development of wind-eroded playas, shallow depressions that are found in several locations across the state, including the southwest, south-central, and east-central areas. These clay-lined depressions collect local runoff, maintaining wetlands that range from large, permanent marshes with considerable open water to small, ephemeral ponds that exist for only a few weeks at a time. Although many of these wetlands have been drained, one playa region still sustains a complex of wetland habitat that is of considerable importance to many migrants. Known as the Rainwater Basin, it covers an area of about 11 000 km^2 (4200 mi^2) in south-central Nebraska (plate 8; see Rainwater Basin subsection also). Scattered south of the Platte River from Gosper County eastward through Phelps, Kearney, Adams, Clay, and Fillmore Counties, and to a lesser extent through Hamilton and York Counties, these wetlands are but remnants of a once very extensive wetlands region, 60–80% of which has been drained for agricultural purposes (Rundquist 1990). Nevertheless, between 11 000 and 13 500 ha (28 000 and 34 000 ac) of wetlands remain in the Rainwater Basin (Raines and others 1990; Smith and Higgins 1990) and are represented largely by extensive marshes of emergent vegetation, some with open water. Some are only ephemeral wetlands, retaining water only in spring.

Numerous impoundments have been developed on Nebraska's waterways since European settlement. Established to provide waters for irrigation and flood control and to control soil erosion, these impoundments have proven to be both ecological boons and busts. It is clearly demonstrated that some Nebraska impoundments and their water-withdrawal systems have significantly altered normal ecological processes associated with stream flow. Species such as Piping Plover, Least Tern, and quite probably Sandhill Crane have been negatively impacted, particularly along the central Platte River.

Since the 1920s water projects have also created more than 58 000 surface ha (145 000 surface ac) of new aquatic habitats that have benefited a number of avian species, including several species of gulls. The largest of these impoundments and one of the earliest and recently most controversial, is that created by Kingsley Dam, located in Keith County on the North Platte River. Known as Lake McConaughy, its surface covers about 14 300 ha (35 700 ac), extends some 35 km (22 mi) from dam to headwaters, and provides about 160 km (100 mi) of shoreline (plate 9). Since construction of the dam in 1941, an extensive marsh has developed in the headwaters of the reservoir, extending upstream for several kilometers from the normal reservoir pool. Several pairs of Western Grebe now regularly nest in this marsh, and the first breeding record for Clark's Grebe comes from this site.

Impoundments may now be found on most other large rivers and tributary systems in the state. Their ability to attract and support avian species depends to a considerable degree on their age, how well the fishery has developed, and the development of vegetation along the banks and in their tailwaters. In the southwest and south-central parts of the state, tributaries to the Republican River support Enders and Swanson Reservoirs (both less than 20 years old and still developing biologically); Harlan County Reservoir, established in the 1930s, is found on the Republican River but supports an important food chain that attracts large numbers of birds, including waterfowl. The Loup River system now has Sherman and Calamus Reservoirs, both less than 30 years old and still maturing. The Salt Creek drainage near Lincoln supports several reservoirs, including Branched Oak, Pawnee, Bluestem, and Wagon Train Lakes

that are less than 50 years old and attract a wide variety of birds. But, according to several Nebraska Game and Parks Commission biologists, some of these impoundments have passed their peak productivity and may need restoration (RSS, pers. comm.). Since 1980 several large reservoirs have been created near Omaha on the Papillion Creek system and include Standing Bear and Cunningham Lakes, which are beginning to attract concentrations of waterfowl.

Few reservoirs are found in the northern part of the state, but they are generally large. Merritt Reservoir, located on the Snake River just before it enters the Niobrara River southwest of Valentine, supports good populations of migratory diving ducks as well as large concentrations of migrant cormorants and pelicans. Downstream, the Niobrara enters the Missouri River in the headwaters of Lewis and Clark Lake, a large reservoir created by Gavin's Point Dam. Developed in the early 1950s, this 13 200 ha (33 000 ac) lake now has a large variety of well-established wetland habitats that attract their associated avifauna.

Climate

Nothing about Nebraska's weather can be considered normal; change and seasonal variation can be extreme. Years of abundant rainfall may alternate with extreme drought. Summers may be dry and hot or cool and moist. Winters may be marked by successive blizzards and long periods of below-zero cold or may be mild and without snow. Daily weather extremes are commonplace as intense high-pressure systems clash with warm, moist air flowing from the south. Temperatures have been known to drop 20 degrees C (about 50 degrees F) in an hour; tornadoes may be spawned from rapidly developing evening storms on a previously cloudless day.

The historic record reflects these yearly and daily changes. Lawson (1976) has demonstrated that the period 1600 to 1850 was considerably more moist than the period 1850 to 1950. Since the last glaciation 10 000 years ago, considerable variation in climate has occurred in the Great Plains, beginning with a cool, moist period, followed by several thousand years of a very warm and dry climate. The present climatic regime is at least a thousand years old but has been interspersed with alternating periods of cool-moist and warm-dry conditions (Flowerday 1986). The results of these relatively long-term climatic shifts have included the slow migration of major plant communities; oak forests have migrated into the prairie of eastern Nebraska during cool, moist periods. Warm, dry periods, accompanied by more frequent fires, have resulted in woodland retreat. Certainly over these long time periods the available habitat for select bird species has expanded and contracted, resulting in distributional shifts for those species.

Today, the state's elevational gradient is paralleled by precipitation. In extreme southeastern Nebraska, annual precipitation exceeds 90 cm (35 in). At Scottsbluff in the extreme western panhandle, less than 38 cm (15 in) of precipitation falls annually. Throughout the state precipitation is greatest during late spring and early summer months in the form of rainfall. Most rainfall is concentrated during relatively brief thunderstorms, which may occur as frequently as twice a week during peak months. Often these storms are intense and are accompanied by strong winds, hail, and lightning. Lightning is still the major cause of prairie fires in the Sandhills, and, propelled by strong winds, these fires may burn thousands of hectares of grassland before being checked.

Although snowfall contributes little moisture to the total annual precipitation, it is an important environmental consideration in the state. Snow cover, while rarely continuous throughout the winter, reduces soil erosion by the strong northwesterly winds that prevail during winter months. Snowfall also aids in the recharge of lowland meadows and temporary ponds. Furthermore, in sufficient quantities snow may bury seeds that most winter passerines feed upon, forcing them to forage elsewhere on the prairies. As a consequence, many winter birds are nomadic. Waterfowl may remain in the state until snow covers winter wheat fields and the residual corn left by mechanical harvesters.

Snowfall, like rain, is frequently accompanied by severe weather. Rapidly dropping temperatures, strong winds, and heavy snow combine to create prairie blizzards. Some of the most severe blizzards occur in late winter and spring, especially in the central and western part of the state, and may account for heavy bird mortality. During the third week of March 1975, the western third of the state experienced one of the most severe spring blizzards on record, a storm that killed numerous waterfowl and Sandhill Cranes along the Platte River. Temperatures fell some 30 degrees C (from 60 degrees to around 0 degrees F) in 12 hours, as an intense storm dropped as much as 53 cm (21 in) of snow in portions of the Sandhills. Winds of 80 km (50 mi) per hour whipped snow into drifts 6 m (20 ft) or more.

On 27 May 1947, heavy, wet snow fell throughout western Nebraska and was followed by below-freezing temperatures and high winds. Accounts tell of thousands of birds, representing many resident and migrant species, that were attracted to the shelter of farm buildings. Surveys of the countryside in the days immediately following the storm revealed that previously abundant breeding species were much reduced in number (NBR 16:23).

Periodic droughts in the plains may also affect birds. Numerous distributional and seasonal anomalies were reported in the Dust Bowl during the 1930s. Western species were often seen well east of their normal range. Magpies, for example, could be found wintering as far east as the Missouri River in Nebraska. Species such as Golden-crowned Kinglets, which nest far north of the state, could be found throughout the summer, and some fall migrants arrived unusually early.

NEBRASKA'S BIOLOGICAL ENVIRONMENTS
Eastern Deciduous Forest
Upland Forest

In pre–European settlement times eastern deciduous forest, dominated by various oak and hickory species, could be found throughout much of southeastern Nebraska (plate 10).

Although this hardwood forest is associated with major rivers and their tributaries, it is predominantly found on the slopes and on the bluffs above the floodplains; thus it is often referred to as "upland" forest. Along the Missouri River bluffs, this forest was continuous, extending into Iowa, Kansas, and Missouri. Farther west, however, the forest becomes increasingly discontinuous, existing on hillsides that were protected from periodic fires and in ravines providing a moisture regime that could support these trees. Now extinct or extirpated, such species as Wild Turkey, Ruffed Grouse, Passenger Pigeon, and Carolina Parakeet were found in this Nebraska forest. Still present are such species as Bluegray Gnatcatcher, Barred Owl, Tufted Titmouse, Whip-poor-will, and Kentucky Warbler.

Remnants of the hardwood eastern deciduous forest still exist in southeastern Nebraska, particularly in those counties bordering the Missouri River, where the topography is so abruptly hilly that crop farming is not profitable. Few of those remaining stands are considered old-growth, however. In early pioneer times hardwood was at a premium for lumber, home fuel, and steamboat fuel, and up until about 1900 a thriving lumber industry existed in eastern Nebraska. Selective harvesting of desirable species altered the diversity, while other stands were clear-cut. Today's forests represent secondary and tertiary regrowth, and many are still undergoing succession to maturity (Garabrandt 1978).

Nebraska's remaining hardwood forest represents the far western extreme of a once vast eastern deciduous forest complex that originally extended to the Atlantic Coast. Eastern Nebraska, however, is at the climatic limits of tolerance for most woody species and in a transition zone between deciduous forest and grassland. Missing in the state's limited representation of forest are the more typical eastern species such as hard maples (*Acer* spp.) and beeches (*Fagus* spp.). Instead, the tree species represented in Nebraska are the more dry-adapted species and include several species of oaks (*Quercus* spp.) and hickories (*Carya* spp.) For example, in extreme southeast

Nebraska where precipitation is highest, forests are dominated by red oak (*Quercus rubra*), basswood (*Tilia americana*), and bitternut hickory (*Carya cordiformis*). Minor constituents include chinquapin oak (*Quercus muehlenbergii*), bur oak (*Quercus macrocarpa*), and shagbark hickory (*Carya ovata*). Canopies may be dense and continuous.

In response to a decreasing moisture gradient, the tree species diversity rapidly diminishes to the north and west along the bluffs adjacent to the Missouri River and its tributaries. Bur oak, a drought- and fire-tolerant species, replaces red oak as a dominant. Red oak may be entirely restricted to moist ravines. Chinquapin oak is totally absent at Omaha. Near the mouth of the Niobrara River in northeast Nebraska, only bur oak remains. The forest canopy also becomes more open, allowing a stratified community to develop beneath the dominant trees. Lower strata constituents in addition to young saplings of dominant species may include smaller-growing trees such as gray dogwood (*Cornus foeming*) and ironwood (*Ostrya virginiana*) and bushes such as hazelnut (*Corylus americana*) and Missouri gooseberry (*Ribes missouriense*).

The eastern hardwood forests in Nebraska are continuously altered by regular, though unpredictable, climatic aberrations. Precipitation and fire play important roles in determining the distribution and tree species composition of these forests. These environmental variables may also rapidly alter what is accepted as the normal distribution for a given tree species. For example, Albertson and Weaver (1945) documented the reduction of the western periphery of red oak and basswood ranges in eastern Nebraska by some 16 km (10 mi) in response to severe drought in the 1930s. These 2 species have slowly expanded their range since that time in response to increased moisture.

Fire, although a well-controlled physical element today, was also an important factor limiting woody plant distribution in Nebraska. Much of today's forest distribution was influenced by historic fire. Garabrandt (1978) documented the impact of historic fires in Fontenelle Forest near Omaha and found that fire was indeed an important variable before and immediately following European settlement times. Some areas of the forest that had been burned differ in species composition and abundance from similar unburned areas.

One can only speculate about the influence of fire before European settlement along the interface of forest and grassland throughout eastern Nebraska. Spring and early summer thunderstorms, spawning lightning, still cause numerous fires in the grasslands of the state. Fueled by the strong northwesterly winds of a passing weather front, grass fires may continue unchecked for many kilometers. Grasses are well adapted to fire, because their growing point is at or near the ground surface, and they may be grazed or burned without major damage to their growth potential. Most woody plants, on the other hand, have their growing point at the tips of stems, where fire may do the most damage.

Some woody species have partial protection against fire. For example, the bark of an adult bur oak is very thick and dense, and occasional grass fires may do little permanent damage. Coupled with its drought tolerance, it is therefore the dominant tree at the forest-grassland border in Nebraska and may also be found in isolated parklands well into the prairies.

Oak Savanna

The boundary between forest and grassland was once an unbroken band of open parkland, consisting of scattered bur oaks, hillsides of sumac (*Rhus* spp.) draws filled with buckbrush (*Symphoricarpos orbiculatus*), and an almost continuous ground cover of grasses and forbs (plate 11). Designated Oak Savanna, much of this community has been converted to agriculture, but in areas where the terrain prohibits farming, particularly in the heavily bisected loess hills of northeastern Nebraska, remnants of Oak Savanna still persist.

Savanna was maintained by periodic fires and occasional periods of drought. Over many years

of increased precipitation, forest would replace parkland, and the parkland would advance into open prairie. During periods of drought, coupled with an increased incidence of fire, the prairie would invade the forest. This continuous interplay along the forest-prairie border has maintained an extensive open-canopy woodland at the northern and western edges of the eastern deciduous forest. Much of what today many consider "forest" in eastern Nebraska is this woodland.

The forest-grassland border creates a diverse habitat that attracts an abundance of bird species. Perhaps the best examples of such edge species are the American Robin and Eastern Bluebird, both of which require trees for nesting and grassland for foraging. Still others, such as the Brown Thrasher and Indigo Bunting, are attracted to the extensive bushy thickets and dense stands of annual weeds created in the transitory woodland.

The great majority of bird species found associated with deciduous vegetation in eastern Nebraska are associated with the edge created by oak savanna and with yet another distinctive deciduous community found across the state: floodplain or riparian forest. More generalists than specialists, they may also be attracted to planted woodlots, windbreaks, and farmsteads that now dot the prairies and plains. Some of the more common breeding birds associated with both oak savanna and floodplain forest include Cooper's Hawk, Great Horned Owl, Northern Flicker, Red-headed Woodpecker, Great Crested Flycatcher, Blue Jay, Black-capped Chickadee, House Wren, Eastern Bluebird, American Robin, Red-eyed Vireo, Yellow Warbler, Baltimore Oriole, and Northern Cardinal.

Floodplain or Riparian Forest

Floodplain woodland is more extensive today than in pre–European settlement times. This community is found adjacent to major streams and rivers throughout the state and is designated woodland rather than forest because the canopy is not continuous, and sunlight frequently penetrates the upper canopy to the woodland floor, allowing the development of distinct vegetative strata. Once a community in continual transition because of frequent changes in river channels, it is currently relatively stable along the Missouri River where bank stabilization, coupled with large impoundments upstream, has significantly reduced flooding and bank undercutting. Since floodplain woodlands now remain undisturbed by the meandering of the Missouri, associated oxbows and backwater areas created by periodic past changes in the river's course are rapidly aging and filling with sediment and will not be re-created naturally. These dwindling habitats are important feeding areas for herons and breeding habitat for the Prothonotary Warbler.

Floodplain woodlands along the Missouri and the lower Platte Rivers are a rich vegetational mixture (plate 12). Willows (*Salix* spp.) and cottonwoods (*Populus deltoides*) are the first to colonize recently exposed soil. In the last stages of cottonwood maturation, when they create a high yet discontinuous canopy, additional species may invade, including American elm (*Ulmus americana*), hackberry (*Celtis occidentalis*), green ash (*Fraxinus pennsylvanica*), and sycamore (*Platanus occidentalis*). An understory dominated by gray dogwood is commonplace. During the 1960s, Dutch elm disease spread rapidly along the floodplains, reducing the elm population to saplings that rarely exceed 4 years of age before succumbing to the persistent disease. As a consequence, floodplain forests have undergone a significant change in structure and are still undergoing changes in response to the loss of the dominant elms.

In some locations where the floodplain woodlands grade into bluffs supporting upland hardwoods, certain upland tree species such as basswood (*Tilia americana*), red oak, and bur oak may migrate into the floodplain and become constituents along the typical floodplain community margin. But in the upper reaches of the Elkhorn River, portions of the Blue River system, and to a lesser degree along several tributaries to the Loup River and in the lower Republican River drainage, small, stunted bur oaks become

important riparian species and exist in a scattered, savanna-like community.

Floodplain woodlands are relatively young communities along Nebraska's interior river systems. Prior to European settlement these rivers meandered through a plain of dense grassland. Only an occasional stand of floodplain woodland was found in more protected locations, because frequent fires in the adjoining grasslands burned to the water's edge. It is estimated that west of a line between Grand Island and Kearney, no woody vegetation existed along the Platte River (Johnson 1994); the South and North Platte Rivers were essentially treeless (Williams 1978). Early accounts by exploring parties also describe treeless conditions along the Loup River and throughout much of the Republican River drainage. With European settlement and the subsequent replacement of floodplain grasslands by agriculture, fire as a natural control was virtually eliminated, and woody vegetation slowly crept westward along the riverways.

A typical floodplain woodland in eastern and central Nebraska includes willow, cottonwood, elm, hackberry, green ash, box elder (*Acer negundo*), and gray dogwood. Along the Platte River, low, ephemeral sandbars may support extensive stands of willow of 5 years old or younger, along with an occasional cottonwood sapling. If these sandbars persist, as is common in the central and lower Platte (and is increasingly common with decreasing annual stream flow), they will continue to succeed to a more mature community, first dominated by older willows and cottonwoods, and then invaded by other tree species such as green ash and box elder.

In the far west, willow and cottonwood prevail somewhat continuously along the stream banks; in some locations a rough woodland develops, the mixture reflecting the available seed source species, including cedar (*Juniperus* spp.). Along the North Platte River, scattered open savanna is achieved away from the stream banks in the nearby floodplain, dominated by older cottonwoods, older willows, and scattered Russian olive (*Elaeagnus angustifolia*). At present in some locations (for example, Keith County), many of these scattered older trees are dying. It has been speculated that the sites that these trees occupy are being elevated by alluvial accumulations which, coupled with a reduced water table, is effectively reducing water availability to existing trees (Williams 1978). What little elm became established in the western floodplains has been lost to Dutch elm disease.

Elements of floodplain woodland may also be found throughout the state in shelterbelts and on farmsteads and ranchsteads. Early European settlers, homesteading a treeless grassland, traveled many miles to rivers for seedlings to plant about their homes. Still today one commonly finds rows of tall, old cottonwoods planted by the first homesteaders near farms and ranches. These early plantings have since provided seed stock for floodplain vegetation found along fencerows, in uncultivated draws, and along small, ephemeral creeks.

The positive impact of the post–European settlement expansion of floodplain forest can be demonstrated for many avian species. Most of the species that extended their ranges into the Great Plains were eastern. Nearly 90% of the contemporary avifauna in northeastern Colorado, 82% of which is associated with riparian habitat, was not present before 1900 (Knopf 1986).

Montane Coniferous Forest

Elements of montane coniferous forest enter the state from the west in 2 locations in the Panhandle and are similar to vegetative conditions in the Black Hills of South Dakota. The Wildcat Hills, south of the North Platte River in Scotts Bluff County, contain numerous stands of ponderosa pine (*Pinus ponderosa*) woodland. Most of this woodland is concentrated in the rough scarps that mark the northern edge of a high plateau region in the southern Panhandle. Farther to the east, the edge of this plateau supports extensive stands of juniper in the scarps above the North Platte River. The plateau continues southward to the valley of Lodgepole

Creek, where junipers again dominate the edge of the plateau.

The most extensive pine woodlands lie in Sioux, Dawes, and Sheridan Counties in what is commonly known as the Pine Ridge region (plates 13, 14). Here the topography is heavily bisected by permanent and intermittent streams, forming canyons, high bluffs, and buttes. A rather extensive line of pine-covered, high bluffs overlook Hat Creek basin in northern Sioux, Dawes, and Sheridan Counties. Another pine-covered scarp forms above the White River in central Sioux and Dawes Counties. Perhaps reflecting the historic impact of fire, in many locations on the plateau above the bluffs and cliffs the woodland opens into pine parkland with considerable grass understory. In Sioux County, on the more shaded north slopes, pine stands are denser and may ultimately yield to deciduous vegetation of Rocky Mountain affinities in the cool, moist canyons that drain the ridge. Thickets of Rocky Mountain western birch (*Betula occidentalis*) and quaking aspen (*Populus tremuloides*) exist along the streams that flow northward from the ridge into Hat Creek.

To the east of Sheridan County, the pine parkland reoccurs in central Cherry County, being distributed along the upper slopes of the canyons formed by the downcutting of the Niobrara River. Stands of pine mixed with juniper continue on the bluffs and hillsides above the river as far east as Holt County. This eastern extremity of pine woodland creates a unique transition zone, as conifers increasingly mingle with residual elements of eastern deciduous forest, principally bur oak. Ultimately, conifers are replaced entirely by deciduous hardwoods in Rock and Holt Counties in the scarps above the Niobrara.

The planted Bessey Division of the Nebraska National Forest in Thomas County contributes an additional 7600 ha (19 000 ac) of coniferous forest (plate 15). Although not reproducing, dense stands of ponderosa pine, jackpine (*Pinus banksiana*), scotchpine (*Pinus sylvestris*), and cedar provide additional needle-leafed evergreen habitat for certain montane bird species in the central Sandhills.

Scarps dominated almost entirely by cedars can be found in several locations in central and western Nebraska, including along the bluffs overlooking the Platte River in Lincoln County, along the North Platte in Keith and Garden Counties, in several isolated locations associated with the Republican drainage system, and on scarps associated with Lodgepole Creek in Kimball and Cheyenne Counties (plate 16). It has been argued that juniper woodlands persist only in these locations because the steep hillsides act as fire refuges, but similar stands may also be found along many floodplains throughout the state and possibly dominate in these locations as a result of long-term overgrazing. Although these cedar woodlands provide cover and nesting habitat for a number of avian species, none are known to be limited to this type of habitat. Most are generalist species that may be found in a variety of woody habitats.

The extension of coniferous habitat into the state and the associated rough terrain of the Pine Ridge and butte country of the Panhandle provide habitat for a number of western species. Some of the species that may be found in the Pine Ridge include Sharp-shinned Hawk, Prairie Falcon, Common Poorwill, Violet-green Swallow, Pinyon Jay, Pygmy Nuthatch, Red-breasted Nuthatch, Mountain Bluebird, Yellow-rumped ("Audubon's") Warbler, Western Tanager, and Dark-eyed ("White-winged") Junco.

Grasslands

Nearly 90% of the state was once covered with grassland (Weaver 1943). Although vast and seemingly continuous, it was also bisected by permanent and ephemeral streams. In many protected locations, woody vegetation made its tenuous stand against fire. In protected draws and on many of the steeper slopes, brushy stands of sumac, buffalo berry (*Shepherdia argentea*), and wild plum (*Prunus americana*) helped break the monotony of the undulating grassland sea

and provided breeding habitat and protection for many breeding birds.

Kaul and Rolfsmeier (1993) have identified 8 distinct prairie types that existed before European settlement. These include tallgrass bluestem prairie, mixed prairie, transition prairie, Kansas mixed prairie, Sandhills prairie, sandsage prairie, Dakota prairie, and shortgrass prairie. Little remains of the tallgrass bluestem, Kansas mixed, mixed, and transition prairies of eastern and southern Nebraska, because they were most suitable for cultivation.

Remnants of the tallgrass bluestem prairie (plate 17) may be found along abandoned railroad rights of way, in old cemeteries, and in some moist meadows in the floodplains of the Platte and Elkhorn Rivers. Many parcels of grassland remain within the tallgrass prairie region in areas where exposed glacial soils are too rocky to permit cultivation. Most, however, have been grazed, and a distinct shift toward mixed and transition prairies has occurred because of grazing intensity (Weaver 1943). One can only speculate about the visual impact of the vast true prairies in pre–European settlement times, but they must have been a sight to behold. Grasses such as big bluestem (*Andropogon gerardi*), Indiangrass (*Sorghastrum nutans*), and switchgrass (*Panicum virgatum*) frequently achieve a height of 2 m (6 ft). Only mounted on a horse could an individual see any distance over the tall, dense grasses.

Rough topography and rocky soils have saved small parcels of the Kansas mixed, mixed, and transition prairies in south-central Nebraska. The Kansas mixed prairie has been least disturbed, and good representations can yet be found along the Kansas border.

Although much of the eastern midgrass and tallgrass prairies is gone, roughly half of the state remains grassland (Williams and Murfield 1977). The greatest representation exists in the Sandhills Region, where some 49 000 km² (19 000 mi²) of ancient sand dunes support a Sandhills prairie community of bluestem grasses (*Andropogon* spp.), sandreed grass (*Calamovilfa longifolia*), needle grass (*Stipa comata*), and soapweed

(*Yucca glauca*). Grasses rarely exceed a height of 1 m (3 ft), except in some depressions and meadows that are moist enough to provide favorable habitat for taller grasses. Except for occasional wild plum thickets in draws, small flats of buckbrush, and scattered wild rose (*Rosa* spp.), woody vegetation is not evident in this prairie. Although cattle grazing is the major agricultural industry, the Sandhills prairie has been altered little. During the drought years of the 1930s, ranchers painfully learned the sensitivities of this grassland to heavy grazing and are careful to protect this resource today.

In the southwestern portion of the Sandhills, the Sandhills prairie association is replaced by sandsage prairie (plate 18). Bluestem grasses and sandreed grass provide an understory for flats of sandsage (*Artemisia filifolia*). Nowhere in Nebraska is sandsage distributed over vast areas as is common in southern and eastern Colorado, southern Kansas, and Oklahoma on unarable arid lands. In Nebraska, sandsage prairie more typically occurs in a patchwork. Each patch, just several hectares or less, persists on sandier exposures surrounded either by grassland or crops on the more favorable soils.

To the west of the Sandhills in the Panhandle, 2 shortgrass prairies constitute the high-plains region of the state, where grasses rarely exceed a height of 30 cm (12 in) (plate 19). These are the true shortgrass prairie of the blue-grama grass–buffalo grass (*Bouteloua gracilis–Buchloe dactyloides*) association and Dakota prairie of the western wheatgrass–threadleaf sedge–needlegrass (*Agropyron smithii–Carex filifolia–Stipa spartea*) association. Few woody plants exist in these prairies except in northern Sioux and Dawes Counties, where some big-leafed sage (*Artemesia tridenta*) flats (most consisting of tens of hectares or less) may be found replacing the Dakota prairie. Both prairie types are reduced from their former distribution. Wheat and sugar beet farming have particularly taken their toll on the shortgrass prairie in the central and southern Panhandle.

There appears to be no close association be-tween a bird species and a given grassland classification. The Upland Sandpiper and Bobolink, for example, are restricted to moist meadows in both tallgrass and midgrass prairies. The Long-billed Curlew, on the other hand, is most often found only in moist meadows of midgrass and shortgrass prairies, irrespective of the plant species (Bicak 1977). Still other birds, such as the Lark Sparrow and Field Sparrow, inhabit grasslands that have woody shrubs distributed throughout or along their edges. Loggerhead Shrikes and Blue Grosbeaks are found in shrub thickets scattered in the grasslands. Meadowlarks and Grasshopper Sparrows may be found in all grassland types, whereas Mountain Plovers and McCown's and Chestnut-collared Longspurs are restricted to wide expanses of high-plains shortgrass prairie, which is found only in the Panhandle.

Aquatic Habitats
Major River Systems

The broad prairie rivers, such as the Republican and Platte, provide considerable resting, feeding, and breeding habitat for many migrant birds. Large concentrations of migrating waterfowl utilize the rivers as resting and feeding places. As a migratory staging area, the central Platte attracts as many as 500 000 Sandhill Cranes and in excess of 1 million geese of several species in the spring (plate 20). However, in the past 50 years, various demands for water in the upper reaches of the Platte have substantially reduced year-round stream flow. Impoundments in Colorado and Wyoming, diversion programs for irrigation, and new well fields in the Platte Valley have had their collective impact, reducing water volumes such that long stretches of the river above Columbus are entirely dry in the summer.

Stream flow reduction has also had an impact on the breadth of the river. It has been strongly argued that the winter capture of waters in Colorado and Wyoming have so reduced spring thaw flows that the normal scouring effect of ice flows, coupled with reduced water volume, have narrowed the channels and allowed for the establishment of significant willow and cottonwood stands in what was previously streambed. Although favorable to woodland nesting species, these ecological changes have had a significant impact on several aquatic birds, including a reduction in breeding habitat for Piping Plovers and Least Terns, both designated threatened species. Concerns have also been expressed for Sandhill Cranes, which use the river as a crucial roost site during their spring migration.

The braided Loup Rivers contain numerous cattail marshes in the great bends they make as they flow to the Platte. Attracting Yellow-headed Blackbirds, Swamp Sparrows, and numerous waterfowl, these streams and their wildlife make an interesting contrast with the rolling seas of Sandhills prairie surrounding them. In the central Sandhills, little woody vegetation grows at the edge of these streams. In their lower reaches, willow and cottonwood woodlands may be present along the broader floodplains.

The Niobrara River is a typical prairie stream for the first half of its course, with grasses growing up to its banks. In Cherry County, it enters a deep valley below the Pine Ridge (plate 21). As the stream flows rapidly through the valley with its steep, pine studded slopes, the environment assumes a montane setting. In the most protected locations along the south bank, relic populations of paper birch (*Betula papyrifera*) contrast with the deep shade. Downstream, as the Niobrara increases in width and sandbars become exposed, suitable breeding habitat is provided for Piping Plover, and family groups of migrating Whooping Cranes regularly utilize this portion of the river for resting and foraging.

The Sandhills Lake Region

Nearly all of the approximately 830 km² (320 mi²) of standing surface water in Nebraska lies in the Sandhills and consists of lakes, small ponds, marshes, and seasonal ponds (Weaver 1965). Although these aquatic habitats are scattered throughout the Sandhills, 2 lake regions may be identified as having the greatest wetland habitat. These are the western Sandhills lake region in northern Garden and southern Sheridan Counties and the central Sandhills lake region in Cherry, Grant, and Arthur Counties.

Some visitors to the Sandhill lakes region question the use of the term "lakes," because most bodies of water cover less than 8 ha (20 ac) and rarely exceed a depth of about 3 m (10 ft). A few are fairly large, however, including the largest, Big Alkali Lake in Cherry County, which covers about 290 ha (720 ac). Their clear waters and sandy bottoms often support excellent fish populations as well as a variety of submergent and floating aquatic vegetation. Along the edges of most lakes, stands of emergent cattails and bulrush grow, while some lakes are surrounded by extensive bulrush marshes. Frequently marshes give way to moist lowland meadows that support dense stands of a variety of grasslike sedges.

Many waterfowl species, including Ruddy Duck, Mallard, Northern Pintail, Redhead, and Canvasback, nest in the Sandhills lake country. Eared Grebe, Marsh Wren, and Yellow-headed Blackbird may nest in the marshes. The adjoining moist meadows are favorable feeding habitat for Long-billed Curlew.

Rainwater Basin

The Rainwater Basin in the south-central and east-central portion of Nebraska, although now a region of widely scattered marshes and permanent and temporary wetlands, has been shown to be a very important, perhaps even critical, habitat for many species of migratory birds (LaGrange 1997). Enormous concentrations of migratory waterfowl, sometimes in the tens of thousands, may pack into these marshes in early March to rest and feed and gather additional energy for the next northward migratory leg and subsequent reproduction (Gersib and others 1992). Spring utilization of these wetlands includes 90% of the midcontinent population of White-fronted Goose, 50% of the midcontinent Mallard population, and 30% of the midcontinent Northern Pintail population (LaGrange 1997). In addition to the waterfowl, occasional

Sandhill Cranes and family groups of Whooping Cranes frequent these wetlands. Centrally located on the migratory route for most arctic-nesting sandpipers, the wetlands also attract and sustain large concentrations of these shorebirds, once including the now likely extinct Eskimo Curlew. Estimates for shorebird utilization are as high as 200 000–300 000 individuals of some 30 different species during the spring migration (LaGrange 1997). During the breeding season, some Rainwater Basin wetlands remain sufficiently moist to maintain extensive cattail and bulrush marshes and thus may support breeding populations of aquatic bird species.

Maps and
Illustrations

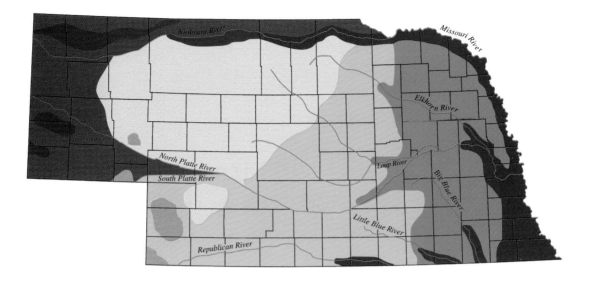

	Short Grass - Mixed Grass Plains		Loess Mixed Grass Prairie
	Ponderosa Pine Forest and Savannah		Tallgrass Prairie, Scattered Oak Savannah
	Sandsage - Mixed Grass Prairie		Oak Forest and Savannah
	Sandhills Mixed Grass Prairie		

MAP 1. Distribution of Native Vegetation before European Settlement (adapted from Kaul and Rolfsmeier 1993)

MAP 2. Nebraska Land Use (adapted from Conservation and Survey Division, University of Nebraska–Lincoln 1973, Land use in Nebraska)

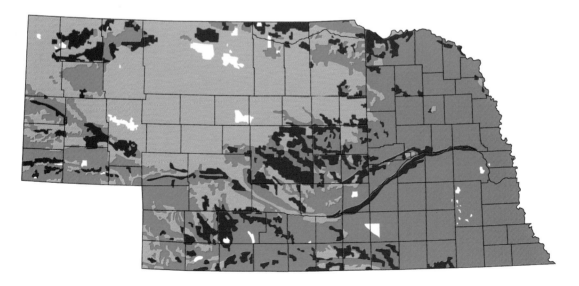

	Range Land grazingland mostly privately owned for seasonal, nonmigratory use		Transitional Land a mixture of cultivated and grazingland
	Cultivated Land 80% or more of total		Special Use Land parks, national forest, government land

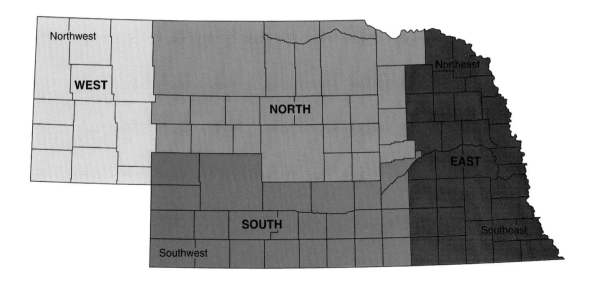

map 3. Nebraska Regions

map 4. Nebraska Counties

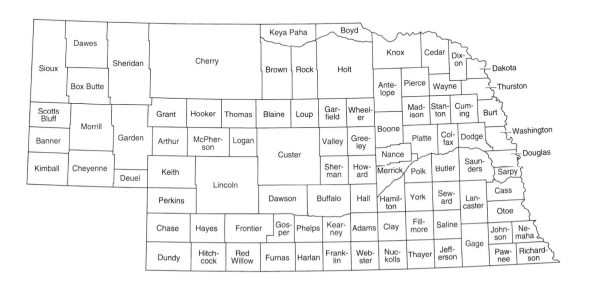

PLATE 1. Buttes in the Pine Ridge, Dawes Co

PLATE 2. Broad, vegetated arroyo, Sioux Co

PLATE 3. Prairie chickens on lek, Johnson Co

PLATE 4. Sandhills prairie, Arthur Co

PLATE 5. Exposed and vegetated sandbars, Platte River, Hall Co

PLATE 6. Shallow playa containing water, foraging avocets in Sandhills, Sheridan Co

PLATE 7. Sandhills lake with emergent edge vegetation, Garden Co

PLATE 8. Snow geese on Harvard Lagoon, Rainwater Basin, Clay Co

PLATE 9. Part of Lake McConaughy, surrounded by Sandhills prairie, Keith Co

PLATE 10. Oak forest, Sarpy Co

PLATE 11. Oak savanna, Otoe Co

PLATE 12. Floodplain forest along Platte River, Sarpy Co

PLATE 13. Pine Ridge with pine-covered escarpment, Sioux Co

PLATE 14. Pine parkland, Sioux Co

PLATE 15. Mature plantings of Austrian pine, Bessey Division, Nebraska National Forest, Thomas Co

PLATE 16. Juniper stands on escarpment, Keith Co

PLATE 17. Tallgrass prairie on Platte River floodplain, Hall Co

PLATE 18. Sandsage prairie, Garden Co

PLATE 19. Shortgrass plains, Sioux Co

PLATE 20. Sandhill Cranes in wet meadow near Platte River, Hall Co

PLATE 21. Niobrara Valley, Cherry Co

History of Nebraska Ornithology

Although small parties of Spanish and French explorers, trappers, and adventurers ventured into the Nebraska region during the 18th century, little record exists of these travels. The documents that do exist provide few insights into the natural history of the region and make no specific reference to its birdlife. It was only after Thomas Jefferson engineered the Louisiana Purchase from the French in 1802 and subsequently sent an expedition into the region in 1803–6 that a record of Nebraska ornithology begins to develop. The chronicle of the Lewis and Clark expedition included descriptions of the resources, people, geography, and natural history that the explorers observed.

The Lewis and Clark Expedition, 1803–6

The Lewis and Clark expedition began in St. Louis and traveled along the Missouri River north into the Dakotas (Thwaites 1904, 1905). Part of the route included the portion of the Missouri River that marks the border of what is now eastern Nebraska and western Iowa. Hindered by the strong current of the river, numerous snags, and piles of flood-deposited trees and debris, the party often traveled no more than 10 km (6 mi) per day. This slow pace offered numerous opportunities for observation and exploration of the environs, including the bluffs that bordered the floodplain. Several of the party, including Lewis and Clark, recorded these observations in their personal journals.

The notes include descriptions of the physical environment as well as occasional descriptions of the natural history encountered, including some of the bird species. In some ways the Lewis and Clark party observed a geomorphology that today can still be glimpsed: broad floodplains covered by expanses of grassland (today replaced by corn and other cultivars); floodplain woodlands bordering the river (albeit greatly diminished); and loess hills covered with oak-hickory woodland (where sheltered on the east facing slopes) and with grasses (on more exposed slopes). Among the wildlife observed were bison (*Bison bison*) and elk (*Cervus americana*), which vanished from the region shortly after pioneer settlement began in the 1840s. Birds most often mentioned in their accounts were primarily those of economic importance to the party, including turkeys, "partridge" (likely *Bonasa umbellus*, or Ruffed Grouse), and waterfowl, but the list is sufficiently varied to give one a flavor of the avifaunal assemblage and species distribution that existed at that time. Included among these records is the first and one of just a few regional records for the Carolina Parakeet, which became extinct later in the century.

Several expeditions explored the Nebraska region subsequent to the travels of Lewis and Clark. Accompanying some of these expeditions were naturalists and observers who gathered scientific information and left to future generations additional views of the avifauna. Among the early naturalists to venture into the region were John Bradbury, Thomas Nuttall, and John Stuart, who accompanied various American Fur Company expeditions sponsored by John Jacob Astor between 1811 and 1813. Their ornithological findings were summarized by Swenk (1937a) and include observations in southeastern Nebraska of species such as Ruffed Grouse and Wild Turkey, which were extirpated later in the century, and the Passenger Pigeon, which eventually became extinct.

Thomas Say and the Long Expedition, 1819–20

In 1819 naturalist Thomas Say accompanied the government-sponsored expedition led by Major Stephen H. Long up the Missouri River to a site known as Engineer's Cantonment (just above present-day Omaha). Say spent the winter there, remaining until June 1820 when the expedition continued its trek, traveling west across the Nebraska plains to the mountains. During the months encamped along the Missouri River in eastern Nebraska, Say recorded 145 species, including Carolina Parakeets and Passenger Pigeons. He also recorded nesting Bald Eagles and Canada Geese (Thwaites 1905: Appendix A). His accounts of the trip with Long across Nebraska to the Rocky Mountains reveal little unique about the birdlife encountered, except for a record of a Marbled Godwit with young, which he observed in floodplain grassland near the mouth of the Loup River near Columbus on 9 June 1820 (James 1972). This remained the only documented breeding record for that species in Nebraska until 1990 (AB 44:1153) and can be accepted with a high degree of confidence because he distinguished the godwits from other grassland sandpipers (curlews, upland sandpipers), which were also present in the area. However, among those species recorded along the Missouri during his winter stay were some whose occurrence, based on subsequent records, seems questionable; among these are Oldsquaw and Northern Hawk Owl (Haecker and Moser 1941).

Similar suspect identifications are commonplace among 19th-century observers, leaving one to question historic geographic distributions of some species or the observational skills of observers or both. It should be noted, however, that some of these naturalists had limited experience in the American West and were often foreign-born and educated. Thus their familiarity with the regional avifauna was limited, and they were also hampered by the absence of suitable reference works and field optics.

The Expedition of Prince Maximilian von Wied, 1833–34

One of the best accounts of the natural history of the Missouri Basin was left to posterity by a European explorer and traveler, Prince Maximilian von Wied. Maximilian, intent on observing and describing the American West before its corruption by European settlement, was accompanied by Swiss artist Karl Bodmer and hunter-taxidermist David Dreidopple. Bodmer's task was to prepare illustrations of the environment, wildlife, and native peoples the expedition encountered. His works (now in the permanent collections of the Joslyn Art Museum, Omaha) along with Maximilian's detailed journal entries and published accounts of the trip, provide a clear and accurate picture of the natural history along the Missouri River at that time (Orr and Porter 1983).

Maximilian, an accomplished explorer, prepared well for the trip, spending part of the winter of 1832–33 traveling with Thomas Say in the eastern states. His travels with Say allowed Maximilian to learn about the conditions that the impending expedition would meet as it repeated Say's route up the Missouri River. The Maximilian party left St. Louis on 10 April 1833, on board the steamer *Yellow Stone* en route to their destination of Fort Mackenzie in Montana. The party traveled along Nebraska's eastern border during the first half of May 1833. After spending the summer exploring the surroundings of Fort Mackenzie, they withdrew downstream to Fort Clark in North Dakota, where they spent the winter. They returned downstream the following spring, arriving at the mouth of the Niobrara River on 5 May. They continued downstream along the Nebraska border at a relatively rapid pace, arriving at Fort Leavenworth (just below the border of Nebraska) on 18 May.

Maximilian was clearly an accomplished naturalist. His particularly detailed descriptions of the flora give one a very good understanding of the predominant vegetation seen by the party. From his accounts one learns that along the Nebraska border upstream to at least the mouth of the Platte River the composition of the dominant woody vegetation has changed little since that time. Maximilian, for example, describes the northernmost distribution of pawpaw (*Asimina* spp.) as near the mouth of the Platte River; he also notes that sycamore were not seen beyond the Council Bluffs. The local geographic distributions of both species are essentially the

same today. Maximilian also notes the common hardwood species (bur oak, hickories, elms) that were found in the hillside forests; an identical list could be constructed today.

Maximilian took great care in recording the animals observed and collected. After he returned to his home in Germany he extracted from his journals his zoological findings for separate publication, including his bird observations. Most of the bird species observed can be encountered today in a similar trek up the Missouri. For example, throughout the Nebraska stretch of the river he makes frequent references to nesting Bald Eagles, nesting Canada Geese, and Wild Turkeys, species whose reproductive populations were all extirpated from the region later in the 19th century.

It was clear to Maximilian that European settlement was already having an impact on the animal life of the Missouri Basin. At a settlement on "Nadaway Island" (near the Nodaway River, which enters the Missouri just above St. Joseph, Missouri) he learned that "Capt. Martin wintered with 20 men 2 years long; they lived from hunting. In one year they shot 1800, in another 1600 head of game (*Cervus* [*Oidicoilous*]*virginianus*) and shot and wounded just as many that they could not get; in addition they still shot elk and bear. The elk is now rare in the area" (Orr and Porter 1983:349). Except for this account about game that once existed along this part of the Missouri Valley, entries of deer and elk are notably absent from his journal until the party reached the Little Sioux River, 270 km (170 mi) above the Nodaway. Upstream from the Little Sioux both species are mentioned with increasing frequency.

Maximilian learned that Native Americans also altered the natural environment, particularly by periodically starting grass fires. Near the mouth of the Little Sioux River he wrote that "the entire prairie was overgrown luxuriantly green with all kinds of grasses and plants, but we also could not find a single flower. Everywhere black, burnt wood was lying about, proof that the Indians had burned off everything here" (Orr and Porter 1983:385). In this instance Maximilian was referring to fires started by Native Americans either to "fertilize" the soil and thereby attract game to the resultant fresh growth or to hide the tracks of the their travel. Both purposes were noted by Maximilian as well as by other early travelers.

Explorations of Zebulon Pike (1829) and John C. Frémont (1842)

Although government-sponsored expeditions led by Pike in 1829 and Frémont in 1842 ventured into the Nebraska territory, little was recorded about the birdlife. Pike barely entered the state as he explored the Republican River valley. Frémont, however, followed the Oregon Trail from Kansas into Nebraska and traveled along the trail as it continued westward up the Platte River valley to a point just west of Chimney Rock in Morrill County. Here he turned southward, crossing the plateau that separates the North Platte River and Lodgepole Creek, continuing down the creek to its mouth. From this point he journeyed up the South Platte River into Colorado. His trip into Nebraska's interior gives us a glimpse of the prairies and plains before they began to be destroyed by European agriculture. Although his observations of the western plains differ little from what can still be encountered, the south-central and eastern grasslands that he traversed now persist as relicts of their vastness. Frémont also noted considerable woody vegetation as he traveled the Oregon Trail from Kansas to the Platte River. While confined to floodplains, sheltered draws, and hillsides, it was sufficiently dense and broadly distributed to sustain such species as turkeys, which he observed in the wooded floodplain of the Little Blue River in Nuckolls County (Frémont 1988). Timber was noted along the Platte from the point of intersection (Grand Island), although with some irregularity and confined principally to islands, as far west as Brady (Lincoln County). From this point westward Frémont noted the increasing scarcity of trees, except for occasional stands of willows in association with the river.

The Missouri River Travels of John James Audubon, 1843

In 1843 the famous artist-naturalist John James Audubon traveled up the Missouri by steamboat. Having already completed his monumental work, *The Birds of America*, Audubon was traveling in the region to gather information, collect specimens, and make sketches of mammals for a new work *The Viviparous Quadrupeds of North America*. Accompanying Audubon were several friends, including John Bell (Bell's Vireo), Edward Harris (Harris's Sparrow), and Isaac Sprague (Sprague's Pipit). He and his party took every opportunity to leave the craft when the party stopped for fuel or trading; the craft was also occasionally hung up on snags or sandbars for periods of several hours or more. These periods of time as well as evenings and early mornings while the boat was moored for the night afforded members of the party opportunities to explore the immediate environment (Audubon and Coues 1960).

The Audubon party spent about 2 weeks traveling the Nebraska stretch of the Missouri River, arriving on the Nebraska-Missouri border on about 6 May and leaving the Nebraska portion of the river on about 24 May, as the boat continued upstream into South Dakota (Audubon and Coues 1960). Audubon and John Harris (McDermott 1951) kept very good journals of the natural history encountered on the trip and included in their notes information on birds seen and collected. Together they recorded more than 150 species, most of which can be seen with regularity during present-day springs. Notably absent today, however, are such species as the Carolina Parakeet, Pileated Woodpecker, Swallow-tailed Kite, and Common Raven, the latter of which was seen with regularity by Audubon and his party upstream from about Rulo, Nebraska. Other noteworthy observations include nesting Bald Eagles near Rulo, caged magpies at Fort Atkinson (near present-day Fort Calhoun) that were trapped in the vicinity during the previous winter, and nesting Canada Geese throughout the Nebraska portion of the trip.

Their general natural history observations provide some insights into the habitat and the changes they encountered en route. Woodland dominated the floodplain and ravines of the loess hills along the lower Missouri River; this habitat was also reflected in the birds observed. However, as the party traveled upstream above the mouth of the Platte River, incremental changes in the biota are evident from their journals. Harris, for instance, noted the beginning of extensive flood-plain prairie just north of Blair and observed that the party began to resort to using drift-wood for fuel, owing to the absence of mature floodplain woodland (McDermott 1951). That prairie habitat was increasing as they traveled upstream is supported by their first observations of a wolf, *Canis lupus*, the principal predator of large grassland ungulates such as bison and elk. Just north of this point, near the mouth of the Little Sioux River, they encountered the first mule deer (*Odocoileus hemionus*), a species limited to shortgrass and midgrass prairie. At this location they also saw elk tracks for the first time. As they traveled upstream from Sioux City, Iowa, they began to encounter with increasing frequency bison carcasses floating in the river, ultimately seeing live bison for the first time near the mouth of the James River, which enters the Missouri just opposite Cedar County in north-east Nebraska. In this general vicinity they saw their first Chestnut-collared Longspurs and Lark Buntings, which suggests that the surrounding environment was indeed dominated by short-grass and midgrass prairie.

Although grassland was the predominant community type in northeast Nebraska, apparently considerable woodland persisted as isolates of floodplain woodland and on sheltered slopes of the loess hills. For example, the party encountered a black bear (presumably *Ursus americanus*), an indicator species for extensive woodland, swimming across the river near Sioux City, and several additional observations were made of bear tracks upstream to about Yankton, South Dakota. Audubon's notes, however, suggest that the extent of woodland habitat was clearly

diminishing upstream from the Sioux City vicinity. Near Yankton Audubon noted (on about 1 June 1843):

The country grows poorer as we ascend. Harris shot a Marsh Hawk, Sprague a Nighthawk, and some small birds, and I saw Martins breeding in Woodpecker's holes in high and large cotton-trees. We passed the "Grand Town'" [prairie dog] today. . . . They [bison] all look extremely poor and shabby; we see them sporting among themselves, butting and tearing up the earth, and when at a gallop they throw up the dust behind them. . . . The Elks, on the contrary, are found on the islands and low bottoms, well covered with timber; the common Deer is found indifferently every-where. . . . We have not seen Parakeets or Squirrels for several days; Partridges have also deserted us, as well as Rabbits; we have seen Barn Swallows, but no more Rough-winged. We have yet plenty of Red-headed Woodpeckers (Audubon and Coues 1960:506–7).

Audubon and his party continued up the Missouri into the Dakotas to Fort Union, just above the point where the Yellowstone enters the Missouri, and spent the summer in this vicinity, exploring, collecting, and hunting. They returned down the river in October, but this rapid, uneventful trip yielded little new in their journals. Audubon and his cohorts were the last ornithological explorers to see the environs of the Missouri River in eastern Nebraska before the first significant waves of European settlement began to populate the eastern portion of the state in the late 1840s and early 1850s. Their ornithological records, along with those of Maximilian and Thomas Say, are clearly the most complete and comprehensive of the pre–European settlement era.

F. V. Hayden, 1852, 1856–57

One additional account, that of F. V. Hayden (1862), offers us some limited insight into the natural history of the state's interior prior to pioneer settlement. Hayden, a geologist for the U.S. Geological Survey, made several trips into the state as part of his responsibility to survey the regional geology. Contained in his various reports are natural history observations, some of which included the avifauna. His 1862 report summarizes his observations of these several explorations and includes a list of 191 bird species. Unfortunately, most of the species accounts are very general and do not clearly identify a species' presence in what would now be the state of Nebraska. His more specific report to Elliot Coues on his avifaunal observations and collections while accompanying the Raynold's expedition in 1852 and as naturalist for the Warren expedition in 1856–57 was incorporated by Coues (1874) in his classic work, *Birds of the Northwest*. In it Coues lists the specimens collected by Hayden in each species account. In a few cases the locations given for Nebraska specimens are detailed enough to determine the exact place of collection. Neither the specimens taken nor his observations suggest that the state's interior avifauna differs from that found today. Some care must also be taken in accepting Hayden's avifaunal impressions, because a careful reading of Coues suggests that Coues did not always agree with Hayden's observations. A list of bird specimens taken by Hayden on the Raynold's and Warren expeditions also appears in Baird et al. (1858).

THE EARLY UNIVERSITY PERIOD (1860–1915)
Samuel Aughey

Shortly after the University of Nebraska was established in 1873, it acquired the services of Samuel Aughey, who served as chair of Natural History until 1883, when he was asked to vacate his position. Aughey, a minister by training, produced 2 works that treat Nebraska avifauna; the first (1878) provides a list of birds that Aughey purportedly observed or collected, along with biological notes of food habits for each species, particularly as they related to insectivory. The second work (1880) is essentially a condensed version of the first. Although Aughey lists some species as having been observed or collected in the state from several locations (Lancaster, Cedar, Red Willow, and Richardson Counties), the absence of specimens and serious questions about the authenticity of his work diminish the reliabil-

ity of his avifaunal list (see Sharpe 1993). Similar questions about his botanical list were raised as early as 1893 (Pound and Clements 1900).

Some early workers used Aughey's list to assist them in determining distributional limits for several species, and subsequent workers have sometimes perpetuated potentially spurious data. For example, in the fifth edition of the *American Ornithologists' Union Check-list of North American Birds* (AOU 1957), the northwestern breeding limits for the Prairie Warbler include southeastern Nebraska. This record was apparently obtained from the work of Bruner et al. (1904), who used Aughey's data but clearly noted that no recent breeding records existed (nor are there any records since).

Bruner, Wolcott, and the Nebraska Ornithologists' Union

The years following Aughey's dismissal from the University of Nebraska saw the addition of two important zoologists to the staff, Robert H. Wolcott to the Department of Zoology and Lawrence Bruner to the Department of Entomology and Ornithology. Both were influential in fostering the development of ornithology in the state by writing about its birdlife, generating public interest in birds, and encouraging students to study birds.

Bruner (1896) published the first comprehensive list of Nebraska birds, entitled *A List of Nebraska Birds*. Several years later (1904) he joined forces with his graduate student Myron Swenk and Wolcott to produce a more definitive work on the birds of the state. Their individual studies may also be found in the first publications of the Nebraska Ornithologists' Union, the *Proceedings*, (1899–1915).

Bruner and Wolcott, along with 12 other individuals, were responsible for helping found the Nebraska Ornithologists' Union (NOU) in December 1899. The NOU not only provided an organization that the university could use to disseminate information about Nebraska birds, but the organization also attracted many amateurs whose contributions over the years have ultimately far exceeded those of the professionals. Among the early amateurs who contributed significantly to the *Proceedings* were M. A. Carriker of Nebraska City, Merritt Cary of Neligh, and F. H. Shoemaker and I. S. Trostler of Omaha. Carriker later became one of the most prolific bird collectors in South America, and several species are named in his honor.

This period of ornithological activity must have been exciting for professionals and amateurs alike. Much of the state remained to be explored ornithologically, and small parties of bird enthusiasts made numerous extended field trips, especially to unexplored locations such as the Pine Ridge and the southwestern plains. These activities attracted many young birders seeking adventure, including 2 students who later became professional ornithologists: Myron Swenk, who remained with the university for his entire career, and John T. Zimmer, who was associated with the American Museum of Natural History for most of his professional life.

THE SWENK YEARS, 1915–40

Myron Swenk grew up near Beatrice, Nebraska, and moved with his family to Lincoln around the time he entered the university in 1901. Already deeply interested in natural history, he was encouraged to pursue these interests by Wolcott, Bruner, and other zoologists at the university. During his student years Swenk developed two major interests, birds and insects, largely through Bruner's influence. Although formally trained as an entomologist, he was equally proficient as an ornithologist, writing 72 articles on birds throughout his career. Following the completion of his master's degree in entomology in 1908, Swenk remained with the university in the Department of Entomology until his death in 1940.

Swenk became an important figure in Nebraska ornithology very early in his career, coauthoring with Wolcott and Bruner a checklist of state birds in 1904, entitled *A Preliminary Review of the Birds of Nebraska*. In 1907 he was president of the NOU and thereafter held the office of

secretary-treasurer until his death. Throughout these years Swenk was the driving force for the NOU and its publications. In 1916, apparently because of financial problems in the NOU, Swenk was instrumental in the NOU joining forces with the Wilson Ornithological Club. From 1916 until 1924 the cover of the *Wilson Bulletin* noted that it was the "Official Organ of the Wilson Ornithological Club and the Nebraska Ornithologists' Union." From 1924 to 1933 Swenk was responsible for publishing the mimeographed "Letters of Information" for the NOU. Swenk ultimately convinced the membership that a more substantial publication was justified, and in 1933 the publication of the *Nebraska Bird Review* (NBR) began under his editorship. He remained editor until his death.

The Swenk years were marked by his significant contributions to Nebraska ornithology, including several systematic reviews of avian taxa and their Nebraska status, an early investigation into the phenomenon of hybridization and its taxonomic implications, and the beginnings of a major work on the state's avifauna. Other professionals were also strongly encouraged by Swenk to publish contributions to the knowledge of Nebraska birds. One of those responding was A. M. Brooking, director of the Hastings Museum, in Hastings, Nebraska, and an active collector of eggs, nests, and birdskins.

Brooking was a very popular museum director and had a strong following in the Hastings area. One outgrowth of this relationship was the development of the Hastings (later Brooking) Bird Club, which remained very active many years after Brooking's death. As a result of Brooking's influence, the Hastings Museum developed an excellent representative collection of the birds of Nebraska. Several significant specimens either reside in or were once part of that collection.

Doubts have arisen in recent years about the validity of some of the significant specimens in the Hastings Museum collection. We have dealt with these cases in the relevant species accounts. These doubts concern a major contributor of specimens to the Brooking collection, Cyrus A.

(Cy) Black of Kearney, a sportsman, hunter, and professional taxidermist, and his association with Miles Maryott of Oshkosh. Black and Maryott not only collected several quite rare specimens but also some that subsequently have proved unique for Nebraska. Black provided taxidermy services to the Hastings Museum and presumably had access to the accession and labeling procedures. It has been alleged by longtime member of the Nebraska Ornithologists' Union John Plank of Council Bluffs, Iowa, now deceased, that Black, or Black and Brooking in collusion, "created" some Nebraska records by obtaining specimens from elsewhere and claiming that they were of Nebraska origin. In discussions with RSS, Plank related that he learned of these activities from Black, who was his mentor while growing up in Kearney (see, for example, the species account on Lesser Prairie-Chicken).

Many others were active contributors to knowledge about the state's avifauna in the Swenk period; they also provided continuity after Swenk's death. These include L. O. Horsky of Omaha, who was the leading figure with the Omaha Nature Study Club and provided many important records from Fontenelle Forest before it was developed as a nature center. Wilson Tout was active through much of his life in North Platte, ultimately authoring and publishing *Lincoln County Birds* in 1947. Dr. T. C. Stephens of Sioux City, Iowa, contributed information about the avifauna of northeast Nebraska and authored *Occasional Paper* No. 3 of the NOU, entitled "The Birds of Dakota County, Nebraska." Bill Youngworth, also of Sioux City, was a significant figure in Nebraska ornithology, particularly in the midcentury; he was an important critic of questionable records that were submitted to the NBR. The Blincoes of Chadron reported on birds of the Panhandle and remained the only birders in that region for more than a quarter century. Harold Turner of Adams County began his birding during the Swenk years and contributed his regular seasonal reports for more than 50 years. Although an anatomist, George Hudson taught ornithology at the University of Nebraska in the

1930s and 1940s and contributed many records to the NBR; he also added several important specimens to the permanent collections in the University of Nebraska State Museum UNSM. Mr. and Mrs. Earl Glandon of Stapleton provided more than 25 years of data from the Stapleton vicinity, as did Charles Ludlow from the Red Cloud vicinity.

THE THIRD QUARTER, PERIOD OF TRANSITION

Myron Swenk's death in 1940 left Nebraska ornithology without professional leadership. However, Nebraska ornithology did not die with Swenk, as many serious amateurs and professional biologists filled the void and insured the continued publication of the NBR. Although the university was without an ornithologist for almost 20 years, individuals such as Mary Lou Hansen Pritchard and George Hudson of the Department of Zoology helped maintain the bird collection. Leadership also arose from the amateur community. In the 1940s and early 1950s individuals such as R. Allyn Moser of Omaha, Earl Glandon of Stapleton, Wilson Tout of North Platte, and Bill Rapp of Crete filled Swenk's void and continued the NBR. In 1958 the *Revised Checklist of Nebraska Birds* by William F. Rapp Jr., Janet Rapp, Henry Baumgarten, and R. Allyn Moser was published. Other figures included Doris Gates who taught ornithology at Chadron State College; Willeta Lueshen of Wisner, who was very active in several national ornithological organizations; and Dr. J. C. W. Bliese, who taught ornithology at Kearney State College. These individuals helped provide momentum and leadership for the NOU and fostered additional interest in Nebraska birds. Other active contributors during this period include C. W. Huntley of Brule, Gail Shickley and Glenn Viehmeyer of North Platte, Harold Turner of Hastings, and Ralph Velich of Omaha.

In 1959 the university reaffirmed its interests in Nebraska ornithology with the addition of Paul Johnsgard to the Department of Zoology staff. Johnsgard has written more than 50 books on various ornithological topics, including an abbreviated checklist of Nebraska birds in 1980, most recently updated in 1996. His presence and influence at the university has also helped guarantee the maintenance of the major state ornithological collection at the UNSM, currently curated by Tom Labedz.

One individual whose efforts have spanned almost half a century is R. G. (Rusty) Cortelyou, who became active with the NOU in the 1950s and served as editor of the NBR for more than 20 years. His leadership had a major impact in reorganizing and modernizing the journal, including the development of systematic procedures for reporting bird records.

Another individual who had a major impact on Nebraska ornithology during this period was James Malkowski of Omaha. Hired by the Fontenelle Forest Association in 1962 as its naturalist, Malkowski's infectious personality moved that organization to become a nature center of national importance. He also became quite involved with other nature organizations, including the NOU, providing leadership and new ideas for its publication, and he helped found the Omaha Bird Club, now the Omaha Audubon Society. As the Fontenelle Forest Association grew under his leadership, many new birders also developed.

Some of those who were influenced by Malkowski became important contributors to the NOU, many in leadership roles. Among those are Ruth Green of Bellevue, who has been a very active force in Nebraska ornithology and conservation. She has served as president several times of both the Omaha Audubon Society and the NOU. During a period of languishing membership, Green took the reins of the NOU. With her enthusiasm and careful guidance the membership increased by more than a third in a year.

RECENT ORNITHOLOGY

During the last quarter of the 20th century, Nebraska ornithology has leaped forward with the addition of many serious amateurs as well as professionals who have pursued birding with a healthy vigor that has vastly improved our un-

derstanding of the distribution and abundance of Nebraska birds. Their involvement has added about 40 species to the state list since 1979. They have not only demanded care and accuracy in identifying and reporting birds, but their leadership has led to even more careful publication of records. One result has been the addition of a Records Committee to the NOU, which carefully reviews all records submitted for publication; another is an evaluation of the entire list of Nebraska birds, *The Birds of Nebraska: A Critically Evaluated List*, authored in 1986 by Tanya Bray, Barbara Padelford, and Ross Silcock. This publication was the basis for the first "Official List of the Birds of Nebraska," published by the NOU (1988).

Several state and regional works have been published in this recent period, including Paul Johnsgard's *Checklist* (1980), Richard Rosche's (1982) summary of avifaunal records for the northern Panhandle and his (1994a) publication on the birds of the Panhandle and the North Platte River Valley, and Jim Ducey's (1988) summary of Nebraska bird breeding records. Rosche's published contributions from the Panhandle for a period of almost 20 years have been especially important to our understanding of the avifauna of that region; without his efforts, the present authors could not comfortably complete this current work. Mark Brogie and Ed Brogie of Norfolk have been very active in northeast Nebraska and spent several seasons surveying the Niobrara Valley Preserve. Gary Lingle produced many important publications about the birds of the central Platte Valley while he was associated with the Platte River Whooping Crane Critical Habitat Maintenance Trust.

Other significant contributors include Fred Zeillemaker, refuge manager at Crescent Lake National Wildlife Refuge from 1978 to 1981, who birded and netted daily during migratory periods in a location that was very short of records. His work has clarified many questions about western migrants and winter occurrences. Wayne Mollhoff of Albion has been especially active in north-central and northeastern Nebraska, providing specimens to Wayne State College. He has more recently been involved in the coordination of the state Breeding Bird Atlas project. Ruth Green has continued her birding activities, adding banding to her activities. Loren and Babs Padelford have contributed many important records; their 5-year fall raptor study has provided important fundamental information about raptor migrations, and for the last few years they have also operated the Nebraska Birdline, sponsored by the Omaha Audubon Society, from their home in Bellevue. Clyde and Emma Johnson of Omaha coordinated the state Breeding Bird Survey for much of its 30-year history. Still others who have been important, regular contributors over a long period include George Brown of Kearney, Lee Morris of Polk County, Tom Labedz of the UNSM, Alan Grenon and Jim Alt of the Fontenelle Forest Association, and Joe Gubanyi of Concordia College.

As we begin a new century, we can say that Nebraska ornithology is in good hands. Many young, energetic, and critical birders are now active in the state. They have skills and equipment far superior to most of what was available throughout much of the 20th century and drive and enthusiasm that rivals that of Bruner, Wolcott, and their colleagues. There is no question that their efforts will help sharpen our understanding of Nebraska birds and eliminate many of the questions that this work reveals.

How to Use the Species Accounts

This chapter explains the sources and analysis of data, defines terms used, and outlines conventions applied in the species accounts. To assist the reader, a list of abbreviations of places, organizations, and observers' initials as well as a gazetteer of locations mentioned in the species accounts can be found in the appendixes.

Throughout the 100-year publishing history of the Nebraska Ornithologists' Union (NOU), several data-gathering activities have been developed, refined, and continued that have allowed members an opportunity to report their bird sightings on a regular basis. These largely annual reports have been the primary sources for the data that we have used to develop the species accounts. While we did not have time to study the fascinating early correspondence carried out by Swenk and now in the NOU archives at the University of Nebraska State Museum, we did make extensive use of notes typed by Myron Swenk regarding the collections of B. J. Olson and C. A. Black at Kearney and notes typed by A. M. Brooking of Hastings on specimens in his collection. The notes by Swenk are in two parts, before 1925 and after 1925, and are cited as such ("Swenk, Notes before 1925" and "Swenk, Notes after 1925) in this work. The notes by Brooking are cited simply as "Brooking, Notes."

Although published reports represent thousands of records of occurrence, most had not been closely checked for accuracy; it was not until 1986 that the NOU Records Committee was established. In summarizing these data, it became clear to us that, in the absence of published criteria regarding accuracy of identification, many published reports suffered from varying observer experience and accuracy. The highly populated eastern part of the state has provided the vast majority of records (Thorson 1950), and less populated regions such as the Panhandle have yielded far fewer reports. Indeed, large regions such as the central Sandhills have provided almost no regular records over this 100-year period.

Clearly there are many problems associated with records that have been contributed by birders with varying levels of experience and published by editors with varying levels of scrutiny. Nevertheless, when nearly 100 years of records are organized temporally and geographically by species, the questionable, out-of-place records begin to stand out. In summarizing these records, we have used our best judgment to separate questionable from likely records and have attempted to portray as accurately as possible an account of the seasonal status, geographic distribution, relative abundance, and ecology for each species. We admit to some creative license in filling the occasional gaping holes in the data, but we have used regional trends and our collective judgment in making those decisions. With few exceptions, we have not included reports after 31 December 1999.

PUBLISHED REPORTS OF SEASONAL OCCURRENCE

"Spring Occurrence Reports" began in 1924 as an annual report by the NOU covering the first half of the year in the mimeographed "Letters of Information"; it was continued with the *Nebraska Bird Review* NBR in 1933 and has been published annually since. Only single dates of occurrence, normally the first date, per location were published until 1964. Beginning in 1965 both first and last dates of occurrence per species and per location were published. In 1992 all seasonal reports began reporting high counts in an effort to approach the problem of understanding relative abundance across the state and to determine peak migration timing.

"Fall Reports" were begun in 1959 (NBR 27:67)

as an annual report covering the second half of the year, first with a single date per locale representing the first date of occurrence. It quickly became clear that this single date did not deal with fall departures, and in 1964 "Fall Reports" included a first date for migrants and arriving winter visitors, and a second date, representing last date of occurrence per locale.

Breeding season reports were begun in 1956 (NBR 24:14) as "Reports of Nests, Nestlings and Fledglings," becoming the "Annual Nesting Report" in 1957 (NBR 25:18) and later the "Nebraska Nesting Survey." Reporting conventions have varied, but the reports usually include evidence of breeding activity per locale, without periodicity; sometimes a species is simply listed as being present, without evidence of breeding activity.

In 1965 the NOU began cooperating with the Cornell University Laboratory of Ornithology Nest Card Program. For about 20 years, results from that survey were incorporated into the breeding season reports. The nest card program also gave us important information as to breeding periodicity and ecology for a number of species. We obtained copies of the nest cards from Cornell for many select species; for several more abundant species, available nest card data had already been summarized (Jacoby 1976).

In 1966 members of the NOU began working with the National Breeding Bird Survey, conducted by the Office of Migratory Bird Management, U.S. Fish and Wildlife Service. Although many of the survey routes have been remote, nearly 100% of the routes have been covered each year. A 24-year summary (1966–89) and a 10-year summary (1980–89) of Nebraska data have provided additional distributional information as well as insights into regional relative abundance and recent population trends.

We also used banding recovery data provided by the National Banding Laboratories, Patuxent Wildlife Research Center, Patuxent, Maryland. In summarizing breeding distribution for a species, we relied most heavily on Ducey's (1988) work, which we consider to be a comprehensive survey of the available literature.

Winter occurrences have not been systematically reported in the state, except by way of the Christmas Bird Count (CBC) sponsored by the National Audubon Society. We have used the useful summary of CBC data published by Root (1988).

In the early 1960s CBCs began to be conducted by several local bird clubs across the state, including Omaha, Lincoln, Kearney, and Scottsbluff. In addition to reports of results of those forays in Audubon *Field Notes* and its successors, in 1964 reports began to be published annually in the NBR for those local organizations submitting them.

We recognize that CBCs have many limitations (Root 1988), including wide yearly variance in weather conditions that influence species' presence and numbers. The late-December period often samples late fall migrants rather than midwinter visitors. However, in cumulative value over a 30-year period, occurrence trends for most reported species, including their relative abundance, are revealed with some reliability.

USE AND TREATMENT OF DATA

Seasonal terms used are "spring," "summer," "fall," and "winter." All denote nonpermanent occurrence.

Spring and Fall

"Spring" is the period in which individuals of a species are moving from winter range to summer range; "fall" is the period of the corresponding return movement to winter range. Such species are usually referred to as "migrants." Species that do not normally migrate are referred to as "residents"; they are the individuals that remain within the same limited area throughout their lives. Few Nebraska species are residents.

Summer

"Summer" is the period when sexually active individuals of a species are engaged in reproduction; such species are referred to as "breeders." In many species there are sexually immature individuals that may not complete migration to a

summer range north of Nebraska during spring and, along with failed breeders, may simply wander during the summer. In addition, recently fledged juvenile individuals of some species may wander considerably during late summer. Species whose summer occurrence in Nebraska fits this description are referred to as "visitors," a term that implies movement during the season rather than occupation of a fixed limited range.

Winter

"Winter" is the period when sexually mature individuals within a species are sexually inactive and usually begins after a fall migratory movement. Most of the species that winter in Nebraska are considered visitors, implying movement during the season in response to ecological conditions, although such movement is normally confined within a recognized winter range.

Arrival and Departure Dates

Using the published seasonal reports, arrival and departure dates from Nebraska were determined by listing chronologically all available dates. Reliable arrival and departure dates were chosen by finding an initial or final cluster of 3 dates, which preferably did not extend over more than 3–5 days. This method was used first, to the best of our knowledge, by Green and Janssen (1975). In some cases we present arrival or departure dates for various regions of the state if significant differences exist. Dates not derived from NBR Occurrence or Seasonal Reports are cited.

We consider undocumented dates that fall outside these clusters to be at least questionable; those that we believe to be reasonable are cited in the text as "earlier" or "later" dates, but all others are provided under the heading "Other Reports," solely in the interest of completeness. We consider dates listed there to be either reporting errors or errors of identification.

Relative Abundance

Using the same array of occurrence dates, we can also develop an estimate of relative abundance. Given the birding activity of the average birder, we have generally assumed that many seasonal reports represented but single sightings or the first of 2 or 3 weekend sightings by weekend birders in any given season. We argue that such birders would make their first sighting at a point at or approaching movements of the largest numbers for a species. Given many years of such records, a cluster of a third to a half of the records becomes apparent, often in a 10-day to 2-week period. We believe that this cluster of records would reflect dates of highest abundance, or what we would call the "peak migration period." This grouping of dates appears to be well correlated with recently accumulated high-count data, mentioned above. High counts reported for a species in this work are used only to reinforce our estimate of the peak migration period and cannot be used as any reasonable estimate of relative abundance for a species because the accumulated data from those reports are insufficient.

EXPLANATION AND ORGANIZATION OF SPECIES ACCOUNTS

Citation of Records

Throughout the species accounts, we have cited all reports of occurrence, whether sight reports or those involving tangible evidence (specimens, photographs, recordings), except for those found in the NBR Occurrence Reports (through 1992) and the successor Seasonal Field Reports. These reports generally (but not always) are in the form "26 Feb 1997 Douglas Co," and some from combined counties appear as "26 Feb 1997 Douglas-Sarpy Cos."

Geographic Conventions

Throughout, we have used the terms "north," "south," "east," and "west" to refer to regions of the state (maps 3, 4). The term "central" combines "north" and "south." We have also on occasion used the terms "southwest," "southeast," "northwest," and "northeast." "Southwest" generally refers to those counties in the south that are in close proximity to both the Col-

orado and Kansas borders and includes at least Dundy, Chase, and Hitchcock Counties. "Southeast" includes those counties in the east that lie south of the Platte River. "Northwest" refers to Sioux, Dawes, Box Butte, and Sheridan Counties. "Northeast" refers to the counties in the east that lie north of the Elkhorn River, except for Washington County.

Elements of the Species Accounts

Each species account is divided into sections:

Species Name

The species name includes the common name and scientific species name as designated by the American Ornithologists' Union (1998).

Status

The Status section briefly and concisely summarizes the seasonal presence, expected frequency of occurrence in that season, relative abundance, and geographic distribution of a species in the state. For example, "common regular migrant west" qualifies seasonal status ("migrant") with descriptors for relative abundance ("common"), frequency of occurrence (in a 10-year period) ("regular"), and geographic distribution ("west").

In contrast with the often-used convention among birders of blending relative abundance and frequency of occurrence descriptors (abundant - common - uncommon - rare - casual - accidental), we have separated this hierarchy into two hierarchies. We believe this will minimize confusion among readers who are not field birders. The hierarchies are:

Relative Abundance
Abundant
Common
Fairly Common
Uncommon
Rare
Frequency of Occurrence
Regular
Casual
Accidental

Seasonal relative abundance descriptors have always been problematic for birders as well as for those writing about birds. Our intent is to use these terms to indicate the level of probability that an active observer will see these species in any given season and in what numbers. These terms are merely a guide and may in a few cases be applied somewhat arbitrarily. "Common" species are those that will be seen in numbers on almost every field trip without special effort but which generally do not occur in large flocks; "abundant" species differ from "common" species in that they occur in large flocks. "Fairly common" species are generally seen on every field trip but only in small numbers and with some effort. "Uncommon" species are those not seen on every field trip but can be expected several times during the season. "Rare" species often require considerable effort to find and generally occur in small numbers, often no more than one in a season, and in 1 or 2 years out of 10 may not be seen at all. "Rare, regular" suggests a species that is present somewhere in that region during that season but in numbers so low that only extreme birding efforts would reveal its presence. We have used the qualifier "local" in combination with numerical terms to indicate that the stated level of abundance is reached only in certain limited locations.

Frequency of occurrence is described as "Regular" when a species can be expected to occur in a region each year during that season or "Casual" when it has been documented for that season at least twice and for which a reasonably predictable pattern of occurrence can be discerned based on a species' entire documented distributional record. It is treated as "Accidental" when the species has occurred only once and when repeat occurrence cannot reliably be predicted. Species whose frequency of occurrence is less than accidental fall into an additional two categories, "Extirpated" and "Extinct." "Extirpated" species are those that at one time in the past were "Regular" but no longer occur in a region or in the state. "Extinct" species are those that no

longer occur anywhere. Species in the frequency of occurrence categories regular, casual, accidental, extirpated, and extinct are considered part of the "State List," or "The Official List of the Birds of Nebraska" according to the NOU Records Committee; their occurrence in Nebraska has been acceptably documented. The most recent State List was published in 1997 (NBR 65:3).

Two additional frequency of occurrence descriptors are used: "Hypothetical" and "Appended." Species in these categories are not considered part of the State List. "Hypothetical" species are those for which evidence for occurrence in Nebraska is suggestive but not acceptable beyond a reasonable doubt. These species are included in taxonomic order within the main body of the species accounts. Appended species, listed in Appendix 3, are those whose occurrence in the state is not at all likely or for which Nebraska records are likely erroneous.

Comments

In the case of species considered Accidental, Extirpated, Extinct, Hypothetical, or Appended, all pertinent information is presented under the heading "Comments," and no other headings are used. For Regular and Casual species, this heading is used only to present information not readily applicable to the other sections described below.

Documentation

The Documentation section cites the first tangible evidence for occurrence of each species in the state. In most cases a specimen is available, usually held in a major collection. For many recent additions to the State List, photographs, sound recordings, or written descriptions are the best evidence available.

Taxonomy

The Taxonomy section is used to discuss items relating to multiple taxa, morphotypes, or hybrids that have been reported in Nebraska. We discuss subspecies when there is evidence that more than one subspecies occurs. Some of these subspecies can be identified in the field, but many cannot. In numerous cases evidence confirming the presence of more than one subspecies can be deduced from disparate migratory arrival dates in different parts of the state, usually eastern Nebraska in comparison with the Panhandle. Several species occur in more than one color morph; these are discussed in this section. Of significant importance in Nebraska is the presence of hybrid zones, where range overlap between western and eastern populations has resulted in hybridization. This phenomenon is of critical importance to taxonomists, and we have attempted to present all available Nebraska information on these cases.

Distribution and Ecology

We consider the Distribution and Ecology section of key importance to this work. In this section we describe the relationship between the distribution of each species within the state as it is related to habitat used. Nebraska has a diverse range of habitats, and this is reflected in the uneven distribution, both spatial and numerical, of virtually every avian species in any given season. In addition to our own experience, supporting evidence is drawn from others and from the published literature.

The occurrence data for each species are presented under seasonal headings in the order Spring, Summer, Fall, and Winter. For Resident species, only Summer and Winter are generally used. The organization of seasonal accounts follows a loose pattern, beginning with dates of occurrence and related information and followed by additional pertinent data. We have included CBC, Breeding Bird Survey (BBS), and U.S. Fish and Wildlife Service (USFWS) bird-banding data where useful in explaining species occurrence and relative abundance phenomena not otherwise revealed in the NBR records. These data are used to indicate general trends or patterns only, and it is not our intent to be completely accurate in a statistical sense. For this reason we have not analyzed the entire database available for Nebraska CBCs and BBSs but have used data from time periods that we believe provide conclusions that would not be significantly changed by use of the entire database. Most egg and nestling date data are derived from Cornell nest cards.

Finding

In the Finding section we include tips that we hope will help birders in their search for specific species. We have noted at least one good location and the time of year to search.

Species Accounts

GAVIIFORMES

Gaviidae (Loons)

Red-throated Loon

Gavia stellata

STATUS

Rare casual spring migrant central. Rare casual fall migrant east, accidental elsewhere.

DOCUMENTATION

Photograph

Fall, about 1899, Frontier Co (Swenk 1933b).

DISTRIBUTION AND ECOLOGY

This species occurs mainly as a fall migrant on lakes and reservoirs, mostly in the east. Most apparently migrate directly from breeding areas to the coasts.

Spring

There are 10 reports, only 2 with details:

8 May 1998 basic adult L McConaughy (Dinsmore 1999c)

16 May 1999 molting to alternate L McConaughy (SJD)

Undocumented reports range 6 Apr–13 May: 6 Apr 1897 Omaha, 17 Apr Omaha, 22 Apr Omaha, 27 Apr Omaha, 29 Apr 1989 Sheridan Co (AB 43:500), 29 Apr–7 May 1961 Douglas Co, and 13 May 1972 Offutt Base L, a bird in "breeding plumage," likely the same bird seen the same day nearby in Iowa (NBR 40:85).

Documented records from surrounding states are few, 14 in the period 7 Mar–18 May.

Fall

There are 8 documented records, these in the period 30 Oct–25 Nov and all but 1 in the east. Fall 1996 saw an unprecedented influx of Red-throated Loons, accounting for 4 of the 8 records.

Fall ca. 1900 (cited above)

30 Oct–3 Nov 1996 Branched Oak L (JS, WRS; Brogie 1997)

31 Oct–3 Nov 1996 Stagecoach L (WRS; Brogie 1997)

31 Oct 1997 Louisville Lakes (TA; Brogie 1998)

3 Nov 1996 Bluestem L (WRS; Brogie 1997)

5 Nov 1995 Branched Oak L (JGJ; Gubanyi 1996c; Brogie 1997)

12 Nov 1996 Wehrspann L (LP, BP, JGJ; Brogie 1998)

19–25 Nov 1993 Standing Bear L (BJR, WRS; NBR 62:3; Gubanyi 1996a; Brogie 1998)

Reports that fit this pattern are 2 Nov 1975 Lancaster Co, 3 Nov 1976 DeSoto Bend NWR, 3 Nov 1996 Johnson L (NBR 64:108), 19–29 Nov 1960 Douglas Co, 24 Nov 1974 Lancaster Co, 27 Nov 1976 Sarpy Co, and 2 Dec 1973 Branched Oak L. Undocumented reports are 22–23 Sep 1930 (5) Douglas Co and 28 Sep 1894 Sarpy Co.

Other Reports

22 Jun 1996 L Ogallala, a report not accepted by the NOURC (Brogie 1997); 30 Aug (reported as 3 Sep) 1980 (8, likely Double-crested Cormorants) DeSoto Bend NWR.

FINDING

Most Red-throated Loons have been observed on large lakes and reservoirs such as those in Lancaster and Douglas Cos. Best time is late Oct–early Nov.

Pacific Loon

Gavia pacifica

STATUS

Rare regular fall migrant statewide. Rare casual spring migrant statewide. Rare casual summer visitor west. Rare casual winter visitor L McConaughy.

DOCUMENTATION

Photograph

9–16 Nov 1971 Sarpy Co (Bray and others 1986).

TAXONOMY

The Pacific Loon was formerly considered a North American subspecies of the Arctic Loon, *Gavia arctica*, but it is now considered to be a distinct species (AOU 1998). We consider Nebraska records to be of *pacifica*, as indicated by Mollhoff (1987), at least for individuals in transitional or alternate plumage, although the slight possibility exists that *arctica* could occur in Nebraska.

DISTRIBUTION AND ECOLOGY

All individuals have been observed on larger lakes and reservoirs, with no apparent regional preference, except in summer, when western reservoirs are preferred.

Spring

There are 2 records.

3 Apr 1998 Zorinsky L (NR)

12–16 May 1999 basic L Ogallala (SJD)

Summer

There are 5 records, all but 1 from the western part of the state and all documented:

10 Jun 1982 L Ogallala, "an excellent description was submitted" (DTW; AB 36:922), although the description is no longer extant (WRS)

14 Jun–6 Aug 1995 L McConaughy, 2 birds, 1 in alternate plumage and the other in first alternate (BP, LP; Gubanyi 1996c; SJD, pers. comm.)

15–22 Jun 1996 L Ogallala, juvenal/first basic (JGJ, WRS, JS; Brogie 1997)

1–4 Jul 1997 Offutt Base L, alternate (BP, LP; NBR 65:103; Brogie 1998)

19 Jul 1979 Box Butte Res, alternate (Rosche 1982; Bray and others 1986)

Fall

8, 16, 19 Oct→27 Nov, 6, 9 Dec

There are about 30 reports, most documented, and most in the short period 2–24 Nov. Earliest documented records are 8–19 Oct 1980 L Minatare (FZ) and 16 Oct 1999 L Alice (SJD) and latest are 6 Dec 1997 Swanson Res (SJD; Brogie 1998) and 9 Dec 1986 L North (AB 41:297). There is an earlier undocumented report 18 Sep 1982 L Minatare.

Fall occurrences in Nebraska are consistent with records in neighboring states, suggesting that some fall migrants drift south inland from their arctic breeding grounds en route to wintering areas along the Pacific Coast. Spring migrants presumably move up the Pacific Coast and fly a more direct route inland to breeding areas.

High counts include 4 in Scotts Bluff Co 18 Nov 1999 and 3 at L Minatare 13 and 18 Nov 1999.

Winter

There are 2 records of birds attempting to overwinter, both at L McConaughy.

2–15 Jan 1998 first basic L McConaughy (Dinsmore 1999c)

9 Jan 1999 first basic L McConaughy (Dinsmore 1999c).

COMMENTS

A specimen collected prior to 1900 in Frontier Co (Bruner and others 1904) and identified as this species was at one time in the Rees Heaton collection. It was later found to be a Common Loon by Oberholser (Swenk 1918b; Bray and others 1986). Another specimen, reportedly collected in Boone Co in the 1930s (NBR 57:86), is also a Common Loon (JGJ).

FINDING

Large lakes and reservoirs anywhere in the state are likely to attract Pacific Loons. Best time is early to mid-Nov.

Common Loon

Gavia immer

STATUS

Fairly common regular spring and fall migrant east, becoming uncommon west. Locally uncommon regular summer visitor west, becoming rare casual east. Rare casual winter visitor statewide.

DOCUMENTATION

Specimen

UNSM ZM7652, 30 Oct 1919 Lancaster Co.

DISTRIBUTION AND ECOLOGY

This diving species prefers large, open bodies of water, including impoundments as well as natural lakes. Increased construction of impoundments in the past 30 years may be responsible for an increased frequency of records. However, most large impoundment development has been limited to the eastern quarter of the state, which may account for a disproportionate number of eastern sightings. Records for the central part of the state are highly irregular, which may be explained by the relative absence of impoundments and natural lakes near populations of birders. One report implies that migrants may utilize rivers (NBR 30:33).

Spring

21, 25, 27 Mar→29, 30, 31 May (east)

5, 7, 9 Apr→summer (west)

Most migrants pass through the east, with a peak in Apr and early May. Arrival is later in the west. Late dates in the west are generally of immatures, which linger and sometimes summer. There are early dates in the west: 16 Mar 1944 Logan Co, 18 Mar 1975 Lincoln Co, 28 Mar 1993 Deuel Co, and 28 Mar 1999 L McConaughy.

High counts include 14 at L McConaughy 26 Apr 1991 (Rosche 1994a), 11 there 14–15 May 1998, and 8 there 12 May 1996.

Summer

In the 1990s this species has become regular in summer on western reservoirs, as in the Keith Co area where there are "many summer records of non-breeding individuals" (Brown and others 1996). In the east, however, it is only casual in summer. Most summering birds are in subadult plumages, suggesting either incomplete spring migration by some nonbreeding birds or some tendency for their wandering away from traditional breeding areas that are farther north.

Fall

16, 17, 20 Sep→5, 7, 8 Dec

The fall migration is apparently leisurely as in spring, with occurrence dates evenly distributed from the latter third of Sep into the first third of Nov. Early dates in the west are of summering birds, but in the east such reports are few, including 7 Sep 1985 Pierce Co and 11 Sep 1982 Douglas-Sarpy Cos. Individuals may linger into Dec in the south and west on open water, especially on reservoirs in the North Platte Valley, with reports as late as 28 Dec 1932 at Harlan Co Res (NBR 1:88), 1 Jan 1952 at L McConaughy (NBR 11:10), and 2 Jan 2000 (2) at L McConaughy (WRS, LR, RH).

High counts include 27 at Harlan Co Res 1 Nov 1998, 22 at L Ogallala 7 Nov 1998, and 19 at Pawnee L 11 Nov 1998.

Winter

Midwinter records are few, presumably due to a lack of open water. These are 8 Feb 1991 Douglas-Sarpy Cos, 15 Feb 1995 North Platte NWR, and 24 Feb 1995 Cass Co (FN 49:162), the latter possibly a very early migrant.

FINDING

Look for migrant Common Loons on large lakes and reservoirs Apr–May and Oct–Nov. They are usually seen as solitary individuals, but occasionally small groups of 3–5 may be seen.

Yellow-billed Loon

Gavia adamsii

STATUS

Rare casual fall visitor east and central.

DOCUMENTATION

Photograph

17–23 Nov 1996 Branched Oak L (JGJ, WRS, JS; Brogie 1997, 1998).

DISTRIBUTION AND ECOLOGY

This species has been regularly reported on larger reservoirs on the Great Plains in recent years as a fall and winter visitor.

Fall

There are 2 records.

17–21 Nov 1996 first basic Branched Oak L (cited above)

8 Aug–18 Oct 1998 first alternate L McConaughy (Brogie 1999; m.ob.)

FINDING

As with other loons, this species should be looked for on larger reservoirs, especially late in the migration period for loons, mid to late Nov.

PODICIPEDIFORMES

Podicipedidae (Grebes)

Pied-billed Grebe

Podilymbus podiceps

STATUS

Common regular spring and fall migrant statewide. Common regular breeder Sandhills, fairly common elsewhere. Rare casual winter visitor Keith Co.

DOCUMENTATION

Specimen

UNSM ZM6004, 20 Apr 1890 Lancaster Co.

DISTRIBUTION AND ECOLOGY

Breeding birds prefer marshes with open water and considerable emergent vegetation. The largest breeding concentration is found in Sandhills marshes, and numbers are fewest in the south and east. Migrants are found on lakes and reservoirs, as well as in natural marshes, on temporary ponds, and in water-filled ditches.

Spring

27, 27, 28 Feb→summer (east)

24, 26, 30 Mar→summer (west)

Although Pied-billed Grebes have been recorded in every month, peak migratory movements are detectable. Good concentrations of migrant grebes may be observed late Mar through Apr; early migrants arriving coincident with the ice thaw on lakes and ponds may occur as early as Feb. There is an early westerly report 3 Mar 1962 Lincoln Co.

High counts are 100 at Funk Lagoon 30 Apr 1994 and 86 at Zorinsky L 24 Apr 1996.

Summer

Breeding records are scattered throughout the state (Ducey 1988). Breeding can be expected in any permanent marsh with emergent vegetation. The largest summer concentrations are found in the extensive Sandhills marshes of Garden, Sheridan, and Cherry Cos. Drainage of wetlands and fluctuating water levels reduce breeding

populations, such as at Keystone L (Brown and others 1996).

High counts include 143 at Crescent L NWR 12 Aug 1997.

Fall

summer→4, 6, 6 Dec

The fall migration tends to be more leisurely, beginning in mid to late Aug and continuing until late Nov, with stragglers until freeze-up. Cold fronts may bring large concentrations, such as the 720 at DeSoto Bend NWR 23 Sep 1994 (see below). There are later reports 13 Dec 1986 Lancaster Co, 18 Dec 1998 Harlan Co Res, 20 Dec 1998 (5) Douglas Co, 21–22 Dec 1995 Sutherland Res, 26 Dec 1998 Omaha, 31 Dec 1984 Dakota Co, 31 Dec 1993 Offutt Base L, "W" (= 31 Dec) 1987 Lancaster Co, 1 Jan 1983 Douglas-Sarpy Cos, and 2 Jan 2000 (4) L McConaughy.

High counts include 860 at L Babcock 20 Sep 1998; 720 at DeSoto Bend NWR 23 Sep 1994, when many flocks of 20–70 birds were noted after a strong cold front; and 165 at L Ogallala 10 Sep 1977 (Rosche 1994a).

Winter

Rosche (1994a) indicated that "during most winter seasons a few remain on the open waters of the Lewellen marshes, the Kingsley Dam spillway, and elsewhere." One was at L Ogallala 21 Dec 1995–20 Jan 1996, and possibly the same individual was there during winter 1997–98 (SJD). Five were on Grove L, Antelope Co, 7 Feb 1999 (MB). In general, individuals may remain in the state until lakes and marshes freeze.

FINDING

This common small grebe may be found on any standing body of water. Migration peaks are Apr and late Sep–early Oct. Numbers of individuals are greatest during the spring migration.

Horned Grebe

Podiceps auritus

STATUS

Fairly common regular spring and fall migrant

statewide. Hypothetical breeder. Rare casual summer visitor Sandhills. Rare casual winter visitor central.

DOCUMENTATION

Specimen

UNK Olson #3, 16 Apr 1910 Alda, Hall Co (Bray and others 1986).

DISTRIBUTION AND ECOLOGY

Migrant Horned Grebes have been sighted on lakes and impoundments and occasionally in large marshes. During the breeding season birds frequent shallow lakes with perimeters that support extensive emergent marsh vegetation.

Spring

12, 13, 13 Mar→3, 4, 5 Jun

Migrants are most commonly first seen during the latter third of Mar and the first third of Apr. There are earlier reports 21 Feb 1970 Adams Co, 28 Feb Keith Co (Rosche 1994a), 3 Mar 1967 Adams Co, and 8 Mar 1999 L McConaughy (SJD).

High counts include 236 at L McConaughy 5 Apr 1998, 47 there 28 Mar 1999, and 15 at Funk Lagoon 1 Apr 1995.

Summer

There are no recent breeding records. Bruner and others (1904) indicated that Trostler had "found it breeding in the alkali lakes of northern Cherry Co," and Wolcott had found a pair with a newly made nest in the same region 6 Jun 1903. A 1963 report for Lincoln Co (Wensien 1964) is undocumented. The contemporary southernmost breeding noted for this species is northeast South Dakota, although it bred south to south-central South Dakota in the 1880s and 1890s (SDOU 1991).

There are several summer reports for the western Sandhills, but no nesting evidence has been noted. These are "s" (= 30 Jun) 1966 McPherson Co, "s" 1967 Garden Co, 17 Jul 1978 Garden Co, 22 Jul 1980 McPherson Co, and 28 Jul 1994 (2) Garden Co.

Fall

3, 4, 5 Sep→19, 19, 20 Dec

Migrants are normally first seen in mid-Sep and depart by late Nov. There is an earlier report 27 Aug 1996 Scotts Bluff Co and later reports at L McConaughy 23 Dec 1995, 31 Dec 1998–2 Jan 1999, and 2 Jan 2000.

Rosche (1994a) stated that this species "occurs sporadically in Western Nebraska," but when it appears it does so in "quite large numbers." For example, Zeillemaker (pers. comm. RSS) related that at Crescent L NWR during 1978 "fifteen appeared 15 Sep, six were present 30 Sep, 50 on 7 Oct, 55 on 23 Oct, and 60 on 7 Nov. At L Minatare 70 had arrived by 22 Sep, 190 were there on 3 Oct, 266 on 20 Oct, 155 on 1 Nov and 4 on 15 Nov."

High counts include 266 at L Minatare 20 Oct 1978, 145 at North Platte NWR 8 Nov 1997, and 133 at L Minatare 6 Nov 1999.

Winter

This species has been reported in midwinter on large reservoirs, as long as they remain ice-free. There are reports 12 Jan 1991 Lincoln Co and 29 Jan 1995 Johnson Res.

COMMENTS

Swenk (Notes after 1925) stated that the 16 Apr 1910 bird cited above was actually collected at Inland 6 Apr 1916 and was labeled HMM 2489 and that Horned Grebe was unrepresented in the Olson and Black collections. Swenk apparently confused these specimens, however, as the 16 Apr 1910 specimen cited above is indeed in the Olson collection (Bray and others 1986).

FINDING

Lakes and reservoirs during the migratory periods, especially Apr and Oct, are the most likely places to see Horned Grebes.

Red-necked Grebe

Podiceps grisegena

STATUS

Hypothetical in spring. Accidental in summer. Rare casual fall migrant statewide.

DOCUMENTATION

Photograph

24–26 Oct 1995 Lewis and Clark L, Knox Co.

DISTRIBUTION AND ECOLOGY

This species nests in northern prairie marshes and potholes from South Dakota northwestward through the prairie provinces into Alaska (Palmer 1962); breeding is irregular in South Dakota (Whitney and others 1978) and is limited to the extreme northeast (SDOU 1991). The species displays chiefly longitudinal migration, overwintering on the Atlantic and Pacific Coasts, and is limited to cold-water zones (Root 1988). These include most of the Pacific Coast south to Baja and the northeast Atlantic Coast, straggling south to Florida.

Spring

There are no documented records. However, the following undocumented reports are within the expected migration period (SDOU 1991): 7 Apr 1927 Omaha (Swenk 1933c), 10 Apr 1963 Lincoln Co, 17 Apr 1984 Dakota Co (Bray and others 1986), 19 Apr 1966 Gage Co (Fiala 1970), 7 May 1967 Scotts Bluff Co, 12 May 1992 Scotts Bluff Co, and 13 May 1979 Douglas Co (NBR 47:42).

Summer

The only documented report is of one that could be judged a late migrant:

16 Jun 1985 Lancaster Co (Garthwright 1985b; Mollhoff 1987)

There is a report without details of a pair near Bingham 9 Jun 1947 (Fichter 1947).

Fall

There are 13 documented records and 10 additional undocumented reports in the period 26 Sep–4 Jan, most 24 Oct–24 Nov, indicating peak occurrence. Earliest are 26–27 Sep 1996 Sutherland Res (Dinsmore 1997b; Brogie 1997); 27 Sep 1997 Willow L, Brown Co (LR, RH; Brogie 1998); and 24 Oct 1917 northeast Nebraska (Bray and others 1986). Latest are 7 Dec 1997–4 Jan 1998 (1–2) L McConaughy (SJD), 21 Dec 1995 L McConaughy (Dinsmore 1996b; Brogie 1997), 1 Jan 1999 L McConaughy (SJD), and 2 Jan 2000 (2) L McConaughy (SJD).

COMMENTS

Earlier literature referred to this species as Holboell's Grebe. A specimen purported to be this species in the Rees Heaton collection was a Horned Grebe (Swenk 1933c).

Other Reports

One reported at Offutt Base L 11 Nov 1981 was identified by others as a Western Grebe (TB, BJR, pers. comm. WRS). A report of 6 juveniles in Hooker Co 9 Sep 1993 was not accepted by the NOURC (Gubanyi 1996b). A 13 Mar report with no details (Johnsgard 1980) may be a typographical error for the 13 May 1979 report above.

FINDING

Red-necked Grebes should be looked for on large lakes and reservoirs; best time is late Oct.

Eared Grebe

Podiceps nigricollis

STATUS

Common regular spring and fall migrant central and west, fairly common east. Common regular breeder north and west. Rare casual summer visitor south and east.

DOCUMENTATION

Specimen

UNSM ZM6006, 25 Sep 1927 Lancaster Co.

DISTRIBUTION AND ECOLOGY

During the breeding season most are concentrated in marshes associated with Sandhills lakes, although singles may also be seen on large reservoirs in the central part of the state. During the migratory period Eared Grebes frequent lakes and reservoirs statewide.

Spring

11, 13, 14 Mar→3, 3, 4 Jun

Last dates above are away from the breeding range. The main migratory movement occurs mid-Apr through early May. Numbers of individuals frequenting lakes are larger to the west of the eastern border of counties. There are early spring reports 18 Feb 1974 Howard-Hall Cos, 21 Feb 1970 Adams Co, 27 Feb(?), and 9 Mar 1972 Lancaster Co and late reports 8 Jun 1967 Adams Co, 12 Jun 1991 Kearney Co, and 19 Jun 1988 Pierce Co.

High counts include 450 at Crescent L NWR 12 May 1999, 230 there 9 May 1997, and 162 at Harvard Marsh 18 Apr 1998.

Summer

Breeding populations may be found on many Sandhills lakes that have extensive emergent vegetation, but it may not be possible to predict that individual lakes maintain breeding colonies. In 1965 Sharpe surveyed Smith L at Crescent L NWR and observed about 60 nests in a loose colony (Sharpe and Payne 1966). Subsequent years revealed far fewer individuals. Following restoration of another Sandhills lake in 1970 (Roundup L), large numbers bred on and utilized that lake (John Wilbrecht, pers. comm. RSS).

The eastern extent of regular breeding appears to be in the central Sandhills, as Ducey (1988) cited no records east of Cherry Co, and Blake and Ducey (1991) gave no records for Holt Co, calling it a "possible summer resident" only.

There are few reports away from the breeding range in summer. Ducey (1988) cited only 3 reports of breeding south of the Platte River and east of the Sandhills: Clay Co in 1915, Douglas Co prior to 1896, and Madison Co prior to 1901. Of most interest are reports from the Rainwater Basin and Lincoln Co, where breeding has not yet been recorded. Birds were present in Phelps Co 1 Jun–8 Aug 1991, in Adams Co 30 Jun 1984 and 23 Jun–4 Jul 1966, at Johnson Lagoon (3) 6 Jul 1997, and at Funk Lagoon 14 Aug 1994. There are also reports from Lincoln Co 14 Jun 1972, 15 Jun 1989, 30 Jun 1981, 24 Jul 1975, and 12 Aug 1988. Rosche (1994a) cited a report without details of breeding on Keystone L in 1977 and the presence of a pair on Oshkosh SL, "a likely breeding location," 17 Jul 1988. There is an undocumented report 28 Jul 1974 Douglas-Sarpy Cos.

Nests are composed of small, floating masses of dead stems of aquatic vegetation. When approached, adults quickly slip into the water and cover the eggs with vegetation. On 5 Jun Sharpe and Payne (1966) found clutches of 2–5 eggs. On occasion large concentrations of nests occur, such as the 90–100 at Lakeside 12–14 Jun 1998.

High counts include 932 at Crescent L NWR 10 Jul 1997 and 436 there 13 Jun 1996.

Fall

5, 5, 9 Sep→7, 8, 9 Dec

Early dates above are away from the breeding range. There are earlier reports 20 Aug 1970 Perkins Co and 23 Aug 1969 Lancaster Co and later reports of 2 at Sutherland Res 12 Dec 1997 (SJD) and 2 at L Ogallala 12–14 Dec 1998, which apparently remained until 2 Jan 1999 (SJD).

Although a good breeding population exists in the state, a clear fall migratory movement is evident, peaking in Oct.

High counts include 487 in the western Panhandle 16 Oct 1999, 69 at Sutherland Res 7 Nov 1994, and 58 at North Platte NWR 9 Oct 1996.

FINDING

Eared Grebes utilize natural lakes and impoundments during migration. During the breeding season Sandhills lakes with scattered and extensive emergent vegetation will often yield results.

Western Grebe

Aechmophorus occidentalis

STATUS

Common, locally abundant, regular spring and fall migrant west, becoming uncommon east. Locally common regular breeder north and west. Rare casual summer visitor south and east. Rare casual winter visitor L McConaughy area.

DOCUMENTATION

Specimen

UNSM ZM14365, summer 1982 Keith Co.

TAXONOMY

This species and Clark's Grebe, *A. clarkii*, were formerly considered color morphs of a single species, but a high degree of assortative mating has been noted (Ratti 1979; Nuechterlein 1981; Nuechterlein and Storer 1982), leading to elevation of the morphs to species status (AOU 1983). The two species are highly sympatric, although Clark's Grebe is more predominant southward (Sibley and Monroe 1990).

DISTRIBUTION AND ECOLOGY

During the breeding season groups are concentrated on lakes with marshy edges, primarily on large, shallow, western Sandhills lakes; individuals may also occasionally utilize nearby reservoirs and lakes, such as the recently established breeding population at the west end of L McConaughy (Rosche 1994a). During the migratory periods individuals may be seen on any large lake or reservoir.

Spring

28, 28, 30 Mar→summer (west, north)

10, 14, 14 Apr→9, 12, 14 Jun (south, east)

There are earlier reports 16 Mar 1962 Lincoln Co and 21 Mar 1972 Brown Co.

Although eastern records for spring and fall migrants may be interpreted as extralimital wandering, the existence of similar records in Kansas (Thompson and Ely 1989) and more rarely in Missouri (Robbins and Easterla 1992), as well as on the southern Great Plains, suggests that some of the eastern records may represent individuals that have drifted south from breeding areas located in northeastern South Dakota and western Minnesota, accounting for what appear to be anomalous eastern and southern records. The increase in reservoirs in eastern Nebraska has also added habitat suitable for wandering individuals. Persistent overwintering records for the Gulf Coast for a very small population of individuals and regular overwintering in western Texas and New Mexico reinforce this hypothesis.

At Crescent L NWR spring migrants arrived nocturnally in small groups as early as 9 Apr and as late as 22 Apr over a 3-year period (Zeillemaker, pers. comm. RSS).

High counts include 14 500 at L McConaughy 19 Apr 1997, about 5000 there 12 May 1996, and about 5000 there 12 May 1999.

Summer

The largest breeding concentrations may be observed on Crescent, Island and on Smith Lakes at or near Crescent L NWR, where the breeding population was 60–90 birds around 1980 (Zeillemaker, pers. comm. RSS), and on larger lakes north of the refuge in Garden Co. Smaller breeding populations may be found on larger lakes in Sheridan, Grant (Wampole and Fichter 1946; Rosche 1982), Arthur (RSS), and extreme southwest Cherry Cos, as well as on L McConaughy and Keystone L in Keith Co, where breeding was first observed in 1993 (Rosche 1994a). At L McConaughy 20–25 adults with young were present 16 Jul 1994. In extreme southeast Sheridan Co 12–15 broods were on a lake just east of Bingham in 1994 and 1995. Ducey (1988) cited reports of nesting from McPherson Co in 1969 and Lincoln Co in 1972 (Bennett 1970, 1973a). Blake and Ducey (1991) stated that breeding occurs at Carson L, northwest Garfield Co, the most easterly breeding location in Nebraska.

There are several reports in late Jun and early Jul away from breeding areas, notably in Scotts Bluff and Lincoln Cos, presumably prebreeders that did not travel to breeding lakes or wandering nonbreeders. There is usually a sizeable number of Western Grebes on L McConaughy during the summer, many of which presumably are not breeding.

Elsewhere there are reports 21 Jun 1991 Holt Co, 12–13 Jul 1996 Antelope Co, 16 Jul 1991 Phelps Co, 22–24 Jul 1989 Knox Co, 29 Jul 1997 Branched Oak L, 8 Aug 1975 Douglas-Sarpy Cos, and 17 Aug 1997 Funk Lagoon.

Fall

summer→6, 6, 7 Dec (west, central)

15, 19, 27 Sep→1, 2, 3 Dec (east)

Immediately as young birds are able to fly, there is a southward drift of family groups, apparently a forerunner to the massive buildup of numbers on L McConaughy, a major fall staging area for the species, where 35 000 were estimated 26 Sep 1999 (SJD). Some 20 000 were present 4–5 Oct 1997, as counted from a boat (SJD, JS, WRS, JGJ), and 1600 were still present 24 Nov (SJD, JS). As many as 700 were present as early as 1 Jul 1996, reaching a peak of 7000+ on 28–29 Sep. As many as 2000 were present there as late as 5 Nov 1994.

At L McConaughy and Sutherland Res a few birds linger late into Dec and may overwinter (see Winter). At North Platte NWR 2 remained until 12 Dec 1997. In the east one lingered at Offutt Base L until 13 Dec 1996.

High counts include 35 000 at L McConaughy 26 Sep 1999, 20 000+ there 4–5 Oct 1997, and 17 600+ there 18 Oct 1998.

Winter

A few birds have recently overwintered in the Kingsley Dam vicinity; first midwinter report there was of 2 at L Ogallala 5 Feb 1995, 1 wintered 1996–97, and in the mild winter of 1997–98 as many as 342 were present 3 Jan, 71 were there 15 Jan, and 14 remained 22 Feb (SJD). As many as 365 were at L McConaughy 1 Jan 1999, but only 1 remained through 19 Feb (SJD). At Sutherland Res first overwintering was in 1997–98, when 2 were noted 14 Jan (SJD).

A report of 2 wintering in Scotts Bluff Co 1994–95 (AB 49:162) is incorrect; 1 was there 3 Dec, but there was no indication that it wintered (WRS).

FINDING

Individuals determined to see breeding Western Grebes should prepare to travel to Crescent L NWR in spring and summer. Crescent L, just outside the extreme southern end of the refuge, has traditionally maintained 10 or more pairs of this large grebe. During late Apr and early to mid-May look for the spectacle of paired courtship dances across the water's surface. Incredible numbers can be seen on L McConaughy in late Apr–early May and especially in late Sep–early Oct.

Clark's Grebe

Aechmophorus clarkii

STATUS

Rare regular spring migrant west. Rare regular breeder west. Local rare regular summer visitor west. Rare regular fall migrant west, becoming rare casual east. Accidental in winter.

DOCUMENTATION

Specimen

UNSM ZM15749, 11 Jun 1986 L Ogallala (Labedz 1987).

TAXONOMY

See Western Grebe. A small *Aechmophorus* grebe (UNSM ZM5913) collected by Jim Tate on 1 Nov 1962 in Otoe Co was in the size range of a female Western or Clark's, but, lacking details as to gender of the specimen, Labedz was unable to determine which one it was. Morphological characters about the head conformed with *clarkii*, however (Bray and others 1986). The presence of white lores and an orange-yellow bill in adults is considered indicative of Clark's Grebe (Storer and Nuechterlein 1985); birds appearing to be intermediate in lore color were thought to be immatures or in winter plumage, as were the few individuals that appeared to be intermediate in bill color. Of interest, therefore, is the report of 2 possible hybrids at L McConaughy 27 Sep 1996 (SJD) that had bright orange bills and dark lores, although Ratti (1981) noted that bill color is "not as discrete as the facial pattern" and "alone should not be used to identify these species." Other reported hybrids were 1 at North Platte NWR 15 Aug 1997 (SJD) and 1–2 at L McConaughy 18 Apr–9 May 1998 (SJD).

DISTRIBUTION AND ECOLOGY

Most reports of this species are of birds in mixed flocks with Western Grebes. The only breeding locations known are the shallow western end of L McConaughy and Willy L in extreme southeast Sheridan Co. Migrants have been reported within the range of migrant Western Grebes, including reports from the east.

Spring

12, 19, 19 Apr→summer

There are earlier reports 27 Mar 1998 at Branched Oak L (js), 28 Mar 1999 at L McConaughy (sjd), and 5 Apr 1998 at L Ogallala (sjd). There are few reports prior to May, when nesting probably begins. The first spring report was 1 May 1987 in Cherry Co (AB 41:455), and it now is reported each year. The easternmost reports are 27 Mar 1998 at Branched Oak L (js) and 19 Apr 1995 at Funk Lagoon.

High counts include 39 at L McConaughy 15 May 1998 and 23 there 26 Apr 1997.

Summer

Breeding was first noted at L McConaughy in 1993, when Rosche and Rosche observed 2 adults feeding a newly hatched downy brood on 31 Jul (Rosche 1994a). In 1994 several adults with young were present Jun–early Aug. The only other breeding location reported is Willy L, near Bingham, where 3 adults with young on their backs were noted 12 Jun 1998 (WRS, js).

There are several additional reports during summer of birds at L McConaughy that are presumably part of the local breeding population: 4 Jun 1992, 9 Jun 1996 (2), 11 Jun 1986 (specimen cited above), 16 Jun 1996–13 Aug 1996 (high count 13), and 26 Aug 1989 L Ogallala (Rosche 1994a). Elsewhere, summer reports are widely scattered throughout the west: 3 Jun 1987 Sheridan Co (not accepted by the NOURC, Mollhoff 1989a), 5 Jun Crescent L NWR (AK), 14 Jun 1993 Garden Co (AB 47:1122), 15 Jun 1997 North Platte SL, 16 Jun–12 Jul 1996 North Platte NWR, 17 Jun 1994 Sheridan Co (with a suspected Western Grebe), 19–20 Jun 1991 Enders Res (NBR 59:93; AB 45:1133), 1 Jul 1988 southeast of Alliance in Morrill Co (Grenon 1991), 12 Jul 1990 Cherry Co (AB 44:1152), 2 Aug 1996 (5) Crescent L NWR, 3 Aug 1996 Dawes Co (FN 50:965), and 3 Sep 1993 Sutherland Res.

The only easterly reports are of singles at Branched Oak L 29 Jul 1997 (js) and Funk Lagoon 14–21 Jun 1998 (LR, RH).

Fall

summer→24, 27, 28 Nov

There are later records 6 Dec 1986 Gavin's Point Dam (Mollhoff 1989a) and 7 Dec 1997 (adult) L McConaughy (sjd; Brogie 1998). Migrants were noted at L Minatare 26 Sep–18 Nov 1999 (sjd).

Fall reports are fewer than in spring, possibly due to less observer persistence in finding these birds in this season. There are 2 eastern reports: 11–19 Oct 1997 Pawnee L (js, jw, jgj, mb) and 6 Dec 1986 (cited above).

High counts include 22 at L McConaughy 18 Oct 1998, 20 there 28 Sep 1996, 11 there 26 Sep 1999, and 11 there as late as 2 Jan 2000.

Winter

The first record of overwintering was of 1–3 present at L McConaughy 1997–98 (js, sjd). Three were on L McConaughy 1 Jan 1999, declining to 1 on 9–10 Jan (sjd). As many as 11 were still at L McConaughy 2 Jan 2000 (sjd).

FINDING

The most consistent location has been the west end of L McConaughy, specifically the Omaha Beach Access. In summer it is usually not difficult to find a Clark's among the many Westerns. Clark's Grebe has a single note call, whereas Western's is doubled (Storer 1965), although this may not be diagnostic in the northern parts of the range where Clark's are not numerous (Storer and Nuechterlein 1985; Eckert 1993).

PELECANIFORMES

Pelecanidae (Pelicans)

American White Pelican

Pelecanus erythrorhynchos

STATUS

Abundant regular spring and fall migrant statewide. Uncommon, locally common, regular summer visitor central and west, rare elsewhere. Locally rare regular winter visitor Keith and Lincoln Cos, rare casual elsewhere.

DOCUMENTATION

Specimen

UNSM ZM12393, 31 May 1959 Lincoln, Lancaster Co.

DISTRIBUTION AND ECOLOGY

American White Pelicans are attracted to large, shallow water bodies that are particularly productive and sustain an abundant fish population, although during migration flocks sometimes rest at grassy wetlands. Normally flocks of 10–100 birds are seen.

Spring

8, 9, 11 Mar→summer

There are earlier reports from locations where wintering is unknown: 15 Feb 1998 along I-80 at mile marker 162 in Lincoln Co, 17 Feb 1999 (48) Harlan Co Res, 21 Feb 1970 Adams Co, 22 Feb 1992 Kearney Co, and 25 Feb 1955 Adams Co.

Typically flocks do not arrive until lakes are entirely clear of ice, usually in early Apr, achieving peak numbers in mid-Apr. Although the major migratory movement is completed by mid-May, small groups or occasionally individuals may remain anywhere in the state throughout summer months.

High counts include 1400 at Wood Duck Area 12 Apr 1998, 1000 at Sacramento-Wilcox Basin 19 Apr 1994, and 1000 at Branched Oak L 12 Apr 1996.

Summer

Breeding colonies of White Pelicans are scattered throughout the northern Great Plains as well as in Wyoming and Colorado, but none have been established in Nebraska. One large colony exists just north of the state line at LaCreek NWR, which may be the source of flocks that utilize many large Sandhills lakes in Nebraska during the summer months; small wandering groups may also utilize large reservoirs such as L McConaughy and Merritt Res. Summer occurrences in the south and east are rare.

High counts include 76 in Otoe Co 21 Jun 1996, 53 at Merritt Res 16 Jul 1994, and 51 at Crescent L NWR 13 Jul 1995.

Fall

summer→23, 27, 31 Dec (away from Sutherland Res)

The fall migration is considerably more leisurely than the spring migration. Detectable migratory movements are first seen in late Aug and early Sep. Migration peaks in Sep–Oct, but small groups may remain until late Nov, such as 12 at Sutherland Res 28 Nov 1998, and there are several reports for late Dec and later (see Winter).

High counts include 2100 at L McConaughy 8 Aug 1998, 1234 at Calamus Res 7 Sep 1998, and 1000 in Buffalo Co 19 Sep 1997.

Winter

In recent years small numbers have lingered into winter; there were no Dec reports until 1977 and no Jan reports until 1991. Some lingering birds are injured, such as one at Johnson Res 19 Jan–4 Feb 1996, and a sick bird was in Cedar Co 6 Jan 1996, but most are not. All overwintering records are at Sutherland Res, which has a cooling pond that remains open throughout the winter. Singles were present there 15 Feb 1992 and 19 Dec 1992, but overwintering began in 1993–94 and has continued each year since, best count of survivors 4 in 1996–97.

Midwinter reports away from Sutherland Res include 1 Jan 1998 Branched Oak L, 21 Dec 1991–4 Jan 1992 Buffalo Co, 12 Jan 1991 Keith Co, and 2 Feb 1997 Harlan Co Res.

COMMENTS

Federal banding returns indicate that bands taken from dead birds found in Nebraska were placed on juveniles produced in breeding colonies from virtually every Great Plains state and Canadian province, including Colorado, Wyoming, Ontario, and Alberta, suggesting that American White Pelicans wander widely during their lives. However, of 101 band returns, 44 were banded in North Dakota.

FINDING

During migration any large lake or reservoir will attract pelicans. Large Sandhills lakes and reservoirs such as L McConaughy and Merritt Res may attract pelicans during the summer months. Flocks of pelicans travel widely, however, in search of an abundant prey source. They may utilize a lake for several days but presumably once having depleted the resource will move on to another lake and may not be seen at the former location again for several weeks. American White Pelicans less commonly will utilize sandbars in the Platte River as resting places during migration.

Brown Pelican

Pelecanus occidentalis

STATUS

Rare casual spring visitor. Hypothetical in fall.

DOCUMENTATION

Photograph

12 May 1991, near Fremont, Dodge Co (Grenon 1991; NBR 59:94).

DISTRIBUTION AND ECOLOGY

Brown Pelican frequents marine coastal habitats; the closest population to Nebraska is found along the Gulf Coast. Inland records of this species are known throughout the southern Great Plains, but occurrences are highly irregular and unpredictable (Palmer 1962).

Spring

There are 15 reports over a period of almost 160 years, beginning with Audubon's record of 1842 (Audubon and Coues 1960). All

sightings since 1912 have been in Apr–May (see Fall).

Documented records are:

spring 1912 Carter L, Omaha, specimen #254 in collection of the Northwest School of Taxidermy, Omaha, and diagnostic measurements published; specimen now lost (Bray and others 1986; Swenk 1934a)

12 May 1991 west of Fremont, Dodge Co (cited above)

21–23 May 1991 Missouri River, Dakota Co (Grenon 1991)

last week Apr–8 May 1992 DeSoto Bend NWR (Green 1992; NBR 60:69)

Undocumented reports are 10 Apr 1955 Keya Paha Co, 18 Apr 1955 Cherry Co, May 1930 (18) North Platte, 6 May 1952 Alexandria, 9 May 1937 (15) North Platte, 28 May 1977 near Milburn (Kieborz 1978).

Other reports include fall 1842 (Audubon and Coues 1960), fall 1819 or spring 1820 Washington Co (Swenk 1934a), and spring early 1890s Omaha (Swenk 1934a).

Fall

The few reports are undocumented. Two birds were killed at Lincoln 10 Jul 1872 (Swenk 1934a), and a male was collected at St. Paul 10 Oct 1885 and placed at the University of Iowa museum (Swenk 1934a). This specimen cannot now be located (Bray and others 1986).

FINDING

This species is possible around reservoirs, lakes, and river oxbows and on sandbars in Apr and May. It does not associate with American White Pelicans.

Phalacrocoracidae (Cormorants)

Neotropic Cormorant

Phalacrocorax brasilianus

STATUS

Rare casual summer visitor statewide.

DOCUMENTATION

Specimen

UNSM ZM14226, 2 Oct 1982 Sutherland Res (Wright 1983).

Summer

There are 6 records, all recent, 5 acceptably documented (NOURC). The sixth was of one at Branched Oak L 26 Mar 1998. Several recent records exist for Kansas (Thompson and Ely 1989), about 15 for Missouri, and other central Great Plains states (Robbins, pers. comm. WRS), suggesting that wandering individuals may drift northward from the Gulf Coast and on rare occasions into Nebraska. Careful observation of small cormorants should yield additional state records.

2 May 1998 first alternate Sutherland Res (Dinsmore 1999b)

20–30 May 1996 Chambers (Brogie 1997)

19 Jul 1995 Valentine NWR (Gubanyi 1996c; NBR 63:71)

4 Sep 1993 Hackberry L (Gubanyi 1996a)

2 Oct 1982 immature (TEL) Sutherland Res (cited above)

FINDING

Any group of cormorants should be checked for this species May–Oct.

Double-crested Cormorant

Phalacrocorax auritus

STATUS

Abundant regular spring and fall migrant statewide. Locally common regular breeder north and west. Rare casual summer visitor south and east. Rare casual winter visitor south and east.

DOCUMENTATION

Specimen

UNSM ZM7651, 1 Nov 1932 North Platte, Lincoln Co.

DISTRIBUTION AND ECOLOGY

During migration cormorants utilize medium-sized to large lakes and reservoirs and may occasionally frequent rivers such as the Platte and Missouri. Breeding colonies have been established on a number of Sandhills lakes, particularly those with nesting platforms erected for Canada Geese. Colonies have also developed in large, dead trees associated with these lakes, as well as dead trees found in the headwaters of reservoirs.

Spring

14, 15, 15 Mar→31, 31, May, 1 Jun

Late dates above are for areas away from the breeding range. Movement normally commences in late Mar and peaks in early to mid-Apr.

There are several late Feb–early Mar reports, which may be early migrants: 20 Feb 1991 Douglas-Sarpy Cos, 23 Feb 1958 Lincoln Co, 27 Feb 1954 Thayer Co, 28 Feb 1974 Douglas-Sarpy Cos, 2 Mar 1985 Polk Co, and 9 Mar 1947 Lincoln Co. Earlier reports are probably wintering birds (see Winter).

High counts include 3400 at Branched Oak L 27 Apr 1995, 1500 there 27 Apr 1997, and 1450 in Lancaster Co 18 Apr 1996.

Summer

Around 1900 breeding was unknown (Bruner and others 1904), but by the 1950s Rapp and others (1958) listed Scotts Bluff, Garden, Keith, Cherry, Grant, and Lincoln Cos as breeding localities. Ducey (1988) presented no breeding records prior to 1921 but added Sheridan Co to the list of Rapp and others. Breeding now occurs in Rock Co, where 29 nests were counted in 1988 (Bennett 1989), and in 1997 nesting was noted at L George (8 nests) and Twin Lakes (2 nests), both sites in Rock Co (JGJ). Probable breeding occurred at Swan L, Holt Co, in 1988 (Blake and Ducey 1991).

Major breeding colonies exist at Crescent L NWR and Valentine NWR. Any of the larger lakes of these refuges will be frequented by individuals during the summer months. Most southerly is a small but growing colony at the west end of Harlan Co Res, where 9 nests were noted in 1995 and at least 26 nests in 1996; as many as 70 birds were on nests in 1998. Nesting no longer occurs in the L McConaughy area, but when Kingsley Dam was under construction, the resulting dead trees attracted a peak of 228 nests and 1000 birds in the 1940s (Collister 1948; Rosche 1994a).

Individuals have been particularly successful in

colonizing the nesting platforms that have been erected on refuges for Canada Geese, displacing the geese in the process. One platform, surveyed by Sharpe on 5 Jun 1965 at Smith L, Crescent L NWR, supported 7 nests containing eggs and young as old as 1 week. Nests were strewn with the remains of bullheads, yellow perch, and tiger salamander tadpoles.

Nonbreeders, mostly immatures, are common throughout the north and west but are only casual in the south and east.

Fall

13, 13, 14 Aug→14, 18, 18 Dec (away from Sutherland Res)

Early dates above are for areas away from the breeding range. Most migratory activity is seen during the month of Oct. Stragglers have been reported in late Dec and Jan since 1990, as in the case of American White Pelican, and overwintering has occurred at Sutherland Res. Later reports include 20 Dec 1991 L McConaughy, 29 Dec 1993 Cass Co, 30 Dec 1994 Harlan Co Res, 2 Jan 2000 L McConaughy, and 1–2 through 8 Jan 1998 at Branched Oak L (see Winter).

High counts include 5000+ at Harlan Co Res 5–16 Oct 1999, 4000 there 20 Oct 1996, and 1500 there 8 Nov 1998.

Winter

Cormorants typically are not present during winter months because of ice cover, although during very mild winters when some lakes remain partially open, occasional individuals may be seen. At Sutherland Res, which has a cooling pond that remains open in winter, wintering occurred 1995–96, when 4 birds present 22 Dec dwindled to 1 by 13 Jan. Other reports from there are 17 Dec 1992, 20 Dec 1991, 1 Jan 1999, 8 Jan 1982, 5 Feb 1963, and 15 Feb 1992. Additional midwinter reports are 12 Jan 1992 Lancaster Co and 30 Jan 1972 at L McConaughy (Rosche 1994a).

FINDING

These large diving birds may be seen with regularity on any large lake or reservoir during the peak migratory periods Apr and Oct. Individuals may also perch on the bare branches of trees adjacent to these lakes, often spreading their wings out to dry. Best locations for seeing this species during summer months include Crescent L NWR and Valentine NWR.

Anhingidae (Darters)

Anhinga

Anhinga anhinga

STATUS

Rare casual spring and fall visitor central and east.

DOCUMENTATION

Specimen

HMM 21719, 20 Sep 1913, Wood River, Buffalo Co (Bray and others 1986; Swenk 1934a).

Spring

Of 4 reports, 2 are documented:

27–30 Apr 1975 near Wolbach (Faeh 1975)

2–5 May 1987 Fontenelle Forest (NBR 55:50; AB 41:455)

Two additional reports may be correct, as they fit the above pattern of occurrence, but are undocumented: 8 Apr 1978 Fontenelle Forest (NBR 46:68) and 4 May 1955 Hamilton Co. Swenk (1934a) noted that one purported to have been shot in the 1890s in the Omaha area may have been shot in Iowa or even Missouri.

Fall

There are 2 reports, 1 documented:

20 Sep 1913 (cited above)

The other report is of 1 present 3 days in late Oct 1976 at the Hamilton-Merrick Cos bridge.

COMMENTS

Scattered records exist for Kansas (Thompson and Ely 1989), where it is also considered a vagrant; it occurs casually during the spring and summer in southeastern Missouri but is accidental in northern and western Missouri (Robbins and Easterla 1992). The nearest breeding areas are in southeast Oklahoma and central Arkansas (AOU 1998).

FINDING

It may be looked for in oxbow areas in the lower Missouri Valley around 1 May.

Fregatidae (Frigatebirds)

Magnificent Frigatebird

Fregata magnificens

STATUS

Hypothetical in spring.

COMMENTS

There is a single report, a sighting by Lawrence Bruner in Cuming Co in spring 1884. Bruner lacked field guides as well as direct experience with this species at the time of the sighting; unfortunately, the observations that led to the identification were never published for what currently would be considered an out-of-season sighting. Swenk, however, accepted the record (Swenk 1934a) and presumably had many opportunities to discuss it with Bruner, as they were associates for many years. Records of this marine species for Kansas (Thompson and Ely 1989), Iowa (Kent and Dinsmore 1996), Missouri (Robbins and Easterla 1992), and Oklahoma (Baumgartner and Baumgartner 1992) lend credence to Bruner's record.

CICONIIFORMES

Ardeidae (Herons, Bitterns)

American Bittern

Botaurus lentiginosus

STATUS

Uncommon regular spring and fall migrant statewide. Uncommon regular breeder north and west, rare casual elsewhere. Hypothetical in winter.

DOCUMENTATION

Specimen

UNSM ZM12409, 19 Sep 1911 Lincoln, Lancaster Co.

DISTRIBUTION AND ECOLOGY

During migration and in summer American Bitterns frequent marshes and waterways with stands of emergent aquatic vegetation. Any suitable habitat throughout the state may attract this species.

Spring

12, 25, 26 Mar→14, 16, 17 May

Late dates above are from areas where breeding was not noted, although in suitable habitat these dates would suggest breeding. A decided migratory movement is evident for this species commencing in early Apr and peaking in late Apr and early May.

High counts include 11 at Crescent L NWR 13 May 1997.

Summer

Although American Bitterns may be found nesting in any appropriate habitat throughout the state, summer numbers are clearly greatest in the Sandhills lakes region. Blake and Ducey (1991) stated that nesting probably occurs in southern Holt Co. Reports in May–Jul since 1995 from Funk Lagoon and Weis Lagoon suggest a breeding population in the Rainwater Basin. Strangely, it is absent as a breeding bird from the extensive Clear Creek Marshes in western Keith Co (Rosche 1994a). There are numerous reports in the period late May–early Aug from elsewhere in the state without evidence of nesting.

There appears to be a decline in numbers of this

species in Nebraska. Ducey (1988) showed nesting reports since 1960 only in Sheridan, Garden, and Cherry Cos, although prior to 1960 there were nesting reports in several eastern counties south to Nemaha Co (Bruner and others 1904). Frequency of reports in NBR for the spring period was 52% lower 1970–93 compared with 1925–49.

The American Bittern is a solitary nester that locates its nest in marshes that contain considerable emergent vegetation. One nest, located at Crescent L NWR, consisted of a platform of old emergent vegetation in a dense stand of bullrush (*Phragmites* sp). The nest contained a single egg on 5 and 14 Jun, representing a less than typical clutch, according to Palmer (1962).

High counts include 19 at Crescent L NWR 8 Jun 1997.

Fall

22, 25, 26 Aug→14, 18, 21 Nov

Early dates above are from areas where breeding was not noted. There is a later report 17 Dec Lancaster Co. The fall migration is quite leisurely, with no peaks evident. Data from areas where summering birds were noted reveal that fall departure is quite variable, ranging from as early as 10 Aug in Howard Co to as late as 24 Oct at Crescent L NWR. Additional last dates of occurrence for Crescent L NWR include 7 and 27 Sep and 1 Oct.

Winter

Mollhoff (NBR 45:34, 36) noted that this species and Great Blue Heron winter along Beaver Creek in Boone Co but provided no further details. Although American Bitterns may linger late in fall, there are no other reports of overwintering.

FINDING

Look for bitterns in marshes with good stands of emergent vegetation, paying careful attention to objects in the vegetation. Bitterns have a habit of "freezing" when near humans or vehicles and will lift their head and bill upright, a behavior that allows them to blend

very well with the dead vegetation of the previous season. Occasionally an observer may also detect a sound emanating from a marsh that closely resembles the sound of an old-fashioned hand-operated water pump. This vocalization, most commonly heard in late spring and early summer, has given the American Bittern its still-heard colloquial name, "thunder pumper." Observers who are intent on finding American Bitterns should travel to either Crescent L NWR or Valentine NWR during late May or Jun for a likely view of this species.

Least Bittern

Ixobrychus exilis

STATUS

Uncommon regular spring migrant east, rare elsewhere. Uncommon regular breeder east, rare elsewhere. Rare regular fall migrant east.

DOCUMENTATION

Specimen

UNSM ZM7658, 18 May 1901 Nebraska City, Otoe Co.

DISTRIBUTION AND ECOLOGY

The Least Bittern is associated with dense marshes with emergent vegetation and waterways with marshy edges. Cink (1971) found a nest of this species in a marsh north of Lincoln: "The nest was in dense cattails about 6 inches [15.4 cm] off the water and contained four eggs." Although this kind of habitat may be found throughout the state and in particular in the Sandhills, recent records suggest that the species is most numerous in the eastern third of the state. However, it is especially secretive, and the relatively low number of observers in western Nebraska may contribute to an inaccurate reflection of this species' true status there.

It is our impression that frequency of sightings in the east is related to seasonal moisture conditions, in that springs and summers with abundant moisture and concomitant standing water in ditches and low-lying areas seem to yield an increase in sightings for this species.

Spring

22, 24, Apr, 1 May→summer

There are earlier reports 8 Apr 1978 Douglas-Sarpy Cos and 15 Apr 1981 Lancaster Co. Peak migration is in mid to late May.

Westernmost reports are 13 May 1961 near Lakeside (Tremaine 1962); a specimen, HMM 7655, taken 19 May 1914 near Oshkosh (Rosche 1994a); and 26 May 1983 at Ash Hollow SHP (Rosche 1994a).

Summer

Breeding records are few, although relatively small areas of cattails can be used for nesting. When the species is present, nesting density can be amazing, such as the 185 nests and 476 birds found at Squaw Creek NWR in extreme northwest Missouri in summer 1996 (Hilsabeck and Bell 1999). It is likely that intensive searches of cattail marshes, particularly in eastern Nebraska, will yield additional breeding records, although depletion of this habitat has contributed to a decline of reports in recent years.

Around 1900 it was a common summer resident in the east (Bruner and others 1904). Ducey (1988) cited breeding records for Cuming, Dakota, Douglas, Lancaster, Nemaha, and Thayer Cos in the south and east, although the only reports since 1960 are a nest in Lancaster Co (Cink 1971), 4–5 in 1984 (AB 38:1035; Bennett 1985), and 1 in 1986 (Bennett 1987). In Sarpy Co a bird was seen carrying nesting material in 1984 (AB 38:1035), 1 was in a small cattail marsh at Fontenelle Forest 7–12 Jun 1992, 3 were at LaPlatte Bottoms 3 Aug 1998, and 2 were there in 1999. There are several summer reports from Knox and Dakota Cos in recent years, including 1 at Crystal Cove L 13 Aug 1998.

Until recently there were few reports from the Rainwater Basin after the 1920s. It was a "very common summer resident" and common breeder at Inland Lagoon 1915–16 (Swenk, Notes before 1925), when Brooking collected a pair in Jun 1915, now HMM 2051, and a male 4 Jun 1916, now HMM 2521. It was reported at Weis Lagoon 23 Jun 1985 (Garthright 1985b) and Rainwater Basin summer 1996. One to 2 were at Funk Lagoon 24 Jul–3 Aug 1997, up to 7 were there 12 Jun–2 Aug 1998, 2 were there 12 Jun 1999, 1 was at Kissinger Basin 6 Jun 1999, and 2 were at a private basin east of Funk Lagoon 11 Aug 1998.

Westernmost breeding reported is a single nesting in 1938 in Garden Co (Tout 1947), although it was reported in Garden Co 20 Jul 1979 and 13 Jun 1997 (WRS, JS) and in Arthur Co 15 Jun 1990 (AB 47:113). It was considered "common in the Garden Co lakes" prior to 1925 (Swenk, Notes before 1925). It may be more numerous in Sandhills marshes than records indicate; Andrews and Righter (1992) considered it a rare migrant and summer resident in eastern Colorado.

Breeding records suggest that periodicity is somewhat variable; nests containing eggs have been found as early as 4 Jun and as late as 13 Jul.

During the 1980s the annual Iowa Breeding Bird Survey regularly revealed nests in the Forney L marshes, Fremont Co, Iowa (Barb Wilson, pers. comm. RSS). Forney L is in the Missouri River floodplain.

High counts include 8 in eastern Otoe Co 5–16 Aug 1997.

Fall

summer→8, 13, 14 Sep

Departure from breeding locations is in Aug. There are late reports 26 Sep 1998 at Black Island Area, 2 Oct 1986 Lancaster Co (AB 41:111), 4 Oct 1999 Walnut Creek L, and 17 Oct 1983 Lancaster Co.

Westernmost reports are 14 Aug 1986 and 30 Aug 1986 at Keystone L (Rosche 1994a). There is a specimen, WSC 914, collected in Boone Co 13 Sep 1982 (Bray and others 1986).

Other Reports

30 Mar 1935 Lincoln Co.

FINDING

Least Bitterns are probably easiest to find in mid to late Jun when adults regularly fly above the cattails while feeding young. A reasonable count can be made by staying in a spot with a good view of dense cattails. There is no particular location in the state that predictably attracts Least Bitterns, as water levels vary from year to year, but dense cattail marshes with a meter or less of water should be checked.

Great Blue Heron
Ardea herodias
STATUS
Common regular spring and fall migrant statewide. Locally common regular breeder west and central, uncommon east. Locally rare regular winter visitor Platte Valley.
DOCUMENTATION
Specimen
UNSM ZM10615, 24 Jun 1912 Halsey, Thomas Co.
TAXONOMY
The breeding subspecies in Nebraska is the northern *herodias*, but *wardi*, the southeastern race, has been reported in Frontier Co, and *treganzai*, the Rocky Mountain race, has been reported in Hall Co (AOU 1957). According to Bent (1926), the breeding ranges of the three may coincide in Nebraska, suggesting that most breeding Great Blue Herons in the state are intergrades. Indeed, it is likely that none of these are valid subspecies but merely represent points on a cline (Robbins, pers. comm.). Brooking (Notes) reported a specimen of *wardi* collected 18 Aug 1914 at Inland, possibly a postbreeding wanderer, and Swenk (Notes before 1925) reported 2 specimens of *herodias* in the Black collection that approached *wardi* in size, suggesting intergrades.
DISTRIBUTION AND ECOLOGY
This large heron is attracted to virtually any water body that sustains a population of fish and other aquatic vertebrates; therefore, it may be seen along any river or stream, farm pond, natural lake, or reservoir statewide. It nests in tall trees, typically in loose colonies. In central and western Nebraska, particularly in the Sandhills, colonies may be located in an isolated clump of tall trees that may be many kilometers from available foraging sites.

Although their numbers decline markedly during the most severe winter periods, some individuals may remain in the state as long as they have access to open water that sustains a food supply.

Spring
Major movements are generally undetected, suggesting that the species is a nocturnal migrant. In early spring groups of 10 or 12 are often found in the morning at sites where the evening before none were present. It is unknown whether these groups represent arriving local breeders or migrants. Although occasional individuals may be seen throughout the winter near open water, a decided increase in sightings corresponds with the spring thaw, which typically occurs in mid to late Mar.

High counts include 50 at Crescent L NWR 10 Apr 1996, 50 around L McConaughy 13 Apr 1997, and 48 at Crescent L NWR 14 May 1995.

Summer
At present about 20 active colonies are distributed across the state (John Dinan, pers. comm. RSS), including several in the Keystone L area (Brown and others 1996; Rosche 1994a). In 1994 Rosche located 9 rookeries with 81 nests in Webster, Harlan, Hitchcock, Cheyenne, Kimball, Sheridan, and Sioux Cos. Ducey (1988) indicated breeding records since 1960 in 30 counties statewide.

Breeding colonies have been observed with as few as 3 to in excess of 100 active nests. Although tall cottonwoods are most often utilized as nest sites, at least one active colony has been maintained for many years in a stand of relatively low-growing hackberry in the Sandhills of Thomas Co; another colony there was situated in a stand of cultivated jackpine.

Nest sites are occupied first by males, coincident with spring arrival. Females apparently arrive somewhat later. Incubation has been observed in May and Jun. Young judged ready to leave the nests have been noted on 29 Jun and

11 and 12 Jul. However, entire colonies of incubating adults have been recorded in late Jul (18, 19, 20), suggesting either late nesting or renesting attempts. One colony of 23 active nests, surveyed on 26 Jun, contained a nest with 3 young; the remainder all contained 2 young.

Fall

In early fall flocks begin to form, probably including birds from farther north and postbreeding wanderers. There is no clear fall movement, although about 50% of departure dates occur in Oct. By mid-Nov most Great Blue Herons have left the state.

High counts include 119 at L McConaughy 20 Sep 1998, 102 there 16 Sep 1983, 85 in the eastern Rainwater Basin 1 Aug 1999, and 80 at Funk Lagoon 2 Aug 1997.

Winter

Winter records exist statewide, but overwintering is essentially limited to the Platte Valley. It winters regularly around springs near Lewellen and below Keystone L along the North Platte River (Rosche 1994a), where 11 were present 4 Feb 1996. It is probable, at least regionally, that during periods of very severe winter weather all individuals may move to warmer climes, where access to food is not limited by ice cover. Banding returns from juveniles banded in Nebraska indicate that individuals migrate directly south into Oklahoma, Texas, and Mexico during winter periods.

FINDING

These large bluish gray birds may be seen silently fishing in the shallows of nearly any body of water, particularly during summer months.

Great Egret

Ardea alba

STATUS

Fairly common regular spring migrant east, becoming rare casual west. Common regular summer visitor east and central, rare west. Accidental breeder. Common regular fall migrant east, becoming rare west.

DOCUMENTATION

Specimen

UNSM ZM7653, 4 Aug 1930 Saunders Co.

DISTRIBUTION AND ECOLOGY

Great Egrets are attracted to shallow waters of rivers, streams, lakes, impoundments, and marshes during migration and in summer months. About 90% of the records are from the eastern quarter of the state. Reports are increasing.

Spring

25, 26, 27 Mar→summer

In the east arrival is in the first third of Apr. There are earlier dates 21 Feb 1970 Adams Co and 10 Mar 1995 Hall Co. It has been suggested (NBR 64:44) that some spring birds, especially flocks, may be postbreeding dispersers from the Gulf Coast, where nesting occurs in Jan–Feb.

There are 15–20 reports from the Panhandle, probably vagrants rather than migrants, in the period 18 Apr–31 May.

High counts include 30+ at Nebraska City 30 May 1996 and 20 at Pawnee L 4 May 1996.

Summer

About 90% of the summer reports are from the eastern quarter of the state. Reports are fewest in Jun. Numerous late Jul and Aug records suggest a tendency for regular movement of postbreeding or nonbreeding transients from centers of greater concentration to the northeast, east, and southeast. At this time Great Egrets become rather common, at least west to the Rainwater Basin.

One nesting record exists for Sarpy Co, where observers noted "a pair or more" nesting in a large colony of Great Blue Herons that once existed on Gifford Peninsula (Cortelyou 1960). Occasional nesting has been observed in Kansas (Thompson and Ely 1989) associated with Great Blues, as well as in Colorado (Andrews and Righter 1992). Steve Duecker (pers. comm. RSS) has observed several pairs nesting in association with Great Blues in Monona Co, Iowa, which borders Thurston and Burt Cos, Nebraska. It is therefore likely that nesting occurs in Nebraska with a frequency greater than the record reflects. Careful observation of heronries, particularly in the Missouri River floodplain, may yield additional nesting records.

Fall

summer→22, 23, 26 Oct

There is a later report 8 Nov 1997 Branched Oak L (BFH, MB, JS).

No clear fall migration peak is evident from the records; numbers build up in Jul and peak in Aug, gradually declining thereafter into Oct. The few Panhandle reports are in the period 24 Jul–22 Oct.

High counts include 132 at Harlan Co Res 9 Aug 1998, 130 there 24 Sep 1999, and 49 there Aug 1997.

FINDING

These large white herons are found in the eastern half of the state in wetlands, most predictably in the Rainwater Basin. Peak numbers are present in Aug. They can be confused with immature Little Blue Herons, which are also white and often wander into the state during summer months.

Snowy Egret

Egretta thula

STATUS

Uncommon regular spring and fall migrant east, becoming rare west. Uncommon regular summer visitor central, rare elsewhere. Hypothetical breeder.

DOCUMENTATION

Photograph

Jun 1965 Crescent L NWR (RSS).

DISTRIBUTION AND ECOLOGY

The Snowy Egret is attracted to typical marsh habitat with extensive emergent reeds and sedges. It is usually seen in shallow waters on the open-water edges of the stands of emergent vegetation. Migrants may appear anywhere, but in late summer best numbers are usually reported from the Rainwater Basin.

Spring

10, 12, 14 Apr→summer

There is an early undocumented report 29 Mar 1992 Lancaster Co. Reports are distributed rather evenly from mid-Apr through mid-May and are fewest in late Jun and early Jul.

Twenty-nine of 66 records are for western counties west of the 100th meridian; given that the preponderance of birders have resided and been active in the eastern half of the state, this suggests that Snowy Egret occurs with far greater frequency in the west.

Another curiosity about the records is that for the first 30 years of NOU record keeping (1933–63), 9 of 12 total reports (75%) were from locations west of the 100th meridian. During the second 30 years (1963–93) only 17 of 54 records (32%) were from the west, suggesting that in recent times Snowy Egrets are seen with increasing frequency in eastern Nebraska. The records also may be interpreted to imply that two different egret populations are contributing to the records, and a review of population trends in neighboring states tends to support this hypothesis. Many western reports may represent individuals associated with a scattered breeding population in southeastern Wyoming (Scott 1993) and northeastern Colorado, where it was recorded nesting as early as 1937 (Andrews and Righter 1992). A single, albeit around 1900, nesting record for L Alice (Bent 1926) suggests a long-term but small population in the central high plains that may have contributed to earlier sightings primarily in the west. The increased frequency of sightings in eastern Nebraska during the past 30 years appears to correlate with an apparent population expansion from the south and east. Prior to 1978 the Snowy Egret was considered a "rare postseason visitant" in South Dakota (Whitney and others 1978), but since the 1980s there have been more reports in the eastern part of the state, and it has established breeding colonies at several eastern South Dakota locations (SDOU 1991).

High counts include 3 at Sacramento-Wilcox

Basin 23 Apr 1994 and 3 at Fahrenholz Ponds 22 Apr 1995.

Summer

Wandering individuals may be found in appropriate habitat statewide during the summer months. Such individuals may remain in a location for up to several weeks, judging from some reports: 30 Jul–23 Sep Garden Co, "summer"–5 Aug Cherry Co, and 4 Aug–7 Sep Polk Co.

Although nesting records exist for Lancaster Co 1895, Scotts Bluff Co about 1900, and Hall Co in 1985, none are well documented (Ducey 1988). The Lancaster Co report (Eiche 1901) involved a female purportedly shot from a nest at Lincoln in Jun and placed in the Eiche collection (Bruner and others 1904); the Eiche specimens were included in the UNSM collection, where there is an immature (white) Little Blue Heron (ZM14585) also "said to be shot from a nest" (UNSM data) in Lincoln Jun 1895. Thus the specimen appears to have been misidentified, and the nesting report of Snowy Egret is therefore in error. No details have been published for the reports of young in Hall Co in 1985 (Bennett 1986) or nesting in Scotts Bluff Co about 1900 (Bent 1926).

Fall

summer→1, 4, 5 Oct

There are later reports 11 Oct 1998 (2) eastern Rainwater Basin, 15 Oct 1997 (4) Branched Oak L (JS), 17 Oct 1989 Lancaster Co, 19 Oct 1993 Lincoln Co, 23 Oct 1993 Cass Co, and 30 Oct 1987 Douglas-Sarpy Cos, and a "very, very late" bird was roosting with gulls in a gale at L McConaughy 6 Nov 1976 (Rosche 1994a).

High counts include 30 at Harlan Co Res Aug 1988 (AB 43:124), 29 at Funk Lagoon 19 Aug 1998, and 22 at Crescent L NWR 4 Oct 1996.

COMMENTS

Three specimens listed by Swenk (Notes before 1925) were at one time in the Brooking collection but apparently are no longer extant. A specimen, HMM 2682, collected 2 Jun 1911 near Gibbon was said to have been photographed by Swenk, but neither it nor the photograph appears to be extant.

FINDING

Shallow marshes with emergent vegetation scattered about expanses of open water will attract this small heron during migration and in summer. The marshes of the Sandhills and the Rainwater Basin are likely places to find this species.

Little Blue Heron

Egretta caerulea

STATUS

Uncommon regular spring and fall migrant east and central, rare casual west. Uncommon regular summer visitor east and central, rare casual west.

DOCUMENTATION

Specimen

UNSM ZM14585, Jun 1895 Lincoln, Lancaster Co (see Snowy Egret).

DISTRIBUTION AND ECOLOGY

Little Blue Herons utilize wet meadows and shallow waters of marshes, lakes, and rivers as foraging sites. Although considerable habitat is available to attract this species, it has been most abundant south and east of Nebraska.

Spring

4, 4, 5 Apr→summer

More than half of 73 reports through 1995 are in the period 8–23 May. There are few reports from the Panhandle, where Rosche (1994a) considered this species "always the rarest" of the "southern herons" reaching Nebraska: 25 Apr 1986 Dawes Co (AB 40:441) and 13 Jun 1936 Garden Co.

High counts include 14 in eastern Otoe Co 4–13 Jun 1996; these birds apparently summered in the area and were still present 24 Jul (FN 50:965).

Summer

This species is recorded with regularity late Jul–Aug in the east and south, areas that are the source of 46 of 47 such records, the exception 17 Aug 1978 Boone Co. There are no Panhandle reports. Reports are fewest in late

Jun and early Jul. Typically sightings are of solitary individuals that may utilize a foraging area for a period of up to several weeks (see Spring).

Until recently, about half the Nebraska records were concentrated in mid to late summer. As many as 50% of those records were of birds in subadult plumage, suggesting that this species' presence in the state could be attributed to postbreeding dispersal. However, since about 1980 it has been found to be a regular but rare breeder in eastern South Dakota (SDOU 1991), suggesting that the increase in recent years of spring sightings in eastern Nebraska may reflect migrants from the South Dakota population.

Fall

summer→14, 14, 19 Sep

There are later reports 7 Oct 1970 Lancaster Co, 8 Oct 1990 Douglas-Sarpy Cos, and 20 Oct 1974 Adams-Clay Cos. There are no reports from the Panhandle; farthest west is 8 Sep 1990 Cherry Co.

High counts include 14 in Clay Co 5–13 Sep 1998, 9 in Otoe Co 5 Aug 1996, and 8 there 22 Aug 1997.

FINDING

Checking Rainwater Basin wetlands in Aug would likely produce a few Little Blues, although most would be in white immature plumage.

Tricolored Heron

Egretta tricolor

STATUS

Hypothetical in spring. Rare casual summer visitor central.

DOCUMENTATION

Specimen

HMM 21718, 14 Oct 1918 Kearney Co (Black 1933, 1941; Bray and others 1986).

Spring

There are 2 reports, neither documented. One reported 13 May 1991 in Sarpy Co may have been the same bird as that seen across the Missouri River in Iowa 4 May 1991 and

acceptably documented there (NBR 59:64, 94; Greer 1992). Another was reported in western Lancaster Co 22 May 1997.

Summer

There are 3 reports, 2 documented, one a very late date for a presumed postbreeding wanderer of this species:

7 Aug 1971 Clay Center (Cink 1973a)

14 Oct 1918 (cited above)

The third report is of one collected 10 Aug about 1918 at Inland; the specimen is now lost (Cink 1973a; Bray and others 1986).

FINDING

This species may appear Apr–May as a spring migrant or, more likely, in late summer as a postbreeding wanderer. The best place to look is probably the Rainwater Basin. It has nested since 1976 in central Kansas (Thompson and Ely 1989).

Cattle Egret

Bubulcus ibis

STATUS

Uncommon, locally common, regular spring and fall migrant east and central, rare west. Uncommon, locally common, regular summer visitor east and central, rare west. Fairly common casual breeder statewide.

DOCUMENTATION

Photograph

7 May 1983 Albion, Boone Co (Molhoff 1989a).

DISTRIBUTION AND ECOLOGY

Since 1972 Cattle Egret has become increasingly regular in central and eastern Nebraska but is still "casual" in the Panhandle (Rosche 1994a). Migrants and postbreeding visitors utilize shallow wetlands and damp pastures, although this species is the least dependent of the herons on the presence of water. Breeding birds, however, usually nest over fairly shallow water, often in willows.

Spring

3, 4, 7 Apr→summer

There is an earlier report 29 Mar 1998 Dodge Co, but most arrive from mid-Apr into the first third of May. The first Panhandle record was 26 May 1979 in Garden Co.

High counts include 140 at Nebraska City 30 May 1996, 50 near Lincoln 10 May 1996, and 50 in Wheeler Co 20 May 1996.

Summer

There are 5 reports of breeding, only 1 involving a site where the colony has continued to be active. The first report was by Ducey (1984b) at Valentine NWR, when 6 nests were found among those of cormorants and night-herons on Marsh Lakes in Jun 1984. This colony has apparently persisted. A colony of at least 7 active nests was established in a clump of willows on a small island at Offutt Base L in 1993 (RSS). In extreme northeast Garfield Co at least 70 nests, all of which had young by 9 Jul, were found in 1994 (NBR 62:128). Also in 1994, 31+ birds with 3 nests and 2 more under construction were at the west end of L McConaughy in willows 6–7 Aug 1994 (Dinsmore and Silcock 1995a). A third 1994 record is of 115 adults carrying nest material into tall vegetation in the center of Lamb's L, Holt Co (NGP).

Fall

summer→8, 9, 11 Nov

Reports are fewest in Jun and early Jul and most numerous Aug–early Sep, gradually declining through Oct.

High counts include 500 in Otoe Co 5 Aug 1996, 450 at East Odessa Area 8 Aug 1994, and 303 at Funk Lagoon 19 Aug 1998.

COMMENTS

This species is an immigrant to the New World from Africa, arriving in South America in the late 19th century. It was first seen in North America in 1941 or 1942 and has been expanding its range north and west since that time (Palmer 1962). This small egret was first observed in Nebraska 11 Sep 1965 in Adams Co (Welch 1966), and until 1972 its occurrence was casual.

FINDING

These small white herons are normally seen foraging in flooded fields and moist meadows often in association with livestock.

Green Heron

Butorides virescens

STATUS

Common regular spring and fall migrant southeast, becoming rare casual northwest. Common regular breeder southeast, becoming rare casual west central.

DOCUMENTATION

Specimen

UNSM ZM12399, 26 Apr 1917 Lincoln, Lancaster Co.

DISTRIBUTION AND ECOLOGY

The Green Heron is most commonly found associated with wooded rivers and streams and less commonly found in marshes and along lake edges. Those regions of the state, such as the Sandhills, that support little appropriate habitat will attract fewer Green Herons.

Spring

16, 19, 20 Mar→summer

There is an earlier report 10 Mar 1962 Gage Co. Most arrival dates are distributed from the latter third of Apr through the middle third of May. Panhandle dates of occurrence tend to be later (late May and Jun), suggesting that it is more of a vagrant there; Rosche (1982, 1994a) listed it as casual in spring in the northwest and as uncommon in the Keith Co area. There are few Panhandle reports: 2 Apr 1935 Sioux Co (Rosche 1982), 25 May 1977 Dawes Co (Rosche 1982), and 16 Jun 1978 Dawes Co (Rosche 1982).

High counts include 12 at Wolf L 27 May 1996 and 6 in southwest Hall Co 11 May 1996.

Summer

Except for the Panhandle, breeding occurs statewide in appropriate habitats (Ducey 1988). This is probably a recent phenomenon, as Bruner and others (1904) considered this species a "very common summer resident in eastern Nebraska" but not reported farther west. Ducey (1988) stated that the present "nesting range is larger than the records indicate," and Rosche (1994a) suggested that nesting probably occurs in the Keith Co area, but there was no definite evidence. The westernmost nesting records cited by Ducey (1988) are for Cherry and Lincoln Cos.

A single was at Rock Creek L 12 Jun 1999 (MB, DH).

Two nests surveyed each contained 3 nearly fledged young on 25 Jun and 1 Jul. Both nests were found in trees at elevations of 2.7 and 5.5 m (9 and 18 ft), respectively.

Fall

summer→31, Oct, 2, 9 Nov

Most depart by mid-Oct. In the Panhandle it was listed as "accidental" by Rosche (1982), but there are a few reports: 24 Jul 1983 Scotts Bluff Co, 12–26 Aug 1926 near Fort Robinson (Rosche 1982), 21 Aug 1971 Scotts Bluff Co, 25 Sep 1994 at North Platte NWR, and 30 Sep 1986 Sioux Co.

High counts include 6 in Otoe Co 14 Aug 1997 and 5 at Harlan Co Res 29 Aug 1997.

FINDING

This small, dark heron is often seen along wooded riverbanks and oxbows. Its unmistakable "keeow" alarm call announces its presence. Usually Green Herons will fly less than several hundred meters after being startled and will continue feeding after a few minutes if undisturbed. The careful, patient observer may approach slowly for a better look. An almost guaranteed observation of the Green Heron can be found along Stream Trail in Fontenelle Forest during May and Jun.

Black-crowned Night-Heron

Nycticorax nycticorax

STATUS

Fairly common regular spring and fall migrant statewide. Locally common regular breeder central, rare elsewhere.

DOCUMENTATION

Specimen

WSC 275, 22 May 1977 Cherry Co.

DISTRIBUTION AND ECOLOGY

During migration these herons may be found in or near marshy vegetation associated with ponds, lakes, reservoirs, and occasionally rivers and streams. They often travel in small groups of 3–15 and may perch in small trees and saplings adjacent to foraging sites. During the summer months they are attracted to marshes with dense emergent vegetation both for foraging and nesting. Thus during the summer most are found around Sandhills lakes and in a few of the larger marshes of the Rainwater Basin.

Spring

29 Mar, 2, 2 Apr→summer

Most arrive in mid-Apr, but there are earlier reports 21 Feb 1970 Adams Co, 19 Mar 1980 Adams Co, and 25 Mar 1982 Douglas-Sarpy Cos.

High counts include 84 in the eastern Rainwater Basin 26 Apr 1997, 47 at Kissinger Basin 27 Apr 1999, and 19 at Lakeside-Antioch 7 May 1994.

Summer

Ducey (1988) indicated that nesting may occur statewide except west of the Sandhills and, in recent years, the eastern part of the state. There are, however, several reports from these areas in summer, indicating that nesting might occur virtually anywhere if conditions are suitable.

This heron may on occasion be a solitary nester, but it most often nests in loose colonies. One colony located in a clump of Siberian elm (*Ulnus*) contained 47 nests, all attended by incubating adults on 19 Jul. Another colony found at Crescent L NWR consisted of more than 12 nests and was located in dense *Phragmites* reeds. One nest contained 4 eggs on 6 Jun.

High counts include 100 at Funk Lagoon 12 Jul 1998 and 13 at Crescent L NWR 13 Jul 1995.

Fall

summer→15, 17, 22 Nov

A first-year bird was very late at L Ogallala 2 Jan 1999 (SJD, DCE). Movement is detectable in late Sep.

High counts include 45 at Funk Lagoon 14 Aug 1994, 17 there 19 Aug 1998, and 16 at L McConaughy 20 Sep 1998.

FINDING
The best time to find this species is during the late spring and summer at Sandhills refuges such as Crescent L NWR and Valentine NWR. These birds are most obvious early and late in the day.

Yellow-crowned Night-Heron
Nyctanassa violacea
STATUS
Rare regular spring migrant southeast, rare casual elsewhere. Rare regular summer visitor southeast and south-central, rare casual elsewhere. Rare casual breeder southeast. Rare casual fall visitor (after Aug) statewide.

DOCUMENTATION
Specimen
HMM 28539, 17 May 1956 Hastings, Adams Co.
DISTRIBUTION AND ECOLOGY
This heron is attracted to small marshes, wooded waterways, and the wooded edges of lakes and reservoirs. There may be an irregular breeding population in the southeast, as indicated by a limited spring movement, but most reports are of presumed postbreeding wanderers in southeast and south-central Nebraska.

Spring
10, 10, 12 Apr→summer
Arrival dates are clustered in the first third of May, although there is an earlier date 2 Apr 1977 Douglas-Sarpy Cos. That reports become increasingly less regular westward is consistent with this species' normal geographic distribution. Panhandle reports are 7–25 May 1969 Dawes Co and 20 May 1991 Scotts Bluff Co. There are northerly reports, one documented, a specimen (UNSM ZM17043) collected southwest of Ainsworth 14 Apr 1992 (Brogie 1997), and undocumented reports 6 May 1955 Antelope Co, 14 May 1960 Garfield Co, and 16 May 1961 Antelope Co.

Although no breeding populations are known in the Dakotas or in western Minnesota, Nebraska records suggest regular arrival in the extreme southeast in early May, even though numbers are very small, typically 2–5 individuals per year.

Summer
There are about 21 reports in the period mid-Jun through mid-Jul, including these away from Douglas-Sarpy and Lancaster Cos: 28 Jun–11 Jul 1998 Funk Lagoon, 2 Jul 1987 Phelps Co, 6 Jul 1986 Kearney Co, and 13 Jul 1999 (an adult) Funk Lagoon.

There is a single record of nesting and 2 additional records indicative of nesting. These few records are surprising in that the species is a regular breeder as close as northwest Missouri and northeast Kansas (Robbins, pers. comm.). In 1963 Sharpe (1964) found a loose nest of sticks in a ashleaf maple about 12 m (40 ft) above the ground in a woodland adjacent to a stream in Fontenelle Forest. Although incubation was initiated, the nest was abandoned. Ducey (1980b) reported on a recently fledged juvenile found 18 Jun 1980 in northern Lancaster Co; it had down feathers on the body and incompletely emerged primaries. Another juvenile was found northeast of Lincoln 9 Jul 1980 that also appeared to have fledged nearby (Ducey 1980b).

More than 50% of the birds that we have observed were in subadult plumage, suggesting that extralimital wandering of prebreeders is the source of most summer reports.

Fall
summer→3, 4, 4 Oct
There are later reports 10 Oct 1979 Douglas-Sarpy Cos (AB 34:176) and 11 Oct 1998 at Glenvil Basin. There are only about 20 reports after Aug, indicating that there is no fall migration per se; rather, summer birds gradually depart the state, most by the end of Aug.

Panhandle reports include 30 Jul–5 Aug 1980 Box Butte Res (Rosche 1982), 22 Aug 1991 Scotts Bluff Co, 30 Aug 1996 Crescent L NWR, and 4 Oct 1991 Scotts Bluff Co.

High counts include 4 at Glenvil Basin 15 Aug 1998, 4 at Funk Lagoon 19 Aug 1998, and 3 in Otoe Co 14 Aug 1997.

FINDING

Look for this species early in summer along wooded streams and oxbows and later, when postbreeding wanderers appear in late Jul and Aug, in marshes in the Rainwater Basin. It is most likely to be seen early in the morning and late in the evening. Many are juveniles and subadults and thus must be carefully observed to obtain proper identification.

Threskiornithidae (Ibises, Spoonbills)

White Ibis

Eudocimus albus

STATUS

Accidental in summer.

DOCUMENTATION

Photograph

5 Jul 1999 Kissinger Basin, Clay Co (JGJ).

COMMENTS

The only documented record is of a first-year bird at Kissinger Basin 4–25 Jul 1999 (JGJ, m.ob.)

There are 2 additional reports, both likely correct but lacking extant documentation. One was observed 12–19 Jun 1916 at Inland Lagoon; the observer "was close enough to be sure" of the identification (Brooking, Notes), but no details were provided. Another was filmed by Karl Menzel during its stay in Rock Co 1–19 Aug 1963; although Paul Johnsgard viewed the film and confirmed Menzel's identification of an individual in juvenal plumage (Menzel 1964), the film cannot now be located (Bray and others 1986).

Scattered summer records for neighboring states suggest that this species is likely to occur on occasion in Nebraska.

Glossy Ibis

Plegadis falcinellus

STATUS

Accidental in spring.

DOCUMENTATION

Description

24 Apr 1999 Wilkins Basin, Fillmore Co (JGJ).

COMMENTS

The record cited above was of an adult with a flock of 28 White-faced Ibis (JGJ).

There are 11 additional reports in the literature, none convincingly documented, but at least 2 are suggestive of this species. A single bird was seen 16 May 1946 in Scotts Bluff Co (Fichter 1946), and 2 were reported in York Co May 1964 (Morris 1965). While both observers were aware of the rarity of Glossy Ibis, neither noted the iris color or bluish facial skin characteristic of adults of this species in May (Kaufman 1990). Adults in breeding plumage in May and Jun are readily identified, but other age groups are difficult to separate from White-faced Ibis. May and Jun flocks of White-faced Ibis often contain both adults and presumed 1-year-old birds; the latter do not exhibit the characteristic white facial markings of adults and may account for the undocumented reports.

Bruner and others (1904) listed 3 specimens attributed to Glossy Ibis, but Swenk (1918a) noted that these immature birds could not be identified to species and dropped Glossy Ibis from the State List. Other reports without sufficient information to allow identification as Glossy Ibis are of 4 seen in Kimball Co about 1946 and 2 seen northwest of North Platte 14 May 1950 (Gates 1950), southeast of Grand Island 6 Oct 1954 (Larson 1956), Clay Co 24 Apr 1975 (NBR 43:47), Lancaster Co about 1983 (NBR 51:89), and Rainwater Basin 26 Apr 1996.

There are several records for northwest Missouri, most recently in 1999, and the species is now regular at Quivira NWR and Cheyenne Bottoms in central Kansas, possibly breeding (Robbins, pers. comm.).

White-faced Ibis

Plegadis chihi

STATUS

Fairly common regular spring and fall migrant west and central, uncommon east. Uncommon casual breeder central.

DOCUMENTATION
Specimen
Olson #43 at UNK, 12 Jun 1916 Island L, Crescent
  L NWR.
DISTRIBUTION AND ECOLOGY
The White-faced Ibis is a western species at the
  eastern edge of its range in Nebraska during
  migration. Small (3–15) migrating groups are
  attracted to flooded meadows, marshes, and
  the marshy edges of lakes. Nesting is limited
  to emergent vegetation in extensive marshes.
Spring
9, 12, 13 Apr→8, 9, 10 Jun
Most move through the state in late Apr and
  early May, but there are several reports
  into Jun, which may reflect nonbreeding
  transients, including later reports from the
  north and west 14 Jun 1995 Sheridan Co, 17
  Jun 1994 Morrill Co, and 22 Jun 1991 Holt Co.
  No details of age or plumage were provided
  with these reports.
High counts include 39 at Weis Lagoon 8–10 May
  1995, 28 at Wilkins Basin 24 Apr 1999, and 27
  at Oliver Res 1 May 1999.
Summer
Historically, Nebraska records suggest that
  White-faced Ibis was once rather common
  in the state but declined substantially in the
  period 1920–60, possibly as marshes were
  drained. Rosche (1982) indicated that there
  were no published records for the northwest
  between 1931 and 1972. Since 1960, and
  especially since 1990, the species has been
  observed with increasing frequency and
  regularity, including nesting at Valentine NWR
  and Crescent L NWR (see below).
Nesting has been recorded only 3 times in
  Nebraska. Ibises were found breeding in
  Clay Co in 1916 when Brooking and Wallace
  located a nest (Swenk 1918b). Ducey (1984b)
  found 4 nests with eggs at Marsh Lakes in
  the northeast corner of Valentine NWR 21 Jun
  1984; young were seen by refuge personnel
  later in the season. In 1988 at Crescent L
  NWR Huber (1987) found 2 nests at Smith
  L that produced at least 5 flying young by

mid-Aug; this was considered "the first known
  production" by this species at the refuge.
  More recently nest building was observed at
  Crescent L NWR in 1998 (SJD).
The species has become a regular breeder
  at several localities in eastern South
  Dakota (SDOU 1991) and is a common and
  regular breeder in Wyoming (Scott 1993) and
  Colorado (Andrews and Righter 1992). Recent,
  regular breeding has been established at 2
  locations in west-central Kansas (Thompson
  and Ely 1989). Given this trend, further
  breeding should be anticipated in Nebraska.
Fall
8, 9, 9 Jul→3, 10, 11 Oct
There are earlier reports 27 Jun 1999 (4) Funk
  Lagoon, 28 Jun 1998 (3) Funk Lagoon, and at
  least 1 near Antioch in late Jun 1999 and later
  reports 17 Oct 1996 Ayr L, 25 Oct 1998 (2)
  Harvard Lagoon, and 12 Nov 1987 Polk Co.
  Since 1994 there have been several reports in
  Jul; there were no Jul reports prior to 1994.
Although most, if not all, fall records would be
  expected to pertain to White-faced Ibis, the
  possibility of some being Glossy Ibis cannot
  be ruled out. Separation of the 2 species in late
  summer and fall is very difficult, and fall birds
  should probably be reported as *Plegadis* sp.
High counts include 47 at Harvard Lagoon 5
  Sep 1998, 43 at North Hultine Basin 19 Sep
  1998, and 31 in the eastern Rainwater Basin 25
  Sep 1999.
FINDING
Most recent records are from the Rainwater Basin
  in Jul and Aug. Migrants may appear anywhere
  in late Apr–early May around wetlands.

Roseate Spoonbill
Ajaia ajaja
STATUS
Rare casual summer visitor south-central and
  southeast.
DOCUMENTATION
Description
16 Sep 1884, Lincoln, Lancaster Co (LOI 66; Bray
  and others 1986)

Summer

There are 4 records, all documented. The 1932
Buffalo Co specimen (see below) was at one
time in the Brooking collection (Brooking
1942b) but cannot now be located (Bray and
others 1986).

5 Jun 1932 Buffalo Co (LOI 66; Bray and others
1986)

5 and 14 Aug 1997 juvenile Otoe Co (Dinsmore
1999d; Brogie 1998)

20 Aug 1966 (2) near Hastings (Maunder 1966)

16 Sep 1884 Lancaster Co (*Lincoln Daily News*;
LOI 66; Bray and others 1986)

COMMENTS

Individuals occasionally wander north in the
interior during the summer months; 3 late
summer and 1 early spring records exist for
Kansas (Thompson and Ely 1989).

FINDING

The best chance to see Roseate Spoonbill appears
to be to check late summer concentrations
of "southern herons," especially Little Blue
Herons and Snowy Egrets, itself a rather rare
phenomenon in Nebraska.

Ciconiidae (Storks)

Wood Stork

Mycteria americana

STATUS

Accidental in summer.

DOCUMENTATION

The only documented record is a description of
one seen near Fontenelle Forest, 29 Jun 1925
Sarpy Co.

COMMENTS

There are at least 2 additional undocumented
reports. One collected Apr 1885 in Hamilton
Co was at one time in the Brooking collection
(Swenk 1918b) but cannot now be located
(Bray and others 1986). Swenk (Notes before
1925) stated that one shot by a farmer near
Cozad about 1912 was fed to hogs. Swenk
(1918a) noted that "there have been several
reports of the seeing or shooting of Wood
Ibises in Nebraska, but most of these have

been alone insufficient to definitely place the
bird on the state list."

Other Reports

A report of one along the Platte River in Cass Co
22 Jun 1996 was not accepted by the NOURC
(Brogie 1997).

COMMENTS

This species occasionally wanders beyond its
normal southeastern range but is only rarely
seen on the Great Plains (Palmer 1962);
currently occurrences seem unlikely, as the
southern population is at a low ebb. South
Dakota has 1 record (SDOU 1991), and Kansas
has 5 (Thompson and Ely 1989).

Cathartidae (Vultures)

Black Vulture

Coragyps atratus

STATUS

Accidental in summer.

DOCUMENTATION

Specimen

HMM 2649, 14 Jul 1916 Hildreth, Franklin Co.

Records

The only acceptably documented record is the
specimen cited above. Bray and others (1986)
accepted a minimal description of 3 birds
seen near Hordville 15 Mar 1963 (NBR 32:66),
but the date of the sighting is rather early
relative to the few confirmed sightings from
neighboring states (see Comments).

There are about 6 additional reports, most
without details. One was collected by Talbot
at Wolf Creek, Pawnee Co, before 1904 and
placed in the University of Iowa collection
(Bruner and others 1904), but none of Talbot's
Black Vulture specimens in that collection
are labeled (Bray and others 1986). Brooking
(Notes) mentions one taken by a boy near
Kearney Oct 1918, mounted by Black, and now
in the Olson collection at UNK; this specimen
could not be located (WRS). A specimen
was described by Swenk (letter dated 11 Oct
1918, NOU Archives) as collected by Black "on
the 18th of last month," thus 18 Sep 1918; a

photograph was sent to Swenk with an offer to purchase the specimen, and it appeared to be "nicely mounted and in good plumage." It seems likely that this is the specimen supposedly in the Olson collection discussed above.

Other reports, with few or no details, are Jan 1950 Keith Co (Johnsgard 1997), 7 Apr 1926 Sioux Co (LOI 23), 27 Apr 1955 Keya Paha Co, and Apr 1951 Logan Co (Johnsgard 1997).

COMMENTS

The occurrence of Black Vulture in Nebraska is poorly documented. The only recent Kansas reports are a few unconfirmed sightings in southeast Kansas (Thompson and Ely 1989), although more recently a group of experienced birders found one in southeast Kansas (Robbins, pers. comm.). Northward vagrancy by this species is limited, even in Missouri, where it occurs regularly in the south (Robbins and Easterla 1992; Kent and Dinsmore 1996). There are about 10 Colorado reports, but none have been accepted (Andrews and Righter 1992).

Turkey Vulture

Cathartes aura

STATUS

Common regular spring and fall migrant statewide. Locally uncommon regular breeder statewide. Hypothetical in winter.

DOCUMENTATION

Specimen

UNSM ZM10880, 24 Jul 1910 Naponee, Franklin Co.

DISTRIBUTION AND ECOLOGY

Turkey Vultures seek out cliff faces, inaccessible ledges, and cavities for nesting, requirements that tend to limit their distribution during midsummer to those parts of the state that provide appropriate nesting habitat, namely the central and lower Niobrara Valley, areas of heavy relief in southern Nebraska along the Republican River and Frenchman Creek, and throughout much of the Panhandle and western Platte Valley. Elsewhere, they

have been known to utilize abandoned farm buildings for nest sites, which explains the presence of small summer populations in southeast Nebraska and along the Missouri Valley.

This carrion feeder locates carcasses by vision and smell; individuals or small groups may be far-ranging in their search for food. They effectively utilize their broad, slotted wings to allow rising thermals to carry them to elevations that not only allow a broad panorama of the landscape but also allow them to soar from higher altitudes across the landscape until they can pick up another thermal. In such a fashion these vultures may remain airborne for hours without exerting much flight energy.

They tend to be concentrated in areas that produce thermals, that is, those that have dramatic relief. Thus individuals and groups are often seen soaring along and just above the edges of bluffs, canyons, and cliff faces where thermals and winds are directed upward. Such behavior is particularly noticeable during the spring.

Spring

25, 26, 26 Feb→summer

Spring arrivals appear in early and mid-Mar, but most arrive in very late Mar and early Apr, a little later in the west. Numbers observed rarely exceed a dozen, and thus no clear migratory peak can be established.

High counts include 15 near Keystone 14 Apr 1979 (Rosche 1994a) and 12 at Gering 7 Apr 1994.

Summer

Individuals or small groups may be seen statewide during the summer months, generally fewer in the heavily cultivated areas. Highest numbers are found along the Niobrara Valley, in the Panhandle, and along the escarpments of the Republican and western Platte Rivers.

It is a regular breeder, with known nest sites concentrated in the west, north, and southeast (Ducey 1988; Rosche 1982; Hatch and Garrels

1971; Mossman and Brogie 1983; Brown and others 1996; NBR 64:47). It typically utilizes cavities and ledges on steep canyon faces or rocky outcroppings (Hatch and Garrels 1971; Mossman and Brogie 1983), but it has also utilized lofts of abandoned farm buildings (Lionberger 1944). Ed and Mark Brogie (pers. comm. RSS) found nests containing young as early as 13 Jun (downy young) and as late as 20 Aug (ready to fledge) in Brown and Keya Paha Cos in 1982.

Roosts form in summer, probably containing nonbreeders or prebreeders; age at first breeding may be several years (Palmer 1988).

Fall

summer→28 Nov, 1, 2 Dec

There is a general southward drift in fall, with high counts indicating significant movement in early Sep. Very few linger after Oct, with only 11 reports in the period 30 Oct–2 Dec. Although occurrence is questionable in midwinter, there are several Nov records from neighboring states (see Winter).

High counts include 100 at Lincoln 1 Oct 1999, 85 near Keystone 3 Sep 1993 (Rosche 1994a), and 66 at Valentine 30 Aug 1997.

Winter

The 17 reports in the period 9 Dec–18 Feb are undocumented and most are likely to be erroneous, although there is a 5 Jan record for Minnesota (Janssen 1987), 7 Iowa records Jan–Feb (Kent and Dinsmore 1996), scattered Kansas records 8 Dec–22 Jan (Thompson and Ely 1989), and Colorado records 21 Dec and 13 Jan (Andrews and Righter 1992). This species is easily confused with large soaring hawks of dark plumage that occur in Nebraska only in winter. The subspecies found in Nebraska, *C. a. teter*, typically overwinters in the subtropics and tropics.

FINDING

In summer look for dark hawklike birds usually singly or in groups of 6 or less that soar with their two-toned wings held in a "V" posture and with spread, fingerlike distal primary feathers. Although they are most likely found in open country or associated with steep hills and canyons, during migration individuals may also be seen passing over towns and cities.

ANSERIFORMES

Anatidae (Ducks, Geese, Swans)

Black-bellied Whistling-Duck

Dendrocygna autumnalis

STATUS

Rare casual summer and fall visitor.

DOCUMENTATION

Specimen

UNSM ZM16079, 29 Oct 1989 Hansen Marsh, Clay
  Co (Labedz 1990a; Grenon 1990; AB 44:115).

Records

There are 3 records, 2 documented:

2–28 Aug 1999 (2 adults) North Hultine Basin
  (JGJ; m.ob.)

29 Oct 1989 Hansen Marsh; shot by a hunter and
  brought in to NGP (notes of Joe Gabig, NGP
  files; cited above)

It is noteworthy that when the specimen cited
  above was collected, a flock of 18 was seen the
  same week in Missouri (Robbins and Easterla
  1992).

The third report was from Stewart Porterfield,
  who reported 6 Black-bellied Whistling
  Ducks on a small wetland just north of Ong
  on 15 Apr 1990, but no further details are
  available (NGP files).

COMMENTS

A question is always raised as to whether
  out-of-range waterfowl are indeed wild
  birds, especially those that appear outside
  an established pattern of vagrancy. In the
  case of Black-bellied Whistling-Duck, there
  has been a general pattern of spring and fall
  vagrancy in recent years, which has resulted
  in this species being added to the state lists
  of most states adjoining Nebraska. There are
  no records for Colorado or Wyoming, but 9
  of 10 reports of apparently wild birds from
  Missouri (Robbins and Easterla 1992), Kansas
  (Thompson and Ely 1989), and Iowa (Kent
  and Dinsmore 1996) are in the migratory
  periods 8–29 May and 20 Sep–11 Nov; the
  exception is 6 Jul. Thus it is unclear whether
  summer reports are of true vagrants or

escaped or released captives, although large
  numbers of undoubtedly wild birds were
  noted in 1999 as far north as Oklahoma.

FINDING

This species should be looked for in the
  Rainwater Basin mid to late May and also in
  Oct.

Bean Goose

Anser fabalis

STATUS

Accidental in spring. Accidental in winter.

DOCUMENTATION

Photograph

29 Dec–10 Jan 1984–85 DeSoto Bend NWR (AG,
  Rose 1985).

Records

An individual was present with Snow Geese at
  DeSoto Bend NWR 29 Dec–10 Jan 1984–85.
  A photograph was published (cited above),
  and many observers had the opportunity
  to see the bird. This bird was assignable to
  the subspecies *middendorfi*. Another was
  observed at Funk Lagoon 4 Apr 1998 with
  Greater White-fronted Geese; it was also
  identified as of the race *middendorfi* (SJD).

COMMENTS

The eastern Siberian race *middendorfi* is rarely
  kept in captivity, which suggests wild origin
  for stragglers to Nebraska (Bray and others
  1986). The only other record of this species in
  the continental United States is from Gray's
  Harbor, Washington, 26 Apr 1993 (ABA 1996).

Greater White-fronted Goose

Anser albifrons

STATUS

Abundant regular spring and fall migrant
  central, common east, rare west. Rare regular
  summer visitor central and east. Rare regular
  winter visitor central and east.

DOCUMENTATION

Specimen

UNSM ZM12412, 1937 or 1938 Lancaster Co.

TAXONOMY

Taxonomy of this species in North America is in

a confused state (for a discussion, see Banks 1983). It is generally agreed that the most numerous subspecies, breeding on arctic tundra from northeastern Siberia eastward to Hudson Bay, is *A. a. frontalis* (AOU 1957; Ely and Dzubin 1994). Most *frontalis* migrate through Nebraska to winter in southeastern Texas and northeastern Mexico, although southwestern Alaska breeders migrate along the Pacific Flyway (Ely and Dzubin 1994).

Some authors (Palmer 1976) use the name *gambelli* (the case for spelling this name *gambeli* is made by Banks 1983) for birds breeding in North America (except for southwestern Alaska) and wintering in Texas and Mexico. This name was first applied by Hartlaub (1852) to large, dark individuals collected during winter in Texas and thought to be representative of interior North American White-fronted Geese. However, Swarth and Bryant (1917) used the name *gambelli* for the large, dark birds wintering in California, since they assumed that, because these birds were rare in California, their main breeding range was eastward in Canada and thus they were the same as the birds described by Hartlaub (Banks 1983). Delacour and Ripley (1975) named the California birds *A. a. elgasi* (Tule Goose) on the assumption that they were the westernmost of two taiga-breeding populations distinct from the vastly more numerous *frontalis*, a tundra breeder, and used *gambelli* for the eastern birds.

The western and eastern taiga-breeding populations use separate breeding, migration, and wintering areas (Palmer 1976), with minimal crossover between flyways (Ely and Dzubin 1994), and a case can be made that they are genetically isolated from each other. Very little is known about the eastern population, but it is thought to breed near the MacKenzie River delta (Palmer 1976).

Some authors (Banks, Ely, pers. comm.) currently believe these birds to be within the range of variation of *frontalis* and thus not separable from them. There have been sightings of large, dark ("chocolate-colored") White-fronted Geese in the North American interior (Nick Lyman, Martin Reid, pers. comm.), which may refer to *gambelli* (per Hartlaub).

Of interest are band recoveries in Kansas and south-central Texas in the early 1980s of Tule Geese banded in the Pacific Flyway; these recoveries (4 out of about 300 birds banded) suggest a crossover of migrants from the Pacific to Central Flyways of about 1% (Ely, pers. comm.).

DISTRIBUTION AND ECOLOGY

As many as 1 million birds, more than half of the North American White-fronted Goose population, migrate through Nebraska between arctic breeding grounds and Gulf Coast wintering areas. The greater portion of the migratory movement is concentrated in the central part of the state. The spring movement includes a major buildup in early Mar in the Rainwater Basin, notably around Funk Lagoon, and along the Platte River between North Platte and Columbus. Far fewer occur along the Missouri River floodplain, but occasionally flocks may number in the several hundreds. The Panhandle population is very small, but on rare occasions flocks of a hundred or more are seen. Like Snow Geese, White-fronted Geese forage in fields in the morning and again late in the day, spending midday and the night on lakes, in marshes, and on sandbars in the Platte Valley. Most of the population remains in the state for just a short time, building up for 2 or 3 weeks and leaving in just a few days.

The fall migration is far less dramatic. White-fronted Geese frequently fly nonstop from resting areas in Alberta and Saskatchewan to wintering grounds in Texas, often overflying the state at altitudes of 2135–3050 m (7000–10 000 ft). Small groups may on occasion settle and rest for a day or two before continuing south.

Spring

Movement typically begins in late Feb and continues through late Apr, with largest concentrations in Mar, although there are many earlier and later reports. Arrival in 1998 was 6–7 Feb at 3 locations. Few occur in the Panhandle; Rosche (1994a) noted that large numbers occur westward in the Platte Valley to the Hershey-Sutherland area in Lincoln Co but cited only a single spring record in the Keith Co area. Rosche (1982) considered it a "very rare" spring transient in the northwest.

High counts include 100 000 at Funk Lagoon 12 Mar 1994, 6000 there 1 Apr 1995, and 5000 there 4–25 Feb 1995. An unusually high count for a westerly location was 2141 on 12 Mar 1994 in Perkins Co.

Summer

Reports are fewest between mid-May and late Sep, probably injured birds unable to migrate. Most reports are from central Nebraska, very few from the east, and it is unrecorded in midsummer in the Panhandle.

Fall

Movement is often protracted, beginning in late Sep and continuing into Dec, with a peak in mid-Oct. There are many earlier and later reports, however. Rosche (1982, 1994a) indicated that west of Lincoln Co in the Platte Valley this species is rare. Up to 8 were in Scotts Bluff Co 22–29 Nov 1997 (SJD).

High counts include "several hundred" at Funk Lagoon 12 Nov 1994, 300 over the eastern Rainwater Basin 10 Oct 1999, 200 at Funk Lagoon 23 Nov 1995, and 125 at Gavin's Point Dam 4 Nov 1995.

Winter

Reports of stragglers into late Dec are not uncommon, and individual birds are sometimes found wintering with flocks of Canada Geese. Unusual in magnitude for the date, however, was a count of 5000+ in the eastern Rainwater Basin 17 Dec 1998. Reports are fewest early Jan through mid-Feb, most from central Nebraska. The only Panhandle records are of one at Scottsbluff SL 19 Jan 1998 and one at L Chappell, Deuel Co, 23 Jan 1999.

FINDING

A trip to marshes of the Rainwater Basin during Mar should yield numerous opportunities to see this sleek, fast-flying goose.

Emperor Goose

Chen canagica

STATUS

Accidental in spring.

Records

A dead adult was picked up during a cholera outbreak at Harvard Lagoon 17 Mar 1997 (NBR 65:77). It showed no signs of captivity and was determined to be a wild bird by the NOURC (Brogie 1998). William Lemburg, of Cairo, Nebraska, has kept this species in captivity and has "never lost one" (pers. comm. WRS).

COMMENTS

This species occurs with some regularity along the Pacific Coast, with the southernmost record in Orange Co, California. About half the records have been inland but not far from the coast, such as in the Lower Klamath Basin; there is only a single record away from the coastal states, in Nevada 11 Dec 1960 (Mlodinow and O'Brien 1996). Emperor Geese found inland tend to occur with "Cackling" Canada Geese or Greater White-fronted Geese (Mlodinow and O'Brien 1996), both of which have significant migratory populations passing through Nebraska that are likely to mix with Pacific Coast populations in the southwestern United States and northwestern Mexico (Bellrose 1976).

Snow Goose

Chen caerulescens

STATUS

Abundant regular spring migrant central and east, common west. Uncommon regular summer visitor central, rare elsewhere. Hypothetical breeder. Abundant regular fall migrant central and east, uncommon west.

Rare regular winter visitor east and central, uncommon casual west.

DOCUMENTATION

Specimen

UNSM ZM10619, 28 Mar 1917 Lancaster Co.

TAXONOMY

Until the late 1960s the blue and white color morphs of Snow Goose were treated as separate species, Snow Goose, *C. hyperborea*, and Blue Goose, *C. caerulescens*. NOU Occurrence Reports maintained the distinction until 1973.

The genetic distribution of color morphs varies between segments of the breeding population of "Lesser" Snow Goose, *C. caerulescens*, the subspecies occurring in Nebraska. The proportion of the blue morph declines westward and may shift over time within segments (Cooke and Cooch 1968; Cooch 1961). Breeding populations tend to migrate as a unit, and currently the white morph overwhelmingly predominates in the Nebraska Panhandle migratory population (Rosche 1994a), whereas the blue morph is predominant in the eastern Nebraska migratory population.

DISTRIBUTION AND ECOLOGY

The bulk of the Snow Goose population migrates through the central Great Plains and Missouri River valley between nesting areas on the arctic tundra and typical overwintering areas from the central Missouri River basin southward to the Gulf Coast.

Before the development of a series of rest refuges in the eastern Dakotas and along the Missouri River in the 1950s and 1960s, Snow Geese were a common migrant throughout the state. However, after these refuges were established, the migrating population has shifted significantly to the east. Where once a regular fall migrant in the Panhandle, it is now irregular, and where once abundant in spring, it is now less common though still regular (Rosche 1982, 1994a).

The largest migrant concentrations are observed in central and eastern Nebraska, although there is some tendency for fall concentrations to be largest along the eastern border, where a buildup in numbers to around 500 000 is often observed at refuges such as DeSoto Bend NWR over a period of several weeks in late fall. Numbers may remain high at favored locations until increasingly inclement weather and a diminishing food supply force the flocks farther south.

During the spring movement flocks may be broadly distributed from central to eastern Nebraska, often developing concentrations in the tens of thousands over a period of weeks along the central Platte Valley, in the Rainwater Basin, and eastward to the Missouri River. Prior to the past 2 decades, the largest spring buildup, as many as 900 000 birds, was traditionally in the Missouri River floodplain from DeSoto Bend NWR south; however, very large flocks are more commonly developing on or near the Platte River west to about Kearney and especially in the Rainwater Basin. Available food supply and water levels in resting areas such as the Rainwater Basin may have contributed to these regional shifts.

Spring

The onset of spring movement is dependent on weather and snow cover. Thus in years with heavy snow cover and a late Mar melt no or few migrants are observed until the beginning of the melt. In years with little or no snow cover early arrivals may be observed in mid to late Feb and concentrations may build into mid-Mar. Unusual were as many as 100 000 in Buffalo Co 14 Feb 1999. The bulk of the movement is typically observed in mid-Mar, after which time concentrations diminish in size. Migrating flocks, however, may be observed throughout Apr and less commonly into May.

High counts in the Rainwater Basin include 1 000 000 in Fillmore and Clay Cos through 2 Mar 1995, 400 000 in Phelps Co 12 Mar 1994, and 400 000 at Funk Lagoon 8 Mar 1997; in the east 10 000 in Dixon Co 21 Mar 1998, 5000 at Wagon Train L 27 Mar 1996, and 3000 in

Dixon Co 13 Mar 1994; and in the west 4700 at
L Minatare 2 Mar 1996.

Summer

Several summer records exist, often involving
injured or diseased individuals, although on
occasion small, apparently healthy, groups
remain through summer, especially in the
Rainwater Basin. Numbers are lowest late
May–late Sep.

Evidence for nesting is scant, although the
propensity for injured birds to summer,
along with their healthy mates, suggests that
occasional nesting is not unexpected. Ducey
(1988) discussed published reports, 2 of which
are likely to be correct. Swenk (Notes before
1925) relates that a nest with eggs along the
Platte River south of Kearney was seen by
C. A. Black, although when Black arrived
the nest was destroyed but still attended by
2 Snow Geese. Four Snow Geese were said
to be nesting at Home L, Sheridan Co, in
1929 (Ducey 1988). Two additional records
of breeding are on file at the NGP in Lincoln:
a pair with young was observed Jun 1990 at
Willow Creek L, and 2 goslings were produced
by a pair, one of which was injured, at a
recharge lake west of York in 1996.

High counts include 42 at County Line Marsh
1 Jun 1997 and 21 at North Lake Basin late
Jun–11 Aug 1995.

Fall

Early migrants are often observed during the
first third of Oct, and the migration continues
into early Nov. However, the fall movement
may vary by as much as 2 or 3 weeks,
depending upon weather and food conditions
farther north. A flock of 95 was flying south
over Seward Co as early as 9 Sep 1999 (JG).
Snow Geese may remain well into Dec and
into Jan in particularly mild winters.

Far fewer occur in the west in fall than in spring
(Rosche 1994a). Fall counts are highest along
the Missouri River; Schilling Refuge often
sustains numbers in excess of 5000 in late Nov.
Late fall counts at DeSoto Bend NWR often
reach 500 000, and, reflecting the current

population explosion, an all-time high of
630 000 birds were there in late Nov 1998.

High counts in the west include 1075 at North
Platte NWR 9 Nov 1997, 800 at Whitney L 19
Nov 1994, and 220 over Ponderosa Area 28
Oct 1995.

Winter

Small numbers may remain throughout the
winter in mild conditions, mostly in southeast
Nebraska. As many as 50 000 have remained
near Clay Center in mild winters (Joe Gabig,
pers. comm. RSS), but scattered birds occur
elsewhere. Small groups occasionally venture
into the state from adjacent overwintering
areas in northeast Kansas and northwest
Missouri during extended warm spells in
midwinter.

FINDING

During spring migration look for large flocks
along the Missouri River floodplain from
DeSoto Bend NWR southward, along the Platte
River floodplain in the Kearney–Grand Island
area, and especially in the Rainwater Basin,
which may support several million birds.

Snow Geese are often seen in cultivated fields,
grazing on spilled grain or young winter
wheat and oats. Grazing periods include early
to midmorning and mid to late afternoon.
Snow Geese prefer to spend midday and
overnight on lakes and in marshes, less often
on river sandbars, and therefore strings of
geese may be seen moving between aquatic
resting habitat and terrestrial feeding grounds.

Ross's Goose

Chen rossii

STATUS

Fairly common regular spring migrant statewide.
Uncommon regular fall migrant central and
east, rare casual west. Accidental in winter.

DOCUMENTATION

Specimen

UNSM ZM11255, 6 Nov 1922 Hooper (Tate 1966).

TAXONOMY

As with Snow Goose, 2 color morphs are
recognized. There are 2 reports of the rare

blue morph Ross's Goose: a single bird near Pierce 19 Mar 1994 (Gubanyi 1996c) and another in Stanton Co 26 Feb 1998 (DH); it is possible these reports are of the same individual.

DISTRIBUTION AND ECOLOGY

In Nebraska this species is generally associated with Snow Geese, utilizing the same habitats. Numbers are probably greater in the Rainwater Basin and Panhandle than in the east, which would be expected based on the essentially western range of Ross's Goose.

In recent years this species has expanded its breeding range eastward as its numbers have increased, thus mingling with Snow Goose populations that migrate into Nebraska. In 1963 it was noted that Ross's Goose was unrecorded in Nebraska (Lueshen 1963). It may have been overlooked, however, as 79 were trapped and banded in Box Butte Co with Snow Geese in spring 1963–68. Huntley (1971) and Tate (1966) published information on a mounted specimen (cited above) taken by a hunter near Hooper on or before 6 Nov 1922, the first Nebraska record. The species has been reported regularly since about 1970.

Most observers do not carefully count Ross's Geese among Snow Geese, a tedious task, but a check of 1200 Snow Geese killed in York Co in a tornado yielded 2% Ross's Geese (Johnsgard 1997), which extrapolates to some 20 000 in the Rainwater Basin with the current population of about 1 000 000 Snow Geese using that area (JGJ). Prevett and MacInnes (1972) found that the ratio of Ross's to Snows in northwest Missouri in fall may be as high as 1:1000.

Spring

19, 22, 22 Feb→8, 9, 9 May

There are earlier reports 4–5 Feb 1995 at Funk Lagoon (SJD, WRS) and 7 Feb 1998 (2) at Grandpa's Steakhouse Pond and later reports, all from the Rainwater Basin: 15 May and 10 Aug–5 Sep 1998, apparently the same bird at a wetland in Clay Co; 16 May 1997 Adams Co;

17 May 1997 Fillmore Co; and 21 May 1997 Clay Co.

Records indicate that spring and fall periodicity parallels that of the Snow Goose, peaking in Mar. A female trapped in Box Butte Co 1 Apr 1963 was neck-banded in Kindersley, Saskatchewan, the previous fall (Sweet and Robertson 1966). The Panhandle spring Snow Goose population winters in north-central Mexico and arrives there in fall via a westerly route through the Rockies from staging areas on the primary Ross's Goose fall migration corridor (Bellrose 1976).

High counts include 500 at Funk Lagoon 29 Mar 1997, 250 at Sutherland Res 29 Mar 1997, 187 at Cochran L 19 Apr 1997, and 30+ out of "possibly hundreds present" at Funk Lagoon 14 Mar 1994.

Fall

4, 6, 10 Nov→4, 6, 6 Jan

Migration appears to peak in mid-Nov. There is an early report 16 Oct 1988 Douglas-Sarpy Cos. Ross's Goose may not be as numerous in fall as it is in spring, even in the east (JGJ). This may have been true in the Panhandle for some time, as Rosche (1982) noted only spring records there, and Bellrose (1976) did not show a fall migration in the west. It is currently casual in the west; the few reports are in the period 10–30 Nov: 9 Nov 1997 North Platte NWR, 10–11 Nov 1977 (88) Scotts Bluff Co, 11 Nov 1994 north of Antioch in Sheridan Co, 12 Nov 1979 Garden Co, 19 Nov 1999 (12) Sutherland Res (SJD), 23 Nov 1999 (2) Kiowa Springs (SJD), and 30 Nov 1977 Garden Co.

High counts include 88 at North Platte NWR 9 Nov 1997, 33 at Harlan Co Res 18 Dec 1998, and 13 at Keystone L 2 Nov 1997.

Winter

The only report of overwintering is of 3 at Grandpa's Steakhouse Pond with 4 Snow Geese 22 Jan 1995.

FINDING

Ross's Geese are usually found with flocks of Snow Geese. Careful observation of settled flocks may reveal individuals, pairs, or family groups of this species. Favored locations for Ross's Geese are Grandpa's Steakhouse Pond just south of Kearney and North Platte SL.

Canada Goose

Branta canadensis

STATUS

Abundant regular spring and fall migrant statewide. Common regular resident statewide. Locally common regular winter visitor statewide.

DOCUMENTATION

Specimen

WSC 705, 22 Mar 1981 Hall Co.

TAXONOMY

Canada Goose taxonomy is complex because this polytypic species has a broad breeding distribution across much of northern North America, with discrete breeding populations that tend to migrate and winter as units, thus maintaining a significant degree of genetic integrity. Size of individuals and extent of dark feathering are reasonably diagnostic; smallest individuals are typically from arctic populations, while large geese, as much as three times the size of their northern relatives, are produced in southerly regions with long growing seasons.

As many as 10 subspecies were described by AOU (1957), 5 of which had been reported in Nebraska (Rapp and others 1958). Since then there have been additional attempts to clarify subspecific relationships. Bellrose (1976) attempted to simplify the taxonomy and designated migratory units. Units that frequented Nebraska were "High Plains" (small and large Panhandle migrants), "Western Prairie" (large geese that are from the south-central interior), and "Tallgrass" (small and large geese using the Missouri Valley). On the other hand, Palmer (1976) attempted to clarify the evolutionary histories and relationships among populations and modified the AOU (1957) classification. Together these works have left considerable taxonomic confusion that also impacts our understanding of the populations found in Nebraska.

Reintroductions during the last 20 or so years have to some extent obscured the taxonomic status of Nebraska Canada Geese, as released birds are often of mixed or unknown heritage. As such birds establish breeding populations, they have in some cases developed new wintering or migrating units. Resident birds in historic times, thought to be the subspecies *maxima*, are generally believed to have been extirpated, and recently reintroduced large geese have been of mixed subspecies (see below) but primarily *maxima* and *moffitti*.

It has been suggested recently (AOU 1998) that at least 2 species may be included in Canada Goose, designated Tundra Goose, *B. hutchinsii*, and Canada Goose, *B. canadensis*. Of currently constituted subspecies occurring in Nebraska, assigned to Tundra Goose are *hutchinsii*, *minima*, and *taverneri* and to Canada Goose *interior*, *maxima*, *moffitti*, and *parvipes*. The taxon *taverneri* has been reported in Nebraska (Rapp and others 1958), but Palmer (1976) considered this taxon a result of interbreeding between *parvipes* and *minima*. Palmer (1976) also considered *maxima* and *moffitti* inseparable and named the combined taxon *moffitti*. The AOU (1998) further notes that the small subspecies may be separate species within the Tundra Goose group, Hutchins's (or Richardson's) Goose, *B. hutchinsii*, and Cackling Goose, *B. minima*.

In summary, we prefer Palmer's (1976) taxonomy, recognizing the following populations that frequent the state: *interior*, a medium-size to large goose that breeds in east-central Canada and migrates primarily along the Mississippi Flyway, with small numbers using the Missouri Valley and wintering from Squaw Creek NWR southward; *moffitti*, a very large goose of the central and northern Great Plains found throughout Nebraska but whose taxonomic integrity has been highly corrupted by reestablished resident birds;

hutchinsii, a small arctic goose that migrates through central and eastern Nebraska, with high concentrations in the Rainwater Basin in spring; and *parvipes*, a small arctic goose that migrates through the Panhandle and the L McConaughy area, sometimes farther east, and winters in small numbers in Nebraska but mostly from southeast Colorado southward.

DISTRIBUTION AND ECOLOGY

Although migrants may be seen statewide, significant concentrations are distributed along the North Platte River, the central Platte River, the Rainwater Basin, the Republican Valley, and the Missouri River floodplain. Migrants may utilize preferred lakes and reservoirs for several days to weeks. Overwintering populations tend to be strongly site-specific although highly subject to weather conditions and food availability.

Canada Geese are grazers, and large flocks are particularly attracted to fields of young winter wheat, rye, and oats during the fall, winter, and spring. They may also forage in harvested cornfields.

During the summer months Canadas are most often seen as pairs or family groups. During the breeding period they are much more secretive and utilize marshes, marshy streams, and river islands and vegetated sandbars, more often restricting their foraging to that available in the immediate habitat. Those that have chosen breeding sites on lakes and reservoirs proximate to parks and housing developments are rapidly becoming semidomesticated and may be seen grazing on lawns and in gardens, a habit that has led them to be viewed as pests elsewhere (Andrews and Righter 1992) and in Nebraska.

Spring

Although Canada Geese are now found throughout the state almost year-round, large spring and fall migratory movements are evident as a result of an influx of smaller forms than those that breed in Nebraska as well as other large geese that breed to the north of Nebraska. The major spring movement peaks in early to mid-Mar. Large concentrations of small-size geese begin to develop in late Feb, particularly in south-central Nebraska, and numbers are greatly reduced by early Apr, although a "Richardson's Goose" (*B. c. hutchinsii*) was at L Ogallala 14–15 May 1998 (SJD), and birds of small races were at Harvard Marsh 14 May 1999 (SJD), Funk Lagoon 16 May 1999 (SJD), and Harvard Marsh as late as 6 Jun 1999 (JGJ). Rosche (1994a) stated that migration peaks consist of some 25 000 birds in the L McConaughy region.

High counts include 100 000 at Funk Lagoon 12 Mar 1994, 100 000 in the Rainwater Basin 5 Mar 1995, 60 000 at Funk Lagoon 25 Feb 1996, 3509 at North Platte NWR 4 Mar 1996, 500 in Dixon Co 13 Mar 1994, and 500 in Cass Co 15 Mar 1994.

Summer

Historically, breeding populations were limited to the Missouri River valley, Sandhills lakes, and Platte River valley (Bruner and others 1904; Ducey 1988), but these were extirpated before the end of the 19th century. Presumably these birds were part of the Western Prairie population, mostly *maxima* or *moffitti*, but no specimens are available for confirmation.

Contemporary breeding populations are scattered statewide and are the result of reintroduction efforts. The earliest-known public reintroduction efforts were in 1936 at Crescent L NWR and in 1951 at Valentine NWR; although problematic in their beginnings, local breeding populations have now been established in the Sandhills (Gabig 1986). Rosche (1994a) noted that nesting was first observed in the L McConaughy area near Lewellen in 1985, and breeding is now common along the North Platte River in Garden Co. Individuals in these reestablished populations are of mid to large size, some approaching the measurements of the largest *moffitti* (= *maxima*), but expansion was at first slow owing to hunter vulnerability (J. Wilbrecht, pers. comm. RSS).

In the late 1960s and throughout much of

the 1970s the NGP conducted a statewide reintroduction effort that has resulted in the establishment of numerous local breeding populations that are rapidly expanding, particularly in eastern Nebraska. These geese are also of mid to large size and were obtained from a captive flock maintained by the NGP (supposedly *maxima*) as well as 100 *moffitti* obtained from the Colorado Division of Wildlife (Ducey 1988).

The resident population continues to grow. Many counties not recorded as having breeding activity by Ducey (1988) now have breeding populations, and "thousands of pairs of large Canada Geese now breed in the state" (Johnsgard 1997).

The earliest brood recorded is 10 Apr at Crescent L NWR (Zeillemaker, pers. comm. RSS). Brood sizes observed range from 1 to 14; however, broods often mix and move between parents.

High counts include 1030 at North Platte NWR 14 Jun 1994 and 619 there 25 Jun 1997.

Fall

The migration of small- to mid-size geese begins in late Sep and reaches its peak in mid-Oct. Flocks may remain into early winter depending upon weather and food availability. Larger geese arrive in mid-Nov or later and may remain throughout the winter, depending on food availability and weather.

High counts include 36 900 in the Kearney area 21 Dec 1996, 30 575 at North Platte NWR 16 Dec 1996, and 20 000 in the Kearney area 16 Dec 1995.

Winter

Major overwintering populations occur along the North Platte Valley, including L McConaughy, where about 12 000 winter (Rosche 1994a); at Sutherland Res; along the central Platte Valley; in the vicinity of Harlan Co Res; in the vicinity of Sherman Res; and along the eastern Platte Valley in Douglas and Saunders Cos. Midwinter waterfowl surveys by the NGP for 1997 and 1998 found overwintering Canada Geese along the North Platte River, South Platte River, central Platte

River, lower Platte River below Fremont, lower reaches of the Loup River system, and numerous lakes and reservoirs scattered across the state.

Some of the overwintering flocks, particularly in the Panhandle, are composed of small arctic forms such as *parvipes* and *hutchinsii*, but the majority are medium-size and large geese, apparently *moffitti*. In the last decade we have seen increasing numbers of larger geese overwintering, particularly during mild periods. These are composed of populations of *moffitti* and *interior*, but some are likely locally and regionally produced and thus of mixed heritage (Gabig 1986; Rosche 1994a).

High counts of wintering birds include 52 000 at Sutherland Res in 1997 (NGP), 46 000 at Grandpa's Steakhouse Pond in 1997 (NGP), 20 000 at Sutherland Res 5 Feb 1995, and 20 000 in Kearney Co 5 Feb 1995.

FINDING

Canada Geese are found near or on nearly every body of water statewide, particularly during migration and in summer months. Largest concentrations may be found along the Platte River and at marshes in the Rainwater Basin in early to mid-Mar.

Brant

Branta bernicla

STATUS

Rare casual spring and fall migrant statewide.

DOCUMENTATION

Specimen

HMM 2557, 10 Nov 1916 Phillips, Hamilton Co.

TAXONOMY

Two races are recognized, the Atlantic, or light-colored, race, *B. b. bernicla*, and the Pacific, or black, race, *B. b. nigricans*. Of 7 Nebraska documented records for the species, 6 are *bernicla*. On geographic grounds, *bernicla* would be most likely to appear in Nebraska, as *nigricans* is essentially a Pacific Coast migrant.

Recently support has grown for reelevating the subspecies to species status (AOU 1998), as constituted previously (AOU 1957) but with an additional split of the eastern North American and European subspecies, *bernicla* and *hrota*, which previously were part of Brant, *B. bernicla*. The North American species would be Black Brant, *B. nigricans*, and Pale-bellied Brent Goose, *B. hrota*.

Spring

There are 6 published reports, 3 documented, all *bernicla*:

22 Feb 1998 Kearney Co (LR, RH)

21 Mar 1988 Funk Lagoon (Fall 1988; Mollhoff 1989a).

22 Apr 1990 Phelps Co (Grenon 1990)

Additional reports without details, all of which fit the above pattern of occurrence, are 18 Feb 1968 (2) between Kearney and Darr (NBR 36:67), 18 Mar 1959 Nemaha Co, 26 Mar 1957 Adams Co, 26 Mar 1960 Webster Co (Turner 1960), 30 Mar 1952 Adams Co, 30 Mar 1958 Buffalo Co, and 20 Apr 1952 Butler Co.

Fall

There are 15 reports, 5 documented, including 2 records of *nigricans*:

7 Oct 1930 Kearney (Brooking 1934)

10 Nov 1916 Phillips (*nigricans*, cited above)

30 Nov 1910 (12) Dawes Co (Zimmer 1911a)

23 Nov 1999 L Minatare (SJD) (*nigricans*)

7 Dec–6 Jan 1988–89 DeSoto Bend NWR (NBR 57:3, 11, 60, 65; AB 43:331; Silcock 1989b)

There are additional reports without details that all fit the pattern of the documented reports: 9 Nov 1895 (3) Omaha (Bruner and others 1904), one killed Nov or Dec 1986 at Clear Creek Marshes (Rosche 1994a), 19 Dec 1967 (3) Clear Creek Marshes (Rosche 1994a), an undated report of one shot on the Platte River near Omaha (Bruner and others 1904), one reported in 1960 in the L McConaughy region (Rosche 1994a), another among a flock of 3 that was killed near Lewellen (Rosche 1994a), and one shot 1960–61 at Kingsley Dam, listed as "Black Brant" (Fuller 1961).

COMMENTS

The preponderance of reports of *bernicla*

suggests that a few birds travel with the populations of Canada Geese or Snow Geese that migrate around or along the west shore of Hudson Bay. The 1988–89 DeSoto Bend NWR Brant was with Snow Geese. The Canada Goose population reaching Nebraska from Hudson Bay is the tallgrass prairie population that migrates through central and eastern Nebraska and winters from Kansas south to Texas (Bellrose 1976).

FINDING

Brant should be looked for among Canada Goose flocks in the North Platte Valley, the Rainwater Basin, and eastward, as well as with Snow Geese (less likely) in the Missouri River valley. Best times are mid to late Mar and Nov.

Barnacle Goose

Branta leucopsis

STATUS

Hypothetical in spring and fall.

COMMENTS

There are 3 reports. A photograph of a mounted specimen taken 2 Nov 1968 in Otoe Co has been published (NBR 37:2). Another was at a sandpit pond south of the Odessa bridge in Phelps Co 9 Mar 1995; the NOURC did not accept this as a wild bird (Gubanyi 1996c). One was with small Canada Geese at Massie Lagoon, Clay Co, 28 Mar–4 Apr 1998 (JS), and possibly the same bird was at Harvard Lagoon 9 May 1998 (JGJ).

In the surrounding states Missouri has 3 reports 25–26 Mar and 28 May in spring and 2 reports 20 Oct–6 Nov in fall; Robbins and Easterla (1992) suggested that these reports remain hypothetical because the species is commonly raised in captivity. A case can be made, however, that wild birds indeed reach North America unaided and could reach the interior (Mlodinow and O'Brien 1996). It is casual Oct–Apr in the northeast United States, with reports south to Colorado, Texas, and the Gulf Coast, and while many may be escaped birds, there are banding recoveries of European birds (ABA 1996).

Mute Swan

Cygnus olor

STATUS

Hypothetical.

COMMENTS

There is no established population breeding
unaided in Nebraska. This species is generally
sedentary, but in severe winters individuals
migrate, especially immatures. An immature
banded in Wisconsin appeared on the
Mississippi River near Lincoln Co, Missouri,
in 1985 (Robbins and Easterla 1992). Thus
unmarked immatures, which would be most
likely to appear in eastern Nebraska, could be
wild birds. Such a bird was at Crystal Cove
L 9 Nov 1991, but the NOURC was unable to
rule out the possibility that it was an escapee
(Gubanyi 1996a). Three adults were on a lake
near Odessa 5–8 Jun 1997 (LR, RH), probably
also escapees. Bray and others (1986) list 2
reports of escaped birds.

Trumpeter Swan

Cygnus buccinator

STATUS

Rare casual spring and fall migrant statewide.
Locally uncommon regular resident Sandhills.
Locally common regular winter visitor
central.

DOCUMENTATION

Specimen

HMM 573, about 15 Mar 1898 Overton, Dawson
Co.

DISTRIBUTION AND ECOLOGY

Trumpeter Swans graze on aquatic submerged
vegetation and thus are attracted to large
marshes with open water that support
abundant food as well as extensive emergent
vegetation that provides cover. Most
appropriate habitat exists in the Sandhills in
Cherry, Sheridan, and Garden Cos. Trumpeter
Swans often remain in the breeding area
throughout winter months if the water
remains open but become more nomadic
as lakes ice over and the food supply is
depleted.

Spring

Most of the breeding birds in the Sandhills
winter at LaCreek NWR (SDOU 1991) or in
open-water streams such as Blue Creek in
Garden Co (Roger Knaggs, pers. comm. WRS)
and the North Platte River below L Ogallala.
Thus migratory movement is minimal in this
population. In recent years, however, perhaps
as a result of expansion of the LaCreek NWR
population, a few birds have begun to migrate
longer distances, such as a female killed in
1978 in Missouri that was banded at LaCreek
NWR in 1972 (Robbins and Easterla 1992).
Palmer (1976) noted that Alberta banding
data since 1954 included fall recoveries in
western Nebraska.

There are a few reports away from the Sandhills
in recent years, which indicate some
migratory tendency: Feb 1996 (7) Hayes Co; 21
Feb 1992 Scotts Bluff Co; 5 Mar 1982 Lincoln
Co; 7–21 Mar 1998 Kiwanis Park, Omaha;
16–21 Mar 1990 Cedar Co; 17 Mar 1995 (5)
Cunningham L; 27 Mar 1993 Douglas-Sarpy
Cos; 29 Mar–12 May 1997 Buffalo Co; 30 Mar
1962 Scotts Bluff Co; 2 Apr 1998 Zorinsky L;
and 14 Apr 1997 Wood Duck Area. Some of
the birds seen in eastern Nebraska apparently
originate from reestablished populations in
southern Minnesota and northern Iowa (John
Dinan, pers. comm. RSS, WRS); many are
marked with patagial tags or neck collars.

Summer

Prior to 1900 Trumpeter Swans apparently bred
in the Sandhill lakes region of Nebraska and
South Dakota in small numbers (Ducey 1988)
but were extirpated before the end of the
19th century. The resurgence of the small
Trumpeter Swan population at Rcd Rocks
NWR in Montana led the USFWS to explore
reintroductions of swans elsewhere in their
former range.

In 1960–63 cygnets from Red Rocks were
introduced to LaCreek NWR in South Dakota,
just north of the Cherry Co, Nebraska,
border. By 1987 descendants of these swans
numbered 268 individuals (SDOU 1991). Some

of these have exploited appropriate habitat in Nebraska, particularly in Cherry and Sheridan Cos, establishing a small but regular breeding population numbering about 150 in 1995 (Johnsgard 1997). The first modern Nebraska nesting record was in 1968 at Hoover L in northeast Sheridan Co (Ducey 1988; Peabody 1974). Since then, the breeding population has expanded into most parts of the Sandhills. To date, nesting has been reported in Cherry, Sheridan, Garden, Morrill (Johnsgard 1979), Arthur (Rosche 1994a), Grant (Rosche 1994a), McPherson (Rosche 1994a), and Holt (Blake and Ducey 1991) Cos. Attempts at reintroduction of swans at Crescent L NWR during the late 1970s were unsuccessful (RSS), but nesting has occurred there more recently.

It should also be noted that at least one rancher has attempted to establish Trumpeters from private game-farm stock. A specimen found by the NGP in Dundy Co 1 Jun 1975 (UNSM ZM13397) may have been associated with that rancher's private attempts, as well as a bird reported along Frenchman Creek in Chase Co in the late 1980s and reports 24 May 1991 and 22 Jun 1991 in Chase Co.

The only summer report away from the breeding range is of an unbanded bird at Hansen Lagoon 13 Jun 1999 (JGJ).

Fall

The increasing incidence of late fall and winter reports for this species away from the Sandhills suggests that local migration is developing. Records include 28 Oct 1999 (4) L Ogallala, 7 Nov 1979 Scotts Bluff Co, 13 Nov 1988 Lancaster Co, 17 Nov 1987 Pierce Co, 23 Nov 1981 Scotts Bluff Co, 23 Nov 1991 Buffalo Co, and 26 Dec 1989 Lancaster Co.

Birds derived from reestablished flocks appear on occasion; these are usually marked with neck collars or patagial tags. Two from Storm Lake, Iowa, were at Wolf L 18 Jan–7 Mar 1997.

Winter

As noted, Trumpeter Swans become somewhat nomadic during winter months when the food supply becomes depleted or inaccessible because of ice cover, although most Sandhills nesters winter in or near the breeding range, notably at LaCreek NWR (SDOU 1991), Blue Creek in Garden Co, the North Platte River below L Ogallala (Rosche 1994a), and in Scotts Bluff Co.

High counts include 130 along the Snake River from Merritt Res to near Nenzel in winter (Ducey 1999), 70+ along Blue Creek 1994–95 (Roger Knaggs, pers. comm. WRS), and 58 along the North Loup River near Mud L, Cherry Co, 6 Mar 1998 (Ducey 1998).

FINDING

The best opportunities for viewing Trumpeter Swans are during summer months at Valentine NWR and the large lakes and marshes found along Highway 20 in western Cherry and northern Sheridan Cos. Wintering birds occur most years on L Ogallala or the nearby North Platte River.

Tundra Swan

Cygnus columbianus

STATUS

Rare regular spring and fall migrant statewide. Uncommon casual winter visitor statewide.

DOCUMENTATION

Specimen

HMM 2822, 27 Oct 1917 Doniphan, Hall Co.

DISTRIBUTION AND ECOLOGY

Most reports of Tundra Swans are from larger bodies of water, where both migrants and wintering birds occur. This species usually is seen in small flocks, probably family groups. They are grazers and thus may also be seen frequenting fields of winter wheat and sometimes harvested cornfields.

This species may be easily confused with the resident Trumpeter Swan, which is about 20% larger but otherwise can be difficult to distinguish.

Spring

25, 25, 27 Feb→6, 6, 6 Apr

Records suggest peak movement during the

latter third of Mar with later reports, none recent, 27 Apr–14 May 1952 Platte Co (Kinch 1968), 30 Apr–3 May 1936 Clay Co (Kinch 1968), 4 May 1933 (14) Sheridan Co, 5 May 1947 Brown Co (Youngworth 1955), 7 May 1930, 1931, or 1932 Sheridan Co (Rosche 1982), and 15 May 1952 Boone Co. It is likely that these reports were indeed of Tundra Swans, as Trumpeter Swans were rare on the Great Plains until about 1960, when the LaCreek NWR population began to increase (Bellrose 1976). About 56% (20 of 36) of reports are in the period 16 Mar–7 Apr.

Midcontinental migrant Tundra Swans are known to overwinter regularly in Yellowstone Park (Scott 1993) and along the mid–Atlantic Coast (Root 1988). The small population that overwinters in the Yellowstone vicinity apparently migrates directly to that area from points north. A much larger mid–Atlantic Coast population regularly utilizes inland migratory pathways to and from nesting areas in northern Canada and Alaska that include northeast South Dakota (SDOU 1991) and all of Minnesota (Janssen 1987), thus flying southward and eastward en route to Atlantic Coast overwintering areas. This pathway passes to the north of Nebraska, but a few birds stray south to the state.

High counts include 15 in Sheridan Co spring 1930, 1931, or 1932 (Rosche 1982) and 14 in Sheridan Co 4 May 1933.

Fall

1, 1, 5 Nov→10, 11, 14 Dec

Most reports are in the period 8–30 Nov. There is an early documented record 29 Sep 1950 Columbus (UNSM ZM10989) and another early report 21 Oct 1970 Lancaster Co.

Winter

Reports of birds at least attempting to winter include 23 Dec 1995–6 Jan 1996 Wolf L (Brogie 1997), 23 Dec 1998–7 Feb 1999 (2) Douglas Co (NR, pers. comm. RSS), 1 Jan 1938 Logan Co, 4 Jan 1986 Washington Co, 14–15 Jan and 7 Feb 1998 (up to 3 immatures) below Keystone Diversion Dam (SJD, DCE, JS, WRS), 21–27 Jan

1980 Washington Co (AB 34:286), and 5 Feb–5 Mar 1972 Lancaster Co.

FINDING

Tundra Swans are most often seen on larger lakes and reservoirs and occasionally on smaller ponds. Best times are Mar and Nov.

Wood Duck

Aix sponsa

STATUS

Common regular spring and fall migrant east, becoming uncommon west. Common regular breeder east, becoming uncommon west. Rare casual winter visitor south and east.

DOCUMENTATION

Specimen

WSC 33, Oct 1971 Antelope Co.

DISTRIBUTION AND ECOLOGY

Migrants are normally attracted to the edges of lakes and reservoirs, marshes, and smaller streams and occasionally to wet meadows. Most common eastward, their numbers decrease westward.

Breeding populations may be found in all sections of the state but are restricted by the absence of tree cavities for nesting, particularly in the Sandhills and the Panhandle. The addition of nest boxes to edges of lakes and reservoirs and along some streams during the past 2 decades has contributed to the spread of the species statewide and a substantial increase in numbers. Breeding birds are most widespread in the east where the species is very common, somewhat more locally distributed in south-central Nebraska, and localized in the Sandhills, mostly along more heavily wooded stretches of the Loup River system. Breeding in the Panhandle is generally limited to the North Platte River, where some trees have matured to develop natural cavities, or on lakes where nest boxes have been placed.

Spring

20, 22, 22 Feb→summer

Early dates cluster around 23 Feb, but earlier

dates in Feb may also be migrants: 12 Feb 1977 Adams Co, 12 Feb 1997 Holmes L, 14 Feb 1988 Polk Co, and 18 Feb 1990 Saunders Co.

Wood Ducks are typically seen as individuals, pairs, or groups of 5 or fewer during migration. Arrivals are concentrated in mid-Mar and are likely coincident with the spring thaw, but numbers peak in mid-Apr.

High counts include 84 at Cunningham L 24 Mar 1996 and 67 at North Platte NWR 16 Apr 1997.

Summer

The Wood Duck is a contemporary success story in Nebraska, partly because of proper management and partly because of habitat changes. Prior to 1900 the Wood Duck was recorded (Ducey 1988) as nesting in only 2 counties (Douglas and Nemaha) bordering the Missouri River, areas where mature trees with natural cavities were found adjacent to appropriate aquatic habitat. From 1900 to 1960 one additional county breeding record, Gage Co, was added (Ducey 1988), although Brooking (Notes) considered it "very common in the summer of 1915" at Inland Lagoon. For the period 1933 to 1948 only 4 migration records exist. However, since the late 1960s reports have increased substantially, including definitive nesting records. Part of this success may be attributed to the maturation of floodplain trees and their development of natural cavities westward along Nebraska streams. Part may also be explained by the placement of nest boxes near appropriate habitat, coupled with severe hunting restrictions imposed by the USFWS for this species during the 1940s and 1950s.

Nesting records are now widely distributed throughout the state and include Scotts Bluff and Garden Cos in the Panhandle, where Rosche (1982) noted that it had increased markedly since 1976, Brown and Boyd Cos in north-central Nebraska, and numerous counties in the east (Ducey 1988).

Nesting has been recorded in natural tree cavities as high as 12.2 m (40 ft); nest boxes are commonly utilized. Downy young have

been observed as early as 27 May and as late as 2 Aug (GH, WH).

High counts include 98 at North Platte NWR 17 Jul 1995, 90 there 25 Jun 1997, and 85 there 14 Jun 1994.

Fall

summer→1, 4, 5 Jan

Flocks form in Aug, such as 60–65 at Box Butte Res 12–22 Aug 1994. Departure dates are clustered in mid-Oct, although individuals linger into Dec and straggle into early Jan.

High counts include 117 at North Platte NWR 30 Oct 1997, 109 there 10 Sep 1998, and 72 there 13 Oct 1994.

Winter

Wood Ducks rarely winter in the state, because their typical pond and marsh habitat usually freezes. There are these midwinter reports: 11 Jan 1994 Fort Kearny SHP, 14 Jan 1981 Lincoln Co, 19 Jan 1975 Lancaster Co, 26 Jan 1997 Sarpy Co, 27 Jan 1980 Washington Co, 29 Jan 1983 Dakota Co, 29 Jan 1989 Douglas-Sarpy Cos, 30 Jan 1988 Lancaster Co, 2 Feb 1997 Harlan Co Res, and 3 Feb 1985 Boone Co.

FINDING

Wood Ducks as singles or pairs may be found along heavily wooded streams, on ponds and lakes, in marshes, and at reservoirs with nearby mature woodland. They are particularly common at Fontenelle Forest. Their whistled call, given when flushed, makes them easy to locate.

Gadwall

Anas strepera

STATUS

Common, locally abundant, regular spring and fall migrant statewide. Common regular breeder north-central, uncommon south-central, rare casual elsewhere. Rare, locally uncommon, regular winter visitor statewide.

DOCUMENTATION

Specimen

KSC Olson #20, 20 Mar 1906 Kearney, Buffalo Co.

DISTRIBUTION AND ECOLOGY

Although a dabbling duck and thus attracted to

temporary and permanent shallow ponds, the Gadwall is more likely than other dabbling ducks to be found on larger lakes and reservoirs during migration and in winter. During the summer months it is more restricted to marshes and shallow lakes with marshy edges, with highest concentrations in Sandhills lakes.

Spring

11, 11, 11 Feb→4, 9, 9 Jun

Late dates cited are from locations where breeding was not known to occur. Earlier dates are 7 Feb 1981 Saunders Co and 7 Feb 1995 (2) Garden Co.

This species typically begins arriving in the state in early Mar, using any open water available. Numbers peak in late Mar and Apr, and migration is usually over by mid-May.

High counts include 2600 at Crescent L NWR 15 Apr 1980 (Zeillemaker, pers. comm. RSS), 1750 there 23 Apr 1996, 1500 there 5 Apr 1994, and 1250 at Funk Lagoon 17 Mar 1995.

Summer

The most significant summer populations are sustained in the Sandhills lakes regions and in the Rainwater Basin. In the Sandhills it is the fourth most common breeder among dabbling ducks, following the Blue-winged Teal, Mallard, and Northern Shoveler, based on yearly inventories by USFWS personnel at Crescent L NWR and Valentine NWR. Harding (1986) found 32 nests in Clay Co 1981–85, the third most common nester after Blue-winged Teal and Mallard.

Elsewhere, there are nesting reports from most parts of the state (Ducey 1988), although there are few away from the Sandhills and the Rainwater Basin: Cuming Co (Wensien 1962), Dawes and Sioux Cos (Rosche 1982; Ducey 1988), Lancaster Co (Cink 1971, 1975b), Saunders Co (Ducey 1988), and Webster Co (Wensien 1962).

There have been several summer reports from counties where breeding is unrecorded, including Douglas-Sarpy, Lancaster, Scotts Bluff, Platte, and Knox. Some of these reports may be of molt migrants (see Fall).

Cornell Nest Record data reveal that nests are located in grasses and sedges on dry ground near a marsh or pothole. Thirty nests at Crescent L NWR and Valentine NWR contained a mean of 8.4 eggs, with clutches ranging from 4 to 12. Nests in incubation stages were found as early as 26 May and as late as 14 Jul. Of 25 nests visited more than once, 16 were destroyed by predators. Nests were found empty, implicating snake predation.

Fall

17, 18, 19 Sep→4, 7, 8 Jan

There are few reports after early Jan (see Winter). Jul and Aug reports from areas where breeding did not occur may be of molt migrants, which disperse in any direction, often for long distances (Palmer 1976). True migration becomes evident by mid-Sep, achieving a peak in early to mid-Oct, although significant numbers may be present well into Dec provided open water is available. As many as 50 were still in Cedar Co as late as 26 Dec 1994, and 61 were on Wehrspann L 20 Dec 1998.

High counts include 2200 at Crescent L NWR 9 Oct 1996, 2030 in Lancaster Co 3 Nov 1998, and 1650 at Crescent L NWR 23 Oct 1995.

Winter

Small numbers may persist into winter, occupying sheltered areas where spring-fed waters maintain ice-free conditions. Small populations may regularly occur in the open tailwaters of large reservoirs, such as L McConaughy, and along sheltered, open, warm-water streams. There are 14 reports for the period 10–23 Jan; Feb reports are noted under Spring.

High counts include 32 at Two Rivers Park 16 Jan 1995 and 30 wintering at L Ogallala 1998–99.

FINDING

This medium-size dark-gray duck is often difficult to identify in poor light conditions, but careful observation will reveal very distinctive elegant plumage, particularly in the male during the spring. Like other dabbling ducks, it selects shallow lakes and ponds during migration and tends to avoid rivers and streams. It can be observed most readily statewide during Apr.

Eurasian Wigeon

Anas penelope

STATUS

Rare regular spring migrant central, rare casual elsewhere. Rare casual fall migrant statewide.

DOCUMENTATION

Specimen

HMM 1782, 26 Sep 1914 Inland, Clay Co (Swenk, Notes before 1925).

DISTRIBUTION AND ECOLOGY

Eurasian Wigeons are usually found with American Wigeons. Edgell (1984) has speculated that small numbers of wild Eurasian Wigeons frequent North America each year, concentrating on both coasts in the fall and winter. Individuals mix into American Wigeon flocks; they return north with American Wigeons and continue north across Alaska into Siberian breeding areas. Edgell (1984) has also speculated that some may return via the North American interior to southern plains overwintering areas (Bray and others 1986).

TAXONOMY

A hybrid Eurasian x American Wigeon was at Funk Lagoon 6–7 Apr 1994 (NBR 62:69). Such hybrids were discussed by Palmer (1976).

Spring

29 Feb, 10, 10 Mar→25, 27 Apr (Brogie 1997), 2 May

There are 20 spring reports, 13 in the period 10 Mar–10 Apr. Of the others, documented reports are 29 Feb 1988 Seward Co (Mollhoff 1989a), 25 Apr 1992 Dawes Co (NBR 60:137), 27 Apr 1996 Clay Co (JGJ), and 2 May 1998 Clear Creek Marshes (SJD).

This species has been recorded each year since 1991 in the Rainwater Basin, usually with American Wigeons.

Fall

There are 5 reports, 3 documented:

26 Sep 1914 Clay Co (cited above)

29 Nov 1966 Lincoln Co (Shickley 1968; NBR 35:71)

20 Dec–8 Jan 1986–87 Scotts Bluff Co (Mollhoff 1989a)

The specimen cited above, taken 26 Sep 1914 in Clay Co, was listed by Bray and others (1986) as taken 14 Sep 1914, but Swenk (Notes before 1925) stated the specimen was collected 26 Sep. It is a male in alternate plumage at a rather early date for molt to be complete.

Undocumented reports are 2–6 Nov 1992 Douglas-Sarpy Cos and a vague "second or third hand" report from Cedar Co in fall 1995.

FINDING

Flocks of American Wigeons in the Rainwater Basin should be checked in the period 10 Mar–8 Apr.

American Wigeon

Anas americana

STATUS

Common, locally abundant, regular spring and fall migrant statewide. Uncommon regular breeder north-central and northwest, rare casual elsewhere. Uncommon casual winter visitor statewide.

DOCUMENTATION

Specimen

UNSM ZM7676, 12 Apr 1913 Lancaster Co.

DISTRIBUTION AND ECOLOGY

Small numbers of American Wigeons are scattered throughout the Sandhills in summer, being most often found on larger lakes edged with a marsh zone. Small potholes and temporary ponds tend to be ignored. Migrants occur throughout on smaller shallow bodies of water, also usually edged with mudflats.

Spring

12, 12, 13 Feb→8, 9, 10 Jun

Late dates above are from locations where breeding was not known to occur.

The initiation of migration is coincident with the thaw of frozen rivers, lakes, and ponds, typically occurring during the first third of Mar. However, early opportunists are common, appearing in mid-Feb on occasion; some of these may be nearby winter residents. Numbers peak during early Apr, but movement continues to be evident into very early May.

High counts include 2581 at North Platte NWR 14 Apr 1995, 1260 at Funk Lagoon 17 Mar 1995, and 1200 at Crescent L NWR 5 Apr 1994.

Summer

American Wigeon is restricted to the northwestern Sandhills lakes, where the largest breeding population is apparently in northern Garden and southern Sheridan Cos. At Crescent L NWR, the breeding population consisted of 45 and 60 individuals, respectively, for 2 different years (Zeillemaker, pers. comm. RSS). A few breeding records exist for Valentine NWR in Cherry Co. Adults with downy young have also been seen in the headwaters of L McConaughy (RSS), and breeding was reported in Lincoln Co in 1961 (Wensien 1962). The only other county with breeding reported is Webster Co (Wensien 1962); no details were provided.

There are multiple breeding season reports, suggestive of nesting, from Scotts Bluff, Sioux, Dawes, and McPherson Cos. There are few additional reports in the breeding season: 12 Jun 1999 Funk Lagoon, 14 Jun 1986 Lancaster Co, 14 Jun 1998 (4) Funk Lagoon, 16 Jun 1995 Kilpatrick L, 30 Jun 1973 Clay Co, 30 Jun 1984 Adams Co, 30 Jun 1985 Polk Co, 4 Jul 1999 Funk Lagoon, 5 Jul 1986 Polk Co, 6 Jul 1999 Harvard Marsh, 19 Jul 1997 (2) Funk Lagoon, 21 Jul 1988 Lancaster Co, and 27 Jul L McConaughy (Rosche 1994a). Aug reports probably are of early migrants or molt migrants; this species may fly long distances prior to molt (Palmer 1976).

A nest card record exists from Sheridan Co for a nest that contained 5 eggs on 11 May. The nest was sited in dry grassland, some 20 m (65 ft) from a marsh.

Fall

17, 21, 27 Aug→6, 8, 8 Jan

The beginning of the fall movement is evident in early Sep; earlier reports may be of molt migrants, which may appear as early as late Jul (Palmer 1976). Numbers peak in Oct, but significant numbers may persist into mid-Nov or until snow cover reduces the food supply as late as early Jan. An example was 133 at Scottsbluff SL 19 Dec 1998.

High counts include 1500 at Palmer L 11 Nov 1994, 1306 at Crescent L NWR 4 Oct 1996, and 1000 at Keystone L 10 Sep 1977, a staging location (Rosche 1994a).

Winter

Small numbers of individuals may remain well into Dec and even Jan where open water exists, together with open fields of winter wheat and barley for grazing. They may also be sustained by aquatic surface vegetation. Large, partially open reservoirs, such as L McConaughy, often sustain small numbers throughout the winter. Up to 30 wintered there and at L Ogallala 1998–99.

There are about 20 reports in the period 9 Jan–9 Feb, mostly from the Platte Valley and Lancaster Co. Exceptional was a flock of 60 in Scotts Bluff Co 7 Feb 1997.

FINDING

Wigeons may be found in largest numbers in south-central Nebraska during the major spring movement in late Mar. Marshes with expanses of open water, potholes, and flooded fields may attract wigeons during this period, but rivers may also be utilized, particularly early in the spring migratory period. Large lakes and reservoirs also support good numbers.

American Black Duck

Anas rubripes

STATUS

Rare regular spring and fall migrant east, rare casual elsewhere. Rare casual summer visitor statewide. Rare regular winter visitor east.

DOCUMENTATION

Specimen

UNK unnumbered, 20 Oct 1912 Kearney, Buffalo Co.

TAXONOMY

Although hybridization with Mallard occurs, it

is not extensive enough to consider the two conspecific (Johnsgard 1975). Nevertheless, American Black Ducks seen in Nebraska should be carefully examined for evidence of hybridization. A bird presumed to be such a hybrid was at Branched Oak L 14 Feb 1999 (JS).

DISTRIBUTION AND ECOLOGY

Nebraska is at the western edge of its range, and individuals are usually associated with the large Mallard flocks that occur in the Missouri Valley during migration and winter. Within its normal range, this species occurs in wooded swamps as a breeding bird.

American Black Duck normally breeds in association with deciduous forest northeast of Nebraska. The nearest breeding populations are in Minnesota and eastern North Dakota, although it is considered a possible rare breeder in South Dakota (SDOU 1991).

Traditional overwintering areas are to the east and include the central and upper Mississippi drainage and the Great Lakes (Root 1988). Thus Nebraska lies west of any established migratory pathways. Although this species' numbers were apparently diminishing prior to 1980 (Johnsgard and DiSilvestro 1976), Nebraska reports increased substantially during the 1980s but began to decline in the 1990s.

Spring

16, 18, 23 Feb→5, 5, 6 Apr

There are several midwinter reports (see Winter). Later reports are 11 Apr 1989 Cedar Co, 25 Apr Oshkosh (HMM 3076), 6 May 1930 Antioch (Rosche 1982), 12 May 1958 Douglas Co, and 12 May 1991 Sarpy Co.

Reports have declined since 1990, although in the 1980s there were several. There are fewer than 100 reports for the state. Almost all reports during migration are of individuals in Mallard flocks.

There are few reports west of Grand Island: 14 Mar 1949 Chadron (Rosche 1982), Apr 1925 Oshkosh (HMM 3076), and 6 May 1930 Antioch (Rosche 1982). During Zeillemaker's 5-year tenure (1978–82) at Crescent L NWR

and North Platte NWR, he recorded no American Black Ducks.

Summer

A few summer reports exist in the period 15 Jun–22 Aug, suggesting postbreeding wandering, although an intriguing source of summer birds, especially near population centers, may be released Easter ducklings (NBR 43:24, 25). Summer reports are 15 Jun–2 Aug 1984 Douglas-Sarpy Cos; one paired with a female Mallard 20 Jun 1993 Sioux Co (WRS; AB 47:1123); 25 Jun 1989 Polk Co; 12–26 Jul 1981 Douglas-Sarpy Cos; 18 Jul–2 Aug 1974 (2) Sarpy Co; 28 Jul 1973 (4) Pawnee L; 11 Aug 1987 Douglas-Sarpy Cos; 12 Aug 1961 Douglas Co; an adult male trapped 14 Aug 1967 Goose L, Crescent L NWR (Hyland 1968); 17 Aug 1985 Polk Co; and 22 Aug 1998 L Babcock.

Fall

1, 3, 4 Nov→2, 3, 4 Jan

There are earlier reports: Sep 1906, a specimen HMM 1960 taken at Funk Lagoon (Swenk, Notes before 1925); Sep 1910, a specimen taken at Sutton (Swenk, Notes before 1925); and 3 undocumented reports 13–15 Sep 1977 Garden Co, 9 Oct 1999 Brown Co, and 10 Oct 1986 Platte Co. Later reports are of presumed wintering birds.

Reports suggest that small numbers arrive with Mallards in very late fall in extreme eastern Nebraska and may remain in the area into Jan and later if Mallard flocks remain to winter (see Winter).

Panhandle reports are few: 13–15 Sep 1977 Garden Co and 29 Oct 1976 Box Butte Res (Rosche 1982). Westerly was one taken on the Dismal River in Thomas Co 10 Nov 1906 (HMM 2562).

Winter

On occasion, if Mallard flocks remain into midwinter, American Black Ducks remain with them. Midwinter reports, all from Boone Co eastward, include 10 Jan 1989 Cedar Co, 12 Jan 1985 Lancaster Co, 17 Jan 1986 Douglas-Sarpy Cos, 18 Jan 1983 Douglas-Sarpy Cos, 18 Jan 1986 Dakota Co, 20 Jan 1980 Washington

Co, 20 Jan 1989 Boone Co, 30 Jan 1983 Lancaster Co, 3 Feb 1979 DeSoto Bend NWR, and 6–13 Feb 1999 (1–2) Branched Oak L.

FINDING

Best chance to find this species is to scrutinize Mallard flocks in the Missouri Valley in Nov–Dec and in Mar.

Mallard

Anas platyrhynchos

STATUS

Abundant regular spring and fall migrant statewide. Common regular breeder statewide. Common, locally abundant, regular winter visitor statewide.

DOCUMENTATION

Specimen

UNSM ZM7674, 11 Apr 1916 Lancaster Co.

TAXONOMY

See Mexican Duck and Mottled Duck.

DISTRIBUTION AND ECOLOGY

Throughout migration and during summer months Mallards are attracted to oxbows, temporary ponds, marshes, and other shallow-water habitats that provide for surface and immediate subsurface foraging. Migrants may also be attracted to fields to forage on spilled and unharvested grains, especially corn.

During the winter months concentrations numbering in the thousands may be maintained on large, open lakes, stretches of open rivers, and tree-protected oxbows, provided that harvested fields remain snow-free and grain is exposed for foraging.

Spring

No discrete patterns are evident, particularly in areas where birds are present throughout the winter. However, during years of heavy snow cover, Mallards begin entering the state coincident with snowmelt, and numbers quickly rise to maxima that are normally achieved in late Mar.

High counts include 23 792 at North Platte NWR 11 Mar 1997, 14 886 there 19 Feb 1996, and 6533 there 28 Mar 1995.

Summer

The Mallard breeds in grassland that is immediately adjacent to aquatic habitat that supports emergent vegetation. This includes marshes and potholes and small prairie rivers with marshy edges, such as the Loup, Dismal, and upper reaches of the Niobrara and Elkhorn Rivers.

Breeding activity has been noted in May, Jun, and Jul. Nests containing 1–10 eggs have been located as early as 12 May and as late as 15 Jun. Nests are depressions in the ground that are constructed of cattails, reeds, and grasses and are lined with down. They are usually concealed in taller vegetation and sometimes in clumps of grass, such as little bluestem (*Andropogon scoparius*).

Harding (1986) found Mallard the second most common nesting duck in Clay Co 1981–85, when 119 nests were found.

Mallard drakes leave nesting sites in May and Jun and wander prior to forming small flocks with other males while molting and flightless. Females join these flocks later, prior to true fall migration (Palmer 1976).

Fall

Major movements are much more discernible in fall than in spring. Populations begin to increase in the state in mid-Oct and achieve their peak in mid to late Nov, particularly concentrating along streams and rivers and at large reservoirs. Peak numbers may be maintained for a month or more, depending upon snow cover.

Bands taken by Nebraska hunters from fall and early winter kills indicate that the majority of individuals were produced in the prairie provinces of Canada.

High counts include 88 765 at North Platte NWR 28 Nov 1995, 73 792 there 25 Nov 1997, 64 300 there 5 Jan 1998, and 60 000 at Sutherland Res 1 Jan 1998.

Winter

Winter concentrations are dependent on available open water and exposed grain in nearby fields and feedlots. Large concentrations may be found statewide, particularly in regions associated with the North Platte, Platte, Republican, and lower Loup Rivers. Some larger reservoirs may also maintain wintering populations.

High counts include 8387 at North Platte NWR 17 Feb 1998 and 8000 at Sutherland Res 11 Feb 1994.

FINDING

During the late spring, summer, and early fall look for these large dabbling ducks in small potholes, in shallow marshes, and along marshy edges of slow-moving streams. During the winter months large concentrations may be seen in early morning and late afternoon foraging in harvested cornfields. During nonforaging periods large concentrations may be seen on protected oxbows, along open stretches of rivers, and on large lakes and reservoirs.

Mexican Duck

Anas diazi

STATUS

Hypothetical.

TAXONOMY

Whether Mexican Duck is a valid species is a matter of debate; it is currently generally considered a part of Mallard, *A. platyrhynchos* (AOU 1983; Sibley and Monroe 1990; ABA 1996). Sibley and Monroe (1990) noted that Mallard and Mexican Duck hybridize extensively and that Mallard characters are swamping those of Mexican Duck. Earlier, however, Bellrose (1976) had cited the observation of Bevill (1970) that apparent assortative mating was occurring in a population of Mexican Ducks near Rodeo, New Mexico.

COMMENTS

There are 2 reports (Bray and others 1986; Silcock and others 1986a). A specimen referred to as Mexican Duck was collected 17 Oct 1921 in Cherry Co (Bent 1923). This is a female #973 in the Chicago Field Museum of Natural History, but it is now considered likely a hybrid between Mallard and American

Black Duck (David Willard, pers. comm. to Bray and others 1986). The second report is of an adult male shot in Rock Co 19 Oct 1969 (Menzel 1970). Although its identity as a Mexican Duck was confirmed by Paul Johnsgard and the bird showed no evidence of captivity, the possibility of its being an escaped bird could not be eliminated (Menzel 1970). Menzel (1970) noted that Mexican Ducks were kept in captivity in Hall Co and that escapes occurred occasionally. There are 3 specimens and 2 sight records listed by Bailey and Neidrach (1967) for Colorado, but this taxon was not mentioned by Andrews and Righter (1992).

Mottled Duck

Anas fulvigula

STATUS

Accidental in fall.

DOCUMENTATION

Photograph

5 Oct 1968 Loup River, Howard Co (Lemburg 1980; Bray and others 1986).

TAXONOMY

Relationships among the "southern Mallards" (Johnsgard 1975) are controversial. Johnsgard (1975) lists 3 groups as subspecies of Mallard: Mottled Mallard, *A. platyrhynchos maculosa*, breeding along the Gulf Coast with interior outlying populations in southeast Colorado, western Kansas, Oklahoma, and northeastern Texas (AOU 1983); Florida Mallard, *A. p. fulvigula*, restricted to Florida; and Mexican Mallard, *A. p. diazi*, breeding in northern Mexico with outlying populations in southeast Arizona, southern New Mexico, and west-central Texas (AOU 1983). Palmer (1976) indicated that *fulvigula* and *maculosa* did not differ enough from each other to be considered separate and listed both as part of Mottled Duck, *A. fulvigula*, considered specifically distinct from Mallard. There is, however, evidence of hybridization between Mottled Duck and Mallard in the Colorado-Oklahoma area (AOU 1983); Andrews and

Righter (1992) noted that specimens listed as Mottled Duck by Bailey and Neidrach (1967) had been found to be "probable hybrids" and did not consider Mottled Duck a valid species on the Colorado list. Thompson and Ely (1989) did not indicate any evidence for hybridization in the small breeding population of Mottled Ducks at Cheyenne Bottoms, Kansas.

Records

The single record consists of a photograph of a female that was injured by a hunter 5 Oct 1968 on the Loup River in Howard Co (Lemburg 1980). Identification from the photograph and the lack of evidence for hybridization were confirmed by Johnsgard (Silcock and others 1986b; Bray and others 1986).

COMMENTS

Primarily a Gulf Coast species, Mottled Duck is considered casual inland to Oklahoma and southern Kansas (AOU 1983). Thompson and Ely (1989) reported that it is a "spring to fall resident" (15 Feb–30 Nov) at Cheyenne Bottoms in central Kansas where it is apparently rare (about 12 pairs were recorded in 1966); it was considered accidental elsewhere in the state. Its presence cannot be anticipated in Nebraska except possibly as a postbreeding wanderer or spring overshoot.

Blue-winged Teal

Anas discors

STATUS

Abundant regular spring and fall migrant statewide. Uncommon, locally common, regular breeder statewide. Accidental in winter.

DOCUMENTATION

Specimen

UNSM ZM12441, 14 Apr 1909 Cass Co.

DISTRIBUTION AND ECOLOGY

Blue-winged Teal may be found statewide in appropriate seasons, normally associated with shallow standing water in marshes, flooded fields, ditches, oxbows, edges of lakes, and reservoirs. It may also frequent small, slow-moving streams.

Spring

28, 28, 28 Feb→summer

Typically arrival is in mid to late Mar, achieving peak numbers around mid-Apr, but large concentrations continue into early May, when breeding commences and individuals become more secretive.

High counts include 1825 at Crescent L NWR 23 Apr 1996, 705 there 9 May 1995, and 600 in the Rainwater Basin 5 May 1996.

Summer

Nesting has been documented for nearly all regions of the state. Summer concentrations are highest in the Sandhills lakes region; elsewhere pairs are scattered throughout the state in appropriate breeding habitat, which is grassland adjacent to permanent water that will sustain a large aquatic insect population.

Nests containing eggs have been found as early as 16 May and as late as 21 Jul. Clutches have varied from 1 to 10 eggs with a mean of 6.8 (N = 19). Harding (1986) found this species to be the most common nesting duck in Clay Co 1981–85 and located 536 nests, 74% of all duck nests found.

Fall

summer→15, 17, 19 Nov

A detectable fall migratory movement is evident at the very end of Aug, achieving a peak by about mid-Sep. Small groups of individuals may remain in the state typically until late Oct or very early Nov. Later reports are few and unexpected for this cold-intolerant species.

High counts include 6000 at Funk Lagoon 19 Sep 1999, 4200 in the eastern Rainwater Basin 22 Aug 1999, and 1700 at Sinninger Lagoon 29 Aug 1997.

Winter

The only documented record of a presumably wintering bird is of one at an open spring-fed pond in Lincoln Co 12 Jan 1991 (SJD, WRS). There are many undocumented reports, however, including 9 in Jan and 42 between 19 Nov and 28 Feb, 25 of these in Feb. There are 2 Jan records for Missouri (Robbins and Easterla 1993), 3 for Iowa (Kent and Dinsmore

1996), and 5 for Jan–Feb in Colorado (Andrews and Righter 1992.

FINDING

This small, warm-weather teal is very commonly seen, often in large numbers, particularly during the spring migration. Look for it in flooded fields, in ditches containing water, and along the marshy edges of lakes and streams. It also frequents large water puddles in feedlots.

Cinnamon Teal

Anas cyanoptera

STATUS

Uncommon regular spring migrant west and central, rare east. Locally uncommon regular breeder west. Rare regular fall migrant west, rare casual elsewhere. Hypothetical in winter.

DOCUMENTATION

Specimen

UNSM ZM7673, 4 Apr 1909 Fillmore Co.

DISTRIBUTION AND ECOLOGY

This western species occurs eastward in migration regularly as far as the Rainwater Basin. Closely related to the Blue-winged Teal, Cinnamon Teal usually occurs with it in shallow, standing-water bodies with marshy and grassy edges and occasionally along small, slow-moving streams. As a breeding bird, it is found regularly in alkaline marshes in the Panhandle such as Kiowa Springs and Facus Springs (Rosche 1994b), as well as on western Sandhills lakes, notably at Crescent L NWR.

Spring

11, 11, 14 Mar→1, 1, 2 Jun

Late dates above are from locations where breeding is unknown. There are earlier reports 24 Feb 1956 Adams Co, 1 Mar 1995 Clay Co, 4 Mar 1988 Antelope Co, 6 Mar 1997 Crescent L NWR, 7 Mar 1984 Lincoln Co, and 8 Mar 1981 Douglas-Sarpy Cos. Normally arrival is in very late Mar, and numbers peak in late Apr and early May. Recently it has been found to be a regular migrant as far east as the Rainwater Basin, where 1–5 are found each year late Mar–Apr (JGJ).

It is rare east of the Rainwater Basin; there are about 31 reports in the period 8 Mar–2 Jun. However, all but 4 are in the period 24 Mar–4 May, the exceptions 8 Mar 1981 Douglas-Sarpy Cos, 13 May 1991 Lancaster Co, 24 May 1952 Lancaster Co, and 2 Jun 1985 Lancaster Co.

High counts include 16 at Kiowa Springs 9 Apr 1999, 10 at Crescent L NWR 9 May 1995, and 8 there 12 Apr 1997.

Summer

It occurs regularly at Crescent L NWR during the summer months but is usually represented by fewer than 12 individuals (Zeillemaker, pers. comm. RSS). A nest containing 8 eggs was found in a grass-sedge meadow at Crescent L NWR 10 Jun 1937. Several additional records for downy young come from that refuge, including 1974 (Bennett 1975), and in 1984 a nest with eggs was found 24 May but was predated on 11 Jun (Helsinger 1985b). Cinnamon Teal continue to summer at Crescent L NWR, including 8 there 9 Jun 1995 and 6 there 24 Aug 1993.

Rosche (1994b) noted that Facus Springs and Kiowa Springs are regular breeding locations also; Cinnamon Teal are regularly reported from Scotts Bluff Co in summer. In recent years there have been multiple reports from Sheridan, Dawes, and Sioux Cos as well as the Rainwater Basin, all locations where breeding may occur.

Elsewhere, summer reports are few: 14 Jun 1995 Whitman, 12–16 Jul 1984 Douglas-Sarpy Cos, 15 Jul 1993 Dawes Co, and 2 Aug 1987 Howard-Hall Cos.

Fall

summer→11, 19, 20 Sep

Later reports are 29 Sep 1990 Scotts Bluff Co, 10 Oct 1968 Brown Co, 10 Oct 1988 Cherry Co, 14 Oct 1995 Crescent L NWR, 4–8 Nov 1987 Pierce Co, 14 Nov 1976 Cass Co (AB 31:194), 20 Dec 1980 Scotts Bluff Co, and one shot Dec 1975 along the Platte River in Cass Co (NBR 44:19).

There are only 21 reports for Sep and later, possibly because males in eclipse plumage are inconspicuous and then leave the state soon after regaining flight. Last dates for summer residents at Crescent L NWR include 12, 17, and 27 Jul (RSS), suggesting that Aug reports are of migrants rather than summering birds.

Winter

There is an undocumented report 9 Jan 1966 Adams Co.

FINDING

Probably the most reliable location for finding this species is Kiowa Springs in Jun. Other good locations at that time are Facus Springs and Crescent L NWR. At Crescent L NWR, ask refuge personnel which lakes are being frequented.

Northern Shoveler

Anas clypeata

STATUS

Common, locally abundant, regular spring migrant statewide. Locally common regular breeder north-central, rare elsewhere. Common regular fall migrant statewide. Rare regular winter visitor Platte Valley, rare casual elsewhere.

DOCUMENTATION

Specimen

UNSM ZM12449, 14 Apr 1909 Cass Co.

DISTRIBUTION AND ECOLOGY

This dabbling duck is primarily a surface feeder, using its broad bill equipped with hairlike strainers to filter plankton, which are typically highly concentrated in shallow, standing-water bodies and the edges of slow-moving streams. Thus any habitat that provides an adequate food source may attract shovelers. During migration larger lakes and reservoirs as well as rivers with sandbars, such as the Platte, can serve as resting places.

Spring

18, 19, 20 Feb→summer

The first arrivals are typically in mid-Mar or immediately following the melt of snow and ice cover. Numbers peak in late Mar and are sustained through late Apr. Reports after mid-May are likely breeders; such reports

from the south and east are few until fall migrants begin to arrive in Sep.

High counts include 1920 at North Platte NWR 25 Apr 1995, 1210 at Crescent L NWR 26 Mar 1996, and 1000 at Johnson Lagoon 20 Apr 1997.

Summer

Breeding pairs may be found associated with marshes throughout the state (Ducey 1988), but the largest concentrations are found in the Sandhills lakes region. Harding (1986) was able to locate only 8 nests in Clay Co 1981–85. Four nests located in early Jun contained 7–11 eggs.

Fall

summer→1, 1, 2 Jan

Migratory movement commences in late Sep and continues without significant peaks until mid-Oct, when peak numbers are attained. Peak numbers are far lower in fall than in spring (Rosche 1994a). Numbers decline until the migration ceases about mid-Nov. Unusual for the late date were as many as 56 at Harlan Co Res 18 Dec 1998.

High counts include 2000 at North Platte SL 8 Oct 1996, 1500 at Gering SL 15 Oct 1997, and 1240 at Crescent L NWR 26 Sep 1996.

Winter

There are several midwinter reports, essentially restricted to the Platte Valley, with few elsewhere: 5 Jan 1990 Cedar Co, 2 Feb 1997 Sarpy Co, 6 Feb 1999 Harlan Co Res, 7 Feb 1954 Harlan Co, and one wintering with Mallards at Jackson, Dakota Co, 1995–96.

FINDING

Largest concentrations are found during the spring migration period statewide; shallow marshes are particularly good locations to find this duck.

Northern Pintail

Anas acuta

STATUS

Common, locally abundant, regular spring and fall migrant statewide. Uncommon, locally common, regular breeder statewide. Locally uncommon regular winter visitor statewide.

DOCUMENTATION

Specimen

UNSM ZM12426, 4 Apr 1909 Grafton, Fillmore Co.

DISTRIBUTION AND ECOLOGY

This medium-size dabbling duck may be found in shallow-water wetlands during spring and fall migration, including sloughs, marshes, and wet fields. They may also utilize larger lakes, reservoirs, and rivers but with lesser frequency and in smaller concentrations. In addition, they may be attracted to harvested fields to forage on grain spillage.

Spring

Typical of most waterfowl species, migration is usually coincident with snowmelt, although in some locations early sightings may be of nearby winter residents. True migrants have a tendency to be among the first duck species to arrive and to depart, numbers achieving maxima in late Feb and early Mar.

Largest concentrations can be seen in south-central Nebraska, where it can be considered abundant; during this period it is common in the east and common to abundant in the Panhandle.

High counts include 75 000 at L McConaughy 10–11 Mar 1979 (Rosche 1994a), 30 000 at Youngson Basin 16 Mar 1998, and 12 000 at Funk Lagoon 2 Mar 1996 and 6 Mar 1999.

Summer

Although breeding has been documented in suitable marsh habitat statewide (Ducey 1988), the largest regularly occurring breeding populations are found in the Sandhills lakes region of Cherry, Sheridan, and Garden Cos.

Nests are located in grass or sedge meadows near marshes or prairie potholes. Nests in various stages of incubation have been found as early as 24 Apr and as late as 23 Jun, with clutch size 2–9 eggs. Six "small juveniles" were at Funk Lagoon 1 Aug 1999 (LR, RH). Harding (1986) found 24 nests in Clay Co 1981–85.

Reports are few away from known breeding areas, especially in Jun, when reports of drakes may involve those that have undergone a molt migration, while the presence of females would suggest nesting.

Fall

Numbers peak during late Sep and the first half of Oct in locations where summer residents are absent. As with Mallard, Northern Pintail drakes leave females early, in late May and Jun, to form molting groups. Some of these groups undertake extensive molt migration (Palmer 1976) and may appear as early as Jun–Jul in areas where breeding did not occur. Representatives of Sandhills breeding populations typically remain well into Nov; last dates for Garden Co, for example, include 15 and 19 Nov and 7 Dec (Zeillemaker, pers. comm. RSS).

High counts include 880 at Crescent L NWR 4 Oct 1996, 424 at Sinninger Lagoon 20 Aug 1998, 200 in Kearney Co 8 Oct 1994, 200 at Funk Lagoon 14 Aug 1997, and 200 at L McConaughy 20 Aug 1978 (Rosche 1994a).

Winter

Individuals may remain in small concentrations throughout the winter months, provided open water and open field foraging are available. Most reports of wintering birds are of males mixed with flocks of wintering Mallards, possibly due to females being overlooked.

High counts include 22 at L Minatare 19 Jan 1998 and 20 at Branched Oak L 20 Jan 1998.

FINDING

Large numbers of Northern Pintails can be found in the Rainwater Basin in early to mid-Mar. When winter runoff is plentiful, some of the marshes in the region support thousands of individuals along with large concentrations of several other species.

Garganey

Anas querquedula

STATUS

Rare casual spring migrant central.

Spring

There are 2 records, possibly of the same individual:

28 Mar 1997 alternate male 4.8 km(3 mi) south of Kearney (NBR 65:77; Brogie 1998)

29 Mar–5 Apr 1998 alternate male Wild Rose Ranch (GL)

COMMENTS

Records from neighboring states are few; 8 spring records from Colorado, Kansas, and Missouri are in the period 21 Mar–11 May, and a single fall record from Kansas is 23 Oct 1988 (Mlodinow and O'Brien 1996). Most records are of males with Blue-winged Teal in spring, but identification of females and immatures is very difficult, and these are probably overlooked.

Green-winged Teal

Anas crecca

STATUS

Common, locally abundant, regular spring and fall migrant statewide. Rare regular breeder north-central, rare casual elsewhere. Rare casual summer visitor statewide. Rare, locally uncommon, regular winter visitor west, south, east.

DOCUMENTATION

Specimen

UNSM ZM12432, 14 Apr 1909 Cass Co.

DISTRIBUTION AND ECOLOGY

Migrants utilize flooded fields, moist meadows, ponds, marshes, lakes, reservoirs, and occasionally streams; thus they are found in appropriate locations statewide during the migratory period. Rosche (1994a) indicated that they prefer flooded fields in spring and mudflats in fall.

During the breeding season Green-winged Teal utilize ponds, lakes, and reservoirs with considerable marshy edge. Such habitats are scattered throughout the state, as are summer records for this species.

Spring

Movement begins in mid to late Feb and peaks in late Mar in the east and mid-Apr in the Panhandle, but migrants may be seen into early May. Late Jan and early Feb reports may represent overwintering individuals.

High counts include 8500 at Harvard Lagoon 3 Apr 1997, 1800 at Kissinger Basin 28 Mar 1999, and 1200 at L McConaughy 27 Mar 1999.

Summer

Regular breeding, as reported in recent years, may occur only in Garden Co, although Green-winged Teal often occur on Sandhills lakes and marshes during the breeding season and thus presumably breed more regularly throughout this region than the records indicate. Published breeding records exist, however, only for Scotts Bluff, Morrill, and Garden Cos in the Panhandle and for Holt, Cherry, and McPherson Cos in the Sandhills (Ducey 1988); young were reported in Hall Co in 1987 (Bennett 1988). A single nest found in Clay Co in 1985 (Harding 1986) is apparently the only nesting record south of the Platte Valley.

There are few summer reports away from the Sandhills; such reports are fewest in Jun and Jul. Reports in Aug are mostly fall migrants (see Fall).

Fall

A detectable movement begins in mid-Aug. Earliest arrivals are probably molt migrants (Palmer 1976). Peak movement is late Sep–early Oct, and scattered small groups of 5–10 individuals may continue to be present into Dec. Occasional individuals or pairs may remain locally through the winter months.

High counts include 4000–5000 at the west end of L McConaughy 17 Sep 1994, 4400+ at L McConaughy 1 Nov 1998, and 3000–4000 there 25 Oct 1991 (Rosche 1994a).

Winter

Although they may not be present during more severe winters, Green-winged Teal are among the most likely dabbling ducks to overwinter. Individuals or small groups have been observed overwintering in Scotts Bluff, Keith, Harlan, Washington, and Cass Cos. Rosche (1994a) considered Green-winged Teal a "locally uncommon to very common" winter visitant at L McConaughy.

It is regularly reported on CBCs, including a high count of 105 at Grand Island 15 Dec 1990. Good midwinter counts include 87 in southern Sioux Co 19 Jan 1998 and 73 below the Keystone Diversion Dam 19 Jan 1998.

High counts include 130 in southwest Sioux Co 3 Jan 1999.

FINDING

Largest concentrations are found during the early spring migratory period, particularly the first half of Mar, and in fall in late Sep and early Oct. Flooded fields, shallow ponds, and lakes with marshy edges are particularly attractive to this small dabbling duck.

Canvasback

Aythya valisineria

STATUS

Common, locally abundant, regular spring and fall migrant statewide. Rare, locally uncommon, regular breeder Sandhills. Rare casual summer visitor statewide. Uncommon casual winter visitor statewide.

DOCUMENTATION

Specimen

UNSM ZM12458, 3 Apr 1932 Red Willow Co.

DISTRIBUTION AND ECOLOGY

This large diving duck may be found on moderate to large lakes and reservoirs during migration and during the breeding season. Very large reservoirs with areas of open water may sustain small numbers during the winter.

Spring

5, 7, 10 Feb→20, 21, 22 May

Late dates above are from locations where summering birds did not usually occur. Arrival is coincident with the ice melt on large lakes and reservoirs. The spring movement typically begins in late Feb and very early Mar and peaks in mid to late Mar. Movement of migrants continues into early May. Late May migrants likely are nonbreeding subadults, such as a report 28 May 1983 Lancaster Co.

The bulk of the spring movement occurs in central and western Nebraska, even though many birds are seen in the east. Rosche (1982) considered this species to be an abundant spring migrant in the Panhandle. Population estimates from Crescent L NWR tend to support that assessment; Zeillemaker (pers. comm. RSS) recorded peak numbers of 1900 on the refuge 15 Apr 1980. Those numbers were exceeded only by Lesser Scaup (2775)

and Gadwall (2600). In eastern Nebraska such numbers are almost never achieved.

High counts include 5000 at L McConaughy 21 Mar 1993 (Rosche 1994a), 1900 at Crescent L NWR 15 Apr 1980 (cited above), and 600 in Grant Co 31 Mar 1996.

Summer

A small summer population is sustained in the Sandhills lakes region, particularly in Garden Co. At Crescent L NWR, for example, the population varies between 10 and 20 breeding pairs (Zeillemaker, pers. comm. RSS). A high count of 25 were there 12 Jul 1995. It is now casual or rare at Valentine NWR (McDaniel, pers. comm. RSS). Johnsgard (1996) considered it a "local and rare summer resident in the Sandhills (Valentine National Wildlife Refuge)," although nesting reports exist for Garden, Cherry, Morrill, and Grant Cos (Ducey 1988).

Rosche (1982) presumed it bred in the sandhills of the northwest Panhandle but had no direct evidence. There are breeding season reports for Scotts Bluff, Dawes, Sheridan, and McPherson Cos, where breeding is likely, as well as 8 Jul 1980 and 28 Jul 1975 Lincoln Co; 20 May–Aug 1971 Brown Co; 19 Jun 1967, 30 Jun 1964, 4 Jul–2 Sep 1973, and 10 Aug 1975 Adams Co; 30 Jun 1993 Hamilton Co; 6 Jul 1997 Funk Lagoon; and 16 Jul 1971 Lancaster Co.

Fall

23, 25, 25 Sep→1, 1, 3 Jan

The fall migratory movement is considerably more leisurely than in spring, beginning in early Oct and continuing until early Dec, provided open water is available. Peak fall populations are evident in late Oct. Late Jul through mid-Sep reports from areas where regular breeding does not occur, of which there are few, may be of molt migrants: 16 Jul 1971 Lancaster Co, 21 Aug 1973 Lancaster Co, 29 Aug 1997 Johnson Lagoon, and 11 Sep 1982 Douglas- Sarpy Cos.

High counts include 3000 at Crescent L NWR 5 Nov 1994, 1200 there 23 Oct 1995, and 515 there 6 Nov 1999.

Winter

Small numbers may remain on larger lakes and reservoirs, provided they remain partially ice-free. L McConaughy regularly sustains a small population at least well into Dec, and there are these midwinter reports: 8 Jan 1961 Scotts Bluff Co, 11 Jan 1987 L McConaughy (Rosche 1994a), 13 Jan 1980 Washington Co, 19 Jan 1996 L Ogallala, and 2 Feb 1958 Scotts Bluff Co.

High counts include 60 at L McConaughy 11 Jan 1987 (Rosche 1994a) and 29 there wintering 1998–99.

FINDING

This largest of Nebraska's divers is most often seen on very large lakes and reservoirs. Early Apr is probably the best time to see Canvasbacks.

Redhead

Aythya americana

STATUS

Common, locally abundant, regular spring and fall migrant statewide. Common regular breeder north-central, uncommon south-central, rare casual elsewhere. Uncommon casual winter visitor statewide.

DOCUMENTATION

Specimen

UNSM ZM7678, 27 Mar 1932 McCook, Red Willow Co.

DISTRIBUTION AND ECOLOGY

A diving duck, it most often frequents larger lakes and reservoirs, but it may also utilize smaller lakes and ponds provided they have sufficient depth to sustain its feeding habits.

During the breeding season this species requires extensive marshes that border expanses of open water. Such habitat is concentrated in Garden and Sheridan Cos, but similar habitat that may occasionally support a pair occurs in other sites scattered throughout the Sandhills and Panhandle.

Spring

4, 7, 8 Feb→22, 23, 24 May

Late dates above are from locations where summering is essentially unknown; there are later reports 27 May 1999 at Alma SL (GH, WH) and 31 May 1997 at Alma SL.

The spring migratory movement is initiated coincident with early snowmelt. Partially open lakes will attract good numbers in late Feb. Peak concentrations are achieved by mid to late Mar and may be sustained into early Apr. The spring movement is usually completed by early May, but stragglers have been observed in late May.

Rosche (1982) considered the species to be a "very abundant" spring migrant (more than 1000 per day) in the Panhandle, noting large concentrations on Box Butte Res and Smith L, Sheridan Co. Rosche (1994a) also noted high concentrations on L McConaughy (see below). Apparently some lakes in or near the Panhandle are particularly attractive to Redheads, because Zeillemaker (pers. comm. RSS) did not find peak concentrations to exceed 500 individuals at Crescent L NWR for the springs of 1979 and 1980. During those same seasons on the refuge Zeillemaker recorded peak concentrations of Canvasback and Lesser Scaup that exceeded Redhead numbers by not less than 50%.

High counts include 15 000 to 20 000 at L McConaughy 21 Mar 1993 (Rosche 1994a), 14 000 there 24 Mar 1989 (Rosche 1994a), and 4640 at Winters Creek L 8 Mar 1994.

Summer

Breeding occurs in the Sandhills and casually in the Rainwater Basin (Johnsgard 1997), although there are few actual records. A regular breeding population of 100–200 individuals is sustained at Crescent L NWR (Zeillemaker, pers. comm. RSS). Since 1960, nesting records exist for Sheridan, Garden, Cherry, and Polk Cos, with earlier records for Box Butte, Morrill, Brown, and Clay Cos (Ducey 1988). There are multiple breeding season (Jun–Aug) reports for Scotts Bluff, McPherson, Grant, Lincoln, Keith, Adams, and Phelps Cos, all of which provide suitable breeding habitat. Excellent numbers apparently summered at Funk Lagoon in 1996, when 22 were present on 22 Jun, and in 1999, when 20 were found on 27 Jun. Scattered reports from additional Rainwater Basin locations are 13 Jun 1999 Sinninger Lagoon, 30 Jun 1964 Hamilton Co, 30 Jun 1964 Platte Co, 11 Jul 1998 (22) Kearney Co, 16 Jul 1995 (8) Theesen Lagoon, 9–10 Aug (7) Rainwater Basin, and 12 Aug 1984 Polk Co.

Elsewhere, there are few reports: 9 Jun 1990 Knox Co, 13 Jun 1984 Douglas-Sarpy Cos, 10 Jul 1992 Cheyenne Co, 16 Jul 1971 Lancaster Co, 28 Jul 1998 (3) Alma SL (GH, WH), 6 Aug 1972 Douglas-Sarpy Cos, and 21 Aug 1999 (a male) Alma SL (GH, WH).

Two nests located near Crescent L NWR 22 Jun contained 8 and 13 eggs, respectively. Eight young hatched from the first nest between 14 and 17 Jul. The second nest was predated. Both nests were located in a moist sedge meadow adjacent to Crescent L (Cornell Nest Record data).

Fall

18, 18, 25 Sep→9, 10, 11 Jan

The migratory movement begins about mid-Sep, although there is an earlier report 3 Sep from Howard-Hall Cos. Concentrations are even across the state, achieving peak numbers in mid to late Oct. Most depart by the end of Nov, but some remain into Dec and early Jan, utilizing large lakes and reservoirs that remain ice-free.

High counts include 3292 at North Platte NWR 10 Nov 1994, 2104 at Crescent L NWR 27 Sep 1995, and 500 at Johnson L 24 Oct 1999.

Winter

Redheads occur casually in midwinter (12 Jan–10 Feb) in the Platte Valley, although 92 at L McConaughy 10 Jan 1999 was an unusually high number. There are a few reports elsewhere: an individual remained at Crystal Cove L 7 Dec 1997–17 Jan 1998, 18 Jan 1997 Branched Oak L, 23 Jan 1995 Offutt Base L, 4 Feb 1996 Branched Oak L, 9 Feb 1957 Adams Co, 9 Feb 1977 Custer Co, and 14 Feb 1966 Dawson Co.

FINDING

Look for this species on larger lakes and reservoirs, particularly during the spring migratory period, Mar–Apr.

Ring-necked Duck

Aythya collaris

STATUS

Common, locally abundant, regular spring and fall migrant statewide. Rare casual summer visitor statewide. Uncommon casual winter visitor statewide.

DOCUMENTATION

Specimen

UNSM ZM12460, 2 Apr 1909 Cass Co.

DISTRIBUTION AND ECOLOGY

This diving duck frequents larger lakes and reservoirs. During migratory periods it is often seen in mixed flocks of diving ducks, especially Redheads and Lesser Scaup.

Spring

8, 12, 14 Feb→5, 6, 7 Jun

First dates of occurrence are typically in early Mar, coincident with the spring thaw. Migration peaks in mid to late Mar, but small numbers may be seen throughout Apr.

High counts include 1450 at Crescent L NWR 5 Apr 1994, 872 at North Platte NWR 14 Apr 1995, and 312 at Kissinger Basin 28 Mar 1999.

Summer

There are no recent breeding records for the state, although it was once (before 1920) a "regular historic nester" in the Sandhills (Ducey 1988). Today, a breeding occurrence in Nebraska would be considered extralimital; the nearest regular breeding is in far northeast South Dakota (SDOU 1991).

Individuals may occur casually during summer months, notably 1–2 summering birds at Crescent L NWR in 1995; it was reported in Garden Co 17 Jul 1978, 30 Aug 1980, and 30 Jun 1981; at Oshkosh SL 24 Jul–12 Aug 1984 (Rosche 1994a); and at Crescent L NWR 16–31 Jul 1996 (FN 50:965) and 30 Aug–7 Sep 1998 (12). One was in Sheridan Co 17 Aug 1996 (FN 51:79). It may have summered at North Platte NWR in 1994, where 2 were noted 14 Jun and 10 Sep; it was reported in Scotts Bluff Co 10 Jun 1991. Multiple reports also exist for McPherson Co: 12 Jun 1975, 30 Jun 1974, 22 Jun 1980, and 4 Jul 1984. Such summer reports may be of wandering nonbreeders or possibly occasional breeding birds.

There are few summer reports away from apparent suitable breeding habitat: 10 Jun 1991 Scotts Bluff Co, 4 Aug 1990 Polk Co, 15 Aug 1980 Douglas-Sarpy Cos, and 17 Aug 1981 Douglas-Sarpy Cos.

Fall

9, 10, 11 Sep→1, 1, 1 Jan

Migration is clearly evident by late Sep, peaking in late Oct and early Nov. A good population may be maintained on larger lakes and reservoirs well into Dec.

High counts include 2100 at Crescent L NWR 27 Sep 1995, 1228 at North Platte NWR 10 Sep 1994, and 260 at Winters Creek L 16 Oct 1999.

Winter

Very small numbers may be sustained on large reservoirs such as L McConaughy throughout the winter months, provided that large expanses of open water persist. The spillway on L Ogallala below the dam may also maintain Ring-necks in winter (Rosche 1994a). Midwinter reports elsewhere are few: 8 Jan 1987 Scotts Bluff Co; 16 Jan 1996 (3) Branched Oak L; 19 Jan 1998 (6) Scottsbluff SL; 20 Jan 1998 Little Salt Creek, Lancaster Co; 21 Jan 1980 Washington Co; 3 Feb 1988 Scotts Bluff Co; 3 Feb 1992 Scotts Bluff Co; and 4 Feb 1999 (8) Branched Oak L.

FINDING

To locate this species go to larger lakes and reservoirs during the spring migration (Mar–Apr) and carefully study the rafts of diving ducks.

Tufted Duck

Aythya fuligula

STATUS

Accidental in fall.

DOCUMENTATION

Description

3–4 Dec 1999 Keystone L, Keith Co (SJD).

Records

The only record is of a male on Keystone L 3 Dec 1999–2 Jan 2000 (SJD; m.ob.).

COMMENTS

This Palearctic species occurs rarely but with some regularity along both the Pacific and Atlantic Coasts, south to about southern California and New Jersey, respectively (AOU 1998). It is very rare in the western interior United States, however (Toochin 1998). The only records from states neighboring Nebraska are from Wyoming 10 Apr 1966 (Scott 1993) and 16 Mar 1994 (Mlodinow and O'Brien 1996), Colorado 21–23 Mar 1997 (FN 51:900; Brandon Percival, pers. comm. WRS), and Kansas 27 Oct 1996 (FN 51:79; Jim Williams, pers. comm. WRS; Lloyd Moore, pers. comm. WRS). Hybrids, with scaup or Ring-necked Duck, are reported also; one was at Denver 22–26 Apr 1997 (FN 51:900; Brandon Percival, pers. comm. WRS).

Greater Scaup

Aythya marila

STATUS

Uncommon regular spring migrant statewide. Rare regular fall migrant statewide. Uncommon, locally fairly common, regular winter visitor statewide.

DOCUMENTATION

Specimen

UNSM ZM14529, 9 Apr 1904 Waco, York Co.

DISTRIBUTION AND ECOLOGY

This species utilizes large lakes and reservoirs during migration. According to Root (1988), wintering populations are concentrated along the northern Pacific and Atlantic seaboards and around the Great Lakes. Smaller numbers may also be found on the Gulf Coast near the lower Mississippi River (AOU 1983). The breeding range extends along the tundra-taiga border of northern North America, and thus migrants may travel north and westward from

Atlantic wintering grounds across the Great Lakes or northward up the Mississippi Flyway again crossing the Great Lakes, where inland migrant concentrations are traditionally known to be greatest. However, as reservoir development has increased along the western edges of the Mississippi Flyway, observations of this species have become increasingly regular where they were once casual. Such is the case in Nebraska, where records have also been increasing.

Recognition by birders also may contribute to an increase in records. This species is morphologically very similar to the Lesser Scaup, but careful observation of rafts of divers during appropriate times often yields records of Greater Scaup that heretofore would have been lumped with their much more common cousin.

Spring

15, 15, 17 Feb→27 May 1, 1 Apr

There are later reports, all but one from northern and western Nebraska: May 1978 Cherry Co, 26 Apr–8 May 1997 Mallard Haven, 18 May 1956 Scotts Bluff Co, and 28 May 1982 Sheridan Co.

This species is an early migrant, utilizing the first open water on large lakes and reservoirs. Numbers achieve maxima in mid-Mar, based on frequency of records, with a few birds occurring through Apr. Numbers are typically very low, rarely exceeding 5 or 6 individuals per sighting.

High counts include 25 at L McConaughy 24 Mar 1978 (Rosche 1994a), 23 there 13 Apr 1997, and 22 at L Ogallala 8 Mar 1999.

Fall

26, 27, 30 Oct→2, 2, 5 Jan

Late dates above do not include Keith Co, where a few winter most years (Rosche 1994a). There are earlier reports 3 Oct 1987 Washington Co and 20 Oct 1981 Douglas-Sarpy Cos.

There are only about 30 reports, the bulk in late Nov and early Dec; it is less common in fall than in spring.

High counts include 63 at Keystone L 2 Jan 2000,

50 in Keith Co 24 Nov 1990 (Rosche 1994a), and 46 there 18–20 Dec 1991 (Rosche 1994a).

Winter

This species has become a rare winter visitor on large, open reservoirs and lakes. In very recent years it has been a regular winter resident in Keith Co, where it was described as a "locally fairly common to common winter visitant" (Rosche 1994a). As many as 32 have been recorded in midwinter (Rosche 1994a); these birds use the spillway of L McConaughy and the open water of associated L Ogallala and the channel of the North Platte River. The only midwinter report elsewhere is of 2–3 at Scottsbluff SL 15–19 Jan 1998 (SJD).

High counts include 32 in Keith Co 13 Jan 1991 (Rosche 1994a).

FINDING

A trip to the L McConaughy area during early winter could be rewarding.

Lesser Scaup

Aythya affinis

STATUS

Common, locally abundant, regular spring and fall migrant statewide. Rare regular breeder north-central. Locally uncommon regular winter visitor central, uncommon casual elsewhere.

DOCUMENTATION

Specimen

UNSM ZM 7679, 11 Apr 1913 Lancaster Co.

DISTRIBUTION AND ECOLOGY

This diver, like others, utilizes large lakes, reservoirs, and large, deep marshes with extensive open water.

Spring

6, 7, 8 Feb→10, 10, 12 Jun

Late dates above are from southern and eastern Nebraska.

Movement begins coincident with the thaw of larger lakes and reservoirs, typically in very late Feb and early Mar. Peak numbers are achieved in Apr, and small numbers persist into late May and even early Jun on occasion.

High counts include 2880 at Crescent L NWR

25 Apr 1995, 2300 there 5 Apr 1994, and 1760 there 23 Apr 1996.

Summer

A small but persistent breeding population may be found in southern Sheridan and northern Garden Cos. An estimated 28 young were produced in 1978 at Crescent L NWR, and in 1979 the refuge maintained a summer population of about 20 adults, not all of which were likely breeding (Zeillemaker, pers. comm. RSS). Recent summer counts include 48 at Crescent L NWR 16 Jun 1996 and 26 there 17 Jun 1995 and 12 Jul 1995.

Ducey (1988) cites only 1 breeding report, in Cherry Co in 1966. Johnsgard (1997) stated that it is an "occasional summer resident in the Sandhills (probably Garden, Morrill, Cherry, and Brown Cos). Known to have nested at Crescent Lake National Wildlife Refuge but not at Valentine National Wildlife Refuge." There are numerous summer reports from Garden Co, but apart from the single nesting report, there are no recent reports from Cherry, Morrill, or Brown Cos.

The only recent summer reports in northern Nebraska away from Garden Co are 10 Jun 1985, 15 Jun 1997, 30 Jun 1967, 19 Jul 1997, and 4 Aug 1997 Lincoln Co;, 12 Jun 1973 Dawes Co; 13 Jun 1992 and 16 Jun 1993 Sheridan Co; 19 Jun 1976 Sioux Co; 21 Jun 1991 Holt Co; 25 Jun 1999 Fort Robinson SP; 30 Jun 1967 Greeley Co; and 17 Jul–18 Aug 1978 L Ogallala (Rosche 1994a). There are several summer reports from Adams Co in the 1960s and as recently as 30 Jun 1984, as well as recent Rainwater Basin reports 22 Jun 1996 and 19 Jul 1997 at Funk Lagoon, 17 Jul 1999 and 7 Aug (2) 1999 eastern Rainwater Basin, and 21–28 Jul 1996 at Harvard Marsh. Elsewhere there are few reports: 12 Jun 1992, 18 Jun 1997, 20 June 1998, and 17 Aug 1974 Lancaster Co; 21 Jun 1978, 9 Jul 1981, 15 Aug 1980, and 25 Aug 1985 Douglas-Sarpy Cos; and 1 Aug 1986 Pierce Co.

That summer birds may well be nonbreeders is indicated by the presence of 20 at Gering SL 24 Jun 1995 and 8 at Gordon SL 8 Jul 1994.

One-year-old birds that do not breed appear at molt sites first, often by late Jun, although most molt migration tends to be northward (Palmer 1976).

Fall

9, 11, 13 Sep→6, 6, 8 Jan

Migration commences in late Sep, peaking in early Nov, although there is an earlier report 2 Sep 1993 of a male at Oshkosh SL (Rosche 1994a). Some last dates of occurrence for Crescent L NWR are 19 Nov, 25 Nov, and 12 Dec (Zeillemaker, pers. comm. RSS).

High counts include 700 at North Platte NWR 9 Nov 1997, 643 there 10 Nov 1994, and 380 at Keystone L 2 Jan 2000.

Winter

Small numbers may persist on the open waters of large lakes and reservoirs. A small group regularly occurs on L Ogallala, utilizing the spillway from L McConaughy; as many as 72 were there 1998–99. There are few midwinter reports away from the L McConaughy area: 1994–95 (2) Bufflehead Pond and 1 Feb 1989 Douglas-Sarpy Cos.

FINDING

Nearly any raft of divers on a medium-size to large lake or reservoir should have representatives of this very common diving duck during either the spring or fall migration.

King Eider

Somateria spectabilis

STATUS

Accidental in fall.

DOCUMENTATION

Photograph

10–24 Nov 1985 DeSoto Bend NWR, Washington Co (NBR 54:81; Mollhoff 1989a; Kent 1987).

Records

There is 1 record of 2 birds that remained at DeSoto Bend NWR 10–24 Nov 1985; identification of the birds as King Eiders is discussed exhaustively by Kent (1987).

COMMENTS

Kent (1987) noted that King Eider might be expected to occur in the upper Midwest with more frequency than Common Eider but that reports of Common Eider were surprisingly numerous. He suggested that some early Common Eider specimens may have been misidentified, as some early records of Common Eider for Iowa later were shown to be in fact King Eiders.

There are 4 Iowa records of King Eider, all in the period 8–24 Nov (Kent and Dinsmore 1996), and 1 from Kansas, a specimen collected in Douglas Co 27 Nov 1947 (Thompson and Ely 1989).

Common Eider
Somateria mollissima
STATUS
Accidental in fall.
DOCUMENTATION
Specimen
UNSM ZM12341, about 29 Nov–2 Dec 1967 Maloney Canal, Lincoln Co (Tate 1969).
Records
There is only 1 record, the specimen cited above, which consists of the head and a foot. Johnsgard (Tate 1969) determined that it belonged to the interior race *S. m. sedentaria*.
COMMENTS
The interior subspecies *sedentaria* overwinters in the open waters of Hudson and James Bays. According to the AOU Check-list (1957), *sedentaria* is casual southward to the Niagara River, based on a single record. The Check-list notes also that the western subspecies *S. m. v-nigrum* has been found to be accidental as far south as Kansas, based on a specimen collected 3 Nov 1891 in northeast Kansas (Thompson and Ely 1989). One present 27 Jan–7 Apr 1995 in southeast Iowa was considered to be *v-nigrum*, and a specimen collected 1 Nov 1901 near Sioux City, Iowa, on the Missouri River was also identified as *v-nigrum* (DuMont 1934; Kent and Dinsmore 1996). Tate (1969) noted that the latter record could as well apply to Nebraska, as its site of collection is not accurately known (Silcock

1992); DuMont (1934) noted that it was "found in market. Brought in by hunter from Missouri River bottoms below city." Based on 3 records, the species is considered a casual fall migrant in South Dakota (SDOU 1991), subspecies not reported. There are no Colorado records (Andrews and Righter 1992).

Harlequin Duck
Histrionicus histrionicus
STATUS
Accidental.
DOCUMENTATION
Specimen
about 1901 Platte River a little west of Grand Island (Swenk, Notes before 1925).
Records
There are 6 reports of this species. The only documented record is a specimen, HMM 2499, on display at the HMM, that "was purchased from Fred Johnson, whose father shot the bird" (HMM data). Swenk (Notes before 1925) stated that "one killed by a farmer on the Platte River a little west of Grand Island, Hall Co, about 1901, and mounted by Mr. J.O. Hasty, the Arapahoe taxidermist, is now #2499 in the A. M. Brooking collection at Hastings." Brooking (Notes) stated that "one obtained of Eric Johnson (male) whose father shot on Platte near Doniphan about 1901." Taken as a whole, despite minor discrepancies, we believe the provenance of this specimen is reasonably proved.

The 5 additional reports are undocumented: specimens were reportedly taken in Douglas Co 16 Sep 1893 and 19 Sep 1895 (Bruner and others 1904), one was among specimens at the Omaha market reportedly shot in Burt Co several years prior to 1904 (Bruner and others 1904), one reported to the NGP was reportedly shot at Calamus Res during the second week of Nov 1992, and one was reported without details to the editor of NBR (NBR 41:79–80).

There is a recent record of one at Barr L, Adams Co, on the northeast Colorado plains 24

Oct 1976; there are no breeding records for Colorado since the 1880s (Andrews and Righter 1992). The first record for South Dakota was 16 Sep–5 Oct 1996 (SDBN 48:19). A female-plumaged bird was shot by a hunter in northwest Missouri 3 Nov 1972, and a female was in St. Charles Co, Missouri, 30 Jan–mid-Feb 1999 (Robbins, pers. comm.); these are the only recent records for the state, although there are 2 others in the period 29 Oct–22 Nov and a spring record 21 Mar 1897 (Robbins and Easterla 1992). There are two records for Iowa, 27 Dec 1932 and 31 Oct 1976 (Kent and Dinsmore 1996). Robbins and Easterla (1992) noted that Harlequin Ducks occasionally associate with Buffleheads in the Midwest.

Surf Scoter

Melanitta perspicillata

STATUS

Rare casual spring migrant statewide. Rare regular fall migrant statewide.

DOCUMENTATION

Specimen

UNSM ZM10300, 7 May 1911 Lancaster Co.

DISTRIBUTION AND ECOLOGY

This species overwinters on both coasts and is a rare but regular winter visitor to the Gulf Coast (AOU 1983). It is possible that individuals that overwinter along the Gulf Coast also migrate through the interior to breeding areas in the Arctic and that these account for Nebraska's migratory population.

Spring

There are 8 reports, 3 documented:

3 May 1991 L Yankton (Grenon 1991)

7 May 1911 Lancaster Co (specimen cited above)

8–15 May 1998 first alternate male Keystone L (SJD)

Other reports are 5 Apr 1986 Hall Co, 12 Apr 1993 Lincoln Co, 21 Apr 1964 Douglas-Sarpy Cos, 27–28 Apr 1990 Keith Co, and 30 Apr 1993 Keith Co.

Fall

5, 9, 10 Oct→25, 27, 27 Nov

There are 30–35 reports, including later reports 18 Dec 1991 Lincoln Co, 20 Dec 1991 Keith Co, and 2 Jan 2000 Keystone L.

High counts include 3 at North Platte NWR 27 Oct 1997.

FINDING

As with other scoters, this species should be looked for among diving ducks on large bodies of water in late Oct and the first half of Nov.

White-winged Scoter

Melanitta fusca

STATUS

Rare casual spring migrant statewide. Rare regular fall migrant statewide. Accidental in winter.

DOCUMENTATION

Specimen

UNSM ZM7684, 14 Feb 1916 Clay Co.

Spring

26, 31, 31 Mar→29 Apr, 1, 4 May

The 17 reports, west to Lincoln Co, include earlier reports 14 Feb 1916 Clay Co and 14 Mar 1981 Washington Co.

Fall

15, 17, 20 Oct→5, 6, 10 Dec

There are about 55 reports, including an earlier report 7–20 Oct 1965 Adams Co and later reports 18 Dec 1900 near Kearney, a specimen, HMM 2531 (Brooking, Notes); and a female or immature at L Ogallala 19–20 Dec 1991 (SJD, WRS).

As is the case with many other waterfowl species, the fall migration is often more leisurely and less direct to wintering areas. Individuals or small groups may first drift south of breeding grounds before moving westward to the Pacific Coast. That numbers and records are more frequent in fall than in spring in South Dakota, where it is considered rare to uncommon (SDOU 1991), and Wyoming, where it is considered an uncommon fall migrant (Scott 1993), supports this concept. Spring migrants are likely more direct and thus may bypass Nebraska.

High counts include 10 in Lancaster Co 2 Nov 1986 and 7 there 18–31 Oct 1984.

Winter

A single bird was at L Ogallala 13 Jan–28 May 1985 (Rosche 1994a); the late date has been incorrectly listed as 25–28 Jun 1985 (NBR 53:50; Loren and Barbara Padelford, pers. comm.).

FINDING

As with other scoters, White-wings are most likely to be found on large lakes and reservoirs, although several early state records are for birds that were observed on the Missouri River.

Black Scoter

Melanitta nigra

STATUS

Rare casual spring migrant statewide. Accidental in summer. Rare regular fall migrant statewide. Accidental in winter.

DOCUMENTATION

Specimen

UNSM ZM7675, 1904 Omaha, Douglas Co.

DISTRIBUTION AND ECOLOGY

This species overwinters on both coasts and on the Great Lakes. Inland records continue to increase, however, particularly during fall. This trend may be attributed to the development of many large inland reservoirs during the last half of the 20th century, including those on the Great Plains, which have provided new islands of habitat among which transients can move.

Spring

There are 7 spring reports, only 2 documented: 25 Feb 1999 Branched Oak L (JS)

1 May 1949 (4) Logan Co (NBR 18:33–34)

Additional reports, without details, are 25 Mar 1953 Boone Co, 27 Apr 1952 (NBR 20:83), 28 Apr 1992 Sarpy Co, 4 May 1958 Scotts Bluff Co, and 17 May and 25 May Adams Co (NBR 35:2).

There are about 19 spring records from

neighboring states in the period 27 Feb–26 May; 13 of these are from Iowa and Missouri. Based on these records, at least some of the undocumented Nebraska reports are likely to be correct.

Summer

There is a single record:

25 Jul 1986, male DeSoto Bend NWR 25 Jul 1986 (NBR 55:21; Mollhoff 1987).

Fall

21, 23, 28 Oct→16, 16, 17 Dec

There are about 35 reports, many documented. There are earlier reports: 1 at Lancaster Co (Bruner and others 1904) 28 Sep 1895, 10 Oct 1975, 3 at Box Butte Res (Rosche 1982) 10 Oct 1975, and 2 at L Minatare 16 Oct 1999 (SJD). Later reports are 2 at L Ogallala 7–18 Dec 1997 with 1 remaining until 4 Jan 1998 (SJD; Brogie 1998), 4 at L Ogallala 12 Dec 1998 with 1 remaining until 1 Jan 1999 (SJD), and 1 at Keystone L 2 Jan 2000.

Specimens mounted by Velich were taken near Blair Dec 1956 (2 females) and near Ashland in Dec (1956?) (NBR 26:9).

High counts include 3 in Sarpy Co 18 Nov 1993 (AB 48:125) and 3 at Pawnee L 13 Nov 1997.

Winter

There is a single midwinter report, 1 of only 3 Jan records (see Fall).

16 Jan 1927, Missouri River, Sarpy Co (Bray and others 1986; LOI 20)

FINDING

This species should be looked for on large bodies of water among other diving ducks in late Oct and early Nov.

Oldsquaw

Clangula hyemalis

STATUS

Rare regular spring and fall migrant statewide. Rare casual winter visitor Keith and Lincoln Cos.

DOCUMENTATION

Specimen

UNSM ZM12300, 11 Nov 1931 east of Tekamah, Burt Co.

DISTRIBUTION AND ECOLOGY

Reports are from larger lakes and reservoirs where Oldsquaw usually occur as single birds with other diving ducks, especially goldeneyes. Johnsgard (1980) argued that with an increase in large reservoirs across the central Great Plains, this species would be seen with increasing regularity in Nebraska. Recent records of occurrence in Nebraska and surrounding states clearly support this prediction, although this may be true only of birds seen at reservoirs in midwinter but not of birds utilizing major rivers; it was considered a "regular, but not common winter visitor" by Bruner and others (1904), and it seems likely that subsequent channelization of the Missouri River has diminished the fall flight of waterfowl there and caused a shift to reservoirs, with wintering birds remaining where water is open (JGJ).

Spring

2, 2, 3 Mar→24, 24, 24 Apr

There are later reports 17 May 1985 and 18 May 1986 Lincoln Co.

Fall

31, 31 Oct, 2 Nov→12, 18, 18 Dec

There is an earlier report 18 Oct 1946 Platte Co. There are virtually no reports after 18 Dec away from Lincoln and Keith Cos (see Winter).

High counts include 11 at Keystone L 29 Nov 1998, 6 in Lancaster Co 7 Dec 1975 (AB 30:735), 6 in Cedar Co 12 Dec 1993 (AB 48:222), and 6 at Keystone L 2 Jan 2000.

Winter

Most of the 13 reports for the period 18 Dec–1 Mar are from Keith and Lincoln Cos, most of single birds with other wintering diving ducks, although 2 were on L McConaughy 15 Jan 1998.

High counts include 9 at L McConaughy 12 Dec 1998–2 Jan 1999.

FINDING

Spillways below dams such as Kingsley and Gavin's Point in Dec are the most likely locations to find this species.

Bufflehead

Bucephala albeola

STATUS

Common, locally abundant, regular spring and fall migrant statewide. Rare casual summer visitor statewide. Rare, locally uncommon, regular winter visitor statewide.

DOCUMENTATION

Specimen

WSC 544, 4 Nov 1979 Gavin's Point Dam, Knox Co.

DISTRIBUTION AND ECOLOGY

This tiny diver may utilize any lake, reservoir, or pond with sufficient depth and fauna to provide for its feeding habits. It may also utilize still backwaters of larger rivers during the spring migration.

Spring

14, 15, 15 Feb→29, 29 May, 1 Jun

There is a later report 6 Jun 1999 at Youngson Basin.

Buffleheads begin arriving with the spring thaw, typically in early Mar. Movement peaks in very late Mar, and numbers remain relatively constant through Apr, when they begin to drop off substantially. Small numbers may be seen into late May, however.

High counts include 1440 at Crescent L NWR 25 Apr 1995, 285 there 8 May 1996, and 229 at L Ogallala 24 Apr 1999.

Summer

This species has not been known to breed in Nebraska. Most summer reports are of individuals occasionally seen for several weeks in midsummer at Crescent L NWR, likely subadults or unpaired individuals. Of about 13 summer reports in the period 7 Jun–17 Aug, most are in Garden Co, including as many as 12 birds at Crescent L NWR 7 Aug 1998. Reports elsewhere are few: 10 Jun 1995 Dakota Co, 12 Jun 1987 Scotts Bluff Co, 25 Jun 1997 North Platte NWR, 30 Jun 1984 Adams Co, late Jun 1999 near Antioch, 1 Jul 1986 Pierce Co, 10 Jul 1916 Inland (Brooking, Notes), and 28 Jul 1991 Keystone L (Rosche 1994a).

Fall

16, 17, 19 Sep→5, 7, 10 Jan

Late dates above exclude those from Keith Co, where wintering is regular.

Movement may be first evident in early Oct, although there is an early report of an immature at Gordon SL 10 Sep 1994, and as many as 52 were at Crescent L NWR 16 Sep 1995. Numbers typically peak in early Nov, but relatively high populations may be maintained until lakes begin to freeze in early Dec or later. Rosche (1994a) indicated that birds seen as early as late Jul may be early fall migrants (see Summer).

High counts include 1200 at Keystone L 9 Nov 1990 (Rosche 1994a), 390 at Branched Oak L 20 Nov 1998, and 310 at Crescent L NWR 23 Oct 1995.

Winter

Small numbers winter at the spillway below Kingsley Dam and possibly at Gavin's Point Dam and Sutherland Res, although reports of overwintering from the latter locations are lacking. Latest reports from Cedar Co are 10 Jan 1990 and 7 Jan 1996 (2), and the only midwinter report from Sutherland Res is 4 Feb 1996. Rosche (1994a) stated that about 100 Buffleheads winter below Kingsley Dam most years; 31 were present 1995–96, and 27 were there 1998–99.

Midwinter reports away from Keith Co are few: 20 Jan 1962 Adams Co and 30 Jan 1983 Douglas-Sarpy Cos.

FINDING

Look for this small diving duck on large lakes and reservoirs during migration. Concentrations tend to be larger during the spring.

Common Goldeneye

Bucephala clangula

STATUS

Common regular spring migrant statewide. Rare casual summer visitor north and west. Common, locally abundant, regular fall migrant statewide. Uncommon, locally common, regular winter visitor statewide.

DOCUMENTATION

Specimen

UNSM ZM7681, 10 Apr 1926 Lincoln, Lancaster Co.

TAXONOMY

This species is known to hybridize with Hooded Merganser. There are 3 records: a male at L Ogallala 23–26 Jan 1999 (SJD), a male at Cunningham L Mar 1997 (JGJ), and a male at Cunningham L 28 Feb 1999 (JGJ). The latter 2 reports may have been the same bird (JGJ).

Complicating separation of this species and Barrow's Goldeneye are hybrids between the two (Martin and Di Labio 1994); one was recently photographed in northwest Missouri (Robbins, pers. comm.).

DISTRIBUTION AND ECOLOGY

Although a diver typically attracted to deep lakes and reservoirs, during the winter and early spring it also may be found in small numbers on large, open rivers, such as the central and lower Platte, lower Niobrara, and the Missouri (see Winter).

Spring

winter→16, 19, 20 May

There are later reports 26 May 1990 Howard-Hall Cos, 27 May 1981 Garden Co, 29 May 1993 Keith Co, and 3 Jun Keith Co (Rosche 1994a).

Although small to moderate populations may be present throughout the winter where open water is available, a spring migratory movement is evident, likely reflecting populations that may be overwintering as far south as the Gulf Coast (Root 1988). Migration begins in late Feb and early Mar and peaks in mid-Mar. Numbers then drop off substantially, a few persisting into late Apr.

Summer

There are only about 10 reports between 3 Jun and 22 Sep: 7 Jun 1981 McPherson Co; 14 Jun 1980 Scotts Bluff Co; 30 Jun 1967 Garden Co; late Jun 1999, a male near Antioch; 6 Jul 1993 Scotts Bluff Co; 14 Jul–21 Aug 1978, a female seen by Johnsgard on L McConaughy and by Rosche on Keystone L (Rosche 1994a); 8 Aug–8 Sep 1998 L Ogallala; 9 Aug 1991 Keystone L (Rosche 1994a); 6 Sep 1993 Sarpy Co; and a specimen collected in Sep 1916 in Clay Co, HMM 2394 (Brooking, Notes).

Fall

18, 20, 20 Oct→winter

Migration becomes evident in late Oct, although there are earlier reports 11 Sep 1997 (28) Crescent L NWR, 23 Sep 1986 Pierce Co, 2 Oct 1965 Adams Co, 11 Oct 1991 Lincoln Co, and 13 Oct 1988 Scotts Bluff Co. Usually there is a peak in mid to late Nov, with good numbers remaining on open water well into the winter (see Winter).

High counts include 3800 at Sutherland Res 17 Dec 1998, 3000 there 18 Dec 1997, and 1500 at L McConaughy 27 Dec 1975 (Rosche 1994a).

Winter

This species will overwinter wherever open water occurs, usually in small numbers, but 300–500 normally winter below Kingsley Dam (Rosche 1994a), and 350 were at Johnson Res 4 Jan 1997.

FINDING

Look for this striking species on large bodies of water in Nov–Dec; a flock winters each year below Kingsley Dam.

Barrow's Goldeneye

Bucephala islandica

STATUS

Rare casual spring and fall migrant statewide. Rare casual winter visitor Keith Co, accidental east.

DOCUMENTATION

Photograph

19 Apr 1987 Keystone L, Keith Co (Mollhoff 1989a).

TAXONOMY

See Common Goldeneye.

DISTRIBUTION AND ECOLOGY

Although historically the large majority of birders have resided in the eastern part of Nebraska, most of the few records of Barrow's Goldeneye are from the Panhandle,

suggesting that individuals from the small eastern Wyoming wintering population (Scott 1993) of Barrow's Goldeneye may wander into Nebraska in late fall, utilizing the North Platte River as a habitat corridor. Most of the Nov reports are from Scotts Bluff Co, while all Dec and Jan reports are from Keith Co, suggesting gradual movement eastward in the North Platte Valley to the winter location at Keystone L. Strangely, however, most of the spring reports are from eastern Nebraska. Most Iowa records are from the west (Kent and Dinsmore 1996), suggesting a liking for the Missouri Valley by a small population of wintering birds; these may be derived from the Great Lakes and may winter with Common Goldeneyes south of Iowa and Nebraska.

Spring

There are 8 spring reports, most from eastern Nebraska: 3 Mar 1997, a female at Cunningham L (JGJ; Brogie 1998); 4–14 Mar 1984 Douglas-Sarpy Cos; 14 Mar 1992 Scotts Bluff Co; 18 Mar Lincoln Co (RSS); 21 Mar 1988 Lancaster Co; 25 Mar 1951 Douglas Co; 28 Mar 1988 Lancaster Co; 30 Mar 1997, an immature male L McConaughy (SJD; Brogie 1998); 1 Apr 1951 Antelope Co; and 5–19 Apr 1987 Keith Co, a male and possibly a female photographed (Rosche 1994a; Brogie and Brogie 1987a).

Fall

There are 9 reports, not including 1 Nov and several Dec sightings in Keith Co, a wintering location. Most are from Scotts Bluff Co: 30 Oct 1990 (Grenon 1991), 5–26 Nov 1982, 19 Nov 1991, 26 Nov–19 Dec 1979, and 17 Dec 1998 (male). Elsewhere, there are these reports: 13 Nov 1986, one shot at Branched Oak L (NGP files); 16 Dec 1963 Lincoln Co (Shickley 1968); 25 Nov 1995 Hall Co; and 1 Jan 1939 Sioux Co, a flock of 6 birds possibly this species (Cook 1939).

Winter

There are about 12 records 19 Nov–Feb in Keith Co of birds wintering with a Common Goldeneye flock below Kingsley Dam, as well as these records elsewhere: 7 Dec 1988–8 Jan 1989 Missouri River valley (Silcock 1989a) and 18 Dec 1997 Sutherland Res, a male and female (SJD).

High counts include 3 at Keystone L 2 Jan 2000.

FINDING

The most consistent location is below Kingsley Dam during Dec–early Feb; a careful check of the wintering Common Goldeneye flock may yield a Barrow's in most years.

Hooded Merganser

Lophodytes cucullatus

STATUS

Fairly common regular spring and fall migrant east, becoming fairly common west. Rare regular summer visitor east and central, rare casual west. Rare, locally uncommon, regular winter visitor statewide.

DOCUMENTATION

Specimen

UNSM ZM13026, 5 Nov 1970 Lancaster Co.

TAXONOMY

Hybrids with Common Goldeneye are listed under that species.

DISTRIBUTION AND ECOLOGY

This small merganser utilizes small to large lakes and reservoirs during migration. Although unrecorded as a breeding bird in Nebraska since 1915 (Ducey 1988), nesting may occur, especially in the Missouri River valley, which is just west of the regular breeding range (Palmer 1976). It is a cavity nester, and during the breeding season it may be found on slow-moving streams, oxbows, lakes, and ponds that are bordered by mature riparian woodland. It has been known to utilize Wood Duck nest boxes in South Dakota (SDOU 1991) and Iowa (SJD).

Spring

21, 21, 21 Feb→6, 6, 7 Jun

Migration commences late Feb to early Mar. Numbers peak in mid to late Mar, quickly waning in early Apr, although small numbers may be seen sporadically into early May or later.

High counts include 35 at Niobrara Marsh, Knox Co, 28 Feb 1999; 31 at Crystal Cove L 27 Mar 1996; and 11 at L Yankton 18 Mar 1995.

Summer

Although no firm breeding records exist since 1915 (Ducey 1988), persistent summer reports have been evident throughout the 20th century. Swenk (Notes before 1925) stated that "it remains all summer at Inland, Clay Co, and probably breeds." The presence of 5 young-of-the-year 19 Jun 1995 in Grant Co "raised questions as to breeding status in western Nebraska" (AB 49:947). Four downy young possibly of this species were noted at Waterloo on or before 3 Aug 1997 (NGP files).

There are about 28 breeding season reports, most, if not all, of immature/female-plumaged birds, probably prebreeders. These reports are statewide, including westerly reports 31 May–5 Sep 1979 Garden Co; 16 Jul 1993 and 18 Jul 1973 Sheridan Co; 31 Jul 1995 (3) Crescent L NWR, possibly early fall migrants; and 14 Jun 1987, 14 Jun 1989, and 29 Jun 1984 in the L McConaughy area, all female-plumaged birds (Rosche 1994a). In 1995 at least 10 immature/female-plumaged birds were observed in the Rainwater Basin 5 Jun–31 Jul.

Recent breeding records of Hooded Merganser utilizing Wood Duck nest boxes in southeast South Dakota (SDOU 1991), bordering Nebraska, suggest that this species may occasionally breed in appropriate habitat in eastern or northeastern Nebraska.

Fall

21, 23, 24 Sep→1, 1, 4 Jan (away from L McConaughy and Sutherland Res)

There are earlier reports 4 Aug 1997 and 15 Aug 1997 and a pair 20 Aug–6 Sep 1998, both reports at North Platte NWR and likely of summering birds; 20 Aug–10 Sep 1995 L Babcock; and 14 Sep 1977 Garden Co. Although there are scattered reports throughout the month of Sep, no clear migratory movement is evident until late Oct. A peak is evident in early to mid-Nov, and individuals may remain well into Dec.

High counts include 112 at L Alice 6 Nov 1999, 70 at North Platte NWR 6 Nov 1998, and 41 at Cunningham L 15 Nov 1996.

Winter

The species may utilize the open waters of large reservoirs until such waters become ice covered. There are about 17 midwinter (5 Jan–20 Feb) reports statewide, including overwintering birds in Keith Co and at Sutherland Res. Westerly reports are 19 Jan 1964, 29 Jan 1987, and 9–19 Feb 1985, all in Scotts Bluff Co.

High counts include 26 at Sutherland Res 14 Jan 1998.

FINDING

Look for this species on medium-size to large lakes and reservoirs, especially in mid to late Mar and around the end of Oct.

Common Merganser

Mergus merganser

STATUS

Common, locally abundant, regular spring and fall migrant statewide. Rare casual summer visitor statewide. Rare casual breeder west and central. Uncommon, locally common, regular winter visitor statewide.

DOCUMENTATION

Specimen

WSC 522, fall 1975 Madison Co.

DISTRIBUTION AND ECOLOGY

This diving fish-eater is most often seen on large lakes and reservoirs, but during the winter and early spring it may also utilize medium-size to large rivers. During migration it may be found statewide where appropriate habitat is available.

A small breeding population exists in the Black Hills of South Dakota (SDOU 1991) and may be responsible for the occasional wandering individual or pair that may be seen in the Panhandle and along the Niobrara River during summer months.

Spring

winter→20, 24, 25 May

Late dates above do not include those from

Scotts Bluff or Keith Cos (see Fall). A later report was 4 Jun 1997 Branched Oak L.

It is difficult to separate winter visitors from early migrants because records are scattered throughout Jan and Feb. However, populations appear to increase in very late Feb, peaking in mid-Mar. Populations wane in very early Apr, but occasional individuals or small groups may be seen throughout the month.

Recently, lingering birds have been observed during the nesting season at L McConaughy, leading Rosche (1994a) to suggest that breeding might occur there; the possibility also exists that at least the later arrivals among these may be molt migrants. Such reports include 19–31 May 1997 (up to 24 birds), 29 May 1993, 3 Jun 1992, 10 Jun–15 Jul 1996 (10–18), and 21 Jul 1991.

High counts include 4000 at L McConaughy 24 Mar 1989 (Rosche 1994a), 3000 there 12 Apr 1997, and 2500 there 21 Mar 1998.

Summer

There are 3 reports of breeding, only 1 with published evidence. It apparently nested at Victoria Springs in 1968; a male and female were present during the spring, and on 27 Jul the female with 6 ducklings was seen by several observers (NBR 37:45). Additionally, there is a "recent nesting record for the North Platte Valley to the west of Lake McConaughy" (Rosche 1994a), and a pair at L McConaughy 31 May 1996 was "suspected as nesting" (FN 50:297) (see Spring; see Fall for late summer birds at North Platte NWR).

Several Jul records exist for Cherry, Rock, and Brown Cos along the Niobrara River (RSS), where good breeding habitat (mature trees, clear river) exists. There are few reports elsewhere: 14 Jun 1985 Lancaster Co, 26 Jun 1991 Lincoln Co, 17 Jul 1978 Garden Co, 19 Jul 1997 Lincoln Co, 25 Jul 1998 Harlan Co Res, 1 Aug 1991 Lincoln Co (AB 41:1456), and 1 Aug 1999 Branched Oak L (LE).

Fall

4, 7, 9 Oct→winter

Early dates above do not include those from

Scotts Bluff Co (see below). There are earlier reports away from Scotts Bluff Co: 21 Aug 1999 Oliver Res (MB), 30 Aug 1998 (2) Harlan Co Res, 4 Sep 1973 McPherson Co, 17 Sep 1994 L McConaughy, 18 Sep 1971 Douglas-Sarpy Cos, 26 Sep 1986 Pierce Co, and 30 Sep 1992 Lincoln Co.

Although small numbers begin to enter the state in early Oct, the major movement is not evident until late Oct and early Nov. Following the peak the fall population may remain high well into Dec, and good numbers sometimes remain into midwinter.

In recent years L Alice has become noted as a molt migration destination for 100–200 Common Mergansers (NBR 63:73). There have been reports of birds arriving in mid-Jul most years since 1973 and as early as 25 Jun in 1997. Recent counts include 170 on 10 Aug 1994 and 156 on 11 Aug 1995. Similarly, 10–25 have appeared at L Ogallala in recent years, such as 21 there in 1998. Few yearling and mature males remain on breeding grounds after mid-Jul (Palmer 1976). Presumably the Nebraska birds move eastward in the North Platte Valley to L Alice from breeding locations in the mountains of Wyoming.

Major central Nebraska reservoirs, L Mc-Conaughy, Sutherland Res, and Harlan Co Res, are important staging areas for this species. Large numbers concentrate on these reservoirs and remain as long as open water allows.

High counts include 48 000 at Harlan Co Res 14 Dec 1997, 40 000 at L McConaughy 23 Dec 1994, and 35 000 at Harlan Co Res 30–31 Dec 1998.

Winter

Small numbers may be found throughout the winter months on large lakes and reservoirs that remain ice-free; occasionally very large numbers overwinter, such as 15 000 at Sutherland Res 11 Feb 1994 and 12 000 at L McConaughy 15 Jan 1998. Likewise, ice-free rivers such as stretches of the Platte, lower Niobrara, and Missouri will sustain small winter populations. Spillways of dams may also attract overwintering birds.

FINDING

This species is most evident at L McConaughy in Dec but may be located almost anywhere on large bodies of water in Nov–Dec and Mar.

Red-breasted Merganser

Mergus serrator

STATUS

Uncommon regular spring migrant east, becoming rare west. Uncommon regular fall migrant east, becoming rare west. Rare casual winter visitor statewide.

DOCUMENTATION

Specimen

HMM 2492, Apr 1900 Funk Lagoon, Phelps Co.

DISTRIBUTION AND ECOLOGY

During migration this species utilizes large lakes and reservoirs and, in contrast with Common Merganser, is almost never found on streams and rivers except in the winter, when it may frequent spillways below dams.

Red-breasted Merganser tends to migrate in flocks, which appear overnight and depart soon afterward. Thus estimates of its abundance in neighboring states vary somewhat from those for Nebraska, presumably dependent on the experience of the writers. South Dakota records (SDOU 1991) indicate that it is found predominantly in the east, where it is judged to be "uncommon to rare." In Kansas it is considered to be a "rare spring and fall transient" statewide (Thompson and Ely 1989). However, in Nebraska, particularly on lakes and reservoirs in the east, such as those in the Salt Creek and Papillion Creek watersheds, Carter L, and DeSoto Bend NWR, spring concentrations may be as large as several hundred individuals (RSS). Palmer (1976) noted that this species is "abundant at favored places" during migration.

Spring

6, 7, 9 Feb→18, 18, 19 May

There are later reports 25 May 1994 at Crescent L NWR and 27 May 1990 Saunders Co.

Although several late Feb reports exist, it is a later migrant than the Common Merganser, typically arriving in mid to late Mar. About a third of spring records are for the last week of Mar, but the species may be seen regularly into mid-Apr. Most reports are from the east; L Ogallala and L Minatare are probably the only westerly locations where the species occurs with any regularity.

High counts include 103 at L Ogallala 18 Apr 1998, 34 at Cunningham L 24 Mar 1997, and 27 at L Yankton 4 Apr 1994.

Fall

20, 24, 24 Oct→18, 19, 23 Dec

There are earlier reports 22 Sep 1978, 23 Sep 1981, 3 Oct 1984, and 11 Oct 1977 Scotts Bluff Co and later reports, undocumented, 29 Dec 1964, 29 Dec 1989, 30 Dec 1990, and 3 Jan 1998 Cass Co.

The fall migration is more leisurely than in spring, without large concentrations or a significant peak. This species is less cold-tolerant than the other merganser species and generally leaves before Dec. There are far fewer fall reports than spring reports, but a greater proportion are from the west.

High counts include 137 in fall 1996, most at eastern reservoirs (FN 51:80); 47 at Summit Res 9 Nov 1997; and 44 at L Minatare 7 Nov 1998.

Winter

There are several reports, most from L McConaughy, where it is likely that a few winter each year with the large Common Merganser flock. There are these midwinter reports in the period 1 Jan–6 Feb: 1–10 Jan 1999 (1–3) L Ogallala (SJD), 1 Jan–4 Feb 1965 Douglas-Sarpy Cos, 13 Jan 1995 Sutherland Res, and 30 Jan Douglas-Sarpy Cos (RSS). A female below Kingsley Dam probably wintered 1996–97, as it was seen there with Common Mergansers 28 Dec and 15 Feb (SJD), and 2–4 were there 1997–98 (SJD). This species may also utilize the spillway at Gavin's Point Dam, where it apparently wintered 1982–83 on the South Dakota side (SDOU 1991).

COMMENTS

Females of this species closely resemble females of the Common Merganser. It is possible that some of Nebraska's more extreme records for this normally cold-sensitive species, particularly in winter, are actually female or immature Common Mergansers.

FINDING

The best location to look for this species would be a larger lake or reservoir in eastern Nebraska in late Mar or early Apr.

Ruddy Duck

Oxyura jamaicensis

STATUS

Common, locally abundant, regular spring and fall migrant statewide. Locally uncommon regular breeder Sandhills, rare Rainwater Basin, rare casual elsewhere. Rare casual winter visitor statewide.

DOCUMENTATION

Specimen

UNSM ZM12481, 26 Dec 1912 Lancaster Co.

DISTRIBUTION AND ECOLOGY

During migration this species is attracted to large lakes and reservoirs statewide, sometimes in very large rafts. During the breeding season it seeks out expanses of emergent marsh vegetation that have open-water corridors interspersed. In summer most are concentrated in the Sandhills, but occasional nesting may occur in the Rainwater Basin and elsewhere in appropriate habitat.

Spring

23, 23, 25 Feb→4, 6, 6 Jun

Migratory populations quickly build up in mid-Mar, coincident with the ice thaw. Peak numbers are usually attained in mid-Apr, but substantial numbers may be maintained until early to mid-May.

High counts include 300 at L Ogallala 12 Apr 1975 (Rosche 1994a), 300 at Funk Lagoon 14 Apr 1996, and 166 at Crescent L NWR 9 May 1997.

Summer

Recent definitive nesting records are scattered throughout the state (Ducey 1988), but the bulk of the breeding population is in Sheridan and Garden Cos, with lesser concentrations in Cherry Co. Summer records are regular for Valentine NWR and Crescent L NWR, where it regularly nests; about 100 were at Crescent L NWR in mid-Jul 1995 and 1996. Rapp and others (1958) stated that it was a common breeder on Sandhills lakes in Garden, Morrill, and Cherry Cos. There are numerous breeding season reports for McPherson Co and fewer for Arthur, Grant, and Brown Cos.

There is a small breeding population in the Rainwater Basin. Evans and Wolfe (1967) found 5 nests in Clay Co, and a Ruddy Duck nested in 1996 in Phelps Co, apparently for the first time. There are also a few breeding season reports from Adams, Polk, York (Ducey 1988), Hall, Platte, and Seward Cos; 4 birds, including a displaying male, were in western Seward Co 23 Jun 1996.

As reservoirs in the eastern part of the state mature and marshy habitat develops, this species may also nest with increasing frequency there. It was considered a regular though rare breeder in the Salt Creek Valley lakes, Lancaster Co (Cink 1975), and there are several breeding season reports from Lancaster and Douglas-Sarpy Cos.

There are a few breeding season reports from Keith and Lincoln Cos, with no evidence of breeding, although Rosche (1994a) observed a downy young bird with a female at Oshkosh SL 24 Jul 1984.

There are numerous breeding season reports for Scotts Bluff and Dawes Cos, although definite evidence of breeding is lacking. There are a few reports from Knox and Dakota Cos, also without evidence of breeding, and 5 were in Otoe Co 4 Jul 1997.

High counts include 46 at Crescent L NWR 8 Jul 1997.

Fall

17, 20, 20 Sep→26, 27, 31 Dec

Movement becomes evident in late Sep in areas where summer residents are absent. Peak numbers are found from mid-Oct into early Nov. A few linger into Dec, but this is not a cold-tolerant species. An unusually late high count was 55 at Wehrspann L 20 Dec 1998.

High counts include 1500 at North Platte SL 28 Oct 1996, 1492 at L Minatare 16 Oct 1999, and "at least 1000" at Johnson Res 29 Oct 1995.

Winter

There are few midwinter reports. As many as 17 wintered on University L 1995–96; 6 at L Ogallala 3 Jan 1998 declined to 1 on 6 Feb; 9 there 9–10 Jan 1999 declined to 6 on 24 Jan; and there are reports 15 Jan 1991 Howard-Hall Cos, 30 Jan 1972 L McConaughy (NBR 40:86), 2 Feb 1997 Wood Duck Area, and 12 Feb 1977 Howard-Hall Cos.

FINDING

Look for rafts of small, dark ducks bunched together on mid-size to large lakes and reservoirs during the spring and fall migratory periods.

FALCONIFORMES

Accipitridae (Osprey, Kites, Hawks, Eagles)

Osprey
Pandion haliaetus
STATUS
Uncommon regular spring and fall migrant
statewide. Rare casual summer visitor
statewide. Hypothetical breeder. Hypothetical
in winter.
DOCUMENTATION
Specimen
UNSM ZM14607, Oct 1938 North Platte, Lincoln
Co.
DISTRIBUTION AND ECOLOGY
During migration this species frequents lakes
and reservoirs as well as the lower Niobrara,
lower Platte, and Missouri Rivers in search of
live fish.
Spring
22, 23, 25 Mar→2, 4, 5 Jun (Rosche 1994a)
Excepting scattered Feb and early Mar reports,
some of which could be misidentifications
(see Winter), regular movement begins in
early Apr and continues into early May,
peaking in late Apr. Numbers quickly wane
after about 10 May, with scattered reports
through late May.
It is possible that migrants represent 2 distinct
breeding populations. Eastern migrants may
be part of the large breeding population in
Minnesota and Manitoba that overwinters
in the lower Mississippi Valley and in south
Texas (Root 1988) and accounts for the higher
frequency of records from eastern Nebraska.
The second breeding population is found
in Wyoming and Montana and is apparently
increasing (Scott 1993), which may account
for the increasing frequency of reports from
western Nebraska during the last few years.
Individuals from this population may drift
eastward through the Panhandle en route to
overwintering areas; their spring return likewise could cross the western part of the state.

High counts include 7 at Branched Oak L 27 Apr
1995 and 5 there 28 Apr 1996.
Summer
There are about 20 reports in the period 17 Jun–
7 Aug, distributed statewide. Most are from
large reservoirs and probably represent 1–3-
year-old birds, many of which only partially
migrate toward the breeding range (Palmer
1988). Although the nearest regular nesting
location to Nebraska is in the Rockies and
northwest Minnesota, a nest has been active
and productive in Pennington Co, South
Dakota, every year since 1991 (Peterson 1995).
According to Bruner and others (1904), it
formerly bred at least once along the Missouri
River near Rockport on the Washington-
Douglas Co boundary (Ducey 1988). A report
of 3 "young Ospreys" in a tree together at
Wellfleet L 8 Aug 1967 (Nielson 1968) may be
correct, although no adults were present and
the age of the young birds was not noted.
Fall
19, 20, 20 Aug→18, 19, 21 Nov
There are earlier reports 12 Aug 1962 Lincoln Co
and 14 Aug 1991 Cass Co and later reports 3
Dec 1988 Lancaster Co and 11 Dec 1991 Scotts
Bluff Co.
Migration is evident beginning in late Aug and
peaks in mid to late Sep. Individuals may
remain in an area for several weeks. There are
scattered records into early Dec.
High counts include 26 at L McConaughy 3 Oct
1998, 9 in Lancaster Co 10 Sep 1995, and 8 at
Branched Oak L 9 Sep 1998.
Winter
There are no documented winter reports among
the 25 or so extant. The reports are in the
period 26 Dec–23 Feb, but there are none since
1985. It is likely that as observers have gained
expertise, misidentification of immature
Bald Eagles, for example, as Ospreys has
been avoided. There are no accepted winter
records for Colorado, despite some 20
reports (Andrews and Righter 1992). The
northernmost midwinter Missouri record was
near St. Louis (Robbins and Easterla 1992).

Of the Nebraska reports, almost half are from Scotts Bluff Co 1973–83; the others are from Lincoln (6), Howard-Hall, Hamilton, Adams, Jefferson, Lancaster, Washington, and Douglas-Sarpy (4) Cos.

FINDING

Look for this species in mid to late Apr or mid-Sep through Oct hunting above the surface of medium-size to large lakes and reservoirs and occasionally larger rivers.

Swallow-tailed Kite

Elanoides forficatus

STATUS

Extirpated.

DOCUMENTATION

Specimen

UNSM ZM14116, 8 Aug 1896 Cass Co (Eiche 1901).

COMMENTS

The above specimen, the male of a pair purportedly nesting near Greenwood (Eiche 1901), appears to be the last tangible Nebraska record for this species, although Bruner and others (1904) considered it "a regular visitor in the eastern third of the state, not uncommon," and William Townsley shot one as it flew over his house in Harvard about 1910 (Swenk, Notes before 1925). Ducey (1988) documented its summer occurrence and apparent breeding activity in the state and in adjacent South Dakota, all occurring in the 19th century. He noted its presence as early as 1833. The 19th-century records suggest an eastern Nebraska distribution. During this same period it occurred casually and with varying abundance in eastern Kansas, where it was known to have bred as late as 1876 (Thompson and Ely 1989). It was formerly found across much of Iowa, but its numbers rapidly declined during the latter part of the 19th century in Iowa as well; the last record prior to its extirpation was in 1931 (Kent and Dinsmore 1996).

The species utilizes riparian woodland for nesting. In Nebraska at this time such woodland is more extensively distributed than it was in the 19th century, and at the end of the 19th century such woodland was increasing its range as grass fires were increasingly controlled. Thus at least in Nebraska loss of breeding habitat is not a likely factor contributing to this species' extirpation. Of course, habitat loss elsewhere in this species' year-round range may have had an effect. In this context, it is curious that a large insect species, the migratory locust *Melanoplus spretus*, abundant in the 19th century, became extinct as this insectivorous kite species dropped in numbers.

Robertson (Palmer 1988) documented the extensive extirpation of this species in the 19th century, its range being reduced by some 80%. No clear factors that led to this loss can be demonstrated. However, Robertson does note a resurgence in numbers that was developing in the existing population prior to that publication and predicted a range expansion.

In recent years Swallow-tailed Kites have indeed reappeared on the northern Great Plains, with recent Apr–Aug reports from Colorado as close to Nebraska as Yuma Co (Andrews and Righter 1992) and a single record from Iowa 14–15 May 1992 (Kent and Dinsmore 1996). It would be expected to occur in Nebraska in the near future.

A report from Douglas Co 31 Mar 1960 (NBR 28:61) was later corrected to Caspian Tern (Swanson 1961).

White-tailed Kite

Elanus leucurus

STATUS

Rare casual summer visitor south and west.

DOCUMENTATION

Photograph

31 Aug–8 Oct 1983 Mormon Island Crane Meadows, Hall Co (Lingle and Lingle 1983).

Summer

There are 4 reports for this primarily neotropic species, 3 documented:

6 May 1995 Polk Co (Morris 1995)

19 Aug 1981 Garden Co (Green 1982).

31 Aug–8 Oct 1983 Hall Co (cited above)

The fourth report, also likely correct, was of one seen 10 May 1999 at Crane Meadows Nature Center in Hall Co (LP, BP).

COMMENTS

The nearest location for regular summer occurrence is southern Oklahoma (AOU 1983). It is unrecorded in Colorado (Andrews and Righter 1992), Kansas (Thompson and Ely 1989), and Iowa (Kent and Dinsmore 1996). However, one was in central South Dakota 9–10 Jul 1978 (SDOU 1991). There are 2 Wyoming records, one near Casper Mountain Jun–Aug 1989 and the other in Grand Teton National Park 20 Aug–14 Sep 1982 (Scott 1993). In Missouri one was near Springfield 15 May 1983 and another in Nodaway Co 14 Jun 1976 (Robbins and Easterla 1992).

The Nebraska records are part of a regional trend. Palmer (1988) noted that this species is experiencing a population resurgence, reoccupying areas where it had been extirpated and in some areas extending its historic range.

FINDING

This species should be looked for in grassland areas.

Mississippi Kite

Ictinia mississippiensis

STATUS

Rare casual spring and fall visitor statewide. Locally uncommon regular breeder Ogallala, Keith Co. Rare casual summer visitor statewide.

DOCUMENTATION

Specimen

UNSM ZM17034, 9 Sep 1991 North Platte, Lincoln Co (Brogie 1997).

DISTRIBUTION AND ECOLOGY

This species occupies prairie-riparian and savanna habitat throughout much of the southern Great Plains (Palmer 1988). In contrast with other kite species its numbers have remained relatively stable during the 20th century. The population in the western part of its range, including that of the southern Great Plains, is the highest in density and most contiguous (Palmer 1988). Apparently this western population is expanding in response to shelterbelt plantings and accounts for the increased frequency of Nebraska sightings, as well as the recent breeding activity in the state. It is now considered a common summer resident in southwestern and south-central Kansas and has recently been found breeding in Kansas along the south-central Nebraska border (Thompson and Ely 1989).

Spring

Other than reports from Keith Co, there are few additional spring reports; only these are documented: 15 Apr 1950 Antelope Co; 24 Apr 1991 Polk Co; 11 May 1986, an adult at Wilderness Park, Lancaster Co (NBR 55:47); 12 May 1948 Webster or Adams Co; 14–21 May 1944 (2) Adams Co; 16–21 May 1980, an adult in Scotts Bluff Co (NBR 48:81); and 19 May 1974 (2) Lincoln Co. A report 16 May 1987 Cherry Co was not accepted by the NOURC (Mollhoff 1989a).

Summer

Dinsmore and Silcock (1995b) documented active nests of this species within the city limits of Ogallala, the first confirmed nesting for Nebraska. Mississippi Kites had first been noted in Ogallala 27 Jul 1991 (Rosche 1994b; Brown and others 1996) and have been seen each year since. As many as 11 adults were present on 6 Aug 1994 (Dinsmore and Silcock 1995b).

Palmer (1988) stated that egg dates range from the latter half of May to mid-Jun, and young fledge about 60 days later, in late Jul through mid-Aug. Reports during this period indicative of nesting are few, most notably from Polk Co, where nesting probably occurred. There are also late May and Aug–Sep reports from Douglas-Sarpy Cos indicative of nesting, as well as 2 early summer Hall Co reports.

Possibly a postfledging wanderer was an immature in Dawes Co 26 Jul 1990 (AB 44:1152).

There are Polk Co reports 7 Jun 1987, 30 Jun 1985, 5–9 Aug 1985, 9 Aug 1988, and 11 Aug 1990. No ages were given for the birds seen, but timing suggests young birds fledged in Aug. Immatures were reported present for "over a week" in Polk Co 3–4 Sep 1983, and it was reported there 5 Sep 1987. Two Hall Co reports 12 Jun 1988 and 10 Jul 1983 were of adults, but there are no late summer reports indicative of breeding from Hall Co; it was also reported there 11 May 1991. One was reported at Kearney 1 Jul 1992 (Chris Hobbs, Tim Barksdale, pers. comm. WRS).

Elsewhere, there are reports from Douglas-Sarpy Cos 16 May 1975, 18 May 1977, 26 May 1975, 27–30 May 1965, 24 Jul 1988, 17 Aug 1984, 4 Sep 1983, 7 Sep 1980, and 27 Sep 1986. One was near Mitchell 30 May–3 Jun 1998 (Steve Kerr, pers. comm. AK), and an adult was in Lincoln 22 Jul 1999 (Cliff Lemen, pers. comm. TEL).

Fall

Several early fall (Sep) reports likely represent postbreeding wandering, not necessarily of birds fledged in Nebraska. Reports away from locations discussed above are few; only documented reports are cited: 27 Aug 1998 Lincoln; 7 Sep 1990, one picked up after being stunned in a hailstorm (NBR 59:17, 34); 9 Sep 1978, a subadult at Wilderness Park, Lancaster Co (NBR 47:18); 9 Sep 1991, a specimen, UNSM ZM17034, found dead in Lincoln Co (Brogie 1997); 11 Sep 1994 Thomas Co (Gubanyi 1996c); 15 Sep 1998, an injured first-year male picked up at Holdrege; 18 Sep 1986 Thayer Co (Mollhoff 1989a); 18 Sep 1988 Lancaster Co; 19 Sep 1974 (9) Lincoln Co (NBR 43:39); and 29 Sep 1989 Buffalo Co.

Two additional reports are questionable: Bruner and others (1904) considered "likely an error" a report of 6–7 seen 12 Oct near Omaha, and no information was provided on a specimen purportedly collected Nov 1912 near Omaha (Bent 1937).

FINDING

At present, Ogallala appears to be the best location to look for this species, especially when young are being fed in Aug.

Bald Eagle

Haliaeetus leucocephalus

STATUS

Common, locally abundant, regular winter visitor statewide. Common regular spring and fall migrant statewide. Locally rare regular breeder (resident?) statewide. Rare casual summer visitor statewide.

DOCUMENTATION

Specimen

UNSM ZM7697, 31 Oct 1921 Hall Co.

DISTRIBUTION AND ECOLOGY

This fish-eating species is distributed along waterways, lakes, and reservoirs during migration and near open water during the winter months. Breeding sites are widely scattered across the state but are associated with lakes, rivers, and reservoirs.

Spring

winter→15, 17, 20 May

Even though a significant wintering population is maintained in the state, migratory movement is evident in Feb–Mar, with over half of the spring records occurring 10–30 Mar. Concentrations begin to develop as ice cover melts, usually in the first third of Mar, and continue to build to the end of Mar; numbers quickly wane in early Apr. Peak counts at the J2 and Kingsley Dam viewing sites on the North Platte River were 23 on 8 Mar and 65 on 22 Feb, respectively (Peyton and Knaggs 1995). A kettle of 13 birds was moving northward in Dawson Co 19 Feb 1996. A few individuals, usually subadults, may remain in an area into May.

Summer

In 1998 the following counties supported active nests (John Dinan, pers. comm. RSS): Boyd, Gage, Garden, Knox, Holt, Loup, Nemaha, Pawnee, Saunders, Scotts Bluff, Sherman, Valley, and Webster. This list has been growing since the first nesting attempt in the 20th century in Cedar Co in 1973 (Lock and Schuckman 1973). A nest site along the North Platte River in Garden Co was attended 1987–98 but was never successful (Rosche 1994a; John Dinan, pers. comm. RSS). The first recent successful nesting in the state was in 1991, when an eaglet was fledged from a nest near Valley (*Omaha World Herald* 29 May 1996); it later died, but 2 were fledged from the site in 1992 (NBR 59:50, 60:59). Birds were reported in Scotts Bluff Co 23 May 1990 and 25 Aug 1992, and nesting first occurred at North Platte NWR in 1993 (McKinney 1993). NGP data show that the nests in Scotts Bluff Co (1993–98) and Sherman Co (1992–98) have produced 14 and 13 young, respectively. In addition, an immature and an adult were at Wolf L 8–14 Aug 1993, adults were present at Crescent L NWR 28 Jul 1994 and at Calamus Res in 1995, and chicks were noted in Holt Co 20 Apr 1996.

Nesting adults may be far-ranging in search of food and thus may appear almost anywhere associated with medium-size to large rivers, lakes, and reservoirs, even though they may not be nesting close by. The few breeding season reports away from known breeding locations are 13 Jun 1946 Logan Co, 14 Jun 1982 Niobrara Valley Preserve (Brogie and Mossman 1983), a series of Jun–Jul reports from Dawes Co 1984–90, 22 Jun 1999 near Valentine, 25 Jun 1953 Lincoln Co, 30 Jun 1967 Garden Co, 30 Jun 1991 Douglas-Sarpy Cos, 22 Jul 1999 Nebraska City, 27 Jul 1981 Scotts Bluff Co, 29 Jul 1970 McPherson Co, Aug 1994 DeSoto Bend NWR, and 8 Aug 1987 Garden Co.

Fall

12, 15, 16 Sep→winter

There are earlier reports from locations prior to or away from known breeding: 1 Sep 1990 Knox Co, 3 Sep 1989 Howard-Hall Cos, and 4 Sep 1991 Cass Co.

Migration begins in mid-Oct and peaks in late Nov. Numbers may be maintained for several weeks or until weather conditions and snow and ice cover reduce food availability. Some individuals may remain well into winter, provided that open water is available (see Winter).

Winter

Early winter distribution may be widespread, provided that lakes and rivers remain open. As ice cover develops, numbers may become more geographically concentrated, and many leave the state. Some favored wintering locations are indicated by a 1–15 Jan 1998 survey conducted by the NGP in conjunction with the National Wildlife Federation (NBR 60:52): North Platte River 145 individuals, South Platte River 11, Platte River 126, Loup River 28, lower Niobrara River 57, Republican River 525, Missouri River 435, major reservoirs 62, other locations 22. The total of 1340 was the highest reported since the survey began in 1980. The 1996 survey, held 3–15 Jan, yielded 719 birds, compared to the 1980–96 average of 743 and the previous 5-year average of 909 (Dinan 1996).

During the most severe winter periods numbers may be concentrated along open, warm waterways, dam spillways, and open areas of larger reservoirs. In late winter as the migration period approaches, numbers may increase to several thousand statewide.

High counts at single locations include 284 at Kingsley Dam 11 Feb 1994, 218 at L McConaughy and Sutherland Res 1 Feb 1997, and 176 at Harlan Co Res 4 Jan 1997.

COMMENTS

Until federal laws were instituted and enforced in the 1930s, Bald Eagles were unprotected in the United States and thus subject to being shot. But perhaps more significant to the welfare of the Bald Eagle and its relationship with people was the development of pesticides and their application following World War II. The use of chlorinated hydrocarbon insecticides (DDT and its relatives) led to their entering food chains. Being persistent and nonbiodegradable, they were subject to biomagnification. Large predator species, such as the Bald Eagle, accumulated high concentrations of these chemicals in their bodies. Acting as hormone mimics, they interfered with the egg-production process. Females produced eggs whose shells were so thin that they were crushed when incubated. Bald Eagle populations plummeted over much of the already reduced range until public concern led to a ban on chlorinated hydrocarbons in 1970. Followed shortly by a Federal Endangered Species designation, the Bald Eagle began its remarkable recovery.

FINDING

Look for this species during late fall and early spring along major rivers and perched along the edge of large lakes and reservoirs. In late winter and early spring, before ice fully melts from lakes, these birds may be seen standing on the ice or picking at dead fish frozen in the ice. Some select locations include the spillway of L McConaughy in midwinter, Sutherland Res in early Dec, along the central Platte River in Mar, and DeSoto Bend NWR in Dec and again in Mar.

Northern Harrier

Circus cyaneus

STATUS

Common regular spring and fall migrant statewide. Uncommon regular winter visitor statewide. Rare, locally uncommon, regular breeder statewide.

DOCUMENTATION

Specimen

UNSM ZM10319, 8 Oct 1960 Gage Co.

DISTRIBUTION AND ECOLOGY

This species feeds primarily on rodents and thus seeks out habitats such as old fields, grasslands, moist meadows, and weedy ditches. Such habitat is scattered throughout the state, and the species therefore may be seen almost anywhere and during any season.

Spring

Following severe winters with extensive snow cover, a clear movement of this species may be seen coincident with the thaw, typically in late

Feb and early Mar. Although the movement may be evident by several to a dozen or more seen in a day, no large concentrations occur. In mid-Mar along the Missouri Valley, which may act as a north-south corridor, numbers as high as 0.6 per linear km (1.0 per linear mi) may be encountered (RSS). Migration is typically completed by mid-Apr, and birds seen in May are indicative of nesting.

High counts include 21 in southern Kimball Co 14 Apr 1997, 12 in southern Cheyenne Co 17 Apr 1998, and 11 there 16 Apr 1997.

Summer

This species tends to be concentrated in the Sandhills during the summer months. In locations such as northern Garden, southern Sheridan, and Cherry Cos, where extensive marshes are present, several may be seen in a day. Smaller numbers may also be seen associated with marshes elsewhere in the Sandhills and in the Rainwater Basin. The species may, however, be found nesting in widely scattered habitat statewide (Ducey 1988).

In Nebraska nesting has not been limited to marshes; a ditch containing dense, tall grasses was the site for one nest; another was built on a pile of brome straw in a brome-grass field (*Bromus* sp.). Nests with eggs have been found in Apr and May; a nest containing 6 eggs was found 7 May in Jefferson Co, and another with 4 eggs was found 17 May 1984 in Clay Co (Robin Harding, pers. comm.).

Fall

No distinct fall movement is evident; in the east migrants are responsible for an increase in sightings in very late Oct and into Nov. In the west migration peaks in early Oct. The earliest arrivals in areas where breeding does not occur are in late Aug and early to mid-Sep. Individuals may remain as long as snow does not accumulate and cover the habitat over which they hunt.

High counts include 54 at L McConaughy 12 Oct 1997 and 34 in the western Panhandle 16 Oct 1999.

Winter

Harriers are present in appropriate habitat statewide during open winters. During severe winters with extensive snow cover, the species may be absent entirely.

FINDING

This species is broadly distributed during its early spring and late fall migration. Look for it over extensive marshlands, along the edges of open farmland, and throughout the grassland areas of the state.

Sharp-shinned Hawk

Accipiter striatus

STATUS

Uncommon, locally common, regular spring and fall migrant statewide. Rare casual breeder statewide. Uncommon regular winter visitor statewide.

DOCUMENTATION

Specimen

UNSM ZM12494, 23 Aug 1908 Monroe Canyon, Sioux Co.

DISTRIBUTION AND ECOLOGY

This small hawk preys chiefly on small birds and is found in open woodland and forest habitat during all seasons. Migrants occur statewide and may travel at heights above 1500 m (about 5000 ft), circling and soaring above many types of habitat.

Spring

winter→28, 30 May, 1 Jun

Because Sharp-shins may be encountered in appropriate habitat during all times of the year and because migration is normally solitary or in low numbers, it is difficult to determine timing of spring migration. Records reveal an increase in frequency during Apr and early May, peaking in mid-Apr. Later reports (late May–early Jun) may be nesting birds or immatures. Palmer (1988) noted that some breed as yearlings; a pair nesting in Brown Co were both in immature plumage (Mossman and Brogie 1983).

There appears to be no geographic variance in distribution of records during the migratory period. Numbers are uniformly low (typically fewer than 1 sighting per day) but evenly spread across the state.

Summer

Until the 1980s and 1990s, summer reports were few and concentrated chiefly in the wooded southeast and had not been noted since Bruner and others (1904) cited 2 reports from Sioux and Douglas Cos. Youngworth (1955), however, considered Sharp-shinned Hawk a "resident in small numbers" in the central Niobrara Valley. Summer reports have increased in more recent years, perhaps because of birder awareness, and include breeding reports for Brown (Mossman and Brogie 1983), Sherman (Ducey 1988), and Saunders Cos (NBR 62:71, 131). There were reports of territorial birds in Lancaster Co in May 1981 and in Johnson Co as late in the season as 2 Jun 1981; observers speculated that nest sites were available, but low numbers of birds available for breeding may have resulted in desertion of nest sites, as only single birds were sighted (Ducey 1981).

Low numbers also occur in the Pine Ridge, where breeding is likely (Rosche 1982). RSS secured a male with enlarged testes on 19 Jun 1968 in typical ponderosa pine habitat at the head of Little Monroe Canyon, Sioux Co. There are breeding season reports from Carter Canyon 11 Jun 1995 and from Scotts Bluff Co Jun 1996 and 4 Jul 1999.

There are about 25 additional reports in the period 5 Jun–9 Aug, at least 19 of which are from southeast Nebraska. The few reports elsewhere are 8 Jun 1974, 12 Jun 1941, and 1 Jul 1994 Lincoln Co; 3 Jul 1986 Pierce Co; 26 Jul 1978 Garden Co; and 29 Jul 1995 Dixon Co.

The Sharp-shinned Hawk is often confused with the slightly larger Cooper's Hawk, but as birders become more familiar with the calls of both species, we believe that the summer records of both species will increase.

Fall

13, 13, 14 Aug→winter

There is an earlier report 4 Aug 1997 Scotts Bluff Co.

Migratory Sharp-shins become noticeable in early Sep and continue into Nov. Numbers peak from mid-Sep to mid-Oct. A record tally of 87 was made during a 7-hour period on 1 Oct 1994 by Barbara and Loren Padelford in Pottawattamie Co, Iowa, as they counted migrating raptors above the Missouri River valley. It may still be numerous in mid to late Dec most years; best CBC totals are 36 in 1994–95 and 26 in 1989–90.

High counts include 41 at Camp Wakonda, Sarpy Co, 21 Sep 1998.

Winter

Winter records are scattered statewide, except for the grasslands of the Sandhills. Reports are fewest in mid-Jan. Some individuals frequent urban neighborhoods during the winter, preying upon passerines that are attracted to bird feeders.

High counts in midwinter include 4 at Gavin's Point Dam 23 Jan 1994.

FINDING

The easiest way to see this species is during fall migration from a hawk watch. Any elevated location with an unobstructed view of the sky should yield several migrating Sharp-shinned Hawks during northerly winds in Sep–Oct.

Cooper's Hawk

Accipiter cooperii

STATUS

Uncommon, locally common, regular spring and fall migrant statewide. Rare regular breeder statewide. Rare regular winter visitor statewide.

DOCUMENTATION

Specimen, UNSM ZM12499, 17 May 1909 Lancaster Co.

DISTRIBUTION AND ECOLOGY

This species occurs in those regions of the state with extensive woodland and forest, including the southeast, the floodplain woodlands associated with major streams, and the Pine Ridge.

Except for migration, it is absent from large areas of the state, including the Sandhills.

Spring

winter→2, 3, 4 Jun

The data do not show a clear peak in movement. Late Feb dates could reflect winter visitors or early migrants. Reports continue through Mar and Apr into early May. May dates may be either nesting birds or late-migrating prebreeders.

Summer

Summer records are widely scattered but are clearly associated with regions that support significant deciduous or coniferous woodland and forest. Although regularly occurring in very small numbers in scattered locations, few nesting records exist.

Pre-1920 nesting records are limited to the eastern quarter of the state and an undocumented report for Sioux Co (Ducey 1988). More recent records cited by Ducey (1988) include Saunders, Lancaster, and Cass Cos in southeast Nebraska and Frontier Co in the southwest. A pair was performing courtship flights in Carter Canyon 27 Apr 1997. Nesting was reported in Thomas Co 1946–48. A nest with recently fledged young was found in 1992 (Bray 1994), and individuals were present 30 Jun–2 Jul 1993. A pair at a nest site in Schramm Park 22 May 1995 probably used the same site in 1994 (NBR 63:40), and a female was on a nest at Platte River SP 8 May 1996 (NBR 64:47) and again in 1997. A pair was nest building in Towle Park, Omaha, late Apr 1997 but abandoned the nest. A nest at Wehrspann L fledged young in 1998 (NGP files). There are some 50 reports in the period 10 Jun–14 Aug, and of these, about half are from the southeast.

Elsewhere, breeding season reports are clustered in the Pine Ridge, in the Niobrara Valley, and in Lincoln Co. There are about 6 recent summer records for the Pine Ridge, where Rosche (1982) considered it casual in summer, with no evidence for breeding. Mossman

and Brogie (1983) believed Cooper's Hawk to
be a probable nester in the Niobrara Valley
Preserve in 1982, with several sightings during
the nesting season. It was reported at Long
Pine 13 Jun 1995 and in Brown Co 4 Jul 1968.
There are a few reports for Lincoln Co, 1
Jul 1964, 14 Aug 1992, and 30 Jun 1990, but
apparently no other summer reports for the
Platte Valley west of Grand Island, where it
was reported 15 Jun 1954. Lingle (1994) listed
no records for May–Aug in the Grand Island
region. Johnsgard (1997) noted an undated
report from Harlan Co of an adult carrying
food; other reports from Harlan Co were 10
Jun 1961 and through the summer in 1999.
The only other breeding season reports are 30
Jun 1966 Logan Co, 30 Jun 1971 Greeley Co, 19
Jul 1993 Cherry Co, 8 Aug 1959 Box Butte Co,
and 13 Aug 1983 Boone Co.

Data on nesting activity include adults
incubating on 4, 8, and 14 May and young in
nests on 15 and 24 Jun.

Fall

19, 22, 22 Aug→winter

This species is a regular but uncommon migrant
statewide. Movement is evident beginning in
mid-Sep and continues into late Oct. Most
activity is concentrated in very late Sep and
early Oct. During a 4-year fall migratory study
along the bluffs overlooking the Missouri
Valley, Barbara and Loren Padelford achieved
a high count of 16 on 23 Sep 1993 during a
7.5-hour period. That year they recorded 49
Cooper's Hawk sightings during 102 hours
of observation. In 1995 they had 19 sightings
during 83 hours of observation (pers. comm.
RSS).

High counts include 7 at Camp Wakonda, Sarpy
Co, 21 Sep 1998.

Winter

Winter reports are scattered geographically.
There are fewer reports than for Sharp-
shinned Hawk; CBC data (1967–95) indicate
a 3:1 ratio of Sharp-shinned to Cooper's.
High CBC totals were 14 in 1994–95 and 11 in
1995–96. Reports are lowest in Jan, when the

species is rare or absent some years. Rosche
(1994a) noted that in the Keith Co region few
are reported after mid-Jan. Rosche (1982)
considered it a very rare winter visitor in the
Panhandle, where extensive forest habitat is
lacking.

FINDING

Cooper's Hawks are encountered under
conditions similar to those of Sharp-shinned
Hawks. Recordings of calls may be useful
in distinguishing between these species,
particularly in late spring and summer.

Northern Goshawk

Accipiter gentilis

STATUS

Rare regular spring and fall visitor statewide.
Rare casual summer visitor northwest. Rare
regular winter visitor statewide.

DOCUMENTATION

Specimen

UNSM ZM12491, Nov 1896 near Maxwell, Lincoln
Co.

DISTRIBUTION AND ECOLOGY

This northern species is typically found
associated with heavy timber, either along
floodplains or with upland forest. Migrants
may only be present in a location for several
days or less. Winter visitors may frequent an
area for several weeks or more.

Although the Sandhills region provides little
habitat to sustain goshawks, select locations
with suitable habitat may attract this species.
It has been observed preying on Ring-necked
Pheasant at Valentine NWR (Ned Peabody,
pers. comm. RSS), and Lawrence Blus, former
biologist for the NGP, noted that the goshawk
was a significant predator of prairie grouse
during the winter at Halsey State Forest (pers.
comm. RSS).

Spring

winter→11, 14, 16 May

There are later reports 24 May in northwest
Nebraska (Rosche 1982) and 1 Jun (RSS).
Reports are most numerous late Mar and Apr
but are too few to establish a clear peak. Later

reports suggest that some birds depart rather late in spring, likely migrants rather than wintering birds. Johnsgard (1997) indicated that there is a concentration of reports 14 Apr–16 May, but many of these would appear to be misidentifications.

Summer

A few scattered observations during summer months in the Pine Ridge (RSS) probably are wandering birds derived from the small resident population in the Black Hills of South Dakota (SDOU 1991). Rosche (1982) listed no summer occurrences.

Fall

29 Sep, 1, 2 Oct→winter

There are earlier reports 8 Sep 1979 Douglas-Sarpy Cos, 16 Sep 1978 Lancaster Co, and 16 Sep 1981 Garden Co.

Fall migration begins in late Sep; Johnsgard (1997) indicated that there is a concentration of reports 21 Sep–17 Oct. A second influx occurs in early to mid-Nov and probably constitutes winter visitors.

Winter

Winter visitors probably arrive in Nov (see Fall) and generally remain in an area for several weeks; however, they may not necessarily remain throughout the winter, often leaving in midwinter. Some reports of long duration, presumably involving the same individual throughout, include 9 Nov–5 Jan, 15 Nov–30 Jan, 11 Dec–26 Feb, and 3 Jan–10 Apr.

FINDING

This large accipiter is most likely to be found from late Nov through early Jan, often occupying high perches with good views in large expanses of woodland. The NNF in central Nebraska and upland woodlands along the Missouri River are good locations for finding this species during appropriate periods.

Harris's Hawk

Parabuteo unicinctus

STATUS

Rare casual fall and winter visitor west and east.

DOCUMENTATION

Specimen

HMM 2895, 28 Oct 1922 Elkhorn, Douglas Co.

Records

There are 2 records. One taken in Douglas Co 28 Oct 1922 is cited above. This bird was shot by a hunter "while he was shooting ducks in a blind at Elkhorn. It was mounted in Omaha and secured by Mr. A. M. Brooking for his collection, where it is #2895" (Swenk, Notes before 1925). Another seen at Stateline Island on 13 and 27 Jan 1995 was documented (LM; Gubanyi 1996c; NBR 63:17).

COMMENTS

This species regularly occurs from central Texas southward. Of the surrounding states, only Kansas has records, 5 in the period 11 Dec–7 Jan, plus a spring breeding record (Thompson and Ely 1989). No fall records exist for Kansas. Lending credibility to the Nebraska record in early 1995 was the appearance during winter 1994–95 in Oklahoma, Kansas, and Colorado of several Harris's Hawks; 3 in Oklahoma and 1 in Kansas were observed beginning 26 Oct and were seen through 21 Jan (AB 49:65, 163). Colorado also reported 2 Harris's Hawks 17 Dec–4 Feb (AB 49:173). A Harris's Hawk in Columbia, Missouri, Feb 1995 was photographed and the record accepted by the Missouri Records Committee; it may have been there for several months (Mark Robbins, pers. comm.).

Red-shouldered Hawk

Buteo lineatus

STATUS

Rare regular spring and fall migrant east, rare casual elsewhere. Locally rare regular resident Fontenelle Forest. Rare regular winter visitor east, rare casual elsewhere.

DOCUMENTATION

Specimen

UNSM ZM12519, 15 Jan 1941 Fontenelle Forest, Sarpy Co.

DISTRIBUTION AND ECOLOGY

During the breeding season this species occupies

floodplain and upland deciduous woodlands, often adjacent to marshes, which support much of its prey, chiefly amphibians (Palmer 1988). Thus the most appropriate breeding habitat for this species is found in the eastern quarter of Nebraska.

Spring

Records are scattered throughout the period, with no clear migratory trends. In areas where this species is migratory, generally at the north edge of its range, most movement is in early Mar (Palmer 1988). Nebraska reports are evenly distributed Feb–Jun, however.

Most reports for Feb–Jun are from the southeast, in the area bounded by Gage, Saunders, and Douglas Cos, but there are more than 10 reports west to Lincoln Co in the Platte Valley, as well as reports 13 Mar 1960 Keya Paha Co, 17 May 1956 Keith Co, and a few Panhandle reports: 29 Mar 1995 Crescent L NWR, 11 May 1980 Garden Co, 13 May 1978 Garden Co, 24 May 1975 Dawes Co (Rosche 1982), and 10 Jun 1973 Sioux Co (Rosche 1982). The Rosche (1982) reports were both of immature birds. There are several reports from the northeastern Colorado plains (Andrews and Righter 1992).

Summer

Breeding records have been limited to extreme southeastern Nebraska in Lancaster, Otoe, Douglas, and Sarpy Cos but only in Douglas and Sarpy Cos in the 20th century (Ducey 1988). The most consistent location for breeding activity has been Fontenelle Forest, where the species has nested somewhat regularly throughout the 20th century. Breeding there recommenced in 1989 after having last been documented in 1964 (NBR 33:10). Most recent activity has been limited to a single pair. Nests have been located on wooded upland slopes but near open marshland. Breeding activity, including courtship, has been observed as early as late Jan and fledging on 6 and 15 Jun.

There are few reports Jun–Aug away from Douglas-Sarpy Cos; most are recent and those

in Cass Co suggest breeding: 6 Jun 1990, 8 Jun 1989, and 18 Aug 1991 Cass Co; 6 Jun 1936 Lincoln Co (Tout 1947); 10 Jun 1973 Sioux Co; 11 Jul 1973 Clay Co; 11 Aug 1990 Dodge Co; 12 Aug 1954 Loomis, Phelps Co (an albino specimen, HMM 28383); 17 Aug 1997 eastern Otoe Co; and 19 Aug 1988 Lancaster Co.

Fall

Fall records are scattered and irregular. There are far fewer fall reports than for spring, although fall distribution resembles that for spring, including westerly reports 11 Jul 1973 Clay Co, 3 Sep 1989 Howard-Hall Cos, 4 Oct 1983 Adams Co, 12 Oct 1934 Lincoln Co (Tout 1947), 24 Oct 1937 Lincoln Co (Tout 1947), 1 Nov 1970 Custer Co, and a single Panhandle report 24 Oct 1981 Garden Co.

Five consecutive falls of daily hawk counting from the bluffs overlooking the Missouri River have only yielded 3 sightings, all in late Sep (Barbara and Loren Padelford, pers. comm. RSS). Similar results have been obtained in late Sep in northwest Missouri (Mark Robbins, pers. comm.).

Winter

There are about 30 reports for Jan, a surprisingly high number; most are undocumented. Westerly reports are 3 Jan 1955 and 18 Jan 1958 Hamilton Co, 7 Jan 1987 Polk Co, and 10 Jan 1974 Hall Co, with northwesternmost reports 11 Jan 1986 Knox Co and 11 Jan 1950 Keith Co.

Individuals may remain in an area for several weeks, such as at Fontenelle Forest 1995–96, although it is not known whether these are the same birds that breed there.

FINDING

The breeding birds at Fontenelle Forest can usually be located in May–Jun, but during winter months pay close attention to medium-size soaring hawks or those that perch at midheight along the edge of woodlands.

Broad-winged Hawk

Buteo platypterus

STATUS

Uncommon regular spring migrant east,

becoming rare west. Rare casual breeder southeast. Fairly common regular fall migrant east, becoming rare casual west.

DOCUMENTATION

Specimen

UNSM ZM7669, 29 Apr 1917 Lancaster Co.

DISTRIBUTION AND ECOLOGY

This small woodland hawk may migrate singly but is sometimes seen at high altitudes in kettles circling on rising thermals. When resting they are usually seen associated with either floodplain or upland woodland. Individuals are most often seen perched on a lateral limb, midway or lower in the tree.

Although this species has been found nesting from Alberta south to Texas and east to the Atlantic Coast and is associated with deciduous or mixed forest and woodland (AOU 1983), it is but a rare summer visitor and breeder on the central and northern plains (Thompson and Ely 1989; SDOU 1991).

Spring

28, 28 Mar, 2 Apr→2, 2, 5 Jun

Late dates above are from locations where summering birds were not reported and may be young birds or possibly breeding birds, but no evidence of breeding was presented. Numbers peak in the last part of April and quickly drop off in the first few days of May. This species is a regular but not numerous migrant in the eastern half of the state, becoming increasingly scarce westward. In the Panhandle it is considered "very rare to rare" (Rosche 1982). There are about 20 Panhandle reports in the period 16 Apr–25 May.

There are numerous Nebraska reports in the period 1 Jan–17 Mar, none documented; we consider these reports highly doubtful. Mosher and Palmer (Palmer 1988) indicate that this species normally arrives in south Texas no earlier than 25 Mar, and breeding adults begin arriving in the northern tier of states on about 18 Apr. Kansas data show an early date of 6 Apr (Thompson and Ely 1989), and in Missouri first arrivals are in early Apr, although a very early date of 10 Mar is cited (Robbins and Easterla 1992).

High counts include "hundreds" moving north along the Missouri River at Omaha 2 Apr 1978 (AB 32:1025).

Summer

This species breeds irregularly in the lower Missouri River valley. Ducey (1988) cited only 5 records, from Douglas, Sarpy, and Nemaha Cos, the most recent in Nemaha Co in 1951 (Fox 1952) and Sarpy Co in 1979 (AB 33:876), the latter report involving 2 juveniles with adults in early Aug. There are numerous reports of pairs of adults, including those exhibiting courtship, from appropriate woodland habitat, particularly in Fontenelle Forest. Breeding occurred in 1995 when 2–3 fledglings were heard 5 Jul and adults were present also. There is a series of summer reports from Douglas and Sarpy Cos 1959–79 and additional reports during the summers of 1983, 1987, and 1992. Three heard calling at Indian Cave SP 2 Jun 1995 may have been nesting (NBR 63:73). There are no other breeding season reports, except for an undocumented report Jun 1970 Custer Co and 20 Jun 1998 Lancaster Co (LE).

Fall

20, 22, 22 Aug→22, 25, 26 Oct

There is a specimen, an immature male, UNSM ZM16658, collected 22 Aug 1992 in Lancaster Co. There are earlier reports where summering birds were not noted 7 Aug 1988 Lancaster Co, 9 Aug 1988 Polk Co, and 15 Aug 1987 Douglas-Sarpy Cos.

During the 5-year Padelford (pers. comm. RSS) fall raptor survey, conducted in extreme eastern Nebraska, early fall sightings were 15, 16, and 17 Sep, generally singles. Numbers rapidly increased to a peak in just a few days, with high counts 178 on 20 Sep 1995 observed during a 2.75-hour period, and 107 on 23 Sep 1994 during a 6-hour period. Preceding and following days were represented by 6 or fewer observations, and observation periods ranged from less than 1 to 4 hours. Although stragglers were noted into early Oct, late dates were 6 and 7 Oct.

There are 4 undocumented Nebraska reports for Nov and 1 for Dec, which we consider highly doubtful; there are no verified Nov records for Missouri (Robbins and Easterla 1992) or Kansas (Thompson and Ely 1989).

In the Panhandle this species is reported far less frequently than in spring; there are 6 reports: 20 Sep 1980 and 21–22 Sep 1979 Garden Co; 24 Sep 1978 Walgren L (Rosche 1982); 28 Sep–1 Oct 1977 Garden Co; 29 Sep 1978, a kettle of 12 over Belmont, Dawes Co (Rosche 1982); and 3 Oct 1991 Scotts Bluff Co. The only other reports away from the east are 20 Sep–8 Oct 1984 Lincoln Co, 27 Sep 1996 Keith Co, and 27 Sep 1997 Halsey.

High counts include 55 at Camp Wakonda, Sarpy Co, 21 Sep 1998.

COMMENTS

The rare dark morph is reported on occasion; a report of a group of 5 "Common Black-Hawks" in Washington Co 26 Apr 1997 (B) probably refers to dark morph Broad-winged Hawks. The dark morph breeds in Alberta, and there are several reports of migrants in the central United States (Palmer 1988).

FINDING

The best chance to see this small, soaring hawk in flight is during the fall migration in late Sep, especially at a hawk watch location. Resting migrants may be encountered in spring along woodland edges, perched on lateral branches.

Swainson's Hawk

Buteo swainsoni

STATUS

Common regular spring and fall migrant statewide. Fairly common regular breeder statewide except southeast, where rare casual.

DOCUMENTATION

Specimen

UNSM ZM12521, 17 Jun 1901 Monroe Canyon, Sioux Co.

TAXONOMY

There is a dark morph (Palmer 1988) that has

been reported in Nebraska (see, for example, Rosche 1994a).

DISTRIBUTION AND ECOLOGY

This species is a social migrant, often traveling in groups of several hundred or more. Larger concentrations are more common during the fall movement than in spring. On the breeding grounds Swainson's Hawk distribution seems to correspond with the presence of grassland ground squirrels (*Spermophilus* sp.) (Palmer 1988); thus most suitable habitat is found in the Sandhills and much of the Panhandle. Occasional pairs may be found in farmland in the eastern third of the state, particularly where considerable pastureland is available to support its preferred prey.

Spring

10, 11, 11 Mar→summer

Arrival dates are difficult to determine based on existing data. There are many undocumented Feb and early Mar reports that should be disregarded. The earliest Nebraska specimen is 7 Apr 1994, UNSM ZM17074. Migrating flocks do not reach south Texas until about 1 Apr (Palmer 1988), although Robbins and Easterla (1992) cite early dates for Missouri, 14 and 18 Mar, and Andrews and Righter (1992) show arrival in Colorado in early Mar. The earliest cluster of Nebraska dates is in Mar, as shown above.

Migration peaks mid to late Apr, and stragglers occur in areas outside the breeding range until early Jun; latest such dates are 8 Jun 1984 Polk Co, 9 Jun 1984 Saunders Co, and 11 Jun 1985 Polk Co.

Spring migratory flocks, when observed, typically number fewer than 100 individuals, more often 12–20. Many spring migrants may be overlooked because they are often soaring and kettling at very high altitudes. A flock of 31 on 9 Apr 1977 was seen in western Douglas Co (RSS). These individuals were following a tractor that was plowing a weedy field, capturing small rodents that were being "exposed" by the plow.

High counts include 93 in the Rainwater Basin 18

Apr 1998, 50+ in Clay Co 21 Apr 1994, and 31 in Douglas Co 9 Apr 1977 (RSS).

Summer

Although breeding records suggest a relatively even distribution across the state, BBS data (NBR 46:38–62) reveal that breeding densities and frequency of occurrence are highest in the Sandhills and the plains habitats of the southwest and Panhandle, regions where considerable appropriate grassland habitat still persists.

The only nesting report from southeast Nebraska (east of the Rainwater Basin) since 1960 (Ducey 1988) was in Gage Co, where it was said to have bred near Blue Springs in Apr (Fiala 1970), but no details were provided for what would be a surprisingly early nesting attempt. In addition, there are few breeding season reports from southeast Nebraska east of the Rainwater Basin and south of the Platte River. Most are from Lancaster Co, where there are late Jun reports 1962–66, 1973, and 1986; these may be merely late spring migrants or nonbreeders, although there are early Aug reports in 1963, 1964, and 1985. Reports from Douglas-Sarpy Cos 26 Jul 1959 and 1 Aug 1964 may also indicate local nesting.

Swainson's Hawks typically choose a tree as a nest site, often the only tree available for some distance. Nests have also been found in shelterbelt trees, and RSS observed a nest that was built on the superstructure of a utility pole alongside Highway 2 in Grant Co.

Active nests have been found 18 Apr (NBR 62:106) through 21 Jun. A nest in Garden Co contained 4 young approximately 2 weeks of age on 21 Jun.

Fall

summer→18, 19, 20 Nov

There are later undocumented reports 26 Nov 1973 Douglas-Sarpy Cos and 28 Nov 1985 Howard-Hall Cos. The latest Nebraska specimen date is 12 Nov 1937, HMM 21764, and Andrews and Righter (1992) cite the latest specimen date for Colorado as 22 Nov 1985.

Migration begins in early Sep, when solitary individuals begin to appear outside the breeding range. As the month progresses, records increase in frequency until late Sep, when a substantial movement is evident, peaking about 20 Sep and extending until about 10 Oct. The peak is usually represented by very large flocks, numbering from the hundreds to as many as 2500 individuals (Menzel 1974). Not only are streams of migrants and large kettles of migrating individuals seen with some regularity, whole fields have been observed to fill as flocks settle at dusk to roost (Menzel 1974). Most of these very large flocks have been observed in central Nebraska, including Jefferson, Adams, Merrick, Cherry, Lancaster, Saunders, and Gage Cos. Outside the typical migration corridor smaller flocks of 100 or more have been observed in Douglas Co (RSS), along the western edge of Pottawattamie Co in Iowa (Barbara and Loren Padelford, pers. comm. RSS), and in Dawes Co (Rosche 1982).

High counts include as many as 2500 at Valentine 29–30 Sep 1973 (Menzel 1974), 150 at Fontanelle 25 Sep 1998, and 49 in Fillmore Co 3 Oct 1998.

COMMENTS

Although there are at least 60 published reports in the period 1 Dec–8 Mar, none are documented, and we consider all highly doubtful. Indeed, reports for late Dec–Feb for temperate North America are virtually nonexistent (Browning 1974; Clark and Wheeler 1987; but see AOU 1998). Swainson's Hawks are transequatorial migrants, spending the winter months in the grasslands of southern South America but casually northward to Central America and extreme southern North America (AOU 1983).

FINDING

A leisurely drive along Highway 2 through the Sandhills during the months of May and Jun should yield several good views of this handsome hawk.

Red-tailed Hawk

Buteo jamaicensis

STATUS

Common regular spring and fall migrant statewide. Fairly common regular breeder (resident?) statewide. Common regular winter visitor south, uncommon north.

DOCUMENTATION

Specimens: *B. j. borealis*, UNSM ZM7666, 9 Sep 1917 Lancaster Co; *B. j. calurus*, UNSM ZM16366, 26 Sep 1991 Butler Co; *B. j. harlani*, UNSM ZM12515 28 Mar 1911 Crete, Saline Co; *B. j. kriderii*, UNSM ZM11452, 16 Jul 1965 Fort Robinson SP.

TAXONOMY

Based on current understanding (Palmer 1988), the breeding subspecies in Nebraska is eastern *borealis*; this is the "normal" type found in eastern North America and is an intermediate color morph for the species, with a typically medium brown back, distinct belly-band against a light breast and abdomen, and rufous tail feathers. There are no confirmed Nebraska records of nesting for western *calurus*, although it has been suggested (Bruner and others 1904), and data in Andrews and Righter (1992) indeed imply that *calurus* nests throughout the Panhandle. Probably most breeding Red-tailed Hawks in the western half of Nebraska are intergrades of the 2 subspecies.

The status of "Harlan's Hawk," *B. j. harlani*, and "Krider's Hawk," *B. j. kriderii*, has been controversial, each variously treated as subspecies (AOU 1957) or more recently (Palmer 1988) as color morphs of Red-tailed Hawk. Palmer (1988) believed that Red-tailed Hawks occur in regionally distinct populations with high percentages of a given color morph rather than in populations of a given morph with clearly identifiable geographic boundaries (Palmer 1988). Mendell (1983), however, made a case for subspecific status of both, conceding varying zones of interbreeding between them and adjacent subspecies but concluding that their ranges were sufficiently well defined and interbreeding not widespread enough to warrant removal of subspecific status.

"Krider's Hawk" nests in extreme northern and northwest Nebraska (Bruner and others 1904; AOU 1957; Palmer 1988), and "Harlan's Hawk" is a regular migrant and winter visitor in Nebraska.

DISTRIBUTION AND ECOLOGY

This is a woodland edge species, commonly seen in summer throughout the east but less generally distributed westward at that season as woodland diminishes. Because tall trees are preferred for nest sites, Red-tails may be absent from large areas of grassland such as the Sandhills, except for valleys of larger streams such as the Loup River that sustain mature riparian trees. Mature deciduous or coniferous trees that provide a good view across open terrain are required; only rarely are ledges on cliff faces utilized. The "Krider's" morph has been noted to nest on cliff faces in the northwest (Carriker 1902; Bruner and others 1904). According to Bent (1937), both *calurus* and *borealis* utilize tall trees for nesting. Nests are commonly reutilized in subsequent seasons.

Grasslands and meadows that support rodent and rabbit populations are preferred foraging habitats. Thus typical habitat includes woodland edge with adjacent meadows for foraging, agricultural land with considerable weedy edge and occasional treed farmsteads, open grasslands with isolated clumps of trees, and vegetationally mature highway rights-of-way. Treed parks in cities may be used also; fledglings were observed at Towle Park in central Omaha 21 May 1998 (NR).

Migrants and winter visitors are found anywhere there are at least a few trees. Wintering birds are scarce in open grasslands.

Spring

winter→31 Mar, 15, 17 Apr ("Harlan's")

winter→18 Apr (dark birds)

It is difficult to differentiate migrants and winter visitors from possible permanent residents, especially among typical "eastern" birds, but presence or absence of dark morphs is a useful indicator of movement, as are increased concentrations along roadsides from late Feb onward. Significant movement occurs in late Feb–early Apr when most, if not all, dark-colored birds depart. Dark birds that are not "Harlan's" are presumably either rufous or dark morphs of *calurus* (Clark and Wheeler 1987), as dark morphs are rare within *borealis* (Palmer 1988; Clark and Wheeler 1987). Rosche (1994a) cited extreme dates for dark birds in the Keith Co area 23 Sep–18 Apr and for "Harlan's" Red-tailed Hawk 26 Oct–24 Mar. There are specimens of "Harlan's" at UNSM for the period 13 Nov–28 Mar and a specimen of "Harlan's," HMM 2806, collected at Tilden 30 Sep 1918. Rosche (1982) listed a sighting of "Harlan's" 15 Apr 1978 in the Panhandle. Late reports of "Krider's" include 18 Apr 1940 Logan Co.

High counts include 19 in Keith Co (Rosche 1994a) 1 Apr 1988, 19 in Dixon Co 28 Mar 1998, and 17 at L McConaughy 1 Apr 1996.

Summer

Nesting has been recorded statewide (Ducey 1988). It is unknown to what extent breeding birds remain to winter, although Rosche (1994a) states that in the Keith Co area "most" of the paler breeding birds arrive in Mar and depart in Oct–Nov. It seems likely that a substantial portion of the summering birds in the state remain in winter, at least in mild conditions, as apparent pairs of light-colored birds, presumed *borealis*, are often seen in the winter months.

Nesting begins in late Feb and Mar, usually in the tallest trees in a woodlot or woodland edge. Copulation has been noted as early as 10 Feb in Lancaster Co (RSS). According to Cornell Nest Record data, incubation has usually commenced in the east by 10 Mar. Downy young have been observed as early as 20 Mar

in Lancaster Co; in Knox Co 2–3-week-old young were found on a nest 9 May; and young have been known to have fledged as early as 25 May in Lancaster Co. Breeding activity is later by as much as 2 or more months in the Panhandle. Downy young have been observed as late as 8 and 21 Jun in Scotts Bluff Co; young with partially developed flight feathers have been observed 1 Jul in Banner Co.

Fall

30 Sep, 27 Oct, 13 Nov→winter ("Harlan's")

23 Sep, 8 Oct→winter (dark birds)

Significant movement occurs in late Oct and Nov, although dark birds are noted as early as late Sep. The study by Padelford and Padelford (pers. comm. RSS) in the Missouri Valley in the Douglas Co area revealed a movement of nonresidents beginning in mid-Sep, a small peak in late Sep, and major movement mid to late Oct; fall data for 1995 revealed an average count of 4.5 per hour (range 1.1–10.6) in the period 20 Sep–7 Oct, but this average increased to 15.6 per hour (range 8.0–30.8) in the period 13 Oct–3 Nov. Counts diminished rapidly thereafter.

High counts include 38 at Fontenelle Forest 21 Oct 1997, 30 in Dixon Co 23 Nov 1995, 19 in Keith Co 15 Oct 1984 (Rosche 1994a), and 19 in Otoe Co 30 Oct 1994.

Winter

In certain winters large numbers may overwinter, especially when snow cover is minimal or absent; such conditions are most likely to occur in the south-central, east, and southeast areas of the state. A roadside survey of wintering hawks in Nemaha Co in 1980–81 and 1981–82 estimated a winter population of 0.28 per square km (0.72 per square mi) (Shupe and Collins 1983). It was considered uncommon to rare in the northwest in winter (Rosche 1982), but small numbers may occur along the North Platte Valley, where Lingle (1989) found it to be the most common *Buteo* along the lower reaches of the river, with 10.8 birds per 100 km (60 mi) of highway. Zeillemaker (pers. comm. RSS) noted that it was uncommon to rare in the western Sandhills in winter, as is the case in the remainder of the Sandhills, except for the Loup drainage, where several winter in less harsh conditions.

Many wintering birds are dark morphs; Rosche (1994a) indicated that in some years these outnumber pale morphs in the Keith Co area.

CBC data (1968–95) show a decline from 0.53 birds per party-hour in the east to 0.08 in the west. Best CBC totals are 125 at Omaha in 1994 and 101 at Lincoln the same year.

High counts include 21 in the Keith Co area 1 Jan 1988 (Rosche 1994a).

FINDING

This species is easy to find in summer. Dark morphs, such as "Harlan's" and rufous and dark morphs of the western subspecies *calurus*, are best looked for during fall migration in Oct–Nov.

Ferruginous Hawk

Buteo regalis

STATUS

Uncommon regular spring and fall migrant west and central, hypothetical east. Uncommon regular breeder north and west. Rare regular winter visitor central and west, hypothetical east.

DOCUMENTATION

Specimen

UNSM ZM12527, 10 Jun 1894 Lancaster Co.

TAXONOMY

The uncommon dark morph has been reported in Nebraska on occasion, including a specimen, UNSM ZM7687, taken at Stapleton 24 Jan 1935 and another bird 1 Nov 1987 near Keystone (Rosche 1994a); 1 out of 10 Ferruginous Hawks found in the winters 1978–79 and 1979–80 along the central Platte Valley were "dark-phase" birds (Lingle 1989). Clark and Wheeler (1987) noted that dark morph birds are not uniformly distributed and represent 1–10% of local populations.

DISTRIBUTION AND ECOLOGY

Ferruginous Hawk is a western species characteristic of extensive grasslands and is thus most often found in summer in the western Sandhills and Panhandle, although it is nowhere numerous. Foraging habitat is usually adjacent to terrain of high relief, consisting of scarplike exposures that have a commanding view of the surrounding landscape. These high points are often used as perches when resting, and even less accessible locations may be chosen as nest sites. Trees are also used as nest sites. Although the species is often considered intolerant of human activity, Podany (1996) found that some agricultural encroachment was tolerated in the Panhandle breeding population but that breeding success may be reduced as lands adjacent to nest sites exceed 50% agricultural use.

In its overall range, its distribution is thought to be determined by food supply (Palmer 1988), largely ground squirrels (*Spermophilus* sp.), although prairie dogs (*Cynomys* sp.) are significant also. Cursory examinations of debris in and around Nebraska nest sites by Podany (1996) suggest a more diverse diet, including ground squirrels (*S. richardsoni, S. tridecumlineatus*), pocket gophers (*Geomys* sp.), and jackrabbits (*Lepus* sp.). Nesting studies by the NGP 1976–85 revealed that pocket gophers represented 30% of total prey.

In winter (as in summer) it is often found in the vicinity of prairie dog towns, where a single bird may lay claim to a town and winter in the area. Wintering birds are most often reported from the extensive grasslands around the Meat Animal Research Center in Clay Co.

Movements of this species are poorly understood in Nebraska, largely due to difficulty of identification of young birds, the group most likely to wander beyond and within the breeding range. Many adults may be only short-distance migrants or merely nomadic in the winter months.

Spring

winter→10, 12, 14 May

Late dates above, none with details, are from areas east of the breeding range, mostly

Jefferson, Webster, Clay, Adams, Lancaster, and Polk Cos. Farther east there are few reports, none with details: 4 Mar 1992 Cass Co, 8 Mar 1981 Nemaha Co (Shupe and Collins 1983), and 25 Mar 1979 Douglas-Sarpy Cos. Most of the above reports would be expected to be of young birds (see Palmer 1988), but no age details are available. Immature Ferruginous Hawks are difficult to identify relative to similar-aged Red-tailed and Rough-legged Hawks (Clark 1987).

According to Palmer (1988), most migration occurs in Mar–Apr, with yearlings (prebreeders) following later, as late as Jun. Rosche (1994a) listed this species as a spring transient in the period 10 Mar–18 May.

Summer

Ducey (1988) cited breeding reports eastward to Cherry, Hooker, Keith, and Lincoln Cos, and, according to Johnsgard (1997), regular breeding occurs west of a line from Dundy to Keya Paha Cos, with recent breeding reports from Banner, Cherry, and Lincoln Cos. Cornell Nest Record data show recent records in Sioux, Dawes, Box Butte, Banner, Kimball, Dundy, Hayes, and Lincoln Cos, although 141 of the 176 records were from Sioux, Dawes, and Box Butte Cos. Nesting was reported, although unsuccessful, in 1995 in Grant Co (NBR 62:106).

There are breeding season reports at the eastern edge of the range outlined above for Keya Paha Co 23 Jun 1962, Brown Co 5 Jun 1982 (Brogie and Mossman 1983), and Polk Co 23 Jul 1985. These may be nonbreeders or prebreeders.

Undocumented reports exist for Aug from Adams Co 1963–67 and for Jul–Aug from Webster Co 1957–58 (Turner 1959), as well as 7 Jul 1980 Lancaster Co. The specimen cited above, which was collected in Lancaster Co 10 Jun 1894, was taken at a time when grassland was probably more extensive than now.

Breeding data suggest that incubation commences in late Apr–early May; downy young have been observed as early as 5 Jun,

but 2–3-week-old young have been noted on 8 Jun. Most are fledged by 15 Jul. Of 28 traditional nest sites monitored by Podany (1996) over a 2-year period, 3 were abandoned after breeding activity was initiated, 9 were utilized in only one of the seasons, and 22 produced 1 or more fledged young. Of the latter, the mean was 2.4 per nest (range 1–4).

Fall

10, 13, 15 Sep→winter

Early dates above, none with details, are from areas east of the breeding range, mostly Webster, Lancaster, and Adams Cos. However, an immature was picked up near Osceola prior to 17 Sep in 1997 and taken to the Lincoln Rehabilitation Center (*Omaha World Herald*). Farther east there are few reports, also without details: 9 Oct 1966 Douglas-Sarpy Cos, 20 Oct 1999 Plattsmouth, 24 Oct 1973 Douglas-Sarpy Cos, and 14 Nov 1965 Cass Co.

Migration begins in late Sep as cold weather arrives and some prey species, such as ground squirrels, begin to hibernate (Palmer 1988). Young birds may wander widely as early as 40 days after fledging (Palmer 1988), which would allow for sightings in early Aug in areas where breeding was not noted. Rosche (1994a) listed Ferruginous Hawk as a fall transient in the period 2 Oct–12 Jan.

High counts include 10 between Merriman and Ainsworth 24 Nov 1994 and 7 at a prairie dog town near Kearney 8 Dec 1996.

Winter

Wintering birds occur virtually statewide in extensive grasslands when snow cover is lacking, often near prairie dog towns. In the northwest Rosche (1982) considered it a "casual winter visitant," but it was not present in winter in Garden Co (Zeillemaker, pers. comm. RSS). The eastward edge of the normal wintering range appears to be a line from Boone Co to Lancaster Co, although it is rare east of Clay Co in the southeast. Reports farther east are few and undocumented: 3 Dec 1990 Cass Co, 29 Dec–6 Jan 1964–65 Cass Co, 5 Jan 1999 Cass Co, 7 Jan 1979 Douglas-Sarpy Cos, 10 Jan

1990 Cass Co, 12 Jan 1956 Nemaha Co, 15 Jan 1988 Cass Co, 19 Jan 1993 Cass Co, 2 Feb 1974 Douglas-Sarpy Cos, 10 Feb 1991 Dakota Co, and 18 Feb 1969 Douglas-Sarpy Cos.

FINDING

Apart from known nest sites, which should not be disturbed, the best chance to locate this species is to drive Sandhills roads in Cherry Co or Garden Co in Jun–Jul. In winter a good location to check is the grasslands in the vicinity of the Meat Animal Research Center in Clay Co. In general, this species is likely to be encountered anywhere in the grasslands of the Panhandle at almost any time of year but especially at prairie dog colonies during migration and winter.

Rough-legged Hawk

Buteo lagopus

STATUS

Fairly common regular spring and fall migrant statewide. Hypothetical in summer. Fairly common regular winter visitor statewide.

DOCUMENTATION

Specimen

UNSM ZM7688, 19 Nov 1920 Lancaster Co.

TAXONOMY

Clark (1987) stated that there are 2 color morphs, light and dark, each with an immature and 2 adult plumages, although Palmer (1988) suggested that "to describe color phases" is "to label a segment of each end of a continuum of variation from darkest to lightest birds." Identification of age class and color morph can be confusing both within and between this species and others. Dark morphs are far more common in Rough-legged Hawk than other *Buteos* (Clark 1987), although none are geographically correlated and thus are not recognized as taxonomically distinct from each other. The entire color range of birds has been observed in Nebraska, although no data exist quantifying the relative proportion of each type.

DISTRIBUTION AND ECOLOGY

It is most numerous in grasslands in central and western Nebraska. At times it may be rather common in these areas, depending on snow cover and availability of prey. During periods of heavy snow cover individuals may concentrate in more sheltered locations, such as shelterbelts and riparian woodland.

Spring

winter→19, 19, 20 May

Most movement is in Mar–Apr, although movement is gradual and young birds may linger into May. Later reports (see Summer) are undocumented, although presumed nonbreeders have a propensity to linger into Jun as far south as Idaho and Wyoming (Palmer 1988) and northern Minnesota and northern Wisconsin (Kent and Dinsmore 1996).

Summer

Although none are documented, there are about 40 reports in the period 21 May–13 Sep distributed statewide. We consider these reports highly doubtful.

Bruner (1901) speculated on its nesting in the Pine Ridge, but Rosche (1982) correctly dispelled this possibility as involving confusion with Ferruginous Hawk.

Fall

14, 16, 16 Sep→winter

In poor prey years Rough-legged Hawks may leave the breeding range early and arrive on the wintering range as early as Aug, but most depart the breeding range Sep–Oct (Palmer 1988). There is a late Aug record for Colorado (Andrews and Righter 1992), and early dates for Iowa are 1, 12, and 23 Sep (Kent and Dinsmore 1996).

High counts include 10 in the Keith Co area 28 Nov 1974 (Rosche 1994a) and 8 in southwest Sioux Co 20 Nov 1998.

Winter

Wintering birds occur statewide, usually in open areas and on extensive grasslands. Presence is somewhat irregular, ranging from common to absent from year to year (Rosche 1982). It is least numerous in the south and east.

High counts include 35 between Neligh and

North Platte 27 Dec 1996, 25 in southwest Sioux Co on both 25 and 27 Dec 1996, and 10 in Dixon Co 6 Feb 1995.

FINDING

Rough-legs are not hard to find in the winter months in open agricultural areas or grasslands without snow cover, where they often perch on fence posts. They are also conspicuous when hovering, a characteristic habit of the species. Typical light morph birds are not hard to identify, but dark morph birds can be confused with other dark *Buteos*.

Golden Eagle

Aquila chrysaetos

STATUS

Uncommon regular spring and fall migrant west and central, rare east. Uncommon regular resident west. Uncommon regular winter visitor west and central, rare casual southeast.

DOCUMENTATION

Specimen

UNSM ZM12988, 8 Nov 1912 Comstock, Custer Co.

DISTRIBUTION AND ECOLOGY

Breeding birds occupy isolated habitats in the western part of the state, requiring inaccessible sites associated with extensive grasslands for nesting. Suitable habitat exists in Sioux, Dawes, and Box Butte Cos in the northwest; Morrill and Kimball Cos in the Panhandle; and southern Keith, southern Lincoln, and Dundy Cos farther east. Home breeding ranges may average as much as 90 km² (35 mi²) (Johnsgard 1979). Nest sites are typically ledges on rock or dirt cliff faces. Only 5 of 63 Nebraska nests reported to Cornell were in trees.

Breeding birds are probably resident, but in winter there is an influx of younger birds, at which time the species occurs almost statewide, although it is most numerous in extensive grassland habitats.

Spring

winter→15, 15, 18 May

There are only about 11 reports in May away

from the breeding range, including the late dates above and a few later reports, most of which probably represent subadults or possibly nonbreeding adults: 20 May 1978 Howard-Hall Cos, 20 May 1994 Lakeside (immature, WRS), 23 May 1994 Crescent L NWR, and 30 May 1957 Brown Co.

Summer

Bruner and others (1904) stated that "years ago" it bred "quite generally over the state" but that by 1904 its range was limited to the Pine Ridge area. Ducey (1988) noted breeding reports from Douglas Co in 1884 and possibly in Nemaha Co in the 1880s.

Currently Golden Eagles breed throughout the Panhandle and possibly farther east (Ducey 1988), although there are no reports more recent than 1979 and 1985 in Chase Co (Busch 1979; Bennett 1986), 1979 and 1982 in Keith Co (Rosche 1994a), 1972 and 1974 in Lincoln Co (Bennett 1973b, 1975), and 1974 in Hayes Co (Bennett 1975). The 1979 nest in Chase Co fledged 2 chicks on 2 Jun (Busch 1979).

There are a few reports in summer from the Panhandle in areas where breeding sites are not known; these are likely immatures. These reports are 2 Jun 1996 Box Butte Co; 30 Jun 1967, 17 Jul 1996, and 25 Jul 1977 Garden Co; 26 Jul 1973 Sheridan Co; and 13 Aug 1992 Morrill Co.

There are few reports in summer away from the Panhandle: it was listed as a "PR" (= permanent resident) in spring 1963 in Antelope Co without details, an adult was just south of Springview 25 Jul 1999 (BP, LP), and it was reported 12 Aug 1983 in McPherson Co.

According to Cornell Nest Record data, in Nebraska incubation begins in late Mar–early Apr, and young hatch in early May. Fifty-two nest sites surveyed in early Jun revealed an average of 1.5 young per nest, most of which were 4–5 weeks of age. One nest site held 4 young, another 3, the remainder 1 or 2. No data are available as to fledging success.

Fall

3, 4, 8 Oct→winter

Early dates above are away from the breeding range. There are a few earlier reports, the more westerly of which are likely to be immatures that may have summered within the breeding range: 12 Sep 1992 Sheridan Co, 16 Sep 1978 Garden Co, 25 Sep 1964 Adams Co, and 25 Sep 1990 Lincoln Co. Padelford and Padelford (pers. comm. RSS) recorded Golden Eagles in the Missouri Valley in 3 of 5 years in the period 28 Sep–9 Nov, with maxima of 4 on 5 Nov 1994 and 9 for the season.

Winter

Breeders are probably resident year-round (Rosche 1982; Palmer 1988), but individuals, mostly immatures, may appear anywhere in the state during the winter months. Numbers are fewest in the east, where there are about 45 reports in the period 6 Oct–30 Apr. Eastern reports are fairly evenly spread throughout the winter period.

National Federation of Wildlife Jan surveys conducted by the NGP in conjunction with midwinter Bald Eagle counts 1992–98 yielded totals ranging from 15 in Jan 1992 to 35 in 1995 (NBR 60:52, 61:22, 63:23, 64:26).

COMMENTS

This species was and continues to be subject to active and passive human exploitation, even in the face of heavy federal protection. Native Americans seek the flight feathers for costumes, and ranchers as part of coyote control programs often employ poison baits that inadvertently kill Golden Eagles.

FINDING

The best time to find this species is during the breeding season. The cliffs at the north (lower) end of Sowbelly Canyon are probably the best site. Also look for them in the upper Niobrara River Valley and White River basin in Sioux, Dawes, and Box Butte Cos and on high bluffs along the North Platte River in Scotts Bluff and Morrill Cos.

Falconidae (Falcons)

American Kestrel

Falco sparverius

STATUS

Common regular spring and fall migrant statewide. Fairly common regular breeder statewide. Uncommon regular winter visitor south, rare north.

DOCUMENTATION

Specimen

UNSM ZM7701, 26 May 1901 Jim Creek Canyon, Sioux Co.

TAXONOMY

Currently only a single subspecies is recognized over most of North America, nominate *sparverius*. The specimen cited above was originally identified as a "nearly typical" representative of the subspecies *phalaena*, the pale "Desert" Sparrowhawk (Bruner and others 1904). Swenk (Notes before 1925, Notes after 1925) listed additional specimens of this form collected near Oshkosh and Kearney, the latter 16 Apr 1908.

DISTRIBUTION AND ECOLOGY

Kestrels are hole nesters and utilize old woodpecker excavations as well as nest boxes. They occur statewide in summer in a variety of open habitats. Extensive open grasslands, as in the Sandhills or Panhandle, are not favored.

There is a noticeable increase in numbers in Mar–Apr and Sep–Oct as migrants pass through, while lowest numbers occur in midwinter.

Spring

Movement peaks around the first week in Apr, often following thawing of winter snow cover.

High counts include 69 in the Keith Co area 1 Apr 1988 (Rosche 1994a), 59 from Scottsbluff to L Ogallala 12 Apr 1997, and 38 in Keith Co 13 Apr 1997.

Summer

Nesting occurs in open areas where dead or dying older trees are present that provide old woodpecker holes. Old cottonwoods are particularly favored. Kestrels will also nest on sheltered ledges of tall buildings as well as in cavities in natural rock or dirt cliffs. In recent years a nest-box program along interstate highways has been successful; nest boxes between Omaha and Lincoln on I-80 fledged 43 birds from 32 boxes in 1994 and a high of 73 birds in 1992 (NBR 62:71).

Nesting activity has been noted as early as late Mar. Eggs have been observed as early as 6 Apr and young by 6 May, and young have been noted still in the nest cavity as late as late Jun. One nest contained 5 eggs on 8 Jun. As many as 6 fledglings have been recorded from one nest.

Fall

Movement peaks in mid-Sep. Padelford and Padelford (pers. comm. RSS) noted a slight increase in frequency of sightings in the Missouri Valley in the latter third of Sep.

High counts include 78 in the southern Panhandle 20 Sep 1999, 74 in Scotts Bluff Co 21 Sep 1997, and 30 around L McConaughy 10 Sep 1988 (Rosche 1994a).

Winter

Numbers are fewest in midwinter, especially in the north and west. Rosche (1982) considered it but a "rare winter visitant" in the northwest, most birds occurring along river valleys. It is fairly numerous and evenly distributed during the CBC period, although data indicate that highest numbers at that time occur in the south. Lingle (1989) found this species the most common raptor in his winter surveys along the lower North Platte and Platte Rivers. It is generally rare or absent from the Sandhills in midwinter.

There is some evidence from banding of birds using nest boxes in Iowa that a few birds remain all year near nesting sites, usually females but also males (Bruce Ehresman, Iowa Department of Natural Resources, pers. comm. WRS). Because females are much less numerous in Nebraska in winter than are

males, any wintering female may thus be nonmigratory.

FINDING

Kestrels are conspicuous during spring and fall migrations, especially in Apr and Sep on roadside utility lines.

Merlin

Falco columbarius

STATUS

Uncommon regular spring and fall migrant west and central, rare east. Rare casual breeder northwest. Uncommon regular winter visitor west and central, rare east.

DOCUMENTATION

Specimen

UNSM ZM12536, 20 Oct 1906 Omaha, Douglas Co.

TAXONOMY

According to Palmer (1988), there are 3 subspecies in North America. Two (*columbarius* and *richarsonii*) occur regularly in Nebraska (Rapp and others 1958) as migrants. The specimen cited above is an example of *richardsonii*, and there is a specimen, UNSM ZM 7702, collected 1 Oct 1940 at Stapleton, referable to *columbarius*. Northern North American breeders are the subspecies *columbarius*, a long-distance migrant that passes through Nebraska, probably mostly in the central and eastern parts of the state as it is rare in Colorado (Andrews and Righter 1992). Some authors (see Palmer 1988) distinguish darker birds in the western part of the range of *columbarius* as the subspecies *bendirei*, noted by Rapp and others (1958) as occurring in Nebraska, but these birds are generally thought to be insufficiently different from *richardsonii* to warrant subspecific status (Bent 1937; Palmer 1988). Prairie breeders, more southerly than *columbarius*, are of the rather pale subspecies *richardsonii*, which migrates much shorter distances than *columbarius*, winters in part in Colorado, and probably accounts for most wintering Merlins in Nebraska. This subspecies nests in the Black Hills of South

Dakota (Pettingill and Whitney 1965) and probably in Nebraska. Most Colorado Merlins are *richardsonii* (Andrews and Righter 1992), but this subspecies is rare in Missouri (Robbins and Easterla 1992), suggesting that it is more numerous westwardly in Nebraska.

The very dark, even blackish, Pacific Northwest subspecies *suckleyi* was reported in Banner Co 15 Jan 1998 (sjd). There have been several sight reports of this race in Colorado, but there are no specimens (Andrews and Righter 1992).

DISTRIBUTION AND ECOLOGY

The few breeding records are in ponderosa pine habitat in the Pine Ridge. Migrants and wintering birds occur virtually anywhere, although wintering birds have a tendency to reside in cities and towns where there is a source of small birds as prey.

Spring

winter→16, 18, 20 May

Late dates above do not include reports from the Pine Ridge. There is a later report 23 May 1959 Scotts Bluff Co. Additional later reports in the east, undocumented, are 29 May 1943 Lancaster Co, 2 Jun 1957 Webster Co, 6 Jun 1979 Brown Co, 12 Jun 1961 Cass Co, 29 Jun 1956 Brown Co, and 4 Jul 1974 Lancaster Co; we consider these reports doubtful.

Data from the east indicate a peak of occurrence in Apr, when 24 of 60 reports in the period Jan–May fall, while in the west and central areas of the state the 65 reports in the same period are fairly evenly distributed. The eastern peak in Apr probably is due to migrants of *columbarius*, a subspecies not known to winter in Nebraska, while other reports from eastern as well as those from the western and central Nebraska probably refer to wintering birds of the race *richardsonii*.

Summer

This species has been suspected of breeding in the Pine Ridge since 1938 (Hudson 1939), but no direct evidence was obtained until 1975.

On 13 Jun 1975 a nest with 2 eggs was found in a ponderosa pine in a sheer-walled canyon in

Fort Robinson sp (Lock and Craig 1975). On 9 Jun 1978 another nest was found, also in Fort Robinson sp; this nest was only 2.4 km (1.5 mi) from the first and contained 1 egg and 4 chicks (Lock 1979). Lock (1979) reported seeing adults in Jun 1977 northwest of Crawford and in Jun 1978 southeast of Crawford. Success of either nest was undetermined. Nest cards were submitted for Sioux Co 1983 (Bennett 1984) and 1984 (Bennett 1985) without details. It was suspected of breeding in Squaw Canyon in 1938 (Hudson 1939), and there are additional breeding season reports for Sioux Co 15 Jun 1986, 28 Jun 1999 (js), 16 Jul 1992, 17 Jul 1994, and 5 Aug 1995. In 1991 ngp surveys found 10 active nests, all in the Pine Ridge of Sioux and Dawes Cos. Five young were observed at one of these nests 14 Jul; another contained 4 young 13 Jul.

Two further reports from the Pine Ridge area may represent early fall migrants, 9 Aug 1994 Box Butte Co and 10 Aug 1994 Sheridan Co, as may a female reported 23 Jul 1994 in Phelps Co.

Fall

11, 14, 17 Aug→winter

Early dates above are away from the Pine Ridge. An earlier report was of one in Deuel Co 3 Aug 1997 (sjd). Most easterly reports are in Sep. The study by Padelford and Padelford (pers. comm. rss) yielded a maximum for any one year of 10 sightings; all occurrences were in the period 10 Sep–25 Oct.

High counts include 4 in the eastern Rainwater Basin 26 Sep 1999 and 3 in the western Panhandle 16 Oct 1999.

Winter

Wintering birds occur statewide, although numbers are fewer eastward. Rosche (1982) listed it as an "uncommon winter visitant" in northwest Nebraska, and single birds are usually reported on cbcs statewide. Most wintering birds are probably of the subspecies *richardsonii*, as discussed above.

High counts include 4 in the Panhandle 24 Jan 1999.

FINDING

Merlins are solitary, fast-moving, and easy to miss. The best chance of seeing one is to check habitats where small birds are plentiful during the winter months. Alternatively, spending a morning on a ridge crest overlooking a pine-studded canyon in the Pine Ridge may yield a sighting as morning thermals develop. However, this species occurs anywhere and is usually chanced upon.

Prairie Falcon

Falco mexicanus

STATUS

Fairly common regular spring and fall visitor west, becoming rare east. Uncommon regular resident west. Fairly common regular winter visitor west, becoming rare east.

DOCUMENTATION

Specimen

UNSM ZM12533, 25 Feb 1896 Warbonnet Canyon, Sioux Co.

DISTRIBUTION AND ECOLOGY

Breeding occurs on cliffs and rocky outcrops in the Panhandle, occasionally farther east. In winter individuals occur statewide, although rarely as far east as the Missouri River. Preferred habitat for foraging is open grassland, although in winter stubble fields are utilized, especially in eastern Nebraska or when snow cover forces birds to river valleys in western Nebraska (Rosche 1982). In winter the usual summer diet of mammals changes largely to birds (Bent 1937).

Spring

winter→30, 30 Apr, 1 May

Late dates above are outside the breeding range; most birds are on the breeding range by late Apr, although there are several later reports, most from central and western Nebraska, likely young birds or wide-ranging foraging breeders. Easternmost late reports are 13 May 1973 Thomas Co, 16 May 1998 Funk Lagoon (JJ), 18 May 1980 Thomas Co, and 20 May 1972 Cherry Co.

There is little information to indicate whether significant migration occurs in Nebraska, although data from daily surveys for 3 years in Loup Co (McClure 1966) showed no increased movement in expected migration periods. It is likely that eastward drift accounts for most wintering birds.

Summer

Nesting occurs regularly in Sioux, Dawes, Sheridan, Scotts Bluff, Banner, and Morrill Cos, all of which have large, perpendicular, rocky cliffs and outcroppings suitable for nesting. There are also reports suggestive of breeding for Cherry Co; Short (1965) observed a pair "probably nesting" on a cliff face along the Niobrara River southwest of Merriman May–Jun 1964, and Ducey (1989) noted a "very defensive" pair 20–21 May 1989 at sandstone cliffs near Eli. Farther east, Brogie and Mossman (1983) reported 3 sightings near Plum Creek Canyon and along the Niobrara River in Brown Co 11 May–28 Jun 1982, but no evidence of nesting was found; it was reported in Brown Co in "summer" 1966. In Keith Co Benckeser (1950) reported nesting, and it was reported there in "summer" 1977 and 16 Jun 1979, although Rosche (1994a) knew of no current nesting sites in the Keith Co area. A nest card was submitted for Garden Co for 1982 (Bennett 1983), where it was also observed "summer" 1967, 31 Jul 1978, and 18 Aug 1981.

There are a few summer reports away from the breeding range, possibly wide-ranging breeders (Rosche 1994a), failed breeders, or prebreeders. Such summer reports, generally undocumented, include a listing as permanent resident in Logan Co in 1963, 30 May–4 Jun Dakota Co, 4 Jun 1981 and 11 Jun 1979 Lancaster Co, 27 Jun 1978 McPherson Co, 30 Jun 1991 Lincoln Co, 1 Jul 1958 Keya Paha Co, 10 Jul 1994 Sherman Co, 16 Jul 1994 Buffalo Co, and 21 Jul 1978 McPherson Co. May and Aug reports are treated under Spring and Fall.

Biologists with the NGP have taken a special interest in this species and surveyed a number of nest sites during the 1980s. Virtually all of the 118 nests visited were on ledges or in holes in cliff faces. Surrounding habitat was commonly midgrass or shortgrass prairie, sometimes pine parkland. Observations of young indicated that hatching occurred late May and early Jun, although nearly fledged young were at one site as late as 23 Jun. As many as 5 young were noted in nests, but most nests had 2–3 young.

Fall

3, 3, 7 Sep→winter (central)

28, 29 Sep, 5 Oct→winter (east)

The early dates above are away from the breeding range; the 28 Sep observation was in the Missouri Valley (Barbara and Loren Padelford, pers. comm. rss). Earlier reports in central Nebraska are Aug 1911 Thomas Co (Bray 1994), 14 Aug 1991 Kearney Co, 17 Aug 1981 Thomas Co, and 23 Aug 1970 and 25 Aug 1972 Perkins Co. Earlier reports in the east are 14 Aug 1991 Stanton Co (specimen wsc 769) and 2 Sep 1965 Adams Co.

Winter

Data from Wyoming and Colorado indicate a strong tendency for young birds to move eastward after fledging, while some adults tend to be resident (Palmer 1988). There is no information on the source of Nebraska wintering birds.

Individuals that occur in winter east of the breeding range may remain in a given area for some time, occupying prominent perches such as the state capitol in Lincoln, a courthouse in Beatrice, and grain elevators.

cbc data show even distribution statewide except for the east, where only about 10% as many occur as compared to elsewhere in the state. Census data in 1979 and 1980 showed variable but even distribution along the Platte River valley eastward to Grand Island (Lingle 1989).

FINDING

This species is easiest to locate in May–Jun in Pine Ridge canyons and other Panhandle escarpments such as Scotts Bluff nm (Cox and Franklin 1989) by scanning sheer cliffs for nest sites. In western Nebraska single birds are often seen in open grasslands.

Peregrine Falcon

Falco peregrinus

STATUS

Uncommon regular spring and fall migrant statewide. Locally uncommon (restocked) regular breeder east. Rare casual winter visitor statewide.

DOCUMENTATION

Specimen

(*anatum*) unsm zm7696, 14 May 1915 Garden Co; (*tundrius*) unsm zm7695, 5 May 1935 near Stapleton, Logan Co.

TAXONOMY

Although published information (Bruner and others 1904; Rapp and others 1958) suggests that only the widespread subspecies *anatum* has been recorded in Nebraska, there are actually 3 specimens of the somewhat paler northern subspecies *tundrius* in the unsm collection (jgj); this subspecies would be expected to occur in migration in Nebraska (Palmer 1988) and is the commonest of the subspecies occurring in Missouri during migration (Mark Robbins, pers. comm.).

The genetic heritage of captive stock that has produced young in Omaha and Lincoln in recent years is unknown, although wild birds that paired with stocked birds are thought to be of the subspecies *anatum*.

DISTRIBUTION AND ECOLOGY

This species is most often reported from the Rainwater Basin, where it follows migrating shorebirds. It may also attack grouse; rss has seen individuals worry prairie grouse on their spring display grounds on several occasions. Recently, breeding success has been achieved by reintroduced birds on the Woodmen Tower in Omaha. ngp studies of prey items at this nesting site have found remains of nearly 50 species of birds.

Spring

6, 7, 9 Apr→22, 23, 24 May

Late dates above do not include reports from Douglas-Sarpy Cos since 1990, when breeding birds became established. There are earlier reports 15 Mar 1983 Kearney Co, 24 Mar 1963 Hamilton Co, 27 Mar 1959 Dawes Co, 31 Mar 1984 Polk Co, and 3 Apr 1974 Lincoln Co. Some early dates may refer to birds that wintered in or near Nebraska (see Winter).

High counts reported statewide include totals of 25 in 1999, 22 in 1997, 21 in 1998, 16 in 1996, 15 in 1995, and 9 in spring 1994. Nine were counted between the Rainwater Basin and Sarpy Co 24 Apr 1998.

Summer

Until recent restocking efforts in Omaha, there was only a single report indicative of nesting; Bruner and others (1904) noted that young and old birds were seen 5–19 Aug 1903 around cliffs 12.9 km (8 mi) west of Fort Robinson in and out of a "recess that may have been the nesting site."

Efforts to establish breeding birds in Omaha began in 1988 when 6 young birds were hacked at the Woodmen Tower in Omaha (NBR 56:81). Two died, and the last bird left in early Oct. The Mutual of Omaha building in Omaha was also used as a hack site. Apparently the first nesting was at the Woodmen Tower in 1992, when a male from the 1988 release returned with a female that had been released in Des Moines, Iowa, in 1991 (Morris 1992). This pair fledged 3 young, which hatched 11–13 Jun. Nesting has taken place at the Woodmen Tower each year since, involving at least 2 successful pairings of a bird fledged in Omaha with birds released from other locations. A total of 22 young were hatched at the two sites 1988–98, and in the period 1992–98 4 of the 8 nesting attempts have been productive, fledging a total of 11 young.

A banded male frequented the state capitol building in Lincoln in the summers of 1990–93 and also 1996–98. It had been released in Omaha in 1989 (NBR 58:63, 59:12). After arriving in spring, it behaved

territorially, courted females without success, remained for the summer, and departed in Sep or early Oct (John Dinan, NGP, pers. comm. RSS).

Apart from Aug reports (see Fall), there are additional summer reports: 7 Jun 1922 Omaha, Douglas Co (specimen HMM 6275, measurements; Swenk, Notes after 1925); 11 Jun and 1 Jul–8 Aug 1980 Garden Co; 26 Jun 1968 McPherson Co; 12 Jul 1987 Sarpy Co (not accepted by the NOURC, Grenon 1990); 14 Jul 1963 Douglas-Sarpy Cos; 20 Jul 1988 Lancaster Co; and 26 Jul 1970 Custer Co.

Fall

11, 13, 15 Aug→31 Oct, 6, 7 Nov

According to Padelford and Padelford (pers. comm. RSS), there is a slight increase in sightings late Sep–early Oct in the Missouri Valley.

There are earlier reports 5 Aug 1988 in south-central Nebraska and 6 Aug 1902 at Badger in the Niobrara River valley (Ducey 1983a). There are several later reports of birds lingering into winter. The few additional Nov reports are 15 Nov 1982 (Rosche 1982), 20 Nov 1982 Washington Co, 23 Nov 1991 Douglas-Sarpy Cos, 25 Nov 1991 Cass Co, 28 Nov 1974 Keith Co (Rosche 1994a), and 30 Nov 1992 Douglas-Sarpy Cos. For Dec reports, see Winter.

Winter

There are at least 44 reports in the period 3 Dec–7 Mar. Most are in Dec–Jan, with only these few Feb reports: 1 Feb 1975 Kearney Co (Turner 1975), 4 Feb 1962 Harlan Co, 4 Feb 1974 Adams Co, 11 Feb 1960 Antelope Co, 16 Feb 1970 Lincoln Co, 16–21 Feb 1981 Douglas-Sarpy Cos, 18 Feb 1963 Dawes Co, and 23 Feb 1985 Saunders Co.

COMMENTS

This species underwent a severe population decline in North America due to pesticides beginning in the 1940s. By 1964 the population was much reduced, but after restrictions on the pesticide DDT in 1972 the population has recovered (Palmer 1988). Restocking efforts have also assisted in increasing the population. Most Nebraska reports have come since 1980.

FINDING

Peregrine Falcons are most likely to be seen during shorebird migration in late Apr–early May in the Rainwater Basin. The breeding birds in Omaha can also be seen by watching the Woodmen Tower in May–Jun.

Gyrfalcon

Falco rusticolus

STATUS

Rare casual winter visitor statewide.

DOCUMENTATION

Specimen

UI 1746923, Feb 1885 Johnson Co (DuMont 1933).

DISTRIBUTION AND ECOLOGY

Reports of this species are during the winter months, primarily from the Sandhills region but also extending to southwest Nebraska (Rosche 1994a). Individual birds do not remain long in any one location (Steve Duecker, pers. comm. RSS). The reports essentially reflect the range of extensive grasslands, although there are sightings from all parts of the state, which may be related to increasing snow cover on the Canadian prairie provinces and the Dakotas as winter progresses. Prey consists of large birds such as grouse and waterfowl.

Winter

4, 5, 11 Nov→23, 23 Feb, 3 Mar

Available reports, about 60 in all, may not reflect the true frequency of occurrence, as most observers avoid open plains habitats in the winter months. The reports are in the period 25 Sep–3 Mar, although most are in Jan. Early reports are 25 Sep 1979 Thomas Co (specimen HMM 36541; Ohlander 1979) and 20 Oct 1978 Hall Co (Ohlander 1979). There is a later report 10 May 1993 in York Co (Morris 1993). There were several reports in the winter of 1984–85; falconers reported 12 in the Sandhills for the season (AB 39:183), and it was reported from Lancaster Co 3–21 Jan (Green 1984a; NBR 52:55), Saunders Co 8 Jan (Green 1984a), Polk

Co 28 Nov (NBR 53:17), and Douglas-Sarpy Cos 7 Jan–14 Feb (Green 1984a).

Falconers indicate that Gyrfalcons are regular in winter; sightings are reported "every year since 1979 in the Sandhills between Norfolk and Gordon" (Rosche 1994a). There are 4 records for the northeast Colorado plains (Andrews and Righter 1992), "several" for Kansas (Thompson and Ely 1989), and 3 for Missouri (Robbins and Easterla 1992; Mark Robbins, pers. comm.).

COMMENTS

All color morphs have been reported in Nebraska (RSS).

FINDING

This species should be looked for in Jan in the northern Sandhills, especially where there are concentrations of prairie grouse or waterfowl. Falconers consider good locations to be Highway 20 in the northern Sandhills and open grasslands in Sioux, Box Butte, and Dawes Cos (Steve Duecker, pers. comm. RSS).

GALLIFORMES

Phasianidae (Grouse, Turkeys, Quail)

Gray Partridge

Perdix perdix

STATUS

Uncommon regular resident northeast.

DOCUMENTATION

Specimen

WSC 683, Nov or Dec 1978 or 1979 Wayne Co.

DISTRIBUTION AND ECOLOGY

A native of much of Eurasia, where it occupies
   open grasslands and cropland borders, it
   has been introduced to North America. In
   Nebraska it appears to prefer agricultural
   fields rather than pure grasslands, although
   edge cover is necessary. Nests are located in
   hay fields and grainfields, with a preference
   for alfalfa (Johnsgard 1979). Currently the
   range is restricted, partridges occurring only
   in northeast Nebraska, although the center
   of abundance, as well as overall numbers,
   changes periodically.

Summer

Although several attempts have been made
   to establish this species in various parts of
   Nebraska, beginning as early as 1925–26, none
   appear to have been successful (Mathisen
   and Mathisen 1960b; Rosche 1982). The
   current Nebraska population was apparently
   self-established by spreading from South
   Dakota and Iowa. The first population to be
   so established is that in the north-central part
   of the state, especially Boyd Co and northern
   Holt Co (Mathisen and Mathisen 1960b), and
   probably also Knox Co (Rapp and others
   1958). Nesting was reported from this general
   area 1957–61 (Pritchard and Pritchard 1958,
   1959, 1960, 1961; Wensien 1962). These birds
   probably came from South Dakota. Reports
   from that area persisted at least until 1992
   in Holt Co and 1994 in Boyd Co, as well as
   neighboring Knox Co through 1999. Coveys
   were seen in southern Holt Co on the Blake
   Ranch 1986–87 but not since (Blake and

Ducey 1991), although a report of 7 birds on
   the 1994–95 Calamus-Loup CBC suggests that
   a population in the Calamus Res area of Loup
   and Garfield Cos may persist.

A report from eastern Cherry Co 11 May 1982
   (Brogie and Mossman 1983), one 22 km (14
   mi) north of Mullen in south-central Cherry
   Co 28 Jan 1983 (Wyman 1983), one northeast
   of O'Neill in Holt Co 18 May 1997, and an early
   report from Antelope Co 10 Feb 1950 may
   have been of stragglers from South Dakota.

Additional reports in the 1990s are mostly from
   Dixon and Wayne Cos, with reports from
   neighboring Cedar Co through 1990 and
   again 8 May 1997 and Dakota Co through
   1991. The most recent population peak was
   in the 1980s, when birds were reported from
   several counties west and south of the current
   range, such as Antelope Co 1988; Boone Co
   1980–89, including the Beaver Valley CBC in
   northwest Boone Co 1987–88 and 1988–89;
   Pierce Co 1986–89; Madison Co, where it was
   reported on the Norfolk CBC 1985–86 and
   1988–89; Stanton Co 1989; and Cuming Co,
   where it nested in 1980 (Einemann 1980)
   and was reported in 1987. One was reported
   northwest of Fremont 29 Mar 1999.

In the Missouri River valley it was seen regularly
   around Blair during the 1980s, when winter
   flocks of up to 20 birds were noted (JGJ), but
   the only sightings in the 1990s are of singles
   in Washington Co near Kennard 6 Aug 1994
   (JGJ) and near Herman 18–20 Jun 1997 (JGJ).
   There are reports from Douglas-Sarpy Cos
   21–23 Feb 1979 (NBR 47:42) and 14 May 1991,
   possibly locally released birds, as may have
   been reports from Polk Co 17 May 1982, 10
   Apr 1988, and 13 Mar 1989 and another from
   Logan Co 3 Apr 1951. A report from near
   Edgar, Adams Co, 25 Jun 1966 was said to
   be from birds introduced there a few years
   previously (Ritchey 1966).

Winter

Flocks of as many as 100 individuals may
   be encountered working the stubble in
   grainfields during the winter months.

Although normally sedentary, when heavy snow cover exists this species may migrate considerable distances (Cramp and Simmons 1980). This may explain entry of a few birds into Nebraska from South Dakota as well some of the more southerly isolated Nebraska reports.

COMMENTS

Reports from Douglas Co 1963–64 were in error, actually referring to Chukar (NBR 33:19). A single was reported at Boyer Chute NWR 7 Sep 1997, but the report was not accepted by the NOURC.

FINDING

Generally the best strategy for locating this species is to drive back roads in northeastern Nebraska in Jul–Aug when broods of young birds are present. Another possibility is to listen in early morning for the 2-noted "crowing," most often heard Apr–Jun.

Ring-necked Pheasant

Phasianus colchicus

STATUS

Fairly common, locally common, regular resident statewide.

DOCUMENTATION

Specimen

WSC 394, 17 Feb 1978 Boone Co.

TAXONOMY

North American birds are mostly from Asian stock, although derivatives of European birds have also been mixed into many North American releases (AOU 1998). Some authors have treated Asian and European birds as separate species, *P. torquatus* and *P. colchicus*, respectively (AOU 1998; Sibley and Monroe 1990).

DISTRIBUTION AND ECOLOGY

This species is a native of the grasslands and brushlands of eastern Asia, including parts of northern China, Mongolia, and the steppe regions of Siberia. Successfully introduced to Europe several centuries ago, the species was considered an ideal candidate for introduction to North America, especially the eastern Great Plains, where habitat was similar to ancestral habitat.

Pheasants occur statewide in varying densities depending on the presence of cover for protection and nesting, although they regularly forage in open agricultural fields. After severe winters, populations may decline significantly, such as in the northwest (Rosche 1982).

Summer

Nesting has been reported in most counties across the state (Ducey 1988), and nesting probably occurs in all. Densities are greatest in the east and south, although the Panhandle also sustains a significant population, mostly limited to river valleys and associated with cropland. Populations tend to be low in the Sandhills, where grass cover appears to be insufficient, although pheasants may become numerous near lakes and marshland. Densities vary significantly from year to year and over a period of years depending on the severity of winters and the amount of farmland idled in government programs.

Nebraska's first pheasants may have been self-introduced from Kansas around 1900, but the first deliberate introduction was around Ord in 1909 (Mathisen and Mathisen 1960b). Since then there have been repeated introductions and restockings around the state. According to reports of hunters and records of the NGP, very high populations were attained by the late 1940s and early 1950s, particularly in south-central and east-central Nebraska. Prospering during this time of limited herbicide use, hundreds were frequently flushed by hunters from cornfields, cropland edges, and weed patches. Shelterbelts often provided cover to additional hundreds during periods of severe winter weather.

Today the Nebraska pheasant population is but a token of its historic highs, but hunting pressure is not to blame. Improved farming techniques that allow for more efficient use of land, coupled with effective herbicides, have reduced weedy cropland edge, habitat important for nesting, brood rearing, and shelter. The net result has been a continuing general decline in population densities. Various government-sponsored set-aside programs have, however, provided the pheasant with some periods of habitat respite (Baxter and Wolfe 1973).

This ground-nester typically chooses weedy fields and cropland edge for nest sites and brood rearing. Nests are well concealed by a clump of grass or other herbaceous vegetation, including dry cattail beds. Eggs have been noted 2 May–12 Jul; of 19 nests surveyed, clutch size was in the range 8–16.

Winter

Pheasants are essentially sedentary. Highest CBC totals are from the east; high counts from various localities are 565 at DeSoto Bend NWR 1983–84; 510 at Norfolk 1994–95; 86 at Lincoln 1990–9; 72 at Omaha 1983–84; 79 at Sioux City, Iowa, 1984–85; 117 at Grand Island 1983–84; 70 at Beaver Valley 1989–90; 22 at North Platte 1981–82; 36 at Scottsbluff 1981–82; and 23 at Crawford 1978–79.

FINDING

Cock pheasants are conspicuous in Apr when they occupy an exposed spot and display to attract hens. Their crowing carries for some distance. Pheasants are often seen along rural roadsides year-round.

Ruffed Grouse

Bonasa umbellus

STATUS

Extirpated.

DOCUMENTATION

Specimen

HMM 1984 (pair), 1916 Dodge Co (Swenk, Notes before 1925).

COMMENTS

Ducey (1988) summarized the little that is known about this species' occurrence. Most had disappeared by the late 1880s, although it was considered a rare resident by about 1900 at Child's Point (Shoemaker, cited by Ducey 1988). According to Bruner and others (1904),

this species still occupied woodlands along the Missouri River valley around 1900 but was by then rare. The specimens cited above were shot by Anthony Machianach near Nickerson and apparently are the specimens examined by Aldrich and Friedmann (1943); one is a red morph and the other gray, and they constitute the last known record for the state. The location of collection in the lower Elkhorn Valley and the date of collection some years after the species was rare in prime habitat raise some questions as to the authenticity of the specimens, however.

Aldrich and Friedmann (1943) indicate on their range maps that the Rocky Mountain subspecies *incanus* may occur in the Pine Ridge, but there are no known reports from that area (Bray and others 1986), even though it is currently an uncommon resident in the Black Hills of South Dakota (SDOU 1991).

An attempt to reintroduce this species in the Nemaha Co portion of Indian Cave SP in 1968 (NBR 38:90) was unsuccessful, none being reported since 1973 (Johnsgard 1997). In recent years releases have been made in Kansas, including areas along the south side of the Missouri River in Doniphan Co; if this population becomes established, a few birds may wander into extreme southeast Nebraska. Similar releases in extreme northwest Missouri have been unsuccessful (Missouri Department of Conservation data).

Two reports 26 Mar 1949 Lincoln Co and 25 Jan 1956 Boyd Co are probably locally released birds or misidentified prairie grouse.

Sage Grouse

Centrocercus urophasianus

STATUS

Uncommon casual winter visitor extreme northwest.

DOCUMENTATION

Description

ca. 6 Sep 1987 Sioux Co (Grenon 1990).

DISTRIBUTION AND ECOLOGY

This species is fairly common very close to

Nebraska in southwest South Dakota (Rosche 1982) and probably is "very rare in extreme nw Nebraska" (Rosche 1982). Its status may be unchanged in the 20th century, as Bruner and others (1904) noted that it was "not common and confined to the extreme northwestern portions of the state in regions where [big sagebrush] *Artemesia tridentata* abounds." Because big sagebrush barely reaches Nebraska and has been eradicated to some extent, especially in the Hat Creek Basin, only limited habitat is currently available to support Sage Grouse.

Winter

Around 1900 it occurred along Hat, Antelope, and Indian Creeks in Sioux County, where it had several times been reported to breed and where Carriker and Cary found old birds with half-grown young in the summer of 1901 some 11 km (7 mi) east of the Wyoming border in the Hat Creek Basin (Bruner and others 1904). There are no reports since the 1987 record cited above, the only documented record for the state (Grenon 1990), although conversations (RSS) with NGP personnel indicate that small flocks are still seen "rarely" in winter in Sioux Co.

There were a few reports in the 1960s and 1970s. At least "two different flocks of Sage Grouse" were seen in the northwest corner of Sioux Co 9–10 Dec 1960 (Mathisen 1961). Rosche (1982) reported that 6–7 were seen by a hunter Sep 1975 in extreme northwest Sioux Co, and an adult male was poached 16 Jan 1979 by a hunter from a flock about 8 km (5 mi) west of Harrison and confiscated by NGP personnel.

There are several old reports, including the collecting of an "old sick hen" (Brooking, Notes) after a blizzard Jan 1917 22 km (14 mi) west of Crescent L NWR in Garden Co; this specimen, at one time HMM 2773, was sold in 1976 to Hastings College but cannot now be located (Bray and others 1986; Swenk, Notes before 1925). In Jul 1901 "two large male birds were secured" in northwest Sioux Co (Carriker 1902), but these specimens also

cannot now be located (Bray and others 1986). Swenk (Notes before 1925) noted that B. J. Olson had killed "Sage Hens" west of Chadron and northwest of Hemingford in western Dawes Co. or Sioux Co in 1888 and also that a Mr. Swanson of Chappell shot "Sage Hens" in Sioux Co in the winter of 1915–16, probably Dec 1915.

Other more recent though undocumented reports follow. A remarkable 87 birds were reported 1935–37 by respondents to a farm shelterbelt questionnaire (Orendurff 1941); these reports are likely misidentifications. It was stated that "local investigation in Sioux and Dawes Counties indicates that a few sage hens may be present in extreme northwestern Nebraska" (Mohler 1944). Sage Grouse was listed as seen in Dawes Co 11–13 May 1945 on NOU field trips (NBR 13:57).

FINDING

The best way to find this species is at a lek (display area) in spring, but no lek sites are known in Nebraska. Sage grouse are conspicuous starting in late Jul and continuing into early winter when broods are present and flocking begins; this might be a good time to check the sage habitat in extreme northwest Sioux Co along Indian and Antelope Creeks. Periods of snow cover may make these birds easier to see; NGP personnel believe that small flocks may still appear in winter.

Sharp-tailed Grouse

Tympanuchus phasianellus

STATUS

Locally common regular resident north and west, accidental south of Platte River. Uncommon casual winter visitor south and east of breeding range.

DOCUMENTATION

Specimen

UNSM ZM12561, 23 Feb 1896 Harrison, Sioux Co.

TAXONOMY

See Greater Prairie-Chicken.

DISTRIBUTION AND ECOLOGY

This species occupies generally treeless native

grasslands throughout the Sandhills and northern Panhandle. It occurs in the same areas as Greater Prairie-Chicken in the central and eastern Sandhills but is far more common than the prairie-chicken in western grasslands (Sharpe and Payne 1966), which tend to be drier. Sharp-tailed Grouse are not as dependent on the presence of croplands in winter as are Greater Prairie-Chickens. Brushy cover is utilized in winter (Johnsgard 1997), when rosehips (*Rosa arkansana*) and buds of woody plants found in shelterbelts and riparian areas provide sustenance.

Like the Greater Prairie-Chicken, male Sharp-tailed Grouse gather on leks known as "dancing grounds," where they display communally in an effort to attract a mate. Although dancing activity may continue from late winter through spring, peak activity occurs during the hen visitation period mid-Apr to early May. Dancing grounds are usually broad hilltops of short grass, but large open valleys may be used in the Sandhills, particularly where hilltops are sharp, choppy, and unstable.

Summer

All counties north of the North Platte and Platte Rivers are occupied, east to Knox Co, where it has been observed on leks near Center (Williams 1987), Antelope Co, Boone Co, and Hall Co, where it occurs on the Taylor Ranch, just northwest of Grand Island (Lingle 1994). It occurs in small but apparently increasing numbers south of the North Platte River in the Panhandle; 16 were seen in central Kimball Co 23 Sep 1995 and 20 in southwest Banner Co 12 Apr 1997. South of the Platte River the only summer report is at Victor L 7 Jul 1994. An undocumented report 24 Apr–30 Jun 1984 in Lancaster Co is apparently in error.

A nesting study in the western Sandhills indicated a nest density of 1 per 70 ha (1 per 174 ac), even though the density of displaying males was somewhat higher, about 1 per 22–49 ha (1 per 60–120 ac) (Blus and Walker 1966). Sixteen nests have been documented, with egg

dates 21 May–5 Aug and clutches in the range of 10–15 eggs with a mean of 11.4. Thirteen nests in Sandhills prairie were concealed in clumps of dense grass, usually little bluestem (*Schizachyrium scoparium*). Three nests were located in shorter grama grass. All nests consisted of a slight depression lined with fine grasses and numerous feathers. Hens are often very secretive and nest-attentive, only flushing when almost stepped upon.

Winter

This species is not as prone as Greater Prairie-Chicken to move southeastward of the breeding range in winter. There are few such reports: 26 Oct 1986 and 21 Dec 1987 in Polk Co, 3 Dec 1982 in Lancaster Co, and 12 Jan 1976 in Adams Co.

Highest CBC totals include 129 at Calamus-Loup in 1994, 104 at North Platte in 1992, 54 at Greeley in 1969, and 53 at Crawford in 1979. Wintering birds often congregate in river or stream valleys or among cedars for shelter.

FINDING

This species is easy to see at leks in Apr, notably those with blinds for the public: Crescent L NWR, McKelvie Division of the NNF, and Valentine NWR. Flocks are often encountered in winter, especially in river valleys.

Greater Prairie-Chicken

Tympanuchus cupido

STATUS

Locally common regular resident north, uncommon south. Uncommon casual winter visitor east of breeding range.

DOCUMENTATION

Specimen

UNSM ZM10301, 25 Nov 1912 Comstock, Custer Co.

TAXONOMY

Some authors (Johnsgard 1979; Sibley and Monroe 1990) have considered this species and Lesser Prairie-Chicken (*T. pallidicinctus*) conspecific.

Greater Prairie-Chicken sometimes hybridizes with Sharp-tailed Grouse (*T. phasianellus*)

(Mathisen and Mathisen 1959); hybrid specimens UNSM ZM13554 and ZM13555 were collected 12 Feb 1959 in Loup Co, and ZM13553 was collected Dec 1965 in Cherry Co. Hybrids have also been reported from Garden, Thomas, Keya Paha, and Rock Cos. Johnsgard and Wood (1968) suggested that hybridization occurred when numbers of each species were about equal, but it could be argued that the species' ranges tend to overlap in areas where there is peripheral habitat for each and where overall numbers are low, limiting the choice of potential mates. In Michigan hybridization was noted to be a temporary phenomenon between these species, occurring when the two were rapidly undergoing a change in relative numbers as Sharp-tailed Grouse replaced Greater Prairie-Chicken (Ammann 1957).

DISTRIBUTION AND ECOLOGY

This species occupies "relatively undisturbed grassland, especially lush Little Bluestem" (Rosche 1994a), probably reaching greatest numbers where such grassland "interdigitates with grain croplands" (Johnsgard 1997) in the eastern Sandhills. The cropland provides important winter forage in the form of seed spillage that was formerly provided by mast such as acorns (Sharpe 1968). In severe winters some southward movement occurs to the south and east of the breeding range.

It is not at all clear what the pre–European settlement distribution was in Nebraska, but judging from habitat requirements and early accounts from Iowa and Illinois (Bent 1932), it is likely that a small population was supported in the midgrass and tallgrass prairies of extreme eastern Nebraska. What is clear, however, is that as European settlement progressed westward beginning in the 1850s, when croplands mixed with grasslands, Greater Prairie-Chicken numbers increased dramatically, a phenomenon experienced earlier and described by pioneers in Illinois and Iowa (Bent 1932). Such explosions in population were short-lived, however,

and numbers decreased as grasslands were almost completely converted to croplands.

Summer

The most extensive area occupied by this species is in the grasslands of north-central Nebraska, between the Platte River and the South Dakota border. The western edge of this area appears to be central Cherry Co, through Grant and Arthur Cos to Keith Co (Rosche 1994a). Reports farther west are few. Rosche (1982) considered it extirpated in Dawes, Sioux, and Sheridan Cos, citing only a few reports and none since 1978, when one was seen Dec 1978 south of Antioch. It was also reported without details as a permanent resident in Dawes Co in 1977. There are few reports from Garden Co; it is not listed on the USFWS bird checklist at Crescent L NWR, and other reports are limited to 11 Jun 1991 and permanent resident in 1981. Formerly it was more common in the west, especially in the Antioch area (Rosche 1982), and possibly in grasslands along and north of the North Platte River as far west as Wyoming (Scott 1993).

The eastern edge of the northern Nebraska range is from Knox Co south through Antelope and Boone Cos to Hall Co, probably including Merrick Co, where it was listed as a permanent resident in Merrick Co in 1963 and 1964. There are no summer reports farther east (north of the Platte River) since 1900, when Bruner and others (1904) noted that it bred over the entire state.

South of the Platte River this species occurs in summer in three discontinuous areas. In the southwest there is a continuation of the range from north of the Platte River in Keith and Lincoln Cos southwest into Perkins, Chase, Dundy, and possibly Hayes and western Hitchcock Cos. Easternmost reports from the southwest population include "rare" status at Camp Opal Springs in southeastern Lincoln Co in 1959 (Huntley 1960) and its occurrence on a BBS route in western Hitchcock Co in the period 1967–77. There are no reports for Frontier, Gosper, Red Willow, or Furnas Cos nor summer reports from Webster Co, but there are several recent reports from Phelps, Kearney, Adams, Harlan, and Franklin Cos, possibly a continuation southward of the populations north of the Platte River in Buffalo and Hall Cos. These south-central birds occur in isolated sandhills remnants; at least half-section expanses of native grassland are required for roosting, display areas, nesting, and brood rearing. Such areas exist, for example, south of Atlanta in Harlan Co, 3.2 km (2 mi) west of Holstein, and southeast of Lowell in Kearney Co.

There are no reports from Clay, Nuckolls, or Fillmore Cos, but in the southeast there is a large and viable population in several counties, including Otoe, Richardson, Johnson, Pawnee, Thayer, Jefferson, and western Gage Cos, where natural grasslands persist. In spring 1995 the NGP initiated a survey system in the southeast and documented about 100 leks in Johnson, Pawnee, and western Gage Cos; survey routes also exist in Jefferson and Thayer Cos. The few reports outside these "survey" counties may reflect lingering winter visitors, except possibly for recent reports from Otoe and Richardson Cos. These reports are from northwest of Cook in Otoe Co May 1974 (Antholz 1975) and also in Otoe Co 12 May 1985, southwest of Auburn in Nemaha Co 24 May 1963 (Wensien 1964), northwest of Stella (McMullen 1989), in Richardson Co spring 1990, and a lek of 7 near Denton in Lancaster Co spring 1965 (RSS). The only report from Jefferson Co is for fall 1963. Two in an Omaha yard near Eppley Airfield 20 Apr 1986 (Fritz 1986) may have been locally released or possibly lingering winter visitors, as may have been one near Olive Creek L 5 Aug 1999.

In late winter and spring males gather at leks (also called "booming grounds"). Here the males go through vocal and visual display rituals in an effort to attract a reproductive partner. Display activity reaches a peak about 10 Apr when females frequent the leks in greatest numbers.

Following copulation females seek out a nest site, typically in or under a clump of grass. A slight depression lined with fine grasses and feathers is constructed and is usually initiated within a few days of mating unless delayed by weather or vegetation conditions. The 4 available nest records likely represent renesting; the earliest was 29 May and the latest 12 Jul. Clutches were in the range of 8–12 eggs. Upon hatching, the young leave the nest site with the female, foraging in surrounding grassland or weedy fields.

Winter

This species has a propensity to move southeastward in severe winters, accounting for several reports from counties south and east of the eastern Sandhills range, usually in the period Oct–Mar. Movement of small flocks flying south at high altitude has been observed in very late fall by duck hunters on the Platte River near Columbus (Bill Erwin, pers. comm. RSS). Rosche (1994a) indicated that in winter it has been reported as far west in the North Platte River valley as Broadwater. At Mormon Island Crane Meadows in Hall Co Lingle and Hay (1982) listed this species as rare from fall through spring. During winter storms in 1979 there were sightings south and east to Fremont and Omaha as a result of "snow, lack of food, and a high population" (NBR 47:19), and in the 1949 blizzard, prairie-chickens left Thomas Co but returned in spring and bred as usual (NBR 18:32). In Polk Co the winter of 1983–84 was marked by an influx mid-Dec through 9 Jan, including a flock of 100 on the York-Polk Co line 4.8 km (3 mi) from Benedict (Morris 1984). In Sarpy Co during the winters of 1934–37 flocks were seen each year, "evidently migrants" (Velich 1958).

Reports of winter visitors from counties without summer records are 15 Mar 1989 Cedar Co; 4 Mar 1994 Dixon Co; fall 1986, 2 Mar 1986, and 22 Mar 1987 Pierce Co; 17–18 Dec 1988 Wayne

Co; 17 Dec 1988 and 6 Mar 1994 Stanton Co; 26 Oct 1986, 3 Nov–14 Mar 1984–85, mid-Dec–13 Jan 1984, and 22 Jan 1989 Polk Co; 21 Oct 1997 Hamilton Co; and 16 Oct 1993, 27 Dec 1967, 6 Jan 1980, and 17 Jan 1993 Lancaster Co.

FINDING

This species is easy to see in Apr and early May at booming grounds such as those with blinds at Burchard L in Pawnee Co and Valentine NWR in Cherry Co or at Taylor Ranch just northwest of Grand Island (Lingle 1994). Finding prairie-chickens at other times is difficult. Best chance in winter is to check at dawn at cornfields that are adjacent to grasslands in the Sandhills for birds flying in to feed.

Lesser Prairie-Chicken

Tympanuchus pallidicinctus

STATUS

Hypothetical.

COMMENTS

The entire history of this species in Nebraska is based on records from Red Willow Co, including 3 extant specimens. A male shot 17 Sep 1924 near Danbury, Red Willow Co, by an H. Frates is HMM 3071 (HMM files; Swenk, Notes after 1925). Swenk (Notes after 1925) noted measurements of a male in the Brooking collection shot near McCook 2 Oct 1925; it was described by Swenk as a "typical light-colored bird, L:14.5, W (R) 8.125, chord 199 mm, W (L) 8.25, chord 200 mm, culmen 15, DBB 9.25, TS 42, and tail 96.5." On 7 Oct 1925 2 females and as many as 5 males were shot by Frates from a single flock (HMM files; UNSM files; Swenk, Notes after 1925). A male, now UNSM ZM14210, presumably the bird given to Swenk (Notes after 1925), and a female, HMM 3330, were from this flock. Apparently other birds from this flock were given to C. A. Black of Kearney and a Mr. Hopkins (Swenk, Notes after 1925).

These records are somewhat enigmatic, as there is no independent indication of this species occurring between Nebraska and

the northernmost extent of its historic range in southeast Colorado and southwest Kansas (Rakestraw 1995). None of the literature (Baker 1953; Jones 1963; Sutton 1967) documents range shifts or population expansion as in the case of the Greater Prairie-Chicken; indeed, Lesser Prairie-Chicken demonstrates strong range contiguity in grasslands associated with shinnery oak (*Quercus mobriana*) or sandsage (in southwest Kansas, southeast Colorado, western Oklahoma, the panhandle of Texas, and northeast New Mexico.

According to a paper Swenk read to the Wilson Ornithological Society in 1931 (unpublished copy in possession of RSS), it was believed in Bruner's time and later that "somewhere in the Nebraska sandhills there might still survive a few of the Lesser species, even though none had been reported by hunters for many years. Mr. A. M. Brooking of Hastings particularly interested himself in this quest. Finally, in the fall of 1924, he was rewarded by receiving 2 fine male specimens of the Lesser Prairie Chicken from Red Willow County, Nebraska." Swenk went on to say that the following fall Brooking enlisted hunters and obtained 5 additional specimens. These are listed above. Brooking's involvement in the collection of the Nebraska specimens raises questions about their provenance.

Of interest is a comment in Bent (1932) that this species had been reported from Nebraska, but that "in the lack of specimen evidence it is thought that the records refer to [Greater Prairie-Chicken]." Evidently Bent was not aware of the specimens discussed above and is referring to additional reports, possibly those mentioned by Bruner and others (1904); the latter involved "a number" shot in Cuming and Washington Cos in the early 1870s and one seen by Bruner near West Point during the winter of 1871–72. It seems that indeed these birds were misidentified Greater Prairie-Chickens. Swenk (Notes before 1925) mentioned that in 1901 Black shot a bird at Cozad that he believed was a Lesser Prairie-Chicken, but it was not preserved.

It is, however, a bird of sandsage prairie, the habitat of the area in which the Nebraska specimens were taken. Thompson and Ely (1989) stated that just prior to the Dust Bowl era of the 1930s this species was especially abundant, at least in Kansas. At one time its range is thought to have extended to southeast Kansas, although reports from Missouri (AOU 1957) have been questioned (Robbins and Easterla 1992).

Wild Turkey

Meleagris gallopavo

STATUS

Common regular resident statewide.

DOCUMENTATION

Specimen

WSC 550, 25 Oct 1980 Holt Co.

TAXONOMY

Prior to its extirpation from the state around 1900, the eastern race *silvestris* occurred in eastern Nebraska. It has been speculated that the southern Great Plains subspecies *intermedia* may have occurred as far north as southern Nebraska (Mathisen 1962).

The currently restocked birds occurring statewide are mostly of the southwestern race *merriami* and the southern Great Plains race *intermedia* (AOU 1957; Mathisen 1962), although some populations may have been established in the 1950s with eastern stock, notably in the Niobrara Valley south of Gordon, along the Dismal River in Hooker Co, and along the Platte River in Lincoln Co (Mathisen and Mathisen 1960a).

DISTRIBUTION AND ECOLOGY

Turkeys are found statewide in riparian woodlands along river and stream valleys; they are probably most numerous in the extensive woodlands of the Missouri River Valley and the Pine Ridge.

Summer

This species was once widely distributed in eastern Nebraska but disappeared as Europeans settled the state. Since about 1960

turkeys have been successfully reintroduced statewide.

Prior to 1900 Wild Turkey was "abundant throughout the wooded portions of the state and even on the adjoining prairies as far west along the Platte and Republican Rivers as McCook and North Platte"; it was also common "in territorial days" along the Elkhorn, Big Blue, and Little Blue Rivers and along the Niobrara to Long Pine (Bruner and others 1904). Swenk (Notes before 1925) noted that turkeys occurred in the Republican Valley in the 1870s near Bloomington in Franklin Co and along Prairie Dog Creek near Orleans. Around 1900 it had become very rare and was limited to heavier woodlands in the Missouri Valley (Bruner and others 1904); it finally disappeared "soon after the advent of white man" (Mathisen and Mathisen 1960b).

Reestablishment began with populations in the Niobrara Valley, along the Dismal River, and near North Platte in the 1950s (Mathisen and Mathisen 1960b). Early releases met with varied success because the birds were at least partially domesticated and of ancestry not well-suited to Nebraska environmental conditions. These birds were likely mixed stock of *merriami* and *gallipavo* (NGP biologists, pers. comm. RSS). In 1958–59 "Merriam's" turkeys, wild-caught in New Mexico, were introduced in the Pine Ridge, and the population grew rapidly to at least 1000 birds by 1962 (Mathisen 1962). During the winters of 1960–61 and 1961–62 additional releases were made in Sheridan, Scotts Bluff, Morrill, Cherry, Blaine, Brown, and Keya Paha Cos, and "Rio Grande" turkeys were released in many other localities from Knox Co westward (Mathisen 1962). Most of these releases were successful.

In eastern Nebraska they were established in the Fontenelle Forest area apparently as a result of a brood released on the neighboring Gifford Farm in 1972, although some birds in the area may have come from various Iowa releases at about that time or even from Missouri birds moving northward (Cortelyou 1967). By the mid-1980s turkeys were widely established in Nebraska and common in many localities.

Little nesting information is available, but females with as many as 14 young have been observed in Jun. A female with 9-week-old chicks was observed 6 Jun in the Pine Ridge (RSS).

High counts include 83 below Keystone Diversion Dam 21 Mar 1998 and 64 in Carter Canyon 29 Jan 1997.

FINDING

Turkeys are easy to locate in deciduous woodlands in Apr–May by the persistent gobbling of toms. During the breeding season these birds are wary and hard to find, but flocks become conspicuous in fall and winter, often feeding in open fields where snow cover is sparse.

Northern Bobwhite

Colinus virginianus

STATUS

Fairly common regular resident southeast, rare northwest and west except for North and South Platte River valleys, where uncommon.

DOCUMENTATION

Specimen

UNSM ZM6093, 19 Jun 1902 Keya Paha Co.

DISTRIBUTION AND ECOLOGY

Bobwhites are most commonly found in brushy areas with adjacent or interspersed grassland or cropland. Nesting requirements include woody cover and weedy areas. As these are cleared through modern farming practices, range extension is limited to riparian corridors; it has been suggested (JGJ; NBR 65:21) that Northern Bobwhites are becoming less common in eastern than in central Nebraska as a result of farming practices.

Since Nebraska is at the northwest edge of the range of this species, numbers are greatest in the southeast, but the species is found throughout the Platte and Republican Valleys in fair numbers. It is uncommon in the central Niobrara Valley (Brogie and Mossman 1983) and irregular in northwest Nebraska (Rosche 1982); the western range limits are

"highly variable" and dependent on weather conditions (Johnsgard 1997), as indicated by increasing numbers northwestward following a succession of mild winters but rapid decline after a single severe winter.

Summer

The range of this species in Nebraska has not apparently changed much in the 20th century; while noting that it formerly was limited to the east, Bruner and others (1904) stated that it was distributed "practically over the entire state, especially along river and creek valleys and about farms where natural shelter occurs." Currently it may be more numerous in parts of the central and western portions of the state, as eastern habitat extent and quality decline (JGJ; NBR 65:21); 6 were at Crescent L NWR 30 Aug 1996, 12 there 31 Oct (NBR 64:112), and another there 9 May 1997.

Ducey (1988) stated that it was a regular nester in the eastern two-thirds of the state, but it is irregular elsewhere. Rosche (1994a) considered it "sporadic" in the Keith Co area, varying from nearly absent in some years to fairly common in others.

Two nests, both found in Jun, contained 16 and 18 eggs. Both were located on the ground, concealed by dense clumps of weeds and grass. One was located in an ungrazed pasture dominated by "ryegrass," and the other was found along the weedy edge of a cornfield.

Winter

CBC data indicate highest populations in the east and south (0.568 and 0.565 birds per party-hour), declining to the north (0.343) and west (0.172). Highest CBC totals include 387 at Norfolk 1989–90, 137 at DeSoto Bend NWR 1981–82, 75 at Kearney 1970–71, 27 at Calamus-Loup 1991–92, and 31 at Scottsbluff 1971–72.

FINDING

Bobwhites are best located in Apr–May by their well-known whistled calls. At other times of the year they can be found by walking brushy riparian areas in southeast Nebraska.

GRUIFORMES

Rallidae (Rails, Gallinules, Coots)

Yellow Rail

Coturnicops noveboracensis

STATUS

Rare casual spring and fall migrant central and east.

DOCUMENTATION

Specimen

UNSM ZM12604, 8 May 1909 South Bend, Cass Co.

DISTRIBUTION AND ECOLOGY

There are few reports, but most are from wet, grassy habitats. Yellow Rails generally do not occur in cattails and tall vegetation but in grasses generally under 0.3 m (1 ft) in height. There are no reports west of Cherry and Keith Cos; there is only one Colorado record, that in Jul 1906 (Andrews and Righter 1992). Yellow Rails are extremely secretive and difficult to flush. Most reports result from the use of dogs, burning, or mowing of grassy habitat in mid-Apr and collision with television towers in fall; these circumstances indicate that this species is more common as a migrant than the few reports suggest.

Spring

Of 14 spring reports, 9 are in the period 26 Apr–12 May, 4 in the period 10–12 Jun (see below), and a rather early report 10 Apr 1974 of 5 in various flooded meadows along Highway 61 in Arthur and Grant Cos (Tremaine 1974).

Documented records are 30 Apr 1973 at Bellevue (Hoffman 1974), a specimen 8 May 1909 (cited above), and one flushed during the first or second week of May 1977 at Pilger Recreation Area (Dave Stage, pers. comm. MB, WRS). Other reports in this period are 26 Apr 1958 L Babcock (NBR 26:55), 30 Apr 1909 (2) Lincoln (Zimmer 1911b), 3 May 1927 Omaha (LOI 23:6), 9 May 1954 Keith Co (Benckeser 1955), 11 May 1979 Gleason Lagoon (Brown 1979), and 12 May 1980 Cherry Co (NBR 48:74).

There are also 4 reports in Jun, leading some to suggest the possibility of breeding in Nebraska (Cink 1973c; Johnsgard 1979), although these may merely be late migrants. The Jun reports are 10 Jun 1915 at Pelican L, Cherry Co (Oberholser and McAtee in Cink 1973c); 10 Jun 1987 south of Bassett in Rock Co, a brief sighting not accepted by the NOURC (Mollhoff 1989a); 11 Jun 1972 Lancaster Co (Cink 1973c); and 12 Jun 1920, a specimen, HMM 1604, collected at Inland (Bray and others 1986). Currently the nearest regular breeding is in northwest Minnesota and North Dakota (AOU 1983).

There are 3 additional old reports noted by Bruner and others (1904) without dates; one was obtained at the Omaha market and another at Bellevue, and a specimen taken at Norfolk was said to be in the Sessions collection.

Fall

Prior to 1982 there were no fall records except for a tentative identification in Frontier Co (see below), but since then there are 8 reports, essentially all in late Sep: tower kill in Boone Co, now WSC 893 16 Sep 1982 (Mollhoff 1983; NBR 51:6); 3 flushed while mowing alfalfa in Dodge Co 19 Sep 1992 (Paseka, pers. comm. NOURC); 3–4 flushed using a dog at Jack Sinn Marsh 21–22 Sep 1986 (Lesick 1987); a tower kill, now UNSM ZM15789, at Waco 21 Sep 1991; Mead 23 Sep 1985; 1 flushed while mowing hay in Dodge Co 27 Sep 1994 (Paseka, pers. comm. NOURC); 3 flushed while mowing native meadow in Dodge Co 29 Sep 1997 (DFP; Brogie 1998); 1 flushed in Wayne Co 1 Oct 1983 (MB, pers. comm. WRS); and a specimen, UNSM ZM15790, collected at Davey "between 15 Sep–6 Oct" 1986. Lingle (1994) noted its occurrence in Aug in the central Platte Valley, but no details were provided. A bird captured by a cat the "first week" of Sep 1935 in Frontier Co was tentatively identified as a Yellow Rail but eaten by the cat before identity was confirmed (NBR 4:13).

FINDING

In late Apr and early May and in late Sep the use

of a dog in grassy areas adjacent to marshes may be successful in locating this species. Occasionally Yellow Rails can be flushed by walking, mowing, or burning these same areas. There is some evidence to suggest (Bent 1926) that Yellow Rails become more active late in the day and early in the evening.

Black Rail

Laterallus jamaicensis

STATUS

Rare casual spring and fall visitor statewide. Hypothetical summer visitor Crescent L NWR.

DOCUMENTATION

Description

22 Apr 1980 Lincoln, Lancaster Co (Ducey 1980a).

DISTRIBUTION AND ECOLOGY

There is little information on this species in Nebraska, although it appears that migrants may occur anywhere in suitable habitat, generally considered to be short sedges or grasses with no more than about 2.56 cm (1 in) of water. All summer reports are from Crescent L NWR, where birds identified as this species were seen in 1995 in short sedges and grass at the outer margins of lakes during a period of an unusually high water table (Larry Malone, pers. comm. WRS).

Spring

In addition to 2 old undated reports, at West Point and at the Omaha Market (Bruner and others 1904), there are 7 spring reports, only 2 documented:

22 Apr 1980 near Lincoln, cited above

25 May 1986 a tape in Knox Co (Brogie and Brogie 1987b)

The other reports are 10 May 1987 near Holstein (Turner 1987); 13 May 1979 Funk Lagoon (Brown 1979); 18 May 1978, a tape that may not be this species, Crescent L NWR (Bray and others 1986); 29 May 1981 Crescent L NWR (NBR 49:42); and 31 May 1995 Crescent L NWR (NBR 63:74).

Summer

Rapp and others (1958) stated, without citing supporting data, that Black Rail "probably breeds in suitable habitat, as Nebraska is well within its breeding range." The only reports for summer are recent ones from Crescent L NWR, where it seems that a breeding population may exist (NBR 63:74), although no birds could be detected in 1997 (WRS, JS). Of 15 total Nebraska reports, 7 are from Crescent L NWR in the period 18 May–6 Sep, including these summer reports: 19 Jun 1995 (NBR 63:70); 11 Jul 1980 (NBR 49:17, 18); 15 Aug 1995, description not accepted by the NOURC although an experienced observer noted a small, black bird with back spots that flew up in front of him (LM; Brogie 1997; NBR 63:70); and 6 Sep 1995 (NBR 63:70).

Nebraska is north of the generally accepted summer range, although there are reports of migrants to the north of this range (AOU 1983).

Fall

The only reports, neither documented, are of 2 birds in Richardson Co in 1873 (Bruner and others 1904) and 20 Sep 1979 at Cunningham L (Cortelyou 1979). In addition, there is a 6 Sep 1995 report at Crescent L NWR (NBR 63:74), possibly a summering bird (see Summer).

FINDING

A likely strategy for locating this species would be to walk shallow sedgy marshes or lake margins at Crescent L NWR (or elsewhere in the Sandhills or Rainwater Basin) during Jun, possibly in conjunction with taped calls, to which this species is somewhat responsive.

Clapper Rail

Rallus longirostris

STATUS

Accidental in winter.

DOCUMENTATION

Specimen

UNSM ZM14120, 30 Jan 1951 near Stapleton, Logan Co (Rapp and Rapp 1951).

Records

This record involved a bird caught in a mink trap about 19 km (12 mi) east of Stapleton in Logan Co. The identification was confirmed by Van Tyne, who determined it to be a "typical Northern Clapper Rail," the Atlantic Coast subspecies *crepitans* (Rapp and Rapp 1951). The specimen has gray cheeks, which do not contrast with the grayish neck coloration (WRS, JGJ).

This is by far the farthest inland this species has been recorded; the closest is a Tennessee record (White 1990).

King Rail
Rallus elegans

STATUS

Rare regular spring visitor statewide. Rare casual breeder east, hypothetical elsewhere. Rare casual summer visitor statewide. Rare casual fall migrant statewide.

DOCUMENTATION

Specimen

UNSM ZM6095, 8 May 1920 Lincoln, Lancaster Co.

DISTRIBUTION AND ECOLOGY

Although there are fewer than 50 reports, available data indicate that this species occurs regularly in spring and early summer and breeds on occasion. Its secretive nature and the lack of concentrated searching for it probably explain the limited number of reports. Most recent reports are from Lancaster Co, the Clear Creek Marshes, and the Rainwater Basin. Habitat loss in the Missouri River valley probably explains the few recent reports there, in contrast to its status in the 1950s and earlier, when Rapp and others (1958) described it as a "rare summer resident in the eastern quarter of the state."

King Rail is a bird of marshes with relatively deep water, where it has a propensity for ditches and waterway edges.

Spring

26, 28, 30 Apr→12, 12, 15 Jun

Of 48 total reports for all seasons, over half are in the period 26 Apr–15 Jun. Arrival is in late Apr, although there are earlier reports 10 Mar 1990 Sheridan Co (AB 44:455) and 2 Apr 1967 Adams Co, both possibly birds that wintered nearby.

Summer

There are no recent breeding records, although there are several reports that indicate that breeding does occur on occasion. Ducey (1988) cited early records from Douglas and Lancaster Cos around 1900, and Swenk (Notes before 1925) noted that it possibly nested at Inland in 1915 and 1916, although it was unclear whether the nest was of King Rail or Virginia Rail. A sighting of 4 near Carter L, Omaha, 28 Jun 1930 is suggestive of breeding (LOI 52:3). An adult with 5–6 young was seen near Fairbury 29 Jul 1940 (NBR 8:94). An adult at a marsh in Lancaster Co 8 Jun 1971 and subsequent sightings there during the summer were suggestive of breeding (Cink 1971), and there were reports from Lancaster Co 20 May and 1 Jul–11 Sep 1971. The presence of up to 5 at Capitol Beach, Lancaster Co, 24 Jul–18 Aug 1984 (Grenon 1990; AB 39:1035, 73) indicated successful breeding. Rosche (1994a) suggested that King Rail may occur year-round in the Clear Creek Marshes, although no midsummer reports later than 6 Jun were listed.

Summer reports without evidence of breeding include a specimen, HMM 2496B (Bray and others 1986), collected at Kearney 5 Aug 1923, 2 at Plattsmouth 5 Aug 1966 (Heineman 1967), 1 in eastern Otoe Co 5–16 Aug 1997 (14 Aug, SJD; Brogie 1998), and 1 that responded to a tape at Facus Springs 4 Sep 1992.

Fall

Reports are few but suggest that departure is in Sep. All reports are: Sep 1898 Ballard's Marsh, Cherry Co (Bruner and others 1904); 9 Sep 1984 Saunders Co; 20 Sep 1900 Nebraska City (Bent 1926); 20 Sep 1992 Clear Creek Marshes (Rosche 1994a); 22 Sep 1890 Lancaster Co (Bent 1926); 23 Sep 1922, a specimen, HMM 2496A, collected at Inland (Swenk, Notes before 1925); 24 Sep 1896 Gresham, York Co (Bent 1926); 1 Oct 1916, a juvenile specimen, HMM 7458, collected at Inland (Swenk, Notes after 1925); "first week of" Oct 1916, an adult, HMM 2496A, collected at Inland (Swenk,

Notes before 1925); 8 Oct 1918, Kearney (Swenk, Notes before 1925); a late sighting at Inland 16 Nov (Brooking, Notes); and a very late sighting at Clear Creek Marshes 17 Dec 1992 (Rosche 1994a). The latter two reports raise the possibility of overwintering (Rosche 1994a).

FINDING

This species is rather easy to attract in May–Jun; best locations are cattail marshes with interspersed grassy waterways or ditches such as Jack Sinn Marsh in Lancaster Co, North Lake Basin in Seward Co, and Deep Well Basin in Hamilton Co.

Virginia Rail

Rallus limicola

STATUS

Uncommon regular spring and fall migrant east and west, common central. Uncommon, locally common, regular breeder north and west, rare south and east. Uncommon local regular winter visitor North Platte Valley.

DOCUMENTATION

Specimen

UNSM ZM6100, 14 Jun 1902 Marsh L, Cherry Co.

DISTRIBUTION AND ECOLOGY

Like other rails, this species' secretive nature probably does not allow an adequate assessment of its abundance. It prefers marshes that possess thick growths of emergent vegetation as well as mudflats where it probes for food. Johnsgard (1979) indicated that Virginia Rail and Sora occupy similar habitats, but Virginia Rail tends to nest in cattails and probe for food while Sora nests in sedges and picks food from the surface.

Virginia Rail probably breeds statewide, although lack of habitat limits occurrence in the east, southwest, and southern Panhandle to a few locations.

Spring

10, 12, 15 Apr→21, 23, 27 May

Late dates above are from locations where breeding was not reported. There are earlier reports 24 Mar 1992 Sioux Co and 6 Apr

1957 Cherry Co. Arrival is in mid to late Apr, while migration peaks in the first half of May.

High counts include 7 at Clear Creek Marshes 23 Apr 1977 (Rosche 1994a) and 6 at Cunningham L 7 May 1995.

Summer

Breeding numbers are highest by far in the Sandhills, where the population at Crescent L NWR was estimated at 2000 in early Jun 1978 (AB 32:1180). Brown and others (1996) noted that up to 8 nests were located in a marsh at Ackley Valley Marsh and another at the junction of Highways 61 and 92 in Keith Co in 1994–95. It apparently breeds regularly east to Knox Co and west to Sioux Co locations such as University L (Rosche 1982). The Rainwater Basin is within its breeding range (Johnsgard 1979; Lingle 1994), but the only published breeding record from there is of 2 adults and 4 young at Funk Lagoon 4 Jul 1998 (LR, RH). There are also recent summer reports from Funk Lagoon 9 Jul 1995 and 17–19 May and 30 Jun 1996. Swenk (Notes before 1925) mentioned a nest possibly of this species at Inland in 1915.

There are a few breeding records in the east: 1972 and 1984 Lancaster Co (Ducey 1988; Cink 1975a), 1982 Douglas Co (Bennett 1983), and 1995 Jack Sinn Marsh (Gubanyi, pers. comm. JGJ). In addition, there are several summer reports: 2 Jun 1999 Lancaster Co, 6 Jun 1998 (3) Crystal Cove L (BFH), 14 Jun 1983 Lancaster Co, 20 Jun 1987 Pierce Co, Jun–Jul 1968 Douglas-Sarpy Cos, 1–20 Jul 1970 Lancaster Co, 6 Jul 1984 Douglas-Sarpy Cos, and 23 Jul 1988 Lancaster Co.

In the southwest it was reported 2 Jun 1992 from Dundy and Chase Cos and 12 Jun 1999 at Rock Creek L, and there is an observation of an adult with downy young 9 Aug 1987 just west of Sutherland Res in Lincoln Co. There are no summer reports from the Panhandle south of Scotts Bluff Co and only these from the latter: 17 Jun 1980 and 11 Jul 1985.

Clutches have been found 15 May–24 Jun.

Clutch size is 4–9 eggs. Young have been seen throughout Jul and downy young as late as 29 Aug at Valentine NWR (JS).

Fall

8, 12, 15 Aug→12, 14, 18 Oct

Early dates above are from locations where breeding was not reported. There are later reports in Nov, most along the North Platte Valley, presumably of birds attempting to winter (see Winter). It is difficult to assess fall arrival and peak migration timing. That large numbers may occur at times is indicated by a postbreeding estimate of 5000 in Aug 1977 at Crescent L NWR (AB 32:225). Most leave by late Sep, as there are few Oct reports.

Winter

In recent years it has become clear that this species winters regularly at cattail seeps in the North Platte Valley (SJD; Rosche 1994a). There are multiple reports from Facus Springs, the Clear Creek Marshes near Lewellen, below Kingsley Dam, and along the north side of L Ogallala. Late reports from other locations are of one at Crescent L NWR 1 Dec 1996 and one on the Boone Co CBC 19 Dec 1997.

High counts include 16 at Facus Springs 10 Jan 1999, 10 near Lewellen 22 Feb 1998, and 8 there in late Dec 1996.

FINDING

Virginia Rails usually respond aggressively to tapes, although they rarely come into the open like King Rails. Calls are similar to those of King Rail, but differences are apparent with practice. Good numbers occur at Crescent L NWR and Valentine NWR.

Sora

Porzana carolina

STATUS

Fairly common regular spring and fall migrant statewide. Fairly common, locally common, regular breeder north-central, uncommon elsewhere.

DOCUMENTATION

Specimen

UNSM ZM6106, 17 Sep 1902 Lancaster Co.

DISTRIBUTION AND ECOLOGY

This species occurs in most marshes in migration, and in the breeding season it is found in more permanent marshes that contain significant emergent vegetation (see Virginia Rail).

Spring

10, 14, 14 Apr→summer

Most arrive in late Apr, and peak movement is in the first 2 weeks of May.

Summer

As with Virginia Rail, breeding may occur statewide, although densities are low in marginal habitat. The highest breeding densities are in the Sandhills; the breeding population at Crescent L NWR was estimated at 2500 in early Jun 1978 (AB 32:1180) and about 1000 in early Jun 1979 (Zeillemaker, pers. comm. RSS). Breeding occurs at certain locations in the Platte River valley; it was considered an "uncommon summer resident" at the Clear Creek Marshes (Rosche 1994a), and it was listed as a nesting species in the Hall Co area (Lingle 1994). The Rainwater Basin was included in its breeding range by Johnsgard (1979), although there are no breeding records from there; there are, however, several summer reports suggestive of breeding: 12 Jun 1991 Kearney Co, 17 Jul 1999 (3) eastern Rainwater Basin, 19 Jul 1997 Tamora Basin, 20 Jul 1986 and 28 Jul 1986 Polk Co, 31 Jul 1994 (7) Funk Lagoon, and Aug 1999 (about 17) eastern Rainwater Basin. Swenk (Notes before 1925) noted that it was a common breeder at Inland.

Eastern breeding populations appear to have declined, except perhaps in the northeast in Knox and Dakota Cos. Bruner and others (1904) considered Sora a common breeder in the east, but while there are no recent breeding records from Missouri River counties, there are a few summer reports: 3 Jun 1986 Dakota Co, 6 Jun 1983 Dakota Co, 12 Jun 1993 Dakota Co, 18 Jun 1999 Cass Co, Jun–Jul 1968 Douglas-Sarpy Cos, Jun–Jul 1987 Dakota Co, 23 Jul 1989 Knox Co, and 8 Aug 1999 Dakota

Co. It bred in Lancaster Co in 1973 and 1981 (Bennett 1974, 1982), and there are several Jun reports from there in recent years.

While there are at least 4 reports for Jun from Scotts Bluff Co, there are none from the southern Panhandle or the southwest.

Fall

summer→23, 24, 31 Oct

Migrants may arrive as early as mid-Aug, but peak movement is mid-Sep through early Oct, after which reports decline markedly. There are later reports 18 Nov 1972 Adams Co and 22 Nov and 26 Nov 1976 Sioux Co (Rosche 1994a), but no Dec reports, in contrast to Virginia Rail. There was a postbreeding estimate of 5000 at Crescent L NWR in 1977 (AB 32:225); in 1978 about 4000 were estimated present in early Sep, but nearly all had departed by 14 Sep (AB 33:192).

FINDING

During early May Soras can be found at almost any marsh. They respond to tapes. Clapping (or slamming the car door) may elicit a few calls, usually the "whinny."

Purple Gallinule

Porphyrula martinica

STATUS

Accidental in spring.

DOCUMENTATION

Description

2 May 1946 Pauline, Adams Co (Ager 1946).

Records

There is 1 documented record, a description of one seen in a wooded backwater of the Little Blue River near Pauline in Adams Co (cited above). There are 2 other reports, neither with details: Bruner reportedly saw one near West Point Jun or Jul 1884 or 1885 (Bruner and others 1904), and one was reportedly seen in Gage Co on the rather early date of 28 Mar 1962 (NBR 38:50).

COMMENTS

This species is a "notorious vagrant," with most Midwest records Apr–Jun and most birds found climbing in the understory of wooded

ponds (Kent and Dinsmore 1996). There are at least 10 records for Kansas in the period 4 Apr–17 Jun (Thompson and Ely 1989). On the other hand, there are only single records for Colorado and Wyoming, both in late summer or fall: one was in southwest Colorado 6–7 Aug 1978 (Andrews and Righter 1992), and another, a juvenile, was near Laramie 24 Sep 1986 (Scott 1993). Thus one might look for this species in the Republican or Little Blue Valleys in May; the single documented Nebraska record fits this pattern.

Common Moorhen

Gallinula chloropus

STATUS

Uncommon casual spring migrant south and east. Uncommon casual breeder east. Uncommon casual summer visitor east. Hypothetical in fall.

DOCUMENTATION

Specimen

UNSM ZM10394, 4 May 1911 Exeter, Fillmore Co.

DISTRIBUTION AND ECOLOGY

This species nests in cattail marshes, where it prefers to walk rather than swim. Most recent reports are from marshes with dense vegetation in southeast Nebraska.

Spring

Of the 31 or so total reports for Nebraska, 12 are breeding or summer reports. Of the remaining 19, 15 are in spring, only 1 is in fall, and 3 are undated. The spring reports are mostly in the period 24 Apr–19 May, with earlier undocumented reports 23 Mar 1975 Lancaster Co and 3 Apr 1938 Webster Co and a later undocumented report 30 May 1935 Logan Co.

Summer

Around 1900 Common Moorhen was "quite a common but locally distributed summer resident in southeast Nebraska" (Bruner and others 1904). There are old breeding records from Dakota, Douglas, and Lincoln Cos (Ducey 1988), and Bruner and others (1904) stated that it bred in Cherry Co. Breeding is documented for Douglas Co by a set of eggs, UNSM ZM8236, collected 4 Jul 1897 at Cutoff L (Carter L), Omaha.

Since these older records of breeding, the most recent of which was prior to 1957 (Youngworth 1957), the only breeding records are 5 clustered in the mid-1980s and records in 1997 and 1998 when wet conditions resulted in suitable cattail habitat. In 1984 Garthright discovered nesting birds in 2 locations in Lancaster Co, Capitol Beach Marsh and a marsh at 27th Street and Bluff Road (Garthright 1984). In 1985 a nest with eggs was reported in Lancaster Co (Bennett 1986), and it was present 12–22 Jun. Also in 1985 Garthright found a nest with eggs at Weis Lagoon (Garthright 1985a). In 1987 Huser found a pair at Crystal Cove L 5 May which bred and were last seen 29 Aug (NBR 55:54–55, 56:14–17; Mollhoff 1989a). An adult was found 5 Aug (SJD), and a juvenile was accompanying it in eastern Otoe Co 10 Aug–13 Sep 1997 (JS, SJD, JGJ, MB). At least 2 broods were noted at LaPlatte Bottoms in Sarpy Co in 1998, where 2 adults and 5 young were seen 6 Aug and an adult with 7 young on 23 Aug. A juvenile was seen in Clay Co 30 Aug 1998.

Most other breeding season reports are from Douglas-Sarpy Cos; no details were published regarding sightings 2–23 Jun and 26 Jul–29 Sep 1959, and a single bird was at the marsh in Fontenelle Forest 15–19 Jun 1994. In addition, one was at North Lake Basin 7 Jun 1996 (Brogie 1997).

Fall

The only fall report not associated with breeding birds is an undocumented report from Cherry Co 29 Aug–3 Sep 1933 (NBR 1:137). Breeding birds in Lancaster Co in 1984 (see above) and Dakota Co in 1987 were last reported 9 Aug and 29 Aug, respectively.

FINDING

When present, Common Moorhens are not difficult to find, as their red bills and habit of walking rather than swimming helps separate them from the ubiquitous American Coots.

Best time to look is late Apr–early May in cattail marshes in southeast Nebraska and the eastern Rainwater Basin.

American Coot

Fulica americana

STATUS

Abundant regular spring and fall migrant statewide. Common, locally abundant, regular breeder north-central, fairly common elsewhere. Uncommon local regular winter visitor.

DOCUMENTATION

Specimen

UNSM ZM12607, 4 Aug 1911 near Henry, Scotts Bluff Co.

DISTRIBUTION AND ECOLOGY

This species breeds in cattail marshes with areas of open water, notably in the Sandhills. It breeds elsewhere in the state where such habitat exists. Migrants concentrate on lakes and reservoirs, where they generally remain close to shore in tightly packed flocks, although any body of water may host migrants.

Spring

20, 20, 21 Feb→summer

Migrants begin to arrive in late Feb, but peak numbers are not achieved until Apr, declining into late May, when essentially summering birds remain.

High counts include 3000 in Dakota Co 28 Mar 1998, 2400 at Crescent L NWR 18 Apr 1998, and 1600 there 23 Apr 1996.

Summer

There are breeding records statewide (Ducey 1988), with lowest numbers probably in Missouri River counties and the southern Panhandle where habitat is sparse. Largest breeding populations occur on Sandhills lakes, but any pond or lake with significant emergent vegetation such as cattails provides breeding habitat. Adults have been seen with young on marshy oxbows of the Loup River system, in large Sandhills marshes, and on very small farm ponds with limited marsh y edge. The Rainwater Basin supports a significant breeding population in wet years, but water levels are rather more variable there than in the Sandhills. Numbers of breeding birds vary considerably from year to year in areas of marginal habitat due to varying water levels.

Nests with clutches have been found during Jun. Thirty-two nests had 4–12 eggs. Nests are bulky, floating platforms of dead and rooted reeds and other aquatic vegetation. They are often located in sedges, cattails, and bulrushes. Young were fledged by 5 May 1997 at Wood Duck Area.

Fall

summer→4, 5, 9 Jan

Migrants begin to appear in late Aug, and flocks become noticeable from then on. Large flocks occur at any time Sep–Nov, but numbers decline rapidly into Dec and Jan (Rosche 1994a).

High counts include 6315 at Crescent L NWR 15 Sep 1978 (Zeillemaker, pers. comm. RSS), 5775 there 4 Oct 1996, 5000 at L Ogallala 10 Sep 1977 (Rosche 1994a), and 5000 at Crescent L NWR 30 Aug 1998.

Winter

Numbers are lowest mid-Jan to early Feb, with most reports in recent years from Cedar and Keith Cos.

High counts include 586 at L Ogallala 23 Jan 1999, 17 there winter 1997–98, and 10 in Cedar Co 12 Feb 1995.

FINDING

Coots are ubiquitous statewide during migrations, Apr–May and Sep–Nov, and in the Sandhills during the summer months.

Gruidae (Cranes)

Sandhill Crane

Grus canadensis

STATUS

Abundant regular spring and fall migrant central, uncommon west, fairly common casual east. Rare casual breeder Rainwater

Basin. Rare casual summer visitor statewide. Rare casual winter visitor central Platte Valley.

DOCUMENTATION

Specimen

UNSM ZM7836, 19 Mar 1959 Overton, Dawson Co.

TAXONOMY

About 80% (Lingle 1994) of the Sandhill Cranes that migrate through Nebraska are of the subspecies *canadensis*, also referred to as Lesser Sandhill Crane, which breeds in arctic Canada and Siberia. Some 15% (Lingle 1994) are the subspecies *rowani*, which breeds in central Canada (Cramp and Simmons 1980; Tremaine 1970) and is also known as Intermediate or Canadian Sandhill Crane. The remaining 5% (Lingle 1994) are of the southern Canadian and United States subspecies *tabida*, also known as Greater Sandhill Crane, which at one time apparently nested in Sandhills marshes (see Summer). Most *rowani* migrate to the east of the population of *canadensis*, with some overlap (Tremaine 1970).

DISTRIBUTION AND ECOLOGY

Lingle (1994) discussed the habitat requirements and habits of this species during migration in central Nebraska. Largest numbers occur between Grand Island and Kearney, a narrow migration corridor very important as a spring "staging area," a practice that may date from the last glaciation if not earlier (Sharpe 1978). At night the birds roost in the shallow Platte River and feed in wet meadows and grainfields generally within 8 km (5 mi) of the Platte River, although some may wander as far as 32 km (20 mi) to forage. On average a crane spends about 28 days in the area, gaining about 0.23 kg (0.5 lb) of fat.

Water-development projects along the Platte and South Platte Rivers during the 20th century have led to as much as a 90% loss in Platte River stream flows. This in turn has led to a narrowing of channels and serious vegetational encroachment through natural succession. The result is considerably less shallow-water habitat for crane roosting and thus concern for Sandhill Crane endangerment. There is no evidence that cranes will modify their nighttime roosting habits.

Spring

24, 24, 25 Jan→15, 18, 20 May

The first arrivals appear in some years in late Jan; in 1994 they were "regular" after 28 Jan, and in 1998 about 10 000 were present 18 Feb (*Omaha World Herald* 18 Feb 1998). These birds often retreat southward after arrival in the face of snowstorms and severe cold, however. There is an earlier report 18 Jan 1999 in the central Platte Valley and a later report 25 May 1997 at Funk Lagoon. Peak numbers are reached 20 Mar–5 Apr, when over 250 000 may be present between Grand Island and Kearney (Lingle 1994). Most depart by the end of Apr.

Johnsgard (1997) noted that a total of about 400 000 cranes pass through Nebraska. A concentration of about 100 000 occupies an area east of Hershey in Lincoln Co, and Rosche (1994a) stated that a population using the meadows in the Clear Creek Marshes area has increased in recent years to about 14 000. Recent years have also seen a population using the Harlan Co Res area (Rapp and others 1958; Johnsgard 1997).

In the northwest it is a "common to very common spring transient" (Rosche 1982), but it is only casual in the east. There are these eastern reports: 7 Feb 1999 Branched Oak L; 28 Feb 1976, 2 Mar 1974, and 20 Apr 1975 Lancaster Co; 18 Feb 1981 and 23 Feb 1974 Douglas-Sarpy Cos; 5 Apr 1980 and 6 Apr 1985 Saunders Co; 3 Apr 1955 Saline Co; 14 Apr 1989 Cedar Co; and 20–23 Apr 1998 Wood Duck Area (DH).

Summer

Sandhill Cranes nested in Nebraska until about 1900, when Bruner and others (1904) stated that it was still breeding "sparingly in the Sandhills" in Cherry and Holt Cos; prior to that time it had been a "common breeder in the marshes of the state." Bruner (1902) related that as a boy in 1883–84 in southern Holt Co he often encountered young Sandhill Cranes, noting that young "only a couple of days old and covered with fluffy reddish down were already almost as large as a common domestic fowl." The draining of the vast marshes that once existed in that region, coupled with hunting, contributed to the species' extirpation as a breeder in Nebraska. Since that time there have been a few reports in the summer months, most relatively recent, raising the possibility that breeding may again occur, as it has in Iowa (Kent and Dinsmore 1996). Breeding was indeed finally confirmed when a pair with 2 downy chicks were found at Harvard Marsh 29 May 1999 (JGJ). Breeding probably occurred in the area in 1996, when a pair was at Harvard Marsh 14–28 Jul, and in 1998, when 2 adults with 2 immatures were seen 16–23 Aug 1998 at a small private basin in eastern Clay Co, where an adult had been observed Aug 1994 (JGJ). An adult was also seen at Harvard Marsh 13 Jul 1997. Activity suggestive of nesting was observed in a meadow along the Loup River near Columbus 13 Jul 1995, when an adult exhibited flight displays and vocalizations (NGP), and 2 adults at Krause Basin, Fillmore Co, 25 Jun–9 Jul 1996 were seen repeatedly flying in wide circles, returning to the basin, and acting quite agitatedly as observers searched unsuccessfully for young (NGP).

Additional summer reports are "summer" 1967 Garden Co, 18 Jun 1991 Hall Co, 27 Jun 1996 Buffalo Co (FN 50:966), 29 Jun 1984 Howard-Hall Cos, 29 Jun 1989 Lincoln Co, 30 Jun 1995 (2) in an area east of Lexington "for weeks" (NGP), and 20 Jul 1999 Massie Lagoon.

Fall

31 Aug, 2, 3 Sep→22, 25, 26 Nov

Fall migration is more leisurely than in spring, with far fewer present at any one time in central Nebraska; according to Lingle (1994), peak numbers are below 10 000, usually noted in the second half of Oct. Some 90% of the reports are in Oct, 60% in the period 5–20 Oct. Most depart by mid-Nov although on occasion large numbers have lingered well into Dec, notably 5000 on the Grand Island CBC 15 Dec 1990 and 200 on the Kearney CBC 21 Dec 1971. There are surprisingly few Dec reports, with these additional: 3 Dec 1962 Webster Co, 6 Dec 1997 (immature) Swanson Res, 10 Dec 1991 Dawes Co, 11 Dec 1989 Howard-Hall Cos, 16 Dec 1978 (2) Scottsbluff CBC, 26 Dec 1977 Lincoln Co, 26 Dec 1994 (20) Buffalo Co, 27 Dec 1986 (1) Kearney CBC, and 26 Nov–"winter" 1964 Lincoln Co.

In the northwest it is "abundant to very abundant" (Rosche 1982) but only casual in the east, with these few reports: Douglas-Sarpy Cos 11 Oct 1977; 7 in Douglas Co and 50 at 2 locations in Lancaster Co 31 Oct. 1996; 77 flying over Blair 2 Nov 1996; Washington Co 22 Nov 1980; Lancaster Co 5 Nov 1991, 6 Nov 1984, and 9 Nov 1986; and, following very strong northeasterly winds, 1200 in eastern Lancaster Co and 336 over Blair 11 Nov 1998.

High counts include 30 000 in Morrill Co 7 Oct 1994 and 5416 in Scotts Bluff Co 16 Oct 1999.

Winter

Wintering has occurred in the central Platte Valley (Lingle 1994). In addition to the few Dec reports (see above), there are these additional reports in the period 1–23 Jan: 7 Jan 1983, 8 Jan 1974, 14 Jan 1962, 17 Jan 1975, and 17 Jan 1979 Lincoln Co; 11 Jan 1994 Buffalo Co; and 15 Jan 1995 (90) Kearney Co. Spring migrants may arrive during the latter half of Jan (see Spring).

FINDING

A trip to Grand Island and a drive along roads to the south of the Platte River between Grand Island and Kearney in Apr or Oct should yield this species; many tours are available, and blinds can be reserved for close observation of cranes roosting in the Platte River (Lingle 1994).

Common Crane

Grus grus

STATUS

Rare casual spring migrant central Platte Valley.

DOCUMENTATION

Photograph

31 Mar–1 Apr 1972 Elm Creek, Phelps Co (Tremaine and others 1972).

Records

There are 5 records, all documented, in the narrow period 6 Mar–1 Apr. One in Phelps Co 31 Mar–1 Apr 1972 was photographed (cited above). Another 16–25 Mar 1974 in northern Kearney and southern Buffalo Cos was photographed (NBR 42:63–64). A bird seen several times 25–31 Mar 1972 between Hershey and North Platte is generally assumed to be a different bird than that seen near Elm Creek at about the same time (Tremaine and others 1972). One was seen in southern Hall and northern Adams Cos 30–31 Mar 1996 (Lingle 1996; Brogie 1997; NBR 64:49; FN 50:297). A pale individual was in northeast Kearney and southeast Buffalo Cos 6–26 Mar 1999 (Silcock and Dinsmore 1999).

COMMENTS

Common Crane is an Eurasian species, breeding within 960 km (600 mi) of (Lesser) Sandhill Cranes in northeast Siberia. The Nebraska records constitute a significant portion of the 12 North American records (Silcock and Dinsmore 1999; ABA 1996). Of the others, 3 are in Alberta, 20 Mar 1958, 19 Sep 1958, and 11–20 Dec 1957; 1 in Alaska 24 Apr–10 May 1958; another in Alaska September 1998; and the other in New Mexico 10 Mar 1961.

FINDING

Flocks of Sandhill Cranes in the central Platte Valley should be scrutinized in late Mar for Common Crane. At this time most of the Sandhill Crane population is concentrated in a small geographic area, and the chance of finding a single Common Crane is enhanced.

Whooping Crane

Grus americana

STATUS

Rare regular spring and fall migrant central, rare casual elsewhere. Rare casual summer visitor central and west.

DOCUMENTATION
Specimen
HMM 3823 (2), spring 1899 Grand Island, Hall Co.
DISTRIBUTION AND ECOLOGY
The entire naturally occurring population, currently about 150 birds, migrates through Nebraska en route between its breeding grounds in northwest Canada's Wood Buffalo Park and the winter area at Aransas NWR in coastal Texas. The historic record indicates that the species has always been rare (NBR 11:1), numbering perhaps no more than several hundred around 1900, decreasing to about 100 in 1934 and to an all-time low of 21 in 1951 (Allen 1952).

Most migrants in Nebraska have been reported within counties adjacent to the central Platte River, essentially from Lexington to Grand Island (Johnsgard and Redfield 1977). In this area resting and foraging birds utilize wetlands, much as they do in Kansas, where it is "primarily a bird of the prairie marshes" (Thompson and Ely 1989). Favored locations include Rainwater Basin marshes, the Table Playa wetlands of Custer Co, Loup River system wetlands, and the Niobrara River. However, roosting sites on the Platte River are also of importance, as indicated by Lingle and others (1986) and Lingle (1994). Lingle (1994) noted that critical habitat for river roosts has been designated in the 85 km (53 mi) stretch between the Shelton and Lexington bridges across the Platte River. River roosts require a wide channel, unobstructed view for some distance, and absence of vegetation and human interference (Lingle and others 1986).

Spring
3, 5, 6 Mar→15, 16, 17 May
Migration peaks 1–20 Apr (Lingle 1994; Johnsgard 1997); some 80% of the reports are in the narrow period 8–19 Apr. There are earlier reports 10 Feb 1975, a young bird that had wintered with Sandhill Cranes in Oklahoma, arrived in Nebraska, but retreated southward as the weather worsened (NBR 63:83); a subadult 14 Feb–28 Apr 1998 Hall Co

(*Omaha World Herald* 18 Feb 1998; B); and 28 Feb–17 Mar 1989 Hall Co. Later reports, in May mostly, generally involve younger birds, which sometimes may not migrate the entire distance to the breeding grounds (see Summer). These reports include: 14–28 May 1988 Cherry Co, 20 May–18 Jun 1950 Phelps Co, 23–24 May 1989 Garden Co, 24–26 May 1984 (8) Hamilton Co, 27 May 1950 Dawson Co, and 28 May 1995 Keya Paha Co.

Of some 270 reports examined, about 60% were from Buffalo, Hall, Adams, Kearney, Phelps, Dawson, Custer, and Lincoln Cos. In all, 55 counties had reports, with fewest in the east and Panhandle. Most easterly are 16 Mar 1890 Richardson Co (Bent 1926), 11 Apr 1982 Nemaha Co, 16 Apr 1980 Cass Co, 7 May 1969 Douglas-Sarpy Cos, 25–27 Mar 1989 Lancaster Co, and 30 Apr 1989 Dodge Co, while most westerly are 20–22 Apr 1987 Sioux Co and 1 Apr 1996, 26 Apr 1978, and 27 Apr 1992 Scotts Bluff Co.

High counts include 56 in Buffalo Co 14 Apr 1920 (this must have been virtually the entire population at the time; Brooking, Notes), 30 in Hall Co 10 Mar 1990, 10–12 in Howard Co 14 Apr 1989, 9 in Keya Paha Co 16 Apr 1993, and 9 in Polk Co 14–17 Apr 1993.

Summer
Occasionally single birds remain in summer, probably immatures that did not migrate to the breeding grounds. Age at first breeding is around 5 years (Terres 1980). Reports include one that remained near Whitney 15 Jul–11 Oct 1991, 16–23 Jul 1950 Morrill Co, and 5 Aug 1990 Perkins Co.

There is no evidence that Whooping Cranes ever bred in Nebraska (Ducey 1988), but both Bruner and others (1904) and Bent (1926) suggested the possibility; it was known to have bred in northern Iowa until 1894 (Kent and Dinsmore 1996).

Fall
22, 24, 25 Sep→18, 21, 23 Nov
Peak movement is 11–31 Oct (Lingle 1994; Johnsgard 1997). There is an earlier report 16

Sep 1990 Kearney Co and a later report 3 Dec (Lingle 1994). Of some 191 reports examined, 81 (42%) are from the counties listed above as most utilized in spring, suggesting that preference for these counties is not as strong in fall. There are fewer easterly reports in fall, the only reports 27 Oct 1988 and 16 Nov 1988 Polk Co and 12 Nov 1998 York Co. In the Panhandle there are about as many fall reports as for spring, most westerly 24 Sep 1979 Sioux Co and 17 Oct 1999, 21 Oct 1990, and 7 Nov 1987 Scotts Bluff Co.

High counts include 14 in Buffalo Co 2 Nov 1989, 11 in Phelps Co 10 Nov 1994, and 10–12 in Dundy Co 12 Oct 1988.

FINDING

Most birds do not stay long in Nebraska, although one remained in spring 1987 for 34 days (Lingle 1994). The best chance is to check favored wetlands such as the Funk Lagoon area during migration peaks in mid-Apr and late Oct. Because these birds are critically endangered, wildlife authorities are reluctant to disclose whereabouts of resting migrants.

CHARADRIIFORMES

Charadriidae (Plovers)

Black-bellied Plover

Pluvialis squatarola

STATUS

Uncommon regular spring and fall migrant statewide.

DOCUMENTATION

Specimen

UNSM ZM6127, 24 Oct 1897 Lancaster Co.

DISTRIBUTION AND ECOLOGY

This species is found on mudflats, in flooded fields, in stubble fields with standing water, and, more often than many other shorebird species, on sandy shorelines and beaches.

Spring

4, 10, 10 Apr→14, 19, 21 Jun

There are few reports during Apr, and peak migration occurs mid-May; 77% of the reports (N = 98) fall in the period 8–27 May. Based on high counts from several years, the largest numbers are found on and near 19 May. There is an early report of "thousands on the Platte River" 21 May 1883 (Bruner and others 1904). Individuals have lingered into Jun on at least 6 occasions.

High counts include 75 at Clear Creek Marshes 19 May 1985 (Rosche 1994a), 73 in the Rainwater Basin 15–16 May 1998, and 68 at Theesen Lagoon 18 May 1996.

Fall

25, 27, 28 Jul→8, 9, 11 Nov

Adults are observed from the end of Jul through Aug, with 60% of the reports (N = 25) 12–20 Aug. There is an earlier report 3 Jul 1980 Lincoln Co. Adults are always scarce during the migration south. Juveniles are more numerous than adults and are observed from Sep through Oct and casually into early Nov.

High counts include 28 juveniles at L North 17 Sep 1995, 19 adults at L Babcock 15 Aug 1998, and 17 juveniles there 2 Oct 1998.

COMMENTS

In Apr observers need to identify *Pluvialis*

plovers carefully. Molting American Golden-Plovers can resemble Black-bellied Plovers, resulting in occasional reports of flocks of the latter in Apr, an occurrence not to be expected.

FINDING

In spring observers should look for Black-bellied Plovers at wetlands in the Rainwater Basin or the L McConaughy area around 19 May. In fall check L North in Sep, where the species has been observed somewhat regularly.

American Golden-Plover

Pluvialis dominica

STATUS

Uncommon regular spring and fall migrant east and central, rare west.

DOCUMENTATION

Specimen

UNSM ZM6121, 23 Sep 1916 Lancaster Co.

TAXONOMY

The two populations of North American golden-plovers, formerly considered one species, are now each accorded specific status (AOU 1995), the American Golden-Plover and the Pacific Golden-Plover. The possibility exists, especially in fall, that a Pacific Golden-Plover could occur in Nebraska.

DISTRIBUTION AND ECOLOGY

This species is found on mudflats, on muddy shorelines, and in wet, plowed, and flooded fields and burned-off lowland grassland. It is generally restricted to eastern Nebraska; west of Phelps Co there are fewer than 20 reports. Prior to the late 1800s the species was "abundant, in flocks of hundreds," and by the early 1900s, although greatly reduced in numbers, it was still considered "rather common" (Bruner and others 1904). The species faced some hunting pressure during this time, but whether hunting was a major factor in reducing the population is not conclusively known. The population may have been reduced by the series of factors that Banks (1977) suggested may have also been responsible for the decimation of its congener, Eskimo Curlew.

Spring

22 Mar, 1, 2 Apr→25, 31 May, 5 Jun

There is a later report, a bird in alternate plumage at Crescent L NWR 13 Jun 1997 (WRS, JS), only the second Jun record.

Arrival is in the first half of Apr, and peak migration occurs toward the end of Apr and into the first half of May. Even though much of the migration occurs before most Black-bellied Plovers have arrived in the state, the largest flocks appear in mid-May, as was noted in spring 1997 (Jorgensen 1997) and for Iowa (Kent and Dinsmore 1996). The majority of the few western reports are in spring.

High counts (modern) include 273 at Freeman Lakes 17 May 1997, 206 at Real Basin 17 May 1997, and 113 in the Rainwater Basin 9–10 May.

Fall

11, 11, 12 Jul→11, 18, 20 Nov

Like the Hudsonian Godwit and White-rumped Sandpiper, American Golden-Plovers migrate through the interior in spring and along the East Coast in fall. Unlike the other two species, however, small numbers of American Golden-Plovers regularly migrate through the interior in fall. There are about 55 such reports, many recent. The majority of fall migrants are juveniles, but adults do occur. There are only 9 reports for Jul and Aug, presumably adults, including an injured bird at Kissinger Basin 1–7 Aug 1999 (JGJ). There are at least 2 observations (JGJ) of pairs of molting adults each accompanied by 1–3 juveniles during Sep, which appear to be family groups, and an adult with a juvenile was near Sinninger Lagoon in a mowed field 5 Sep 1998. Almost all reports are in Sep–Oct, especially Sep.

High counts include 116 in the eastern Rainwater Basin 25 Sep 1999, 79 at L Babcock 20 Sep 1998, and 29 at Tekamah SL 28 Sep 1997.

FINDING

The best chance of finding this species is during spring. Observers should search wetlands in the Rainwater Basin, Jack Sinn Marsh, and flooded fields (when present) during the latter half of Apr or early May. In fall the same areas mentioned above should be checked, primarily during Sep.

Snowy Plover
Charadrius alexandrinus
STATUS
Rare regular spring visitor central, rare casual elsewhere. Accidental breeder northeast. Rare casual summer visitor central and west. Rare casual fall migrant statewide.
DOCUMENTATION
Specimen
UNSM ZM6114, 16 May 1903 Lincoln, Lancaster Co.
DISTRIBUTION AND ECOLOGY
In Nebraska this species has been observed on mudflats, river sandbars, and sandy beaches. Its range on the Great Plains is generally south of Nebraska.
Spring
28 Mar, 3, 6 Apr→26, 28, 30 May
Through 1999 there are about 26 reports, 17 of which are documented. The reports suggest a pattern of overmigration; almost all reports are in May. Careful coverage of the Rainwater Basin during the 1990s has produced 14 of the 17 documented records, all of which involve single individuals observed late Apr through mid-May (JGJ; Dinsmore 1996c). An additional undocumented report from Adams Co 25 Apr 1957 and the specimen cited above fit this late Apr through mid-May pattern.
Easterly reports are 28 Mar 1976 Lancaster Co, 3 Apr 1993 Cass Co, and 16 May 1903 Lancaster Co (specimen cited above), while westerly reports are 21 May 1980 Sioux Co, 26 May 1990 Scotts Bluff Co (Grenon 1991), and 28 May 1976 Dawes Co (Rosche 1982).
Summer
Its occurrence in summer in Nebraska is peripheral, but there is a single record of breeding: a nest with 3 eggs attended by 2 adults was found adjacent to the Missouri River near Ponca 10 Jun 1998 but later destroyed by high water (Greg Pavelka, pers. comm. John Dinan, NPG). The normal North American summer distribution is primarily southwestern, with regular breeding in central and southwestern Kansas (Thompson and Ely 1989) and southern Colorado (Andrews and Righter 1992). A recent breeding record exists for southwestern Wyoming (Scott 1993).
There are 7 additional summer reports, only 2 of which are documented, but all are probably correct. The documented records are of 2 observed in Buffalo Co 25 Jul 1983 (Lingle and Labedz 1984) and 1 at Funk Lagoon 23 Jun 1997 (BS; Brogie 1998). In addition, 1 or 2 were observed at Funk Lagoon 20–29 Jun 1974 (Bliese 1975b). The remaining 4 reports are from L McConaughy during a period when the water level was very low, exposing extensive sandy shoreline. These reports are 3–4 Jun 1991 (Peyton 1991) and 4–5 Jun 1992, 11–25 Jun 1992, 8–22 Jul 1992, and 1 Jun 1993 (Rosche 1994a).
Fall
There are 4 reports, 3 documented, in the period 7 Aug–2 Sep. The undocumented observation occurred at L McConaughy 7 Aug 1977 (Rosche 1994a).
10 Aug 1952 Dawson Co (Wycoff 1953)
13–19 Aug 1972 Lancaster Co (Hoffman 1973)
2 Sep 1932 Lancaster Co (Hudson 1933c)
FINDING
Recent observations suggest the best chance of finding a Snowy Plover is to search wetlands in the Rainwater Basin during late Apr and early May.

Semipalmated Plover
Charadrius semipalmatus
STATUS
Fairly common regular spring and fall migrant east, becoming uncommon west.
DOCUMENTATION
Specimen
UNSM ZM6155, 5 Oct 1901 Lincoln, Lancaster Co.
DISTRIBUTION AND ECOLOGY
In Nebraska this species favors mudflats and muddy shorelines but is occasionally found in flooded fields and on sandy beaches. It

is regular throughout the state, but higher densities pass through eastern sections.

Spring

24, 24, 31 Mar→9, 10, 10 Jun

Most arrive in mid-Apr, and peak migration occurs around the first of May. Individuals have lingered into Jun on 8 occasions. Pairs or small groups are usually found, but sometimes large flocks are seen.

High counts include 210 in the Rainwater Basin 1–2 May 1998, 77 at North Hultine Basin 3 May 1994, and 52 at Rolland Basin 3 May 1997.

Fall

3, 9, 13 Jul→14, 15, 20 Oct

Adults arrive by the end of Jul, and most pass through during Aug. By late Aug juveniles arrive, and by the end of Sep most birds are gone, as there are few Oct reports. Smaller numbers are found in fall than in spring.

High counts include 17 at Pearson Basin 26 August 1995 and 14 at L McConaughy 3 Aug 1978 (Rosche 1994a).

Other Reports

7 Mar 1950, 11 Mar 1928, and 14 Mar 1944 Adams Co.

FINDING

Semipalmated Plovers are most easily found during May at wetlands in the Rainwater Basin, the L McConaughy area, or eastern wetlands such as Jack Sinn Marsh.

Piping Plover

Charadrius melodus

STATUS

Locally common regular breeder central and east. Uncommon regular spring and fall migrant statewide.

DOCUMENTATION

Specimen

UNSM ZM6102, 17 Jun 1902 Keya Paha Co.

DISTRIBUTION AND ECOLOGY

Piping Plovers in Nebraska breed on barren, midstream sandbars in wide channel beds of large rivers, including the Platte, Niobrara, Missouri, and Loup. Nesting also occurs along the shores of L McConaughy and on exposed expanses of gravel and sand that are produced in the process of sand-and-gravel mining. Typically, nesting areas consist of exposed sand that supports little or no vegetation, and what vegetation does exist is usually less than 15 cm (6 in) tall and often scattered willow and cottonwood seedlings. Migrants, most numerous in eastern Nebraska, are found on mudflats as well as in sandy areas.

The Piping Plover is currently classified as a threatened species by both state and federal agencies. In Nebraska one of the major contributors to its decline is loss of breeding habitat. Along the North and South Platte Rivers and the central Platte River water-withdrawal projects have reduced stream flow, allowing vegetative encroachment, which has precluded the production of new sandbars. Along the Missouri River channelization and containment structures have nearly eliminated shifting of the channel, and therefore the natural creation of exposed sandbanks is now negligible. Naturally functioning channels such as the Niobrara and lower Platte Rivers still provide adequate sandbar habitat for nesting plovers. Recently it has been shown that sandpits prepared especially for breeding of Piping Plovers and Least Terns (NBR 62:107) have promise as a mitigation measure, although more research is required on these alternatives to natural breeding habitat.

Spring

27 Mar, 6, 7 Apr→summer

Migration occurs during the latter half of Apr and early May. Rarely more than 2 birds are seen together.

Summer

Regular breeding is restricted to the Missouri, Platte, lower Niobrara, Elkhorn, and lower Loup River systems. In the Missouri River valley breeding occurs from Gavin's Point Dam downstream to near Ponca SP (Ducey 1984a). South of Dixon Co the river has been channelized, and suitable nesting habitat is now eliminated. In the Platte River valley breeding extends west to L McConaughy, where the plover was first observed in 1978. Since the initial discovery, the high count was 142 birds in 1993, found during a period when the lake was at record low levels (John Dinan, pers. comm.). In the Niobrara River valley breeding barely reaches Brown Co (Mossman and Brogie 1983; Ducey 1989; Plissner and Haig 1997). In the Loup River system breeding extends west to Valley and Howard Cos (Sidle and others 1991).

Of interest was a pair courting and digging nest scrapes at Swanson Res 13 May 1999 (SJD); this is the only report away from the regular breeding range for many years.

There appears to have been a breeding population in the Sandhills at least until 1917. There is a breeding record from Clear L, Cherry Co, which is now within Valentine NWR, 17 Jun 1902, when 4 adults (UNSM ZM6108–11) and 2 chicks (UNSM ZM6112–13) were collected 17 Jun 1902. Swenk (Notes after 1925) noted that in the Black collection at Kearney was a pair of adults and 4 eggs collected 23 Jun 1917 at Crescent L NWR. The specimen cited above was collected 17 Jun 1902 in Keya Paha Co. Tout (1938) reported a small flock 22–27 Aug 1937 at Crescent L NWR.

In 1996 a comprehensive Piping Plover breeding census was conducted in Nebraska by the NPG (Plissner and Haig 1997). Excluding individuals along the Missouri River, 366 plovers were counted. They were found at 63 sandbar sites, at 36 sandpit sites, and at several shoreline locations. It was calculated that the population had declined 18% when compared to the 10-year mean of 447. Previously, censuses had been conducted on the Missouri River by Ducey (1984) and other river systems by Sidle and others (1991), who found 673 plovers along the Missouri, Platte, Niobrara, Loup, and Elkhorn Rivers; excluding the Missouri River total of 275 birds, the total of 398 (Sidle and others 1991) is similar to that found by the NPG in 1996.

Fall

summer→9, 13, 20 Sep

Migrants are first noted in mid-Jul, and most depart by early Sep. There are these later reports: 3 Oct 1985 Platte Co (AB 40:135), 5 Oct 1991 Saunders Co, and 24 Oct 1992 Saunders Co. Singles or pairs are usually found.

FINDING

One of the most accessible locations to find the species is L McConaughy, where breeding birds can often be closely observed at protected, fenced-off areas near the lake or on sandy beaches. Postbreeders usually remain for a time along the sandy shoreline. High, dry sandbars along the lower Platte River may also yield views in early to midsummer, and the area around the Niobrara River bridge at Niobrara usually has several birds.

Killdeer

Charadrius vociferus

STATUS

Common regular breeder statewide. Common regular spring and fall migrant statewide. Locally rare regular winter visitor west, rare casual east and south.

DOCUMENTATION

Specimen

UNSM ZM6117, 30 Mar 1901 Lincoln, Lancaster Co.

DISTRIBUTION AND ECOLOGY

Killdeer are commonly seen in wetlands, in fields, and along rural roadsides throughout the state and also can occasionally be found in urban areas where flat, gravelly roofs are sometimes utilized as nesting sites. Gravelly roadsides may also be utilized as nesting sites. Because it is a noisy and conspicuous bird, it is often reported, even though its numbers rarely exceed 1 pair per hectare in the most suitable habitat. Wintering birds require open, shallow water such as occurs along the Platte River and in mild winters in the southeast.

Spring

10, 11, 11 Feb→summer

Migrants arrive by the end of Feb or in early

Mar, and the species is ubiquitous through Apr. Killdeer are less frequently encountered during May, as migrants move north and summer residents commence breeding.

High counts include 112 in Dixon Co 1 Apr 1998, 56 at Crescent L NWR 4 May 1994, and 54 there 23 Apr 1994. "Giant numbers" were present near Gibbon 15 Mar 1995.

Summer

This species breeds throughout Nebraska; BBS data (1967–77) show an even distribution statewide. Data from 1966 to 1989 indicate a gradual increase.

Fall

summer→1, 1, 1 Jan

Postbreeders and young birds begin flocking in Jul, and such groups are common through early Oct. Numbers decrease in late Oct through Nov. CBC data show few remaining into Dec; best Platte River valley counts are 21 at L McConaughy 23 Dec 1995, 20 in Lincoln Co 17 Dec 1994, and 20 at Scottsbluff 18 Dec 1976, while elsewhere only single birds have been recorded.

High counts include 456 in the eastern Rainwater Basin 10 Oct 1999, 324 at L Minatare 6 Sep 1998, and 138 in Lancaster Co 12 Aug 1995.

Winter

Small numbers winter regularly in the North Platte River valley downstream from the Keystone Diversion Dam, Keith Co, where the high count is 21 individuals observed on 8 Jan 1982 (Rosche 1994a). In Scotts Bluff Co it has been reported as a permanent resident in 1965, 1968, 1974, and 1981. There are additional reports from Scotts Bluff Co 2–6 Jan 1995, 8 Jan 1967, 18 Jan 1969, and 26 Jan 1969. One was at North Platte NWR 14 Jan–7 Feb 1997, and 3 were there 19 Jan 1998. Wintering may also occur on occasion in the southeast; Killdeer was reported as a permanent resident in Adams Co 1966 and 1969–73 and in Clay Co in 1973, and there are reports from Jefferson Co 13 Jan 1949 and 3 Feb 1948.

There are few additional reports in the period 2 Jan–9 Feb: permanent resident 1980 Douglas-Sarpy Cos and 1981 Boone Co, 11 Jan 1986 Antelope Co, 15 Jan 1966 Douglas-Sarpy Cos, 17 Jan 1987 Madison Co (Wolff 1987), and 2 Feb 1935 Lincoln Co.

FINDING

The Killdeer is easily found during warmer months in wetlands and fields and along country roadsides that are adjacent to wet ditches.

Mountain Plover

Charadrius montanus

STATUS

Rare regular breeder Kimball Co. Rare casual spring visitor west, hypothetical elsewhere. Accidental in fall.

DOCUMENTATION

Specimen

HMM 3053, 7 Apr 1917 Whitman, Grant Co.

DISTRIBUTION AND ECOLOGY

Mountain Plovers typically nest in tracts of high-plains grassland dominated by blue grama at sites that have at least 30% bare ground (Knopf and Rupert 1996), a habitat that is present to a limited extent in Kimball Co. Significant for Nebraska is the observation by Knopf and Rupert (1996) that "population declines of Mountain Plovers appear independent of any recent landscape fragmentation within this [Weld Co, Colorado] breeding stronghold of the species." Prairie dog colonies are often utilized as nest sites in some parts of the breeding range (Stephen Dinsmore, pers. comm. WRS). In recent years Mountain Plovers have been found nesting in Jun in Kimball Co in fallow wheat fields, even those recently disked. Soon after hatching, young may be moved a significant distance, as much as 800 m (2625 ft), from the nest site to areas where there are forbs or other objects providing shelter from the sun; such areas may include fallow agricultural ground (Graul 1973; Knopf and Rupert 1996).

Spring

Migrants are usually not detected in Nebraska,

as the few birds reaching the state are presumably prospective local breeders at the eastern edge of the breeding range. Breeding birds arrive in Weld Co, Colorado, around 1 Apr (Knopf and Rupert 1996). There are only about 20 Nebraska reports in the period 4 Apr–15 May. Those from the current breeding range in Kimball Co are 4 Apr 1999 (6) (SJD), 12 Apr 1982 (RCR; *Birding* 14(3/4) insert), 25 Apr 1997, 27 Apr 1992 (NBR 60:138), 1 May 1993 (NBR 61:95), 4 May 1997 (3 separate sightings), and 8 May 1985 (NBR 53:71). That migrants may appear somewhat east of the expected range is suggested by the following: a specimen collected at Whitman 7 Apr 1917 (cited above); 2 seen southeast of Sidney, Cheyenne Co, 17 Apr 1998 (SJD); a specimen whose current location is unknown said to have been collected by Harris at Capitol Beach, Lancaster Co, in 1912 (Brooking, Notes); and a sighting at Clear Creek Marshes 21 Apr 1989 (Rosche 1994a). Undocumented reports include 8 May 1949 and 15 May 1950 Dawes Co (Rosche 1982), 15 Apr 1949 Antelope Co, and 9 Apr 1967 Douglas-Sarpy Cos. At least the latter 2 reports should be discounted, as separation from basic-plumaged American Golden-Plover is difficult.

Summer

Mountain Plover has suffered due to agricultural practices that have eliminated large areas of its native shortgrass habitat. Before the development of wheat and alfalfa ranching on the plateau lands of the Panhandle, it is likely that this species bred over large areas of the Panhandle, including Cheyenne, Dawes, and Sioux Cos (Bruner and others 1904). Still earlier, when bison grazing maintained shortgrass prairies considerably to the east, at least one breeding occurrence was recorded from Lincoln Co 8 Jul 1859 (Coues 1874; Ducey 1988), and specimens were collected in 1856–57 in the Loup Valley (USNM 9043, 9044). Other old reports during summer include 23 Jun 1927 north of Morrill in southern Sioux Co (LOI 24; Rosche 1982), early observations by

Bruner near Harrison and Marsland in Sioux Co and near Sidney in Cheyenne Co (Bruner and others 1904), probable occurrence at Kearney (Bent 1927), and a report of 4 birds in Antelope Co 6 Jun 1900 (Bates 1901). When bison were present the species also bred in South Dakota (SDOU 1991) and in western Kansas (Thompson and Ely 1989), locations where it is now virtually absent.

At present a small summer population exists in Kimball Co, which is adjacent to a substantial breeding population in Weld Co, Colorado (Andrews and Righter 1992). Recent nesting records include an observation of young 7 Jun 1975 in an area of native prairie about 11.2 km (7 mi) west of Bushnell (Lock 1975; Bennett 1976), a sighting of 3 "less than half-grown young" 4.8–6.4 km (3–4 mi) north of Highway 30 and 1.6 km (1 mi) east of the Wyoming border 13 Jul 1983 (Cox 1983), the finding by Knopf of a nest with eggs 8 km (5 mi) west and 2.4 km (1.5 mi) north of Bushnell in native prairie 21–23 May 1990 (Clausen 1990; Grenon 1991; Johnson-Mueller and Morris 1992; Dinsmore 1997a), a nest with eggs in the same location northwest of Bushnell 18 May 1995 (Dinsmore 1997a; Gubanyi 1996c; NBR 63:42), a nest southwest of Kimball 18 May 1997, a bird that appeared to be incubating in the same general area 12 Jun 1997, one making a nest scrape 11.2 km (7 mi) southwest of Kimball 16 May 1998 (SJD, JS), and a nest with 3 eggs 11–17 May 1999 near the Kimball airport (SJD). In addition, a single bird was in a disked fallow wheat field 10.4 km (6.5 mi) south and 1.6 km (1 mi) west of Kimball 24 Jun 1995, and 2 were in similar habitat 16 km (10 mi) south and 3.2 km (2 mi) west of Kimball (about 6.4 km [4 mi] from the previous sighting) 15 Jun 1996 (Brogie 1997). Further sightings in this area were made in spring 1997, including about 8 birds occupying both stony, freshly disked fallow wheat fields and overgrazed pasture. Together with sightings by Shackford (letters to Knopf and Clausen, provided by SJD to WRS), these reports indicate that breeding may occur somewhat regularly in an area bounded by I-80 on the north, Highway 71 on the east, and a line starting 19 km (12 mi) south of Kimball on Highway 71, passing west 6.4 km (4 mi) and north about 16 km (10 mi) back to I-80.

Fall

The few birds that breed in Nebraska apparently vacate the state soon after young are fledged, as there are no recent reports after Jul (see Summer). Knopf and Rupert (1996) stated that most plovers leave the breeding grounds in mid to late Jul. There is a single report later than Jul, a specimen taken by Mickel in Dawes Co 27 Sep 1920 (Brooking, Notes).

FINDING

Best chance of finding the species is to search "dry, stony, short-grass prairie that is overgrazed" (Rosche 1994b) or fallow wheat fields, especially those recently disked, in western Kimball Co May–Jun.

Recurvirostridae (Stilts, Avocets)

Black-necked Stilt

Himantopus mexicanus

STATUS

Rare regular spring and fall migrant west, rare casual elsewhere. Locally uncommon casual to regular breeder west.

DOCUMENTATION

Specimen

HMM 28534, 12 May 1956 Adams Co.

DISTRIBUTION AND ECOLOGY

Migrants are found in wetlands or flooded fields and grasslands. Breeding birds have been restricted to wetlands in the Panhandle, primarily sandhill marshes.

Spring

17, 20, 21 Apr→summer

Spring migration is difficult to define or predict, as Black-necked Stilts have a propensity for showing up at unexpected locations. Most reports are from a known breeding colony in Sheridan Co, Garden Co, and elsewhere in the Panhandle. Observations away from the

Panhandle not cited below (see Summer) are 21–23 Apr 1999 at Harlan Co Res (GH, WH), 30 Apr 1996 near Creighton (Brogie 1997), 2–3 May 1996 Keith Co (FN 50:297), 3 May 1986 extreme southeast Garden or Keith Co (Rosche 1994a), 4 May 1996 Knox Co, 9 May 1998 Weis Lagoon (JGJ), ca. 11 May 1990 Hall Co, 2–10 Jun 1996 Seward Co, 5 and 10 Jun 1997 Funk Lagoon, and 11 Jun 1996 Phelps Co.

Summer

Prior to 1970 this species was casual in occurrence with only about 8 reports, including 4 from Omaha in the 1890s (Bruner and others 1904), 2 from Crescent L NWR (1933, 1942) (NBR 48:76), 12 May 1956 Adams Co, and an Aug report in the early 1960s in Cass Co (NBR 33:16). During the 1970s and early 1980s reports slowly began to increase, primarily in western areas in spring, including 30 Apr 1979 Scotts Bluff Co, 14–28 May 1977 Lincoln Co, 20–21 May 1980 Garden Co, 28 May–5 Jun 1972 (Tremaine 1972), 29–30 Jun 1985 Sioux Co, and 28 Jul–8 Aug 1978, and more easterly reports 28 Apr 1981 Lancaster Co, 11 May 1983 Dawson Co, and 15–17 May 1987 at Valentine NWR. Beginning in 1988 this species was reported regularly.

In 1985 breeding was reported for the first time, as young were observed at Crescent L NWR (Helsinger 1985a) and again in 1987 (Huber 1987). Since 1987 (Mollhoff 1989a) there has been a stable colony, varying in size from year to year, along Highway 2 at mile marker 106 between Lakeside and Antioch in Sheridan Co. A high count of 27 individuals was observed there 7 May 1994, and 21 were seen in spring 1996 (FN 50:297). An adult with a juvenile was at Paterson L, Garden Co, 7 Aug 1998, and there was an unsuccessful breeding attempt in Dawes Co in 1994 (NBR 62:73). Nesting was noted at the Grand Island SL in 1998, but the outcome was not reported (GL).

The breeding events beginning in 1985 coincided with heavy flooding over much of the normal breeding range in northwest Utah and southwest Wyoming, leading to speculation that individuals colonizing Nebraska were displaced. It is possible also that the cycle of breeding activity in Nebraska is waning, because attempts to locate birds along Highway 2 in 1997 and 1998 were unsuccessful. However, because many marshes and potholes in Sheridan and Garden Cos are quite remote, it is possible that part of the existing breeding population has so far been overlooked.

Fall

summer→8, 12, 18 Jul

Summering and breeding birds tend to disappear in Aug (Thompson and Ely 1989); there are only 3 reports after Jul, both away from known breeding locations: an Aug Cass Co report cited above, 15–16 Aug 1998 Funk Lagoon, and 21 Sep 1989 (8) L Alice.

FINDING

Look for the species May–Jun between Lakeside and Antioch near mile marker 106 along Highway 2.

American Avocet

Recurvirostra americana

STATUS

Common regular spring and fall migrant west and central, rare east. Common, locally abundant, regular breeder west and north-central, rare casual elsewhere.

DOCUMENTATION

Specimen

UNSM ZM6614, 17 Jun 1902 Clear L, Cherry Co.

DISTRIBUTION AND ECOLOGY

In summer American Avocet occurs regularly in the western Sandhills from Cherry Co west and casually in the eastern Sandhills and Rainwater Basin. It is attracted to shallow, alkaline playa wetlands and mudflats along lake margins where it commonly nests. Emergent aquatic insects, including blackflies, are significant attractants in these habitats.

During migration American Avocets may be found in marshes and along shorelines, and they occasionally swim in open, shallow water. Although a regular migrant statewide, it is far less common in the east.

Spring

20, 21, 22 Mar→summer

Migrants usually arrive early to mid-Apr, and peak migration soon follows during the latter half of that month. Migrants are still in evidence well into Jun.

High counts include 151 at L McConaughy 24 Apr 1999, 120 at Harlan Co Res 23 Apr 1999, and 117 in the Rainwater Basin 1–2 May 1998.

Summer

This species breeds regularly in the western Sandhills east to Cherry Co (Ducey 1988) and west of the Sandhills, at least in Scotts Bluff Co, where it has been reported in summer most years since 1971, and 12 adults with at least 2 chicks were at Mansfield Marsh 13 Jun 1998 (WRS, JS). Breeding may occur sparingly east of the Sandhills; it was considered an uncommon summer resident in Holt Co, with a report 21 Jun 1991 (Blake and Ducey 1991), and there is a breeding season report from Pierce Co 17 Jun 1988.

There is a single nesting record from the Rainwater Basin, where 2 nests and young were reported in 1974 (Bennett 1975). There are a few additional reports during the breeding season suggestive of local breeding or postbreeding flocking. Up to 30 birds were at Funk Lagoon 8 Jun–6 Jul 1997, including 30 there 19–20 Jun, somewhat early for migrants. Additional Rainwater Basin reports are "summer" 1973 Clay Co, 6 Jun 1991 Polk Co, 14–30 Jun 1985 Adams Co, 30 Jun 1986 Adams Co, 22 Jun 1988 Howard-Hall Cos, and 30 Jun 1991 Kearney Co. Two birds near Valley 29 Jun 1998 may have been failed breeders that had moved south.

Nests have been located on mudflats and along lake margins. The nest consists of a small depression with a mat of thin grasses where 3 and sometimes 4 eggs are laid. Incubation has been observed as early as 25 May and as late as 20 Jun.

Downy young have been recorded as early as 5 Jun. Adults are very aggressive late in the incubation period and when protecting downy young. Alarm calls and aggressive flights toward intruders may be joined by neighboring adults in a mobbinglike action.

High counts include 270 at Crescent L NWR 16 Jul 1996 (FN 50:966) and 56 there 8 Jul 1997.

Fall

summer→7, 7, 8 Nov

There are later reports 28 Nov and 5 Dec 1998 L McConaughy, possibly an injured bird (SJD).

Despite considerably fewer reports, numbers are similar to those found during spring. Migrants are noted in Jul where breeding does not occur, while numbers begin to decrease during Sep and more so during Oct. Counts of 8–10 at Kissinger Basin 17–25 Jul 1999 were likely early fall migrants.

High counts include 600 at Crescent L NWR 30 Aug 1998, 150 there 2 Aug 1996, and 100 at L McConaughy 3 Oct 1976 (Rosche 1994a).

FINDING

Avocets can usually be found during prime migration periods in the Rainwater Basin and Sandhills marshes. In summer look for the species at Crescent L NWR. The drive between Crescent L and Lakeside can be especially rewarding in late May and early Jun.

Scolopacidae (Sandpipers, Phalaropes)

Greater Yellowlegs

Tringa melanoleuca

STATUS

Fairly common regular spring and fall migrant statewide. Rare casual winter visitor Platte Valley.

DOCUMENTATION

Specimen

UNSM ZM12650, 23 Apr 1910 Lancaster Co.

DISTRIBUTION AND ECOLOGY

This species is found in marshes, along shorelines, and on mudflats. Both species of yellowlegs prefer areas of shallow water with scattered emergent vegetation, although Greater Yellowlegs are usually outnumbered by Lesser Yellowlegs.

Spring

3, 6, 9 Mar→29, 30, 31 May

An early migrant, this species arrives in mid to late Mar, peaking during Apr. A few are observed in early May, but numbers decline rapidly thereafter.

High counts include 152 in the eastern Rainwater Basin 3 Apr 1999, 65 in the Rainwater Basin 11 Apr 1998, and 65 near Minden 16 Apr 1998.

Fall

10, 11, 13 Jun→16, 17, 18 Nov

Adults begin their return migration as early as mid-Jun, peaking in Jul. As migration progresses into Aug, juveniles begin to outnumber adults, and by mid to late Aug most birds seen are juveniles. The main movement of juveniles, however, is evident in mid-Sep.

Greater Yellowlegs is one of a handful of shorebird species expected throughout Oct and into early Nov. Excluding Common Snipe, the Greater Yellowlegs has been reported more times (about 30) in Nov than any other sandpiper.

High counts include 60 at Mallard Haven 17 Sep 1995, 50 in the eastern Rainwater Basin 25 Sep 1999, 30 at Crescent L NWR 12 Sep 1995, and 30 at Pawnee L 25 Oct 1997.

Winter

This species may remain well into Dec. There are these reports of lingering birds, which may have been attempting to winter: 17 Dec 1993 Keith Co, 28 Dec–1 Jan 1961–62 Adams Co, and 14 Jan 1997 Stateline Island.

COMMENTS

Ducey (1988) cited old reports of nesting in Holt and Nemaha Cos prior to 1900. In the absence of substantiating evidence, we consider these reports highly doubtful. This species breeds no nearer to Nebraska than central Canada (AOU 1998).

FINDING

Greater Yellowlegs prefer marshes rather than shorelines and are found in such areas throughout during migration. Separation from Lesser Yellowlegs, especially of single birds, can be difficult.

Lesser Yellowlegs

Tringa flavipes

STATUS

Fairly common regular spring migrant statewide. Common regular fall migrant statewide.

DOCUMENTATION

Specimen

UNSM ZM12652, 12 Apr 1898 Lancaster Co.

DISTRIBUTION AND ECOLOGY

Lesser Yellowlegs are found in marshes, along shorelines, and on mudflats. Both species of yellowlegs prefer areas of shallow water with scattered emergent vegetation. Lesser Yellowlegs usually outnumber Greater Yellowlegs.

Spring

10, 13, 14 Mar→4, 5, 6 Jun

There are 4 reports in the period 9–16 Jun, which are probably birds that did not complete migration: 9 Jun 1984 Douglas-Sarpy Cos, 13 Jun 1992 Sheridan Co, 14 Jun 1961 Lincoln Co, and 16 Jun 1979 Garden Co (see Comments).

Arrival is generally at the end of Mar or the beginning of Apr. Numbers peak during the latter half of Apr and decrease during May until the last week when the species is scarce. Observations in spring usually are of small groups.

High counts include 812 in the Rainwater Basin 24–25 Apr 1998, 454 in the eastern Rainwater Basin 17–18 Apr 1999, and 211 at Hansen Marsh 25 Apr 1997.

Fall

19, 21, 22 Jun→16, 19, 23 Nov

Adults arrive by the end of Jun and are observed through Jul. Juveniles arrive at least by mid-Aug and are the majority age group by and through Sep. Juveniles often flock in larger groups than do adults. The high counts listed below presumably include a large majority of juveniles based on the dates. Numbers dwindle in Oct, and the species is only casually observed during the first half of Nov.

High counts include 400 at L McConaughy

25–27 Aug 1989 (Rosche 1994a), 271 there 20 Aug 1998, and 175 at Mallard Haven 17 Sep 1995.

COMMENTS

Reports of breeding in Cherry Co and of young in Jun 1931 in Sheridan Co (Ducey 1988) are highly unlikely since this species is unknown as a breeder outside its typical tundra-muskeg habitat of far northern North America; it breeds regularly as far south as central Saskatchewan and southeast Manitoba (AOU 1998).

FINDING

Lesser Yellowlegs can usually be found in marshes and along shorelines during Apr and through the first half of May and Jul through Sep.

Solitary Sandpiper

Tringa solitaria

STATUS

Fairly common regular spring and fall migrant statewide. Rare casual summer visitor statewide.

DOCUMENTATION

Specimen

UNSM ZM6144, 7 Jul 1902 Carns, Keya Paha Co.

TAXONOMY

Swenk and Fichter (1942) examined 10 specimens and found that both *solitaria*, the eastern subspecies, and *cinnamomea*, the western subspecies, migrate throughout the state. Each subspecies was represented by 5 specimens. Specimens of *solitaria* were taken in Cass, Lancaster, Thomas, Sioux, and Sarpy Cos, and those of *cinnamomea* were taken in Lancaster (2), Sioux, Clay, and Keya Paha Cos. Juveniles of *solitaria* were noted in eastern Nebraska late Aug 1996 and of *cinnamomea* in early Sep (JGJ). Field observers should be aware that juveniles in autumn, studied at close range, are the only individuals that can be reliably identified to subspecies in the field (Paulson 1993).

DISTRIBUTION AND ECOLOGY

This species is found at small, shallow wetlands

and marshes with some vegetation and along wooded streams or creeks.

Spring

27, 29 Mar, 1 Apr→25, 25, 29 May

This species does not usually arrive until the latter half of Apr, and peak migration occurs during the first half of May. There are earlier reports 16 Mar 1989 Lancaster Co and 17 Mar 1963 Lincoln Co. Almost without exception, singles and less often pairs are found at a single locale.

Summer

This species has been considered in the past a "hypothetical breeder" based on Jun sightings prior to 1900 (Ducey 1988); however, breeding in Nebraska is an extremely remote possibility, as the nearest regular breeding is in central Saskatchewan (AOU 1983). There are indeed a few reports in the period 30 May–23 Jun, probably late spring and early fall migrants: 5 Jun 1964 Webster Co, 5 Jun 1994 Scotts Bluff Co, 7–10 Jun 1966 Cass Co, 15 Jun 1994 Otoe Co, and 20 Jun 1968 McPherson Co.

Fall

24, 25, 26 Jun→28, 28 Oct, 2 Nov

Adults arrive by the end of Jun and are present through Jul. Juveniles arrive in mid-Aug and tend to outnumber adults shortly thereafter. Most have moved south by mid to late Sep. There are only 18 reports after 23 Sep, several undocumented.

High counts include 20 in Lancaster Co 30 Jul 1995, 16 east of Gering SL 3 Sep 1999, and 14 in southeast Garden Co 16 Jul 1994.

FINDING

Solitary Sandpipers live up to their name in spring, usually only single birds being observed. Jul may be the best time to find the species, when small numbers may occur together. Search small wetlands with vegetation and along slow-moving wooded streams.

Willet

Catoptrophorus semipalmatus

STATUS

Common regular spring migrant central and

west, uncommon east. Uncommon, locally common, regular breeder north-central and west. Uncommon regular fall migrant central and west, rare east.

DOCUMENTATION

Specimen

UNSM ZM6611, 7 Jul 1902 Carns, Keya Paha Co.

DISTRIBUTION AND ECOLOGY

Migrants utilize mudflats, shorelines, and flooded grassy areas. Breeding birds are limited almost entirely to the Sandhills, where they are associated with playas, marshy edges of lakes, flooded meadows, and other wetland sites. Occasionally they occur in valleys of high-plains prairies. In the northern Great Plains, Willets set up large territories averaging about 40 ha (100 ac) adjacent to shallow wetlands (Ryan and Renken 1987).

Spring

22, 22, 23 Mar→3 May, 1, 2 Jun

Late dates above are away from the breeding range.

In spring the Willet is found throughout the state, although it appears to be least common in the eastern quarter. It arrives in early and mid-Apr. The Willet has an exceptionally clearly defined peak migration period, the last week of Apr; 37% of all reports (N = 348) fall within the very short period 24–30 Apr.

High counts include 87 in the eastern Rainwater Basin 24 Apr 1999, 70 in the Rainwater Basin 25–26 Apr 1998, and 51 at Harlan Co Res 23 Apr 1999.

Summer

Willets breed primarily in the western Sandhills region. Ducey (1988) cited breeding records north of the Platte River east to Cherry Co. Rosche (1982) apparently did not find Willets breeding in the high plains of the western Panhandle, remarking that breeding was "primarily confined to the sandhills region." There is evidence that Willets breed sparingly farther east, to the eastern edge of the Sandhills. Ducey (1988) cited an early breeding record, and there are recent summer observations from Holt Co (Blake and Ducey 1991). A copulating pair was observed in early Jun at Pony L, Rock Co (Ducey and Schoenenberger 1991), and 2 were in suitable breeding habitat in extreme western Knox Co 25 May 1998 (wrs, js).

Fall

26, 27, 28 Jun→26, 27, 29 Sep

Early dates above are away from the breeding range. There are later reports 10 Oct 1985 Lancaster Co, 20 Oct 1991 Lancaster Co, 22 Oct 1992 Scotts Bluff Co, and 25 Oct–8 Nov 1959 Platte Co.

There are dramatically fewer fall reports than in spring, and Willets are rare in the east in fall. Willets likely vacate Nebraska soon after breeding. Most leave before Sep, as there are only about 15 reports after Aug.

High counts include 10 at Funk Lagoon 5 Jul 1999, 8 at Sinninger Lagoon 14 Aug 1995, and 6 in the eastern Rainwater Basin 1 Aug 1999.

Other Reports

4 Mar 1951 Lincoln Co.

FINDING

Willets are most easily found during the last week of Apr at good shorebird areas throughout Nebraska. In summer territorial birds may be most conspicuous as they exhibit flight aggression toward intruders during peak breeding activity in late May and early Jun. Playas with mudflats along the road between Crescent L nwr and Lakeside support especially high concentrations.

Spotted Sandpiper

Actitis macularia

STATUS

Fairly common regular spring and fall migrant statewide. Uncommon regular breeder statewide.

DOCUMENTATION

Specimen

UNSM ZM12643, 28 May 1910 Lancaster Co.

DISTRIBUTION AND ECOLOGY

Spotted Sandpipers are found on sandy shorelines, mudflats, or mud shoreline near at least sparse vegetation. The species is found throughout the state, but it does not associate with mixed groups of shorebirds.

Spring

24, 27, 29 Mar→summer

Migrants arrive during the latter half of Apr, and peak migration occurs during the first half of May; 50% of all reports (n = 471) are in the period 1–15 May.

High counts include 25 in the Rainwater Basin 9–10 May 1998, 21 at "various sandbars" in Hall Co 14 May 1995, and 19 at L McConaughy 12 May 1999.

Summer

This species breeds in suitable habitat throughout the state (Ducey 1988) but appears to be a low-density breeder. It is often seen individually or in pairs foraging along stream- and riverbanks and along lake shores that are frequently heavily vegetated, often with dense, woody vegetation.

A nest containing 4 eggs was found in Hall Co in an old field near a sandpit 27 May 1974; it consisted of a depression in the ground. On 10 Jun the eggs had not yet hatched, but on 13 Jun the nest was empty, the young by then presumably hatched. Adults with young have been observed in Jun and early Jul.

Fall

summer→17, 19, 20 Oct

The majority of reports are in Aug and Sep, with few into early Oct. Later reports are 26 Oct 1966 Lincoln Co and 1 Nov 1980 and 15 Nov 1991 Douglas-Sarpy Cos.

High counts include 21 at North Platte nwr 10 Sep 1998, 14 at L Ogallala–Keystone L 3 Aug 1978 (Rosche 1994a), and 12 at Scottsbluff sl 3 Sep 1999.

FINDING

During migration, and less reliably during breeding season, Spotted Sandpipers can be found along shorelines of calm bodies of water. Often birds will flush before being seen and are easily identified if the observer is familiar with the species' distinctive flight.

Upland Sandpiper

Bartramia longicauda

STATUS

Fairly common regular spring and fall migrant

statewide. Fairly common regular breeder statewide.

DOCUMENTATION

Specimen

UNSM ZM11419, 18 Jun 1931 Atkinson, Holt Co.

DISTRIBUTION AND ECOLOGY

A grassland species that utilizes tallgrass, midgrass, and shortgrass prairies, Upland Sandpiper also may be found in mixed agricultural lands with meadows interspersed as well as freshly mowed alfalfa fields in fall. Breeding birds prefer tracts of grassland, but the species is somewhat tolerant of limited farming near its preferred grassland habitat.

It would seem obvious that numbers have declined since the 19th century as they have in Iowa (Kent and Dinsmore 1996), although serious declines may have occurred only in areas of the state where grasslands have been converted to cropland, as has been the case in most of Iowa and eastern Nebraska. Numbers in relatively untouched grasslands in Nebraska are still significant.

Spring

23, 25, 30 Mar→summer

There is an earlier undocumented report 9 Mar 1977 Lancaster Co. Migrants usually arrive during the latter half of Apr, with peak migration occurring in early May.

High counts include 20 at Clear Creek Marshes 10 May 1987 (Rosche 1994a) and 17 in the Rainwater Basin 15 May 1998.

Summer

Upland Sandpiper breeds statewide with the possible exception of the extreme southwest and the southern Panhandle, although it was reported in Scotts Bluff Co during summer 1970–72, 1975–76, 1978, and 1983; Perkins Co 1974; and 13 Jun 1990 in Chase Co. Breeding densities are greatest in the Sandhills and northern Panhandle, where the species is common. It is less common in the most intensely cultivated areas of eastern and south-central Nebraska. Even small grassland tracts may attract a breeding pair, however; a pair raised a brood on an isolated tract of

lowland meadow of less than 6 ha (15 ac) (RSS).

Nests containing eggs have been found in the period 21 May–19 Jun. Clutch size varies from 2 to 4 eggs.

High counts include 35 at Calamus Res 27 Jun 1997.

Fall

summer→18, 20, 28 Oct

Most have moved out of Nebraska by the first week of Sep; there are fewer than 20 reports for Oct.

High counts include 67 in an alfalfa field in Buffalo Co 12 Aug 1995, 45 in the eastern Rainwater Basin 22 Aug 1999, and 32 at Clear Creek Marshes 31 Jul 1988 (Rosche 1994a).

FINDING

A drive through meadows and prairies in eastern Nebraska or throughout the Sandhills in late spring and summer should yield a few of these birds, often perched on fence posts. They also can be easily located by their haunting "wolf-whistle" calls.

Eskimo Curlew

Numenius borealis

STATUS

Extirpated. Formerly abundant regular spring migrant central. Formerly rare regular fall migrant central.

DOCUMENTATION

Specimen

HMM 2469 (2), Apr 1880 Hamilton Co.

DISTRIBUTION AND ECOLOGY

This species is virtually extinct and may be so; there are no accepted records for North America since 1962 (ABA 1996). This curlew's former abundance and the unregulated hunting that may have played an important role in its decline are described by Swenk (1915b):

They [flocks] contained thousands of individuals, and would often form dense masses of birds extending for a quarter to a half mile in length and a hundred yards or more in width. When the flock would alight the birds would cover forty to fifty acres of ground. During such flights the slaughter of these poor birds was appalling and almost unbelievable. Hunters would drive out from Omaha and shoot the birds without mercy until they literally slaughtered a wagon load of them; the wagon being actually filled and often with the sideboards on at that. Sometimes . . . their wagons were too quickly and easily filled, so whole loads would be dumped out on the prairie, their bodies forming piles as large as a couple tons of coal.

It took only about 50 years for this species essentially to disappear. The curlews' tameness, habit of flying in dense flocks, and the propensity of a flock to circle over the same area again and again as hunters shot to their content were all liabilities that made the species particularly vulnerable to hunting. However, the notion that hunting was the sole factor leading to the species' near extinction has been challenged by Banks (1977). Banks suggested that a series of other factors (including climate) may have increased mortality during the fall migration; more specifically, a minor change in prevailing wind patterns may have changed the flight from Labrador to South America just enough that many birds may have perished in the South Atlantic Ocean. Consecutive years of breeding failure may have also played a role (Banks 1977). During the 1860s the curlew was still numerous, but "year by year the birds decreased in numbers, until 1878," after which "they were seen only in small flocks or individually here and there" (Swenk 1915b). By the 1890s the curlew was rarely seen, and after 1900 records are few. In 1911, 2 females were shot in York Co, and 7 of a flock of 8 were taken in Merrick Co (Swenk 1915b), one of which is now a specimen, UNSM ZM14114. The last accepted occurrence of the species in Nebraska was a flock of 8 near Hastings 8 Apr 1926 (Brooking 1942a; Bray and others 1986).

Spring

Generally migration was from early Apr through the first half of May, with earliest dates 22 Mar and 1 and 2 Apr, but the species was primarily a mid to late Apr migrant, similar to the American Golden-Plover. The curlew was found in eastern Nebraska in large numbers, but like the Buff-breasted Sandpiper, "the chief feeding grounds of these curlews at the time (1877) was in York, Fillmore and Hamilton Counties, and their heaviest lines of northward migration was [*sic*] between the 97th and 98th meridians" (Swenk 1915b). Those counties make up a majority of the eastern portion of the Rainwater Basin, and it is likely that the species' area of preference encompassed Clay Co as well. Burned over prairies were favored by the Eskimo Curlew, but with onset of agriculture, the species also foraged in newly plowed fields.

Fall

Records are few, but it appears that the curlew did migrate through Nebraska in fall, primarily during the first half of Oct (Swenk 1915b).

COMMENTS

Any small curlew found in Nebraska, especially during Apr in the eastern Rainwater Basin counties mentioned above, has the potential of being an Eskimo Curlew. Any observation thought to be this species should be thoroughly documented, especially keeping in mind the (remote) possibility of Little Curlew, the Eurasian counterpart of Eskimo Curlew.

Other Reports

A sighting of a single bird at Mormon Island Crane Meadows 16 Apr 1987 (Faanes 1990) was not accepted by the NOURC (Grenon 1991).

Whimbrel

Numenius phaeopus

STATUS

Rare regular spring migrant central and west, rare casual east. Rare casual fall migrant central and east.

DOCUMENTATION

Specimen

HMM 2498 (2), 12 Oct 1901 Hall Co.

DISTRIBUTION AND ECOLOGY

Migrants may be found in flooded fields and pastures and on moist prairies and mudflats. They are more often reported in central and western areas. In the eastern half of the state there are fewer reports than in the western half, even though there are fewer observers in the west. Lincoln Co has several reports.

Spring

10, 12, 27 Apr→24, 27, 30 May

Of about 35 reports, only 4 are in Apr (Dinsmore 1996d). Some 67% of all reports are in the short period 10–24 May. Singles or pairs are most often discovered, but large flocks sometimes occur at peak migration, as shown below.

High counts include 62 at Bronco L 15 May 1948 (NBR 16:54), 45 at Branched Oak L 18 May 1997, and 29 at Funk Lagoon 18 May 1996.

Fall

There are only 5 reports, 1 documented:

12 Oct 1901 Hall Co, specimen cited above

The others, likely correct, are 19 Jul 1979 Dawes Co (Rosche 1982), 25 Jul 1984 Buffalo Co (NBR 52:70–71), 8 Oct 1898 Lancaster Co, and 17 Oct 1934 Douglas Co. The Jul reports are likely adults, while the 3 Oct reports are likely juveniles.

FINDING

The best chance of encountering this species appears to be searching wetlands in the western areas of the state during the peak spring migration period, 10–24 May.

Long-billed Curlew

Numenius americanus

STATUS

Fairly common regular spring and fall migrant central and west, rare casual east. Fairly common regular breeder north-central and west.

DOCUMENTATION

Specimen

UNSM ZM6609, 20 Jun 1901 Indian Creek, Sioux Co.

DISTRIBUTION AND ECOLOGY

Breeding birds are found on native prairie, away from water, most commonly in the western Sandhills and Panhandle. The species was at one time more widespread and was an "abundant migrant throughout Nebraska" (Bruner and others 1904). Just before 1900 Cary (1900) noted that the Long-billed Curlew was "formerly common, but is now quite rare" in Antelope Co. The species became increasingly less common in the east and south of the Sandhills as the prairies were converted to agriculture. Since 1925, there are fewer than 20 reports east of the Sandhills and south of the Platte River.

Spring

20, 22, 25 Mar→summer

Arrival is relatively early, generally by the last week of Mar, with a peak in Apr. Some 61% of all reports (N = 233) fall within the period 1–23 Apr, including 9 of the 16 reports from east of the Sandhills and south of the Platte River.

High counts include 26 at L Ogallala 2 May 1998, 15 at Harlan Co Res 3 May 1999, and 13 near Keystone 18–19 Apr (Rosche 1994a).

Summer

In the early 1900s Long-billed Curlew was "a common breeder locally from Cherry County westward" (Bruner and others 1904). The breeding range has remained stable since. Breeding densities appear to be highest in the western Sandhills, primarily the "lake country" of Sheridan and Garden Cos (Rosche 1982). It also breeds commonly on the Ogallala National Grasslands in Sioux Co (Rosche 1982), but there is little information on breeding in the southern Panhandle. A pair was noted west of Long Canyon in Banner Co 3 May 1997, and one was calling in suitable breeding habitat in southern Kimball Co 9 May 1998 (SJD). Eastern and southern limits of the breeding range are not well known. It has bred as far east as Brown Co (Mossman and Brogie 1983) and is "thought to breed a few miles west of Bruner Lake" in Holt Co (Blake and Ducey 1991). Territorial birds were noted in extreme southwest Rock Co

24–25 May 1997. Breeding occurs as far south as northern Lincoln Co (Tout 1947; Ducey 1988) and just north of the Platte River in the Keystone area (Mohler 1946; Rosche 1994a).

There are a few breeding season reports from the southwest but no evidence of breeding: records from Perkins Co include Jun or Jul 1970, 1972, 1974, and 2 Jun 1992; there is one from Chase Co 23 May–30 Jun 1988. Breeding is unknown in adjacent Colorado (Andrews and Righter 1992).

Nests are hollow, grass-lined depressions on the ground and are typically located near ridgetops in upland sandhills within several hundred meters of a meadow foraging area (Bicak 1977). Three eggs are normally laid, but clutches of 2 and 4 have been found. Eggs hatch late May–early Jun (Brown and others 1996). Females leave attending males and young several weeks after hatching and roam in large, all-female flocks. Throughout spring and summer small groups of nonbreeding, perhaps yearling, birds may also be found roaming throughout the Sandhills.

Fall

summer→6, 13, 18 Sep

By late Jun curlews begin flocking (Brown and others 1996) and migrating out of Nebraska. Flocks of as many as 30 birds may occur as early as 10 Jun, as was the case at Crescent L NWR in 1997. During Jul they grow increasingly less common, and most are gone before Aug. Tout's (1947) latest date from Lincoln Co is 22 Jul. Most of the 30 or so Aug reports are from the first half of the month, and there are only 4 Sep reports.

Easterly reports are very rare; one was at Funk Lagoon 13 Sep 1997 (LR, RH).

High counts include 67 at North Platte NWR 24 Jul 1997, 52 there 2 Aug 1997, and 37 at Crescent L NWR 16 Aug 1997.

FINDING

Long-billed Curlews can usually be found on any trip to the central and western Sandhills or northern Panhandle in May or Jun. Migrants can be found along the shoreline

at such locations as L McConaughy during migration.

Hudsonian Godwit

Limosa haemastica

STATUS

Uncommon regular spring migrant east and central, rare casual west. Accidental in fall.

DOCUMENTATION

Specimen

UNSM ZM12692, 12 Jun 1910 Ceresco, Saunders Co.

DISTRIBUTION AND ECOLOGY

Hudsonian Godwits are often found on mudflats that have puddles of standing water, sometimes on muddy shorelines of lakes, reservoirs, and less often rivers, and may also frequent flooded fields of stubble. They are less often attracted to flooded fields and wetlands with fresh emergent vegetation. Hudsonian Godwits occur occasionally in mixed flocks with Marbled Godwits.

Spring

13, 14, 14 Apr→4, 5, 6 Jun

The specimen cited above, 12 Jun 1910, also provides the latest spring record. Migration begins in mid-Apr and is ending by mid-May. In contrast with other migrant shorebird species, peak numbers are generally coincident with the species' arrival. Typically flocks rather than solitary individuals or pairs are seen, with flocks arriving during the third week of Apr, about a week after Marbled Godwits. After the initial wave, successive flocks occur during the rest of Apr and about the first two-thirds of May, after which numbers dwindle. Hudsonian Godwits are most numerous in the Rainwater Basin, common west to Phelps Co, but farther west there are only about 11 reports. Four of these are from the L McConaughy area (Rosche 1994a; FN 50:966).

Westerly reports are 25 Apr 1984 Lincoln Co, 28 Apr 1938 Lincoln Co, 28 Apr 1991 Box Butte Co (AB 45:467), 1 May 1935 Lincoln Co, 11 May 1947 Lincoln Co, 15 May 1977 Cherry Co, and spring 1984 Dawes Co (AB 38:930). There is also a very late undocumented report 18–28 Jun 1982 McPherson Co.

High counts include 93 in the Rainwater Basin 18–19 Apr 1998, 85 at Sacramento-Wilcox Basin 20 Apr 1994, and 54 at Sinninger Lagoon 16 Apr 1995.

Fall

There are 3 fall reports, only 1 documented:

30 Aug 1998 juvenile Sinninger Lagoon (JGJ)

Additional undocumented reports are 6 Sep 1982 McPherson Co and 22 Sep 1979 Lancaster Co.

Although it is generally accepted that Hudsonian Godwits do not return in fall through the interior but migrate south off the Atlantic Coast (Hayman and others 1986), records from the northwestern United States suggest that a few juveniles occur in the interior in Sep (Paulson 1993). There are 5 records of single birds for Missouri: 30 Aug, 21 Sep, 4 Oct (specimen), 9 Oct, and 24 Oct (Robbins and Easterla 1992) and 5 records for the northeastern plains of Colorado (Andrews and Righter 1992).

FINDING

Finding Hudsonian Godwits can be a difficult task at times. Best chances of encountering this species is to visit as many wetland locales as possible on a single outing to the Rainwater Basin during the latter half of Apr.

Marbled Godwit

Limosa fedoa

STATUS

Uncommon regular spring migrant central and west, rare east. Accidental breeder west. Rare regular fall migrant west, rare casual elsewhere.

DOCUMENTATION

Specimen

UNSM ZM15300, 10 May 1908 Dorey L, Cherry Co.

DISTRIBUTION AND ECOLOGY

Marbled Godwits are found wading in shallow water on mudflats and muddy shorelines and in flooded fields. In North Dakota

postbreeders showed a clear preference for foraging in feedlots (Ryan and others 1984). Marbled Godwits occasionally flock with Hudsonian Godwits in central and eastern Nebraska.

Spring

29 Mar, 5, 9 Apr→26, 26, 30 May

Migration begins about the second week of Apr and peaks mid to late Apr. Some 60% of the reports (N = 198) fall within the period 8–30 Apr. Arrival on North Dakota breeding grounds is tightly timed in the period 15–22 Apr (Higgins and others 1979), indicating that most Marbled Godwits indeed "rush through the state in mid-Apr" (JGJ). The smallest numbers are observed in the eastern third of the state, although 25 were at Jack Sinn Marsh, Lancaster Co, 18 May 1997.

High counts include 134 in Keith Co 24 Apr 1999, 78 at Harvard Lagoon 16 Apr 1995, and 50 at Harlan Co Res 5 May 1999.

Summer

The only recent breeding record occurred in 1990 when adults where observed defending young in Dawes Co (AB 44:1153). Marbled Godwits breed regularly in the LaCreek NWR area in south-central South Dakota (SDOU 1991), very near the Nebraska border.

There is another report, likely correct, of a nesting near present-day Columbus 9 Jun 1820 reported by Thomas Say (James 1972). The only other mention of possible breeding was "a godwit involved in homemaking" at a pond in Madison Co in 1876 (Ducey 1988). Bruner and others (1904) thought it "almost certain" to breed in the state but provided no evidence.

Fall

14, 16, 17 Jun→4, 21, 24 Oct

Migrants are found as early as late Jun, and most have moved through by the end of Jul. In contrast to spring, fall migrants are essentially restricted to the Panhandle; there are only 12 reports east of there.

High counts include 40 at L McConaughy 13 Aug 1986 (Rosche 1994a).

FINDING

The species is most likely to be encountered in mid-Apr in wetlands in the Rainwater Basin and at Clear Creek Marshes.

Ruddy Turnstone

Arenaria interpres

STATUS

Rare regular spring migrant east and central, rare casual west. Rare casual fall migrant statewide.

DOCUMENTATION

Specimen

UNSM ZM12626, 16 May 1895 Lancaster Co.

DISTRIBUTION AND ECOLOGY

Most of the year the Ruddy Turnstone is found along rocky or stony shorelines. Migrants observed in Nebraska are usually found on mudflats, on muddy shorelines, and often on wet Platte River sandbars.

Spring

29, 30, 30 Apr→2, 4, 4 Jun

There is a later report 14 Jun 1977 Cherry Co. Migrants are usually only found in May, with a peak in the third week. Some 41% of the reports (N = 56) fall within the very short period 14–19 May. There are only about 12 reports from the western half of the state, most from the L McConaughy region, where Rosche (1994a) considered the species "very rare." Singles or pairs are most often encountered, but occasionally small flocks are found.

High counts include 21 at Sinninger Lagoon 17 May 1997; 18 on a sandbar in the Platte River, Sarpy Co, 24 May 1994; and 18 at Sinninger Lagoon 8 May 1999. An unprecedented total of 76 birds was reported at various locations 7–19 May 1997.

Fall

There are only 10 fall reports, 4 of which are in the last part of Jul, corresponding with the primary southward movement of adults (Paulson 1993). These are 19–20 Jul 1997 Funk Lagoon, 25 Jul 1984 Buffalo Co, 27 Jul 1989 Howard-Hall Cos, and 31 Jul 1989 Pierce Co.

Five reports are clumped during the period mid-Aug through early Sep and are likely all juveniles. These are 15 Aug 1998, a juvenile at L North; 3–4 Sep 1999, a juvenile at Gering SL (SJD); 4 Sep 1986 Douglas-Sarpy Cos; 4 Sep 1986 L North; 8 Sep 1996 Cass Co; and 9 Sep 1939 Lincoln Co. These two periods represent the best time to look for the species in fall. The other report is 28 Oct 1986 L North. Note that 3 reports are from L North, a reservoir that possesses extensive stony shoreline, and the 1996 Cass Co report is from a Platte River sandbar. Such habitat may be utilized more often than mudflats during the more leisurely fall migration and may be overlooked by observers searching for shorebirds.

FINDING

Observers hoping to find the species should search areas such as Jack Sinn Marsh, Platte River sandbars, and Rainwater Basin wetlands during the third week of May. Chances of finding the species in autumn are bleak, but one may try L North or possibly Platte River sandbars.

Red Knot

Calidris canutus

STATUS

Rare casual spring and fall migrant statewide.

DOCUMENTATION

Photograph

5 Sep 1986 L Babcock, Platte Co (Mollhoff 1987).

DISTRIBUTION AND ECOLOGY

This species inhabits mudflats and shorelines; the records are somewhat scattered throughout the state. That about a third of the reports are from Lancaster Co may reflect observer concentration.

Spring

The 11 reports are in the period 30 Apr–26 May, 7 of them in the period 15–26 May. Only 3 are documented:

18 May 1974 Lancaster Co (Bray and others 1986)

20 May 1999 Ayr L (27) (JGJ)

26 May 1991 Platte Co (Gubanyi 1996a)

Other reports, most of which are likely correct, are 30 Apr 1994 Dixon Co, 7 May 1967 Pawnee Co, 10 May 1943 Douglas Co (NBR 10:20), 15 May 1975 Douglas-Sarpy Cos, 16 May 1896 Lancaster Co (Bruner and others 1904), 18 May 1986 Lincoln Co, 19 May 1963 Lancaster Co, 22 May 1997 Garden Co, and 23 May 1993 Keith Co (Rosche 1994a).

Fall

There are 20 reports, 12 documented. Most are in the period Aug–early Sep. Documented records are:

4 Aug 1978 Box Butte Co (Rosche 1982; Bray and others 1986)

7 Aug 1998 molting juvenile L McConaughy (SJD)

11 Aug 1974 Lancaster Co (NBR 43:24)

13 Aug 1995 York Co (Gubanyi 1996c)

15 Aug 1998 adult L Babcock (JGJ)

17 Aug 1985 Douglas Co (Bray and others 1986)

19 Aug 1979 Dundy Co (Bray and others 1986)

5 Sep 1986 Platte Co (Mollhoff 1987)

11 Sep 1975 Dawes Co (Rosche 1982; Bray and others 1986)

12 Sep 1998 juvenile L Minatare (SJD)

23 Sep 1996 juvenile Cass Co (WRS; Brogie 1997)

1 Nov 1963 Lincoln Co (Shickley 1964)

Undocumented reports, most of which are likely correct, are 25–27 Aug 1989 Keith Co (Rosche 1994a), 27 Aug 1896 Lancaster Co (Bruner and others 1904), 31 Aug 1986 Polk Co, 7 Sep 1975 Lancaster Co, 18 Sep 1976 Lancaster Co (AB 31:195), 30 Sep 1893 Douglas Co (Bruner and others 1904), Oct 1874 Nemaha Co (Bruner and others 1904), and 30 Oct 1977 Lancaster Co.

FINDING

Best time to look for the Red Knot is probably Aug–early Sep. Good locations are L McConaughy, L North, Branched Oak L, and wetlands in the Rainwater Basin.

Sanderling

Calidris alba

STATUS

Fairly common regular spring migrant statewide. Common regular fall migrant statewide.

DOCUMENTATION

Specimen

UNSM ZM12694, 21 May 1895 Lancaster Co.

DISTRIBUTION AND ECOLOGY

The Sanderling is usually found along sandy shores, a habitat that is limited in Nebraska to certain reservoirs, and on river sandbars. As a result of recent availability of preferable habitat at reservoirs, stopovers by Sanderlings appear to be increasing. Tout (1947) did not have any Lincoln Co records until the construction of L Maloney. Prior to 1980 there were only 4 reports from Scotts Bluff Co, which now boasts the state's highest count of 703 individuals (NBR 63:101). Sanderlings are consistently observed only at certain reservoirs such as L McConaughy, Sutherland Res, North Platte NWR, and L North. River sandbars are also utilized, such as those in the Platte River, but to what degree is not fully known due limited coverage. Sanderlings also utilize mudflats but usually briefly and in very small numbers.

Spring

26, 26, 29 Mar→3, 3, 4 Jun

There are earlier reports 28 Feb 1981 Washington Co (AB 35:314) and 4 Mar 1992 Scotts Bluff Co and a later report 10 Jun 1972 Brown Co.

Individuals occasionally are seen during the first half of Apr or before, although most do not arrive until the latter half of Apr. Peak migration occurs during May, which has 66% of the reports (N = 95).

High counts include 104 at Crescent L NWR 28 May 1996, 53 at L McConaughy 12 May 1996, and 37 there 12 May 1999.

Fall

1, 4, 4 Jul→7, 9, 10 Nov

Adults tend to arrive during the third week of Jul, although earliest dates are at the beginning of Jul (see above); however, adults are rather uncommon and usually do not remain for extended periods. Juveniles arrive during the latter part of Aug and generally represent the majority of birds reported thereafter.

High counts include 157 at North Platte NWR

10 Aug 1994, 75 there 28 Aug 1996, and 67
juveniles at L North 10 Sep 1995.

FINDING

Sanderlings are more difficult to find in spring
than in fall. In spring good locations to search
are the sandy shorelines at the reservoirs
mention above. One should not expect to see
Sanderlings in alternate (breeding) plumage
in spring, as molt is late (Paulson 1993).
During fall the odds of finding Sanderlings
greatly improve; the same places should
be searched late Aug through early or
mid-Sep.

Semipalmated Sandpiper

Calidris pusilla

STATUS

Abundant regular spring migrant east and
central, fairly common west. Common regular
fall migrant east and central, fairly common
west.

DOCUMENTATION

Specimen

UNSM ZM6190, 13 May 1901 Gage Co.

TAXONOMY

Although no subspecies are recognized, bill
and wing lengths exhibit a cline from
longest in the east to shortest in the west.
Within this cline 3 breeding populations
are distinguished, each with a characteristic
migration route. In spring migrants through
Nebraska appear to include all populations,
although probably fewer of the Eastern
Canadian (eastern) population occur. Of
8 spring UNSM specimens, 4 are from the
Alaskan (western) population, 3 are from
the central Canadian (central) population,
and one, a female, by bill length of 21 mm
(0.8 in), appears to be from the eastern
population. In fall the central and eastern
populations migrate east to the Atlantic Coast,
and Nebraska migrants are of the western
population; the absence of eastern and central
birds results in reduced numbers compared
with spring. All of the 4 UNSM fall specimens
are of the western population.

DISTRIBUTION AND ECOLOGY

This species is found on mudflats and along
muddy shorelines throughout Nebraska,
although larger numbers are found in eastern
areas.

Spring

21, 23, 26 Mar→7, 8, 10 Jun

There is a later undocumented report of one at
Branched Oak L 18 Jun 1997.

Migrants usually arrive mid-Apr, and peak
migration occurs late Apr–early May, a period
in which 55% of all reports (N = 288) occur.
Semipalmated Sandpipers are still common
in late May, when they often flock with
White-rumped Sandpipers. Higher densities
apparently move through eastern sections of
Nebraska, where the highest counts have been
recorded in the eastern Rainwater Basin. West
of Clay Co concentrations do not reach the
same levels.

High counts include 1354 in the Rainwater Basin
1–2 May 1998, 1000 at North Hultine Basin 3
May 1994, 800 at County Line Marsh 21 May
1997, and 400 at Freeman Lakes 18 May 1997.
In the west there were 227 north of Antioch
15 May 1998, 140 at Clear Creek Marshes 2
May 1998, and 65 there 10 May 1985 (Rosche
1994a).

Fall

4, 5, 6 Jul→2, 3, 10 Oct

Determination of late fall dates is difficult due
to the problem of separation from Western
Sandpiper, generally a later fall migrant than
Semipalmated Sandpiper. The late dates
above are considered reliable by us; there are,
however, several undocumented Oct reports,
the latest 25 Oct.

Adults arrive in early Jul and tend not to remain
for extended periods. Juveniles begin to
outnumber adults by mid to late Aug. Nearly
75% of reports are from the latter half of Jul
through the first week of Sep. Abundance
is much reduced in fall relative to spring,
no doubt a result of the absence of the
central and eastern Canada populations (see
Taxonomy).

High counts include 330 at L Babcock 22 Aug
1998, 150 at Pearson Basin 19–26 Aug 1995, and
150 at Funk Lagoon 27 Jul 1997.

FINDING

Semipalmated Sandpipers should be found
wherever mixed flocks of shorebirds occur.
Best chances of finding sizable groups is to
search wetlands in the eastern portion of the
Rainwater Basin in May.

Western Sandpiper

Calidris mauri

STATUS

Rare regular spring migrant statewide. Fairly
common regular fall migrant statewide.

DOCUMENTATION

Specimen

UNSM ZM6195, 31 Aug 1913 Lancaster Co.

DISTRIBUTION AND ECOLOGY

Western Sandpiper frequents shorelines and
especially mudflats, where it often wades in
shallow puddles.

Spring

12, 14, 18 Apr→13, 16, 19 May

Early and late dates above are considered
reliable by us, although there are several
undocumented reports both earlier and
later. Identification problems and the lack of
documented records in Nebraska have made
the status of the Western Sandpiper difficult
to define. It is likely that some if not many
reports are misidentifications of, for example,
Semipalmated Sandpiper. Some 58% of the
reports (N = 76) are from the first 2 weeks
of May, coinciding with peak migration of
Semipalmated Sandpiper. It has been stated
(Abbott 1998) that "if [one is] struggling with
identification, it's not Western . . . a breeding
Western is something like a breeding Dunlin
without the black belly."

Peak migration of Western Sandpiper is in
late Apr. Most Westerns, including those
that winter on the Gulf and Atlantic Coasts,
migrate up the Pacific Coast in spring,
peaking in the Pacific Northwest in late Apr
(Paulson 1993). Very small numbers migrate

through the central Great Plains in spring;
even at Cheyenne Bottoms in Kansas, a major
shorebird stopover point, at best groups
composed of about 10 are found on any
given day (Senner and Martinez 1982). A
similar situation appears to exist in Nebraska,
recent observations (JGJ) suggesting that
small numbers occur some years and none
in others. When present, Westerns would be
expected in late Apr; Rosche (1994a) stated
that they occur in the Keith Co area 22–23 Apr.
Despite speculation that the species is more
common in western areas in spring, there is
little evidence to support this.

High counts include 50 at Clear Creek Marshes
22 Apr 1989 (Rosche 1994a).

Fall

4, 4, 6 Jul→18, 28, 29 Oct

There is an earlier report 28 Jun 1916 of a "flock"
near Hastings (Brooking, Notes).

Western Sandpipers are definitely more common
in fall than in spring. This is in accordance
with the elliptical migration route, proposed
by Senner and Martinez (1982) and supported
by Butler and others (1996), in which fall
migrants through Nebraska are part of
the population that winters on the Gulf
and Atlantic Coasts and which in spring
moves north along the Pacific Coast to its
breeding areas. It appears that Nebraska is
on the eastern edge of this fall route. Largest
numbers pass south and west of Nebraska,
especially through central Kansas, where
several thousand are regularly recorded at
Cheyenne Bottoms (Senner and Martinez
1982) and high counts of several hundred have
been recorded at Quivira NWR in consecutive
years (Skagen and Knopf 1994). The species
is also considered a common to abundant fall
migrant in Colorado (Andrews and Righter
1992). In contrast, no count has exceeded 75
in fall in Nebraska. Eckert (1996) considers
Western Sandpiper less than regular in
Minnesota, with as few as 4 valid records.

Adults arrive in mid-Jul. There are 2 adult
female specimens collected in Lancaster

Co 13 Jul 1919 (Mickel and Dawson 1920). Stopovers by adults are apparently brief, although more data are needed. In contrast, at Cheyenne Bottoms the average length of stay of 75 banded birds was 34 days (Senner and Martinez 1982). Adults seen in Nebraska may be en route to Cheyenne Bottoms. Juveniles arrive as early as 13 Aug (JGJ) and probably account for most northerly and easterly Nebraska reports. They can be fairly common in the eastern Rainwater Basin, where adults occur in generally low numbers as in spring (JGJ). High count of juveniles is 35 at Pearson Basin 19–26 Aug 1995. Juveniles are mostly gone by mid-Sep, although a few remain with the normally late-migrating Dunlin flocks and migrate with them. This has been noted regularly in the Pacific Northwest by Paulson (1993) and at least once in Nebraska, when a juvenile Western Sandpiper was at North Hultine Basin with 2 Dunlin as late as 28 Oct 1996 (WRS). Western Sandpipers are more common to the south and west in fall in Nebraska.

High counts include 100 at Funk Lagoon 24 Jul 1997, 60 at L McConaughy 28 Jul 1991, 47 at Crescent L NWR 13 Aug 1996, and 35 juveniles at Pearson Basin 19–26 Aug 1995.

COMMENTS

The length of time adults stop over in the state needs further analysis. How long are their stays? Do they remain longer in years when conditions are ideal at western locations such as at L McConaughy?

FINDING

The best chance of finding a Western Sandpiper is to search good shorebird areas during the latter half of Aug, when juveniles are most likely to be encountered. Juveniles are the easiest age class to identify.

Least Sandpiper
Calidris minutilla
STATUS

Common regular spring and fall migrant statewide.

DOCUMENTATION
Specimen
UNSM ZM12672, 8 May 1890 Lancaster Co.

DISTRIBUTION AND ECOLOGY

This species frequents mudflats, flooded fields, and shorelines throughout the state.

Spring

7, 9, 18 Mar→3, 10, 10 Jun

Migrants usually arrive early to mid-Apr, and peak migration occurs from the last week of Apr through the first 2 weeks of May, a period in which 61% of all reports (N = 374) occur. By the end of May most have moved on, as there are relatively few reports after 25 May.

High counts include 361 in the Rainwater Basin 9–10 May 1998, 300 at North Hultine Basin 3 May 1994, and 300 at Johnson Basin 1 May 1999.

Fall

23, 30 Jun, 2 Jul→11, 13, 23 Nov

There is a very late report of 2 at L McConaughy 19 Dec 1998 (SJD).

Adults arrive in early Jul. Juveniles follow and are present at least by mid-Aug. The species is widespread and easily found into the first half of Sep, although large flocks are uncommon.

High counts include 250 at L McConaughy 3 Aug 1978 (Rosche 1994a), 137 at Sinninger Lagoon 29 Aug 1997, and 133 in the eastern Rainwater Basin 17 Jul 1999.

FINDING

Least Sandpiper is usually easily found during peak migration periods in early May and Aug throughout the state wherever shorebirds occur.

White-rumped Sandpiper
Calidris fuscicollis
STATUS

Common regular spring migrant statewide. Hypothetical in fall.

DOCUMENTATION
Specimen
UNSM ZM6159, 17 Jun 1902 Cherry Co.

DISTRIBUTION AND ECOLOGY

White-rumped Sandpipers are found on

mudflats and shorelines and in flooded fields throughout Nebraska.

Spring

19, 23, 24 Apr→18, 20, 21 Jun

Early dates above are documented reports; two of the latest dates are specimens, 17 Jun 1902 cited above and 21 Jun 1902 Keya Paha Co, UNSM ZM6160. There are at least 17 undocumented reports in the period 28 Mar–23 Apr, which we have discounted. In the L McConaughy area, Rosche (1994a) lists the earliest date as 6 May, and Robbins and Easterla (1992) cite the earliest date for Missouri as 27 Apr.

The first birds generally arrive by the very end of Apr or early in May but are not numerous until mid-May. During the latter half of May White-rumped Sandpipers often flock with Semipalmated Sandpipers. Unlike other sandpipers, the migration period of White-rumped Sandpiper extends well into Jun.

High counts include 3600 at Freeman Lakes 17 May 1997, 2816 in the eastern Rainwater Basin 30 May 1999, and 1725 in the Rainwater Basin 9–10 May 1998.

COMMENTS

Although there are about 34 fall reports, none are documented, and we are not aware of any fall specimens taken in Nebraska or in any of the neighboring states. The White-rumped Sandpiper, like the Hudsonian Godwit, migrates through the interior in spring and along the Atlantic Coast in fall. Therefore, until there is a substantiated record we are taking a conservative stance regarding fall reports of this species. Andrews and Righter (1992) have taken a similar position, and Rosche (1994a) does not list any records for fall, although Paulson (1993), Robbins and Easterla (1992), Thompson and Ely (1989), and AOU (1998) all suggest that some migrants occur in the interior in fall, but they offer little or no substantiation.

FINDING

White-rumps should be looked for in mid to late May wherever shorebirds are found.

Baird's Sandpiper

Calidris bairdii

STATUS

Common regular spring and fall migrant statewide.

DOCUMENTATION

Specimen

UNSM ZM6163, 28 Mar 1902 Lancaster Co.

DISTRIBUTION AND ECOLOGY

This species occurs on mudflats and muddy shorelines and in flooded fields, mowed moist meadows, and even drier areas such as plowed fields.

Spring

8, 9, 10 Mar→5, 6, 7 Jun

There are later reports 11 Jun 1951 Logan Co, 12 Jun 1999 Funk Lagoon (LR, RH), 13 Jun 1970 Lancaster Co, and 15 Jun 1919 Lincoln Co (Tout 1947). Arrival is in late Mar; it consistently arrives in numbers earlier than other "peep" species. It is common through Apr; numbers begin to diminish by early May, and by mid to late May most are gone.

High counts include 557 in the Rainwater Basin 1–2 May 1998, 506 in the eastern Rainwater Basin 11 Apr 1999, and 400 in the Sacramento-Wilcox Basin 20 Apr 1994.

Fall

9, 10, 10 Jul→7, 7, 8 Nov

There are earlier reports 24 Jun 1998 (JS) and 4 Jul 1996 (FN 50:966), both Funk Lagoon. An extremely late bird was below L Ogallala on the North Platte River 23 Dec 1994 (Dinsmore 1996a; Gubanyi 1996c; Brogie 1998), and another was at L McConaughy 19 Dec 1998 (SJD). There are additional late reports 17 Nov 1988 Lancaster Co, 20 Nov 1980 Scotts Bluff Co, 21 Nov 1982 Douglas-Sarpy Cos, and 5 Dec 1998 Sutherland Res (SJD).

Adults arrive about the second week of Jul, and juveniles follow in mid to late Aug. The species becomes increasingly less common during Sep and more so in Oct, but individuals have lingered into Nov and beyond on a few occasions. It is often found in smaller numbers than in spring in central and eastern

Nebraska; large numbers appear to migrate through the western Great Plains in fall. L McConaughy has hosted very large numbers during 3 different years, including high counts of 2650 in 1990, 4000–5000 in 1991, and 3000 in 1992 (Rosche 1994a). On the eastern plains of Colorado the species is considered a common to abundant migrant in fall but only an uncommon to fairly common migrant in spring (Andrews and Righter 1992).

FINDING

This species usually can be found wherever flocks of shorebirds occur throughout Nebraska during Apr. In fall it is harder to find in eastern areas but is not uncommon. Western areas, such as the west end of L McConaughy, offer the best opportunity of observing this species in fall.

Pectoral Sandpiper

Calidris melanotos

STATUS

Fairly common regular spring migrant east, becoming rare west. Common regular fall migrant east, becoming fairly common west.

DOCUMENTATION

Specimen

UNSM ZM12658, 22 Apr 1890 Lancaster Co.

DISTRIBUTION AND ECOLOGY

This species frequents mudflats and muddy shorelines, especially areas with some vegetation. Pectoral Sandpipers will also forage in drier areas of a wetland that are grown over with short vegetation, which may lead to some individuals going unnoticed by the less than careful observer, as well as giving rise to its nickname "grass snipe." Pectoral Sandpipers are much less common in the western half of the state.

Spring

4, 8, 13 Mar→6, 6, 8 Jun

There are later reports 12 Jun 1910, a specimen, UNSM ZM12663, collected in Lancaster Co; 12 Jun 1997 Pintail Marsh; 12 Jun 1999 (3) Funk Lagoon; 17 Jun 1988 Howard-Hall Cos; 19 Jun 1978 Lancaster Co; and 30 Jun 1967 Garden

Co. Arrival is in early Apr, and peak migration is from the last week of Apr through the first 2 weeks of May, a period in which 50% of all reports (N = 325) occur. The species is a low-density spring migrant, usually found only in small groups.

Pectoral Sandpiper is increasingly rare westward; there are only about 30 reports west of Phelps Co. Rosche (1994a) considered the species casual in spring and listed only 2 records from the L McConaughy region. It is also considered a rare spring migrant on the eastern plains of Colorado (Andrews and Righter 1992).

High counts include 506 in the Rainwater Basin 1–2 May 1998, 182 in the eastern Rainwater Basin 1 May 1999, and 100 at County Line Marsh 21 May 1997.

Fall

5, 6, 8 Jul→9, 9, 9 Nov

There are later reports 20 Nov 1976 and 21 Nov 1982 Douglas-Sarpy Cos and 20 Nov 1996 Richardson Co. A specimen, UNSM ZM12664, was collected 4 Nov 1911 in Lancaster Co. Southbound migrants arrive in Jul, build to a peak in Aug, when high counts suggest a major movement occurs, and then decrease gradually during Sep and Oct. There are about 10 reports for Nov. There are considerably more reports from the west during fall than in spring.

High counts include 433 at L Babcock 10 Aug 1998; 327 at Koenig Basin, Clay Co, 15 Aug 1998; and 300 at Kissinger Basin 20 Aug 1995.

FINDING

Pectorals should be looked for on mudflats with scattered vegetation in eastern Nebraska during peak migration periods, Apr and Aug–Sep.

Sharp-tailed Sandpiper

Calidris acuminata

STATUS

Rare casual fall migrant east and west.

DOCUMENTATION

Description

12 Oct 1986 Octavia, Butler Co (Kovanda and Kovanda 1986).

DISTRIBUTION AND ECOLOGY

In Nebraska, the species has been observed on mudflats.

Fall

There are 2 records, both of juveniles:

8 Sep 1994 Sheridan Co (Brogie 1998; NBR 62:134)

12 Oct 1986 Butler Co (cited above)

COMMENTS

The only records from neighboring states are 1 from Colorado, 26 Oct–6 Nov 1975 in Boulder Co (Andrews and Righter 1992), and 4 from Iowa, including juveniles 30 Sep 1988, 3 Oct 1974, and 14 Oct 1990 and an adult in spring, one of very few interior North American records in spring, identified by B. J. Rose in Fremont Co 15 May 1994 (Kent and Dinsmore 1996). Identification of adults is difficult, and they are easily overlooked among Pectoral Sandpipers.

FINDING

Juveniles of this species are the easiest age class to separate from Pectoral Sandpiper and should be looked for in late Sep and Oct.

Dunlin

Calidris alpina

STATUS

Uncommon regular spring migrant east, becoming uncommon casual west. Uncommon regular fall migrant statewide. Rare casual winter visitor east and central.

DOCUMENTATION

Specimen

UNSM ZM12675, 14 Oct 1890 Lancaster Co.

DISTRIBUTION AND ECOLOGY

Dunlin occur on mudflats, on muddy shorelines, and in flooded fields. They are found primarily in the eastern half of Nebraska, with as few as 8 reports in all from the western half of the state.

Spring

3, 5, 6 Apr→2, 3, 9 Jun

A few Dunlin arrive in Apr, but most move through the state in May. There is an earlier report 31 Mar 1997. Peak migration occurs 8–23 May, when 65% of the reports (N = 87)

occur. A specimen, HMM 2699, was taken 24 May 1916 at Lincoln. Numbers of Dunlin vary from year to year, likely in response to the availability of favorable conditions.

There are few reports from the west: 18 Apr 1986 Keith Co (Rosche 1994a), 19 Apr 1987 Garden Co, 24 Apr 1994 L McConaughy (SJD, BP, LP), 18–19 May 1983 Dawes Co, 18 May 1985 Garden Co (Rosche 1994a), and 23 May 1939 Logan-Lincoln Cos.

High counts include 62 at Branched Oak L 21 May 1997, 46 in the Rainwater Basin 9 May 1998, and 40 in Saunders Co 15 May 1997 (Poague and Dinan 1997).

Fall

4, 6, 8 Sep→20, 21, 26 Nov

Although there are earlier reports 15 Aug 1971 Lancaster Co, 18 Aug 1973 Lancaster Co, and 25 Aug 1978 Garden Co, reports before Sep are unexpected, as, unlike other long-distance migrants of the genus, there is no early movement of adults; adults and young birds migrate together after all have completed molting (Holmes 1966). The major movement of Dunlin is during the latter half of Oct; 42% of the reports fall 14–27 Oct. Two of the reports during the peak period are specimens taken in Lancaster Co, one a molting adult taken 14 Oct 1890 (cited above) and the other a juvenile molting to first basic taken 20 Oct 1917 (UNSM ZM6170).

There are these reports from the west: 25 Aug 1978 Garden Co, 4 Sep 1979 Garden Co, 7 Nov 1998 L McConaughy, and 16 Nov 1997 Sutherland Res.

High counts include 25 in Clay Co 11 Oct 1998, 10 at Cunningham L 25 Oct 1996, and 6 in the eastern Rainwater Basin 7 Nov 1999.

Winter

There are 2 reports. A single bird was at North Platte NWR, 6–23 Jan 1981, and another in basic plumage was at Snyder's Bend 28 Feb 1998 (BFH).

Dec reports are not unusual (Kent and Dinsmore 1996; Robbins and Easterla 1992; Andrews and Righter 1992; Thompson and Ely 1989), but the Nebraska records, one from Kansas, and one from Colorado are the only midwinter reports from Nebraska and neighboring states.

Other Reports

26 Jul 1989 Cedar Co.

FINDING

Dunlin should be looked for in typical shorebird habitat such as mudflats, shorelines, and flooded fields in mid-May in eastern Nebraska.

Curlew Sandpiper

Calidris ferruginea

STATUS

Accidental in fall.

DOCUMENTATION

Photograph

19 Jul 1997 Funk Lagoon (JGJ; Brogie 1998).

Records

The lone record is of an individual molting into basic plumage found at Funk Lagoon 19–21 Jul 1997 (JGJ, WRS; NBR 65:107).

COMMENTS

Records from neighboring states include 15 May 1971, 18 Jul–3 Aug 1975, 4 Aug 1972, and 8 Aug 1969 in Kansas (Thompson and Ely 1989); 17 Jul 1998 in Missouri (Mark Robbins, pers. comm.); and 10 May 1985 and 13 May 1988 in Iowa (Kent and Dinsmore 1996). The fall dates, all in the period 17 Jul–8 Aug, suggest that later-migrating postmolt basic-plumaged adults and juveniles are being overlooked.

Stilt Sandpiper

Calidris himantopus

STATUS

Common regular spring and fall migrant central, fairly common east and west.

DOCUMENTATION

Specimen

UNSM ZM12678, 29 Sep 1890 Lancaster Co.

DISTRIBUTION AND ECOLOGY

Stilt Sandpiper is often found wading in shallow water, usually near mudflats. It is least common in both the eastern and western

portions of Nebraska, being concentrated in a central north-south migratory corridor through the state.

Spring

14, 15, 15 Apr→30, 31 May, 1 Jun

There is an earlier undocumented report 3 Apr 1979 Lincoln Co and later reports 5 Jun 1996 Garden Co, 7 Jun 1995 Garden Co, and undocumented reports 14 Jun 1964 Lancaster Co and "summer" (= Jun 30?) 1967 Garden Co. Arrival takes place by the end of Apr, and peak migration occurs mid-May. Some 65% of the reports (N = 186) are in the period 8–23 May.

High counts include 927 in the Rainwater Basin 15–16 May 1997, 625 at Mallard Haven 16 May 1997, and 605 in the eastern Rainwater Basin 20 May 1999.

Fall

5, 5, 6 Jul→25, 26, 27 Oct

Adults arrive about the second week of Jul, and small groups are observable through mid-Aug. Large flocks of juveniles, most molting into first basic plumage, begin arriving in late Aug and may continue their migration sporadically through Sep.

High counts include 440 at L Babcock 20 Sep 1998, 372 at L McConaughy 6 Sep 1998, and 275 at the west end of L McConaughy 24 Sep 1989 (Rosche 1994a).

FINDING

Stilt Sandpipers are usually not too difficult to find in appropriate shorebird habitat during peak migration periods, May and Aug–Sep.

Buff-breasted Sandpiper

Tryngites subruficollis

STATUS

Fairly common regular spring and fall migrant Rainwater Basin, uncommon east, rare west.

DOCUMENTATION

Specimen

UNSM ZM6199, 12 Sep 1904 Lancaster Co.

DISTRIBUTION AND ECOLOGY

The enigmatic Buff-breasted Sandpiper has for much of the history of ornithology in

Nebraska escaped detection, even though it has throughout its history probably migrated with predictable regularity and in conspicuous abundance, albeit through a very narrow spring migratory corridor. This, along with its association with agricultural areas, a habitat not usually visited by birders, has contributed to its enigmatic status. Like Swainson's Hawk and Franklin's Gull, Buff-breasted Sandpiper has discovered the value of following farm equipment in spring as fields are being worked and planted. In fall, birds often forage in freshly mowed hay and alfalfa fields. When this species frequents mudflats and muddy or sandy shorelines it is more conspicuous, but its use of such habitats is uncommon.

Spring

23, 23, 26 Apr→19, 24, 25 May

During the northward migration Buff-breasted Sandpipers apparently migrate through a very narrow corridor in east-central Nebraska and only stop over in sizable numbers in the eastern section of the Rainwater Basin. York Co birder and farmer Lee Morris likely has more experience with the species than anyone in the state. Morris has routinely observed Buff-breasted Sandpipers following behind him as he is working a field. He describes the species as "very common" and notes "lots of days there are several hundred" (NBR 46:77). Morris has observed Buff-breasted Sandpipers as early as 1 May but claims that the species does not usually arrive until about 10 May.

The region and habitat used in spring have often been neglected by observers, and the actual number of birds using this area is not fully known, but others have observed large numbers in soybean stubble fields during this period (WRS). There are few reports from counties in the extreme east and even fewer from areas west of the eastern Rainwater Basin. Indeed, there are fewer than 30 reports for the state in all. There are these few reports west of a line from Antelope Co to Hamilton

Co: 23 April 1935 Lincoln (Tout 1947), 11 May 1985 Clear Creek Marshes (Rosche 1994a), 22 May 1939 Logan Co, and an undocumented report 12 Jun 1972 Brown Co.

High counts in the York Co area may be in the hundreds (see above), but no specific data have been published. Published high counts include 162 near Freeman Lakes 10 May 1997, 139 in the eastern Rainwater Basin 6–7 May 1999, and 116 south of Shickley 17 May 1997.

Fall

17, 22, 26 Jul→23, 24, 26 Sep

The 17 Jul report above is a specimen, UNSM ZM6198, taken in 1919 in Lancaster Co (Mickel and Dawson 1920). Migration is under way by the beginning of Aug and continues into mid-Sep. The migratory corridor is somewhat wider in fall than in spring. Buff-breasted Sandpipers are, however, still most common in the eastern section of the Rainwater Basin, often foraging in cut alfalfa fields and dried areas of basins with short vegetation. The few reports well away from east-central Nebraska are during Sep and are certainly juveniles. Reports are as far west as Dawes Co 4 Sep 1993, Scotts Bluff Co 14 Sep 1974, and L Minatare 6–12 Sep 1998 (2–3).

High counts include 90 in an alfalfa field in Buffalo Co 3 Aug 1984 (NBR 52:70), 62 in mowed grass at Branched Oak L 1 Sep 1973 (NBR 42:29), and 45 in Clay Co 16 Aug 1998.

Other Reports

12 Jun 1972 Brown Co.

FINDING

This species should be looked for in mid-May and again Aug–early Sep in the eastern portion of the Rainwater Basin.

Ruff

Philomachus pugnax

STATUS

Rare casual spring and fall migrant central and east.

DOCUMENTATION

Photograph

19 Apr 1994 Phelps Co (JGJ; Gubanyi 1996b).

DISTRIBUTION AND ECOLOGY

Ruffs utilize mudflats and are found with other shorebirds.

Spring

There are 3 documented records, including a red-maned male at Sacramento-Wilcox Basin and a black-maned male at Funk Lagoon:

9 Apr 1998 female northeast Antelope Co (MB)

19 Apr 1994 Sacramento-Wilcox Basin (JGJ; Gubanyi 1996b)

24 May 1997 Funk Lagoon (LR, RH; NBR 65:82; Brogie 1998)

Fall

There are 2 documented records, both of juvenile males in flooded fields:

22–23 Sep 1993 near Axtell (JGJ; Gubanyi 1996b)

27 Sep 1998 Kirkpatrick North Basin (JGJ)

FINDING

Ruffs may appear anywhere other shorebirds are concentrated. Reports from neighboring states include about 16 spring records in the period 31 Mar–1 Jun, a specimen from Kansas 22 Jun, and about 8 fall records in the period 12 Aug–10 Oct (SDOU 1991; Kent and Dinsmore 1996; Robbins and Easterla 1992; Thompson and Ely 1989; Andrews and Righter 1992).

Short-billed Dowitcher

Limnodromus griseus

STATUS

Uncommon regular spring and fall migrant statewide (Jorgensen 1996).

DOCUMENTATION

Specimen

USNM 088176, 9 May 1878 Omaha, Douglas Co.

TAXONOMY

All 4 specimens in the UNSM are, in our opinion, *L. g. hendersoni*, the subspecies expected in Nebraska (Jorgensen 1996), as is the specimen cited above. The first and to date only record of the eastern subspecies *L. g. griseus* was of one photographed 2 May 1998 north of Bradshaw (JGJ); *griseus* normally migrates along the Atlantic Coast, and its status on the Great Plains is poorly understood.

We have considered only documented records

here, essentially following Jorgensen (1996), who summarized the status of this species in Nebraska.

DISTRIBUTION AND ECOLOGY

Short-billed Dowitchers are usually found occupying similar habitat as that of Long-billed Dowitchers. Records are scattered throughout the state, but the majority are from the eastern half.

Spring

25, 30 Apr, 4 May→18, 19, 23 May

There is an earlier documented record 20 Apr 1996 Knox Co (MB; Jorgensen 1996). Jorgensen (1996) cited 9 documented records, and there is an additional specimen in the Smithsonian Museum, USNM 088176, taken at Omaha 9 May 1878 (cited above). Short-billed Dowitcher does not usually arrive in Nebraska until May. Most of the records fall within the short period 9–19 May, which corresponds with peaks noted elsewhere (Robbins and Easterla 1992). Small groups only have been reported in Nebraska.

High counts include 31 at Mallard Haven 16 May 1997 (JGJ), 22 at Father Hupp Marsh 16 May 1997 (JGJ), and 18 at Sinninger Lagoon 20 May 1999 (JGJ). An undocumented report, but at a credible date, was of 125 in Otoe Co 19 May 1997.

Fall

5, 6, 7 Aug→8, 10, 11 Sep

Jorgensen (1996) cited 13 documented records. In general, adult Short-billed Dowitchers migrate during Jul and very early Aug (Jehl 1963), somewhat earlier than Long-billed Dowitchers. The only documented Jul record for Nebraska, however, is of 4 at Jack Sinn Marsh 20 Jul 1997 (NBR 65:107). Another Jul record, 21 Jul 1988 in Boone Co, was accepted by the NOURC but considered inconclusive by Jorgensen (1996). Two records are of large flocks of adults during the first week of Aug: 47 were at Clear Creek Marshes 7 Aug 1994 and 40–45 at Funk Lagoon 6 Aug 1995. Jorgensen (1996) discussed the probability that interior adults are males, as all specimens to date are indeed males; females are known to migrate first and are thought to travel nonstop to the Atlantic Coast.

Most fall records are of juveniles; there are 13 documented reports of juveniles in the period 16 Aug–10 Sep, all from the east. The latest fall record of an adult is a specimen taken 28 Aug 1896, UNSM ZM15281, a surprisingly late date for an adult to be found in Nebraska.

Although Short-billed Dowitcher would be the expected species in Jul, a few Long-billed Dowitchers have been noted in late Jul in recent years, and all dowitchers seen in Jun–Jul should be carefully identified.

FINDING

Recent evidence is accumulating that suggests that Short-billed Dowitcher may be more likely eastward. In spring, mid-May is the best time. In summer, Jul should be the best time to search for adults, but considering the lack of documented Jul records, a search for juveniles in Aug and very early Sep during the southward migration may offer the best chance of observing the species. Any observation of a Short-billed Dowitcher should be made with care and with a thorough knowledge of field marks and should be documented.

Long-billed Dowitcher

Limnodromus scolopaceus

STATUS

Common regular spring and fall migrant statewide.

DOCUMENTATION

Specimen

UNSM ZM6176, 10 May 1890 Lancaster Co.

DISTRIBUTION AND ECOLOGY

This species is found on mudflats with shallow water. It is a low-density migrant in the Missouri River valley, but its numbers tend to increase dramatically westward, achieving peak concentrations in the Rainwater Basin, especially in the Clay Co area. Westward from there numbers diminish, and it is common to uncommon in the Panhandle (Rosche 1982).

Spring

18, 21, 22 Mar→2, 6, 13 Jun

Long-billed Dowitchers arrive by the very end of Mar or in early Apr. Peak migration occurs from the last week of Apr through the first 2 weeks of May, a period in which 66% of all reports (N = 353) occur. Numbers decrease quickly, and the species is scarce by late May.

High counts include 1363 in the eastern Rainwater Basin 1–2 May 1999, 1152 in the Rainwater Basin 1 May 1998, 500 at Funk Lagoon 28 Apr 1994, and 500 at Hultine North Basin 3 May 1994.

Fall

19, 19, 20 Jul→10, 13, 19 Nov

There are earlier undocumented reports 6 Jul 1997, 13 Jul 1998, and 18 Jul 1999 at Funk Lagoon and later reports 3 Dec 1976 Cherry Co and 18 Dec 1998 Branched Oak L (JS).

Adults are observed Aug–Sep, and juveniles follow in late Sep and Oct. Determining early arrival dates for Long-billed Dowitcher is difficult because, even though both it and Short-billed Dowitcher should occur in early Aug, few reports have been documented. The relative abundance of the 2 species in late Jul is unknown, although documented reports favor Short-billed Dowitcher. At Funk Lagoon 1 Aug 1999, observers (LR, RH) reported 1 Long-billed, 9 Short-billed, and 12 unidentified to species. There are about 30 reports of Long-billed Dowitcher or dowitcher species during Jul; at least some of these reports are considered questionable, as there is little evidence to support occurrence of Long-billed Dowitcher in Jul. Long-billed Dowitchers do not outnumber Short-billed Dowitchers until late in Aug in Missouri (Robbins and Easterla 1992; Jorgensen 1996). Any dowitcher observed in Jul should be carefully identified. The earliest Nebraska date for Long-billed Dowitcher, confirmed by calls, is 19 Jul (JGJ, WRS).

After molt, dowitchers are virtually impossible to identify except by call, although late reports almost certainly involve juveniles of this species: 8 Nov 1997 (2) Merrick Co "probably this species" (Brogie 1997), 10 Nov 1996 Sarpy Co, and 13 Nov 1994 Holt Co.

High counts include 558 at L McConaughy 3 Oct 1998, 381 in the eastern Rainwater Basin 10 Oct 1999, and 303 there 11 Oct 1998.

FINDING

Long-bills can be found wherever numbers of shorebirds occur in Apr–May and Aug–Oct. Good locations to search are the Rainwater Basin, Jack Sinn Marsh, and Clear Creek Marshes.

Common Snipe

Gallinago gallinago

STATUS

Fairly common regular spring and fall migrant statewide. Uncommon regular breeder north and west, rare casual elsewhere. Rare regular winter visitor statewide.

DOCUMENTATION

Specimen

UNSM ZM12634, 3 Nov 1910 Lancaster Co.

DISTRIBUTION AND ECOLOGY

Although breeding records are few (Ducey 1988), it is likely that this species is a widespread breeder since summer sightings are frequent in appropriate habitat. It is found in moist hay meadows, periodic wetlands, shallow-water marshes, and ditches with standing water. Extensive areas of appropriate habitat may be found in the upper Elkhorn drainage, including Holt Co, and in meadows adjacent to many Sandhills lakes.

Migrants inhabit mudflats or soft substrate often with various types of scattered vegetation, notably smartweed.

Spring

21, 21, 21 Mar→11, 14, 17 May

Dates above are from areas where summering or wintering birds are not reported. Migrants arrive by the end of Mar, and peak migration occurs during the first half of Apr. Singles or small groups are usually found.

High counts include 87 in the eastern Rainwater Basin 3 May 1999, 38 at Bluewing Marsh

4 Apr 1997, and 24 at Funk Lagoon 30 Mar 1997.

Summer

Breeding records are few. Ducey (1988) cited records from Lancaster, Hall, Howard, Sheridan, and Garden Cos since 1960, although summering birds are most often encountered north of the Platte River in the Sandhills and Panhandle, where there are at least 30 Jun reports without evidence of breeding. There are multiple summer reports from Sioux, Scotts Bluff, Garden, Dawes, and Holt Cos. It has bred in Rock Co (Wood 1970; NPG files) and was considered a summer resident at Pony L, Rock Co (Ducey and Schoenenberger 1991) and probably an uncommon nester in Holt Co (Blake and Ducey 1991). Additional counties with summer reports in this region are Boone, Pierce, Brown, Cherry, Loup, Morrill, and Box Butte.

Apart from breeding records cited by Ducey (1988) for Howard, Hall, and Lancaster Cos, there are summer reports in the Platte River valley and southward; these are from Lincoln, Phelps, Lancaster, and Douglas-Sarpy Cos, where breeding apparently occurs on occasion, although there is definite evidence only for Lancaster Co.

Egg dates are from about mid-May through Jun, with laying preceded by several weeks of territorial winnowing by males (see below) (Johnsgard 1979). Downy young have been observed 10 May 1961 in Rock Co and 12 Jul 1987 in Sheridan Co (NGP files).

Fall

15, 20, 25 Aug→winter

Apparent migrants become evident in nonbreeding locations by late Aug. Johnsgard (1979) indicated that young birds gather into flocks (known as "wisps") after fledging and may begin moving south before adults as early as early Aug. Peak migration occurs during late Sep and Oct. Late dates are difficult to determine due to wintering birds.

High counts include 73 at Koenig Basin, Clay Co,

13 Sep 1998; 47 at Kiowa Springs 2 Oct 1999; and 18 in Colfax Co 21 Sep 1996.

Winter

This species winters locally, especially in areas with warm springs that provide soft mud for probing. There are multiple Jan–Feb reports from Scotts Bluff, Keith, Lincoln, and Lancaster Cos, and it has been reported in midwinter as far north as Dawes, Cherry, and Antelope Cos. Rosche (1982) listed Common Snipe as an uncommon winter visitant in northwest Nebraska. There are at least 90 reports Dec–Feb with at least 40 in Jan. Reports are fewest in Feb.

FINDING

Migrants should be looked for on mudflats that possess scattered vegetation, which often allows the snipe to blend into its surroundings. By walking through these areas snipe can frequently be flushed, usually emitting the "whusk" call. Best times are Apr and Oct.

Observers seeking evidence for breeding should post themselves adjacent to moist meadows at twilight during late May and Jun and listen for the eerie buzzing sounds produced by the wings of a territorial male as part of its breeding and territorial displays. A warm, still night could produce an hour or more of this activity, usually referred to as "winnowing."

American Woodcock

Scolopax minor

STATUS

Uncommon regular spring and fall migrant east and central, rare casual west. Uncommon regular breeder east and central.

DOCUMENTATION

Specimen

UNSM ZM6132, 19 Apr 1900 Lincoln, Lancaster Co.

DISTRIBUTION AND ECOLOGY

American Woodcocks are found in moist lowland woodlands as well as in marshy areas with stands of young willows, where soft substrate allows probing for food. Migrants are found in the east and with diminishing frequency west to Kearney Co in the Platte River valley. Nesting has been documented west to Merrick Co but probably occurs west to Kearney Co.

Spring

22, 27 Feb, 4 Mar→summer

Woodcocks arrive and begin displaying by early to mid-Mar. Reports are few west of Kearney Co: 2 reports without dates Lincoln Co (Tout 1936) and 16 May 1936 Lincoln Co. Other reports toward the western edge of the range are 12 Mar 1975 Clay Co and 26 Mar 1988 Polk Co.

High counts include 15 at Branched Oak L 25 Mar 1997.

Summer

Historically, woodcocks were restricted to the Missouri River and lower Elkhorn River valleys (Bruner and others 1904; Lingle 1981), but during the 20th century the species moved west as habitat became suitable, primarily along the Platte River valley. Ducey (1988) cited breeding records from Lancaster, Sarpy, and Merrick Cos since 1960. It nested in Dodge Co in timber near the Platte River in 1978 (RSS), and displaying birds have been noted in western Douglas Co over a moist meadow since 1975 (RSS). An adult with 4 young was seen late Jun 1998 at Fremont Lakes, Dodge Co. A specimen was reported taken near present-day Columbus in 1856 or 1857 (Coues 1874). Adults were found breeding in Merrick Co 6 Jul 1978 and probably at the same place in 1979 (Lingle 1981; Lemburg 1979). As recently as 1994, woodcocks have been found displaying just south of Kearney (NBR 62:75) and have been reported regularly there since, although nesting evidence is still lacking. It was reported from Howard-Hall Cos 27 Jun 1987 and 27 Jun 1988 and from Knox Co 18 May 1997, 31 May 1988, 8 Jun 1991, and 20 Jun 1980. A displaying male was recorded along the Niobrara River in Brown Co 1 May 1998 (NGP files).

The species is sometimes difficult to detect,

likely the reason for the relatively few reports. The true extent of the breeding range is not well known and may extend farther west along the Platte.

Egg dates are Mar–May in Kansas and late Apr–Jun in Minnesota, with young present in Kansas Apr–May (Johnsgard 1979).

Fall

summer→14, 18, 20 Nov

The few reports are scattered throughout the period. There is a later report 8 Dec 1994 Otoe Co, which consisted of a woodcock emerging from a Wood Duck nest box near a warm spring during a spell of subzero weather (Bergman and Bergman 1995). Westerly reports are limited to a report of 3 birds in Lincoln Co 23 Oct 1928 (Tout 1947) and 1 at Crescent L NWR headquarters 6 Nov 1999 (SJD). Andrews and Righter (1992) list 5 records for the northeast Colorado plains in the period 5 Jul–30 Nov.

FINDING

Look for this species at dusk in Apr–May at areas where it is known to display, such as Branched Oak L and the Kearney Hike-Bike Trail at Fort Kearny.

Wilson's Phalarope

Phalaropus tricolor

STATUS

Abundant regular spring migrant central and west, common east. Common regular fall migrant statewide. Uncommon regular breeder north-central and west, rare casual south-central and east.

DOCUMENTATION

Specimen

UNSM ZM12705, 16 May 1895 Lancaster Co.

DISTRIBUTION AND ECOLOGY

Wilson's Phalarope is found in shallow wetlands, in marshes, and on mudflats, both during migration and the breeding season, although there is a preference for alkaline marshes in the Sandhills and Panhandle during the breeding season. Numbers are fewest in eastern Nebraska.

Spring

23, 30 Mar, 1 Apr→1, 2, 4 Jun

Late dates above are from locations where summering was not reported. Usually arrival is in mid-Apr, and peak migration occurs during the last week of Apr and the first week of May, a period in which 50% of the reports (N = 536) occur. The lowest numbers are found in the eastern quarter of the state.

High counts include 10 000 at North Platte SL 6 May 1996, 2000 in Grant Co 30 Apr 1994, and 1534 at L Ogallala 1 May 1999.

Summer

The species breeds, presumably regularly, in the Sandhills and northern Panhandle (Johnsgard 1979; Ducey 1988). Despite a paucity of records, the species may also occasionally breed in suitable marshes of the Rainwater Basin, such as at Harvard Lagoon, where a downy young bird was found 21 Jul 1996 (JGJ). There are several additional Jun–early Jul observations from Adams, Phelps, Fillmore, Howard-Hall, and Clay Cos.

There are rather recent breeding records from eastern Nebraska, including Lancaster Co in 1985 (NBR 53:80) and 1991 (NBR 59:59–61), Nance Co in 1974 (Ducey 1988), and Saunders Co in 1987 (Ducey 1988). There are additional breeding season reports from Lancaster Co 4 Jun 1982, 18 Jun 1991, 19 Jun 1978, 25 Jun 1983, 5 Jul 1972, and 12–14 Jul 1973.

Midsummer reports away from known breeding locations are few; while some may indicate local breeding, at least 1 bird, at North Platte SL 15 Jun 1997, was in basic plumage (WRS). Other reports are 9 Jun 1984 Douglas-Sarpy Cos, 30 Jun 1985 Pierce Co, and 5 Jul 1997 Pintail Marsh.

Fall

19, 22, 22 Jul→20, 21, 22 Oct

Early dates above are from locations where summering was not reported. Migrants can usually be detected away from breeding areas by late Jul, but most movement is primarily during Aug and early Sep. Large concentrations occur at Crescent L NWR in

Jul, possibly flocking of breeding birds from nearby areas; 1321 were there 8 Jul 1997. By the end of Sep most are gone, as there are few Oct reports.

High counts include 2000 at Crescent L NWR 24 Aug 1997, 1600 there 13 Aug 1996, and 1500 in Morrill Co 28 Jul 1994.

FINDING

The best time to find Wilson's Phalarope is late Apr and early May. During this period the species is usually widespread and numerous, even in extreme eastern Nebraska. During the summer potholes along the road between Crescent L NWR and Lakeside and along Highway 2 between Hyannis and Lakeside should provide numerous glimpses of this species.

Red-necked Phalarope

Phalaropus lobatus

STATUS

Fairly common regular spring migrant west, becoming rare casual east. Fairly common regular fall migrant west, becoming uncommon east.

DOCUMENTATION

Specimen

UNSM ZM6221, 23 Jun 1916 Mitchell, Scotts Bluff Co.

DISTRIBUTION AND ECOLOGY

This species inhabits marshes, sewage lagoons, shallow lakes, and prairie potholes during migration. About two-thirds of all reports are from the northern Panhandle south to Scotts Bluff Co. The remaining records are rather evenly distributed across the remainder of the state, although numbers are generally low, leading some observers to consider it rare or casual (Rosche 1994a; Tout 1947). In eastern Nebraska, Lancaster Co claims at least 11 reports and Douglas-Sarpy Cos 6 reports.

Spring

19, 24, 24 Apr→28, 30 May, 1 Jun

There are later reports 6 Jun 1908 Lancaster Co, 10 Jun Garden-Sheridan Cos (RSS), 12 Jun 1992 Sioux Co, 13 Jun 1992 Sheridan Co, and 23 Jun 1916 (specimen cited above).

A relatively late migrant, the Red-necked Phalarope generally does not arrive until the second week of May, and peak migration occurs during about the third week. All but 16 of 83 reports fall within the short period 16–23 May. Small numbers linger into Jun, being seen regularly as late as 10 Jun in potholes of Garden and Sheridan Cos (RSS). There are about 20 reports from the eastern half of the state.

High counts include 250 at Crescent L NWR 22 May 1997, 54 north of Antioch 15 May 1998, and 52 at Swan L, southwest Holt Co, 20 May 1996.

Fall

25, 28, 30 Jul→11, 13, 14 Oct

There are about 65 reports, including these earlier reports: 23 Jun 1910 Scotts Bluff Co (specimen cited above), 28–30 Jun 1985 Sioux Co, 11 Jul 1973 Sheridan Co, and 17 Jul 1979 Garden Co. Reports prior to mid-Aug are from western Nebraska, presumably adults, with a single exception, 30–31 Jul 1998 Schilling Refuge (BP, LP, GW). Later reports are more widespread and presumably are mostly juveniles, which wander more than adults. In eastern and southern Nebraska juveniles are the most likely age to be encountered; 3 sight records in 1995 and 2 specimens, all in Sep, were juveniles.

High counts include 45 in Sheridan Co 21 Sep 1996 (FN 51:80), 16 at Crescent L NWR 16 Sep 1996, and 10 at L McConaughy 11 Sep 1988 (Rosche 1994a).

FINDING

Red-necked Phalaropes should be looked for in shallower waters of Sandhills marshes, sewage ponds, potholes, and playas in the northern Panhandle during the third week of May.

Red Phalarope

Phalaropus fulicaria

STATUS

Rare casual fall migrant central and east.

DOCUMENTATION

Specimen

FMNH, 15 Oct 1921 Dad's L, Cherry Co (Conover 1934; Bray and others 1986).

DISTRIBUTION AND ECOLOGY

In Nebraska observations have been on shallow
water, shorelines, and mudflats. Its occurrence
statewide in fall is consistent with records
from adjacent states (SDOU 1991; Thompson
and Ely 1989; Andrews and Righter 1992),
although there are a few spring records from
neighboring states (see Comments).

Fall

All but 1 of the 8 records are of birds in juvenal
or first basic plumage. The exception is a 1
Aug record of a female in alternate (breeding)
plumage. All records are documented.

1 Aug 1993 Phelps Co (Gubanyi 1996a)

24 Aug 1993 Phelps Co (JGJ; Gubanyi 1996a)

13 Sep 1998 2 molting juveniles L Babcock (JGJ)

23 Sep 1985 basic Pierce Co (Brogie and Brogie
1985; Mollhoff 1987)

26 Sep 1987 molting juvenile Pierce Co (Brogie
and Brogie 1988; Grenon 1990)

1 Oct 1995 molting juvenile Lancaster Co (JGJ;
Gubanyi 1996c)

10–13 Oct 1986 juvenile Platte Co (NBR 55:64;
Mollhoff 1987)

15 Oct 1921 Cherry Co, specimen cited above

COMMENTS

Although there are no spring records for
Nebraska, records from neighboring states
include at least 10 in the period 10 May–5 Jun
but most in late May. An earlier record (Swenk
1915a) was subsequently withdrawn, as the
specimen involved was collected in South
Dakota (Swenk 1918a; Bray and others 1986).

FINDING

Chances are slim, but one may stumble onto this
species in Sep or early Oct at any wetland.

Laridae (Jaegers, Gulls, Terns)

Jaeger sp.

Stercorarius sp.

STATUS

Rare casual fall migrant statewide. Hypothetical
in spring.

DISTRIBUTION AND ECOLOGY

Jaegers are found at larger lakes and
reservoirs. Most reports are of juveniles,

and identification can be extremely difficult
when dealing with a lone bird at some
distance. Based on records from Nebraska and
surrounding states, Pomarine and Parasitic
are the most likely species to pass through the
interior.

There are about 24 reports of jaegers for the state.
Of these, only 9 are documented; these are
treated under the respective species accounts.

Spring

There are few reports, none documented. Bruner
and others (1904) noted reports May 1869 in
Dakota Co and May 1873 near Fremont of
birds thought to be Pomarine Jaegers, but no
details exist, and a jaeger seen by Brooking
and Black in May 1919 at Crescent L was not
identified; Brooking thought it a Long-tailed
and Black a Parasitic (Swenk, Notes before
1925).

Fall

There are several reports unidentified to
species. One was seen 8 Jul 1975 at Overton L
(Robertson 1977). Single sightings of probably
the same bird unidentified to species were
at L McConaughy 19 Aug 1998 (JT) and 31
Aug 1998 (BP, LP). One collected by Eiche
in Lincoln 13 Sep 1898 was identified as a
Parasitic Jaeger (Bruner and others 1904),
but the location of the specimen is unknown.
An unidentified jaeger was at Pawnee L 22
Sep 1997 (JS). One identified as a Parasitic
Jaeger was collected by Black near Kearney 2
Oct 1926 (LOI), the location of the specimen
unknown. Rosche (1982) had a "quick look" at
a jaeger, thought to be probably a Pomarine,
at Whitney L 7 Oct 1973. A dark juvenile
was seen at a distance in bad weather at
L McConaughy 12 Oct 1997 but was not
identifiable to species (SJD; Brogie 1998). A
light phase bird stated to be a Pomarine Jaeger
was in the Brooking collection, collected Oct
1927 or 1928 (Swenk, Notes after 1925) or 1920
(Brooking, Notes), although no locality was
given and the specimen may now be lost.
Swenk noted measurements of a Pomarine
Jaeger in the Black collection that was taken
Oct 1918 (Swenk, Notes after 1925), but no

locality of collection is noted and the report thus is not acceptable (Bray and others 1986). One was seen at a distance at L McConaughy 3 Oct 1998 (SJD). A bird identified as a Parasitic Jaeger but not accepted by the NOURC was at L McConaughy 8 Nov 1997 (Brogie 1998). One shot near North Platte by Barnum 11 Nov 1895 was identified as a Pomarine Jaeger (Bruner and others 1904), but the location of the specimen is unknown. A dark bird stated to be a Parasitic Jaeger was collected 18 Nov 1926 near Kearney and was at one time HMM 3099a (Swenk, Notes after 1925), but the specimen is not now extant (Bray and others 1986). A dark bird identified as a Pomarine Jaeger was collected near Kearney Nov 1919 and seen in the Olson collection (Swenk, Notes before 1925), but the specimen may be lost.

FINDING

Best time to look for jaegers is during Sep and Oct at larger lakes and reservoirs. Identification is difficult if views are at a distance and other birds are unavailable for size comparison. Any sighting identified to species should be well documented.

Pomarine Jaeger

Stercorarius pomarinus

STATUS

Rare casual fall migrant statewide.

DOCUMENTATION

Specimen

Olson collection at UNK, 8 Oct 1917 Buffalo Co (Swenk, Notes after 1925; Bray and others 1986).

DISTRIBUTION AND ECOLOGY

Reports are from larger lakes and reservoirs. Jaegers are usually associated with gulls or terns, which they harass for food.

Fall

Among several reports of birds purported to be this species (see Jaeger sp.), only 7 are documented:

30 Jun 1990 light phase adult photographed Lewis and Clark L (Grenon 1991; Brogie 1997)

28–29 Aug 1999 light phase adult Merritt Res (JS)

8 Sep 1973 description Branched Oak L (DG, MG; NBR 42:29–30)

8 Oct 1917 specimen cited above

14–19 Nov 1997 juvenile Pawnee L (JGJ, BP, LP, JS; Brogie 1998)

18–19 Nov 1999 3 juveniles, 2 dark and 1 intermediate phase L McConaughy (SJD)

15–17 Dec 1991 descriptions Conestoga L (Gubanyi 1996a)

COMMENTS

See Jaeger Sp. Kent and Dinsmore (1996) indicate that this species occurs in the interior mostly Sep–Nov, but there are records May–Aug and Dec also. There are about 15 documented records from neighboring states in the periods 12–16 May (2), 30 Jun–12 Jul (2), 28 Sep–10 Oct (5), and 7 Nov–5 Dec (6).

Parasitic Jaeger

Stercorarius parasiticus

STATUS

Rare casual fall migrant statewide.

DOCUMENTATION

Specimen

UNSM ZM12309, 23–24 Aug 1968 Sheridan Co (Gates 1969).

DISTRIBUTION AND ECOLOGY

This species is found at larger lakes and reservoirs.

Fall

Of several reports that may have been this species (see Jaeger sp.), there are but 2 documented records. An immature bird found alive 23 Aug 1968 subsequently died and was collected 24 Aug. It was initially identified as a Skua sp. (Gates 1968) but later reidentified as a Parasitic Jaeger by James Tate at UNSM (Gates 1969), where it is a specimen (cited above). Another, a juvenile, was followed in a boat for some 42 km (26 mi) on L McConaughy 5 Oct 1997 (Dinsmore and Silcock 1998); the same bird may have been present 4 Oct (SJD; Brogie 1998).

23–24 Aug 1968 near Hay Springs, specimen cited above

5 Oct 1997 juvenile L McConaughy (Dinsmore and Silcock 1998; JGJ, JS; Brogie 1998)

COMMENTS

See Jaeger sp. Kent and Dinsmore (1996) indicate that this species occurs in the interior more frequently than the other species, with peak records in Sep–Oct. There are about 16 documented records from neighboring states, including 23 Apr (1) and between 13 Aug and 3 Nov, with most (7) in Sep.

Long-tailed Jaeger

Stercorarius longicaudus

STATUS

Rare casual fall migrant statewide.

DOCUMENTATION

Specimen

UNSM ZM13093, 1 Sep 1952 Lancaster Co (Baumgarten and Rapp 1953).

Fall

Of about 4 reports that may be this species (see Jaeger sp.), only 2 are documented. The specimen cited above was collected at Salt L, Lancaster Co and identified as this species by R. C. Murphy; it is an immature (Baumgarten and Rapp 1953).

1 Sep 1952 Lancaster Co, specimen cited above

3 Oct 1998 light or intermediate morph juvenile L McConaughy (SJD)

COMMENTS

See Jaeger sp. Kent and Dinsmore (1996) indicate that this species occurs in the interior mostly in late Aug–Oct but also in late May–late Jun. There are 9 documented records from surrounding states in the periods 29 May–23 Jun (4) and 4 Sep–8 Oct (5). Although apparently the least likely jaeger species to occur in Nebraska, Long-tailed seems to be the most likely in spring.

Laughing Gull

Larus atricilla

STATUS

Rare casual spring, summer, and fall visitor. Accidental in winter.

DOCUMENTATION

Photograph

27–29 Apr 1990 Gavin's Point Dam, Cedar Co (Brogie 1990; Grenon 1991).

DISTRIBUTION AND ECOLOGY

This species is found around larger lakes and reservoirs. Records from most adjacent states are increasing rapidly partly because of an increase in numbers of reservoirs, because existing reservoirs become more productive as they mature ecologically, and because birders are becoming more informed. We predict that records for Nebraska will increase for similar reasons. Most records are during the warmer months.

Spring

There are about 13 reports; most are undocumented sight reports, 7 from the period 1947–55, but there are 3 documented records:

2 Apr 1915 Inland, Clay Co, female, measurements (Brooking 1933a), at one time HMM 2631 but no longer extant (Bray and others 1986)

4 Apr 1997 adult, Father Hupp Marsh, Thayer Co (JGJ; Brogie 1998)

27–29 Apr 1990 adult, Gavin's Point Dam, Cedar Co (cited above)

Additional reports, undocumented and most probably Franklin's Gulls, are 5 Apr 1953 Platte Co, 10 Apr 1947 Keith Co, 12 Apr 1952 Platte Co, 14 Apr 1952 Keith Co, 19 Apr 1992 Phelps Co (NBR 60:138), 28 Apr 1950 Keith Co, 28 Apr 1952 Boyd Co, 1 May 1955 Dawson Co, 19–21 May 1977 Lancaster Co (NBR 45:31), and 26 May 1986 Lancaster Co (Mollhoff 1987).

Summer

There are 4 reports, 3 documented:

14 Jun 1998 second alternate Keystone L (WRS, JS)

30 Jun 1992 second alternate Keith Co (Dinsmore and Silcock 1995c)

3 Aug 1998 juvenile Branched Oak L (BP, LP)

There is an old report without details Jul 1880 at Alda (Bruner and others 1904).

Fall

There are 3 reports, 2 documented:

28 Oct 1996 second basic Lincoln Co (WRS; Brogie 1997)

22 Dec 1994 first basic Lincoln Co (Dinsmore and Silcock 1995c; Gubanyi 1996c; Brogie 1998)

Johnsgard (1980) cited a report 5–7 Dec without further details.

Winter

There is a single documented winter report, highly unusual for this species, which winters along the Gulf and Atlantic Coasts. This bird apparently was present in the area 10 Jan–28 Apr (SDOU 1991).

10 Jan 1990 Gavin's Point Dam (MB; Grenon 1990)

FINDING

Careful observation of gulls at large reservoirs during the summer or in the period Apr–Dec is probably the best strategy for finding this species in Nebraska.

Franklin's Gull

Larus pipixcan

STATUS

Abundant regular spring and fall migrant statewide. Accidental breeder Sandhills. Uncommon regular summer visitor statewide. Rare casual winter visitor statewide.

DOCUMENTATION

Specimen

UNSM ZM12719, 22 Apr 1892 Lancaster Co.

DISTRIBUTION AND ECOLOGY

Flocks of Franklin's Gulls are found in a variety of habitats during migration, including lakes, marshes, and fields. Flocks will often hover behind farm implements that are being used to work fields in spring.

Spring

22, 23, 23 Feb→summer

Usually arrival is in the latter half of Mar, although occasionally individuals arrive earlier. Peak migration occurs during the second half of Apr, and most migrants have passed through by the end of May.

High counts include "thousands" in the Rainwater Basin 23 Apr 1995, 1000 at Syracuse 26 Apr 1995, and 1000 at Lincoln 7 May 1995.

Summer

There are 2 breeding records. A nest was reported for Garden Co in 1965 (Sharpe 1966), probably 1 of the 2 found at Smith L, Garden Co, 7–11 Jun that year, when 30 nesting pairs were present (Sharpe and Payne 1966). An additional 2 nests were reported from Garden Co in 1966 (Sharpe 1967). Nearest regular breeding is in northeast South Dakota (Johnsgard 1979; SDOU 1991).

There are at least 60 reports of Franklin's Gulls during Jun, including as many as 130 birds at L Ogallala 10 Jun 1998, and numerous reports in Jul, a majority of which likely represent nonbreeding subadults.

Fall

summer→26, 28, 30 Nov

Small numbers begin appearing during Jul, numbers building into Aug and achieving their peak during Sep and early Oct. Numbers decline rapidly into Nov. High counts at L McConaughy as early as 6 Aug (NBR 62:15) suggest use of that reservoir as a staging area for migration.

High counts include 85 000+ at Sutherland Res 28 Sep 1996, 35 000 there 29 Aug 1997, and 22 000 at Calamus Res 7 Sep 1998.

Winter

Although this species normally winters in South America, there are at least 10 winter reports for Nebraska. There are 10 reports in the period 1 Dec–21 Feb, only 2 of these prior to 1988, including successful overwintering of a single alternate-plumaged bird at L McConaughy 14 Dec–21 Feb 1993–94 (NBR 62:22, 56). The other reports are 13–17 Dec 1988 Cedar Co, 16 Dec 1993 L Ogallala (Gubanyi 1996c), 20 Dec 1975 Scotts Bluff Co, 25–26 Dec 1988 Lancaster Co, 6 Jan 1990 Keith Co, 14 Jan 1988 Lincoln Co, 19 Jan 1955 Lincoln Co, 8 Feb 1992 Phelps Co, 16 Feb 1992 Douglas-Sarpy Cos, and Feb 1992 Sioux Co.

FINDING

Franklin's Gulls are easily found at reservoirs, on lakes, and in fields. During peak migration periods, in late Apr and Sep, the species can often be found by simply looking into the sky.

Little Gull

Larus minutus

STATUS

Rare casual spring and fall migrant statewide.

DOCUMENTATION

Photograph

26 Apr 1995 Wehrspann L (JGJ; Gubanyi 1996c).

DISTRIBUTION AND ECOLOGY

Little Gull reports are from larger lakes and reservoirs.

Spring

There are 2 records. The alternate-plumaged individual at Wehrspann L 26 Apr 1995 (cited above) was with a large flock of Bonaparte's Gulls.

19–20 Apr 1997 third alternate adult L McConaughy (SJD; Brogie 1998)

26 Apr 1995 Sarpy Co (cited above)

Fall

There are 8 records, including 5 in 1998:

6 Sep 1997 juvenile L Minatare (SJD, JS, WRS; Brogie 1998)

8 Sep 1998 juvenile molting to first basic L Ogallala (SJD, JS)

8 Sep 1998 juvenile L McConaughy (different bird from above; SJD, JS)

20 Sep 1998 first basic L McConaughy (SJD)

3–5 Oct 1996 juvenile Pawnee L (JGJ; Brogie 1997)

17–18 Oct 1998 juvenile/first basic L Minatare (SJD)

19 Oct 1997 second or third basic Summit Res (JGJ)

1 Nov 1998 first basic Sutherland Res (SJD)

FINDING

While the chances of finding a Little Gull are still slight, observers visiting reservoirs could stumble upon this species, especially among Bonaparte's Gulls; all small gulls should be carefully checked.

Black-headed Gull

Larus ridibundus

STATUS

Accidental in fall.

DOCUMENTATION

Description

12 Aug 1979 Walgren L, Sheridan Co (Rosche 1982; Bray and others 1986).

DISTRIBUTION AND ECOLOGY

This species is found on lakes and reservoirs, usually in the company of Bonaparte's Gulls.

Fall

There are 2 reports, one cited above and the other by an experienced observer but reported secondhand and not accepted by the NOURC. The latter was a bird seen near Venice, Douglas Co, 18 Nov 1985 (NBR 54:28).

COMMENTS

This gull is rare on the western Great Plains, although it has become fairly common in recent years along the East Coast. There are at least 12 records from states neighboring Nebraska, indicating spring and fall migrant status. The records are in the periods 4 Apr–25 May and 5 Sep–23 Nov, with 2 northwest Iowa records suggestive of breeding 2–12 Aug 1994 and 23 Jun–9 Jul 1996 (Kent and Dinsmore 1996).

FINDING

This species, like Little Gull, should be looked for among flocks of Bonaparte's Gulls.

Bonaparte's Gull

Larus philadelphia

STATUS

Fairly common regular spring and fall migrant statewide. Accidental in summer.

DOCUMENTATION

Specimen

UNSM ZM6622, 31 Mar 1893 Lancaster Co.

DISTRIBUTION AND ECOLOGY

Bonaparte's Gulls prefer larger lakes and reservoirs, which until the last quarter of the 20th century were rare in Nebraska. Prior to construction of large reservoirs, it is likely that most Bonaparte's Gulls passed over the state. Bruner and others (1904) had records only from Salt L, now Capitol Beach, a fairly large natural lake in Lancaster Co. Even in the mid–20th century records were scarce, particularly in the west. Tout (1947), for example, did not have any such Lincoln Co records. Currently, however, the species is found routinely in large numbers, especially at reservoirs in the west such as Sutherland Res and L McConaughy. Rosche (1994a) noted that at L McConaughy the species' abundance "seems directly correlated with the presence or absence of storms," with more found during stormy years and fewer during mild ones. However, it is still most common in the east, from whence approximately 75% of all reports are derived.

Spring

8, 10, 14 Mar→29, 31 May, 1 Jun

There is an earlier report 22 Feb 1998 (4) at Niobrara and later reports 8 Jun 1919 Lancaster Co (Mickel and Dawson 1920), 10 Jun 1989 Lancaster Co, and "early Jun" 1996 at L McConaughy (Brown and others 1996).

Bonaparte's Gulls usually arrive in early Apr, and peak numbers pass through during mid and late Apr.

High counts include 465 at L McConaughy 18 Apr 1998, 322 there 26 Apr 1997, and 277 at L McConaughy 24 Apr 1999.

Summer

The only report is of 2 present at Inland (now Harvard) Lagoon "all summer" in 1915. One was collected by Brooking 28 Aug 1915 and was seen by Swenk in the Brooking collection (Swenk, Notes before 1925).

Fall

1, 2, 2 Sep→19, 19, 24 Dec

There are earlier reports 19 Jul 1919 Lancaster Co (Mickel and Dawson 1920), 20 Jul 1988 Lancaster Co, 9 Aug 1989 Lancaster Co, 18 Aug 1977 Douglas Co, 20 Aug 1998 L Alice, and 25 Aug 1991 Douglas-Sarpy Cos; later reports 2 Jan 1998 (34) Sutherland Res and 3 Jan 1998

(3) L McConaughy; and an undocumented report 12 Jan 1978 Lincoln Co.

Migrants usually do not arrive until the latter half of Oct, and peak numbers pass through in early to middle Nov. There are only 12 reports before Oct 1 and about 12 reports in Dec. As many as 91 were still at Sutherland Res 18 Dec 1997.

High counts include 675 at L McConaughy 6 Nov 1975 (Rosche 1994a), 550 at Branched Oak L 1 Oct 1995, and 340 at Gavin's Point Dam 22 Nov 1998.

FINDING

Bonaparte's Gulls are most easily found during Apr and again in late Oct and Nov at larger reservoirs such as Branched Oak L, Pawnee L, L McConaughy, and Cunningham L.

Mew Gull

Larus canus

STATUS

Rare casual spring visitor west and central. Accidental in fall.

DOCUMENTATION

Photograph

11 May 1996 Keith Co (JGJ; Brogie 1997).

TAXONOMY

All records to date have been of the western North American subspecies *brachyrhynchus*.

DISTRIBUTION AND ECOLOGY

This species is found on large lakes and reservoirs.

Spring

There are 8 records involving 11 birds, all recent, all documented, and all but 1 from L McConaughy:

17 Feb 1996 second basic Keith Co (WRS; Brogie 1997)

7–22 Feb 1998 first basic, second basic, 2 adults L McConaughy (SJD, WRS, JS)

21 Feb 1999 adult Johnson L (JGJ, WRS)

21 Mar 1998 adult basic L McConaughy (SJD)

12–13 Apr 1997 adult Keith Co (SJD; Brogie 1998)

18–19 Apr 1998 adult L McConaughy (SJD)

19 Apr 1998 adult L McConaughy (SJD)

11 May 1996 adult Keith Co (JGJ; Brogie 1997)

Fall

The only record is from the east. It is of interest that most Iowa records are also in Dec (Kent and Dinsmore 1996).

1–5 Dec 1996 adult Branched Oak L (JGJ; NBR 65:21; Brogie 1998)

COMMENTS

In recent years this species is being recorded more often on the Great Plains. Records from states neighboring Nebraska are in the period 28 Aug–28 Apr.

FINDING

Careful checking of flocks of Ring-billed Gulls may yield this species, especially during late winter and spring at western reservoirs.

Ring-billed Gull

Larus delawarensis

STATUS

Abundant regular spring and fall migrant statewide. Fairly common regular summer visitor statewide. Uncommon regular winter visitor statewide.

DOCUMENTATION

Specimen

UNSM ZM6619, 22 Sep 1917 Lancaster Co.

DISTRIBUTION AND ECOLOGY

Ring-billed Gulls frequent lakes, marshes, and fields during migration. During winter and also in summer the species is found mainly around reservoirs, where birds scavenge and pilfer food. Individuals may be found at any season at the largest reservoirs, although different age classes predominate in different seasons. During migration individuals and small groups are seen with increasing frequency around commercial or urban areas, particularly on parking lots near fast-food establishments.

Spring

Because of wintering birds, the arrival of genuine migrants is indicated only by an increase in numbers by late Feb or early Mar, although to what degree is often decided by the severity of weather, with birds retreating south in adverse conditions. The largest numbers pass through during Mar, with smaller flocks, often consisting of high numbers of subadults, following in Apr and May.

High counts include 19 000 at L McConaughy 14 Mar 1994, 4000 there 7 Feb 1998, and 2550 there 30 Mar 1997.

Summer

Small numbers of nonbreeders, primarily subadults, can be found at larger lakes and reservoirs throughout the state. The nearest regular breeding is in North Dakota (Johnsgard 1979).

High counts include 700 at L McConaughy 9 Aug 1991 (Rosche 1994a) and 245 at North Platte NWR 24 Jul 1997.

Fall

Migrants begin to trickle into Nebraska by Aug or Sep, with numbers peaking in Nov. Most Ring-billed Gulls leave the state with the onset of colder weather.

High counts include 6000–8000 at L McConaughy 1 Dec 1974 (Rosche 1994a), 7000 there 2 Jan 2000, and 5103 at Harlan Co Res 18 Dec 1998.

Winter

Wintering numbers are dictated by the severity of the winter. In general, small numbers overwinter at larger reservoirs that possess open water. Smallest numbers are usually found during Jan, but numbers slowly begin to build at certain reservoirs, such as L McConaughy, during Feb. In the northwest they apparently depart in winter, as Rosche (1982) cited extreme occurrence dates 3 Mar–27 Nov.

COMMENTS

There is a single egg that may be of this species obtained from L McConaughy (Roger Knaggs, pers. comm. SJD, WRS; TEL). A technician working on Piping Plovers at the west end of L McConaughy found a nest with 2 eggs in dead cattails and an adult bird "coming and going" on 22 Jul 1992. The nest was washed out by high water and only 1 egg recovered and sent to UNSM. A photograph of the "nest" that accompanies the single egg at UNSM apparently was taken after the original nest was washed out; Labedz doubted from this photo that there was a real nest and suspected that the egg may have been "dumped."

FINDING

During migration and summer the species is easily found at larger reservoirs. In winter the species is found at reservoirs that possess open water.

California Gull

Larus californicus

STATUS

Locally fairly common regular spring migrant west, rare casual elsewhere. Locally rare regular summer visitor west. Locally common regular fall migrant west, rare casual elsewhere. Locally uncommon regular winter visitor west.

DOCUMENTATION

Photograph

14 Aug 1986 L McConaughy (Brogie 1986; Mollhoff 1989a).

TAXONOMY

There are 2 subspecies of California Gull (Jehl 1987, 1992). While there is no information on the subspecies occurring in Nebraska, it seems that either subspecies may occur. The Canadian Great Plains subspecies, *L. c. albertaensis*, breeds on the Canadian Plains and more recently in eastern North Dakota and northeast South Dakota. Migrants should pass through western Nebraska. However, California Gull numbers at L McConaughy are highest in late summer and early fall, when postbreeding dispersers from Colorado colonies of *californicus* are also likely. At least 2 birds of the nominate race have been seen in Missouri, and both subspecies have been collected in Louisiana (Mark Robbins, pers. comm.).

Albertaensis has a mantle "approaching or matching the paleness of [Herring Gull]" (Jehl 1987) and is larger, paler, and larger billed than the nominate subspecies *californicus*, which breeds mostly in the Great Basin of the United States and also in Wyoming and northern Colorado and has a mantle "much darker" than that of Ring-billed Gull. It has been stated (Pittaway 1997; Leukering 1997; King 1998) that the 2 subspecies are recognizable in the field given good comparison with other gulls for size and mantle color. Opposing views have been expressed, however (Sibley 1997; Jaramillo 1997). Complicating matters is reported hybridization with Herring Gull in Colorado (Andrews and Righter 1992).

DISTRIBUTION AND ECOLOGY

The state's first record (see Comments) is of a bird captured in Howard Co Nov 1934 that had been banded in Montana 17 Jun 1934. Through the early 1980s, there were very few reports: Shickley (1966, 1968) reported birds in Lincoln Co 25 Oct 1965 and 19–20 Oct 1967. Rosche (1982) noted 4 sightings at Whitney L 19 Mar–10 Apr 1974–76 and 1 at Box Butte Res 4 Oct 1975. During the 1980s there were additional reports from Dawes and Keith Cos in the period 18 May–19 Aug (AB 38:930, 1036; 39:73; 41:1456; 44:455); reports 11 Jul and 25 Jul 1984 were considered "astonishing" due to the species' rarity (AB 38:1036).

Most of the history of this species in Nebraska has been focused on L McConaughy, where it was first reported 25 Jul 1984 (AB 38:1036). By 1989 there were 14 observations, and since 1990 the California Gull has become "almost a permanent resident" (Rosche 1994a). Largest numbers are present there from late summer through early winter, either postbreeders from breeding locales in Colorado or migrants from colonies to the north on the Great Plains, although banding data suggest the latter. One bird banded in Montana 8 Jul 1940 and 2 banded in North Dakota 9 Aug 1964 were recovered at Valentine NWR 4 Oct 1940, Hooker Co 14 Sep 1964, and Deuel Co 26 Sep 1964, respectively. A few have been found in recent years at nearby Sutherland Res, and since 1992 there are 3 records from Douglas Co in extreme eastern Nebraska. Knowledge of the distribution and movements of this species in Nebraska is still evolving. California Gull has increased dramatically throughout its range in the past 50 years (Andrews and Righter 1992), an increase coincident with the increase in Nebraska reports.

Spring

19, 19, 21 Mar→27, 27, 30 May

Early dates above are at locations where wintering birds were not reported. There is an earlier report 9 Mar 1999 at Johnson L (SJD).

The lowest numbers occur in the period 31 Dec–18 Mar, although in recent years a few have wintered at L McConaughy and Sutherland Res (see Winter).

Reports away from L McConaughy are more numerous in spring than in fall, with 12 reports in the period 21 Mar–14 May. In addition, there are a few easterly reports: an adult at Cunningham L 19 Mar 1997 (JGJ; Brogie 1998); 2 adults at Cunningham L 27–28 Mar 1996 (JGJ; Brogie 1997); a second alternate bird at Kiwanis Park, Omaha, 18 Apr 1999 (WRS); and a bird at Branched Oak L 30 Apr 1998 (JS).

Prior to the species becoming a regular migrant in the state there were observations from Dawes Co 23 Mar 1974, Cherry Co 21 Mar 1980, and Lincoln Co 18 May 1986.

High counts include 82, mostly adults, at L McConaughy 22 Apr 1994 and 27 there 17 Apr 1997.

Summer

Breeding occurs close to Nebraska in northeastern Colorado, with up to 1500 pairs of the nominate race (Mark Robbins, pers. comm.) nesting at Riverside Res in Weld Co, where islands are used for nesting (Andrews and Righter 1992). Although single birds have occurred recently at L McConaughy, the lack of islands there may obviate breeding. Breeding also occurs in northeastern South Dakota (SDOU 1991).

Reports in Nebraska in the period 1 Jun–23 Jul are few and recent: 12 Jun 1998 (2) L McConaughy (WRS, JS); 13 Jun 1987, an adult in Keith Co (AB 41:1456); 15 Jun 1991,

mixed-age flock of 10 at L McConaughy; 16 Jun 1996, mixed-age flock of 23 flying north at Harrison (WRS); 22 Jun 1996 L McConaughy; 26 Jun 1999 L Ogallala (JS); and 5 Jul 1995 L McConaughy.

Fall

18, 19, 24 Jul→21, 23, 30 Dec

Numbers of California Gulls increase at L McConaughy during late summer, with the first arrivals noted in Jul, and continue to build into fall and then decrease in early to midwinter, with sizeable numbers still present into late Dec. In recent years a few have lingered through winter (see Winter).

There are several reports away from L McConaughy in the period 18 Jul–30 Dec, including easterly reports 9 Nov 1997 Branched Oak L (JGJ), 21–23 Nov 1994 Douglas Co (AB 48:125), 22 Nov 1997 Willow Creek L (JS), and 2 reported without details at Calamus Res 30 Dec 1995.

High counts include 148 at L McConaughy 4–11 Aug 1996, 148 there 5 Nov 1994, and 105 there 4 Oct 1997.

Winter

CBC data at L McConaughy show the species continuing to increase. Counts were 46 in 1991, 68 in 1992, 72 in 1993, 143 in 1994, but only 23 in 1995 and 3 in 1996, when the count was held in late Dec and Jan. Overwintering first occurred 1991–92 at both L McConaughy and Sutherland Res, when 1 adult was at L McConaughy 16 Feb and 3 were at Sutherland Res 15 Feb (NBR 60:138). Overwintering numbers appear to be increasing, as 26 were at L McConaughy 28 Jan 1995 and 21 were still present 5 Feb 1996, but only 3 remained 5 Mar. The prior winter, numbers dwindled to 1–2 on 19 Feb 1995. Twenty were at L McConaughy 1 Feb 1997 but only 1 on 15 Feb. In 1997–98, 78 were present 1 Jan but none 14–15 Jan, although 38 had reappeared by 7 Feb (SJD). A report 10 Mar 1993 in Lincoln Co may have been of a wintering bird.

COMMENTS

The first report of this species for the state was

a specimen collected by Hudson in Lancaster Co 19 Mar 1933 (Hudson 1933b), now specimen UNSM ZM11152, an axial skeleton. However, Hudson (1956) later retracted the report, concluding that the measurements did not eliminate the possibility of the specimen being a small Herring Gull.

FINDING

The species is most easily found at L McConaughy Aug–Dec, but can be difficult to separate from Ring-billed and Herring Gulls without practice.

Herring Gull

Larus argentatus

STATUS

Locally abundant regular spring migrant statewide. Rare casual summer visitor statewide. Common regular fall migrant statewide. Locally fairly common regular winter visitor statewide.

DOCUMENTATION

Specimen

UNK mount, 8 Oct 1917 Kearney, Buffalo Co.

DISTRIBUTION AND ECOLOGY

Bruner and others (1904) considered it a "rather rare migrant." In Lincoln Co Tout (1947) noted, "I have no records most years." Rapp and others (1958) considered the species uncommon. However, in recent years numbers have been increasing throughout its range and, coupled with an increasing number of ecologically mature reservoirs in the region that attract Herring Gulls, there have been an increasing number of reports on the Great Plains and in Nebraska.

Spring

winter→28, 30, 31 May

Later reports are 5 Jun in northwest Nebraska (Rosche 1982), 8 Jun 1957 Gage Co, and 14 Jun in the Keith Co area (Rosche 1994a).

Numbers begin building in late Feb, although to what degree is dictated by the severity of weather. Most Herring Gulls move through the state during Mar, with fewer following

in the first half of Apr. Individuals are scarce thereafter and are almost always immatures.

High counts include 1000 at L McConaughy 21 Feb 1994, 900 at Johnson L 21 Feb 1999, and 500 at L McConaughy 24 Mar 1978 (Rosche 1994a).

Summer

There are about 14 reports 15–30 Jun, mostly from Lincoln Co but also including 30 Jun 1966 Harlan Co, 30 Jun 1966 Brown Co, and 30 Jun 1988 Lancaster Co. There are at least 20 reports in the period 1 Jul–7 Sep, most from Scotts Bluff, McPherson, and Lincoln Cos since 1970. It seems likely that many of these mostly undocumented westerly reports may be in fact California Gulls, which arrive in late Jul and Aug in western Nebraska. There is a single Jul–Aug Herring Gull record for Colorado (Andrews and Righter 1992), none in South Dakota (SDOU 1991), and only about 8 Missouri records Jun–Aug (Robbins and Easterla (1992). Nebraska reports from the east and documented westerly reports in this period are 25 Jul L McConaughy (Rosche 1994a), 2–11 Aug 1996 L McConaughy, 4 Aug 1991 Lancaster Co, 6 Aug 1994 L McConaughy, 8 Aug 1976 Lancaster Co, 10 Aug 1998 Branched Oak L, 10 Aug 1998 Johnson L, 11 Aug 1995 North Platte NWR, 15–16 Aug 1999 (2 immatures) L McConaughy (SJD), 22 Aug 1989 Lancaster Co, 28 Aug 1994 North Platte NWR, 31 Aug 1961 Cass Co, 1 Sep 1992 Dakota Co, 4 Sep 1997 Harlan Co Res (LF, CF), 6 Sep 1986 Platte Co, and 7 Sep 1971 Douglas-Sarpy Cos.

Most, if not all, summer Herring Gulls are immatures, singles of which may summer locally at the largest reservoirs.

Fall

13, 17, 19 Sep→winter

Most migrants arrive mid to late Oct, with numbers building in Nov. Large concentrations of Herring Gulls are not found until after the end of Nov. See Summer for reports in the period 15 Jun–7 Sep.

High counts include 446 at Harlan Co Res 15 Dec 1996, 416 at L McConaughy 23 Dec 1994, and 260 at North Platte 25 Dec 1979.

Winter

Numbers of Herring Gulls wintering in Nebraska are dictated by the severity of the winter. Numbers continue building in Dec, and large numbers can be found locally just prior to the coldest winter weather. Smallest numbers are found during Jan, although high midwinter counts were 1500 at L McConaughy 15 Jan 1998 and 1100 there 6 Feb 1999.

FINDING

The Herring Gull is easy to find in mixed flocks of gulls Nov–Mar at larger reservoirs.

Thayer's Gull

Larus thayeri

STATUS

Rare casual spring migrant statewide. Rare regular fall migrant statewide. Rare regular winter visitor L McConaughy, rare casual elsewhere.

DOCUMENTATION

Photograph

5 Dec 1995 Douglas Co (JGJ).

TAXONOMY

Thayer's Gull was first described in 1915, when Brooks noted it breeding in the western Canadian arctic (DeBenedictis 1990). Subsequently, its taxonomy has been a subject of debate. Until about 1960 it was at various times considered a subspecies of Herring Gull, as accepted by the AOU (1957), or a dark, western form of Iceland Gull, *L. glaucoides*, interbreeding with the variably colored Kumlien's Iceland Gull, *L. glaucoides kumlieni,* to its east, which in turn interbred with the very pale nominate Iceland Gull, *L. g. glaucoides*, in Greenland (DeBenedictis 1990). The work of Smith (1966), however, indicated that Thayer's and Kumlien's Gulls did not interbreed, suggesting species status for Thayer's Gull, a position accepted by the AOU (1983) and supported by Chu (1998),

who found that Thayer's Gull is actually most closely related to Glaucous-winged Gull, *L. glaucescens*, and that Iceland Gull is specifically distinct from both Thayer's Gull and Kumlien's Iceland Gull and is actually most closely related to Glaucous Gull.

Smith's results have been questioned in recent years; they have not been duplicated by other workers, and indeed Thayer's and Kumlien's Gulls have been found interbreeding freely at several locations (DeBenedictis 1990; Zimmer 1991). It is generally believed that Thayer's and Kumlien's Gulls are conspecific, distinct from Iceland Gull, and form a cline from the lightest Kumlien's Gulls to the darkest Thayer's Gulls (Banks, in Sibley and Monroe 1990). It is currently a matter of debate whether this cline is continuous and whether intermediate birds occur; a possible Nebraska example is provided by the Cedar Co Iceland Gull (Brogie and Brogie 1989; AB 43:331), a very dark bird that structurally resembles a Thayer's Gull but lacks a noticeable secondary bar, a feature considered characteristic of Thayer's Gull (Zimmer 1991).

If Kumlien's Gull is indeed distinct from Iceland Gull, as suggested by Chu (1998), the latter, currently *L. g. glaucoides*, becomes monotypic; reports of *glaucoides* in North America are few and "supported by only a few specimens (mostly from northeastern Canada) and some of these are disputed" (Zimmer 1991).

DISTRIBUTION AND ECOLOGY

Thayer's Gull was first reported in 1981 but not documented until 1992, the seventh report of the species. As with other large gull species, reports of Thayer's Gull have dramatically increased in the 1990s. The increase can be attributed to both an actual increase in numbers present as well as an increase in the number of ecologically mature reservoirs and of informed field observers.

As with other large gulls, Thayer's Gulls occur at large reservoirs and lakes.

Spring

winter→18, 19, 23 Apr

Reports of birds that were not noted during the preceding winter months are few and recent and suggest an incipient spring movement late Feb through mid-Apr:

20 Feb 1999 (3) L Maloney (JGJ, WRS)

21 Feb 1999 (3) Johnson L (JGJ, WRS)

26 Feb–4 Apr 1994 up to 2 first basic, adult (Gubanyi 1996b)

5 Mar–23 Apr 1995 first basic Keith Co (Gubanyi 1996c; WRS)

24 Mar 1996 first basic Douglas Co (JGJ)

26 Mar 1996 adult Dakota Co (NBR 64:53)

29 Mar 1997 adult L McConaughy (SJD)

13 Apr 1997 2 first basic and 1 third basic L McConaughy (SJD; Brogie 1998)

15 Apr 1996 first basic Cunningham L (JGJ; Brogie 1997)

19 Apr 1997 first basic L McConaughy (SJD)

23 Apr 1995 first basic L McConaughy (WRS)

Fall

6, 6, 10 Nov→winter

An earlier documented report was of an adult at L Ogallala 4 Oct 1997 (photographed SJD, JS, WRS, JGJ). There is an undocumented report 25 Oct 1995 Cedar Co.

Winter

Overwintering is unknown away from L McConaughy except for 6 birds at Sutherland Res 14 Jan 1998 (SJD) and 5 there 31 Dec 1998 (SJD); recent reports at L McConaughy indicate wintering beginning to occur involving significant numbers.

23 Dec 1995–31 Mar 1996 2 adults, 1 second basic, up to 4 first basic Keith Co (JGJ, SJD, WRS; FN 50:298)

15 Dec 1996–15 Feb 1997 adult, 2 first basic (WRS, SJD, JS)

6 Dec 1997–18 Apr 1998 maxima 11 first basic, 1 third basic, 8 adults (SJD, WRS, JS, DCE)

6 Feb 1999 (10) (SJD)

FINDING

Best chance of finding a Thayer's Gull is to search groups of gulls at L McConaughy Dec–Mar.

Iceland Gull

Larus glaucoides

STATUS

Rare casual spring and fall migrant statewide. Rare casual winter visitor statewide.

DOCUMENTATION

Photograph

4–13 Dec 1988 Cedar Co (Brogie and Brogie 1989; AB 43:331; Grenon 1990; see Comments)

TAXONOMY

See Comments under Thayer's Gull, with which some authors believe the North American race of Iceland Gull, *L. g. kumlieni* (Kumlien's or Kumlien's Iceland Gull), is conspecific.

DISTRIBUTION AND ECOLOGY

Iceland Gull occurs at large lakes and reservoirs. Unlike Thayer's Gull, most records of Iceland Gull are in midwinter, although individuals that were not present in winter appear in spring and fall. Records from nearby states are also few and recent. The 1907 record is significant since it occurred before the explosion in numbers of large gulls in recent years. As is the case elsewhere in the interior, records of Iceland Gull are likely to increase.

Spring

There are 3 documented records of birds that were not noted during winter:

26 Mar 1999 adult L Ogallala (KE, BR)

30 Mar 1997 first basic L Ogallala (SJD; Brogie 1998)

19–20 Apr 1997 first basic L Minatare (SJD, WRS; Brogie 1998)

Fall

There are 2 records of birds that did not remain into winter, both documented. The 1988 Cedar Co bird is discussed under Thayer's Gull; it was a very dark bird. No photographs are available of the 1991 Cedar Co bird.

10 Nov–1 Dec 1991 first basic Cedar Co (BFH; Gubanyi 1996a; NBR 60:32)

4–13 Dec 1988 first basic Cedar Co (cited above)

Winter

There are 10 records that could be considered winter records; all are documented. The initial record consists of a first basic bird shot by a boy from a flock of crows; the specimen cannot be located with certainty, but measurements were published (Swenk 1907; Bray and others 1986). The remaining records are recent.

10 Jan 1999 first basic L McConaughy (SJD)

11–12 Jan 1992 first basic Lancaster Co (BP, LP; NBR 60:139 gives correct date; Gubanyi 1996a has incorrect date)

15 Jan 1907 first basic Saline Co (Swenk 1907; Bray and others 1986)

20 Jan–12 Mar 1994 1–2 first basic L Ogallala (WRS, RCR; Gubanyi 1996c)

27 Jan–5 Feb 1995 adult L McConaughy (SJD, JGJ, WRS; Gubanyi 1996c)

6 Feb 1999 second basic L McConaughy (SJD)

7 Feb 1998 first basic L McConaughy (WRS, SJD, JS)

7 Feb–4 Mar 1998 adult L McConaughy (SJD, WRS, JS)

18 Feb–31 Mar 1996 first basic L McConaughy (BP, LP, KN, LN; Brogie 1997; FN 50:298)

COMMENTS

Separation of Thayer's and Kumlien's Iceland Gull is difficult. There has arisen in recent years an additional problem of separation of Kumlien's Iceland Gull from the small western North American race of Glaucous Gull, *L. hyperboreus barrovianus*. The status of *barrovianus* as the smallest race of Glaucous Gull is controversial; there is evidence that it has a reddish orbital ring as does Kumlien's Iceland Gull, whereas Glaucous Gull's is yellow (Blom 1996). Structurally, however, it resembles Glaucous Gull. This situation has been a problem on the West Coast for some time and has manifested itself at least once in Nebraska, where the identity of a bird thought to be an Iceland and a Glaucous by different observers at Branched Oak L 1–7 Dec 1996 was unresolved (JGJ, BP, LP, WRS; NBR 65:22; Brogie 1997, 1998).

FINDING

The chances of finding an Iceland Gull are slight, and identification is fraught with pitfalls. Look for it in midwinter among groups of large gulls at reservoirs such as L McConaughy.

Lesser Black-backed Gull

Larus fuscus

STATUS

Rare casual spring and fall migrant statewide.

DOCUMENTATION

Photograph

28 Feb 1994 L McConaughy, Keith Co (Gubanyi 1996b).

DISTRIBUTION AND ECOLOGY

This species has been found exclusively at large reservoirs. First recorded in 1992, it has been found each year since except for 1993, and there are now numerous records. Unlike other species of rare large gulls, the majority of records are of adults rather than immatures.

Spring

Most of the records are of adults in the period 7 Feb–19 Apr, although one appeared at L McConaughy as early as 14 Jan in 1998; as many as 6 were there later. As many as 6 individuals were observed at 3 different locales in 1996, and 6 were present at L McConaughy 7 Feb 1998. There is an additional report, not accepted by the NOURC, of a first basic bird at L McConaughy 18–19 Feb 1996 (NBR 64:12; Brogie 1997).

14 Jan–5 Apr 1998 maxima 4 adults, 2 third basic, 1 second basic L McConaughy; 2 adults until 5 Apr, 1 second basic until 16 Mar (SJD, WRS, JS, DCE)

16–26 Feb 1992 adult Pawnee L, Lancaster Co (JJH, AG; Gubanyi 1996a)

17–24 Feb 1996 adult Keith Co (WRS, BP, LP; Brogie 1997)

21 Feb–14 Mar 1994 1–2 adults Keith Co (BP, LP, WRS, JGJ, PL; Gubanyi 1996b; NBR 62:56)

23 Feb–23 Mar 1996 1–2 adults Branched Oak L, Lancaster Co (JGJ; NBR 64:12; Brogie 1997)

5–12 Mar 1995 adult Lancaster Co (JM; Gubanyi 1996c)

8–9 Mar 1999 fourth basic L Ogallala (SJD)

18 Mar 1996 first basic L McConaughy (KN, LN; Brogie 1997)

24 Mar 1996 first basic Douglas Co (JGJ; Brogie 1997)

9–10 Apr 1995 adult Douglas Co (JGJ; Gubanyi 1996c)

19 Apr 1996 Keith Co (FN 50:298)

Fall

There are 6 records, all subadults, in contrast to the spring records of mostly adults. All but one of the records of first and second basic birds are from Branched Oak L.

24 Aug and 26 Sep 1999 first alternate–second basic L McConaughy (SJD)

17–19 Nov 1996 second basic Branched Oak L (JGJ; NBR 64:115; Brogie 1998)

29–30 Nov 1997 first basic Branched Oak L (JGJ, JS, BP, LP; Brogie 1998)

3–6 Dec 1997 second basic Branched Oak L (JS)

3–20 Dec 1997 first basic Branched Oak L (JS)

17–19 Dec 1992 third basic Sutherland Res (Dinsmore and Silcock 1993b; Gubanyi 1996a)

FINDING

This species should be looked for at larger reservoirs during early spring, especially L McConaughy and Branched Oak L.

Glaucous-winged Gull

Larus glaucescens

STATUS

Accidental in spring.

DOCUMENTATION

Photograph

12 Apr 1995 L McConaughy, Keith Co (Roberson and Carratello 1996; Gubanyi 1996c).

Records

The single record consists of a first basic bird observed at L McConaughy (cited above).

COMMENTS

Two essentially identical first basic white-winged gulls that may have been introgressants of this species and Herring Gull were observed and photographed at L Ogallala 16 Feb 1996 (JGJ, WRS; NBR 65:22). It was also suggested that these birds were "Nelson's Gulls," introgressants between Herring and Glaucous Gulls, but the absence of checkering on the tertials reduces the likelihood of this possibility.

This species has occurred in northeast Colorado 3 times, 31 Mar–6 Apr, 11–19 Apr, and 25 Jul (Andrews and Righter 1992), and is likely to occur again at L McConaughy.

Glaucous Gull

Larus hyperboreus

STATUS

Uncommon regular spring and fall migrant statewide. Locally uncommon regular winter visitor statewide.

DOCUMENTATION

Photograph

4 Apr 1986 L North, Platte Co (Mollhoff 1987).

TAXONOMY

The possible occurrence in Nebraska of the small western Canadian race *L. h. barrovianus* is discussed under Iceland Gull. There are no specimens from the Great Plains of "pure" *barrovianus*, but probable intergrades with the somewhat larger *L. h. leuceretes* have been reported in Colorado, a first-year male 1 Apr 1938 and a first-year female 28 Mar 1938 (Banks 1986), and on the Texas-Oklahoma border a probable female 17 Dec 1880 (Banks 1986). Another reported *barrovianus*, from Mustang Island, Texas, was found to be an albinistic Herring Gull (Banks 1986). Banks (1986) concluded that a few small Great Plains Glaucous Gulls are apparent introgressants of *barrovianus* and *leuceretes*.

An introgressant of Glaucous Gull and Herring Gull ("Nelson's Gull") was reported 26 Feb 1994 at L McConaughy (WRS; Gubanyi 1996b; Brogie 1998).

DISTRIBUTION AND ECOLOGY

Glaucous Gull is found on large lakes and reservoirs. The species was considered accidental by Rapp and others (1958) and, based on 12 reports, mostly juveniles, Johnsgard (1980) considered the species a rare vagrant. Records have increased dramatically since about 1980, with significant numbers now being found, including regular reports of adults.

Spring

winter→4, 4, 5 Apr

There are later reports 12 Apr 1997 L
McConaughy, 22 Apr 1995 L McConaughy,
29 Apr 1975 Lancaster Co, and a worn first
alternate individual 2 May 1998 at Sutherland
Res (SJD). Migrants pass through during Mar
and the first week of Apr.

High counts include 11, including 3 adults, at L
McConaughy 14 Mar 1994 and 6 there 31 Mar
1996 (FN 50:298).

Fall

16, 22, 23 Nov→winter

Migrants appear by the end of Nov and become
more numerous in Dec. Individuals found
during this period are usually immatures.

High counts include up to 5 at Gavin's Point Dam
4–19 Dec 1993 and up to 5 at Harlan Co Res 15
Dec 1996.

Winter

Prior to 1989 there was a single midwinter
report. During the 1990s impressive and
unprecedented numbers have been found
during midwinter at reservoirs in the western
Platte Valley.

High counts include 16, including 4 adults, at
L McConaughy 22 Feb 1996; 12, including
2 adults, there 7 Feb 1998; 9, including 2
adults, there 15 Feb 1997; and 7 at Johnson and
Elwood Res 13 Jan 1994.

FINDING

The easiest place to find the species is L
McConaughy Dec–Mar. In eastern Nebraska
larger reservoirs should be checked late
Nov–Dec and in Mar.

Great Black-backed Gull

Larus marinus

STATUS

Rare casual spring migrant statewide. Accidental
in summer. Rare casual fall migrant east. Rare
casual winter visitor Keith and Lincoln Cos.

DOCUMENTATION

Photograph

20 Jan–18 Mar 1996, first basic L McConaughy
(JGJ, WRS; Brogie 1997).

DISTRIBUTION AND ECOLOGY

As with other large gulls, reports are from
large reservoirs and lakes. There are no
documented records prior to 1988, reflecting
the increase in numbers seen on the Great
Plains in recent years. The first of at least 8
Colorado records was in 1980 (Andrews and
Righter 1992).

Spring

There are 5 records, 3 documented, of birds that
were not present in winter:

21 Feb 1999 first basic Johnson L (JGJ, WRS)

3 Apr 1996 adult Wagontrain L (LE; NBR 64:53; FN
50:298)

12–13 Apr 1997 third year L McConaughy (SJD;
not accepted by the NOURC; Brogie 1998)

Undocumented reports are a first basic bird at
L McConaughy 16 Mar 1998 and a bird near
Niobrara 25 Mar 1998.

Summer

There is a single report of one at L McConaughy
in first alternate plumage. This may have been
the same bird reported in first basic plumage
earlier and second basic plumage later.

4 Aug 1996 first alternate L McConaughy (SJD;
Brogie 1997)

Fall

There are 2 records:

19–20 Nov 1998 first basic Conestoga L (JGJ)

2 Nov 1997 first basic Gavin's Point Dam (MB;
Brogie 1998)

Winter

All documented records are of immatures from
either Sutherland Res or L McConaughy.
This species tends to return to or remain at
favored winter localities; the 1988 and 1989
records are probably the same individual, as
may be a first basic bird at L McConaughy 20
Jan–18 Mar 1996, one there in first alternate
plumage 4 Aug 1996 (see Summer), and
one in second basic plumage 15–17 Feb 1997
(cited below).

14 Jan 1989 immature Sutherland Res (RCR;
Grenon 1990)

20 Jan–18 Mar 1996 first basic L McConaughy
(JGJ, BP, LP, KN, LN; Brogie 1997)

15–17 Feb 1997 second basic L McConaughy (JGJ,
WRS; FN 50:298)

21 Feb 1988 immature Sutherland Res (RCR; Mollhoff 1989a)

21 Feb–14 Mar 1994 first basic L McConaughy (PL; Gubanyi 1996b)

There are 2 additional reports, neither acceptably documented. A reported adult was briefly seen 5 Apr 1952 in Hamilton Co (Swanson 1953), and there is an old report without details of a specimen examined by Aughey that was shot by the Winnebago in May 1887 on the Missouri River (Bruner and others 1904). The latter report has been considered highly suspect (Sharpe 1993).

FINDING

The increasing frequency of occurrence of this species on the Great Plains suggests that finding it may become easier. Look for it at L McConaughy.

Sabine's Gull

Xema sabini

STATUS

Rare regular fall migrant west, rare casual elsewhere.

DOCUMENTATION

Specimen

UNSM ZM6624, early Sep 1899 Lincoln, Lancaster Co (Swenk 1902).

DISTRIBUTION AND ECOLOGY

This species most commonly migrates along the Pacific Coast, is uncommon along the Atlantic Coast (Harrison 1983), and is considered a casual migrant in interior North America (AOU 1983). Predictably, therefore, inland records are more numerous westward, which is consistent with Nebraska reports. That Sabine's Gull occurs mainly in fall likely reflects the tendency of many latitudinal migrants to be more leisurely and less direct in their fall movement than in spring. Sabine's Gull is found on large lakes and reservoirs.

Fall

3, 6, 6 Sep→12, 14, 15 Oct

There are later documented records 20 Oct 1998, a juvenile at Branched Oak L (JS, RS); 22–26 Oct 1995, a juvenile at Pawnee L (JGJ; Brogie 1997); and 26 Oct–6 Nov 1996, a juvenile at Gavin's Point Dam (BFH; Brogie 1998). An additional report is 30 Oct 1981 Garden Co.

Few adults are reported, the only records 26 Sep 1996 L Maloney (JGJ, SJD; Brogie 1997), 13 Sep 1997 L Minatare (SJD), 21 Sep 1997 (2) L Minatare (SJD), and 30 Sep–7 Oct Branched Oak L (JS).

Unprecedented influxes occurred in 1997 and 1998, with about 35 birds reported in the period 6 Sep–15 Oct 1997 and 32 birds 12 Sep–20 Oct 1998.

Documented reports east of L Maloney are few: early Sep 1899, specimen cited above; 22 Sep–15 Oct 1997, up to 6 juveniles at Branched Oak L (JGJ, JS, WRS); 10–14 Oct 1987, juvenile at Branched Oak L (Mollhoff 1989a); 20 Oct 1998, juvenile at Branched Oak L (JS, WRS); and 22–26 Oct 1995, juvenile at Pawnee L (JGJ; NBR 63:104).

High counts include 9 at North Platte NWR 21 Sep 1997 (SJD, DCE), 6 at Branched Oak L 5 Oct 1997 (JGJ, JS, WRS), and 5 at L McConaughy 21 Sep 1997 (SJD).

COMMENTS

There is a single spring report, a "flock" of adults on the Missouri River in Douglas Co or Sarpy Co 15 Apr 1928 (LOI 1928). It seems unlikely that a flock could occur in Nebraska, especially in spring.

There are very few spring records from neighboring states, interestingly none from Iowa or Missouri (Robbins and Easterla 1992; Kent and Dinsmore 1996), suggesting movement of only very small numbers in spring on the western Great Plains; documented records are in the period 17 Apr–15 May (Andrews and Righter 1992; SDOU 1991).

FINDING

The best chance of finding this species is to search large lakes and reservoirs, especially in the west, during Sep–Oct after the passage of a cold front.

Black-legged Kittiwake

Rissa tridactyla

STATUS

Rare casual spring and fall migrant statewide.

DOCUMENTATION

Specimen

UNSM ZM16143, 20 or 21 May 1990 adult L
McConaughy, Keith Co (Johnsgard 1990b;
Grenon 1991).

DISTRIBUTION AND ECOLOGY

Kittiwakes are found on large lakes and
reservoirs. Most records are from eastern
Nebraska, but it is unclear whether
this is due to observer concentration;
there are about 30 Colorado records
(Andrews and Righter 1992). All but 2
of the Nebraska records are subsequent
to 1981.

Spring

Of 7 reports, only 2 are documented:

20–21 Apr 1981 immature Lancaster Co (Schrad
1981)

20 or 21 May 1990 adult (specimen cited above)

Additional undocumented reports are Mar
1932 Lancaster Co (Hudson 1937b), Apr 1976
Lancaster Co, 23–24 Apr 1993 Douglas-Sarpy
Cos, 4 May 1991 Hall Co, and 21 Jun 1991
Holt Co.

Fall

Most reports from Nebraska and surrounding
states are of first winter birds found during
late fall and early winter. The Nebraska
reports, most documented, fall within the
period 30 Oct–23 Dec, with 8 in the short
period 21 Nov–5 Dec.

30 Oct 1988 first basic Sarpy Co (DB; Grenon
1990)

5–9 Nov 1981 first basic North Platte NWR (FZ;
Bray and others 1986)

21 Nov 1995 first basic Cunningham L (JGJ;
Brogie 1997)

22 Nov–16 Dec 1998 1–4 first basic Gavin's Point
Dam (JS, WRS)

26 Nov–5 Dec 1988 first basic Cedar Co (Brogie
1989a; Grenon 1990)

27 Nov 1995 first basic Summit Res (JGJ; Brogie
1997)

1 Dec 1984 Lancaster Co (JG; Bray and others
1986)

1 Dec 1991 first basic Cedar Co (BFH; Gubanyi
1996a)

3 Dec 1995 first basic Branched Oak L (JGJ;
Brogie 1997)

17 Dec 1999 first basic Harlan Co Res (SJD, m.ob.)

23 Dec 1991 first basic Johnson L (PK; Gubanyi
1996a)

There are several additional undocumented
reports in the same period as the documented
reports: 3 Nov 1988 Sarpy Co, 4 Nov 1990
Douglas Co, 24 Nov 1988 Lancaster Co, 1 Dec
1983 Douglas-Sarpy Cos, and 5–9 Dec 1981
Scotts Bluff Co.

High counts include 4 at Gavin's Point Dam 22
Nov 1998 and 3 there 5 Dec 1988.

FINDING

One should look for kittiwakes in late Nov–early
Dec at Gavin's Point Dam in Cedar Co and
other eastern reservoirs, such as those in
Lancaster Co.

Ross's Gull

Rhodostethia rosea

STATUS

Accidental in winter.

DOCUMENTATION

Photograph

17–23 Dec 1992 Sutherland Res (Dinsmore and
Silcock 1993a; Rosche 1993; Gubanyi 1996a).

Records

A bird in second basic plumage found at
Sutherland Res represents Nebraska's only
record.

COMMENTS

There are about 10 interior records of this
species, including 5 from neighboring states:
Colorado 28 Apr–7 May 1983 (Andrews and
Righter 1992), Iowa 31 Oct–1 Nov 1993 (Kent
and Dinsmore 1996) and 13–18 Dec and
27–31 Dec at 2 different localities in 1998,
and Missouri 31 Dec–11 Jan 1991–92 (*Bluebird*

59:84–85). Vagrants tend to appear Nov–Dec and Apr–May (Bledsoe and Sibley 1985; Kent and Dinsmore 1996).

Ivory Gull
Pagophila eburnea
STATUS
Hypothetical.
COMMENTS
A report of an adult seen briefly 12 Oct 1986 in Antelope Co (Marsh and Marsh 1986) was not accepted by the NOURC (Mollhoff 1987). Although there are records from Colorado 2 Jan 1926 (Andrews and Righter 1992) and Iowa 20 Dec 1975 and 20 Dec–1 Jan 1990–91 (Kent and Dinsmore 1996), both habitat and date of occurrence seem unlikely for the Nebraska report.

Caspian Tern
Sterna caspia
STATUS
Uncommon regular spring and fall migrant statewide. Rare regular summer visitor statewide.
DOCUMENTATION
Specimen
UNSM ZM14209, 5 May 1893 Salt L, Lancaster Co (Hudson 1934a).
DISTRIBUTION AND ECOLOGY
In Nebraska the species is found at large lakes and reservoirs and occasionally along rivers, primarily in the eastern third of the state. Although only 12% of the reports (N = 162) are west of a line from Knox Co to Lancaster Co, the species appears to be increasing in western Nebraska (Rosche 1994a). Indeed, most of the western records are after 1984. Most reports come from eastern counties with numerous reservoirs, including Douglas-Sarpy Cos (56) and Lancaster Co (14).
Spring
22, 23, 26 Apr→3, 4, 4 Jun
There are undocumented earlier reports 23 Mar 1961, 31 Mar 1960 (NBR 29:40), 31 Mar 1976, and 16 Apr 1978 Douglas-Sarpy Cos and 3 Apr 1993 Cass Co. There is a later report 8 Jun 1993 Douglas-Sarpy Cos.

Caspian Tern is primarily a May migrant, with 50% of the reports (N = 70) in the period 8–28 May.

Westerly reports are 26 Apr 1962 Lincoln Co, 28 Apr 1990 Keith Co (Rosche 1994a), 8 May 1950 Lincoln Co, 8 May 1955 Lincoln Co, 11 May 1996 Keith Co, 12 May 1997 L Ogallala, 14 May 1998 Sutherland Res (SJD), and 28 May 1995 Scotts Bluff Co.

High counts include 28 at Branched Oak L 21 May 1997, 14 at 2 locations in Douglas Co 16 May 1996, and 9 at Branched Oak L 27 May 1995.
Summer
There are about 17 reports in the period 9 Jun–17 Jul. In recent years reports from L McConaughy have increased, including seasonal high counts of 12 on 24 Jun 1996 (Brown and others 1996) and 6 on both 20 Jun 1989 (Rosche 1994a) and 22 Jun 1996.

Westerly reports are Jun–Jul 1986 Keith Co (Rosche 1994a); Jun–Jul 1989 Keith Co (Rosche 1994a); Jun–Aug 1990 Keith Co (Rosche 1994a); 13 Jun 1998 Mansfield Marsh, Dawes Co (WRS, JS); 20 Jun 1997 Keystone L; 26 Jun 1991 Lincoln Co; 26 Jun 1999 L Ogallala; and 30 Jun 1999 (4) L Alice, North Platte NWR.
Fall
18, 19, 20 Jul→7, 7, 9 Oct
There are undocumented later reports 4–14 Oct 1963 Douglas-Sarpy Cos, 6–13 Oct 1963 Lancaster Co, and 20 Oct 1986 Pierce Co.

Fall reports are distributed fairly evenly from the latter half of Jul through Sep, with about 6 reports from Oct.

The first report from the Panhandle occurred in Dawes Co 18 Jul 1990 (AB 44:1153), and there are these additional westerly reports: 18 Jul 1990 Dawes Co, 27 Jul–2 Aug 1991 Keith Co (Rosche 1994a), 30 Jul 1993 Keith Co, and 5 Sep 1996 Dawes Co (FN 51:81).

High counts include 10 at L McConaughy 30 Jul 1993, 7 at Pawnee L 21 Sep 1998, and 5 at Branched Oak L 8 Sep 1995.

FINDING

The best chance of finding the species is to search large reservoirs in eastern Nebraska in May and Sep. Good locations to check include Branched Oak, Pawnee, and Cunningham Lakes.

Royal Tern

Sterna maxima

STATUS

Hypothetical.

COMMENTS

A sighting in Sarpy Co of a purported immature with Caspian Terns 13 Sep 1993 was not accepted by the NOURC due to the possibility that the bird was an immature Caspian Tern (Brogie 1997). The only report from a neighboring state is one from Colorado 15–16 Jul 1998; it has occurred in Illinois 7 Sep 1985 and 13 Jul 1988 (Bohlen 1989) and in Oklahoma 1–6 Jan 1989 (AB 43:332).

Common Tern

Sterna hirundo

STATUS

Uncommon regular spring and fall migrant statewide.

DOCUMENTATION

Specimen

UNSM ZM6625, Sep 1892 Lincoln, Lancaster Co.

DISTRIBUTION AND ECOLOGY

This species occurs on lakes and reservoirs but less often in marshes than Forster's Tern, although Forster's Tern occurs on larger bodies of water also during migration.

Few reports of Common Tern are documented, and the species can be difficult to distinguish from the more numerous Forster's Tern. In Oklahoma Sutton (1967) collected 30 terns with the general appearance of the Common Tern in an effort to obtain a specimen, but all were Forster's. The species is considered rare in Iowa (Kent and Dinsmore 1996), Colorado (Andrews and Righter 1992), and Kansas (Thompson and Ely 1989) and uncommon in Missouri (Robbins and

Easterla 1992). Breeding occurs as close as northeastern South Dakota, where it is considered a fairly common migrant (SDOU 1991).

Spring

10, 12, 12 May→15, 16, 18 Jun

In the upper Midwest the species appears to migrate primarily in mid to late May, peaking in mid-May, with a few still moving through in early to mid-Jun. Presumed nonbreeders occur on occasion throughout the summer months (Robbins and Easterla 1992; Kent and Dinsmore 1996), but there are no documented Nebraska records for this period. There are no Apr records for Missouri or Colorado (Robbins and Easterla 1992; Andrews and Righter 1992). Documented records for Nebraska are in the period 10 May–18 Jun, with an additional early record 1 May 1999 at Oliver Res (SJD). The only spring specimens for Nebraska are a pair, HMM 2976, collected 14 May 1924 at Kearney.

High counts include 14 at L McConaughy 18–19 May 1997 (SJD) and 6 adults at Crescent L 13 Jun 1997.

Fall

27 Jul, 7, 7 Aug→1, 2, 5 Oct

Peak migration probably occurs in mid-Sep. There are later records 13 Oct 1997 Gavin's Point Dam (MB), 17 Oct 1998 L McConaughy (SJD), 25 Oct 1997 Conestoga L (JS), 1 Nov 1997 North Platte NWR (SJD), and 2 Nov 1997 L Alice (SJD; Brogie 1998).

Documented Nebraska records fall in the period 27 Jul–2 Nov, including several of juveniles with and without adults; Rosche has noted that adults accompanied by dependent juveniles have appeared at Whitney L and Box Butte Res. The only fall specimen for the state, cited above, was taken Sep 1892.

High counts include 18 at L McConaughy 7 Sep 1998 and 6 at Oliver Res 20 Sep 1999.

FINDING

The best chance of finding the species is to search larger bodies of water during mid-May and again in Aug–Sep. Any record of the species should be well documented so that the true status of the species can be determined.

Arctic Tern
Sterna paradisaea

STATUS

Hypothetical.

COMMENTS

There are 2 reports of this species. A first basic bird was reported with minimal details at Gavin's Point Dam 6 Dec 1998 (DLS). This is a very unusual date for this species (see below). Aughey mentions that "a few were seen in Dixon Co in May 1866" (Bruner 1896), a report discounted by Bruner and others (1904), who noted that no specimen was taken and that "the chances for a misidentification are too great to warrant the continual inclusion of *paradisaea* in our list."

This species is quite likely to be recorded in Nebraska, however, as there is a record for the eastern Colorado plains 12 Jun 1991 (Andrews and Righter 1992) and 6 records for the Duluth area in Minnesota 18 May–17 Jun (Janssen 1987), and an adult and a juvenile were found in central Iowa 18 and 19 Aug 1999 (SJD). There are also 2 fall records for the eastern Colorado plains, 11–12 Sep 1979 and 16 Sep 1912 (Andrews and Righter 1992). Any "non-Forster's" terns seen in Nebraska in late May and early Jun should be carefully scrutinized.

Forster's Tern
Sterna forsteri

STATUS

Common regular spring and fall migrant statewide. Locally common regular breeder Sandhills. Rare casual summer visitor statewide.

DOCUMENTATION

Specimen

UNSM ZM6627, 16 Jun 1902 Dewey L, Cherry Co.

DISTRIBUTION AND ECOLOGY

Migrant Forster's Terns utilize both small and large lakes, marshes, reservoirs, and occasionally rivers throughout the state. Breeding birds are most often found associated with large Sandhills lakes, where they forage over open water and nest in shallower waters with emergent marsh vegetation.

Spring

28 Mar, 1, 4 Apr→9, 10, 10 Jun

Late dates above are from locations where breeding is not reported. See Summer for reports after 10 Jun.

Individuals arrive mostly by mid-Apr, and peak migration occurs during the end of that month into May.

High counts include 132 at L McConaughy 18 May 1995, 87 at Branched Oak L 27 Apr 1996, and 85 at L Ogallala 9 May 1998.

Summer

Ducey (1988) cited recent breeding records from Cherry and Garden Cos but considered a report from Douglas Co (Bennett 1970) questionable. Also listed was an old report from Holt Co, cited by Bruner and others (1904) as well. Johnsgard (1996) noted that breeding is "highly localized" in Cherry and Garden Cos and speculated that the species may also breed in the Sandhills lakes of Sheridan Co. Brown and others (1996) indicated that breeding occurs at Crescent L NWR and also at "sandhills marshes in Arthur Co" but provided no further details. About 60 birds were at a nesting site at Goose L, Crescent L NWR, 13 Jun 1997 (WRS, JS).

There are about 16 reports away from breeding locations in the period 14 Jun–10 Jul, most from the north and west; reports from the south and east are 14 Jun 1998 Stagecoach L, 18 Jun 1997 Branched Oak L, 20 Jun 1995 Buffalo Co, 20 Jun 1999 (2) Harlan Co Res, 30 Jun 1985 Lancaster Co, 1 Jul 1991 Phelps Co, 4 Jul 1984 Douglas-Sarpy Cos, and 4 Jul 1988 Howard-Hall Cos.

Fall

11, 11, 13 Jul→27, 28 Oct, 1 Nov

Early dates above are from locations where breeding is not reported. The late date above was a basic-plumaged adult at Sutherland Res (WRS; Brogie 1997); a later report was one at Sherman Res 8 Nov 1998 (JGJ). Most Forster's Terns have departed by mid to late Sep, as there are only about 17 reports later than 23 Sep.

High counts include 69 at L McConaughy 3 Oct 1998, 63 there 3 Aug 1997, 50 at Crescent L NWR 16 Sep 1995, 50 at Branched Oak L 9 Aug 1995, and 50 at L McConaughy 4 Aug 1996.

FINDING

The best chance of finding the species during migration is to search large reservoirs. In summer the best locations are certain lakes at Valentine NWR and Goose L at Crescent L NWR.

Least Tern

Sterna antillarum

STATUS

Fairly common regular spring and fall migrant statewide. Locally fairly common regular breeder east and south-central.

DOCUMENTATION

Specimen

UNSM ZM6222, 4 Aug 1902 Badger, Holt Co.

DISTRIBUTION AND ECOLOGY

The Least Tern is primarily found along larger river systems that provide high and dry exposed sandbars for nesting; they are also attracted to the large sandpiles produced in sand-mining operations adjacent to the Platte River. Because of the ephemeral nature of their breeding habitat, this species could be considered a successional specialist, and thus its breeding locations cannot be predicted with certainty from year to year. However, river development and water-withdrawal projects over the years have had a negative impact on natural sandbar formation. Contemporary stream flows along the Platte River and bank stabilization and channelization along the Missouri River have contributed to a reduction of suitable breeding habitat and a gradual decrease in Least Tern numbers. Consequently, the

Nebraska subspecies, *S. a. athalassos*, has been designated endangered by the USFWS.

During migration Least Terns may also be found associated with large lakes and reservoirs.

Spring

19, 19, 21 Apr→summer

The first migrants arrive by the end of Apr, although the majority of reports are during the latter half of May. There are 2 early undocumented reports 4 Apr 1964 Cass Co and 8 Apr 1991 Douglas-Sarpy Cos.

Summer

Breeding birds are restricted to the Missouri, Platte, lower Niobrara, lower Loup, and lower Elkhorn River systems. In the Missouri River valley breeding regularly occurs along the unchannelized portion of the river above Sioux City, Iowa, but no longer south of Dixon Co because the river is channelized and preferable habitat has been eliminated. Breeding populations on the Platte River extend west to L McConaughy, where the species was first discovered breeding in 1989 (Czaplewski 1989), and it now breeds each year with Piping Plovers at protected sites. In the Loup River system, breeding birds occur as far west as Valley and Howard Cos, along the Niobrara as far west as Rock and Keya Paha Cos, and on the Elkhorn upstream to Madison Co. Breeding has also occurred 1991–94 along Papillion Creek (NBR 62:76), a smaller tributary of the Missouri that empties just above the mouth of the Platte. The most recent comprehensive censuses of breeding birds were made by Ducey (1984) and Sidle and others (1991). Results are shown below.

Least Tern Breeding Populations

Missouri River	216
Platte River	472
Niobrara River	291
Loup River	87
Elkhorn River	26
Total	1092

Fall

summer→14, 15, 18 Sep

There are later reports 30 Sep 1973 Lancaster

Co, 3 Oct 1989 Lancaster Co, and 6 Oct 1971 Lincoln Co and an undocumented report 28 Oct 1990 Polk Co. Most birds depart by early Sep.

High counts include 10 at L Babcock 22 Aug 1998.

FINDING

Look for a small, dainty tern that forages over open water along the rivers mentioned above. The lower Platte River from Columbus to the mouth during late May and Jun is especially productive; good access to the river is available at Schramm Park and Two Rivers Park. Other good areas are the overlook at Ponca SP and the parking area at the north end of Kingsley Dam at L McConaughy.

Black Tern

Chlidonias niger

STATUS

Common regular spring and fall migrant statewide. Common regular breeder north and west.

DOCUMENTATION

Specimen

UNSM ZM12725, 21 May 1895 Lancaster Co.

DISTRIBUTION AND ECOLOGY

The Black Tern is an insectivore and most commonly feeds on aquatic and other insects resting on the water surface. Aquatic insects do not achieve significant numbers until the waters warm, so Black Tern is the last tern species to arrive in spring and first to depart in fall. It is most often found foraging along the edge of open water in marshes but may also frequent quite small potholes and wetlands that produce insects but are too small to sustain a fish population suitable for other tern species. Breeding birds are found on lakes that possess large areas of marsh, primarily the numerous lakes in the Sandhills.

Spring

10, 10, 17 Apr→summer

Individuals usually arrive by the very end of Apr or early May. Only 2% of all spring reports (N = 456) are before 24 Apr. Peak migration

occurs mid-May. Late migrants are moving through well into Jun.

High counts include 1300 at Sutherland Res 17 May 1995, 700 at L Ogallala 16 May 1999, and 300 at North Platte SL 16 May 1999.

Summer

This species formerly bred throughout the state (Bruner and others 1904), but degradation of suitable habitat has left the species restricted to the marshes of the Sandhills. It could breed and should be looked for at larger marshes in the Rainwater Basin, especially Funk Lagoon, where as many as 100 were observed as late as 11 Jun 1996. There are numerous summer reports south of the Platte River, although reports are fewest mid-Jun through early Jul.

Fall

summer→29 Sep, 2, 5 Oct

Returning birds appear in mid-Jul, and peak migration occurs in Aug and very early Sep. Only 9% of the reports (N = 227) are after 15 Sep.

High counts are 428 at L McConaughy 20 Aug 1998, 275 at Funk Lagoon 14 Aug 1997, and 181 at L McConaughy 4 Aug 1979 (Rosche 1994a).

FINDING

During migration the species should be found in marshes and wetlands across the state. During summer marshes in the Sandhills offer the best opportunities for viewing the species. The several marshes and potholes along Highway 2 in Grant and Sheridan Cos are good, easily accessible habitats that regularly support Black Terns late May–Aug.

Alcidae (Murrelets)

Ancient Murrelet

Synthliboramphus antiquus

STATUS

Accidental in fall.

DOCUMENTATION

Specimen

UNSM ZM14801, 27 Oct 1929 Burt Co (Swenk 1933a).

Records
The only record is a mounted specimen collected 27 Oct 1929 along the Missouri River 8 km (5 mi) east of Tekamah in Burt Co by Max Salsman; it is an immature winter-plumaged bird (Swenk 1933a).

COMMENTS

There are 10 records from states neighboring Nebraska (South Dakota, Wyoming, Colorado, Iowa, and Minnesota). Eight of these records are in the period 27 Oct–22 Nov, with other records 22 Aug and 28 Feb. Apparently, when this species moves south along the Pacific Coast a few straggle inland.

COLUMBIFORMES

Columbidae (Pigeons, Doves)

Rock Dove
Columba livia
STATUS
Common regular resident statewide.
DOCUMENTATION
Specimen
UNSM 13458, 26 May 1977 Lincoln, Lancaster Co.
TAXONOMY
North American Rock Doves are an amalgamation of several wild subspecies from the native range in central Asia (AOU 1957).
DISTRIBUTION AND ECOLOGY
Along the Mediterranean Coast the Rock Dove nests on cliff faces, choosing small ledges as nest sites. Domesticated centuries ago in Europe, European immigrants brought this species with them to the New World (Long 1981). Those settling Nebraska apparently brought this species with them, and many Rock Doves have become feral. Today it occurs statewide, essentially around human habitation where nesting sites and food are available. It is most common in larger towns and cities in the east, where it may be found on ledgelike sites on multistory buildings and beneath bridges. In the west feral nesting on cliffs and in canyons is rare but does occur on certain buttes in the northwest (Rosche 1982) and possibly the Wildcat Hills in Scotts Bluff Co (Rapp and others 1958).
Summer
BBS data indicate that numbers are greatest in the east, declining toward the north and west; this is probably a reflection of human population distribution.
Winter
CBC data show the same distribution as BBS data, although numbers recorded in the west, essentially at Scottsbluff, are relatively higher than shown in BBS data. Highest CBC totals are from Lincoln, 685 in 1992 and 480 in 1993,

although some localities probably do not make a major effort to count these birds.

FINDING

Flocks are often seen in flight in cities and around rural farmsteads or perched on roofs of grain elevators and city buildings.

Band-tailed Pigeon

Columba fasciata

STATUS

Accidental in fall.

DOCUMENTATION

Description

15 Oct 1996 Morrill (ECT; Brogie 1997).

Records

The single record is of a bird seen at Morrill 15 Oct 1996 (cited above).

COMMENTS

This species shows no tendency to vagrancy on the northeast Colorado plains (Andrews and Righter 1992), but it occurs just across the Wyoming border in the Sierra Madre in late Aug–early Sep on rare occasions (Scott 1993). There are records from neighboring states: Kansas 19 Jul 1963 and 9 Oct 1969, both in the southwest (Thompson and Ely 1989); South Dakota 20 Apr 1979, 20 Apr 1981, 3 May 1981, 15 May–1 Aug 1969, 13 Jun 1964, and 30 Jul 1983, all in the western half of the state (SDOU 1991); Missouri late Nov–20 Feb 1983–84 (Robbins and Easterla 1992); and Iowa, a hypothetical record summer–late Sep (Kent and Dinsmore 1996). In summary, this species is a low-density vagrant, with most records Apr–Oct and a single winter record in Missouri.

Eurasian Collared-Dove

Streptopelia decaocto

STATUS

Locally uncommon casual resident central, rare casual elsewhere.

DOCUMENTATION

Description

29 Nov 1997–31 Mar 1998 Shelton, Buffalo Co (RH, BP, LP; Brogie 1998).

COMMENTS

This species seems destined to become a common bird in Nebraska, although as it is just becoming established in the state.

The individual cited above, Nebraska's first, appeared at a feeder in Shelton in late Nov 1997 and remained until mid-Apr 1998. It appeared to be a wild bird both by behavior and physical characteristics. At about the time the Shelton bird departed, a pair appeared at the Roger Newcomb residence in Kearney 22 Apr 1998 (LR, RH). These birds had fledged 2 young by 19 May and then renested, and by 24 Sep as many as 9 birds were present in the neighborhood (LR, RH). Nesting was attempted in Oct and Dec but was unsuccessful (LR, RH). Eight birds were still present Dec 1998–Feb 1999.

Records continue to accumulate. Breeding apparently occurred in Pierce, where a pair was present most of the summer in 1999 and 2 young were noted around 23 Jul (MB). A group of at least 3 were present in the Prinz Street neighborhood of West Point during 1998 (David Mlnarik, pers. comm. BP, LP, WRS). A single bird was noted 11 Jul 1998 on a farm about 4.8 km (3 mi) east of Superior (DCE), and another was 11.3 km (7 mi) north of Bradshaw 22 Aug 1998 (JGJ).

The provenance of these birds is unknown, but the species appears to be expanding its range from its Florida base, where it was introduced (DeBenedictis 1994). Range expansion seems to involve appearance at locations some distance from the previous range limit, with subsequent filling in of the gaps.

White-winged Dove

Zenaida asiatica

STATUS

Rare casual summer visitor statewide.

DOCUMENTATION

Photograph

11–16 May 1994 Lincoln, Lancaster Co (Gubanyi 1996b).

DISTRIBUTION AND ECOLOGY

Generally a species of the southwest, preferring arid habitats, in Nebraska it may appear almost anywhere.

Summer

There are 8 reports, 4 documented, all since 1988 and all but one since 1994:

15–21 Apr 1995 Scottsbluff (LM; Gubanyi 1996c)

22–27 Apr 1999 Creighton (MB)

11–16 May 1994 Malcom (KM; Gubanyi 1996b; Brogie 1997, 1998)

8 Jul 1998–21 May 1999 Kearney, one with Eurasian Collared-Doves (Roger Newcomb, pers. comm. LR, RH; m.ob.)

Possibly the same bird as that in Scottsbluff 15–21 Apr 1995 was reported a few kilometers to the west 1 May 1995 but was undocumented (NBR 63:47). One seen from a car near Cozad on 3 Jul 1988 was described but not accepted by the NOURC (Grenon 1990). Another was reported at a Lincoln feeder 7–10 Oct 1999 (Dearald Kohles, pers. comm. JS). One was reported in Gering about 10–24 Aug 1999 (AK).

COMMENTS

This species appears to be expanding its range northward on the Great Plains, with numerous records in recent years. The total number of records for Colorado, Kansas, and Missouri is about 25, but there are fewer than 5 in all for Iowa, Wyoming, and South Dakota.

FINDING

This species should be kept in mind in summer in the Panhandle or southwest. It occurs in habitats similar to those of Mourning Dove.

Mourning Dove

Zenaida macroura

STATUS

Abundant regular spring and fall migrant statewide. Abundant regular breeder statewide. Uncommon regular winter visitor south, rare north.

DOCUMENTATION

Specimen

UNSM 6232, 30 May 1914 Haigler, Dundy Co.

DISTRIBUTION AND ECOLOGY

One of Nebraska's most common and conspicuous species, Mourning Dove occurs statewide in edge and shrubby habitats and even in open treeless country. In winter it is generally found in the southeast and along the Platte Valley in woodland edge near a food source such as harvested grainfields, semiopen corncribs, or grain elevators.

Spring

10, 11, 11 Mar→summer (north, west)

Only in the north and west is this species essentially absent in winter (see Winter). In those areas spring arrival is in mid-Mar.

Summer

BBS data indicate that densities are similar statewide, although somewhat lower in the Panhandle. One of the most numerous species reported on BBS routes, routes in the north, south, and east average 79.7 birds and those in the Panhandle 50.7. Mourning Doves may raise as many as 5 broods per year Apr–Sep (Terres 1980).

Fall

summer→ 24, 26, 26 Oct (west, away from Platte Valley)

summer→8, 10, 11 Nov (north, west of Boone and Antelope Cos)

Juveniles from early broods may begin migration as early as Jul, but adults generally molt before departure (Terres 1980). Of hatching-year birds banded in Nebraska, 43.6% were recovered in Texas and 44.4% in Mexico (Terres 1980). Flocks form in Aug–Sep, prior to migration; 450 were in Buffalo Co 7 Sep 1997.

Departure in the Panhandle (away from the Platte Valley) is in late Oct, although there are later reports 11 Nov 1994 Dawes Co, 25 Nov 1992 Morrill Co, 29 Nov 1986 Sioux Co, 8 Dec 1981 Garden Co, 23–25 Dec 1978 Dawes Co (Rosche 1982), and 28–29 Dec 1974 Dawes Co (Rosche 1982).

Departure in the north is in early Nov, although there are later reports (west of Boone and Antelope Cos) 3 Dec 1967 Custer Co and 11 Dec 1966 McPherson Co.

In the southeast many remain into Dec, as indicated by CBC data; highest CBC totals are 339 at Lincoln in 1979, 303 there in 1978, and 244 there in 1970. Birds per party-hour for the east are 0.725, south 0.222, north 0.023, and west 0.038.

Winter

Wintering occurs mainly in the southeast but also along the Platte Valley, including Scotts Bluff Co, and in the northeast west to Boone and Antelope Cos, although numbers are low away from the southeast. Rosche (1994a) considered it a casual winter visitor in the Keith Co area 25 Nov–15 Jan, and in Webster Co it was recorded in winter only twice in 25 years (Ludlow 1935). There is a single midwinter report for the Panhandle away from Scotts Bluff Co, 11 Jan 1976 Dawes Co (Rosche 1982), and in the north only 2 reports west of Boone and Antelope Cos: 1 Jan 1955 Logan Co and Feb 1971 Custer Co.

FINDING

This species is usually one of the first encountered when observers venture outside in rural areas statewide May–Sep. It is particularly abundant along back roads with scattered brush.

Passenger Pigeon

Ectopistes migratorius

STATUS

Extinct.

DOCUMENTATION

Specimen

HMM 2914, Cook, Johnson Co.

Records

The Nebraska specimen cited above is part of a mounted group on display at the HMM. The collector and date of collection are unknown, but the specimen was recorded as taken at Cook, Johnson Co (Bray and others 1986).

There is very little Nebraska information published. Ducey (1988) cited reports by Maximilian near the mouth of the Niobrara River in May 1833 and a possible nesting "across from Sioux City" noted by Coues prior

to 1874. It was apparently a common although probably irregular migrant and summer resident in the eastern parts of Nebraska (Bruner and others 1904; Rapp and others 1958), although Nebraska was at the western edge of its range.

In Kansas it nested occasionally, only in small groups, unlike the huge nesting concentrations recorded in the Great Lakes states (Thompson and Ely 1989).

The last wild birds of the species were taken in Ohio in 1900, and the last individual died in the Cincinnati zoo 1 Sep 1914 (AOU 1998; Bent 1932).

Inca Dove

Columbina inca

STATUS

Rare casual winter visitor central and west.

DOCUMENTATION

Photograph

28 Oct 1987–7 Mar 1988 Kearney, Buffalo Co (Paine 1988).

DISTRIBUTION AND ECOLOGY

A species of the southwest, like White-winged Dove it appears to be expanding its range northward on the Great Plains. It is now breeding in Kansas (Mark Robbins, pers. comm.). Most Nebraska records are of birds that arrive in fall and attempt to overwinter at feeders.

Winter

Inca Dove has been reported 4 times, all of the reports documented:

11 Sep–23 Nov 1996, apparently a single bird at 2 Chadron feeders (RCR; Brogie 1998; FN 51:81; NBR 64:116, 65:22)

28 Oct 1987–7 Mar 1988 Kearney (Grenon 1990; Mollhoff 1989a; Paine 1988; NBR 56:3; AB 42:284)

14–30 Nov 1989 north of Bassett (Grenon 1990; AB 44:116)

from early Dec 1990 for about 6 weeks Orleans (Grenon 1991; NBR 59:96)

FINDING

Any smallish dove seen in late fall should be

carefully observed in order to eliminate the possibility of Inca Dove. Most records are westerly at a time when Mourning Doves have generally departed. A small dove in flight that shows a flash of rufous coloration is likely to be this species.

Common Ground-Dove
Columbina passerina
STATUS
Rare casual fall visitor east and west.
DOCUMENTATION
Description
18 Nov 1979 Washington Co (Padelford and Padelford 1980a).
Fall
There are 5 reports, only 2 documented. The Sioux Co record is rather early as far north as Nebraska; the earliest Kansas record is 6 Oct (Thompson and Ely 1989).
21 Jun 1996 Sioux Co (RCR; FN 50:966)
18 Nov 1979 DeSoto Bend NWR (cited above)
Additional reports are undocumented but generally fit the pattern of the documented reports. Two birds were noted by an observer familiar with the species in Sarpy Co about 1969, but the observer was unwilling to report the birds (NBR 48:22). One was reported in Lincoln 13 and 15 Nov 1976 (AB 31:96; NBR 45:19). The NOURC did not accept a report based on unclear photos of one near Valentine 16 May 1987 (Grenon 1990).
COMMENTS
Common Ground-Doves have a propensity to wander north from the regular range in the southern United States in late fall. There are about 10 Kansas records, all in the period 6 Oct–20 Dec (Thompson and Ely 1989). There are 2 reports for Colorado (Andrews and Righter 1992) and none for South Dakota (SDOU 1991) or Wyoming (Scott 1993).
FINDING
During Nov any dove that appears small should be carefully checked. Young Mourning Doves appear small after fledging but by Oct–Nov should have attained adult size.

PSITTACIFORMES

Psittacidea (Parakeets)

Monk Parakeet
Myiopsitta monachus
STATUS
Hypothetical.
COMMENTS
On 10 May 1998 2 were seen at Freedom Park, Omaha (JK, SK, JP, WRS); at the time there was a nest under construction on the mast of a minesweeper on display. The birds were said to have been present for several years (Freedom Park personnel, pers. comm. WRS); they disappeared in fall when the weather was cold and reappeared each year in late Apr or May. It remains to be seen whether these birds will establish themselves on a permanent basis.
This introduced or released cage-bird was widely reported in the wild in North America in the 1970s after becoming popular in the wild bird trade in the 1960s (DeBenedictis 1994). There were occasional reports in Nebraska of free-flying birds in the period 1972–76. The first involved a nesting attempt (Padelford and Padelford 1973) in Omaha. This pair had arrived 28 Oct 1972 and were constructing a nest in a spruce 29 Nov 1972. They departed in mid-Apr 1973, however (NBR 41:17). Subsequent reports include a pair in Bellevue in mid-Mar 1973 and a single bird in Omaha 13 Apr 1973 (NBR 41:17), a pair in Omaha 29 Aug 1974 (NBR 47:58; AB 29:78), one near Kearney 6 May 1975 (Bliese 1975a), and a pair in Lincoln mid-Aug 1975 (Di Silvestro 1975) at a location around which 2–4 were reported at intervals through mid-Mar 1976 (NBR 44:35).
This species is considered established by ABA at localities in southeastern Florida and the Austin and Houston areas of Texas (DeBenedictis 1994).

Carolina Parakeet

Conuropsis carolinensis

STATUS

Extinct.

DOCUMENTATION

Specimen

NMNH 4614, 24–25 Apr 1856 McKissock Island, Nemaha Co (Bray and others 1986).

Records

This species is extinct, the last substantiated records of wild birds in 1913 and the last captive bird dying in 1918; there were, however, unverified reports until 1936 (ABA 1996).

Swenk (1934b) has summarized Nebraska information on this species, with updated information published by McKinley (1965, 1978). Ducey (1988) discussed its status also, indicating that it was extirpated about 1875. Prior to 1867 flocks of parakeets wandered throughout the wooded Missouri Valley, and it apparently nested on McKissock Island (Swenk 1934b). It apparently occurred on occasion at least as far west as the Platte River at the mouth of the Loup River (McKinley 1965).

There are 3 extant Nebraska specimens (McKinley 1965). The presence of the one cited above in the Smithsonian Institution and its status as the only specimen there from Nebraska have been confirmed (Bonnie Farmer, pers. comm. to Bray and others 1986). It was 1 of 12 taken by Hayden and Warren on "Bald Island," now known as McKissock Island, Nemaha Co, 24–25 Apr 1856 (Swenk 1934b). The only other extant Nebraska specimens are a female (4613) now in the Academy of Natural Sciences in Philadelphia and a female (4609) sent to a Henry Bryant (McKinley 1965).

CUCULIFORMES

Cuculidae (Cuckoos, Anis)

Black-billed Cuckoo

Coccyzus erythropthalmus

STATUS

Fairly common regular spring and fall migrant east, becoming uncommon west. Uncommon regular breeder statewide.

DOCUMENTATION

Specimen

UNSM 6243, 6 Jul 1901 Warbonnet Canyon, Sioux Co.

DISTRIBUTION AND ECOLOGY

This species is distributed statewide, but because it prefers deeper or denser woodlands than Yellow-billed Cuckoo, it is less common than that species, especially westward. At times it becomes common, however, usually in response to cyclic increases in caterpillar populations.

Spring

1, 1, 2 May→summer

Migrants appear in early May, although there are earlier reports 22 Apr 1964 Platte Co, 23 Apr 1951 Gage Co, and 26 Apr 1978 Cherry Co. Half of the spring arrival dates considered by Johnsgard (1980) were in the period 16–30 May.

Summer

BBS data (1967–77) indicate that numbers are greatest in the east and north and lowest in the Panhandle. In the west it is found mainly along watercourses where denser woodland occurs. There are breeding records statewide (Ducey 1988). It was listed as uncommon at Bessey Division, NNF, by Bray (1994), who suggested that this species had become less common than Yellow-billed Cuckoo since Zimmer noticed the opposite in 1912 (Bray 1994). Short (1961) suggested that it may be "frequently overlooked," as he found it common at Big Springs, a part of the state where it has been thought to be least numerous (Johnsgard 1979,

1980). Short's observation may have been during a caterpillar outbreak, a phenomenon that may determine abundance of this species to a greater extent than that of Yellow-billed Cuckoo (Rosche 1982).

Fall

summer→8, 9, 9 Oct

Departure is in Sep, although late dates are in early Oct, with a later date 15 Oct 1999 Otoe Co (LF, CF). Half the late fall dates noted by Johnsgard (1980) are in the period 25 Aug–9 Sep.

Other Reports

10 Mar 1953 Platte Co, 1 Apr 1967 Scotts Bluff Co, 2 Apr 1993 Box Butte Co.

FINDING

This species is best located by its calls in early Jun. Likely locations are extensive woodlands in the Missouri Valley, where it tends to occur in interior forest with significant vines and dense thickets for nesting. In Missouri it is often found in dense stands of willow (Mark Robbins, pers. comm.).

Yellow-billed Cuckoo

Coccyzus americanus

STATUS

Common regular breeder and spring and fall migrant east, becoming uncommon west.

DOCUMENTATION

Specimen

UNSM 6233, 27 May 1892 Jamaica, Lancaster Co.

DISTRIBUTION AND ECOLOGY

This species occurs statewide in its habitat of open woodland edge, regenerating woodland, and shrublands, especially riparian. It is most common in the east and north and least common toward the west away from riparian habitat. It avoids forest interior.

Spring

24, 25, 26 Apr→summer

19, 20, 20 May→summer (west)

Migrants arrive earliest in the southeast, in mid-May, and latest in the northwest, about 3 weeks later. Half of the arrival dates were in the period 15–29 May (Johnsgard 1980). There are earlier reports in the west 30 Apr 1956 Scotts Bluff Co, 12 May 1959 Scotts Bluff Co, and 9 May Dawes Co (Rosche 1982).

High counts include 16 at Indian Cave SP 23 May 1999.

Summer

BBS data show lowest numbers in the Panhandle, a result of diminished habitat there. Rosche (1982) considered it uncommon in the northwest, while Short (1961) found it to be common westward to Colorado in the South Platte Valley, and there are numerous reports from Scotts Bluff Co. In the Niobrara Valley Brogie and Mossman (1983) listed it as uncommon in the Niobrara Valley Preserve, and it was present in central Cherry Co at Anderson Bridge in summer 1982 (Ducey 1982). Short (1961) found it common at Valentine in 1957 but uncommon at Chadron. Bray (1994) considered it uncommon at Bessey Division, NNF.

A nest with 3 eggs was found at Schramm Park 31 May 1998, nestlings were reported in Sioux Co 14 Jun 1997, and a nest with 4 chicks was noted in Thurston Co 11 Jul 1997.

Fall

summer→30 Oct, 3, 5 Nov

Departure is mainly in Sep with late dates in late Oct, although there is a later report 15 Nov 1979 Adams Co. Half the fall departure dates were in the period 1–27 Sep (Johnsgard 1980).

High counts include 5 at Fontenelle Forest 23 Aug 1996.

Other Reports

1 Apr 1982 Lancaster Co, 9 Apr 1951 Gage Co (Fiala 1970), 14 Apr 1940 Adams Co, 16 Apr 1944 Jefferson Co. A report 12 Apr 1963 Logan Co was incorrect (Glandon 1964).

FINDING

This species can be located in edge habitats in the Missouri Valley most easily by its call in Jun. It can be difficult to see, as it moves slowly and sits still for long periods.

Groove-billed Ani
Crotophaga sulcirostris
STATUS
Rare casual fall visitor central and east.
DOCUMENTATION
Photograph
15 Oct–1 Nov 1975 Hall Co (Stoppkotte 1975b).
Fall
There are 5 Nebraska reports, 2 documented
  with photographs and 1 by description. Dated
  reports are in the period 22 Sep–1 Nov.
late Sep–4 Oct 1952 Elgin (Baumgarten and
  Rapp 1953)
22 Sep 1985 Beaver L, Cass Co (Kiser 1985;
  Mollhoff 1987)
15 Oct–1 Nov 1975 Grand Island (cited above)
A minimal description was published of one
  near West Point 2 Oct 1976 (Einemann 1977).
  Ducey (1989) considered a report from the
  Panhandle questionable; no details were given
  other than that the report was post-1960.
COMMENTS
This species demonstrates a propensity to
  disperse northward from its usual Texas
  and Louisiana range during Sep–Nov in
  most years, although few birds travel as far
  as Nebraska. There are about 12 reports for
  Kansas in the period 15 Jul–8 Dec (Thompson
  and Ely 1989).
FINDING
Observers should be aware of the possibility of
  an ani appearing in fall; it would be easy to
  overlook as a grackle, although anis usually
  stay in brushy undergrowth rather than in the
  open. Any gracklelike bird in a brushy thicket
  should be scrutinized.

STRIGIFORMES

Tytonidae (Barn Owls)

Barn Owl
Tyto alba
STATUS
Uncommon regular spring and fall migrant
  statewide. Common regular breeder south,
  rare north. Rare casual winter visitor east.
DOCUMENTATION
Specimen
UNSM 7719, 30 Apr 1903 Elm Creek, Buffalo Co.
DISTRIBUTION AND ECOLOGY
Barn Owl occurs statewide but in recent years
  has been reported most often in summer in
  the southwest and westward in the North
  Platte Valley. Winter reports in recent years
  are few. It occurs in open areas, preferably
  grasslands, where small rodents are plentiful
  (Epperson 1976). Nesting is usually in crevices
  in rocky cliffs or mammal holes in dirt banks
  or Sandhills blowouts (Bray 1994) but often in
  abandoned buildings and grain bins close to
  human habitation.
Spring
10, 11, 13 Mar→summer
Arrival is in late Mar, although there are earlier
  reports, which are probably of wintering
  birds. In Nebraska, which is near the north
  edge of the species' breeding range, most
  individuals depart prior to winter. Bent (1938)
  indicated that banding data show extensive
  movements that suggest migration; most
  Nebraska reports are in the period Apr–Oct.
Summer
Barn Owl has been reported in summer
  statewide, but there are no nesting records in
  the 20th century from north and east of Holt,
  Boone, and Platte Cos (Ducey 1988), although
  a brood of 7 in a grain bin was photographed
  at the edge of this area 25–32 km (15–20 mi)
  east of Chambers during Aug 1999 (Bub
  Blake, pers. comm. WRS). The only reports at
  any time of year from northeast of this area
  are 29 Apr 1990 Cedar Co and 18 May 1995

Wayne Co. It was considered a rare summer resident at Bessey Division, NNF, by Bray (1994), and one was at Calamus Res 24 May 1997. In recent years most summer reports are from the southwest; the North Platte Valley, including Crescent L NWR; and eastward, less commonly, south of the Platte Valley. As many as 11 were counted at Crescent L NWR 6 Jun 1995, and 5 were roosting in a willow thicket there 8 Sep 1997.

Nesting data for the 5-year period 1985–89 (Bennett 1986, 1987, 1988, 1989, 1990) indicate that this species reaches its peak abundance as a breeding bird in the southwest; 221 nests were reported, including 78 in Lincoln Co, 48 in Frontier Co, and 14 in Hitchcock Co. Seven young ranging in age from less than a week to almost 4 weeks and near-fledged were found in a shallow burrow located on the face of a near-vertical clay bank, about 2.44 m (8 ft) above a road cut in Keith Co 12 Jun 1982 (RSS).

A captive breeding and reintroduction program was initiated in 1980 (Hancock and others 1981) under a management plan of the NPG (Lock 1980). Nest boxes have been set out and used successfully at Crescent L NWR (Huber 1987).

Fall

summer→16, 18, 18 Oct

Departure is in Oct; later reports are probably of birds attempting to overwinter, although there are few late Oct–Nov dates: 9 Nov 1965 Adams Co and 29 Nov 1975 Sherman Co (see Winter).

Winter

While Barn Owl is often considered to be a permanent resident in Nebraska, Dec–Feb reports are few since 1980: 15 Dec 1984 Sioux City, Iowa, CBC; 31 Dec–1 Jan 1982–87 Hall Co; and 20 Feb 1981 Colfax Co, 2 found dead (Peters and Schmidt 1981). Prior to 1980 there are 22 reports, most northerly 20 Feb 1954 Dawes Co, 2 Feb 1958 Scotts Bluff Co, 25 Feb 1950 Logan Co, 4 Jan 1967 Lincoln Co, and 27 Jan 1950 Dawson Co.

FINDING

This species is not hard to find in the southwest,

especially Hitchcock and Frontier Cos, where nest holes can be readily located by driving back roads and looking for whitewash at holes in dirt road cuts. Barn Owls begin incubating after the last egg is laid, and the youngest birds of a clutch may not leave the nest until as late as Aug.

Strigidae (Typical Owls)

Flammulated Owl

Otus flammeolus

STATUS

Hypothetical.

COMMENTS

There are 2 reports. Calls believed to be of this species were heard in ponderosa pines at Chadron SP 30 Jun 1978 by an observer familiar with calls of Northern Pygmy Owl, *Glaucidium gnoma*, and Northern Saw-whet Owl, *Aegolius acadicus* (Egger 1978). One was reportedly shot in spring 1891 near Kearney by C. A. Black, but the specimen was not kept as its importance was not realized; it had buffy plumage and blue eyes but may have been a young screech-owl (Bray and others 1986).

This migratory species breeds in mature ponderosa pine woodlands that occur at lower elevations in the Rocky Mountains from British Columbia south as far as central Mexico (AOU 1998). It has been recorded migrating through similar habitat as far east as Montana, central Colorado, and eastern New Mexico (AOU 1998), and there are 2 fall records Oct–Nov for Wyoming (Scott 1993).

Eastern Screech-Owl

Otus asio

STATUS

Common regular resident statewide.

DOCUMENTATION

Specimen, UNSM 12746, Beatrice, 22 Mar 1894.

TAXONOMY

The taxonomy of North American screech-owls

has undergone considerable revision during the 20th century, yielding as many as 12 races of *O. asio* and recent division into 2 species, Eastern Screech-Owl, *O. asio*, and Western Screech-Owl, *O. kennicottii* (AOU 1983). The geographic range of Eastern Screech-Owl covers most of temperate eastern North America and includes Nebraska at its western range limit. This species is represented by several subspecies, including the Great Plains race *maxwelliae*, which includes the Nebraska breeding population (Marshall 1967; DeBenedictis 1977). Three morphologically distinct color morphs—gray, red, and brown—are represented. These morphs are genetically determined and independent of age, gender, or season but may be distributed to some extent geographically. DeBenedictis (1977) noted that only 7% of *maxwelliae* specimens were red morphs and 6% brown and that red morphs are scarcer toward the north and west. Indeed, Rosche (1982) recorded no red morph birds in northwestern Nebraska, and elsewhere in the state red morph birds are rarely reported.

Since the split of *O. asio* into 2 species, speculation has arisen as to the possible occurrence in Nebraska of Western Screech-Owl, especially as this species is known to occur in riparian habitat in states adjacent to and west of Nebraska. The closest records are in extreme southwest Kansas along the Cimarron River (Thompson and Ely 1989; FN 51:766). Also of interest is the suggestion that long-distance dispersal of screech-owls leads to occurrence of the 2 species in close proximity and in mixed pairs in western Texas (Marshall 1967; AOU 1983). Such dispersal may account for the presence of both species on occasion in the Colorado Springs area in winter (Marshall 1967) and may lead to the occurrence of Western Screech-Owl in western Nebraska.

Mollhoff (1985) found that screech-owls at Rock Creek L in Dundy Co responded to calls of both Eastern and Western Screech-Owls but only gave the Eastern Screech-Owl song type. A bird that responded aggressively to an Eastern Screech-Owl tape along the Arikaree River in extreme southwest Dundy Co in early Jun 1989 was closely examined and determined also to be an Eastern Screech-Owl (WRS, TB, DR).

DISTRIBUTION AND ECOLOGY

This species occurs statewide but becomes local in the west due to its habitat preference of edge woodlands that include mature or dead larger trees that have developed natural cavities suitable for nest sites. Such habitat is more common and widely distributed in eastern and central Nebraska, including towns and cities, but the species occurs only rarely in deep woods, where it tends to become prey for larger owls (Johnsgard 1980). Some pine forests may also support screech-owls (Rosche 1982).

Summer

Accurate estimates of numbers are difficult as this owl is strictly nocturnal. BBS data do not detect this species, and it is apparent that there may be "more Screech-owls around than most people realize" (NBR 40:17). It is common in the Platte Valley at least as far west as Hall Co (Lingle 1994), occurs in numbers in the Niobrara Valley west at least to the Niobrara Valley Preserve (Mossman and Brogie 1983), and occurs throughout the Republican Valley (Johnsgard 1979; Andrews and Righter 1992). It is probably least numerous in the Sandhills due to scarcity of habitat; Rosche (1982) cited no summer records from Sheridan Co, but Ducey (1983) found it in suitable habitat at Anderson Bridge in central Cherry Co in summer 1982.

Nesting begins in Mar; in Seward Co nesting was under way by 12 Mar 1994.

Winter

CBC totals indicate that numbers are highest in the east, diminishing steadily westward. High counts are at Lincoln: 40 on 23 Dec 1979, 33 on 17 Dec 1978, and 32 on 18 Dec 1975. As many as 7 were counted at Crawford 27 Dec 1977.

It has been suggested that because there appear to be more sightings in winter there may be migrants coming from the north and west (Rapp and others 1958; Ludlow 1935), but this has yet to be verified (see Taxonomy).

FINDING

Screech-owls readily respond to tapes, especially in late fall and winter. Woodlots are preferred habitat, particularly where isolated older trees or dead trees are present.

Great Horned Owl

Bubo virginianus

STATUS

Common regular resident statewide. Rare regular winter visitor statewide.

DOCUMENTATION

Specimen

UNSM ZM12752, 4 Nov 1902 Lancaster Co.

TAXONOMY

There are 2 resident breeding subspecies. Eastern *virginianus* occupies the southeastern quarter of the state (Swenk 1937b; AOU 1957) and western *occidentalis* the remainder (Swenk 1937b; AOU 1957; Rapp and others 1958). *Occidentalis* has exhibited some tendency to disperse eastward in fall and winter, beginning when young are abandoned by adults (Swenk 1937b; AOU 1957; Haecker and others 1945).

The pale, occasionally almost white (Bent 1938), northern race *wapacuthu* is an occasional winter visitor (Swenk 1937b). Swenk (1937b) cited 11 specimens 11 Nov–1 Mar from all parts of the state. Two of these, 2542 and 10686, are in the HMM. The northwestern race *lagophonus* also appears casually in winter (AOU 1957), as does the southerly race *pallescens* (Swenk 1937b). Swenk (1937b) cited 8 records of *logophonus* statewide in the period 28 Oct–mid-Feb, including a specimen, 2679, in the HMM and 3 specimens of *pallescens,* from Adams Co 7 Dec 1933 and 19 Oct 1934 and from Saunders Co 9 Dec 1934, although one of the Adams Co specimens was later redescribed as a small example of *occidentalis* (Swenk 1938).

DISTRIBUTION AND ECOLOGY

This species is a statewide resident, occupying the edge of both deciduous and coniferous forest and open savanna. Mature planted woodland habitat may also support this large owl, including woodlots at farmsteads, shelterbelts, and larger parks in towns and cities. Although highly adaptable as a predator and therefore broadly distributed, one limiting factor to its distribution may be the availability of appropriate nest sites that include large, abandoned nests of other species in mature trees, cavities in stumps of fallen trees, and ledges on cliffs.

There is some tendency toward regional winter dispersal, which may reflect local food availability. Nebraska specimen records reveal almost annual winter occurrences of northern birds, and western and even southern subspecies have been recorded (see Taxonomy).

Summer

Peak egg laying in Kansas is around 10 Feb (Johnsgard 1979). Of 45 nests observed by Rosche in 1994, earliest was 4 Mar (NBR 62:110). Swenk (1937b) cited data indicating egg dates 6 Feb–10 Mar, with most young fledged by early May. Incubation regularly begins around 15 Feb in eastern Nebraska, and hatching occurs around 15 Mar (RSS). Nests (N = 5) with young have been reported in the period 4 Apr–9 May in recent years, including young with primaries emerging in a nest near Redington 24 May (WM).

Winter

CBC totals indicate the abundance of this species; high counts are 31 at Norfolk 20 Dec 1986, 30 at Omaha 26 Dec 1987, and 30 at Lincoln 16 Dec 1990.

FINDING

Easily heard just before dawn and at dusk Nov–Apr, this species can often be seen perched prominently in the open at dusk; driving open areas with large trees or telephone poles should yield a sighting.

Snowy Owl

Nyctea scandiaca

STATUS

Rare regular winter visitor statewide.

DOCUMENTATION

Specimen

UNSM 7729, 6 Mar 1918 Harvard, Clay Co.

DISTRIBUTION AND ECOLOGY

Snowy Owls prefer open areas, usually grasslands, but are found in agricultural areas, as well as on frozen lakes. They have a preference for perching on elevated sites such as haystacks (Johnsgard 1980), exposed ridges (Rosche 1982), and fence posts in agricultural lands. Numbers may be larger than reported, as there are vast areas of northern Nebraska that are often snow-free in winter. Reports are statewide but fewer in the south.

This species has a tendency to wander southward in most years in varying numbers. Occasionally an invasion occurs when northern food supplies are severely limited. At least one individual is reported in the state most years.

Winter

11, 12, 15 Nov→13, 22, 23 Mar

Arrival is in mid-Nov, although there is an earlier report 6 Nov (Johnsgard 1980). Most depart by mid-Mar, although there are later reports 30 Mar 1985 Lancaster Co, 7 Apr 1936 Logan Co (Glandon and Glandon 1937; listed as Lincoln Co in NBR), 15 Apr 1981 Douglas Co, 20 Apr 1971 Brown Co, and 30 Apr (Johnsgard 1980, without documentation).

FINDING

This species should be looked for in invasion years in open areas, where it usually perches on power poles, haystacks, or exposed ridges. It often remains in a single location for an extended time.

Northern Hawk Owl

Surnia ulula

STATUS

Accidental in winter.

DOCUMENTATION

Specimen

HMM 1231, 8 Dec 1912 Sheridan Co.

Records

There is a single documented record, the specimen cited above, a male collected by Walter Moore in Sheridan Co 8 Dec 1912 (Swenk, Notes after 1925; Bray and others 1986). Another specimen, reportedly shot near Raymond in Nov 1891 by E. R. Mockett and examined by Bruner and Eiche (Bruner and others 1904), cannot now be located (Bray and others 1986).

COMMENTS

There are single records for South Dakota 12–26 Jan 1978 (SDOU 1991) and Iowa 25 Dec 1981–25 Feb 1982 (Kent and Dinsmore 1996), and this species occurs accidentally during invasion years in southern Minnesota (Janssen 1987).

Other Reports

One was reportedly in an Omaha yard 29 Aug 1974 (Green 1974).

Burrowing Owl

Athene cunicularia

STATUS

Common regular spring and fall migrant west, becoming rare casual east. Common regular breeder west, becoming uncommon east. Rare casual winter visitor central and west.

DOCUMENTATION

Specimen

UNSM ZM11338, 24 Mar 1895 Beatrice, Gage Co.

DISTRIBUTION AND ECOLOGY

Burrowing Owls occur in the western two-thirds of the state, generally in shortgrass prairie or grazed grasslands, a habitat commonly found in prairie dog towns, although it may occupy other burrows and holes. They have declined as prairie dogs have been eradicated, such as in Thomas Co (Bray 1994). On occasion individuals attempt to overwinter on the breeding grounds. Its occurrences in the east are rare, usually during migration periods.

Spring

21, 21, 21 Mar→summer

Arrival is in early Apr, although there are earlier reports 8 Mar 1942 Jefferson Co, 10 Mar 1990 Lincoln Co, 12 Mar 1955 Thomas Co, and 16 Mar 1967 Adams Co.

High counts include 41 in southwest Scotts Bluff Co 19 Apr 1999, 27 near Lyman 27 Apr 1997, and 11 in southeast Cheyenne Co 17 Apr 1998.

Summer

This species was formerly a common breeder throughout the state (Bruner and others 1904) but is now essentially absent from the settled and farmed eastern third (Ducey 1988), occurring there only occasionally and locally. In the absence of prairie dog towns the species typically utilizes the large burrows of carnivores such as badger and fox. Fiala (1970) stated that it no longer occurred in Gage Co. It continues to be regularly reported in summer as far east as Holt, Howard, Hall, and Adams Cos, although it is probably regular in the few remaining prairie dog towns as far east as the eastern edge of the Rainwater Basin, where 2 were seen south of Geneva 27 Apr 1995 (NBR 63:47). Breeding season reports farther east are few: it was seen 30 May–4 Jun 1965 Dakota Co (Wagner 1966), 3 Aug 1989 Stanton Co, 17 May 1986 Merrick Co, and 17 Apr 1985, 15 May 1990, and 16 May 1991 Polk Co.

It is most numerous in the Panhandle, where prairie dog towns are still commonplace and widely distributed. A study in Sioux, Box Butte, Scotts Bluff, Morrill, and Banner Cos in 1989 (Desmond and Savidge 1990) comparing fledging success of colony nesters (in prairie dog towns), solitary nesters (in badger holes), and nesters in abandoned prairie dog towns showed that of 85 nests, success was 3.12 ($N = 60$), 5.0 ($N = 16$), and 0.88 ($N = 9$) fledged young per nest, respectively.

High counts include 19 south of Lyman 8 Jul 1994, 16 north of Alliance 14 Jul 1994, and 12 at Fort Niobrara NWR 21 Jun 1999.

Fall

summer→23, 27, 28 Oct

Departure is in late Oct, although there are later reports 7 Nov 1980 Sioux Co, 9 Nov 1965

Adams Co, 23 Nov–9 Dec 1989 Hall Co, and 30 Nov 1975 Lancaster Co.

Most of the few recent reports from the east are of fall migrants: 25 Sep 1990 Douglas Co, 28 Sep 1985 Polk Co, 29 Sep 1986 Douglas Co, 18 Oct 1975 Douglas Co, 23–27 Oct 1984 Wayne Co, and 30 Nov 1975 Lancaster Co.

High counts include "20 or more" in Cherry Co 16 Sep 1997, 11 in Scotts Bluff Co 4 Sep 1999, and 9 there 23 Aug 1996.

Winter

Rapp and others (1958) noted that "a few winter in mild winters," and Bent (1938) stated that a few remain on the northerly breeding grounds through winter. Curiously, however, there are several Jan and Feb reports but virtually none for Dec, which may indicate that rather than overwintering, a few individuals begin northward movement very early in spring. Reports in midwinter are Jan 1931 Webster Co (Ludlow 1935), 1 Jan 1925 Webster Co (Ludlow 1935), 1 Jan 1941 Webster Co, 6 Jan 1965 Scotts Bluff Co, 8 Jan 1931 Webster Co (Ludlow 1935), 10 Jan 1947 Dawes Co, 10 Jan 1952 Holt Co, 27 Jan 1956 Antelope Co, 3 Feb 1953 Holt Co, and 3 Feb 1985 Scotts Bluff Co.

FINDING

This species is not hard to locate in prairie dog towns in the Panhandle. Scanning the area within a town usually yields at least one owl among the prairie dogs.

Barred Owl
Strix varia

STATUS

Uncommon regular resident east, rare casual elsewhere.

DOCUMENTATION

Specimen

UNSM ZM7741, 28 Nov 1937 Fontenelle Forest, Sarpy Co.

DISTRIBUTION AND ECOLOGY

The range of this species corresponds to that of extensive tracts of mature upland deciduous forest, essentially reaching western limits in Knox, Madison, Seward, and Jefferson Cos. It is rare or absent where such forest is restricted in extent, although in the Missouri River floodplain of southeast Nebraska, Barred Owls may utilize mature riparian forest where such tracts adjoin upland forest.

As early as 1900 it was recognized that the selective cutting of old, mature hardwood trees, many containing natural nesting cavities, was also a serious limiting factor for this species in Nebraska (Carriker 1900). This is essentially the same problem facing the Spotted Owl, *Strix occidentalis*, a closely related species, as it loses old-growth forest habitat in the Pacific Northwest.

Summer

Although nowhere numerous, it is most common in the lower Missouri River valley where extensive hardwood forest is found (Shupe 1985). It may occur, however, in small numbers westward in isolated upland deciduous forest remnants, such as in Knox Co (see below), where it has been reported from and near Bohemia Prairie since 1985 (Schreier 1986; NBR 62:88); in the Verdigre area 1962–63 (Svoboda 1963); and near Niobrara 7 Jan 1998. Also at the western edge of the range are reports from Madison Co of single birds on CBCS 15 Dec 1979 and 17 Dec 1994; from Seward Co, where it was reported from Oak Glen 25 Jun 1995; and from Jefferson Co, where it was reported 10 May 1950 and seen in the Fairbury city park 9 May 1983 (Hoge and Hoge 1984). Its status in Gage Co was considered "uncertain" by Fiala (1970), who cited only 6 reports for the area; there is a specimen, UNSM ZM16954, from Gage Co collected 5 Jan 1994. In Lancaster Co it was noted to have established residency in Wilderness Park in the early 1970s, after the area became a park and hunting was disallowed (NBR 44:7). It was reported from the Ames area 14 Jul 1997.

West of these locations there a few scattered reports. Most intriguing are those cited by Rosche (1982): it was "heard every night" 18 May–1 Jun 1900 in Monroe Canyon (Crawford 1901) and reported in southern Dawes Co 21

May 1965 without details. Barred Owl occurs in very low numbers in the Black Hills of South Dakota (Peterson 1990; SDOU 1991), but there are no other reports from the Pine Ridge area.

Mossman and Brogie (1983), despite careful searching, could not find this species in the Niobrara Valley Preserve in 1982, although Youngworth (1955) had noted it there earlier, and one was reported from Brown Co 17 Sep 1997 (RG). These minimal data suggest that a small population may occur in the lower Niobrara Valley, including Knox Co.

The only other reports are from the Platte Valley. Tout (1947) listed Barred Owl as a "rare straggler" in Lincoln Co, citing 2 reports, neither with descriptions: 4 were flushed from a shrub in a meadow 19 Jan 1925, and another was killed by a train 1 May 1927 near North Platte. Additional Platte Valley reports are York Co (Tout 1902; Ducey 1988), 2 Jun 1991 Hall Co (Lingle 1994), and 3 Oct 1962 Dawson Co, when one was "chased out of cottonwoods" near Lexington (Kinch 1963).

The only Colorado record was of a nesting pair in 1897 at Holyoke, in the extreme northeast, suggesting that on occasion stragglers may occur westward in the Platte Valley. Bent (1938) noted that northerly breeders may move south in severe winters; it is possible such birds may straggle to Nebraska and provide some of the few reports from central and western Nebraska.

Egg laying apparently peaks in early Mar, at least in Kansas (Thompson and Ely 1989). A nest containing 3-week-old young was found by RSS in the cavity of a sheared-off tree trunk 16 Apr 1972 in Fontenelle Forest. A juvenile was in Wilderness Park, Lincoln, 15 May 1999 (LE).

Winter

CBC totals are generally low. High counts include 9 at Omaha 1989–90 and 8 at Lincoln 1988–89.

FINDING

Barred Owl is best searched for in late Mar and Apr in extensive Missouri Valley forests, such as at Indian Cave SP and Fontenelle Forest.

It is very vocal and can usually be heard just before sunrise if present. It typically does not vocalize in response to tapes but often comes in to check out the noise.

Great Gray Owl
Strix nebulosa
STATUS
Accidental in winter.
DOCUMENTATION
Specimen
Mount at wsc, 1 Jan 1978 Dixon, Dixon Co.
Records
The above specimen is the only documented record. The bird was first seen 31 Dec 1977 and illegally shot 1 Jan 1978 near Dixon; it was a female in underweight condition (Lock 1978; Mollhoff 1979).

There are additional undocumented reports. A specimen reportedly taken near Omaha 17 Dec 1893 (Bruner and others 1904) cannot now be located, although it may be the specimen purchased from the Northwest School of Taxidermy in Omaha in 1912 and now #2537 at HMM. Two were said to have been shot about 1896 near Maywood, Frontier Co, and exhibited there for some time (Brookings, Notes; Bray and others 1986). According to Bruner and others (1904), a mounted specimen in a saloon at Long Pine was said to have been taken in the vicinity. It was reported from Omaha 4 Apr 1948 (NBR 16:86) and from Brady 12 Jan 1950 (NBR 18:83).
COMMENTS
This species is a resident in northwest Wyoming (Scott 1993) and in northern Minnesota (Janssen 1987). There are no records for Colorado, Kansas, or Missouri and only 2 from southeast South Dakota, 25 Jan 1984 (SDOU 1991) and 29 Jan–2 Feb 1997 near Sioux Falls (FN 51:763). There are several from Iowa, including 2 during winter 1996–97 (Cecil 1997) and as many as 3 during winter 1995–96, the most recent of a few large invasions since that of 1968–69, including 1977–78, when the Nebraska bird appeared, and 1983–84,

when the first South Dakota bird appeared (Kent and Dinsmore 1996). In Minnesota, the probable source of Nebraska and Iowa birds, fall movement becomes noticeable some years in Oct, sometimes developing into invasions, with most birds departing by mid-Mar (Janssen 1987). Even in Minnesota the species is accidental in the south (Janssen 1987).
FINDING
As with Boreal Owl, this species should be looked for in the northeast in major Minnesota invasion years. It is somewhat crepuscular, so early morning and late afternoon are good times to search. Open woodlands are probably the best habitats.

Long-eared Owl
Asio otus
STATUS
Uncommon regular spring, fall, and winter visitor statewide. Rare regular resident statewide.
DOCUMENTATION
Specimen
UNSM ZM12786, 9 May 1892 Waverly, Lancaster Co.
DISTRIBUTION AND ECOLOGY
Long-eared Owl occurs statewide, occupying dense woodland thickets, including shelterbelts. Nesting usually takes place in abandoned crow or hawk nests. The resident population is probably augmented by winter visitors from north of the state. Wintering birds usually congregate in dense stands of conifers but also occur in riparian thickets and shelterbelts.
Spring
winter→18, 19, 20 Apr
Departure of wintering birds occurs in mid-Apr; based on egg dates (see Summer) some spring reports as well as the following later reports may be of summering birds: 22 Apr 1962 Douglas Co, 24 Apr 1971 Lancaster Co, 28 Apr 1957 Adams Co, and 28 Apr 1957 Webster Co.
Summer
Even though there are reports of nesting from

all parts of the state, this species is not
a predictable repeat nester at any known
location. Nesting has been documented from
Cass (Ducey 1988), Otoe (Carriker 1900),
Douglas (Ducey 1988), Gage-Johnson (Ducey
1988), Lancaster (NBR 39:15, 45:38; Bennett
1971, 1972, 1974, 1978, 1980a, 1982, 1986; Ducey
1988), Dodge (LOI 38:1; Holcomb 1967), Boone
(NBR 54:55; Bennett 1987), Antelope (Cary
1900), Thomas (Bennett 1988; Bray 1994),
Cherry (Held 1958; Brogie and Mossman
1983), Hall (Bennett 1972), Lincoln (Bennett
1974), Keith (Brown and others 1996), Logan
(Glandon 1956; Ducey 1988), Box Butte
(Mathisen and Mathisen 1960a), and Sheridan
(Ducey 1988) Cos.

These owls utilize old crow and magpie nests
as platforms, typically located in conifers,
especially junipers (Mossman and Brogie
1983; Holcomb 1967; Held 1958; Bray 1994),
but also in mixed junipers and hardwoods
(Mossman and Brogie 1983) and in an elm in
a shelterbelt (Mathisen and Mathisen 1960a).

There are many additional reports in the period
May–Aug that probably indicate nesting but
without details. Reports from counties not
listed above are 2 May 1957 and 12 Jul 1997
Nemaha Co, 7 Jul 1986 Pierce Co, 19 May 1957
and 8 May 1971 Brown Co, 4 May 1964 and 30
Jun 1959 Webster Co, 20 May 1957 Buffalo Co,
5 May 1974 and 22 Apr–summer 1970 Perkins
Co, 11 Jun 1999 Chase Co, 1 May 1999 Garden
Co, and 30 Jun 1967 and permanent resident
1970 Scotts Bluff Co.

Bent (1938) listed 8 egg dates for Indiana, Illinois,
and Iowa in the period 20 Mar–28 Apr. A nest
near Alliance had 4 eggs 25 Apr (Mathisen
and Mathisen 1960a). Young were fledged by
28 May (Holcomb 1967), 23 Jun (Brogie and
Mossman 1983), and Aug (Held 1958).

Fall

21, 22, 26 Oct→winter

Numbers increase in the period Nov–Mar,
probably as birds arrive from north of
Nebraska. Birds banded as juveniles in
Alberta and Manitoba have been recovered

in Nebraska in early Dec and early Mar,
respectively. Some Nebraska breeders may
remain in winter congregations as long as prey
is accessible. Although it has been suggested
that some Nebraska birds move south after
breeding (Haecker and others 1945), there is
no evidence to support this.

Arrival is in late Oct, although there are several
Sep and early Oct dates, possibly summering
birds: 9 Sep 1968 Brown Co, 15 Sep 1959
Webster Co, 25 Sep 1962 Webster Co, 25–26
Sep 1973 Perkins Co, 3 Oct 1967 Adams Co,
and 5 Oct 1968 Adams Co.

High counts, usually CBC totals, include 39 at
Lincoln 18 Dec 1977, 22 there 21 Dec 1980, and
19 there 19 Dec 1982.

Winter

Wintering birds occur statewide and consist
of congregations of 25 or more. Bent (1938)
suggested that these winter congregations
consist of family groups, which may account
for roost sites developing some degree
of permanency, as demonstrated by use
in successive winters. A roost site near
Cunningham L has been used for at least 6
winters. Keith Geluso (pers. comm. RSS) has
noted that individuals arrive at this site as
early as 21 Nov when snow cover begins to
develop in South Dakota and that as many as
25 birds may be present, although some or all
may leave if snow reduces prey accessibility.
Geluso has a record of 4 birds at the site as late
as 28 Mar.

Roost sites are typically found in dense stands
of vegetation, frequently junipers, that may
develop in shelterbelts or on overgrown
grazing land. Chosen sites are usually near
a foraging area of open habitat such as
grassland or fallow cropland that supports
small rodents such as voles (*Microtus spp.*).

FINDING

Long-eared Owls are not hard to find in Jan–Feb
by searching dense conifer groves for pellets
and whitewash. Roosting birds usually flush
fairly readily and can often be seen well.

Short-eared Owl

Asio flammeus

STATUS

Rare regular resident statewide. Uncommon
regular winter visitor statewide.

DOCUMENTATION

Specimen

UNSM ZM7740, 4 Dec 1902 Kirkwood, Lancaster
Co.

DISTRIBUTION AND ECOLOGY

This species is essentially restricted to open
areas, especially ungrazed and undisturbed
native grasslands (Dobkin 1994), showing
some preference for moist meadows and
marsh habitats, particularly in spring,
summer, and fall. Wintering birds occur
statewide, generally south of the snowline,
where their rodent diet remains available.
Numbers vary considerably from winter
to winter depending on snow cover, and
the species may be absent in some winters.
Breeding birds may occur in grassland areas
anywhere in the state, particularly in damper
situations, but in recent years reports have
been scarce. It roosts in conifers in winter
when snow cover precludes ground roosting
(Dobkin 1994).

Summer

Breeding populations fluctuate widely
depending on rodent populations, and
breeding may be colonial at times and
nonexistent at others. Most authors consider
this species to have declined significantly in
Nebraska in the 20th century, partly due to
habitat loss (Blake and Ducey 1991; Brogie
and Mossman 1983; Ducey 1988).

Currently it appears to be a rare nester, although
its summer status is poorly known (Ducey
1988); there are only these few reports of
nesting since 1960: a nest was photographed
in Brown Co in the late 1960s (Mossman
and Brogie 1983); another was found in a dry
wheat field in Banner Co in 1966 (Sharpe
1967); a nest card was submitted for Hall
Co in 1967 (Bennett 1968); 2 nests were
reported in Garden Co in 1967 (Bennett 1968);

and young were noted in Holt Co in 1973
(Bennett 1974).

There are numerous reports May–Sep without
evidence of nesting; these reports are from
all parts of the state but are most numerous
in native grasslands, including those in the
southeast.

Egg dates are in Apr and May (Bent 1938; SDOU
1991), and reports at that time are indicative
of nesting, although it should be kept in
mind that this species may be present but not
nesting if rodent populations are low.

Winter

While evidence is scarce, Nebraska birds likely
remain on their summer ranges if weather and
food conditions allow (Rosche 1982). In severe
winters with extensive snow cover, most birds
apparently move south below the snowline,
sometimes resulting in increased numbers
in the state. Reports are fewest in Feb. This
species is probably best considered nomadic
rather than migratory, often appearing in
numbers wherever the rodent population
is high. However, in most years, numbers
increase beginning in late Oct and decrease
by late Apr.

High counts include 9 near Creighton 21 Mar
1999, 8 on the Omaha CBC 26 Dec 1964, 8 at
Branched Oak L 11 Jan 1978 (Lock 1978), and
7–8 at Jack Sinn Marsh 25 Nov 1985 (Lock
1987).

FINDING

Short-eared Owls are most likely to be found in
Nov and Dec in Sandhills or Rainwater Basin
marsh areas; flying birds most often appear
at dusk, although some day flying is evident.
They hunt low to the ground, like Northern
Harriers. Birds seen in summer should be
carefully observed in order to locate any nest
that might be present.

Boreal Owl

Aegolius funereus

STATUS

Accidental in fall.

DOCUMENTATION

Specimen

HMM 2710, 5 Oct 1916 Inland, Clay Co.

Records

There are 4 reports, only 1 documented: a specimen on display at the HMM collected by Clyde Ley near Inland 5 Oct 1916 was described by Swenk (Bray and others 1986). Another report was of one seen in a barn in Webster Co 16–17 Dec 1963 (Turner 1964); it was described as being the size of a screech owl but without ear tufts and with a pale bill. One captured live in Lincoln 10 Dec 1892 (Bruner and others 1904) was said to be in the UNSM collection but cannot now be located (Bray and others 1986). A fourth report was of one perched on a "shock of grass" near Salt Creek in Lancaster Co 19 Oct 1907 (Swenk 1907).

COMMENTS

There are no acceptable records from Missouri, Iowa, or Kansas, but there are 3 from South Dakota, all near the eastern border: Mar 1913, 16 Aug 1949, and 6 Dec 1957 (SDOU 1991). In Minnesota this species stages occasional invasions, during which birds appear in mid-Jan through Mar and are usually emaciated (Janssen 1987). Since 1976 a fall migration has been detected in Minnesota beginning in Oct (Janssen 1987).

Northern Saw-whet Owl

Aegolius acadicus

STATUS

Hypothetical breeder west. Casual summer visitor central and east. Uncommon regular winter visitor east, becoming rare west.

DOCUMENTATION

Specimen

UNSM ZM12801, 8 Dec 1895, Lincoln, Lancaster Co.

DISTRIBUTION AND ECOLOGY

Summering birds, although rare, may be found regularly in ponderosa pine woodland from Chadron SP westward (Rosche 1982). Although most reports are from the east,

winter visitors and migrants occur statewide and may be found in either coniferous or riparian woodland. Wintering birds usually establish a roost site in dense conifers, especially cedars near water, but may be found in any dense growth. They are also found in towns and cities in winter and occasionally become trapped in chimneys as they seek shelter.

Summer

In recent years there has been no indication that summering birds occur anywhere but in the Pine Ridge, where singing birds on territory have been heard 18 Apr–1 Jun from Chadron SP westward (Rosche 1982; Johnsgard 1979). There is, however, no documented evidence for nesting there (Ducey 1988). It is a permanent resident in the Black Hills of South Dakota (SDOU 1991).

There are old reports of nesting in the Missouri Valley, but none since Carriker reported collecting a nest with eggs near Nebraska City prior to 1900 (Carriker 1900; Silcock 1979; Ducey 1988); this is the only documented nesting for the state. However, there was a successful nesting in Doniphan Co, Kansas, in 1951 (Thompson and Ely 1989).

There are a few summer reports away from the Pine Ridge, which raises the possibility of occasional nesting elsewhere in the state. Because the last wintering birds leave by the end of Apr, territorial birds in May–Jun are of interest. One such bird was at Ash Hollow SHP 16–17 May 1986 (Rosche 1994a); one was near Halsey 10 May 1988 (NBR 56:99); one was reported in Douglas-Sarpy Cos 19–23 Jun 1964; and one was singing at Champion Mill SHP 23 Jun 1996 (MB). There are also several Jul and Aug reports: one was at Riverside Park in Scottsbluff in Jul 1985 (NBR 54:15); one was in Lancaster Co 15 Jul 1974; one was calling in Lincoln 28 Jul and 20 Aug 1969 (NBR 38:39); one was seen on a stop sign in Omaha 8 Aug 1979 (Bellinghiere 1980);, and one was in a Scottsbluff yard 17 Aug 1985 (NBR 54:15). Fall migrants do not usually appear until Oct.

Winter

22, 28, 30 Oct→12, 22, 22 Apr

Wintering birds probably occur statewide in small numbers, but most reports are from the east and may be a result of the presence of more observers there. Arrival is in Oct, although there is an earlier report 29 Sep 1972 of a calling bird in Fontenelle Forest; this bird may have summered in the area, as there are no Sep records for Iowa or Missouri (Kent and Dinsmore 1996; Robbins and Easterla 1992). Another earlier record is a specimen, HMM 28234, a female collected at Hastings 12 Oct 1953. Early fall dates for Iowa are 5 and 6 Oct (Kent and Dinsmore 1996). There are several CBC reports, all of single birds, mostly from Lincoln. Departure is in Apr.

FINDING

Efforts to locate this bird involve searching its preferred winter roosting habitat, which includes isolated, medium-size, dense junipers on slopes near water. Its presence is often betrayed by whitewash on the trees. In spring and summer territorial birds will respond to taped calls.

CAPRIMULGIFORMES

Caprimulgidae (Goatsuckers)

Common Nighthawk

Chordeiles minor

STATUS

Common regular spring migrant statewide. Common regular breeder statewide. Abundant regular fall migrant statewide.

DOCUMENTATION

Specimen

UNSM ZM6253, 15 Jul 1901 Indian Creek, Sioux Co.

TAXONOMY

Three subspecies have been recorded in the state as summer residents, but their contemporary distribution is not clearly known. The eastern subspecies *C. m. minor* is known to occur westward to central Nebraska (AOU 1957) but may currently occur farther westward as an urban dweller nesting on flat, gravelly rooftops. The state's open grasslands are occupied by the smaller and paler *C. m. sennetti* (Haecker and others 1945), and entering the state from the west in ponderosa pine parkland in Sioux Co is the rufous Rocky Mountain form *C. m. howelli* (Bruner and others 1904). It is likely that habitat changes during the last 50–60 years have significantly altered subspecific distribution.

DISTRIBUTION AND ECOLOGY

This species is a common migrant statewide, especially so in fall when large flocks, sometimes 200 or more, often pass by at dusk. Breeding birds occur throughout the state but are most abundant in larger towns and cities, where they nest on flat, graveled roofs. Although numbers are lower in open grasslands, they are broadly distributed in this habitat and nest directly on the ground. Urban dwellers are crepuscular and nocturnal, feeding at elevations of 30 m (100 ft) or more, usually above urban structures; they are often seen hawking about large light sources, such as stadium lights, which attract large numbers of insects. Birds of open grasslands forage much closer to the ground, usually 15 m (50 ft) or less, and often forage until midmorning and again in late afternoon. It is not known whether these differences in feeding habits coincide with subspecific differences or merely reflect differing prey availability.

Spring

10, 10, 11 Apr→summer (southeast)

26, 29 Apr, 2 May→summer (west)

Arrival is somewhat earlier in the east than in the west; this interval has been estimated at about 2 weeks (Gates 1966). Early dates in the east are in the south-central counties in mid to late Apr. In the Panhandle arrival is in early May, although there is an earlier report 18 Apr 1995 at Crescent L NWR. Concentrations of spring migrants are rarely seen.

Summer

BBS data indicate that greatest numbers occur in grasslands of the north and north-central parts of the state, and fewest occur in the east. The latter may be an artifact resulting from the tendency for BBS routes to undercount urban habitats, the preferred habitat for this species in the east, or the tendency for grassland birds to fly during daylight hours.

Nesting may take place rather late, as in the case of a nest with eggs near Keystone in early Aug 1981 (Rosche 1994a). Brogie and Mossman (1983) reported 7 egg dates in the Niobrara Valley Preserve 4–23 Jun.

Fall

summer→20, 22, 24 Oct (Johnsgard 1980)

Large aggregations are often seen in fall. Departure is generally in late Aug and Sep, and Oct records are few. There is no clear difference in departure dates between the east and west.

High counts include "several hundred" at Scottsbluff 23 Aug 1994, 200 at Chadron 27 Aug 1993, and 150 in Dixon Co 6 Sep 1997.

Other Reports

10 Mar Gage Co (Fiala 1970), 14 Mar 1958 Gage Co, 23 Mar 1949 Saline Co, 26 Mar 1959 Adams Co.

FINDING

This species is easy to locate by its call at dusk in summer virtually anywhere in the state, although larger towns and cities are best.

Common Poorwill

Phalaenoptilus nuttallii

STATUS

Common regular spring and fall migrant west, becoming casual east. Common regular breeder west, uncommon central, rare casual east.

DOCUMENTATION

Specimen

UNSM ZM6247, 24 May 1900 Monroe Canyon, Sioux Co.

DISTRIBUTION AND ECOLOGY

Nebraska is at the extreme eastern edge of the breeding range for the species, and while the range boundary in Nebraska is not well documented, it apparently encompasses any areas where pines or brushy woodlands are associated with rocky terrain. Such habitat occurs primarily in the northern Panhandle (Rosche 1982) but extends along the Niobrara Valley east at least to Brown Co (Brogie and Mossman 1983; Mossman and Brogie 1983), in the North Platte Valley at least to the Keystone area (Johnsgard 1990a; Rosche 1994a), and in the southwest, including scarps above Frenchman Creek in Chase, Hayes, and Hitchcock Cos. It may occupy mixed grassland-coniferous habitat in the upper Loup drainage (Bray 1994; Smith 1957; Ford 1959) and other native grasslands associated with rough, rocky exposures, notably those south of the Platte River. Migrants occur regularly somewhat farther east, probably at least as far as the central parts of the state.

Spring

24, 25, 25 Apr→summer

The presence of this species in the western Loup drainage in summer, especially Thomas Co, suggests migrants should occur virtually statewide, but there are few such reports. These include a small number from Lincoln

Co; 1 fall record 30 Sep–2 Oct 1973 from Perkins Co; 2 reports from Logan Co, 4 May 1964 and 6 May 1935 (Glandon and Glandon 1964); and farther to the east, specimens HMM 26993 collected at Bladen 26 Apr 1951 and a male collected 2 May 1901 at Kearney (Swenk, Notes before 1925). Arrival is in early May, although earliest reports are in Apr, mostly from south-central Nebraska, and there is an earlier report 18 Apr Keith Co (Johnsgard 1990a).

High counts include 6 at Ash Hollow SHP 17 May 1995.

Summer

Although it has not been recorded breeding in open areas of the Sandhills (Johnsgard 1980), it may indeed nest rarely in native grasslands. Various authors (Johnsgard 1980; Haecker and others 1945) have indicated that breeding may occur farther to the east than outlined above. There are several summer reports for Knox and Holt Cos, suggesting that breeding may occur throughout the Niobrara Valley. It is a "common summer resident and nester" in the Bessey Division, NNF, although it was not recorded there until 1956 (Bray 1994); 12 were counted there 28 Aug 1999 (JS). In the Platte Valley there is a single Lincoln Co summer report 6 Jun 1954. Although documentation is lacking, as are summer reports, the distribution in Kansas (Thompson and Ely 1989) suggests that at least occasional nesting may occur in Nebraska south of the Platte River, especially in rocky areas in native grasslands east perhaps to Gage Co. Thompson and Ely (1989) suggest that this species may indeed be more common in the Flint Hills of eastern Kansas than in western Kansas. Immediately north of the Flint Hills in Nebraska there are in fact 2 nesting records. A successful nest was observed in Lancaster Co in 1976 (NBR 45:42), and another, with eggs, was found near Powell in 1963 (Hoge 1964).

Incubated eggs have been found 27 May–4 Jun in open grassland in Jefferson Co and 9–11 Jun and 13 Jun (with 2 eggs) in grassland among ponderosa pines in Dawes Co. Young, typically 2, have been observed as early as about 20 May in Lancaster Co and as late as 9 Jul in Dawes Co. A female with a soft-shelled egg in her oviduct was taken as a specimen 8 Jul 1957 in Thomas Co (Ford 1959).

High counts include 12 in Sowbelly Canyon 13 Jun 1998.

Fall

summer→3, 3, 4 Oct

Departure is difficult to determine, as these birds usually stop calling in Jul. While most probably leave in Sep, there are later records, mainly road-killed specimens in early Oct. Later records include a specimen captured alive near Valparaiso and banded, photographed, and released in Lincoln 7 Oct 1974 (NBR 45:42); one seen in Thomas Co by many observers through 9 Oct 1995 (NBR 63:105); one south of Gering 11 Oct 1999 (AK); one reported from Crescent L NWR 20 Oct 1996; and another extremely late record of a road-killed specimen in Lancaster Co 1 Nov 1970 that had been dead several days and is now UNSM ZM12946 (Cink and Fiala 1971b).

FINDING

This species is easy to locate by its call at dusk in the Pine Ridge or in the Wildcat Hills in Scotts Bluff Co; Jun is the best time. Surveys at dusk and early evening in central Nebraska and even the east would be of interest also; peak spring migration is in May.

Chuck-will's-widow

Caprimulgus carolinensis

STATUS

Uncommon regular breeder east. Casual summer visitor central.

DOCUMENTATION

Photograph

4 Jun 1983 Camp Merrill, Saunders Co (Lingle 1983).

DISTRIBUTION AND ECOLOGY

This species was first noted in 1963 in western Douglas Co, where it had been heard by residents and was eventually verified by RSS in 1966 (Sharpe 1967).

The contemporary distribution of this species is the Missouri River valley oak-hickory woodlands of extreme southeastern Nebraska, northward in the river bluffs to Dakota Co, and westward in the Platte Valley to the Morse Bluff area. It also occurs a short way up the Elkhorn River. Although Johnsgard (1979) indicates it prefers pine-oak woodlands near streams, the only location of such habitat is along the Niobrara River, where the species is absent. In Nebraska it is more often located along or near ridgetops of oak-hickory parkland.

Spring

2, 4, 5 May→summer

Arrival is in early May.

Summer

Following the first record for the state (see above), another was heard for about a month near Peru from mid-May to mid-Jun 1965 (Gates 1965). At about the same time 2–3 were heard in Indian Cave SP 24 May 1966 (Sharpe 1967), and at least 8 were there 3 May 1969, when it was considered "clearly more common" than in 1966 (Tate 1969).

It has been recorded as far west as Camp Merrill in northwest Saunders Co, some 9 km (5–6 mi) west of Morse Bluff; the only documented nesting was there 4 Jun 1983, where birds had been present since 1978 (Lingle 1983). Hoffman (NBR 62:110) has noted up to 3 birds each year since 1991 in the Morse Bluff area along the Platte River. In recent years it has apparently established itself as far west as the Minden-Kearney bridges stretch of the Platte River, where it was heard singing in summer 1997 (NBR 65:108). It had been reported in that area as early as 19–29 Jun 1983, when one was reported 14.5 km (9 mi) southeast of Kearney on the south bank of the Platte River (Kimball 1984).

Northward in the Missouri Valley it has occurred with some regularity in Dakota Co, where the earliest report was 13 Jun 1988, and it has been reported there each year since 1990 in the period 6 May–14 Jun. There is a single Dixon Co report, 14 May 1993. Surprisingly, the only reports from the Blue and Nemaha Valleys are 21 Jun 1987 York Co, 18 Jun 1985 Gage Co, and 18 Jun 1985 Johnson Co.

A probable vagrant was one in mixed conifers and hardwoods in Thomas Co 13 Jun 1983 (Dwyer 1988), a report considered hypothetical by Bray (1994).

Fall

Departure is probably in Sep based on data from nearby states, although the only Nebraska reports after Jun are 17 Jul 1992 Saunders Co, 19 Jul 1990 Douglas Co, 8 Aug 1983 Cass Co (Green 1984a), and 15 Aug (Johnsgard 1980). The latest Kansas date is 16 Sep (Thompson and Ely 1989). Detection is difficult after singing ceases in early Jul.

Other Reports

A report of one in Sioux Co 18 Sep 1992 (NBR 60:78) is an error (NBR 60:152).

FINDING

This species can be found at Indian Cave SP, where it is vocal in May and Jun, although the numerous Whip-poor-wills make it difficult to hear the few Chuck-will's-widows.

Whip-poor-will

Caprimulgus vociferus

STATUS

Common regular breeder and spring and fall migrant east, rare central.

DOCUMENTATION

Specimen

HMM 2733, 2 May 1896 Giltner, Hamilton Co.

DISTRIBUTION AND ECOLOGY

As a breeding bird, this species occurs in the oak-hickory forest of the southeast, west to Pawnee Co, in similar habitat throughout the Missouri Valley, in the Platte Valley where oak-dominated woodland occupies hillsides west to the Morse Bluff area, and in the Niobrara Valley west to extreme eastern Cherry Co in similar oak habitat. Where such hardwood habitat is heavily grazed or where it becomes open savanna the species is absent. In migration it also occurs in riparian forest.

Spring

15, 19, 22 Apr→summer

Arrival is in late Apr. Few are reported in the western half of the state; reports include one in Perkins Co 6 May 1970 identified by size and wing length (NBR 38:87); one in Dawson Co 28 Apr 1935, the only local record (Kinch 1964); and one in McPherson Co 17 Jun 1972. This species is a casual migrant in eastern Colorado 13–26 May and 14–27 Sep (Andrews and Righter 1992).

Summer

The general distribution outlined above did not apparently change much during the 20th century, as Bruner and others (1904) listed this species as occurring in the eastern part of the state and also rarely in Thomas Co. Rapp and others (1958), however, stated that while it was locally common in the Missouri River region, it was not known elsewhere. Mossman and Brogie (1983) suggested that this species was more common along the central Niobrara River than in prior years but no more so perhaps than around 1900. Its dislike of heavily grazed woodland suggests temporary absence from certain areas, depending on grazing status. Youngworth (1958) noted its occurrence near Homer in 1957, its first appearance there for about 30 years; Youngworth blamed its disappearance on heavy grazing.

In the Niobrara Valley it occurs west to extreme eastern Cherry Co (Brogie and Mossman 1983). It was unrecorded in Hall Co by Lingle (1994) and considered rare in Hamilton Co (Swanson 1962). Currently it occurs regularly in the Platte Valley west to the Morse Bluff area of northern Saunders Co (NBR 63:47) and was reported from Oak Glen 25 Jun 1995. It apparently does not occur on the Republican River but has bred west along the Big Nemaha Valley to Pawnee Co in the extreme southeast (Bennett 1974; Johnsgard 1980).

High counts include 14 in Sarpy Co 23 Jun

1996 and 9 along Trail 9 at Indian Cave SP 13 Jul 1995.

Fall

summer→11, 17, 18 Sep

Migration occurs in late Aug and Sep, although there is a later report 2 Oct 1982 Boone Co. "Scores" were heard at Platte River SP 3 Sep 1983, where none were heard 8 Aug; although these birds were considered migrants (Green 1984a), a period of regular fall singing (albeit weak and nonrepetitive) in early to mid-Sep has been noted by RSS in eastern Nebraska and western Iowa.

FINDING

This species is probably easiest to hear at Indian Cave SP in late May and Jun. Often birds sit on the paved roadway just after dark and can be seen in car headlights.

APODIFORMES

Apodidae (Swifts)

Chimney Swift
Chaetura pelagica
STATUS
Common regular spring and fall migrant east, becoming uncommon west. Common regular breeder east, becoming uncommon west.
DOCUMENTATION
Specimen
UNSM ZM12811, 23 Jun 1892 Lancaster Co.
DISTRIBUTION AND ECOLOGY
This species is abundant at times in summer and as a migrant in the east in cities and towns, becoming less common westwardly. Most nesting currently occurs in brick or masonry structures, including chimneys, but at one time this species nested in holes in trees of mature woodlands.
Spring
20, 25, 25 Mar→summer (southeast)
17, 20, 22 Apr→summer (west)
Arrival is mainly in Apr, although earliest reports are in late Mar in the extreme southeast and late Apr in the west. There are earlier reports 7 Mar 1973 Adams Co, 10 Mar 1955 Gage Co, 15 Mar 1957 Gage Co, and 16 Mar 1956 Gage Co.
High counts include 118 in Sarpy Co 11 May 1996.
Summer
The breeding range has expanded westward during the 20th century. Around 1900 the species was rare west of a line connecting Knox and Nuckolls Cos (Bruner and others 1904). Swenk (Notes before 1925) cited a specimen, #2626 in the Brooking collection, taken at Crescent L 18 Jun 1917; Tout (1947) listed only 1 record for Lincoln Co, a dead bird found while cleaning his chimney in North Platte 22 Sep 1908; and Ludlow (1935) considered it uncommon "this far west," in Webster Co, where it was not known to summer. By 1980 it was considered rare west of a line connecting Dawes and Hitchcock Cos (Johnsgard 1980). Ducey (1988) cited no breeding records west of North Platte, which was the westward limit noted by Rapp and others (1958).

Currently it occurs westward throughout the North and South Platte Valleys, having been noted around Sutherland in 1956 (Short 1961), a first-time summer resident at Brule on the South Platte in 1957 (Huntley 1958), and first reported as a summer resident in Scotts Bluff Co in 1966.

Rosche (1972, 1982) observed Chimney Swifts at Crawford from 1972 onward; there is only a single record from the northern Panhandle prior to 1972, that of birds seen at Alliance in 1956 (Mathisen and Mathisen 1958). It is now established in towns of the southern Panhandle, notably Chappell, Sidney, and Kimball.

Fall
summer→28, 29, 29 Sep (west)
summer→20, 21, 24 Oct (south, east)
Departure is in early Oct, although there is a later report 5 Nov 1990 Douglas Co. In the west departure is in late Sep, although there is a later report 5 Oct 1996 in Lincoln Co (FN 51:81). During migration large numbers may build up, especially in urban areas. One roost site, a church chimney near downtown Omaha, was used for several weeks by hundreds, perhaps the low thousands; these were observed entering the chimney at dusk 16 Sep 1983 (RSS).
High counts include 100 in Otoe Co 3 Sep 1995, 100 at Kearney 9 Sep 1996, 100 there 19 Sep 1999, and 100 at Bellevue 24 Sep 1997.
FINDING
This species is easy to find flying over Missouri Valley cities and towns, especially in the later hours of daylight when it is very noisy.

White-throated Swift
Aeronautes saxatilis
STATUS
Fairly common regular spring and fall migrant west. Uncommon regular breeder west.

DOCUMENTATION
Specimen
UNSM ZM12814, 1892 Squaw Canyon, Sioux Co.
DISTRIBUTION AND ECOLOGY
Essentially restricted to nesting sites on cliffs
in the Panhandle, this species is unrecorded
elsewhere.
Spring
17, 19, 20 Apr→summer
Most arrive in late Apr, although there is an
earlier report 10 Apr (Rosche 1982).
The only report away from Sioux, Dawes, and
Scotts Bluff Cos is a sighting in Sheridan Co
14 May 1993.
High counts include 30 at Scotts Bluff NM 12 May
1999, 20 there 17 May 1998, and 12 there 12
May 1995.
Summer
Rosche (1982) has noted this species in the
breeding season as far east as Beaver Creek
valley, north of Hay Springs in extreme
western Sheridan Co, although more recently
easternmost sightings are from Jailhouse
Rock, Morrill Co (NPG files). It occurs in
the rest of the Pine Ridge, most commonly
westward, and in the Wildcat Hills and at
Scotts Bluff NM. In recent years it has been
reported most often from Scotts Bluff NM and
in Sowbelly Canyon.
Ford (1959) recorded young still in nests in
early Aug 1957 some 12 km (7.5 mi) west of
Crawford, a rather late date, as egg dates are
mostly in Jun and early Jul (Johnsgard 1979).
High counts include 6 at Scotts Bluff NM 1 Jun
1995.
Fall
summer→29 Sep, 2, 3 Oct
Most depart in Sep. Migrants are rarely seen
away from nesting locations; there are only
2 such reports: Rosche (1982) recorded one
3 Aug 1980 at Whitney L, and there is an
undocumented report 13 Oct 1984 in Lincoln
Co, the latest and easternmost report for the
state. There is, however, a record of one 2
Nov 1978 in Manhattan, Kansas (Thompson
and Ely 1989), and a specimen was found 7

Nov 1988 in southeast Missouri (Robbins and
Easterla 1992).
High counts include 150 at Scotts Bluff NM 18
Aug 1995, 53 there 20 Sep 1999, and 29 there
12–13 Sep 1998.
FINDING
This species is best located by checking Scotts
Bluff NM May–Aug, especially the north and
west faces.

Trochilidae (Hummingbirds)

Ruby-throated Hummingbird
Archilochus colubris
STATUS
Uncommon regular spring migrant east, rare
casual central, hypothetical west. Uncommon
regular breeder east, rare casual central.
Common regular fall migrant east, rare casual
central.
DOCUMENTATION
Specimen
UNSM ZM10638, 25 May 1913 Lancaster Co.
DISTRIBUTION AND ECOLOGY
This hummingbird is essentially restricted
as a breeding species to woodlands of the
Missouri Valley and lower Platte Valley,
where a variety of flowering plants provide
suitable habitat for the breeding season. It
normally places its nest in trees over water
(Dobkin 1994). During migration it occurs
in decreasing numbers westward to about
North Platte, where it is only casual. Migrants
are most common among flowering plants in
woodland and gardens, as well as at feeders.
Fall migrants are more often seen than spring
migrants, probably because they are generally
more leisurely, remaining in an area for
several days to a week or more foraging on an
especially productive clump of flowers, often
in a town garden.
Spring
6, 8, 9 Apr→10, 11, 12 Jun
Migration occurs essentially in May, although
extreme dates are early to mid-Apr and early
Jun; there are later reports, possibly nesting

birds (see Summer), 21 Jun 1938 Lincoln Co and 24 Jun 1964 Brown Co.

The westernmost documented reports are of an adult male netted at Cedar Point Biological Station 23 May 1996 (Brown and others 1996) and a female in Dawes Co 29 Sep 1996 (FN 51:81). A female *Archilochus* hummingbird was reported at Kimball 4 Aug 1996; at this location either Black-chinned or Ruby-throated Hummingbirds would be exceptional (FN 51:81). Other westerly reports, none documented, are 27 May 1936 Logan Co, 30 May 1983 Sioux Co, 31 May 1964 Lincoln Co, 8 Jun 1947 Lincoln Co, 8 Jun 1957 Logan Co, 21 Jun 1938 Lincoln Co, and 29 Jun 1982 Sioux Co. There are no Colorado records of Ruby-throated Hummingbird (Andrews and Righter 1992), and the first documented for Wyoming was 7–8 Sep 1995 (AFN 50:87).

Summer

Egg laying probably peaks in Jun (Johnsgard 1979); a nest was located in Schramm Park 22 May 1995. Records at this time away from Missouri Valley counties are few. Fiala (1970) considered it "at least an occasional summer resident" in Gage Co, and it was reported from Pawnee and Johnson Cos in Jun 1985; these reports suggest rare nesting in the oak-hickory woodlands of the extreme southeast. It may also nest rarely in the Platte Valley as far west as Kearney, where it nested for at least a few years prior to and including 1956 (Ludden 1956). There is a recent North Platte report 10–11 Jun 1990 and older ones 8 Jun 1947 and 21 Jun 1938, a Dawson Co report 9 Jun 1956, and a Logan Co report 8 Jun 1957; all may be merely late spring migrants, although it was considered an "uncommon or casual summer visitor" in Logan Co prior to 1934 (Glandon and Glandon 1934a), and Tout (1947) listed it as a "summer resident," probably breeding, in Lincoln Co. Lingle (1994) listed it as a migrant only, uncommon or rare, in the Grand Island vicinity. It does not appear to nest currently in the Niobrara River valley west of the confluence with the

Missouri River (Johnsgard 1979; Brogie and Mossman 1983), although it summered in Brown Co at least in 1970 and was reported there 24 Jun 1964. The only indication of summer occurrence in the western Loup drainage is from the Bessey Division, NNF, where it is a "rare summer visitor" (Bray 1994).

Fall

7, 9, 9 Aug→10, 10, 11 Oct

Migration extends somewhat farther west than in spring and occurs from early Aug through Sep, although there are earlier reports 16 Jul 1988 Howard-Hall Cos, 20 Jul 1957 Brown Co, and 29 Jul 1974 Adams Co, possibly nesting birds, and a later report 23 Oct 1909 Webster Co (Ludlow 1935).

The most westerly documented records are 3 Sep 1912 Thomas Co (specimen, UNSM ZM10637) and 15 Sep 1985 Chase Co (specimen, UNSM ZM15512). There are no Colorado records (Andrews and Righter 1992). Undocumented westerly reports are 7–17 Aug 1970 Perkins Co, an immature that had a red feather on its throat but could not be identified with certainty (NBR 39:23); 18 Aug–8 Sep 1972 Scotts Bluff Co; 28 Aug 1995 Sheridan Co (FN 50:76); and 20 Sep 1966 McPherson Co. In addition, there are about 10 fall records from Lincoln Co.

High counts include 10–12 at Fontenelle Forest 10 Sep 1998, 10 there 3 Sep 1995, and 6 there 29 Aug 1997.

FINDING

This species is common in Missouri Valley woodlands in early Sep, when it frequents beds of blooming Turk's cap, *Lilium canadense*. It also can be attracted to feeders in the Missouri Valley and farther west, especially in fall. Backyard plantings of the vine trumpet honeysuckle, *Lonicera sempervirens*, are also known to attract fall migrants. In spring hummingbirds frequent clumps of wild columbine, *Aquilegia canadensis*, an early bloomer found along the forest edge; some garden cultivars may also attract Ruby-throated Hummingbirds.

Calliope Hummingbird

Stellula calliope

STATUS

Rare casual spring and fall migrant west.

DOCUMENTATION

Specimen

display mount at Wildcat Hills Nature Center, Scotts Bluff Co, 23 Jun 1994 central Sioux Co (Gubanyi 1996c; NBR 62:137; WRS).

Spring

There are 2 records, both adequately documented. A female was found dead 8 Apr 1962 40 km (25 mi) northwest of North Platte; the specimen was identified by Emmet R. Blake at the FMNH (Shickley 1965) but is now lost (Bray and others 1986). The second was a male captured by a dog in central Sioux Co on or about 23 Jun 1994; the specimen was recovered and is on display at the Wildcat Hills Nature Center (NBR 62:137; WRS).

Fall

There are 2 documented records. First was a male seen 8–10 Aug 1960 3.2 km (2 mi) north of North Platte feeding on *Penstemon*; it was accompanied by a similar-size bird, possibly a female (Viehmeyer 1961b). A male was seen 2–8 Aug 1980 in Scotts Bluff Co; it was seen several times at about 1.8 m (6 ft) distance and also may have been accompanied by a female (NBR 49:19).

COMMENTS

Two of the Nebraska records are at the expected time, early Aug, based on records from neighboring states. There are 2 records from the northeastern Colorado plains (Andrews and Righter 1992), 4 from the Black Hills of South Dakota (SDOU 1991), and 1 from southwest Kansas (Thompson and Ely 1989), all in the period 23 Jul–3 Sep. The Apr and Jun Nebraska records are difficult to explain, as most Calliope Hummingbirds migrate west of the Rocky Mountains in spring. There are, however, 5 Colorado spring records and 1 for Jun (Andrews and Righter 1992).

FINDING

Feeders in the extreme west should be carefully watched for this species in late Jul and early Aug.

Broad-tailed Hummingbird

Selasphorus platycercus

STATUS

Rare regular fall migrant west (NBR 64:117), becoming accidental east. Hypothetical in spring.

DOCUMENTATION

Photograph

22–30 Aug 1987 Hastings, Adams Co (Grenon 1990).

DISTRIBUTION AND ECOLOGY

Migrants pass through the western Panhandle in fall, most often at feeders and flowering plants. Occasional individuals occur farther east.

Spring

There are few reports, none documented, even though this species breeds commonly in the Rocky Mountains of Colorado and Wyoming; indeed, Rosche (1982) "strongly suspected" that a female/immature reported 10 Jun 1969 in Dawes Co was in fact a Ruby-throated Hummingbird. Reports are 2 Jun 1987 Sioux Co, 2 Jun 1999 Scottsbluff, 10 Jun 1969 Dawes Co, and 15 Jun 1983 Sioux Co. Also, Rapp and others (1958) noted a "small flock" at Scottsbluff Jun 1956 without further details. There is a Kansas specimen record 18 May 1978 in Morton Co (Thompson and Ely 1992) and a Black Hills record from South Dakota 8–13 Jun 1990 (SDOU 1991).

Fall

18, 20, 20 Jul→14, 17 (FN 51:81), 18 Sep

Migrants are most often reported from late Jul through mid-Sep in Sioux, Dawes, and Scotts Bluff Cos, although there is a later report 5 Oct 1981 Dawes Co. Migrants have been reported on occasion east to McPherson Co 27–29 Aug 1967 and Lincoln Co 11–14 Aug 1959 and 2–7 Sep 1960, and there are a few reports farther east: Bent (1940) cited a bird collected at Kearney 22 Jul 1914, a male was photographed at Hastings 21–30 Aug 1987 (NBR 56:14; Grenon 1990), a female/immature was photographed at Seward 17–24 Oct 1990 (Grenon 1991), Bates reported one at Bassett 10 Sep 1899 (Bruner and others 1904), and Huntley reported it in Keith Co but without date (Rapp and others 1958). Bray and others (1986) noted that one was collected at Kearney in Aug 1903; it was a female, identified by Oberholser (Swenk 1918b; Swenk, Notes before 1925). Swenk (Notes before 1925) also noted that a female in the Brooking collection was 1 of 2 birds that hit a Kearney window Aug 1921 (Swenk, Notes before 1925; Black 1922). None of the specimens mentioned by Swenk can now be located (WRS).

COMMENTS

Bruner (1901) suggested that this species might breed in the Pine Ridge, as he had seen several there in summer 1891 (Bruner and others 1904). Although there is no other indication of breeding in Nebraska, the species was apparently a common breeder in the Black Hills of South Dakota in 1874; no nests have been reported in the Black Hills since 1929 (SDOU 1991).

FINDING

This species should be looked for in the western Panhandle at feeders and in flower gardens during Aug.

Rufous Hummingbird

Selasphorus rufus

STATUS

Rare regular fall migrant west, becoming rare casual east.

DOCUMENTATION

Specimen

UNO B218, late Aug 1971 Scotts Bluff Co (Bray and others 1986).

DISTRIBUTION AND ECOLOGY

This species occurs regularly only in the west, but, especially in late fall may appear as far east as Douglas and Sarpy Cos. It is usually seen at feeders and gardens with red flowers.

Fall

17 (Rosche 1982), 20, 31 Jul→1, 12 (Rosche 1982), 14 Sep

Although it is a regular fall migrant in the Black Hills of South Dakota with extreme dates 26 Jun–8 Sep (SDOU 1991), there are only about 35 Nebraska records. In the west migrants occur from late Jul through late Aug; the latest reports, however, are easterly: a male was seen at a feeder at Hastings 13–14 Sep 1987 (NBR 56:14); another was seen in Sarpy Co 10–17 Nov 1985, an immature/female plumaged bird at a Bellevue feeder (Douglas 1985), which was conservatively judged a *Selasphorus* sp. but "most likely" a Rufous by the NOURC (Mollhoff 1987); no details were provided for one in Omaha 20 Nov 1994 (Green 1994). There is also a Douglas Co report of a "brilliant male" at a feeder 8 Aug 1992 (NBR 60:166; Green 1994), a more usual date of occurrence.

Migrants have occurred during the normal migration period east to Logan Co 26–29 Jul 1936 (Glandon 1936), McPherson Co 15 Aug 1985, Lincoln Co 15 Aug 1992, and Perkins Co 30 Jul–1 Aug 1974 and 31 Jul–9 Aug 1972.

COMMENTS

Identification of this species, particularly in relation to Allen's Hummingbird, has been discussed by Stiles (1972) and Newfield (1983). While the possibility of female/immature Allen's cannot be eliminated in the field, the fact that Rufous Hummingbird is a regular migrant in western Nebraska suggests that the overwhelming majority of birds seen are indeed Rufous Hummingbirds (NBR 64:117).

FINDING

The best chance to see this species is during Aug in Panhandle towns with feeders set out or extensive flowerbeds.

CORACIIFORMES

Alcedinidae (Kingfishers)

Belted Kingfisher

Ceryle alcyon

STATUS

Fairly common regular breeder (resident?) and migrant statewide. Uncommon winter visitor south and east, rare elsewhere.

DOCUMENTATION

Specimen

UNSM ZM6277, 23 Jun 1916 Mitchell, Scotts Bluff Co.

DISTRIBUTION AND ECOLOGY

Breeding birds occur statewide around water areas providing small fish and sandy dirt banks for nesting. Most move south in winter, although individuals remain wherever open water can be found.

Spring

The presence of wintering birds and their early movement north in spring do not allow accurate determination of early and late migration dates. Peak migration probably occurs in Apr, while freezing and thawing of water areas determine actual early arrival. Rosche (1994a) cited peak counts of 7 on 12 Apr and 5 on 14 Apr in Keith Co, indicating peak spring migration, while the first arrivals at James Ranch, Sioux Co, in 1995 were noted 27 Apr.

Summer

Individuals and pairs can be found throughout the state along bodies of water, from streams to reservoirs. Preferred habitat includes woody vegetation for perches and clay or solid sand banks for nesting sites.

Most nesting is in May and Jun (Johnsgard 1979). A nest in Sarpy Co had young 1–2 weeks old 1 Jun 1998 (WM). Possibly a family group or groups of 8 were at Crystal Cove L 30 Jul 1997; this species does not appear to form migrant flocks.

Fall

Departure is dependent on weather conditions, especially freezing of small bodies of water. CBC data show fairly wide distribution of small numbers (see Winter). Most migrants probably pass through in late Sep–early Oct; Rosche (1994a) noted a peak fall count of 6 on 23 Sep 1989 .

Winter

Overwintering occurs wherever open water remains; numbers are greatest along the Platte Valley and in the southeast, areas where open water exists during most winters. CBC data indicate rather even distribution statewide, again probably reflecting distribution of open water from year to year. Highest CBC totals are 11 at Grand Island 1995–96; 11 at Omaha 1988–89; 10 there 1981–82, 1990–91, and 1991–92; and 10 at Grand Island 1988–89.

FINDING

Kingfishers are generally easy to locate along streams and lake edges by their rattling calls when disturbed. They also perch prominently on wires over streams.

PICIFORMES

Picidae (Woodpeckers)

Lewis's Woodpecker

Melanerpes lewis

STATUS

Uncommon regular spring and fall migrant west, rare casual central, accidental east. Rare casual breeder west. Rare casual winter visitor west.

DOCUMENTATION

Specimen

UNSM ZM6299, 30 May 1900 Warbonnet Canyon, Sioux Co.

DISTRIBUTION AND ECOLOGY

This species is most often encountered in the northwest, especially in the Pine Ridge and Wildcat Hills, most often in spring and in late summer and early fall (Rosche 1982). It appears to be only an irregular breeder in Nebraska.

Rosche (1982) observed that it is often found in Aug and Sep around chokecherry or wild plum thickets. It may prefer riparian situations, especially cottonwoods, for breeding (Vierling 1997), although this may simply reflect a requirement for a shrub understory (Dobkin 1994). Linder and Anderson (1998) found that in the Laramie Mountains of southeast Wyoming there was a decided preference in nest siting at openings in the canopy of the ponderosa pine forest resulting from crown burns; these openings allowed development of understory vegetation important for insect production. It has a preference for burned areas in the Black Hills of South Dakota (Peterson 1990) and has nested in the burned area of Deadhorse Canyon, Dawes Co (Rosche 1982). Linder and Anderson (1998) found that a 1986 burned area had significantly more nests than a 1974 burned area.

Spring

1, 2, 5 May→15, 22, 30 Jun

This species is migratory in northern parts of its range (Bent 1939), including the Black Hills

of South Dakota, where it may occasionally linger into winter (SDOU 1991; Peterson 1990). It typically occurs in the Black Hills 7 May–15 Sep (Peterson 1990) and in eastern Wyoming May–Sep (Scott 1993).

It is of interest that prior to the Deadhorse Canyon burn in 1973, after which a summer presence was established, this species was as often recorded away from the northwestern ponderosa pine areas as not; these data suggest that this species was essentially a casual or accidental vagrant for some time prior to 1973, although around 1900 Bruner and others (1904) considered it a common summer resident and breeder in the Pine Ridge and Wildcat Hills. Thus status has varied, probably in response to the availability of burned woodland habitat and competition with its congener, Red-headed Woodpecker. In southeast Colorado, where Lewis's Woodpecker has become more common since the 1950s, it prefers isolated large cottonwoods, while Red-headed Woodpeckers occupy most contiguous riparian habitat (Andrews and Righter 1992).

Most reports are in spring. There are few reports away from ponderosa pine areas: 1 May 1944 Logan Co; 5 May 1959 Loup Co; 16 May 1900 Kearney (specimen in Olson collection; Swenk, Notes before 1925); 17 May 1987 Cherry Co; about 20 May 1999 8 km (5 mi) west of Bushnell, Kimball Co; 27 May 1950 Logan Co; 30 May 1943 Adams Co; 5 Jun 1940 Logan Co; and 7 Jun 1987 Loup Co.

Summer

Reports of nesting are few, although it has been observed in Jun most years since the 1973 Deadhorse Canyon burn in Dawes Co and has bred there possibly since 1978 (Rosche 1982); documentation of nesting there is lacking, however, since 1986 (Bennett 1987). Following the major 1989 Fort Robinson burn, Lewis's Woodpecker had not been noted there in summer until 1996, when 2 were found in Smiley Canyon 1 Aug 1996. There is a Sioux Co report 10–11 Jun 1975 and another from

Loup Co 7 Jun 1987. There are no Jul reports, suggesting that successful nesting birds depart soon after young are fledged.

Ducey (1988) described it as an occasional nester, especially in the Pine Ridge area, but cited only Dawes Co reports since 1960. The only other nesting reports since 1900 are from Logan Co. In 1940 several observers, including Wilson Tout and Earl Glandon, observed a Lewis's Woodpecker in a grove of dead cottonwoods 14 Jul; it was entering a hole and "appeared to be nesting," although no evidence was noted and a second bird was not mentioned. This record was cited as a "possible nesting record" by Rapp and others (1958). In 1944 young were reported in a dead cottonwood southeast of Stapleton; one of the adults disappeared in mid-Jul (Glandon 1948).

Prior to 1920 it apparently occurred with some regularity in the central Niobrara Valley, with nesting cited for Brown Co (Ducey 1988) and observations from Cherry Co (Coues 1874; Bruner 1896) and at least in winter east to Long Pine and Valentine in ponderosa pine areas (Bruner and others 1904).

Fall

7, 8, 15 Aug→23, 28 (specimen UNSM ZM6361), 29 Sep (Dawson 1921)

There is a later report, a specimen taken at Oshkosh 14 Nov 1916 in the Black collection (Swenk, Notes before 1925).

Although irregular, at times this species becomes conspicuous in late Aug and early Sep, when it forms flocks (Bent 1939; Scott 1993; Rosche 1982); Gates (1981) reported 25 near Chadron SP 30 Aug–24 Sep 1980, and 6–7 were in pines at the head of Monroe Canyon 29 Sep 1920 (Dawson 1921). There are no published reports for Jul.

The only report away from ponderosa pine habitat is 23 Sep 1989 Knox Co, the most easterly report, although Bray (1994) mentions a report without details from the Bessey Division, NNF.

Winter

There are few reports: 26 Nov–10 Dec 1998

Verdigre (LS), one present during CBC week at Scottsbluff in 1971, one collected Dec 1916 at Oshkosh by Maryott and placed in the Brooking collection (Brooking, Notes), 10 Dec 1950 Sioux Co (Mohler 1951), 3 Jan 1955 Scotts Bluff Co, 20 Jan 1954 Adams Co, and 30 Jan 1966 Scotts Bluff Co. There are no Oct, Feb, Mar, or Apr reports, and regular overwintering would seem unlikely; Bent (1939), however, cited wintering at Long Pine in 1898–99, and Glandon and Glandon (1934a) cited one wintering south of Stapleton in 1933–34.

FINDING

This species had been fairly easy to find in the 1980s in the Deadhorse Canyon burn area, although it may have become less regular there in the 1990s. It is still expected in the Fort Robinson area because of the 1989 burn there. Otherwise, it is most likely in the northwest in late May and late Aug–early Sep, but numbers are low.

Red-headed Woodpecker

Melanerpes erythrocephalus

STATUS

Common regular spring and fall migrant east, becoming fairly common west. Common regular breeder east, becoming uncommon west. Uncommon regular winter visitor southeast, becoming accidental west.

DOCUMENTATION

Specimen

UNSM ZM6290, 5 Jun 1981 Gage Co.

DISTRIBUTION AND ECOLOGY

During the breeding season nesting birds occur statewide in areas of open woodland, along roadsides near large isolated trees or clumps of trees, at the edge of farmstead woodlots, and occasionally in urban parks with large trees. During late spring, summer, and early fall the species is largely insectivorous, occupying its time either hawking insects on wing or pecking and probing bark of larger trees for adult and larval insects. In fall as the insect supply begins to decrease, red-heads are forced to leave their summer habitat, shift their food habits, and migrate to areas that provide a winter food supply, predominantly mast. The Nebraska woodpeckers that summer in riparian habitat that lacks mast may migrate south into Kansas, Oklahoma, and Missouri to overwinter in oak woodlands with adequate mast supplies. Some southeastern Nebraska red-heads, however, may remain in years that acorn mast is available (Parker 1982).

Parker's (1982) study of competition among bark-foragers in Fontenelle Forest showed that as the fall progressed, red-heads devoted increasingly more time to caching acorns on firmly defended wintering territories and used this stored food reserve during the insect-free winter. Parker also found that when red-heads establish feeding territories they could openly displace other mast-feeding birds, including the slightly larger Red-bellied Woodpecker. During winters that follow growing seasons of local mast failure, presumably a result of late spring freezes, red-heads are absent from southeastern Nebraska.

Spring

12, 15, 16 Apr→summer (south)

21, 24, 24 Apr→summer (west)

This species is migratory in parts of its range, reflecting its food habits (see above).

Arrival in most of the state is in mid to late Apr. Earlier reports away from the east are few; in the west 11 Apr 1989 Dawes Co and 17 Apr 1995 Sioux Co and in the north 7 Apr 1954 Boyd Co. Such reports are more numerous in the south, with about 8 reports from 14 Mar 1949 Lincoln Co and later, possibly reflecting nearby wintering.

High counts include 142 in Sarpy Co 11 May 1996, 91 of which were at Fontenelle Forest.

Summer

BBS data show a noticeable decline in density from east to west. Rosche (1982) pointed out that large numbers occurred in the Deadhorse Creek drainage near Chadron SP after the 1973 burn, presumably due to an abundance of insects typical of burned woodland.

Young were fledged 1 Aug 1999 in Dixon Co (JJ). A rather late family group with 2 juveniles was in Seward Co 29 Oct 1999 (JG).

Fall

summer→30 Sep, 2, 2 Oct (north)

summer→10, 14, 18 Oct (west)

summer→18, 20, 21 Oct (Rosche 1994a) (south)

Most depart by late Sep, although there are later reports in the north 9 Oct 1999 in Blaine and Thomas Cos, 16 Oct 1966 Custer Co, and 31 Oct 1981 Thomas Co and in the south 27 Oct 1990 Buffalo Co, 25 Nov 1979 Hall Co, 28 Nov 1991 Phelps Co, and 15 Dec 1990 Lincoln Co.

Winter

Wintering occurs most years in the extreme southeast, with stragglers north to Dakota Co, but is strongly linked to availability of acorns (Bent 1939; Parker 1982) and possibly fruits such as Russian olive, *Elaeaguus angustifolia*, which is used by lingering individuals, usually immatures, in central Nebraska (Rosche 1994a). CBC data indicate that even by Dec, very few birds remain north and west of Lincoln and Omaha, and none were found in 4 years out of 25 (1967–68 through 1991–92). Single birds were at Overton 30 Dec 1998 (LR, RH) and Verdigre 13 Dec 1998 (WRS, JS). The westernmost CBC reports are of single birds at North Platte (1990–91), Grand Island (1983–84), and Kearney (1984–85, 1988–89, and 1989–90). High CBC totals are 143 at Omaha 16 Dec 1995, 142 at Indian Cave SP 1971–72, and 122 at Omaha 1985–86.

Away from the southeast, midwinter reports are few. In the west there is a single report 19 Jan 1952 Scotts Bluff Co, and there are none in the north. In the south there are these few: 31 Dec 1966 Adams Co, 1 Jan 1980 Hall Co, 10 Jan 1979 Hall Co, 8 Feb 1992 Phelps Co, and 2 Mar 1974 Perkins Co. Two wintered at Wyuka Cem, Lancaster Co, 1994–95.

FINDING

This species is conspicuous in edge habitats and urban areas in the Missouri Valley during the summer months because of its flashy plumage and loud calls.

Acorn Woodpecker

Melanerpes formicivorus

STATUS

Accidental in spring.

DOCUMENTATION

Photograph

19 May 1996 Chambers, Holt Co (Brogie 1997).

Records

The only record is of a single bird that appeared
at a feeder stocked with oranges at Chambers
19–22 May 1996 (Brogie 1997).

COMMENTS

This nonmigratory species occurs in Arizona,
New Mexico, and along the West Coast. There
are, however, 2 Wyoming records, 20 Jun 1975
at Grand Teton National Park and 15 Jul 1989
south of Rock Springs (Scott 1993), and a
single sight record for Colorado, 5 Sep 1980
in Ouray Co, considered hypothetical by
Andrews and Righter (1992).

Red-bellied Woodpecker

Melanerpes carolinus

STATUS

Common regular resident east, becoming rare
west of central. Rare regular winter visitor
west.

DOCUMENTATION

Specimen

UNSM ZM6285, 9 Apr 1903 Roca, Lancaster Co.

DISTRIBUTION AND ECOLOGY

This species is most common in the valleys of
the Missouri River and its major tributaries,
where it inhabits open woodland, including
mixed coniferous woodland in the Niobrara
River valley and cottonwoods westerly. It is
found throughout the Republican Valley, west
in the North Platte Valley to Kingsley Dam,
in the South Platte Valley into Colorado, and
west in the Niobrara Valley to Valentine. It also
occurs in the Elkhorn and Loup drainages,
the Nemaha and Blue drainages, and most
of the area south of the Platte River west to
Adams Co.

Summer

Summering birds now are found over most of

the state except for the Panhandle, but this
species is far more numerous in the east,
where BBS data indicate 95% of all birds are
reported.

This species has spread gradually west during
the 20th century. Ducey (1988) cited breeding
records prior to 1920 only for Otoe and Cass
Cos, while the breeding range outlined by
Bruner and others (1904) was southeast of
West Point, Lincoln, and Beatrice. It has long
occurred in the lower Missouri Valley but was
first recorded in Dakota Co in the winter of
1932–33 (Stephens 1957). In the Republican
Valley it was first recorded in Webster Co 20
Apr 1935 and in Harlan Co 31 Jan 1953. Since
then it has spread westward to Dundy Co
and now is found at Bonny Res in Yuma Co,
Colorado, where breeding has occurred since
the late 1960s (Andrews and Righter 1992).

In the Platte Valley there has been a breeding
population on the Little Blue River south
of Hastings since the 1950s (Rapp 1953);
the species has been reported in that area
at least since 7 May 1933. It had not arrived
at North Platte by 1947 (Tout 1947); first
reports there were 17 Jun 1950 and Oct 1954
(Shickley 1968), although it was recorded at
Brady 28 Mar 1949. Short (1961) noted that
Red-bellied Woodpeckers were common as
far west as Sutherland in 1956–57. Rosche
(1994a) indicated that Kingsley Dam may
be a barrier to westward spread of breeding
birds, as there are no summer reports west
of there and only a single summer report at
Cedar Point Biological Station, in Jun 1989.
There have been, however, breeding birds
along the South Platte River in northeast
Colorado since the late 1960s, where it is rare
to uncommon (Andrews and Righter 1992).
The first Keith Co reports were on the South
Platte near Brule 19 Mar 1955 (Benckeser 1956)
and 3 Mar 1957 (Huntley 1958).

Since the 1930s, when it was first noted in Dakota
Co, it has spread westward in the Niobrara
Valley to the Valentine area. Youngworth
(1955) indicated that it had been seen near

Fort Niobrara by Dille, presumably about 1950 (Mossman and Brogie 1983), but it was apparently still rare, as Short (1965) did not see it in northwest Brown Co in 1955. Ducey (1989) noted that this species occurred west to central Cherry Co in 1988; it was present at Anderson Bridge 14 Jun 1985 (NBR 53:50). There is an isolated record of breeding far to the west on the Niobrara south of Gordon, where copulation was observed 20 May 1959 (NBR 27:50; Rosche 1982). Nesting was reported, but without details, from either Chadron or Alliance in the Panhandle in 1959 (Pritchard and Pritchard 1960). These somewhat confusing reports may have been of the same sighting. Rosche saw one south of Rushville Nov 1980 and one again there 29 Mar 1981 (Mark Brogie, pers. comm.), and one was at a feeder in Crawford Feb or Mar 1989 (Doug Kapke, pers. comm. WM).

It occurred throughout the Loup drainage by the 1950s, although it is rare as far west as Bessey Division, NNF (Bray 1994). It was first reported as far west as Logan Co 11 Apr 1954 and 12 Dec 1959 (Glandon 1960) and north to Boone Co 30 Oct 1956 (Anderson 1957). In the Elkhorn drainage it was first reported in Antelope Co 3 May 1953 and 5 Feb 1954, and it is currently considered uncommon in southeast Cherry Co (Blake and Ducey 1991).

Winter

CBC data indicate that in Dec about 95% of birds recorded were in the east, a distribution like that found in BBS data, suggesting that winter and summer distributions are similar. Bent (1939) indicated that there may be some southward movement in winter, however. In Nebraska during fall, winter, and spring, birds are sometimes seen in areas where summering does not occur, although such sightings are rare. Rosche (1994a) listed dates of occurrence in the Keith Co area 27 Sep–30 Apr. There are these reports west of L McConaughy in the North Platte Valley: one was at a suet feeder in Bridgeport during winter 1985–86 (Rosche 1994a), one was in Scottsbluff 17 Dec 1978–4

Jan 1979, and another was in a yard near Mitchell 2–30 Nov 1999 (AK). In the northwest there are these reports west of Valentine: 16 Dec 1993 Garden Co (NBR 62:26, 50) and a male south of Rushville on the Niobrara River 27 Nov 1980 and (probably the same bird) 29 Mar 1981 (Rosche 1982); Bruner and others (1904) cited without a date a "straggler" seen by Bruner in Monroe Canyon.

FINDING

This species is not hard to find, especially in winter, in lower Missouri River valley woodlands. It often appears at suet feeders in winter.

Williamson's Sapsucker
Sphyrapicus thyroideus

STATUS

Rare casual spring visitor east and central.

DOCUMENTATION

Specimen

UNSM ZM15986, 13 May 1988 Omaha, Douglas Co.

Records

There are 5 reports, 3 documented. A description was published of a male seen at Hastings 24 Mar 1939 (Jones 1939); a male was described near Grand Island 5 May 1959 (NBR 27:51, 58); and a male found dead at Creighton University, Omaha, 13 May 1988 is now ZM15986 in the UNSM collection (Wilson 1989; Grenon 1990).

Undocumented reports are of a "female" in Omaha 18 Feb 1945 (Haecker and others 1945; NBR 13:61) and a single bird visiting feeders in Omaha 4 Dec 1976 (Green 1977).

COMMENTS

This species is a summer resident in lower elevation ponderosa pine forests in the eastern Rocky Mountains of Colorado and Wyoming, arriving in Apr and leaving in Oct (Andrews and Righter 1992; Scott 1993). There are 4 records for the eastern Colorado plains, 2 in fall and 2 in Feb–Mar (Andrews and Righter 1992). The single Kansas and South Dakota records are in Apr (Thompson and Ely 1989; SDOU 1991). Minnesota has 2 records, 1 each for Apr and May (Janssen 1987). The preponderance of Apr and May records suggests eastward wandering during spring migration, but the lack of reports from western Nebraska is puzzling, probably a reflection of low observer concentration.

FINDING

This species should be looked for in Apr and May in pine or riparian habitats in the western Panhandle.

Yellow-bellied Sapsucker

Sphyrapicus varius

STATUS

Uncommon regular spring and fall migrant east, becoming rare casual west. Uncommon regular winter visitor southeast, casual elsewhere.

DOCUMENTATION

Specimen

UNSM ZM6307, 9 Oct 1921 Lancaster Co.

DISTRIBUTION AND ECOLOGY

Migrants occur in woodlands statewide but are rare west of the Missouri Valley. Wintering birds are restricted to the extreme southeast, where they often appear at suet feeders or on planted conifers, especially Austrian pine (*Pinus nigra*) and blue spruce (*Picea pungens*)(Hiatt and Bliese 1971)

Spring

winter→29 Apr, 1, 1, May

Migration peaks in late Mar and early Apr. Arrival is difficult to discern due to the presence of wintering birds, as is departure due to several May and Jun reports, although few are seen after Apr. There are about 20 May reports, with later reports 6 Jun 1972 Lancaster Co, 9 Jun 1957 Hall Co, and 1 Jul 1993 Sarpy Co. There is no evidence for summering or nesting in Nebraska, even though this species nests in extreme eastern South Dakota (SDOU 1991).

Spring migrants are far less numerous westward than in fall. In the Panhandle there are only about 10 spring reports, in the period 7 Mar–21 May. It should be noted that undocumented reports in the west, and especially the western Panhandle, may refer to Red-naped Sapsucker, recently split from Yellow-bellied Sapsucker.

Fall

18, 19, 22 Sep→winter

Migrants appear in late Sep, reported both in the east and west. Some of these western reports probably refer to Red-naped Sapsuckers; see that species for a discussion of a juvenile male specimen, UNSM ZM6312, probably a Yellow-bellied (WRS, JGJ), collected at Chadron 23 Sep 1919, and the likelihood that in fall a majority of western sapsuckers are indeed Yellow-bellieds and not Red-napeds.

Peak movement is in the first half of Oct. Individuals regularly linger into Dec, although at that time only 5 birds of a total of 96 counted on CBCs 1967–68 through 1992–93 were north and west of Kearney and DeSoto Bend NWR. A late westerly report was of one at Scottsbluff 16 Nov 1999 (AK). There are earlier reports, none documented, 23 Jul 1987 McPherson Co, 17 Aug 1974 Howard-Hall Cos, 25 Aug 1981 Lancaster Co, and 1 Sep 1967 Cass Co.

High counts include 3 at Wyuka Cem, Lincoln, 14 Oct 1996.

Winter

Wintering birds are most numerous in the extreme southeast, but there are several Dec reports westward in the Platte Valley as far as the Scottsbluff area. The only Jan reports for the Panhandle, however, are 1 Jan 1977 Sioux Co, 4 Jan 1982 Scotts Bluff Co, and 10 Jan 1976 Scotts Bluff Co. In the north there are no reports between 1 Feb 1981 Boone Co and 11 Mar 1981 Boone Co, and only these few Dec–Jan reports: 3 Dec 1967 Custer Co, 27 Dec 1997 Calamus Res, 27 Dec 1998 Calamus-Loup CBC, 30–31 Dec 1979 Boone Co, 1 Jan 1957 Brown Co, 24 Jan 1993 Thomas Co (Bray 1994), and 1 Feb 1981 Boone Co.

Highest CBC totals are 8 at Tristate (at Arbor Lodge SHP) 1981–82 and 7 in Omaha 1979–80.

COMMENTS

Ducey (1988) cited old reports of nesting in

Cuming and Douglas Cos, but these were discounted by Bruner and others (1904). While this species has been recorded nesting in northwest Iowa in recent years (Kent and Dinsmore 1996), there is no evidence for nesting in Nebraska in the 20th century.

FINDING

This species may be easiest to locate in southeastern cemeteries or at Arbor Lodge SHP in winter, when it forages on conifers, especially Austrian pines. Migrants should be looked for in early Apr and in Oct in Missouri Valley woodlands and cemeteries such as Forest Lawn in Omaha; hardwoods are generally avoided (Johnsgard 1980).

Red-naped Sapsucker

Sphyrapicus nuchalis

STATUS

Rare casual fall migrant west. Accidental in spring.

DOCUMENTATION

Specimen

UNSM ZM6310, 13 Sep 1919 Monroe Canyon, Sioux Co (Mickel and Dawson 1920).

TAXONOMY

This species was treated as a subspecies of Yellow-bellied Sapsucker until recently, when Sibley and Monroe (1990) argued for its specific status. This Rocky Mountain taxon breeds in coniferous forest and woodland, including the Black Hills of South Dakota, where it is uncommon (SDOU 1991) but not difficult to find (Peterson 1990). It is migratory (AOU 1998).

Spring

The only record is of a male photographed in ponderosa pines in Sowbelly Canyon 17 Apr 1987 (Brogie and Brogie 1987c; Mollhoff 1989a).

Fall

There are 4 documented records and 2 reports without details. There are 3 specimens in the UNSM collection taken by C. E. Mickel in Monroe Canyon: ZM6311, taken 13 Sep 1919; ZM6310, a juvenile male, taken the same day;

and ZM6313, taken 29 Sep 1920 (Mickel and Dawson 1920; Dawson 1921; Bray and others 1986). An adult male was photographed at Oliver Res 27 Sep 1998 (SJD).

Undocumented reports are 4 Oct 1929 Monroe Canyon (Dawson 1921) and 24–25 Sep 1995 Dawes Co (FN 50:76).

COMMENTS

Given the presence of a breeding population in the Black Hills of South Dakota, it seems likely that migrants should occur regularly in the Panhandle, albeit in small numbers. On the other hand, there are few reports of sapsuckers of either species in the west, especially in spring, and it seems that if Red-naped Sapsucker occurred with any regularity it would have been noted. Nevertheless it is likely that some of the earlier sapsucker reports from the Panhandle indeed refer to this species, especially reports in fall, when the 2 species can be difficult to distinguish (Kaufman 1990).

Red-naped Sapsucker is present in the Black Hills of South Dakota from mid-Apr through mid-Sep (Peterson 1990), suggesting that migrants pass through Nebraska in mid-Apr and late Sep, as the few records confirm. There is a single record from the eastern plains of Colorado 3 Oct 1936 (Andrews and Righter 1992), and the latest Black Hills record is 5 Oct (SDOU 1991).

Prior to the species' split there are 3 Nebraska spring reports that may be Red-naped Sapsuckers: 13 and 21 Apr 1980 Garden Co and 17 Apr 1981 Sioux Co. In fall most Red-naped Sapsuckers depart before the arrival of Yellow-bellied Sapsuckers (Andrews and Righter 1992; SDOU 1991), and fall reports from the west are in 2 groups, 22 Sep–2 Oct and 11 Oct into winter. It is likely that many, if not all, of the earlier reports are Red-naped Sapsuckers: 22 Sep 1990 Sioux Co, 23 Sep 1979 Dawes Co (Rosche 1982), 26 Sep 1982 Scotts Bluff Co, 28 Sep 1991 Sioux Co, 30 Sep and 1 Oct 1989 Scotts Bluff Co, and 2 Oct 1976 Scotts Bluff Co.

A specimen, UNSM ZM6312, collected at Chadron 23 Sep 1919 (Mickel and Dawson 1920) is a juvenile male, but its identity to species is difficult to determine. At that date, Red-naped Sapsuckers would have molted from juvenal plumage, and indeed there are scattered red feathers in the throat and crown but not on the nape, in contrast with the juvenile male Red-naped Sapsucker cited above, UNSM ZM6310. This specimen (ZM6312) is probably a juvenile Yellow-bellied Sapsucker.

FINDING

This species may be most likely located in mid to late Sep in woodland, including urban habitats, in the Panhandle. In spring it is most likely found in mid-Apr.

It is useful to remember, especially in the west, that juvenal-plumaged birds seen after Aug are Yellow-bellieds, as Red-naped juveniles molt into adult plumage in Aug, prior to migration (AB 48:133; Kaufman 1990).

Downy Woodpecker

Picoides pubescens

STATUS

Common regular resident statewide.

DOCUMENTATION

Specimen

UNSM ZM12374, 25 Nov 1893 Lancaster Co.

TAXONOMY

Two subspecies occur: eastern *medianus* in the east and now breeding over most of the state, albeit more uncommonly to the west, and the Rocky Mountain subspecies *leucurus*, breeding at least uncommonly (Haecker and others 1945) in the northwest. Eastern *medianus* may be interbreeding with *leucurus* (Ford 1959) to the extent that few pure *leucurus* may be present.

These forms are identifiable in the field; *leucurus*, like Rocky Mountain Hairy Woodpecker, has little or no white spotting on the wings, especially the secondaries and outer tail feathers.

DISTRIBUTION AND ECOLOGY

This species is most common in the east, where wooded habitats are most extensive. Numbers decrease westward, and it is far less numerous in the west, except in the Platte Valley and Pine Ridge. Winter distribution is similar.

A range of woody habitats is utilized, including coniferous and deciduous types, upland, riparian, early successional willow and cottonwood, as well as plantations, farmsteads, and urban yards and parks. It is also known to forage on very small shrubs and in weedy areas as well as in corn stubble.

Summer

BBS data indicate highest numbers in the east, with numbers very low in the west. Around 1900 Bruner and others (1904) considered this species very common in the eastern third of Nebraska but rare and local in the western half of the state. Ducey (1988) shows no breeding records in central Nebraska prior to 1920. Thus, coincident with the increase in woody habitat in central and western Nebraska along streams and around farmsteads, this species (in particular, its eastern race *medianus*; see Taxonomy) appears to have spread westward in the 20th century to occupy most of central Nebraska and farther west in river valleys; it is now a fairly common resident in the Keith Co area (Rosche 1994a) but occurs in eastern Colorado only in major river valleys (Andrews and Righter 1992). There has been a breeding population of the Rocky Mountains race *leucurus* in the Pine Ridge region in historic times (Ducey 1988; Bruner and others 1904). This western spread has resulted in contact of the 2 races; Short collected an intermediate specimen (KU 34088) 8 km (5 mi) north of Harrison on 18 Jul 1957 (Ford 1959).

Winter

Winter distribution resembles that in summer, according to CBC data. Apparently less wandering occurs than with Hairy Woodpecker; Bent (1939) and Browning (1995) noted that there is little or no evidence for movement of Downy Woodpeckers. Some wandering and possible influx of *leucurus* may occur in the west, however, as Andrews and

Righter (1992) noted that this race appears on the plains in winter. Rosche (1982) considered Downy Woodpecker more common in winter than in summer. Prior to establishment of woodland at Bessey Division, NNF, this species was considered a "migrant," whereas it is now resident (Bray 1994), suggesting that indeed some movement occurs.

Highest CBC totals include 191 at Omaha in 1977–78 and 175 there in 1981–82. Highest CBC totals in the west are 29 at Crawford in 1977–78 and 20 at Scottsbluff in 1976–77.

FINDING

This species is easily located in the east year-round and is perhaps most visible at feeders in winter.

Hairy Woodpecker

Picoides villosus

STATUS

Common regular resident statewide. Rare regular winter visitor statewide.

DOCUMENTATION

Specimen

UNSM ZM6321, 12 Dec 1895 Fort Robinson, Sioux Co.

TAXONOMY

The breeding subspecies throughout is eastern *villosus* (AOU 1957). The Rocky Mountain race *monticola* occurs in the west, at least in winter, although it may be resident in northwest Nebraska, as it is in the Black Hills of South Dakota (Bruner and others 1904; AOU 1957). Thus winter birds in the west may come from the Black Hills or from mountain ranges to the west.

The northern race *septentrionalis* has been considered a winter visitor based apparently on early specimens taken in the east (Bruner 1896; AOU 1957), but it was pointed out by Bruner and others (1904) that these specimens were probably large examples of *villosus*, a conclusion reached by Oberholser regarding Iowa specimens similarly identified (DuMont 1934). *Septentrionalis* has not been confirmed in Iowa (DuMont 1934) and probably does

not occur in Nebraska, although Short (1959) listed a specimen considered to be this subspecies (KU 34087) taken by Tordoff near Valentine on the unexpected date of 15 Jul 1957; the date perhaps confirms the statement of Bruner and others (1904), in that this may have been merely a large example of *villosus*. *Septentrionalis* breeds no closer to Nebraska than northern North Dakota (AOU 1957).

DISTRIBUTION AND ECOLOGY

The preferred habitat for this species is mature forest, and in summer it is most numerous in the east but is distributed evenly over the rest of the state in lesser numbers. In winter and during migration periods it occurs statewide in virtually all arboreal habitats; at this season it is often found in cities and towns.

It appears that Hairy Woodpecker was more numerous in the 1950s than at present in the mature riparian forests of the lower Platte, Missouri, and lower Blue Rivers (RSS). This change may be a result of the extensive loss of elms (*Ulnus americanus*) during the outbreak of Dutch elm disease in the 1960s in those forests.

Summer

BBS data (1967–77) indicate that this species is far more common in the east than elsewhere, presumably a result of the presence of favorable deciduous woodland habitat in the lower Platte River and Missouri River valleys.

Winter

Although this species is usually considered to be resident, there is "evidence to indicate a general southward movement in fall; the individuals seen in winter are probably not the same as those seen in summer" (Bent 1939). Definite southward migrational movements have been observed in fall, at least in the eastern United States, primarily from late Aug through Oct (Bent 1939). Altitudinal movement of the Rocky Mountain (and Black Hills) birds, *monticola*, occurs in fall and spring, with these birds appearing on the Colorado plains in winter (Andrews and Righter 1992) and presumably also in western

Nebraska. There are few if any specimens of the northern race *septentrionalis* to indicate southward movement to Nebraska of at least this race.

CBC data indicate even distribution across the state, in contrast to the summer concentration in the east indicated by BBS data, suggesting either an influx into Nebraska in winter or significant movement of Nebraska breeding birds. At least in the northwest, this species is more numerous in winter (Rosche 1982); this may be due to an influx of the western race *monticola*, as no winter influx was noted farther east in the Keith Co area (Rosche 1994a).

FINDING

This species is most easily seen at suet feeders in towns in winter but can be found almost anywhere year-round, usually by its sharp call note.

Three-toed Woodpecker

Picoides tridactylus

STATUS

Rare casual summer visitor west.

DOCUMENTATION

Specimen

UNSM ZM6345, 16 Jun 1916 Scotts Bluff Co.

Records

There are 2 documented records, both somewhat surprisingly in summer; the specimen cited above is a male collected 11.3 km (7 mi) west of Scottsbluff 16 Jun 1916 and confirmed as this species by H. C. Oberholser (Swenk 1918b), and a probable pair was seen 18 Jul 1994 (JJ; Gubanyi 1996a; NBR 62:111). The latter report included a description of a female that was associated with a second, undescribed bird in burned ponderosa pine woodland in Smiley Canyon, Sioux Co.

Two additional reports are inadequately documented. These are of one coming to a feeder at Hardy, Nuckolls Co, for several weeks through 15 Jan 1926 (LOI 1926). This bird was described as being the size of a Hairy Woodpecker but with the "glossy black

and white of the Hairy Woodpecker replaced by dull black and gray." This may have been a nutritional consequence of the bird's "mutilated" lower mandible. No other features were described. The second report was of one seen 8 Jan 1957 at Saint Edward (Anderson 1957). The only description was of a "blackish dull striped bird" with a "black cap."

COMMENTS

This species tends to be erratic and opportunistic in its movements, especially when related to burn areas (Holt 1997). There is a limited population in the Black Hills of South Dakota, with some movement to lower elevations in winter (SDOU 1991), and it occurs in "all the mountain ranges of Wyoming except the Black Hills" (Scott 1993), presumably an indication that none occur in the Wyoming portion of the Black Hills. It occurs rarely in the Colorado foothills, usually in winter (Andrews and Righter 1992). In Minnesota, where it is a casual summer resident and regular winter visitor in the north, fall migrants appear in Oct and depart by mid-Apr (Janssen 1987).

FINDING

This species should be looked for in burned areas in the Pine Ridge, especially in winter, although the documented records are for summer. Scott (1993) noted that in Wyoming it is "almost guaranteed" in recent burns, with such habitat good for at least 10 years after the burn.

Black-backed Woodpecker

Picoides arcticus

STATUS

Hypothetical.

COMMENTS

This species has been reported 5 times, but none of the reports are acceptably documented. This is somewhat surprising in that it is rather common in the Black Hills of South Dakota (SDOU 1991). All Nebraska reports, however, are from the southeast: one was collected at Omaha 15 Dec 1895 by I. S. Trostler (Bruner and others 1904); another was collected there

by F. J. Frazee, without date (Bruner and
others 1904); one was collected at Dakota
City by Wallace Bruner, without date (Bruner
and others (1904); one was seen in Fontenelle
Forest 6 Mar 1927 by Dr. Mitchell, without
details (LOI 19); and one described as a female
was seen on the unlikely date of 30 Aug 1970
in Custer Co (Kieborz 1971). None of the
collected specimens are known to be extant.

It is surprising that this species has not been
recorded from burns in the Pine Ridge, as it
is opportunistic and mobile in search of such
habitat, at least in the Black Hills of South
Dakota. The next nearest population to Nebraska is in northern Minnesota, presumably
the source of the several Illinois and few Iowa
records, where it occurs during rare invasion
years mostly in the period Dec–Feb and in
May (Bohlen 1989; Kent and Dinsmore 1996).
That the Nebraska reports tend to be from
the east suggests provenance in Minnesota for
these reports if indeed they are correct.

FINDING

A careful winter search of burns such as the
recent Fort Robinson burn in the Pine Ridge
may yield a Black-backed Woodpecker.

Northern Flicker

Colaptes auratus

STATUS

Common regular spring and fall migrant
statewide. Uncommon regular resident
statewide. Common regular breeder
statewide. Common regular winter visitor
statewide.

DOCUMENTATION

Specimen

UNSM ZM6281, 6 Jul 1892 Lincoln, Lancaster Co.

TAXONOMY

The center of the hybrid zone between the
western "red-shafted" subspecies *cafer* and
the eastern "yellow-shafted" *auratus* passes
from the Black Hills of South Dakota south
through Chadron to near Crook in northeast
Colorado (Short 1961). There are probably no
"pure" *cafer* in Nebraska in summer (Short

1961) and possibly only very few "pure"
auratus in extreme eastern Nebraska, with
most summer birds being hybrids (Short
1961); few parental types occur within the
rather wide hybrid zone, where mating is
random (Sibley and Monroe 1990). Lingle
(1994) stated that "red-shafted" types occur
rarely in south-central Nebraska in summer,
and Glandon and Glandon (1934a) listed this
form as an uncommon summer resident in
Logan Co, although not known to nest.

In winter most birds in western Nebraska are
at least "salmon-colored" (Rosche 1994a),
with some occurring east to the Missouri
River; perhaps some fairly "pure" *cafer* appear
also in the west (Andrews and Righter 1992),
although all collected specimens are hybrids
(Rapp and others 1958). Many migrants and
wintering birds are northern "yellow-shafted"
birds formerly described as *borealis* (AOU
1957); these birds show very little hybrid
character (Andrews and Righter 1992). Two
birds of the race *borealis* in the HMM (HMM
26960) were collected in Adams Co 9 Feb 1951.

There is a large series of intergrade specimens in
the USNM and a representative series in the
UNSM.

DISTRIBUTION AND ECOLOGY

At all seasons this species is rather evenly
distributed over the state. In summer it is
least common in the extensive grassland
areas of the Sandhills and Panhandle but
does occur in the few widely scattered groves
that are present there. In winter it becomes
uncommon or even absent in the north and
throughout most of the Sandhills in severe
weather. During late spring, summer, and fall
it is primarily a ground forager, preferring
ants, and thus is found most commonly in
open woodland areas, particularly riparian,
but it also may be found in windbreaks
and woodlots associated with farmsteads
in rural areas. In winter it becomes a bark
forager and thus occurs in any appropriate
woodland habitat, including urban. During
late winter and early spring it may become

especially conspicuous in treed residential areas and parks but may retreat to more favorable habitat in larger parks and adjacent countryside when breeding territories are established.

Spring

Most migrants are probably northern United States and Canadian breeders, some of which winter in Nebraska. Peak migration occurs in late Mar–early Apr. Rosche (1994a) listed a high count of 20 on 28 Mar. Ludlow (1935) listed 8 reports of birds identified as "*collaris*," now considered part of *cafer*, in the period 1 Nov–7 Apr in Webster Co; other late dates of "red-shafted" birds in the east are 5 Apr 1997 Douglas Co, 10 Apr 1998 at Laurel, and 13 Apr 1999 Otoe Co.

Summer

Breeding birds occur statewide, although numbers are somewhat lower in the Panhandle and over large areas of the Sandhills, where, except for the Pine Ridge (Rosche 1982), suitable habitat is restricted. Early arrival dates of "red-shafted" birds include 21 Sep 1995 at James Ranch, 22 Sep 1996 in Merrick Co, and 25 Sep 1999 in southwest Hall Co (LR, RH). Ludlow (1935) noted birds of red-shafted forms in Webster Co 1 Nov–7 Apr. The last "yellow-shafted" birds left James Ranch 25 Sep 1995.

Fall

Peak migration occurs in early Oct.

There appears to be a significant eastward movement of "red-shafted" birds in fall, possibly a continuation of altitudinal movement out of the mountains (Bent 1939); this movement may be another example of "altitudinal" migration in Nebraska (see Mountain Bluebird).

High counts include 50+ at Red Willow Res 3 Oct 1998, 40 in Buffalo Co 24 Sep 1998, and 30 at Calamus Res 7 Sep 1997.

Winter

CBC data indicate a generally even winter distribution statewide, with somewhat lower numbers in the north. Many Nebraska

breeders remain (Bruner and others 1904; Bent 1939), but in severe winters most move south. Winter numbers in the northwest fluctuate widely depending on food supply (Rosche 1982). Bent (1939) cited information that suggested that winter survival can be difficult due to a paucity of animal-based food; vegetable food may not be enough to support winter survival in some years.

FINDING

This species is easily located in hedgerows or riparian woodland statewide; it commonly feeds on the ground and occurs in small flocks during migration.

Pileated Woodpecker

Dryocopus pileatus

STATUS

Rare casual winter and summer visitor southeast. Accidental breeder Fontenelle Forest.

DOCUMENTATION

Photograph

24 Jan 1991 Bellevue, Sarpy Co (Grenon 1991).

DISTRIBUTION AND ECOLOGY

This species was extirpated from Nebraska prior to 1900 but appears to be reoccupying parts of its former range (Thompson and Ely 1989; SDOU 1991; Rapp and others 1958). Repopulation appears to be coming from northwest Missouri, where a small but persistent population exists (Robbins and Easterla 1992).

Reports in the 20th century are since 1950, mostly from heavy riparian forest and adjacent upland woodland in the lower Missouri Valley, north to Washington Co.

Summer

Prior to 1950 the last report of this species was from near Omaha 20 May 1895; it was "formerly not uncommon in the more heavily wooded portions of the Missouri bottoms" and may have bred (Bruner and others 1904). More recently there has been an increase in reports, culminating in a nesting pair seen in 1999 in Fontenelle Forest.

Calls and holes attributed to this species were

noted at Indian Cave SP 3–4 May 1969 (Tate 1969), although no birds were seen. One was seen along the Republican River near Franklin in summer 1974 (Brown 1975). During the first week of Jun 1983 one was seen near Cris Lake in southeast Sarpy Co (NBR 51:71). One was reported along the Platte River in Sarpy Co 4 Jul 1996 (RG).

The first documented nesting in the 20th century was of a pair along North Stream Trail in Fontenelle Forest, first seen in the area 25 April 1999 (WRS) and subsequently seen by many observers; the birds' activities were carefully studied 30 Apr (CNK). Copulation was observed 4 May (BFH). Two young were successfully fledged by 10 Jul. On 25 Apr at least one other bird was seen southeast of Hidden Lake in Fontenelle Forest (WRS). Prior to these 3 sightings, singles had been reported in the Great Marsh area 16 Jan (JK, SK), 7 Apr (PW), and 18 Apr (JT).

Winter

There was a report of one at a suet feeder near Fort Calhoun 26 Oct 1952 (NBR 21:3). There were 3 reports 1970–75: one was seen in a farm woodlot near Cook Feb 1970 (Fiala 1971); there was an undocumented report of one in Omaha in 1970 (Fiala 1971); and another came to a suet feeder in Omaha 8 and 10 Feb 1975 (NBR 43:20, 55).

There were no further reports until a series of sightings from Fontenelle Forest 1983–86, possibly involving the same bird. These reports, most undocumented, were in the period 7 Dec–early Jun. Another was seen repeatedly and photographed in Fontenelle Forest 24–27 Mar 1991 (Grenon 1991; Wright 1991; NBR 52:54, 59). More recently, there was an undocumented report of one near Union 7 Nov 1998 (Rod Edwards, pers. comm. LF).

FINDING

This bird could be looked for in heavy riparian woodland in the lower Missouri Valley at any time of year but is most conspicuous in late Apr and rather secretive at other times.

PASSERIFORMES

Tyrannidae (Tyrant Flycatchers)

Olive-sided Flycatcher

Contopus cooperi

STATUS

Uncommon regular spring migrant east, becoming rare casual west. Uncommon regular fall migrant east, becoming rare west.

DOCUMENTATION

Specimen

UNSM 6401, 14 Jun 1901 Warbonnet Canyon, Sioux Co.

DISTRIBUTION AND ECOLOGY

Olive-sided Flycatcher occurs in deciduous and coniferous woodland, where it habitually perches on exposed twigs and branches at the tops of trees. It is most numerous during spring and fall in the Missouri Valley and is probably regular in small numbers in the Panhandle, especially in fall (Rosche 1982).

Spring

29 Apr, 1, 2 May→12, 14, 14 Jun

Most reports are from the Missouri Valley, although migrants occur statewide in declining numbers westward. Migration is from early May through early Jun, with a peak mid to late May, although there are earlier undocumented reports 22 Apr 1943 Logan Co and 25 Apr 1972 Hall Co and a later undocumented report 27 Jun 1967 Brown Co.

There are a few Panhandle reports in the period 13 May–14 Jun. Andrews and Righter (1992) considered it a regular though rare migrant in eastern Colorado. Rosche (1982) noted that late spring birds in the Pine Ridge should be monitored for evidence of breeding.

Fall

6, 6, 7 Aug→2, 4, 5 Oct (FN 51:81)

Migration is from mid-Aug through late Sep, although there are earlier reports 1 Aug 1996 Sioux Co and 2 Aug 1999 Harlan Co (GH, WH) and a late record 25 Oct 1926, a specimen from Inland, Clay Co, now HMM 3667.

Reports from the Panhandle are few, in the

period 1 Aug–17 Sep; it is a regular though rare migrant in eastern Colorado late Jul–early Oct (Andrews and Righter 1992).

COMMENTS

Although currently no subspecies are recognized (AOU 1957; Bent 1942), Mickel and Dawson (1920) indicated that the specimen cited above and another collected at Kimball 8 Jun 1919 were representative of the slightly larger (DuMont 1934) western subspecies *majorinus*. Swenk and Dawson (1921) indicated that a specimen, UNSM ZM10564, taken at Roca, Lancaster Co, 4 Sep 1909 was intermediate between eastern *borealis* and *majorinus*.

Other Reports

3 Apr 1979 Howard-Hall Cos, 6 Apr 1948 Logan Co, 14 Apr 1963 Lincoln Co, 14 Apr 1946 Logan Co; the earliest Kansas specimen is 20 Apr (Thompson and Ely 1992).

FINDING

This species should be looked for in Missouri River valley woodlands in mid to late May; at this season it may be located by its song, although at any time its habit of perching on an exposed twig and its large bill are clues to its identity.

Western Wood-Pewee

Contopus sordidulus

STATUS

Common regular spring and fall migrant west, hypothetical elsewhere. Common regular breeder west. Rare casual summer visitor central.

DOCUMENTATION

Specimen

UNSM ZM6396, 24 May 1900 Monroe Canyon, Sioux Co.

TAXONOMY

Determination of the range limits of Western Wood-Pewee and of Eastern Wood-Pewee is difficult because of the problems of identification of the two away from their regular breeding grounds. The two are very similar in appearance, even sometimes considered conspecific (Mayr and Short

1970), and may hybridize, although probably only to a limited extent (Rising and Schueler 1980; Ducey 1989). Plumage characteristics are at best suggestive, and vocalizations may be misleading, although generally species-specific if carefully heard (Rising and Schueler 1980). Short (1961) studied the pewees in Nebraska and noted that while songs usually identify these species, "occasional songs of Western Wood-Pewee are virtually indistinguishable from Eastern Wood-Pewee singing abbreviated [rather than full] songs."

The problem may have been simpler earlier in the 20th century, as Swenk and Dawson (1921) indicated that Western Wood-Pewee occurred only west of the 100th meridian and its range and that of Eastern Wood-Pewee "do not anywhere meet." The lone exception appears to be a specimen of Western Wood-Pewee taken at the "Loup Fork of the Platte" (vicinity of Columbus) in 1856 (USNM 8892; Coues 1874). During the 20th century, however, Eastern Wood-Pewee has extended its range westward, resulting in contact zones in the Platte Valley in Keith and Garden Cos (Brown and others 1996) and probably in Deuel Co along the South Platte River (Short 1961) and the Niobrara River in Sheridan Co (Short 1961; AB 42:1309).

DISTRIBUTION AND ECOLOGY

In the northwest this species is fairly common in open, mature pine woodland and riparian woodland. It also occurs in cottonwoods and other deciduous trees around reservoirs and in city parks (Rosche 1982), but its distribution is essentially restricted to the Panhandle. Short (1961) found no difference in habitat occupied by Eastern and Western Wood-pewees.

Spring

6, 6, 7 May→summer

Arrival is in early May, although there are earlier reports 20 Apr 1988 Dawes Co, 27 Apr 1987 Dawes Co, 1 May 1990 Dawes Co, and 3 May 1964 Scotts Bluff Co. There are no Kansas

specimens earlier than 29 Apr (Thompson
and Ely 1992).

The few easterly reports are undocumented,
although migrants may indeed occur
somewhat east of the summer range (SDOU
1991). These reports are 26 Apr 1957 Brown Co,
9 May 1954 Cherry Co, 10 May 1955 Cherry Co,
12 May 1964 Brown Co, 30 May 1995 Cherry
Co, and an undated report from Lincoln Co
(Tout 1947).

High counts include 6 at Crescent L NWR 6 May
1995.

Summer

Significant numbers of breeding Western
Wood-Pewees are found only in the Pine
Ridge and Niobrara Valley in Sioux, Dawes,
and Sheridan Cos and in the Platte Valley
and Wildcat Hills in Scotts Bluff Co (Ducey
1988). Rosche (1994a) placed the easternmost
site of regular breeding in canyons south of
Redington in Morrill Co. Away from these
areas, riparian habitats, primarily cottonwood
groves, and possibly city parks may support
breeding birds. Examples are reports from
Alliance 7 Aug 1988 and 17 Jun 1989, Cheyenne
Co 11 Jul 1992, and several from southern
Garden Co, particularly in the Lewellen
area, where Rosche (1994a) noted that most
observations have been made since the late
1980s from "lakeside cottonwood forest."
Rosche (1994a) cited 2 nesting records in
southeast Garden Co in 1987 and 1993 and 1
near Lisco, Morrill Co, in 1979; 8–9 singing
males were at the west end of L McConaughy
27 Jul 1997 (RCR). Brown and others (1996)
noted that singles were banded in the L
Ogallala area 16 May, 30 May, and 5 Jun,
including a female with a brood patch. It is
normally resident in summer at Crescent
L NWR, inhabiting the large cottonwoods
around the older headquarters buildings
(RSS), and was reported from Crescent L NWR
7 Jun 1995 and 5 Jun 1999.

East of the Panhandle there are only scattered
summer reports, all west of the 100th
meridian. The present eastward limit of

regular breeding in the Niobrara Valley
is uncertain, but SDOU (1991) shows the
breeding range in southern South Dakota
extending east to about the 100th meridian,
with migrants occurring farther east. Short
(1961) suggested that contact between the 2
pewee species may occur along the Niobrara
River between Valentine and the Pine Ridge,
and a Jun 1988 Sheridan Co report was of
the two "singing side by side" (AB 42:1309).
Ducey (1988) cited an old breeding record for
Cherry Co (Bates 1900), and songs with the
"song pattern of Eastern Wood-Pewee and
tone quality of Western Wood-Pewee" were
heard in Cherry Co 21 May 1989 (AB 43:501).
There is a single Brown Co summer report,
undocumented, 30 Jun 1964, and a singing
male was in Keya Paha Co 7 Jun 1985 (RCR).

Andrews and Righter (1992) and Short (1961)
indicated that this species breeds in Colorado
northeastward to Deuel Co, Nebraska, where
Short (1961) suggested that contact may occur
with Eastern Wood-Pewee. Other summer
reports from this general area are 9 Jun 1990
in cottonwood forest at Swanson Res (RCR;
AB 44:1153), Jul–12 Aug 1974 Perkins Co, 6–7
Jul 1989 Chase Co, 16 Jul 1987 Keith Co (AB
41:1457), and 31 Jul 1993 Keith Co.

Western Wood-Pewee apparently once occurred
in summer east to Thomas Co on the Dismal
River (Bruner and others 1904) and may still
breed occasionally in the extreme western
Loup drainage, as there are these late spring
reports that raise the possibility (but see
Spring): 28 May 1980, a territorial male at
NNF, Thomas Co (RCR, DJR); 4 Jun 1986 Grant
Co; 4 Jun 1993 Thomas Co (Bray 1994); 31
May–7 Jun 1982 McPherson Co; and 27 Jun
1956 Logan Co.

A nest with chicks was found in Monroe Canyon
on the rather late date 20 Aug 1995 (NBR
63:105).

Fall

summer→17, 18, 19 Sep

Departure is in mid-Sep, although there are
later reports 27 Sep 1984, 28 Sep 1979, 9 Oct

1965, and 18 Oct 1981, all from Scotts Bluff Co but undocumented. There are no Kansas specimens later than 17 Sep (Thompson and Ely 1992).

Reports east of the breeding range are few, although these are from likely breeding locations: 22 Aug 1962 Lincoln Co, 3 Sep 1975 Lincoln Co, and 12 Sep 1966 McPherson Co. There are additional undocumented reports 26 May 1984 Howard-Hall Cos and 6 Sep 1987 Polk Co; it should be noted that Eastern Wood-Pewee was reported simultaneously at the latter location.

COMMENTS

Early egg data suggesting breeding in Douglas Co in 1899 and Saline Co in 1941 are incorrect (Ducey 1988).

FINDING

Western Wood-Pewee is best looked for in Pine Ridge canyons, where it can be located by its song. The entire song must be heard, including the burry downward slur of the second part, for positive identification.

Eastern Wood-Pewee

Contopus virens

STATUS

Common regular spring and fall migrant east, becoming rare casual west. Common regular breeder east, becoming rare casual west.

DOCUMENTATION

Specimen

UNSM ZM10565, 4 Sep 1909 Roca, Lancaster Co.

TAXONOMY

See Western Wood-Pewee.

DISTRIBUTION AND ECOLOGY

Eastern Wood-Pewee shows a preference for deciduous forest during the breeding season, utilizing both upland hardwoods and mature floodplain forests. It is most common in forests associated with the Missouri, lower Platte, and lower Elkhorn Valleys but is only casual west of North Platte and in Cherry Co.

The substantial loss to disease of mature American elms in the major eastern Nebraska floodplains during the 1960s has been a factor in the reduction of numbers of Eastern Wood-Pewees in these woodlands during the 1960s and 1970s.

Spring

24, 24, 26 Apr→summer

Arrival is difficult to determine from published data, which are poorly documented, although there is a cluster of early dates beginning 24 Apr. The earliest South Dakota and Iowa records are also in late Apr, but the earliest Kansas specimen is 9 May (Thompson and Ely 1992), and the earliest Nebraska reports in the 1990s are 4, 7, and 8 May. Published Nebraska reports prior to Apr 24 and at least some of those in late Apr and early May are likely to be of Eastern Phoebe, an early migrant.

High counts include 12 in Sarpy Co 11 May 1996 and 8 at Indian Cave SP 23 May 1999.

Summer

Eastern Wood-Pewee is most numerous where deciduous forest occurs along the Missouri, Elkhorn, and Platte Rivers and within the oak-hickory region of extreme southeastern Nebraska. Elsewhere, it is found westward along the Niobrara, Platte, and Republican Valleys where deciduous woodland remains reasonably extensive.

It occurs west along the Niobrara Valley to the Valentine area in eastern Cherry Co, where it was considered uncommon by Brogie and Mossman (1983). Nesting was observed in western Keya Paha Co (Mossman and Brogie 1983). Seven were recorded on the Niobrara in Brown Co north of Johnston, close to the above nesting location, 6–7 Jul 1991 (NBR 60:3).

Short (1961) suggested that there may be contact between this species and Western Wood-Pewee in the Niobrara Valley between Valentine and the Pine Ridge area, although numbers of either species in this area are limited. Songs with the pattern of Eastern Wood-Pewee but with the tonal quality of Western Wood-Pewee were heard in eastern Cherry Co 21 May 1989 (AB 43:501; Ducey

1989), and singing birds of both species were found "side-by-side" in Sheridan Co in 1988 (AB 42:1309). Rosche (1982) considered Eastern Wood-Pewee a "very rare summer visitor" in the northwest, with dates for singing birds in the period 29 Jun–28 Aug only and all sightings but one from Box Butte Res; one was reported from Dawes Co 7 Jun 1996 (FN 50:966).

Zeillemaker netted birds identified as this species near Crescent L in Garden Co 1–3 Jun 1980 and 12 Jul–24 Aug 1979, and singing birds were there 12–13 Jul 1982 (AB 33:877). Additional undocumented reports from the Panhandle are 30 May 1952 Scotts Bluff Co (the only spring report from the west), 29 Jun 1966 Dawes Co, 1 Jul 1968 Dawes Co, 9 Jul 1968 Scotts Bluff Co, and 14 Jul 1987 Scotts Bluff Co (AB 41:1457).

In the Platte Valley Eastern Wood-Pewee breeds regularly west to the North Platte area in Lincoln Co (Johnsgard 1979), although as recently as 1947 (Tout 1947) there were only 3 Lincoln Co reports of "Wood Pewee sp.," none documented, and Tout considered it "very rare." There are 2 reports from the L McConaughy area in Keith Co, 16 Jul 1987 (AB 41:1457) nesting within 200 m (656 ft) of a Western Wood-Pewee pair carrying food (Rosche 1994a) and 4 Jun 1992 in Eagle Canyon (AB 46:1152; NBR 60:141; Rosche 1994a). Rosche (1994a) considered Eastern Wood-Pewee a "casual summer visitant" in this area, and Brown and others (1996) noted that singing males were heard below the Keystone Diversion Dam in 1993 but that a juvenile netted there 12 Aug may have hatched elsewhere. Short (1961) noted Eastern Wood-Pewee breeding along the South Platte to the Colorado border, where it was thought to make contact with Western Wood-Pewee.

It has been stated that this species breeds west in the Republican Valley to Dundy Co (Johnsgard 1979, 1980), but recent fieldwork (JGJ, WRS) indicates that it is virtually absent west of Harlan Co. In north-central Kansas

it breeds regularly west only to Jewell Co (Thompson and Ely 1992).

There are a few summer reports from the Loup drainage, west to Loup Co (BBS) and 6 Jun 1989 Custer Co, but the species is apparently irregular in this area; there are no published reports from Thomas Co, where Bray (1994) listed it as a "rare migrant" only.

Nest card data (N = 4) show egg dates in the period 15–29 Jun.

Fall

summer→5, 6, 7 Oct

Departure is generally in Sep, although there are later undocumented reports 12 Oct 1962 Douglas Co, 14 Oct 1986 Pierce Co, and 20 Oct 1989 Douglas Co. The latest Kansas specimen date is 23 Sep (Thompson and Ely 1992)

High counts include 6 at Indian Cave SP 19 Sep 1999.

FINDING

Eastern Wood-Pewee can be located most easily by its song in Jun in deciduous forest in the Missouri Valley, such as at Indian Cave SP and Fontenelle Forest.

Yellow-bellied Flycatcher

Empidonax flaviventris

STATUS

Rare regular spring migrant east, rare casual elsewhere. Hypothetical in summer. Rare casual fall migrant east and central.

DOCUMENTATION

Specimen

WSC 1027, 13 Sep 1982 Albion, Boone Co.

DISTRIBUTION AND ECOLOGY

This species occurs in dense, shaded underbrush, usually deciduous, but often also in conifers (Robbins and Easterla 1992). It typically perches near the ground. Migrants are rarely reported away from the Missouri Valley.

Spring

8, 8, 8 May→27, 27, 30 May (Gubanyi 1996c)

Migration peaks in late May. The early dates listed above are undocumented but may be accurate, as the early date in Iowa is 4 May (Kent and Dinsmore 1996). Documented

Kansas records are in the period 24 May–5 Jun (Thompson and Ely 1992), and it is rarely recorded before mid-May in Missouri (Robbins and Easterla 1992). The earliest documented Nebraska record is of one at Schramm Park 30 May 1995 (Gubanyi 1996c).

There are later reports 1 Jun 1991 Cherry Co (AB 45:1134) and 7 Jun 1972 Lancaster Co, which appear to be within the expected migration period.

The westernmost reports within the expected migration period, none documented, are 12–13 May 1973 Thomas Co (Bray 1994), 14 May 1954 Cherry Co, 15 May 1949 Lincoln Co, 15 May 1955 Keith Co, 16 May 1911 Lincoln Co (Tout 1947), 20–27 May 1975 McPherson Co, 21 May 1967 Custer Co, and individuals banded at Cedar Point Biological Station, Keith Co, 18 May 1996 and 3 Jun 1992 (Brown and others 1996). Such westerly reports may as likely be Cordilleran Flycatchers, although the Lincoln Co report was of a bird closely examined by Tout, who explicitly eliminated "Western" Flycatcher as a possibility.

Summer

There are a few reports for Jun and Jul that were thought to indicate breeding, although this would be most unexpected. Yellow-bellied Flycatchers were reported in summer from Lancaster and Sarpy Cos in the years 1978–81, including 3 nest cards for Sarpy Co submitted in 1978 (Bennett 1980a). No identification details were published. There are undocumented reports for Cass Co 18 Jun 1966 and Saline Co 18 Jun 1950. One was reported in Omaha 16 Jun 1998 (Phipps 2000), but it was singing a 2-note song with the first note "wheezy" and the second clear (Kristine Phipps, pers. comm. WRS), suggestive of an odd Willow Flycatcher. The most intriguing evidence for summering was a report of a female netted in Sarpy Co 1 Jun 1978 that had a "large brood patch" considered to "indicate nesting in the area" (Green 1978; AB 32:1181). Some, if not all, of these birds may have been Acadian Flycatchers, as suggested by Swenk

and Dawson (1921) for specimens purportedly taken by Aughey in Jul (Bruner and others 1904; Sharpe 1993).

Fall

16, 17, 19 Aug→18, 19, 2 Sep

Migration is from Aug through early Oct, as indicated by documented Kansas and Missouri records, which span the rather protracted period 1 Aug–11 Oct (Thompson and Ely 1992; Robbins and Easterla 1992). Nebraska reports, most undocumented, are in the period mid-Aug through mid-Sep, although there are earlier reports 29 Jul 1981 Boone Co and 8 Aug 1962 Webster Co and later reports 24–25 Sep 1983 Thomas Co (Bray 1994), 1 Oct 1990 Cass Co, 4 Oct 1993 Cass Co, 4 Oct 1999 Cass Co, 5 Oct 1982 Douglas-Sarpy Cos, and 9 Oct 1993, a bird banded but undescribed, Sarpy Co (NBR 60:167), all apparently within the expected migration period.

Identification of fall birds is difficult, as several species of *Empidonax* flycatcher, especially Acadian, may be yellowish in fall (Kaufman 1990).

Other Reports

30 Apr 1952 Cherry Co, 1 May 1975 Douglas-Sarpy Cos.

FINDING

This species is difficult to identify, as it is generally silent in migration (Bent 1942) and frequents shady undergrowth where it is difficult to study well enough to discern structural characteristics. Nevertheless, it should be looked for in late May and late Aug–early Sep in the Missouri Valley.

Acadian Flycatcher

Empidonax virescens

STATUS

Rare regular spring migrant southeast, rare casual elsewhere. Rare regular breeder southeast. Hypothetical in fall.

DOCUMENTATION

Specimen

UNSM ZM10562, 30 May 1910 Lancaster Co.

DISTRIBUTION AND ECOLOGY

This species generally occurs in shady wooded ravines in upland forest, usually near water. It is essentially restricted to southeast Nebraska, north and west to Douglas and Lancaster Cos.

Spring

1, 1, 2 May→summer

Data indicate that Acadian Flycatcher is most often reported in spring, suggesting that the few birds that reach Nebraska may be overshooting the regular breeding range. Most reports are in May.

Arrival is in early May. Half the available spring reports are in the period 8–21 May (Johnsgard 1980).

Away from the east there are few reports; there are no accepted Colorado records (Andrews and Righter 1992). A specimen, UNSM ZM9389, was collected in Cherry Co 29 May 1932. Four birds were netted, banded, and identified as this species by wing formula at Cedar Point Biological Station, Keith Co, 21–29 May 1996 (Brown and others 1996). An adult male specimen assumed to have been collected by Cary in Sioux Co 26 May 1900 (Cary 1901; Crawford 1901; Rosche 1982) cannot now be located; Swenk and Dawson (1921), who also had not seen the specimen, suggested that Cary's report may have been based on a misidentification, as a specimen taken in the same place 2 days later by Hunter and said to be identical in appearance to that of Cary was identified as "*E. traillii brewsteri*" (apparently a Willow Flycatcher) by Oberholser.

Additional westerly reports, none documented, are 5 May 1947 Adams Co, 11 May 1948 Keith Co, 14 May 1944 Adams Co, 16 May 1937 Adams Co, 24 May 1959 Antelope Co, 25 May 1974 Howard-Hall Cos, 25 May 1957 Lincoln Co, 28 May 1976 Howard-Hall Cos, 29 May–4 Jun 1970 McPherson Co, and 10 Jun 1976 McPherson Co.

Summer

Ducey (1988) listed Acadian Flycatcher as a hypothetical breeder along the Missouri River and cited about 6 reports of breeding, none since 1952. Egg dates (N = 6) 10 Jun–6 Jul for Douglas-Sarpy Cos 1898–99 were cited by Hanson (1952), and 6 nests were found in Sarpy Co in 1944 (Garrett 1944).

Recent Jun and Jul reports from extreme southeast Nebraska and 2 as far north as extreme northern Thurston Co may indicate that the species is expanding its breeding range in Nebraska or that observers are better at finding these birds: 13 Jun 1998 and 3 Jul 1999 Ashford BSC, Thurston Co; 20 Jul 1988 and 4 Jun 1992 Lancaster Co; 14–21 Jul 1990 Douglas-Sarpy Cos; 25 Jun 1991 Douglas Co (AB 45:468); 3 Jun 1994 Sarpy Co; 6 Jun and 4 Jul 1997 and 12–19 Jun 1998 Neale Woods; 15 Jun 1994 Nemaha Co; 18 Jun 1991 Cass Co; and up to 3 birds on Trail 9 in Indian Cave SP through 25 Jun 1995 and at least 1 there 9 Jun 1997. The 1990 reports from Douglas Co were described as "perhaps the first since the 1960s" (AB 44:1153), and summering birds were present until 3 Sep (AB 45:124). In 1991 nesting occurred in Douglas Co, but a storm 22 Jun 1991 destroyed the nest (AB 45:1134).

Acadian Flycatcher is unreported elsewhere in summer except for a single undocumented BBS report from northern Holt Co in the period 1967–77.

Fall

summer–→17, 19, 22 Sep

Departure is in mid-Sep, although there is a later undocumented report 1 Oct 1965 Douglas Co. There are no documented fall records for Nebraska, but substantiated records in Kansas extend to 15 Sep (Thompson and Ely 1992) and to 20 Sep in Missouri (Mark Robbins, pers. comm.).

The few reports away from the east, none documented, are 17 Aug 1973 Howard-Hall Cos and 26–28 Aug 1967 Adams Co. There are no records for Colorado (Andrews and Righter 1992) or Wyoming (Scott 1993).

Other Reports

21 Apr 1980 Lancaster Co, 19 Aug 1964
McPherson Co, 31 Aug 1986 McPherson Co.

FINDING

Acadian Flycatcher is best located by its explosive
call in wooded ravines at Indian Cave SP
in early Jun when the birds are strongly
territorial. One technique is to walk brushy
ravines with running water; Trail 9 at Indian
Cave SP is good.

Alder Flycatcher

Empidonax alnorum

STATUS

Fairly common regular spring migrant east, rare
casual central. Rare casual fall migrant east.

DOCUMENTATION

Specimen

PSM 17452, 4 Jun 1992 Lancaster Co.

TAXONOMY

Alder Flycatcher, *Empidonax alnorum*, and
Willow Flycatcher, *E. traillii*, were considered
a single species, Traill's Flycatcher, *E. traillii*,
until 1973, when a long period of considerable
confusion over taxonomy and distribution of
the 2 taxa ended. We have only used data since
1973 in the accounts for these species.

Alder Flycatcher can only be safely separated
in the field from Willow Flycatcher by
vocalizations, and so Nebraska data,
especially for fall, are limited. Kansas
specimens of Alder Flycatcher examined by
Phillips were in the periods 13–29 May and
29 Jul–12 Sep (Thompson and Ely 1992);
undocumented reports outside of these
ranges are questionable.

DISTRIBUTION AND ECOLOGY

Migrants are found in shrubs and small
trees, usually in edge situations, including
backyards. The species is most numerous in
the Missouri Valley.

Spring

12, 12, 13 May→4, 4, 5 Jun

Most reports are from the east and north, but
Alder Flycatcher occurs west at least to Keith

Co. Brown and others (1996) considered
Alder Flycatcher a common migrant in the
Keith Co area 22 May–2 Jun; 62 were banded
there spring and fall. A specimen collected
in Keith Co 27 May 1993 is UNSM ZM20522.
It is very rare in eastern Colorado, where
there are 10 specimen records for the eastern
Colorado plains in very late May and early Jun
(Andrews and Righter 1992). The earliest and
latest Nebraska specimens are 12 May 1993
Lancaster Co (UNSM ZM20498) and 4 Jun 1992
(cited above).

Nebraska reports are generally undocumented
and are in the period 6 May–4 Jun. The
earliest reports are 6 May 1985 Lancaster
Co, 6 May 1993 Lancaster Co, and 8 May
1977 Adams Co and latest are 13 Jun 1975
McPherson Co and 18 Jun 1977 Howard-Hall
Cos; none are documented.

A bird banded in Keith Co 28 May 1993 was
recovered at Fairbanks, Alaska, 7 Aug 1993
(Brown and others 1996).

Fall

The few documented records are in the period
21 Jul–4 Sep. These records are 10 Aug 1988
"identified by song" in Sarpy Co (AB 43:126),
10 Aug 1999 "singing" at Omaha, 29 Aug 1982
Sarpy Co (Bray and others 1986), and 4 Sept
at Fontenelle Forest "identified by call notes"
(NBR 63:105). In addition, banding data from
Cedar Point Biological Station in Keith Co
(Brown and others 1996) indicate that Alder
Flycatcher is a common migrant in the period
21 Jul–17 Aug.

There are undocumented early reports 11 Jul 1988
Lancaster Co and 16 Jul 1973 Douglas-Sarpy
Cos and a later undocumented report 23 Sep
1982 McPherson Co.

COMMENTS

Breeding records cited by Ducey (1988) probably
pertain to Willow Flycatcher.

FINDING

Alder Flycatcher should be listened for in the
Missouri Valley during the last week in May
in shrubby edge habitats. If heard at the

beginning and end of migration periods, recordings should be made to document the occurrence.

Willow Flycatcher
Empidonax traillii

STATUS
Fairly common regular spring migrant east, becoming uncommon west. Uncommon regular breeder statewide, except rare southwest. Uncommon regular fall migrant statewide.

DOCUMENTATION
Recording: 1 Jun 1968 Fontenelle Forest, Sarpy Co (Bray and others 1986).

TAXONOMY
This summary uses reports of Willow Flycatcher published since 1973, when Willow Flycatcher and Alder Flycatcher were recognized as different species. Summer reports of "*Empidonax* species" are included here also. As egg dates are generally in Jun (Bent 1942; SDOU 1991), undocumented Nebraska reports of *Empidonax* flycatchers in the period 15 Jun–15 Jul are most likely to be Willow Flycatchers.

DISTRIBUTION AND ECOLOGY
Willow Flycatcher occurs statewide in shrubby edge habitats, with a preference during the breeding season for willows. Reduction of grazing in riparian habitats and cessation of willow control led to increases in numbers of this species (Taylor and Littlefield 1986). Most reports are from the Missouri Valley, especially during migration.

Spring
2, 2, 3 May→summer (east)
9 (Brown and others 1996), 10, 12 May→summer (elsewhere)
Most migrants pass through Nebraska during May and early Jun, although numerous Jun and Jul reports do not allow easy determination of the end of migration. In Kansas, where this species is essentially a migrant, specimens and banded birds are in the period 10 May–15 Jun (Thompson and Ely

1992; Cink 1998), with 87% of the records at Hays in the period 20–25 May (Ely 1970) and 80% of the records in northeast Kansas 20–28 May (Cink 1998). One of the 2 May dates cited above was for a singing bird in Dixon Co in 1999 (JJ).
Willow Flycatcher is a regular spring migrant in Nebraska only in the east; arrival is in early May. There is an earlier undocumented report 15–28 Apr 1989 Cedar Co, and another, identified as "Willow/Alder," was reported 30 Apr 1994 in northern Saunders Co.
Away from the east, arrival is in mid-May, suggesting that most birds passing to the north of Nebraska use the Missouri Valley and that reports elsewhere are of potential residents.

Summer
Willow Flycatcher breeds virtually statewide in small numbers, as Jun and Jul reports exist for most parts of the state, although it is least numerous in the southwest. The preferred breeding habitat of shrubs and small trees is somewhat transitory, especially in the Panhandle and southwest.
Rosche (1982) stated that Willow Flycatchers are present every summer at Smith L in Sheridan Co and cited a report for Whitney L in Dawes Co; all birds were in willows near water. There is a report 20 Jun 1973 Dawes Co (AB 27:888).
Breeding is regular in the Platte Valley in the Lewellen area (Rosche 1994a) and in Keith Co (Brown and others 1996), such as in 1979 (Rosche and Johnsgard 1984), 29 Jun–2 Jul 1991 (AB 45:1134), and 31 Jul 1993, when a recently fledged brood of 4 was seen (AB 47:1123; Rosche 1994a). A juvenile was netted in Keith Co 27 Jul 1978 (Rosche and Johnsgard 1984); migrant juveniles tend to arrive later (Mark Robbins, pers. comm.). The only other summer reports for the western Platte Valley are from the Mitchell area of Scotts Bluff Co in 1975 and 1976 and Lincoln Co 6 Jun–1 Jul 1968, 10 Jun 1978, and 8 Jul 1964.
The only reports from the southwest are from Perkins Co 12 Jun–26 Jul 1974 and 30 Jul 1971.

Summering birds in northeast Colorado are rare, and breeding has not been proven there (Andrews and Righter 1992).

Egg dates for Sarpy, Douglas, Lancaster, Dodge (2), and Cuming (1) Cos are in the range 16 Jun–4 Jul. Fledged young with attendant parents have been observed as late as 8 Jul in Hall Co. Where known, nests were less than 1.5 m (5 ft) above ground in early successional habitat, dominated by willow.

Fall

summer→11, 11, 12 Sep

While identification of silent fall birds is questionable, most published data are from locations where this species was present during the summer. Kansas specimen dates are in the period 16 Aug–15 Sep (Thompson and Ely 1992), and banding dates from northeast Kansas are 28 Jul–18 Sep, peaking in Aug (Cink 1998).

Migration probably takes place during Aug, ending in mid-Sep. Zeillemaker netted and banded individuals in Garden Co 1, 9, and 11 Sep.

FINDING

Willow Flycatcher is best located in Jun by its song in riparian areas with small willows. This habitat occurs throughout most of the state but is most prevalent in major eastern river valleys.

Least Flycatcher

Empidonax minimus

STATUS

Common regular spring and fall migrant statewide. Rare casual breeder north and west.

DOCUMENTATION

Specimen

UNSM ZM6391, 9 May 1903 Havelock, Lancaster Co.

DISTRIBUTION AND ECOLOGY

Migrants occur in open woodland and edge habitats of all types, including those around human habitation. Summering birds are generally found in riparian stands of deciduous trees such as cottonwoods (Short 1965; Rosche 1982).

Spring

22, 23, 24 Apr→summer

Among the *Empidonax* group, Least Flycatcher is the earliest spring migrant. Kansas specimen data show occurrence 26 Apr–28 May (Thompson and Ely 1992), and banding data in northeast Kansas show occurrence 2 May–8 Jun (Cink 1998). At Cedar Point Biological Station, Keith Co, 94 were banded 10 May–3 Jun (Brown and others 1996).

Spring migration begins in early May, although there are undocumented earlier reports 10 Apr 1953 Gage Co, 12 Apr 1975 Adams Co, 16 Apr 1941 Jefferson Co, and 19 Apr 1985 Lincoln Co. Migration probably peaks in mid-May and ends in early to mid-Jun, but this is difficult to determine due to the many undocumented Jun and Jul reports from areas of the state where nesting would not be expected. These reports may be Willow Flycatchers (but see Summer). In the southeast, where nesting probably does not occur, late reports are at the end of May.

High counts include "about 50" in Garden Co 14 May 1979 (AB 33:786), 54 in Sarpy Co 11 May 1996, and 53 there 13 May 1995 (NBR 62:63).

Summer

Least Flycatcher regularly breeds close to Nebraska in the Black Hills (Peterson 1990) and along the Missouri River in South Dakota (SDOU 1991), as well as along watercourses in eastern Wyoming (Scott 1993). There is 1 report of "probable breeding" for extreme northeast Colorado (Andrews and Righter 1992). As might be expected, therefore, the available evidence for breeding in Nebraska comes from areas in the north and west.

Rosche (1994a) indicated that Least Flycatcher nests "in the tall cottonwoods" at Smith L, Sheridan Co. Rosche (1982) cited a record for Dawes Co 5–10 Jul 1973 (AB 27:888) and suspected as early as 1982 (Rosche 1982) that this might be a breeding species in the northwest, as did Ducey (1988). It was reported in Monroe Canyon 21 Jun 1994. Short (1965) reported 2 pairs, also in "tall cottonwoods," along the Niobrara River in Brown and Keya Paha Cos, apparently in mid to late Jun 1964, although nesting was not confirmed by Short or by Mossman and Brogie (1983) in the same area. There is a summer report of 2 birds in 2 locations in the L McConaughy area 11 Jun 1991 (AB 45:1132; Rosche 1994a).

Territorial Least Flycatchers in early to mid-Jun, especially in cottonwoods, should be monitored for breeding.

There are old reports of breeding, without documentation, for Dakota and Douglas Cos (Ducey 1988) and Richardson Co (Bent 1942).

Fall

summer→29 Sep, 1, 2 Oct

This species apparently begins migrating statewide very early, possibly in early Jul, although the many undocumented Jun and Jul reports make this difficult to determine. Adults apparently move through first, probably early; Cink (1998) noted this in northeast Kansas, where migrants were banded 25 Jul–28 Sep. Kansas fall specimens are in the range 3 Jul–24 Sep (Thompson and Ely 1992), and Cedar Point Biological Station banding data show fall migration in the period 22 Jul–17 Aug, when banding operations ceased (Brown and others 1996). Leukering (pers. comm. WRS) suggested that most Least Flycatchers move through the Barr Lake area in eastern Colorado prior to 20 Aug in fall; his banding operation starts around 20 Aug, and only 5% of *Empidonax* flycatchers captured are Leasts. In Nebraska available data suggest that migration ends in late Sep.

Other Reports

19 Oct 1984 Lancaster Co.

FINDING

Migrants are easy to locate by song in wooded edge habitats in mid-May and are most numerous in the Missouri Valley. A search for nesting birds seems likely to be most successful in Jun in cottonwoods along the Niobrara River, in the Pine Ridge, and along the Platte and South Platte Rivers west of North Platte.

Hammond's Flycatcher

Empidonax hammondii

STATUS

Accidental in spring. Rare casual fall migrant west and central.

DOCUMENTATION

Specimen

UNSM ZM10557, 17 Sep 1911 Dawes Co (Zimmer 1945; Rosche 1982).

DISTRIBUTION AND ECOLOGY

Migrants in Nebraska have been seen mainly in riparian habitats, especially willows.

Spring

There are 2 records. A specimen, UNSM ZM10559, was collected 14 May 1917 by Zimmer at Lincoln, Lancaster Co; it appears identical to the specimen (cited above) collected in 1911, notably a long primary extension and a small, dark bill. Another documented report was of a single at Scotts Bluff NM 11–12 May 1999 (SJD).

Fall

There are 7 records, all documented and all in the period 2–21 Sep:

2 Sep 1995 Adams Co (Gubanyi 1996c)

6 Sep 1998 Oliver Res (SJD)

8 Sep 1997 Garden Co (WRS, JS; Brogie 1998)

12 Sep 1997 Cheyenne Co (JGJ)

17 Sep 1911 Dawes Co, specimen cited above

21 Sep 1997 Scotts Bluff Co (SJD, DCE)

21 Sep 1997 Oliver Res (SJD, DCE)

The specimen collected by Zimmer 17 Sep 1911 was an adult male in fresh fall plumage (Zimmer 1945), as were the 1997 and 1998 records; the species is the only *Empidonax* flycatcher expected in Nebraska that molts into fresh, brightly colored plumage prior to fall migration. Leukering (pers. comm. WRS) found that about 20% of *Empidonax* flycatchers banded after 20 Aug at his station in eastern Colorado were Hammond's.

COMMENTS

Data from Kansas suggest that, at least in fall,

migrants stray east as far as western Kansas and may be expected in western and even central Nebraska. There are 5 fall specimens for Ellis Co in northwest Kansas in the period 8 Sep–4 Oct (Thompson and Ely 1992).

FINDING

This species should be looked for in Sep in the Panhandle and southwest Nebraska. It is noteworthy that Hammond's Flycatcher molts prior to fall migration, when it is in fresh plumage, in contrast to other *Empidonaces*, which are usually in worn plumage in fall. This characteristic can be an aid in identification (Kaufman 1990).

Gray Flycatcher

Empidonax wrightii

STATUS

Accidental in spring.

DOCUMENTATION

Description

17 May 1999 Oliver Res, Kimball Co (SJD).

Records

One in fresh spring plumage was carefully studied at Oliver Res 17 May 1999 (SJD, BP, LP).

COMMENTS

Gray Flycatcher breeds north and east to southeastern and western Colorado and southwestern Wyoming, but there are a few records east of the breeding range, especially in spring. In Colorado there are 8 spring records and 2 for fall; none are far from the foothills, however. There are several Kansas reports but only 1 confirmed record, a specimen collected in Morton Co 29 Apr 1967, and several unconfirmed reports for Sep (Thompson and Ely 1992).

FINDING

This species should be looked for in arid habitats in late May in the southwestern Panhandle.

Dusky Flycatcher

Empidonax oberholseri

STATUS

Hypothetical in spring.

COMMENTS

One was reported 17 May 1992 in Kimball Co (WRS); the documentation provided the NOURC was judged Class IV (Hypothetical) due to the difficulty of identifying silent *Empidonax* flycatchers.

This species breeds in the Black Hills of South Dakota (Peterson 1990; Johnsgard 1979) and would be expected to migrate through the Nebraska Panhandle. The earliest South Dakota records are 11, 20, and 25 May (SDOU 1991). In Kansas it is a "low-density transient" in the west, with specimens collected 29 Apr–17 May; it was considered probably the "most regular of the western *Empidonaces*" in Kansas (Thompson and Ely 1992). There are, however, no specimens from the eastern plains of Colorado (Andrews and Righter 1992), although Leukering (pers. comm. WRS) found that as many as 55% of fall *Empidonax* flycatchers banded at his station at Barr Lake in eastern Colorado after 20 Aug were Dusky Flycatchers.

Cordilleran Flycatcher

Empidonax occidentalis

STATUS

Rare casual spring and fall migrant west. Uncommon regular breeder northwest.

DOCUMENTATION

Recording: 9 Jul 1991 Sowbelly Canyon, Sioux Co (WRS; Grenon 1991).

TAXONOMY

This account includes reports of "Western" Flycatcher prior to 1989, when it was concluded that 2 species were involved, Pacific-slope Flycatcher (*E. difficilis*) and Cordilleran Flycatcher (Johnson and Marten 1988). These species are difficult to identify in the field, as even their vocalizations may be equivocal. As Leukering (pers. comm. WRS) stated, "On the Colorado plains, I'm not even willing to identify silent migrant 'Westerns.' This is because I believe it is too likely that Pacific-slope could be regular over here, where, 30 miles [50 km] east of the foothills,

Cassin's Vireo is more common in fall than is Plumbeous!"

DISTRIBUTION AND ECOLOGY

Cordilleran Flycatcher is known to occur only in Pine Ridge canyons, where it is found in shaded, steep-sided locations with riparian deciduous shrubs.

Cordilleran Flycatcher probably migrates through at least the Panhandle and possibly eastward to the 100th meridian in the periods 10–31 May and 28 Aug–9 Sep. Peterson (1990) stated that it is resident in the Black Hills of South Dakota 15 May–1 Sep.

Spring

There are no documented records of migrant Cordilleran Flycatchers for Nebraska. There are a few reports that may be correct, and, as Johnsgard (1980) suggested, the species "is probably more common as a migrant than these few records suggest."

Reports away from the Sioux Co breeding range are 6 May 1994, 10 May 1986, 11 May 1995, and 12 May 1994 Scotts Bluff Co; 31 May 1982 Garden Co; and 11 Jun 1966 McPherson Co.

Summer

There are few confirmed nesting records. Rosche (1982) discovered a nest with 4 young in Sowbelly Canyon 17 Jul 1974 (NBR 43:18; AB 28:920). The nest was on a cliff face in the upper canyon, at a location where a singing male was found for the first time 4 Jul 1973 (Rosche 1982). At this location an adult was feeding 3 fledged young 5 Aug 1995 (NBR 63:106; FN 50:76).

In recent years Cordilleran Flycatcher has been reported from Sowbelly Canyon most summers in the period 28 May (1988) through 5 Aug (1995). Grzybowski (AB 46:1152) surmised that because this "isolated population" has been reported "so persistently over the years, it (or they) are more likely representatives of a larger breeding group." Indeed, in recent years there have been several reports elsewhere in western Pine Ridge canyons: 2 Jul 1987 Dawes Co (AB 41:1457),

31 May 1997 in East Hat Creek Canyon, and singing birds in side canyons of Monroe Canyon 20 Jun 1993 (AB 47:1123), 8 Jun 1994 (NBR 64:98), May–Jun 1996, and 31 May 1997. It was reported in the Wood Reserve, Sioux Co, 24 June 1994 (NBR 64:98); in Smiley Canyon 31 May 1997; and in West Ash Canyon 22 Jun 1999.

Fall

Fall reports away from Sioux Co are 9 Aug 1995 Garden Co (NBR 63:106; FN 50:76), 28 Aug–8 Sep 1979 Garden Co (banded FZ), 9 Sep 1978 Garden Co (banded FZ), 10 Sep 1994 Scotts Bluff Co, and what was apparently Nebraska's first report of this species, 6 Sep 1961, in the Wildcat Hills of Scotts Bluff Co or Banner Co (NBR 30:14, 42). No identification details were published for this latter report, but the date and location are appropriate. There is an undocumented report Sep 1967 McPherson Co (Rapp and others 1958). Three fall Kansas specimens are in the period 3–5 Sep; one was collected as far east as Ellis Co (Thompson and Ely 1992).

FINDING

Cordilleran Flycatcher is not hard to find when singing in its restricted breeding range in upper Sowbelly and Monroe Canyons; best time is probably late Jun. Migrants are possible anywhere in the Panhandle in the second half of May and the first week of Sep.

Eastern Phoebe

Sayornis phoebe

STATUS

Common regular spring and fall migrant east, becoming uncommon west. Fairly common regular breeder north and east, becoming rare south and west.

DOCUMENTATION

Specimen

UNSM ZM6367, 21 Mar 1903 Lancaster Co.

DISTRIBUTION AND ECOLOGY

This species nests statewide but is least numerous in the Panhandle. Nests are typically constructed on a substructural feature of a bridge that has water beneath it, either a pond or stream. Abandoned or little-used buildings may be used as well. In the Niobrara Valley and at Valentine it breeds exclusively under bridges, which seems to be true in the northwest also (Rosche 1982). Short (1961) noted that where Say's Phoebe occurred with Eastern Phoebe at Valentine, Say's Phoebe utilized buildings only, while Eastern Phoebe was found at bridges.

Spring

4, 4, 9 Mar→summer (south, east)

20, 24, 24 Mar→summer (north, west)

Eastern Phoebe is the earliest flycatcher in the state, arriving in mid-Mar. Arrival in the north and west is in late Mar.

Summer

bbs data indicate that Eastern Phoebe is most common in the north and east. Although there are summer records south of the Platte River, densities are low, especially as far west as Dundy and Chase Cos. Lingle and Hay (1982) and Lingle (1994) did not list this species as a breeding bird in the central Platte Valley, and it is but a rare breeder along the South Platte River in northeast Colorado (Andrews and Righter 1992).

In the north it is common westward at least to eastern Cherry Co (Brogie and Mossman 1983) and Valentine (Short 1961). It is very rare in the western Sandhills (Rosche 1982), and there are few Jun or Jul reports for the Sandhills west of Boone Co: 10 Jun 1974 and 14 Jun 1984 McPherson Co, 30 Jun 1964 Logan Co, and 30 Jun 1971 Custer Co.

In the Panhandle it breeds in small numbers in Pine Ridge canyons and along tributaries of the White and Niobrara Rivers (Rosche 1982) but is rare along the North Platte Valley west of Keith Co. While it is a rare summer resident in Keith Co (Rosche and Johnsgard 1984; Rosche 1994), there are only these Jun and Jul reports farther west: 10 Jun 1990 Garden Co, 1 Jul 1968 Scotts Bluff Co, and 1 Jul 1981 Garden

Co. It has not bred successfully in Wyoming (Scott 1993).

Nebraska nest card data (N = 14) show egg dates in the period 20 Apr–11 Jul; adults feeding fledged young have been observed as late as 24 Jul, and nesting birds had departed territories in Lancaster Co by 30 Jul in 1995.

Fall

summer→19, 20, 20 Oct

Departure is in late Sep, with later departure dates from the east. Eastern Phoebes may linger rather late: 29 Oct 1989 Sarpy Co, 9 Nov 1989 Sarpy Co (AB 44:116), 4 Nov 1928 Webster Co (Ludlow 1935), and 17 Nov 1991 Lancaster Co (AB 46:116).

High counts include 5 in southeast Washington Co 18 Sep 1999.

Other Reports

Feb 1931 Webster Co (Ludlow 1935), 1 Feb 1953 Holt Co.

FINDING

Eastern Phoebe is best looked for along country roads, especially in the east and north, where older wooden bridges cross small streams. Nesting birds in May and Jun usually flush, while spring migrants can be located by their distinctive song.

Say's Phoebe

Sayornis saya

STATUS

Common regular spring migrant west, uncommon central, rare casual elsewhere. Common regular breeder west, uncommon north, rare casual elsewhere. Common regular fall migrant west, rare casual elsewhere.

DOCUMENTATION

Specimen

UNSM ZM6369, 20 May 1900 Sioux Co.

DISTRIBUTION AND ECOLOGY

Say's Phoebe prefers arid habitats where it nests on abandoned or little-used structures, beneath eaves, under bridges, on ledges on cliff faces, and occasionally on rocky outcrops. Although it is broadly distributed in the Panhandle, there are isolated breeding locations elsewhere, especially the northeast. Distribution in the northeast has fluctuated considerably during the 20th century, extending as far east as Cedar, Polk, and Cuming Cos and south to Adams and Hamilton Cos. During periods of population decline it has been limited essentially to the Panhandle. At present its range seems to be expanding.

Migrants occur in open habitats, often somewhat east of the breeding range, but are casual south and east of Polk Co.

Spring

18, 20, 20 Mar→summer

Migrants occur east of the breeding range with some regularity, except for the southeast, but are not numerous. Arrival is in late Mar, although there is an earlier report 15 Mar 1987 Polk Co. There are few reports south and east of Polk Co: 28 Apr 1955 Lancaster Co, 28 Apr 1988 Douglas-Sarpy Cos, 2 May 1952 Saline Co, 15 May 1966 Lancaster Co, and 23 May 1991 Sarpy Co (NBR 59:96).

High counts include 18 in western Banner Co 4 May 1997 and 11 in southern Kimball Co 8 May 1998.

Summer

Say's Phoebe breeds commonly throughout the Panhandle, although Rosche (1982) noted that records in the sandhills of Sheridan and Garden Cos are "scarce" and limited to migrants. There was an unsuccessful nesting attempt at Crescent L NWR in 1980 (AB 34:909).

Since 1977 the only Jun and Jul reports in the east are from Polk Co, where Say's Phoebes were present each year in the period 1984–89 (see also AB 39:932); sightings in 2 summers from Cedar Co in 1988–89; and nesting reports from Hamilton Co in 1985 (Bennett 1986) and York Co in 1986 (Bennett 1987). At least 1 bird has summered in southwest Dixon Co 1996–99 (JJ). Say's Phoebe has not bred south and east of Polk Co.

Other regular summer reports in recent years

away from the Panhandle are limited to Lincoln, Keith (Rosche 1994a), and Chase Cos, although it "apparently summers each year" in Adams Co (AB 39:821). There is a Jul 1981 record from Howard-Hall Cos; an immature that may have hatched locally was in Phelps Co 26 Jun 1994; and a single was in Phelps Co 6 Jul 1996 (FN 50:966). In the north the only recent reports are 2 sightings in 1982 in Brown and Cherry Cos (Brogie and Mossman 1983).

There is some evidence that easterly nesting is sensitive to wet and dry 20-year weather cycles. There were many summer records in the 1960s, even as far east as Cuming Co (Ducey 1988; Short 1961), and again in the 1980s (see above), but few in the 1970s and 1990s.

Nebraska nest card data (N = 4) show egg dates in the period 18 May–4 Jul.

Fall

summer→28, 28, 29 Sep

Departure is in late Sep; the only Oct reports are 3 Oct 1959 Webster Co, 8 Oct 1995 Thomas Co, 10 Oct 1962 Webster Co, 19 Oct 1965 Lincoln Co, and 29 Oct 1964 Lincoln Co. Say's Phoebe is very rare in fall away from the Panhandle, with most of the few records from known breeding locations. This is especially true in the east, where there few reports: 5 Aug 1984, 6 Aug 1988, 18 Aug 1985, 6 Sep 1987, and 22 Sep 1991 Polk Co; 20 Sep 1999 Branched Oak L (LE); and 21 Sep 1989 and 23 Sep 1988 Cedar Co (AB 43:126). There are very few additional fall reports away from the Panhandle and Lincoln Co: 1 Sep 1971 Brown Co, 2–22 Sep 1985 McPherson Co, 10–11 Sep 1994 Thomas Co, 14 Sep 1989 Hall Co, and 30 Sep 1986 McPherson Co.

Other Reports

2 Mar 1950 Adams Co.

FINDING

Say's Phoebe is best looked for in early May in the Panhandle, where migrants are commonly seen in open areas. Breeding birds are somewhat scattered, but territories are established by early Jun at most sets of abandoned buildings.

Vermilion Flycatcher
Pyrocephalus rubinus
STATUS
Rare casual spring and fall visitor central and
  east.
DOCUMENTATION
Photograph
16 May 1995 Lancaster Co (Gubanyi 1996c).
Spring
There are 3 reports, 1 documented. A single bird
  was photographed (cited above) in Lancaster
  Co 16 May 1995. Another was reported in
  Lincoln Co 27 May 1976 (AB 30:861), and a
  pair was seen by veteran out-of-state birders
  at the fish hatchery at Valentine 29 Jun 1980
  (BP, LP).
Fall
There are 3 reports, including a documented
  record; all are in late fall, Oct–11 Dec. A
  newspaper clipping adequately described an
  adult male that was found near Carter L in
  Douglas Co 21 Nov 1954 (Grenon 1991; Rapp
  and others 1958). An immature male that had
  been present on the grounds of the University
  of Nebraska Agricultural Experiment Station
  near North Platte since Oct was collected 11
  Dec 1954 (Rapp and others 1958; Bray and
  others 1986). The specimen was mounted and
  deposited at the HMM, where it was observed
  by Burton Nelson, but the specimen cannot
  now be located (Bray and others 1986). A third
  Nov report was of headless remains found at
  North Platte 4 Nov 1960 and identified as this
  species (Viehmeyer 1961a).
COMMENTS
Vermilion Flycatcher has a tendency for
  wandering north on the western Great
  Plains, usually early in spring and late in fall
  (Andrews and Righter 1992; Robbins and
  Easterla 1992; Thompson and Ely 1992; Kent
  and Dinsmore 1996).
FINDING
This species should be looked for in riparian

situations in the southwest, such as the
State Fish Hatchery in Dundy Co, in May
and possibly in Nov. Any brown-plumaged
flycatcher seen in Nov should be carefully
checked.

Ash-throated Flycatcher
Myiarchus cinerascens
STATUS
Accidental in fall.
DOCUMENTATION
Description
29 Aug 1987 Dawes Co (Mollhoff 1989a).
Records
The single record is of a bird seen near the White
  River north of Chadron 29 Aug 1987 (RCR,
  DJR; Mollhoff 1989a; AB 42:99).
COMMENTS
Ash-throated Flycatcher has occurred several
  times on the northeastern Colorado plains in
  May and Sep (Andrews and Righter 1992). In
  southwestern Kansas there are several records
  in the period 29 Apr–16 Sep, including some
  nesting records (Thompson and Ely 1992).
Bray and others (1986) noted that one was
  reported to the editor of NBR in 1957, but
  the editor cautiously did not publish details.
  The report cited by Ducey (1988) of a set
  of eggs collected in Otoe Co in 1907 is, as
  suggested by Ducey, an obvious error; these
  eggs presumably are those of Great Crested
  Flycatcher.

Great Crested Flycatcher
Myiarchus crinitus
STATUS
Common regular spring and fall migrant and
  breeder east, becoming uncommon west.
DOCUMENTATION
Specimen
UNSM ZM6363, 19 Jul 1901 Beatrice, Gage Co.
DISTRIBUTION AND ECOLOGY
This species occurs in mature deciduous
  woodland, both riverbottom, usually mature
  cottonwoods, and upland. It is a cavity nester,
  using woodpecker holes or natural cavities

(Dobkin 1994). While it is most numerous in the east, it occurs in decreasing numbers westward, occurring only in small numbers in the Panhandle (Rosche 1982; Short 1965).

Spring

20, 21, 23 Apr→summer (except west)

9, 10, 11 May→summer (west)

Migrants are most numerous in the Missouri Valley. Arrival is in late Apr, although peak migration probably is around 2–15 May (Johnsgard 1980). There are undocumented earlier reports 12 Apr 1959 Gage Co, 14 Apr 1962 Hamilton Co, and 16 Apr 1977 Douglas Co. It arrives much later in the Panhandle, in mid-May, although there is an earlier report 1 May 1959 Scotts Bluff Co.

High counts include 39 in Sarpy Co 11 May 1996 and 29 in Sarpy Co 13 May 1995.

Summer

The 20th century has seen an expansion of the breeding range westward along major river valleys as riparian woodlands have spread west and matured and component trees have developed natural cavities. Bruner and others (1904) considered Great Crested Flycatcher a common summer resident only along the Missouri and its larger tributaries in the east; Rapp and others (1958) noted that it occurred as far west as Keith and Logan Cos. Short (1961) stated that it was "apparently spreading westward" and found it in the South Platte Valley west to Crook, Colorado. In Colorado it has been recorded in summer along the South Platte since the 1970s, with breeding first recorded in 1988, and is "definitely increasing in numbers" (Andrews and Righter 1992).

Currently, it is common in the Niobrara Valley west to eastern Cherry Co (Brogie and Mossman 1983), occurring only in small numbers farther west. Short (1965) collected a male in breeding condition 5 Jul 1964 south of Rushville. It was considered a rare migrant and local summer resident in the northwest (Rosche 1982), where it has occurred as a migrant since 10 May 1953 and in summer since 18 Jun 1961. It appears to be increasing

in Pine Ridge canyons; 2 were in Sowbelly Canyon 8 Jun 1996, and 3 were there 2 Aug 1996. There is a report 12 Jun 1992 Sheridan.

It occurs regularly in the Platte Valley west at least to Keith Co, where it is a "local" summer resident (Rosche 1994a; Brown and others 1996), and along the South Platte to Colorado (Short 1961). West of Keith Co there are few reports. There were 2 pairs at Crescent L NWR in 1981, although nesting was not confirmed (AB 35:956); single birds at Oshkosh 16 Jul 1987 (AB 41:1457) and in Scotts Bluff Co 12 Jul 1981; and 2 at Stateline Island 6 Jun 1995. Scott (1993) cites only 2 records for Wyoming.

While it occurs regularly in the eastern Loup drainage, including Boone and Custer Cos, in recent years it has become more common in the western parts of the drainage. Indeed, at Bessey Division, NNF, it is now a "common summer resident" (Bray 1994). Previously, western Loup drainage reports had been few: 8 Jun 1957, 12 Jun 1963, and 30 Jun 1965 Logan Co and 30 Jun–Jul 1987 and 8 Jun–1 Jul 1988 McPherson Co.

It summers throughout the Republican Valley (Johnsgard 1980; Rosche 1994b), although numbers decline west of Harlan Co, with few reports: 3 Jun 1992 Frontier Co; 24 May 1987, 17 May 1992, and 3 Jun 1992 Red Willow Co; and 3 Jun 1989 and 25 Jun 1989 Dundy Co.

Fall

summer→3, 4, 6 Oct

Migrants depart by the end of Sep, with late reports in early Oct, although there are later reports 9 Oct 1985 (Rosche 1982) and 22 Oct 1985 Lancaster Co.

This species is rare in the Panhandle in fall, with few reports: 16 Aug 1989 Sioux Co, 22–24 Aug 1990 Sioux Co, one netted 24 Aug 1978 Garden Co (AB 33:193), 7 Sep 1981 Garden Co, 11–13 Sep 1985 Scotts Bluff Co, and 9 Oct (Rosche 1982).

Other Reports

21 Jan 1958 Nemaha Co, 24 Mar 1959 Thayer Co, 31 Mar 1975 Douglas-Sarpy Cos, 4 Apr 1948 Webster Co, 7 Apr 1957 Hamilton Co.

FINDING

Great Crested Flycatcher can be located in the Missouri Valley at Fontenelle Forest or Indian Cave SP by its calls; it is most vocal in May and Jun and is easily attracted by squeaking.

Cassin's Kingbird

Tyrannus vociferans

STATUS

Fairly common regular spring migrant west, rare casual central. Fairly common regular breeder west. Fairly common regular fall migrant west, rare casual central.

DOCUMENTATION

Specimen

UNSM 6360, 19 Sep 1919 Glen, Sioux Co. This specimen was stated by the collectors to have been taken 9 Sep (Mickel and Dawson 1920).

DISTRIBUTION AND ECOLOGY

Migrants occur in open grasslands and open areas with widely scattered low trees and shrubs, mainly in the Panhandle. They are most numerous in fall. Summer birds are found in widely scattered pine parkland or at the edge of pine woodland that abuts expanses of open grassland. A preference for "narrow fingerlike projections of ponderosa pine woodland into the surrounding prairie" has been noted (Huser 1986). Although this habitat is broadly distributed in the Panhandle and along the Niobrara Valley in north-central Nebraska, Cassin's Kingbird populations have so far been restricted in summer to areas of moderate relief near canyons in the Pine Ridge and in Banner, Scotts Bluff, and Morrill Cos (Rosche 1982, 1994b).

Spring

27 Apr, 4, 4 May AB 39:321)→summer

This species is not often seen away from known breeding locations, with these few reports: 8 May 1953 Keith Co, 11 May 1985 Garden Co (Rosche 1994a; AB 39:321), 12 May 1999 (4) southwest Kimball Co (SJD), 17 May 1964 Keith Co, and 21 May 1961 Lincoln Co.

Summer

Summer reports are from the Panhandle, east to Sheridan Co in open ponderosa pine woodland. It is considered a "fairly common" breeder in canyons with scattered pines south of Redington in Morrill Co (Rosche 1994b). Two birds in Smiley Canyon 8–10 Aug 1996 probably summered there. Adults were feeding young in Kimball Co 21 Jul 1994. In Sheridan Co two were seen north of Hay Springs 4–15 Jun 1986 (Huser 1986) and 12 Jun 1992. The 1986 Sheridan Co birds, apparently a pair, were carrying nesting material, but breeding was not confirmed (Huser 1986). The only 2 documented nestings for South Dakota are from Shannon and Bennett Cos (SDOU 1991), immediately north of Sheridan Co.

This species was not recorded by Bruner and others (1904); the first indication of its presence apparently was a report by Mickel and Dawson (1920) that it was "common in northwestern Nebraska in September 1919." Subsequently it was reported in Scotts Bluff Co 27 Apr 1952 (undocumented) and Keith Co 8 May 1953 (Huntley 1953) but was not regularly reported until the 1970s. Brown and others (1996) cited an observation 3 Jul 1996 in Keith Co and suggested that this species may be overlooked among the many Western Kingbirds in the area.

High counts include 10 in canyons south of Redington 3 Aug 1996, 7 there 11 Jul 1998, 6 there 22 Jul 1995, and 6 in Smiley Canyon 25 Jun 1999.

Fall

15, 19, 21 Aug→25, 26, 27 Sep

Migrants may be numerous at times (see below). Rosche (1982) noted that the Cassin's Kingbird migration peak in fall is rather later than that of Eastern and Western Kingbirds; most "yellow-breasted kingbirds seen after about 10 Sep tend to be Cassin's" (Rosche 1994b).

Away from the Panhandle there are few reports: 11 Sep 1963 and 26 Sep 1963 Lincoln Co.

High counts include 43 in Banner Co 7 Sep 1975 (Rosche 1982), 43 there 13 Sep 1991 (AB 46:116), and 27 in southwest Kimball Co 20 Sep 1998.

FINDING

To find this species, the most dependable location in Jun and Jul is the ponderosa pine canyon directly south of Redington, where there is a consistent breeding population. Check for migrants in open areas with scattered trees in the western Panhandle during the first half of Sep.

Western Kingbird

Tyrannus verticalis

STATUS

Common regular spring and fall migrant and breeder west, becoming fairly common east.

DOCUMENTATION

Specimen

UNSM ZM13464, 10 May 1903 Lancaster Co.

TAXONOMY

Hybridization between Western Kingbird and Eastern Kingbird may have occurred in Keith Co in 1994, when adults of each species were tending a nest; the nest was destroyed by grackles before the plumages of the young could be determined (Brown and others 1996).

Matings with Scissor-tailed Flycatcher are discussed under that species.

DISTRIBUTION AND ECOLOGY

Western Kingbird is found statewide in open edge situations, including pines (Rosche 1982), but requires trees for nesting sites. Rosche (1982) noted that it occupies drier areas than does Eastern Kingbird, "including interiors of cities and towns," and at Bessey Division, NNF, it occurs primarily in prairie areas rather than in riparian habitats as does the more common Eastern Kingbird (Bray 1994). In the east, in addition to its more natural habitat, it may be found in many open areas of towns and cities, including the edges of freeway rights-of-way, large parking lots, and suburban parks with sparse woody vegetation.

Spring

8, 8, 8 Apr→summer (except west)

25, 25, 26 Apr→summer (west)

Migrants occur statewide. Data in Johnsgard (1980) indicate that peak migration occurs 1–10 May. While arrival is generally in mid-Apr, there are earlier reports 31 Mar 1971 Greeley Co, 1 Apr 1974 Adams Co, 2 Apr 1965 Scotts Bluff Co, 4 Apr 1939 Lincoln Co, and 5 Apr 1954 Boone Co. There are 3 records in Mar–early Apr for Colorado (Andrews and Righter 1992).

Arrival in the Panhandle is in late Apr, although there are earlier reports 2 Apr 1965 Scotts Bluff Co and 8 Apr 1954 Dawes Co.

High counts include 49 in Morrill and Box Butte Cos 12 May 1995 and 37 in Dixon Co 22 May 1994.

Summer

Western Kingbird has apparently spread east across Nebraska in the 20th century as a nesting species (Swenk and Dawson 1921; Squires 1928). Bruner and others (1904) stated that it occurred east to about the 100th meridian and to the mouth of the Niobrara River in the north, and farther east it was but a rare migrant. Rapp and others (1958) and Johnsgard (1980) indicated that while it was a common summer resident throughout the state, it was more common in the west. BBS data suggest, however, that numbers of this species are now relatively even across the state (NBR 46:38).

Nest card data (N = 8) show egg dates in the period 4 May–25 Jun.

Fall

summer→>30 Sep, 1, 1 Oct (except east)

summer→22, 27, 27 Sep (east)

Data in Johnsgard (1980) indicate peak migration is around 24 Aug–10 Sep. Most have departed by mid-Sep, about a week earlier in the east, although this may be an artifact resulting from late fall departure in the west of Cassin's Kingbird, which can easily be passed off as a Western Kingbird. Rosche (1994b) stated that "most yellow-breasted kingbirds seen after about 10 Sep tend to be Cassin's." There is a later report 6 Oct 1981 McPherson Co.

High counts include 55 southwest of Lowell 22

Aug 1999, 50 in Scotts Bluff Co 27 Aug 1995, and 35 at North Platte NWR 1 Sep 1996.

Other Reports

10 Mar 1972 Brown Co, 10 Nov 1967 Dawes Co (NBR 36:66), 11 Nov 1993 Sarpy Co.

FINDING

Western Kingbird is easily found statewide in summer, usually on fence wires.

Eastern Kingbird

Tyrannus tyrannus

STATUS

Common regular spring and fall migrant and breeder statewide.

DOCUMENTATION

Specimen

UNSM ZM11592, 10 May 1892 Jamaica, Lancaster Co.

TAXONOMY

See Western Kingbird.

DISTRIBUTION AND ECOLOGY

Eastern Kingbird occurs statewide, with breeding birds found in most types of edge habitats, including roadsides, windbreaks, fence lines, and early successional vegetation along water edges. They avoid ponderosa pine woodlands in the Panhandle but may be associated with nearby deciduous habitat. Rosche (1982), in comparing the habitats of the kingbirds, noted that in the northwest Eastern Kingbirds prefer "wetter, more eastern-like habitats" than do Western Kingbirds.

Spring

2, 2, 2 Apr→summer

Arrival is in early Apr, although there are earlier reports 24 Mar 1959 Thayer Co, 25 Mar 1972 Hall Co, and 31 Mar 1971 Greeley Co. There are very early undocumented reports 10 Feb 1971 and 20 Feb 1972 Brown Co.

High counts include 45 in Dixon Co 21 May 1994, 49 in Pierce Co 8 May 1999, and 45 in Sarpy Co 13 May 1995.

Summer

This species is generally more numerous than Western Kingbird, especially in the north and east, and is least numerous in the Panhandle.

Nebraska egg dates are in the range 3 Jun–26 Jul (N = 33).

Fall

summer→2, 3, 3 Oct

Data in Johnsgard (1980) suggest peak migration around 1–16 Sep. Departure is in late Sep, although there are later reports 7 Oct 1980 Adams Co, 12 Oct 1965 Adams Co, 15 Oct 1978 Douglas Co, 21 Oct 1986 Pierce Co, 31 Oct 1981 Thomas Co, and, undocumented, 21 Nov 1993 Sarpy Co.

High counts include 150 near Gibbon 23 Aug 1999, 60 at Crescent L NWR 30 Aug 1998, and 60 at Harlan Co Res 30 Aug 1998.

FINDING

Eastern Kingbird is conspicuous and easily found anywhere in the state during the summer months, most often on fence wires.

Scissor-tailed Flycatcher

Tyrannus forficatus

STATUS

Rare casual breeder southeast. Rare casual summer visitor statewide.

DOCUMENTATION

Specimen

KU 34091, 9 Jul 1957 Thomas Co.

TAXONOMY

There are 2 reports of this species mating with Western Kingbirds. A female mated with a male Western Kingbird in Lincoln in 1921, but the nests were abandoned and taken over by House Sparrows, and a male that had been present in the summers of 1926 and 1927 in Logan Co mated with a female Western Kingbird one of the years (Glandon and Glandon 1937).

DISTRIBUTION AND ECOLOGY

This species occurs in open grasslands, usually native, with scattered trees. Most reports are from the southeast, but it may appear anywhere. Nesting is usually in isolated trees or tall structures such as power poles and windmills (Johnsgard 1980).

"Migrant" usually indicates passage through an area, but since the breeding range of this species is essentially south of Nebraska, spring birds are probably overshooting spring migrants, and fall birds are likely postbreeding wanderers from within or south of Nebraska.

Spring

There are about 25 reports in the period 20 Apr–31 May, including a northerly report 13 May 1982 Brown Co (Brogie and Mossman 1983).

Summer

The few reports of nesting are from the area south of Cass, York, and Adams Cos, except for a single Logan Co report for 1926–27 cited by Ducey (1988), who listed reports prior to 1945 for Lancaster (2), Logan, and Adams Cos.

Nesting at the northern edge of the breeding range may be related to dry and wet weather cycles, as the breeding occurrences since 1945 are in 2 periods, 1958–65 and 1989–91. Reports for the earlier period were of a pair in Gage Co observed 28 Jul–19 Aug 1958 which fledged young (Sturmer 1959), a nest with young found in York Co (not Platte Co as cited by Ducey 1988) 16 Aug 1959 (Armstrong 1960), and a nest with eggshells below it found in Clay Co with birds observed in the area 1–20 Jun 1965 (Evans and Wolfe 1965). After a 30-year gap, more recent breeding reports are of a pair that nested unsuccessfully on a 15 m (50 ft) tall ballpark light standard in Cass Co in 1989 (AB 43:1335) and adults seen with young in Cass Co 13–25 Aug 1991 (NBR 60:33).

There are several Jun and Jul reports without breeding evidence, including these since 1980: Jun 1984 Lincoln (AB 33:897), "summer" 1984 Keith Co (Johnsgard 1990a), and Jul 1984 Bellevue (AB 33:897). Earlier summer reports are 1 Jun 1979 Lancaster Co, 4 Jun 1967 Gage Co, 5 Jun 1999 4.8 km (3 mi) south of Sumner, 10 Jun 1959 Gage Co, 11–12 Jun 1979 Keith Co (Rosche 1994a), 20 Jun 1957 Hamilton Co, 30 Jun 1964 Gage Co, 5 Jul 1979 Douglas-Sarpy Cos, 5–6 Jul 1965 Dawes Co

(Richardson 1965), 9 Jul 1957 Thomas Co (specimen cited above; Ford 1959), and 24 Jul 1963 Douglas Co.

Fall

There are about 25 reports in the period 13 Aug–14 Oct, including westerly reports 3 Sep 1970 Sheridan Co (NPG), 10 Sep 1988 Keith Co (AB 43:126; Rosche 1994a), 15–24 Sep 1980 Scotts Bluff Co, and 14 Oct 1981 Garden Co (Silcock and others 1986c).

FINDING

The most likely areas to search for this species are grasslands with scattered trees along the southern row of counties, including Pawnee, Gage, Jefferson, and Thayer Cos. Most reports are in May and Sep.

Laniidae (Shrikes)

Northern Shrike

Lanius excubitor

STATUS

Fairly common regular winter visitor north and west, rare south and east.

DOCUMENTATION

Specimen

UNSM ZM6747, Mar 1900 Lancaster Co.

TAXONOMY

Two subspecies have been recorded (Rapp and others 1958). *Invictus* is the western race and winters throughout Nebraska, while eastern *borealis* winters westward to eastern Nebraska and Kansas (AOU 1957). Phillips (1986) stated, however, that AOU's "recognition of *L. e. 'invictus'* . . . amusingly illustrates the disdain of zoo-politicians for mere facts"; Ridgway "long ago correctly synonymized *invictus*" with *borealis*.

Two specimens from Hall Co were *invictus*, while one from Dakota Co was an intergrade (NBR 1:64). A specimen, UNSM ZM11413, collected in Logan Co 28 Feb 1937 was thought to be the westernmost record of *borealis*, although Swenk considered it an intergrade, albeit closer to *borealis* than *invictus* (Glandon and Glandon 1937).

DISTRIBUTION AND ECOLOGY

This species is most numerous in the west and north, becoming rare in the southeast. It occupies open grassland and cropland where perching sites are available, such as scattered trees, windbreaks, and fence lines. Numbers vary somewhat from year to year.

Because Nebraska is near the southern limit of the usual wintering range, there is no noticeable movement through the state.

Spring

winter→11, 14, 15 May

Most depart by late Apr, but there are several May reports, including later undocumented reports 20 May 1957 Brown Co and 23 May 1982 Adams Co. Brownish immature Loggerhead Shrikes can be a source of confusion in shrike identification (Andrews and Righter 1992) and may account for undocumented midsummer reports of Northern Shrike 6 Jun 1959 Douglas Co, 30 Jun 1989 Dawes Co, and 6 July 1986 Lincoln Co.

Fall

12, 12, 14 Sep→winter

Arrival is mainly in mid-Oct, although there are earlier undocumented reports 28 Aug 1974 Scotts Bluff Co, 2 Sep 1976 Cherry Co, and 7 Sep 1982 Adams Co.

Winter

Northern Shrikes are found statewide but are rare in the southeast. The highest CBC single count total is 11 at both Crawford in 1976–77 and Norfolk in 1988–89. The highest overall total is 35 in 1988–89, and the lowest is 1 in 1980–81.

High counts include 15 near Scottsbluff 2 Jan 1999, 12 on the Calamus-Loup CBC 27 Dec 1998, and 9 in the Panhandle 1 Nov 1998.

FINDING

Northern Shrike is most easily found in open areas by driving roads and checking isolated trees and structures such as fences. The best time is probably Jan and Feb, when the population is largest, and the best areas to check are the western Sandhills and the Panhandle.

Loggerhead Shrike

Lanius ludovicianus

STATUS

Fairly common regular spring and fall migrant statewide. Fairly common regular breeder statewide. Rare regular winter visitor southeast, rare casual elsewhere.

DOCUMENTATION

Specimen

UNSM ZM6750, 5 Jun 1901 Warbonnet Canyon, Sioux Co.

TAXONOMY

Most of the breeding Loggerhead Shrikes in Nebraska are probably intergrades of the western *excubitorides* and the eastern *migrans* (Rapp and others 1958). Those in the Panhandle may be fairly pure *excubitorides*, because migration timing and propensity to winter differ from eastern birds (see below).

DISTRIBUTION AND ECOLOGY

Optimum breeding habitat is open areas with scattered small trees or hedgerows that provide nesting sites. While Loggerhead Shrikes are found statewide in summer, breeding habitat is probably less disturbed where agriculture is less intensive, so there are somewhat more Loggerhead Shrikes in the north and west. Similar habitat is occupied in winter; the few wintering birds are found mostly in the southeast.

Spring

11, 12, 15 Mar→summer (west)

9, 9, 10 Feb→summer (elsewhere)

Arrival dates in most of the state differ somewhat from those of the Panhandle, possibly due to subspecific differences in breeding populations (see Taxonomy). The presence of wintering birds tends to obscure arrival dates. There is, however, a concentration of arrival dates in mid to late Feb in areas where no wintering was reported.

In the Panhandle, where it is "best to treat all wintering shrikes as Northerns, unless convincing published details or collected specimens prove otherwise" (Rosche 1982), Loggerhead Shrikes arrive in mid-Mar, with earlier reports 1 Mar (Rosche 1982) and 3 Mar 1957 Scotts Bluff Co. Earlier undocumented reports, 14 Feb 1963 Dawes Co and 8 reports 1 Jan mostly from Dawes Co, are likely Northerns; there are no Dec records of Loggerhead Shrike for the Panhandle.

High counts include 29 in Keith Co 24 Apr 1999, 29 in the Panhandle 25 Apr 1999, 11 in southern Cheyenne Co 17 Apr 1998, and 11 in southern Kimball Co 8 May 1998.

Summer

Loggerhead Shrikes are evenly distributed throughout the state as breeding birds, although BBS data indicate slightly higher numbers in the north and west. Robbins and Easterla (1992) discussed the decline in Loggerhead Shrike numbers in some parts of the United States over the past 30 years or so, possibly due to habitat destruction and pesticide use. That these causes may be operative in intensively cropped eastern and southern Nebraska is suggested by the fact that the lowest breeding numbers are in that area. On the other hand, this species is holding its own in the extensive grassland areas of the state.

Nest building was under way for this early breeder 29 Mar 1998 in northern Lancaster Co.

High counts include 16 south of exit 1 on I-80 on 11 Jul 1998, 14 in southwest Kimball Co 8 Aug 1998, and 11 in Box Butte Co 16 Jun 1995.

Fall

summer ›28, 29 Sep, 1 Oct (Rosche 1982) (west)

Departure is gradual, with many records Dec–Feb. Most leave, however, in Oct and Nov, especially from the north, where there are very few documented reports after mid-Oct.

Departure in the Panhandle occurs in late Sep, although there are later undocumented reports 10 Oct 1970 Scotts Bluff Co, 19 Oct 1968 Scotts Bluff Co, 26 Oct 1994 Garden Co, 3 Nov 1995 Scotts Bluff Co, 11 Nov 1967 Scotts Bluff Co, 11 Nov 1996 Scotts Bluff Co, and 25 Nov 1996 Crescent L NWR. There are also reports of 5 birds seen on 3 CBCs at Whitney

and Chadron, with no details. These are probably Northerns, as none of the latter were listed for these counts.

Winter

This species can be found most winters but is generally restricted to the southeast. The only documented winter reports elsewhere are a specimen, UNSM ZM17545, taken 10 Jan 1997 in Butler Co (WM) and descriptions of one in Sherman Co 1 Jan 1996 (LR, RH; Brogie 1997) and another at Sutherland Res 13 Dec 1997 (JS). CBC data yield only 1 report from the north, at Loup City 18 Dec 1994, and a few reports from the Panhandle, none considered acceptable by Rosche (1982). A report from Carter Canyon 18 Dec 1993 was also undocumented. There is only 1 CBC report from north of Douglas Co in the east, an individual at DeSoto Bend NWR 22 Dec 1979. CBC reports in the south are from Grand Island eastward, so the primary wintering area appears to be south and east of Grand Island and Omaha. Even there, numbers of wintering birds are low; CBC counts held in the first or second week of the count period recorded only 72 birds in 7488 party-hours, about 1 bird per 100 party-hours.

FINDING

Loggerhead Shrikes are most easily found by driving back roads in relatively undisturbed areas with scattered small trees in May–Jun. At that time birds are actively feeding young and are usually seen flying across the road or perched on fencerows.

Vireonidae (Vireos)

White-eyed Vireo

Vireo griseus

STATUS

Rare casual spring migrant east, becoming accidental west. Rare casual summer visitor east. Rare casual fall visitor east.

DOCUMENTATION

Specimen

UNSM ZM11885, 19 May 1910 Sarpy Co.

DISTRIBUTION AND ECOLOGY

This species prefers dense undergrowth and thickets, usually adjacent to woodland and often near water. Recent reports are from the Missouri Valley and extreme southeast.

Spring

19, 21, 23 Apr→summer

There is an earlier record, an adult, banded in Bellevue 6 Apr 1992 (RG).

Since 1981 there have been 19 reports, most from Douglas and Sarpy Cos, but with these exceptions: 7 May 1986 Lancaster Co, 8 May 1983 Boone Co, 8 May 1987 Dakota Co, 9 May 1994 Keith Co (Brown and others 1996), and 10 May 1996 Lancaster Co.

Prior to 1982 White-eyed Vireos were reported more often, including reports for summer and fall, discussed below. The only reports away from the east (in addition to those above) are 6 May 1957 Adams Co; 20 May 1917, a specimen, HMM 2623, taken at Inland, Clay Co (Swenk, Notes before 1925); 26 May 1976 Dawes Co, the only Panhandle report (Rosche 1982); and 31 May 1974 Adams Co.

Summer

This species probably no longer breeds in Nebraska. Ducey (1988) cited no records of breeding after 1920, and the most recent summer reports are 5 Jun–4 Jul 1978 Lancaster Co, 20 Jun–24 Jul 1979 Douglas-Sarpy Cos, and 24 Jun 1981 Lincoln Co. These reports suggest nesting at that time, but no evidence was published.

Prior to 1978 there are numerous Jun and Jul reports from Douglas-Sarpy and Lancaster Cos, as well as one from Cass Co 1 Jun 1965. There is also an undocumented report from Brown Co 1 Jul–2 Aug 1964. A report of a pair in Thomas Co 16 Jun 1993 (Bright 1993) was subsequently withdrawn (Bright 1994).

Fall

summer→19, 22, 23 Sep

The most recent reports are 23 Sep 1982 Lancaster Co and 2 Oct 1994 Sarpy Co, neither documented. There were only 3 reports during the 1970s: 28 Aug–22 Sep 1976 Douglas-Sarpy Cos, 30 Aug 1974 Douglas-Sarpy Cos, and 28 Aug 1973 Lancaster Co.

FINDING

The best time to look for this species is the first half of May. It is usually easily detected by its distinctive song and is best looked for in dense riparian growth and thickets in the southern Missouri Valley. Likely places are Indian Cave SP and Fontenelle Forest.

Bell's Vireo

Vireo bellii

STATUS

Uncommon regular spring and fall migrant statewide. Uncommon, locally common, regular breeder statewide.

DOCUMENTATION

Specimen

UNSM ZM10541, 12 Aug 1907 (2 birds) Lincoln, Lancaster Co.

DISTRIBUTION AND ECOLOGY

Bell's Vireo is found statewide in dense thickets and hedgerows, often associated with river or stream valleys. Clearing of brush along watercourses has reduced breeding habitat in many areas, although Rosche (1994a) noted that "inadequate stream flow" in the North Platte channel in Keith Co has allowed dense riparian brush to return, benefiting this species in recent years. While there has been concern over a decline in numbers in recent years in the central United States (Robbins and others 1989), this species appears to be holding its own in Nebraska.

Spring

20, 20, 20 Apr→summer (east)

24, 26, 27 Apr→summer (central)

11, 15, 19 May→summer (west)

Arrival begins in the east, becoming progressively later westward. In the west earlier reports are in mid to late May, although there are reports 22–29 Apr 1967 Sioux Co and 2 May 1964 Scotts Bluff Co.

High counts include 34 at Calamus Res 24 May 1997 and 23 in Chase Co around 30 May 1999.

Summer

Although this species breeds statewide, it is most numerous in the Loup drainage and southeast and least numerous in the Panhandle, where it may have appeared as a breeding bird only recently. Rapp and others (1958) described it as a "summer resident in the eastern half of the state, reaching as far west as Thedford and north to Valentine." Rosche (1982) considered it a "rare to uncommon . . . summer visitant" with "no nesting evidence" in the northwest. Ducey (1988), however, cited breeding records for Scotts Bluff Co in 1964 and Dawes Co in 1966.

In the north the lower Niobrara Valley is a favored breeding area, as is the Loup drainage (Bray 1994), notably the Calamus Res area. Rosche (1994a) noted a "good nesting population" in the L Ogallala area, and Lingle (1994) considered it common in the central Platte Valley. It may have increased in numbers recently in the Platte Valley, as Lingle and Hay (1982) listed Bell's Vireo as "rare" at their study area in Hall Co. It is also numerous at Rock Creek L, Dundy Co.

In the Panhandle numbers are low due to scarcity of habitat. Johnsgard (1979) suggested that it is absent in most of the Panhandle except for the White River valley, and Rosche (1982) considered it rare to uncommon in the northwest. Birds were found at 4 Garden Co locations 15–17 Jul 1987 (AB 41:1457).

Tout (1947) listed egg dates for Lincoln Co (N = 20) in the period 28 May–19 Jun.

Fall

summer→5, 7, 7 Sep (north, west)

summer→1, 3, 4 Oct (south, east)

There are very few Panhandle reports: 1 Aug 1966, 7 Aug 1976, and 15 Aug 1992 Garden Co; 23 Aug 1991, 28 Aug 1978, and 17 Sep 1973 Sioux Co; and 28 Sep 1961 Scotts Bluff Co. Departure in the north is in late Aug and in the south and east late Sep, although there are later undocumented reports 8 Oct 1996 Otoe Co, 11 Oct 1983 Lincoln Co, and 13 Oct 1987 Lincoln Co.

High counts include 30 at Calamus Res 20 Aug 1995.

Other Reports

8 Mar 1953 Boone Co, 20 Mar 1975 Adams Co, 2 Apr 1967 Custer Co, 10 Apr 1972 Brown Co, 10 Apr 1978 Lancaster Co, 14 Nov 1966 Lancaster Co.

FINDING

This species is most easily found in May and Jun in the Loup drainage and in the southeast in dense brush. It responds to squeaking at this time of year and readily betrays its presence with its distinctive song.

Black-capped Vireo

Vireo atricapillus

STATUS

Accidental in spring.

COMMENTS

There are only 3 published references to the appearance of this species in Nebraska. Skow was said to have recorded it in the Bellevue area (Bruner 1896), and Trostler and Skow "thoroughly identified" a single bird near Bellevue 19 Jun 1894 (Bruner and others 1904). These reports appear to be of the same sighting; neither was documented.

The only record accompanied by tangible evidence is that of a specimen, HMM 2866, collected by Wallace and Brooking near Meadow in Sarpy Co 19 May 1921 and now on display at the HMM (Bray and others 1986). We have some reservations about the provenance of this specimen, given the existence of other unique specimens in the Brooking collection. Although Black-capped Vireo apparently was "quite common" and bred in south-central Kansas in the 1880s, it has not been acceptably recorded there since (Thompson and Ely 1992). A case could thus be made for a spring overshoot migrant during the 1880s, at about the time of the Omaha report cited above, but a spring overshoot seems extremely unlikely as late as 1921.

Yellow-throated Vireo
Vireo flavifrons
STATUS
Fairly common regular spring migrant east, becoming rare casual west. Fairly common regular breeder east. Fairly common regular fall migrant east, rare casual central.
DOCUMENTATION
Specimen
UNSM ZM10537, 28 Apr 1888 (2 birds) Peru, Nemaha Co.
DISTRIBUTION AND ECOLOGY
This species occupies mature deciduous forest in the Missouri Valley and to some extent along major tributaries such as the Platte and Niobrara Rivers.
Spring
15, 15, 19 Apr→summer
Arrival is in mid-Apr, with earlier reports from Douglas-Sarpy Cos 1 Apr 1991, 5 Apr 1977, and 8 Apr 1981. There is a very early undocumented report 24 Mar 1967 Custer Co.
Most of the few reports away from the breeding range, other than breeding season reports discussed under Summer, are of spring migrants, including these from the Panhandle: 25 May 1980 Dawes Co (Rosche 1982) and 29 May 1949 Dawes Co.
High counts include 17 in Sarpy Co 13 May 1995.
Summer
Breeding birds occur throughout the Missouri Valley. Ducey (1988) cited breeding in Dakota Co, and there are summer reports for Washington, Sarpy, Johnson, Nemaha, Pawnee, and Richardson Cos, where breeding presumably occurs also.
There is no evidence that it currently breeds anywhere outside the Missouri Valley, although singing males have been regularly noted in mature floodplain forest of the Platte Valley as far west as Saunders Co and in the Elkhorn Valley as far west as Cuming Co.

There are a few breeding season reports farther west in the Platte Valley, although with no details: 8 Jul–20 Aug 1988 Howard-Hall Cos, 30 Jun 1965 and 30 May–3 Aug 1986 Lincoln Co, and 30 Jun 1965 and 19 Aug 1965 Adams Co. Similar reports exist for the Loup drainage in Custer and Thomas Cos: 28 May 1968, 3 Jun 1978, and 1 Jul 1972 Custer Co and 22–24 May 1990 Thomas Co. The last record involved 2 birds thought to be a territorial pair foraging for 3 days in a cottonwood along the Middle Loup River near the entrance to Bessey Division, NNF; no evidence of nesting was noted (NBR 58:72). Bray (1994) listed this report as "hypothetical."
There is some indication that breeding might occur in Brown Co, where Short (1965) found 3 scattered singing males and a female during late May and Jun 1964 along Plum Creek northwest of Ainsworth. A male with enlarged testes was collected 17 Jun in open second-growth woods. There are subsequent summer records from Brown Co with no details: 30 Jun–1 Jul 1969, 30 Jun 1972, and 30 Jun 1973. Brogie and Mossman (1983), however, found no Yellow-throated Vireos in the area in 1982 and suggested that "breeding populations are unstable this far west."
The only summer report for the Republican Valley is 6 Jun 1925 Webster Co (Ludlow 1935), probably a migrant; it apparently does not breed in this drainage in Kansas (Thompson and Ely 1992).
Fall
summer→29, 30 Sep, 4 Oct
Departure occurs in late Sep, although there are undocumented later reports 14–27 Oct 1963 Lincoln Co and 17 Oct 1976 Douglas-Sarpy Cos.
Fall reports away from the breeding range are few: 1 Sep 1973 Howard-Hall Cos, 5 Sep 1967 Adams Co, 15 Sep 1982 Boone Co, 15–16 Sep 1974 Perkins Co, and 16–29 Sep 1972 McPherson Co.

FINDING

Yellow-throated Vireos are not hard to find due to their distinctive song in May and Jun along the Missouri Valley. Good places to look are Fontenelle Forest and Indian Cave SP.

Plumbeous Vireo

Vireo plumbeus

STATUS

Uncommon regular spring migrant west. Uncommon regular breeder west. Uncommon regular fall migrant west, accidental east.

DOCUMENTATION

Specimen

UNSM ZM6773, 21 May 1900 Sioux Co.

TAXONOMY

See Blue-headed Vireo.

DISTRIBUTION AND ECOLOGY

This species occurs in summer in small numbers in northwestern Sioux Co primarily in ponderosa pine woodlands with riparian deciduous trees and shrubs adjacent (Rosche 1982; Huser 1986). Migrants occur in any wooded habitat but are rarely noted away from breeding locations.

Spring

9, 9, 10 May→summer

Dates listed above are reports of "Solitary" Vireo from the west and are almost certainly of Plumbeous Vireo, although Cassin's Vireo is a very slight possibility in spring (Andrews and Righter 1992). There is an earlier report 23 Apr 1993 Sioux Co (AB 47:428). Migrants are uncommon (Rosche 1982), but 3 were at Oliver Res 17 May 1999 (SJD).

Summer

Small numbers breed in the western Pine Ridge, from the vicinity of Sowbelly Canyon westward (Rosche 1982). A photograph has been published of an incubating bird in a small elm in the transition zone between ponderosa pine and riparian deciduous woodland in upper Sowbelly Canyon 16 Jun 1986 (Huser 1986). A nest was being built in East Hat Canyon 21 May 1996, probably the most easterly nesting reported to date (BP, LP).

Fall

summer→13, 17, 21 Sep

Breeding birds apparently depart by early Sep, but few migrants are seen away from the breeding locations. Some Sep reports prior to 1997 were probably of Cassin's Vireo, which is more common in fall than in spring. Later reports are 24 Sep 1993 Sheridan Co and 12 Oct 1990 Keith Co, the latter reported as "*plumbeus*" (Rosche 1994a). A specimen, UNSM ZM6776, was taken 27 Aug 1921 in Sioux Co. A later report, 30 Oct 1977 Garden Co, may be of any of the 3 recently split species but is most likely Blue-headed, which tends to be an Oct migrant on the eastern Colorado plains (Andrews and Righter 1992).

There is a single record away from the Panhandle; one was documented at Wood Duck Area 21 Sep 1997 (DH; Brogie 1998).

FINDING

Plumbeous Vireo is not hard to find in Sowbelly Canyon in Jun; its burry song, similar to that of a Yellow-throated Vireo, indicates its presence.

Cassin's Vireo

Vireo cassinii

STATUS

Rare casual fall migrant west.

DOCUMENTATION

Specimen

UNSM ZM11878, 8 Sep 1911 northwest of Crawford, Sioux Co.

TAXONOMY

See Blue-headed Vireo.

DISTRIBUTION AND ECOLOGY

The minimal amount of data available indicates that this species migrates through the Panhandle in riparian woodland in fall.

Fall

To date there are 12 records, all in the period 15 Aug–21 Sep.

There are 2 specimens, 1 cited above and the other taken at the same location 13 Sep 1911, UNSM ZM11877 (JGJ, WRS; Brogie 1998).

In fall 1997 several sightings of the newly split

species were made: 24 Aug L Minatare (JS, SJD; Brogie 1998); 1 Sep Oliver Res (SJD; Brogie 1998); 6 Sep L Minatare (SJD, WRS, JS; Brogie 1998); 7 Sep Riverside Park, Scotts Bluff Co (SJD, WRS, JS; Brogie 1998); and 21 Sep Oliver Res (SJD; Brogie 1998). In 1998 there were again 4 sightings, all at Oliver Res: 29 Aug, 30 Aug, 6 Sep, and 12 Sep. Another was at Oliver Res 15 Aug 1999 (SJD), the earliest record to date. These sightings suggest that Cassin's Vireo may be a regular migrant, despite the paucity of prior Panhandle reports of vireos in the "Solitary" Vireo complex. Undocumented reports that may be this species (or Plumbeous, or remotely, Blue-headed) include 22 Aug 1994 Dawes Co, 26 Aug 1994 Dawes Co, 5 Sep 1973 Sioux Co, 11 Sep 1979 Garden Co, 11 Sep 1977 Garden Co, 17 Sep (Rosche 1982), and 23 Sep 1993 Sheridan Co.

Andrews and Righter (1992) believe that Cassin's Vireo, while rare, is more numerous than Blue-headed in eastern Colorado, based on specimen records, and that Cassin's tends to migrate in Sep and Blue-headed in Oct. There is only a single spring specimen for Colorado (Andrews and Righter 1992) but several sight reports for May.

FINDING

Cassin's Vireo should be looked for in riparian deciduous habitat, especially willows or willow-cottonwood mix, in the Panhandle in the period 1–20 Sep.

Blue-headed Vireo

Vireo solitarius

STATUS

Uncommon regular spring migrant east, rare casual central. Uncommon regular fall migrant east, accidental central and west.

DOCUMENTATION

Specimen

UNSM ZM6767, 13 May 1917 Lancaster Co.

TAXONOMY

"Solitary" Vireo was recently split into 3 species by the AOU (1997), based on studies by Murray and others (1994) and Johnson (1995).

All occur in Nebraska: Blue-headed Vireo, Plumbeous Vireo (*V. plumbeus*), and Cassin's Vireo (*V. cassinii*). See separate accounts for the latter 2 species. Rosche (1982) had no records of Blue-headed Vireo in the northwest; all, including migrants, were Plumbeous Vireos. Andrews and Righter (1992), however, cited a few reports of Blue-headed Vireo on the eastern plains of Colorado; most reports are in Oct, in contrast to reports there of Cassin's Vireo, which tend to be in Sep.

Field identification of the 3 species was discussed by Heindel (1996).

DISTRIBUTION AND ECOLOGY

During migration this species occurs in the east, where it is found in mature woodland and forest.

Spring

21, 22, 23 Apr→6, 10, 12 Jun

The above dates are reports of "Solitary" Vireo from the east, where Plumbeous and Cassin's Vireos are not expected to occur.

There are later undocumented reports 21 Jun 1973, 29 Jun 1972, and 30 Jun 1964 Douglas-Sarpy Cos.

The only documented record of Blue-headed Vireo away from the east is a specimen, #2708 in the Brooking collection, taken at Kearney 10 May 1914 (Brooking, Notes). It occurs with some regularity, although rarely, west to about Grand Island (Lingle 1994). A westerly report of "Solitary" Vireo, which was not specified as Plumbeous Vireo and thus may be a Blue-headed Vireo, was 18 May 1986 Keith Co (Rosche 1994a).

High counts include 23 in Sarpy Co 11 May 1996 and 18 there 13 May 1995.

Fall

28, 29, 29 Aug (specimen, NBR 62:142)→17, 17, 18 Oct

Migrants pass through the east between late Aug and mid-Oct, although there is an earlier report, 20 Aug 1983 Douglas-Sarpy Cos, and a few undocumented Jul reports, 1 Jul 1973, 4 Jul 1984, 10 July 1971, and 22 July 1965

Douglas-Sarpy Cos. Blue-headed Vireos may linger rather late in fall: 22 Oct 1973 Lancaster Co, 28 Oct 1997 Douglas-Sarpy Cos (BP, LP), 1 Nov 1987 Lancaster Co, 3 Nov 1972 Lancaster Co, and 7 Nov 1983 Lancaster Co.

There is only 1 Panhandle record, an individual photographed 27 Sep 1998 at Oliver Res (SJD), although a report 30 Oct 1977 in Garden Co may have been this species. Most of the few documented records on the eastern Colorado plains are in Oct (Andrews and Righter 1992). Lingle (1994) did not list any occurrences for "Solitary" Vireo in the Hall Co area in fall; apparently the only documented record in central Nebraska is one in Hall Co 25 Sep 1999 (LR, RH).

High counts include 6 at Wehrspann L 20 Sep 1999.

COMMENTS

Ducey (1988) characterized as "extralimital" 2 nest reports noted in Stipsky's field notes (in NOU Archives) for 1934 along the Elkhorn River in Dodge Co. Without documentation, these reports should be considered hypothetical. Blue-headed Vireo breeds no closer to Nebraska than north-central Minnesota.

Other Reports

2 Apr 1991 Douglas-Sarpy Cos, 30 Jun 1964 Logan Co, 30 Jun 1965 Logan Co, 30 Jun 1967 Logan Co.

FINDING

Migrant Blue-headed Vireos are easiest to find in early May in deciduous forest such as at Fontenelle Forest and Indian Cave SP, usually associated with flocks of migrant warblers.

Warbling Vireo

Vireo gilvus

STATUS

Common regular breeder and spring and fall migrant statewide.

DOCUMENTATION

Specimen

UNSM ZM6788, 23 May 1900 Monroe Canyon, Sioux Co.

TAXONOMY

Two subspecies occur in Nebraska. It has been suggested that the two may be separate species, differing "morphologically, vocally, and genetically" (Sibley and Monroe 1990). Western *swainsonii* breeds in the Panhandle, including Sioux Co and "Crescent Lake" (AOU 1957), and eastern *gilvus* breeds "west to eastern South Dakota, eastern Nebraska, Kansas, and southeast Colorado" (AOU 1957). The 2 forms would thus meet in northwestern Nebraska; delineation of the ranges and characteristics of birds in the putative contact zone, including recordings of vocalizations, would be of interest.

According to Bent (1950), *swainsonii* has a crown darker than the back and a relatively shorter bill and more olivaceous sides than *gilvus*. A very yellow bird at Winters Creek L 6 Sep 1997 (SJD, WRS) may have been a *swainsonii* or a freshly plumaged very yellow *gilvus*.

DISTRIBUTION AND ECOLOGY

Breeding birds occur statewide in riparian woodland, most commonly in cottonwoods. Migrants are found statewide in most types of woodland and forest edge, not necessarily riparian.

Spring

1, 2, 2 May→summer (west)

14, 16, 19 Apr (Ludlow 1935)→summer (elsewhere)

Migration data indicate that arrival in the Panhandle is later than elsewhere, perhaps due to differing schedules of the 2 subspecies discussed above. Away from the Panhandle, spring arrival is in late Apr, while in the Panhandle arrival is in early May.

High counts include 30 in Douglas Co 25 May 1996; 15 near Fremont 20 May 1995; 15 at Wilderness Park, Lincoln, 17 May 1996; and 15 in Dakota Co 17 May 1997.

Summer

Bruner and others (1904) listed Warbling Vireo as occurring in the "eastern portion" of the state, west to Cherry Co, Neligh, West Point, Lincoln, and Beatrice. A specimen from Sioux Co was identified as *swainsonii*, which was a common breeder there but unrecorded elsewhere (Bruner and others 1904). It seems clear that this species currently breeds statewide, however; Lingle (1994) listed it as a common nesting species in Hall Co, and Rosche (1994a) considered it a "fairly common to common summer resident" in the Keith Co area. BBS data indicate that breeding birds are evenly distributed throughout the state except for the Panhandle, where density is lower. The only author noting an absence of breeding birds is Bray (1994), who stated that it was not known to breed at Bessey Division, NNF, Thomas Co, where it was a migrant only.

Fall

summer→8, 8, 13 Sep (west)

summer→5, 5, 5 Oct (elsewhere)

Timing of fall departure differs between the Panhandle and elsewhere. Departure in the Panhandle is completed by early Sep, elsewhere by early Oct. There are 2 later reports from the Panhandle, 30 Sep 1994 Scotts Bluff Co and 8 Oct 1978 Garden Co. Elsewhere there are later reports 13 Oct 1934 Webster Co (Ludlow 1935) and 20 Oct 1977 Howard-Hall Cos.

There is a banding record: a Warbling Vireo banded 20 May 1979 at Norfolk was recovered 8 Oct 1982 in Guatemala, where this species winters.

High counts include 12 at Pibel L 7 Sep 1998.

Other Reports

20 March 1972 Howard-Hall Cos.

FINDING

Almost any stand of cottonwoods and their immediate understory will have Warbling Vireos in May and Jun. The distinctive song locates these birds, although it is often difficult to see them well due to the large leaves of cottonwood trees.

Philadelphia Vireo

Vireo philadelphicus

STATUS

Uncommon regular spring and fall migrant east, becoming rare casual west.

DOCUMENTATION

Specimen

UNSM ZM10533, 16 May 1917 Lancaster Co.

DISTRIBUTION AND ECOLOGY

This species occurs primarily in Missouri Valley woodland and has a preference for deciduous brush. There are few Panhandle reports.

Spring

23, 23, 26 Apr→31 May, 1, 4 Jun

Migrants arrive in late Apr and depart by late May.

Away from the east, there are few reports since 1983: 4 May 1986 Brown Co (AB 40:493), 11 May 1991 Hall Co, mid-May 1996 Thomas Co, 14 and 16 May 1999 Geneva Cem (SJD, JGJ), 16 and 18 May 1996 Adams Co, 18 May 1986 Lincoln Co, and 24 May 1997 Calamus Res.

Panhandle reports are few: 30 Apr 1974 Scotts Bluff Co, 15 May 1971 Scotts Bluff Co, and 29 May 1965 Sheridan Co (Short 1966).

Fall

11, 11, 12 Aug→29 Sep, 1, 3 Oct

Migrants occur between late Aug and late Sep, although there are later reports 10 Oct 1997 Lancaster Co (JGJ), 12 Oct 1973 Douglas-Sarpy Cos, and 12 Oct 1981 Douglas-Sarpy Cos.

There are earlier undocumented reports 30 July 1973 Douglas Co, 1 Aug 1990 Saunders Co, and 5 Aug 1972 Lancaster Co. Aug dates for this species have been questioned (Thompson and Ely 1992); the earliest fall record for South Dakota is 22 Aug (SDOU 1991). However, for Iowa there are early records 4, 13, and 14 Aug (Kent and Dinsmore 1996).

There were no fall reports from 1979 until 1996, but numbers have been relatively high since then; for example, 7 were found in 1997, all in the east, 6 in the period 6–30 Sep.

There are few reports from the Panhandle, all documented: 3 Sep 1997 Riverside Park, Scottsbluff (NK); 28 Sep 1974 Dawes Co (Rosche 1982); and a very late record of one netted and photographed 5 Nov 1977 Garden Co (AB 32:225).

High counts include 3 at Fontenelle Forest 8 Sep 1998.

Other Reports

10 Jun 1967 Douglas-Sarpy Cos, 26 Jun 1975 Douglas-Sarpy Cos, a series of late dates given by Aughey (Sharpe 1993), 3 July 1979 Lancaster Co.

FINDING

The best chance to find this species is in mid-May or early Sep in brushy areas in the Missouri Valley. Song is not very helpful, as it is very similar to that of Red-eyed Vireo.

Red-eyed Vireo

Vireo olivaceus

STATUS

Common regular spring and fall migrant east, becoming uncommon west. Common regular breeder north and east, becoming uncommon southwest.

DOCUMENTATION

Specimen

UNSM ZM6782, 22 May 1900 Monroe Canyon, Sioux Co.

DISTRIBUTION AND ECOLOGY

This species is most numerous in mature deciduous woodland in the Missouri Valley, but it occurs in deciduous woodland elsewhere, although less commonly southwestward.

In the northwest it is limited to riparian habitats in canyons and along river valleys (Rosche 1982). In recent years it appears to be extending its range west in major river valleys in the southwest due to an increase in second-growth trees (Rosche 1994a).

Spring

20, 23, 23 Apr→summer

Migrants arrive in late Apr, with an earlier report 11 Apr 1951 Antelope Co.

High counts include 18 at Fontenelle Forest 28 May 1997 and 8 at Arbor Day Farm 21 May 1995.

Summer

While breeding probably occurs throughout the state, BBS data indicate that the species is far more numerous in the east and north than elsewhere. Red-eyed Vireo is probably least common in the southwest and southern Panhandle, although it may be increasing

there (Rosche 1994a). Rosche (1994a) cited only 1 summer record for Keith Co, 27 Jul 1990 in Eagle Canyon, and a singing male was at Clear Creek Marshes 16 Jul 1994. In the same area Brown and others (1996) noted the capture of 3 females with brood patches 30 May–18 Jun as well as 2 juveniles.

Rosche (1994a) also noted that it "apparently nests farther west" in Morrill Co, where singing males were found 4 Jul 1987 at Pumpkin Creek south of Redington and near Facus Springs. Two were at Carter Canyon 5 Aug 1995. A singing male was reported in Hitchcock Co 9 Jun 1990 (AB 44:1153), and 2 birds were in Dundy Co 22 Jun 1994.

In the northwest it was considered an uncommon summer resident by Rosche (1982), while Brogie and Mossman (1983) found it common in the Niobrara Valley Preserve in 1982.

High counts include 20 at Niobrara Valley Preserve, Brown Co, 28 Jun 1997.

Fall

summer→6, 8, 11 Oct

Peak migration is in early to mid-Sep. Departure generally occurs in late Sep, although there are Oct reports from all regions, including later dates 16 Oct 1979 Douglas-Sarpy Cos and 20 Oct 1967 Brown Co.

High counts include 14 at Fontenelle Forest 5 Sep 1997, 12 in Dakota Co 2 Sep 1994, and 12 in Dakota Co 7 Sep 1997.

Other Reports

13 Mar 1950 Lancaster Co, 20 Mar 1972 Hall Co, 27 Mar 1978 Lancaster Co.

FINDING

Red-eyed Vireos are easiest to find in May and Jun in Missouri Valley woodlands. Like most vireos, their movements are deliberate, which allows easy observation.

Corvidae (Crows, Jays)

Gray Jay

Perisoreus canadensis

STATUS

Accidental in winter.

DOCUMENTATION
Description
2–26 Feb 1930 Antioch, Sheridan Co (LOI 49).
Records
Gray Jay has been reported 11 times, but there are only 2 documented records. The reports are in the period 12 Nov–7 May and are concentrated in the northwest. Eight of the reports are north and west of Sheridan and Scotts Bluff Cos, with additional reports from Thomas Co and far to the east in Cuming and Douglas Cos.

The 2 eastern reports involved 1 bird at West Point in late Feb and early Mar 1886 (Bruner and others 1904), and 1 of the documented records was of another seen at Two Rivers Park in Douglas Co 12 Nov 1968 (Holcomb 1970). It is likely that these 2 reports, if correctly identified, are of the eastern race *canadensis*, which wanders south on occasion in fall from resident breeding populations northeast of Nebraska (Janssen 1987). The West Point bird was believed to be *canadensis* by Bruner and others (1904).

Other Nebraska reports, if indeed correctly identified, are likely to involve birds of the Rocky Mountain race *capitalis* from the Black Hills of South Dakota, although Phillips (1986) lists northwest Nebraska as a location visited casually by *canadensis*. In the Black Hills Gray Jays are common residents above 1220 m (4000 ft) (SDOU 1991), but the species is rarely seen at lower elevations (Rosche 1982; Andrews and Righter 1992). The Nebraska reports are of birds seen spring 1889 at Belmont, Dawes Co (Bruner and others 1904); April 1891 on the Timber Reserve west of Fort Robinson (Bruner and others 1904); Feb 1896 in Monroe Canyon (Bruner and others 1904); 2–4 Feb and 26 Feb 1930 at Antioch (Bray and others 1986); 11–16 Dec 1939 (4) between Whitney and Crawford on the north slope of the Pine Ridge in Dawes Co (Weakly 1940); unspecified winter sighting(s) in the period 1946–50 in Chadron SP (Gates 1951); winter reports from 1946–47 (NBR 14:11, 15:11), 1933–34, and fall 1944 in the NNF near Halsey

(Bray 1994); and 7 May 1956 and 29 Dec 1966 in Scotts Bluff Co. All but the Antioch sighting are undocumented.

Steller's Jay
Cyanocitta stelleri
STATUS
Uncommon casual winter visitor statewide.
DOCUMENTATION
Specimen
UNSM ZM7630, Oct 1916 Mitchell, Scotts Bluff Co.
TAXONOMY
Bruner (Bruner and others 1904) reported that his Apr 1891 sighting was a bird of the northern Rocky Mountain race *annectens*, but Swenk (1918b) discounted this determination based on the collecting of a specimen of *diademata*, the southern Rocky Mountain race, near Mitchell Oct 1916 (specimen cited above). One in Lincoln Co 15 Oct 1936 was identified as *diademata* (Tout 1947), as was another taken at Oshkosh 4 Dec 1918, HMM 2906. Phillips (1986), however, assigned Nebraska birds to *annectens*.
DISTRIBUTION AND ECOLOGY
This species apparently enters Nebraska via riparian corridors, primarily the Platte Valley, and is usually seen in towns or at feeders. Few of the Nebraska reports are from ponderosa pine woodland, its normal summer habitat in the Rocky Mountains.
Winter
There are some 30 reports of Steller's Jay, about 20 of which are from the Platte Valley, east to Lincoln Co, in the period 11 Aug–30 Apr. Of the remainder, 2 are from the east, Lancaster and Nuckolls Cos, in years when the species was recorded in the west, and 3 are from the Pine Ridge in Sioux and Dawes Cos. There are also 3 reports from the Sandhills.

The Platte Valley reports occurred during in the winters of 1916–17, 1937–38, 1945–46, 1946–47, 1948–49, 1950–51, 1962–63, 1969–70, 1971–72, 1976–77, 1977–78, 1983–84, 1985–86, and 1996–97, except for a report 15 Oct 1936 in Lincoln Co (Tout 1947) and 2 specimens, HMM 2906 collected at Oshkosh 4 Dec 1918 and HMM 2799 collected at Kearney 13 Sep 1914 (Swenk, Notes before 1925).

The largest number reported was a flock of about 35 birds in Scotts Bluff Co 1969–70, 12 of which were counted on the CBC 20 Dec 1969 (NBR 38:15). Multiple reports came in the winter of 1946–47, when flocks of up to 7 birds were seen in Lincoln Co (Middleton 1947; Collister 1947) and 1 wintered as far east as Superior (Day 1947); the winter of 1985–86, when several sightings were made in the Scottsbluff area (NBR 54:15); and 11 Aug 1996–13 Apr 1997, when up to 8 birds were reported at various locations in Scotts Bluff Co. In the winter of 1962–63 this species was reported in Lincoln Co in the period Nov–19 Apr and as far east as Lancaster Co 6 Jan 1963.

There are 2 records of birds wintering at Pine Ridge feeders and a third sight report there. One was at a Chadron feeder 9 Nov 1973–8 Jan 1974 (Rosche 1982; AB 28:74, 659), and another wintered at a feeder in Sowbelly Canyon 24 Nov 1989–13 Apr 1990 (AB 44:116, 456). One was seen Apr 1891 in the Timber Reserve west of Fort Robinson by Bruner (Bruner and others 1904). Steller's Jay does not occur in the Black Hills of South Dakota; there is only 1 South Dakota record (SDOU 1991).

There are 3 additional reports, all from the Sandhills. These are from Antioch 15 Feb and 26 May 1931 (LOI 60, 61; Rosche 1982), Sheridan (or Grant) Co 26 Apr 1936 (NBR 4:58), and Logan Co with no date or details (Johnsgard 1980).

FINDING

Steller's Jay should be looked for along the North Platte Valley in Scotts Bluff Co in Nov and Dec.

Blue Jay

Cyanocitta cristata

STATUS

Common regular resident and spring and fall migrant east, becoming fairly common west.

DOCUMENTATION

Specimen

UNSM ZM12259, 30 Dec 1908 Lancaster Co.

TAXONOMY

The northeastern North American subspecies *bromia* breeds in the northeastern corner, including Niobrara, and presumably the eastern edge of Nebraska, and the southwestern race *cyanotephra* occupies the remainder of the state "so far as known" (AOU 1957; Phillips 1986). However, if indeed Blue Jay has "spread westward" (Bruner and others 1904), western birds would be expected to be intergrades or possibly northeastern *bromia*.

DISTRIBUTION AND ECOLOGY

This species breeds statewide but is least numerous in the Panhandle. It requires the presence of trees but apparently avoids coniferous woodland (Rosche 1982) and occurs in cities and towns as well as open areas. Migrants may occur anywhere, including grasslands without trees. Wintering birds are most numerous in the east and are especially conspicuous at feeders in cities and towns.

Spring

Because birds are present all year, migration dates are difficult to determine. Spring migrants tend to be inconspicuous, but most move through in late Apr and the first half of May.

High counts in the west include 13 in Scotts Bluff Co 25 May 1996 and 10 at Morrill 13 May 1994. Elsewhere, there were 372 in Sarpy Co 11 May 1996.

Summer

Blue Jay was spreading "rapidly" westward around 1900 (Bruner and others 1904), apparently in response to an increase in riparian woodland and maturing plantations associated with western farmsteads, ranchsteads, and towns. Even then, however, it occurred over the "entire state," although it was "uncommon westward" (Bruner and others 1904). Tout (1947) considered it a "common summer resident" in Lincoln Co. The earliest specimen from the Panhandle is 7 Oct 1920 (UNSM 7628), taken in Monroe Canyon, although it was first recorded in

Colorado in Yuma Co in 1903 (Andrews and Righter 1992). Numbers are still increasing in the Platte Valley of Wyoming (Scott 1993), as in the Keith Co area, where Brown and others (1996) suggested that its increase may be coincident with a decline in numbers of Black-billed Magpies. Currently it is considered an uncommon summer resident in the northwest, even absent in some summers (Rosche 1982).

BBS data indicate much lower density in the Panhandle than elsewhere (NBR 46:38).

Egg dates are in the period 1 May–9 Jun (N = 6); Tout (1947) cited "fresh eggs" 11 May–18 Jun in Lincoln Co.

Fall

Fall migrants are conspicuous, often occurring in large flocks. Peak movement is in late Sep and early Oct (Bent 1946). Bruner and others (1904) noted that Blue Jays return in mid-Apr and depart in late Oct. A series of fall departure dates from Garden Co in the years 1977–81 covered the period 6–17 Oct.

High counts include 110 in Otoe Co 29 Sep 1995, 100 in Lancaster Co 1 Oct 1995, and 45 at Arbor Day Farm 5 Oct 1999.

Winter

Many Blue Jays are resident, but the species is sensitive to adverse winter weather and may move southward or merely "wander about" (Tout 1947) if necessary, even regularly in some parts of the state. Around 1900 most Blue Jays left the state in winter (Bruner and others 1904); Ludlow (1935) listed this species as a "common summer resident" in Webster Co, with no mention of records Dec–Mar, and Tout (1947) listed it as a "common summer resident" in Lincoln Co, occurring 20 Apr–27 Nov. CBC data indicate that by late Dec, few Blue Jays remain in the north; in the west a good count of 35 was made on the Scottsbluff CBC 16 Dec 1995. In the northwest it is absent in some winters (Rosche 1982), and it tends to be absent in winter in parts of Holt Co where it breeds, reappearing in early May (Blake and Ducey 1991). Brogie and Mossman

(1983) did not see Blue Jays until 30 Apr in the Niobrara Valley Preserve, and the first for the year appeared at James Ranch, Sioux Co, 11 May 1995.

There is no evidence that birds from north of Nebraska winter in Nebraska; most, if not all, apparently pass through the state as migrants. There is also no information on the destination of Nebraska breeding birds that depart in winter.

FINDING

One of Nebraska's most conspicuous birds, especially in fall and winter, Blue Jays can easily be located by their loud calls and presence at feeders in winter.

Western Scrub-Jay

Aphelocoma californica

STATUS

Hypothetical.

COMMENTS

There is 1 report of this species, without details. The NOURC has not accepted this report, so the species is not on the State List. One was reported seen in Bull Canyon, Banner Co, in early Jun 1978 (Bray and others 1986; NBR 48:89) by an observer familiar with the species who did not realize its rarity in Nebraska. The habitat in this canyon and nearby Long Canyon seems suitable for this species.

There is an old report cited by Bruner (1896) of it being a "common transient visitor" at North Platte, but the observer apparently was referring to Pinyon Jay (NBR 48:89). Phillips (1986) listed this species (*A. c. suttoni*) as wintering "irregularly to Nebraska," but no evidence is presented for this statement.

There are no records for this nonmigratory species closer than southeast Colorado (Andrews and Righter 1992) and extreme southwest Kansas (Thompson and Ely 1992). Most vagrant records for this species from Colorado and Kansas are in fall and winter. The likelihood of this species appearing in Nebraska is slight; the 1 report is from a time of year when vagrancy is not expected. Bull and Long Canyons in Banner Co are likely locations, however, and the species might occur from late Sep through Dec.

Pinyon Jay
Gymnorhinus cyanocephalus
STATUS
Fairly common spring and fall visitor northwest. Uncommon summer visitor northwest. Hypothetical breeder northwest. Common regular winter visitor west, rare casual central, accidental east.

DOCUMENTATION
Specimen
UNSM ZM7644, 13 Dec 1895 Fort Robinson, Dawes Co.

TAXONOMY
Phillips (1986) indicated that birds reaching northern and northwestern Nebraska after breeding were examples of the eastern Rocky Mountain subspecies *cyanocephalus*, the breeding subspecies in the Black Hills of South Dakota.

DISTRIBUTION AND ECOLOGY
During the summer this species is found singly or in small groups in open ponderosa pine woodland in the Pine Ridge and similar habitat in Scotts Bluff Co. At other times of the year flocks wander throughout the same areas, often descending to adjacent lowlands (Rosche 1982). Occasionally it appears in winter some distance from its usual range.

Summer
In the Pine Ridge Pinyon Jay occurs in summer, but since about 1900 evidence for breeding is lacking. Numbers are lower in summer than during other seasons (Rosche 1982). Around 1900 it apparently bred east into Sheridan Co almost to the South Dakota border (Bates 1900; Bruner and others 1904), but the only nesting season report from Sheridan Co since then is a listing as permanent resident in 1992 without details. The range may have contracted in modern times.

Nesting takes place in the Black Hills of South Dakota in Apr and May, with dependent young seen as early as mid-Apr (Johnsgard 1979). While there are several reports from Scotts Bluff Co at that time, as well as reports from Banner Co 21 May 1988 and of about 10 birds seen in Morrill Co in Apr and May 1987 (AB 41:456), evidence for nesting in these areas, where suitable habitat exists, has not been published (Ducey 1988), although Rosche (1994b) noted that the species "reaches eastern limits" in a Morrill Co canyon south of Redington, albeit with no mention of breeding. At this location in 1998 flocks of 5–75 birds were seen with young of the year 16 May 1998 (WM). These young birds were rather mobile, however. The observer (WM) also noted finding 2 nests presumably of this species but later withdrew this finding, as he was unable to identify the nests with certainty (Wayne Mollhoff, pers. comm. WRS).

Rapp and others (1958) noted that the species' "status in yellow pine along the Niobrara River is not understood," presumably a reference to old summer reports cited by Bent (1946) for Valentine and Holly.

Ducey (1988) considered a report of "young" in Webster Co in 1935 questionable; there is no apparently suitable breeding habitat at this easterly location. Based on this, a series of about 10 reports, mostly of winter observations, from there at about the same time appear questionable also (Ludlow 1935). A winter notation for Red Cloud (Bent 1946) may be based on these reports.

Winter
This species is highly gregarious most of the year, and flocks may cover considerable distances in search of food (Bent 1946). Large numbers are sometimes seen at feeders in towns, such as the "invasion" in Chadron in Sep 1985 (AB 40:136) and in Scotts Bluff Co, where 350 were counted on the 18 Dec 1971 Scottsbluff CBC. High CBC totals of 170 were also made at Crawford in 1977 and 1978. Rosche (1982) stated that numbers are higher in fall, winter, and spring than in summer and suggested that local populations may be

augmented by birds from "presumably farther west" (see also Phillips 1986).

Pinyon Jays occur irregularly east of the breeding range, mostly in the Platte Valley, but rarely east of Lincoln Co, where Tout (1947) considered it an irregular winter visitor, and Logan Co, where Glandon and Glandon (1934b) noted that a small flock appeared almost every winter. Easterly reports for Nebraska include one photographed at Norfolk 26 Nov 1985–20 Jan 1986 (AB 40:298); a specimen, HMM 2517, collected at Hastings 20 Oct 1926; and a single at a Kearney feeder during the winter of 1996–97 (CD, JD; Brogie 1998), the same winter one appeared in central Iowa. There are additional undocumented reports for Webster Co (see Summer), Lancaster and Clay Cos (Bent 1946), and 28 Feb 1936 Jefferson Co.

Curiously, there were no Nebraska reports for 1972–73, when a major corvid invasion on the Great Plains pushed 2 Pinyon Jays to southwest Iowa, providing that state with its first record of Pinyon Jay (Dinsmore and others 1984).

FINDING

Pinyon Jay can be difficult to find during the summer but is conspicuous in fall and winter in the Pine Ridge, when flocks can be located by their raucous calls.

Clark's Nutcracker
Nucifraga columbiana
STATUS
Uncommon casual irruptive winter visitor statewide. Rare casual breeder northwest.
DOCUMENTATION
Specimen
UNSM ZM10406, 29 Nov 1910 Crawford, Dawes Co.
DISTRIBUTION AND ECOLOGY
This species normally occupies montane coniferous forests, including spruce (*Picea* sp.), Douglas fir (*Pseudotsuga* sp.), and limber pine (*Pinus flexilis*), but is known to wander at least locally altitudinally,

latitudinally (Andrews and Righter 1992), and longitudinally onto the Great Plains during winter. Eruptions, generalized movements of significant numbers of birds from the breeding range, apparently occur when population densities are high and food availability is limited (Tomback 1995). The best Nebraska example of an irruption, a major influx, occurred in fall and winter 1986 when breeding ensued, but the population later dwindled and eventually disappeared.

Clark's Nutcracker may appear virtually anywhere in Nebraska during irruption years, although many reports are from the ponderosa pine habitat in the Pine Ridge. In nonirruptive years there is a tendency for individuals to appear in the North Platte Valley. Most reports are from the Panhandle.
Summer
After winter invasions, breeding may occur. Clark's Nutcrackers have remained into summer after the 1968–69, 1972–73, 1986–87, and 1996–97 irruptions, although evidence of breeding has been noted after only 2 of these influxes. Ducey (1988) cited information regarding a begging juvenile in Sowbelly Canyon in May 1987, reported later by the original observer (Tomback 1995). Although breeding could not be confirmed by another observer (AB 41:1457), the large number of birds present and a nesting record during the same summer in the Black Hills of South Dakota (SDOU 1991) support the likelihood that breeding occurred. A single bird first noted in Morrill 8–9 Jan 1997 was paired during the second week of Feb, and a young bird with brown feathering in its wings was noted at a feeder 7 May; at least 1 bird remained until 4 Jun 1997 (ECT; Mollhoff 1997a). It was stated that breeding apparently occurred also in early 1997 at 2 locations in the Pine Ridge (NBR 65:87), but this was incorrect (Mollhoff 1997a).

Other reports, including 4 in Monroe Canyon 25 Jun 1973 (Rosche 1982; AB 27:889), birds present in the Wildcat Hills of Scotts Bluff

Co in summer 1969 (Gates 1970), and reports
from Sowbelly Canyon 24 Jun 1999 and 16 Jul
1994, were unaccompanied by any evidence of
breeding. Bruner (1896) listed this species as a
"possible nester" in Sioux Co.

Winter

22, 26 Aug, 4 Sep→25, 30 May, 1 Jun

This irruptive montane species has been
reported at least 50 times. Many of the reports
are from ponderosa pine habitat in the Pine
Ridge and in Scotts Bluff Co, and the major
invasion of 1972–73 provided most of the
remaining reports. Single birds appear in
noninvasion years on rare occasions, such as
one at L McConaughy 1 Nov 1998 (Dinsmore
1999a). Significant irruptions occurred in the
winters of 1919–20, 1933–34, 1950–51, 1968–69,
1972–73, 1973–74, 1986–87, and 1996–97.

Easterly reports include one at Wayne 30 Oct
1972 which died 4 Nov (wsc 67; Rasmussen
1973), Merrick Co 6 Oct 1955 (Galloway
and Galloway 1956), Sarpy Co 29 Sep 1972,
Louisville during early Nov 1972 (AB 27:81),
Cass Co 18 Feb 1951 (Heineman 1951),
and Superior 10 Oct 1933 (Johnston 1933).
Curiously, none were noted in the east in
winter 1996–97 when 3 appeared in Iowa.

High counts include 71 in Sowbelly Canyon 7
Feb 1987 (AB 41:299), a flock of 50 just east of
Sowbelly Canyon 19 Sep 1979 (RSS), and 42 in
Monroe Canyon 1 Jun 1997.

FINDING

Pine Ridge canyons, especially Monroe and
Sowbelly, should be checked in fall and winter
for wandering nutcrackers.

Black-billed Magpie

Pica pica

STATUS

Common regular resident west, becoming
locally uncommon east, absent extreme
southeast. Rare casual winter visitor east of
breeding range.

DOCUMENTATION

Specimen

UNSM ZM10305, 14 Jul 1910 Sioux Co.

DISTRIBUTION AND ECOLOGY

This species occurs in open country wherever
small to medium-size bushy trees are available
for nest sites; usually these are riparian
situations, especially westward. Magpies are
more numerous in the west, but breeding has
been documented virtually statewide, with
the exception of the extreme southeast. There
is some wandering in fall, when birds may
appear in areas, notably in the east, where
breeding does not regularly occur.

Magpies often choose shrubby trees with dense
interior branching as sites for their large,
ball-like nests, built of numerous small
sticks. Drought-stunted hackberry (*Celtis
occidentalis*), hawthorn (*Crataegus* sp.), and
mulberry (*Morus rubra*) are frequently used
trees, but large wild plum (*Prunus americana*)
bushes have also been used.

Spring

Occasional migrants are recorded in spring
from Mar to May in areas of the east where
breeding does not usually occur. Such reports
since 1982 are 14 Mar 1987 Douglas-Sarpy Cos,
15 May 1999 on the Stanton-Cuming Co line,
and 25 May 1987 Cuming Co.

Summer

Breeding birds are probably most numerous in
drier areas in central Nebraska where suitable
habitat is most plentiful, although Rosche
(1994a) considered it an uncommon summer
resident in the L McConaughy area. Numbers
decline sharply toward the southeast.

Prior to 1900 Black-billed Magpie bred virtually
statewide, including Cuming, Cass, and Gage
Cos (Bruner and others 1904), but the range
contracted so that after 1900 breeding was
confined to the west and northwest only. Rapp
and others (1958) considered it a common
breeder in the western half of the state. Since
then, this species has expanded eastward
again.

In recent years the easternmost regular breeding
has occurred in Dakota, Saunders, and Clay
Cos (Ducey 1988), and in the 1990s it became
established in Cuming and Dodge Cos (LE).

Breeding has occurred in Lancaster Co but is irregular, with no reports since 1987. A pair was nesting near Raymond 5–6 years up to 1987, 1–2 pairs were nesting in northern Lancaster Co 3–4 years up to 1983 (NGP), and it was listed as a permanent resident in Lancaster Co in 1992. There are few summer reports in the last 60 years southeast of these counties: 18 Jun 1935 Douglas Co, 25 Jun 1993 Gage Co, and 9 Aug 1999 Otoe Co (LF, CF).

High counts include 107 in Scotts Bluff Co 23 Aug 1996 and 65 at L Ogallala 2 Sep 1993, possibly postbreeding flocks on the move.

Fall

This species is not migratory but occasionally exhibits postbreeding dispersal to areas where the species does not breed. Blake and Ducey (1991) noted that at Holt Co locations where it does not summer, "occasional wanderers" appeared in fall, an observation made also by Rosche (1994a) for the L McConaughy area. At locations on the east edge of the breeding range, postbreeding dispersal is noticeable in Sep and later. There are reports from Greenwood 18 Sep 1981 (AB 36:193), Jefferson Co 5 Oct 1996 and 5 Nov 1998, and Washington Co 25 Oct 1998, "the first modern record" (JGJ).

Fall wanderers apparently winter east of the breeding range (see Winter).

Winter

CBC data indicate numbers decreasing eastward, with very few reports from the southeast, although in recent winters the species has moved into the Lincoln area (NBR 61:12). The only recent winter reports east of the breeding range are 17 Jan 1999 in northern Dodge Co and 4.8 km (3 mi) north of Bennington 3 Dec 1998. The only CBC reports from the east are from Norfolk, within the summer range, where as many as 222 were counted 1985–89, and at Lincoln, where 1–2 have been reported on CBCs from 1990 to date. Highest CBC totals are from the Panhandle: 98 at Crawford 1977–78, 87 at Scottsbluff 1987–88, and 81 at Scottsbluff 1989–90.

FINDING

Magpies are easy to find in the western half of the state along river valleys. They are best located by their calls but are often seen in flight. Perhaps the best time to look for this species is in fall, especially Oct–Dec, when they are most conspicuous.

American Crow

Corvus brachyrhynchos

STATUS

Common regular resident statewide. Abundant regular spring and fall migrant statewide.

DOCUMENTATION

Specimen

UNSM ZM7635, 18 May 1907 Cass Co.

DISTRIBUTION AND ECOLOGY

In the breeding season American Crow is most numerous in the east and least common in the southern Panhandle. Nests are placed high in trees, and thus riparian and upland forests and woodlots in agricultural areas are favored. In the last 25 years crows have become common urban dwellers, nesting in neighborhoods where mature trees provide appropriate nest sites. Until this recent adaptation, crows were essentially rural during spring and summer.

Spring

Banding studies of birds wintering in Oklahoma showed that many of these birds bred in the prairie provinces of Canada and that crows of the Great Plains exhibit marked migratory tendencies (Bent 1946). Migration is characterized by movements of flocks; peak migration in Nebraska is probably Mar–Apr (Rosche 1994a; Scott 1993; SDOU 1991).

High counts include 200 in Howard Co 2 Mar 1997.

Summer

BBS data (NBR 46:38) indicate that summering birds are most numerous in the east, with fewer birds distributed rather evenly over the rest of the state. Around 1900 this species was rare west of the 98th meridian but was spreading westward (Bruner and others 1904). An early southern Panhandle specimen was taken in Kimball Co 8 Jun 1919 (UNSM 7639). It is still rare to uncommon on the northeastern Colorado plains (Andrews and Righter 1992), and in Nebraska it is probably scarcest in the southern Panhandle.

Fall

Peak migration is probably Oct–Nov. A high count for the Panhandle was 86 at Chadron 8 Oct 1994, where such concentrations are uncommon (NBR 62:139). Tout (1947) counted 2000 flying over in a string in Lincoln Co 7 Oct 1934, and 3000 were in the Rainwater Basin 28 Oct 1996.

Winter

This species is most common in winter when resident populations are augmented, presumably by birds from farther north. During late fall and winter large flocks form and often roost in cities and towns throughout the state. In the earlier part of the 20th century these large concentrations, often numbering in the tens of thousands, were considered a nuisance in more isolated towns on the Great Plains, leading to the winter practice of "crow shoots," often conducted at night at the roost site. Roosts were present near Kearney in 1988–89 and 1989–90, when 5000 birds were counted on CBCs each year, and over 100 000 have roosted there during the winters of the mid-1990s. A roost at Holdrege for about 25 years prior to 1996 has held more than 100 000 birds.

CBC data show fairly even distribution statewide in late Dec, but occasional roosts in the south, especially the central Platte Valley, boost numbers there. High CBC counts were 3000 at Kearney in 1986–87 and 2767 at Grand Island in 1990–91.

FINDING

This species is easy to locate year-round but is most conspicuous in fall and early winter in fields in agricultural areas. Black corvids in the southern Panhandle and southwest at any time of year should be carefully identified, as ravens of either species may occur there also.

Chihuahuan Raven

Corvus cryptoleucus

STATUS

Rare casual breeder southwest and south-central.

DOCUMENTATION

Specimen

UNSM ZM12283, 23 Jun 1951 no specific location, Nebraska.

Records

The occurrence of this species in Nebraska is enigmatic. Reports in the 20th century are from south-central and southwest Nebraska and have been cyclic, concentrated around 1927, the late 1940s, and the 1970s.

Apparently Chihuahuan Ravens occurred north to eastern Colorado and western Nebraska when the bison herds were being slaughtered, then retreated to the current range in southeastern Colorado and southwest Kansas sometime around 1900 (Andrews and Righter 1992; Bent 1946). Bent (1946) cited 2 early reports from this period: Aughey reported one from the Republican River near the west edge of Nebraska in April 1877, although it is questionable that Aughey actually visited that part of the state (RSS), and Bruner found one near Sidney prior to 1904 (Bruner and others 1904).

There were no further reports until 1927, when 2 were shot and left in a ditch around 25 Apr 1927 near Holstein, Adams Co; the "very large size" and "the fact that the bases of the feathers of the throat were white" were noted by the shooter (LOI 23; NBR 10:25). A nest in the vicinity composed of cedar bark was considered to be a crow's nest, however. The discarded birds were recovered 3 weeks later, examined by Brooking and Swenk, and were apparently preserved as skeletons in the HMM, although they can no longer be located (Bray and others 1986).

There followed 4 additional reports from the Adams-Kearney Cos area. One was seen near Brickton 9 May 1942 and was "approached to such a short distance that the fringe of white feathers surrounding its throat was distinctly seen by everyone" (NBR 10:25). Two birds that had been reared in a cage were ascertained to have been taken (along with a third, which died) from a nest between Axtell and Wilcox on 11 Jun 1944 (Brooking 1944). This nest was also made of cedar bark but had considerable wire woven into it, a characteristic of Chihuahuan Raven nests (Bent 1946). On 13 Apr 1947 another nest was found near Axtell; the young birds were taken as pets (Brown 1947). A fourth report, undocumented, is 13 Feb 1954 Adams Co.

There are 2 undocumented reports at about the same time as the Adams-Kearney Cos reports, 6 Jun 1949 Lincoln Co and 1 Jan 1949 Keith Co.

There were 4 reports in the 1970s. A nest was reported from Dundy Co in 1971 (Cink 1977), but no details have been published. A call purportedly of this species was heard 15 May 1975 in Lancaster Co but could not be confirmed (NBR 43:52). A pair of adults considered to be White-necked Ravens based on size, the "caw" call, and the wedge-shaped tails were seen in Hall Co 24 Apr–28 May 1976 (Lemburg 1976; Coons 1976). Whether a nest with 6 young nearby was that of this species is questionable (Lemburg 1976). The last report, undocumented, is 17 Jun 1979 Keith Co (AB 33:877).

Despite several searches in the Dundy Co area by WRS and others, this species was not reported from Nebraska between 1979 and 1997. Similarly, in northwest Kansas, Scott Seltman (pers. comm. WRS), who has birded the region extensively since 1979, has no records of this species.

The only recent report was a sighting near Sparks 29 Mar 1997; the bird was clearly a raven and thought to be this species based on its small size, nonmassive bill, and range. The NOURC accepted this report only as "Raven sp." (Brogie 1998).

FINDING

Based on the Kansas range (Thompson and Ely 1992) and Colorado information (Andrews and Righter 1992), this species should be looked for in the Dundy and Hitchcock Cos area and possibly in Keith Co and the southern Panhandle. Spring is the most likely time to find this species, when birds return to the northern parts of the range, although some northward movement in fall is evidenced by occasional appearance of groups in southwest Kansas of birds presumed to originate in Texas (Scott Seltman, pers. comm. WRS).

Common Raven

Corvus corax

STATUS

Accidental in fall.

DOCUMENTATION

Description

12 Nov 1900 Kearney, Buffalo Co (Bray and others 1986).

Records

There are only 2 documented records:

7 Nov 1936 Peru (Hudson 1937a)

12 Nov 1900 Kearney (cited above)

Common Raven apparently occurred statewide (although there is only anecdotal evidence) and was fairly common until about 1877, when the disappearance of bison led to its rapid decline (Bruner and others 1904; Johnsgard 1980). Ducey (1988) lists sightings, including an 1820 breeding record, for the Missouri Valley prior to 1900.

Among the last reports of the declining population subsequent to the demise of the bison can probably be counted the following, the latest of which was in 1915. The specimen cited above, HMM 21721 collected 12 Nov 1900 at Kearney and in the Black collection at the HMM, was examined by Brooking and measured by Swenk; the measurements were wing 423 mm (16.9 in), culmen 65 mm (2.6 in), and tarsus 50 mm (2.0 in) (Bray and others 1986). This specimen was sold in 1952, and its current whereabouts are unknown (Bray and others 1986).

There are additional undocumented reports. Swenk (Notes before 1925; see Bray and others

1986) noted that one was killed by Black in the 1890s, presumably in south-central Nebraska, and that 3 were seen near Kearney in 1915 by Black. A specimen at one time in the HMM (HMM 2759) was reportedly collected near Arapahoe in 1899. It cannot now be located (Bray and others 1986).

Since about 1900 there have been few reports, all but 1 undocumented. There is evidence for a small incursion into the Missouri Valley during the winter of 1936–37, when 2 in Nemaha Co 7 Nov 1936 were described well enough to be acceptably documented (Hudson 1937a), and others were reported without details from Washington Co (Haecker 1937). There is an earlier report from Omaha without details 25 Apr 1929, and a female brought to an experienced Omaha taxidermist was shot 13 Feb 1967 9.7 km (6 mi) northwest of Norfolk, probably in Pierce Co (Velich 1970). A brief description of one near Wakefield 16 Dec 1971 (Mollhoff 1986b) was not accepted by the NOURC (Mollhoff 1987). These eastern reports seem plausible because birds from the Minnesota population are known to migrate south in fall, providing eastern South Dakota and northern Iowa records (SDOU 1991; Dinsmore and others 1984).

There are 2 reports from the Sandhills, neither documented: Short (1961) noted "small numbers" near Halsey 12–13 Jun 1955, north of Halsey 14 Jun 1955, and near Valentine 14–15 Jun 1955, and Rosche (1972) saw a single bird 23 Jun 1972 at Valentine NWR. Short's observations have been considered "almost certainly a misidentification" (Bray 1994).

There are 4 undocumented reports from south-central Nebraska: 21 May 1952 Adams Co, 5 May 1956 Thayer Co, 1958(?) Thayer Co, and 21 Mar 1962 Adams Co.

The only other reports are from Scotts Bluff Co and Sioux-Dawes Cos, perhaps the most likely locations for ravens to occur. A few pairs apparently still breed on the plains of northeastern Colorado (Andrews and Righter

1992) and might be expected to wander in winter. There are 4 Scotts Bluff Co reports, none accompanied by identification details: 26 May 1958, 1 Apr 1962, 1 in a yard at Gering 2 Feb 1983, and at about the same time 4 reported from a canyon southwest of Gering (NBR 51:78). One was reported based on calls heard in the "cliff forests" of Fort Robinson SP 30 May 1999.

FINDING

Probably the most likely location to find Common Raven is the southern Panhandle in fall and winter. Migratory ravens in Minnesota are on the move Sep–Nov (Janssen 1987), so late fall might be the best time in eastern Nebraska. Isolated rock outcroppings, dumps, and cattle feedlots should be checked.

Alaudidae (Larks)

Horned Lark
Eremophila alpestris

STATUS

Abundant regular spring and fall migrant and winter visitor statewide. Common regular breeder statewide.

DOCUMENTATION

Specimens: UNSM ZM6423, *praticola*, 4 Feb 1894 Omaha, Douglas Co; UNSM ZM10207, *leucolaema*, 12 Nov 1897 Alliance, Box Butte Co; UNSM ZM6416, *alpestris*, Feb 1899 Lincoln, Lancaster Co; UNSM ZM6442, *hoyti*, 20 Jan 1936 Stapleton, Logan Co.

TAXONOMY

Four subspecies occur. Two or 3 breed, depending on which races are recognized. According to Bruner and others (1904), *leucolaema* breeds in the west, eastward to about the 99th meridian (near Kearney), and *praticola* in the east, westward to about the 97th meridian (near Lincoln). Presumably intermediates occur where the two races meet. The race *enthymia*, included in *leucolaema* by Bruner and others (1904), comprises the easternmost populations of *leucolaema*, breeding from eastern Wyoming and eastern Colorado eastward; Ludlow (1935) considered nesting birds in Webster Co to be of the race *enthymia*. At least some individuals of all 3 races remain in winter (AOU 1957; Glandon and Glandon 1937).

Two additional races occur as migrants and in winter: *hoyti*, which breeds in central Canada, and *alpestris*, which breeds in northeastern Canada (AOU 1957). Bruner and others (1904) considered *hoyti* to be less common than the resident races, appearing mostly in midwinter, and placed "most if not all" reports previously considered to be *alpestris* in *hoyti*. Bruner and others (1904) list only 1 specimen of *alpestris*, stating that it was "far west of its usual range"; *alpestris* is considered a regular winter resident in Kansas, however (Johnston 1965).

DISTRIBUTION AND ECOLOGY

Breeding birds occur statewide in open, sparsely vegetated areas, especially grasslands, where bare soil is present. Johnsgard (1980) suggests that the "sparse grasslands of the Sandhills are probably a nearly optimum habitat." At Bessey Division, NNF, it occupies areas near cattle tanks, which are devoid of vegetation (Bray 1994). In the east, where croplands are the norm, this species breeds commonly also, usually fledging young before cultivation begins, although later nests are often destroyed.

In winter and migration flocks are found throughout the state in open situations, both grasslands and agricultural areas, including roadsides, where they feed on spilled grain. Large concentrations may be found along roadsides during periods of heavy snow cover, notably during migrations.

Migration dates are difficult to determine, as birds are present all year statewide. However, sizeable flocks in Nov–Dec, such as 5275 around L McConaughy 28 Dec 1996 and 3250 in southern Nebraska 17 Nov 1996, and again in Feb–Apr are likely to be migrants, such as "thousands" 23–24 Feb 1924 in Webster Co

(Ludlow 1935) and 6750 along 16 km (10 mi) of road in Scotts Bluff Co 12 Apr 1997.

It is not known whether Nebraska breeding birds are resident. Birds breeding north of Nebraska enter the state during winter and may move through depending on weather conditions. Data from central Ohio (Bent 1942) show that the race *hoyti* occurs there 26 Nov–17 Mar, and peak numbers of the race *alpestris* occur in the period Dec–Feb. Both races occur in Nebraska in winter (see Taxonomy).

Summer

BBS data indicate a strong preference of Horned Larks for grasslands of the west and north as breeding habitat (NBR 46:38). By Jun, however, when BBS routes are run, cultivation practices may have reduced breeding densities considerably in the east and south.

A small sample of egg dates are 4 May–3 Jul (n = 6), but pairs appear to be on territory as early as late Feb (NBR 62:58), and young are observed in Mar; nearly fledged young (of the race *enthymia*) were noted in Webster Co 20 Mar 1930, when snow was on the ground (Ludlow 1935). A flying fledgling was seen in Keith Co 24 Apr 1999 (SJD).

Winter

CBC data indicate a more even distribution in late Dec than in summer, although numbers are highest in the south and are extremely variable from year to year. Highest CBC totals are 4810 from Hastings in 1975–76, 3238 from Lincoln in 1978–79, and 3200 from Hastings in 1973–74. At these same 2 locations, however, none were found at Lincoln twice, and at Hastings counts as low as 6 and 10 have been recorded. Many Dec birds may be migrants that pass through the state. High Dec counts include 11 000 from Garfield Co to Saunders Co 31 Dec 1994, while highest midwinter (Jan–Feb) concentrations may be reached in stubble fields and around feedlots in the west. A cattle feedlot in Box Butte Co had 6000 Horned Larks 13 Jan 1974 (AB 28:659), and 8000 were near Sutherland Res 17 Jan 1994 (WRS).

Breeding birds are abundant in the short grasslands of the Panhandle but are more difficult to find in the east, where they are best located by song in Apr. Wintering birds are easily found in open areas, especially after snowstorms, when they concentrate along roadsides.

Hirundinidae (Swallows)

Purple Martin

Progne subis

STATUS

Common regular spring and fall migrant east, becoming rare casual west. Common regular breeder east, becoming rare casual west.

DOCUMENTATION

Specimen

UNSM ZM10234, 30 May 1895 Lancaster Co.

DISTRIBUTION AND ECOLOGY

Purple Martin is a common breeding bird in the east, becoming less common westward and virtually absent in the Panhandle. Most nesting is colonial, utilizing artificial nest boxes around human habitations, although occasionally nests are found on cliffs or in old woodpecker holes (Johnsgard 1980). Colonies are often located near water and in areas where open perches are available, such as utility wires. Migrants are common in the east, usually around water and especially over ponds and reservoirs.

Spring

12, 14, 15 Mar→summer

This species arrives early in spring, with male "scouts" arriving first, often to perish in inclement weather. Data in Johnsgard (1980) indicate that migration peaks in early Apr, although arrival is in late Mar. There are earlier reports 14 Feb 1987 Lincoln Co, 2 Mar 1933 Jefferson Co, and 8 Mar 1966 Cass Co.

Away from Lewellen in Garden Co, an occasional breeding location, there are few Panhandle reports: 21 Apr 1953 Dawes Co, 7 May 1996 Garden Co, 10 May 1949 Dawes Co, 11 May

1996 (6) Sheridan Co (FN 50:298), 20 May 1961 Scotts Bluff Co, and 23 May 1959 Scotts Bluff Co, as well as 10 Jun 1990 Cheyenne Co (see Summer).

High counts include 40 at Bellevue 31 May 1994, 31 at Gretna 11 May 1996, and 22 in Sarpy Co 13 May 1995.

Summer

This species is common only in the east. While found throughout the south-central part of the state, it is uncommon and local as far west as Chase and Perkins Cos, where there are Jun or Jul reports from Chase Co for 1982, 1986, and 1988–90, and 1 report from Perkins Co 29 Jul 1973, possibly a migrant. Maps in Ducey (1988) suggest westward spread of this species in the 20th century, particularly along the Platte Valley, an observation also made by Johnsgard (1980). Tout (1947) indicated that erection of martin houses had resulted in some nesting in North Platte, although no dates were given. Tout observed Purple Martins nesting on a downtown North Platte building "for several summers around 1910" before House Sparrows had become common. Summering birds have been reported from Lincoln Co at least since 1944, with "a few nesting" in 1969 (AB 23:602). Small populations breed sporadically in the North Platte Valley west as far as Lewellen in Garden Co, around 1977–79 considered the "westernmost regular nesting site in Nebraska" (Rosche and Johnsgard 1984), although there are no currently active colonies in that area. As of the mid-1990s the westernmost regular breeding was at Ogallala (Rosche 1994a; Brown and others 1996). The only summer report from the Panhandle away from Garden Co is 10 Jun 1990 Cheyenne Co.

Reports from the Loup drainage are few, although Purple Martins occur in Boone Co and were at one time regularly reported as far west as Logan Co, with Jun reports from there for as long ago as 1934 (Glandon and Glandon 1934a) and for 1953–54, 1961, and 1968.

Along the Niobrara Valley the westernmost

regularly summering birds are at Ainsworth, where Short (1965) found them common over the town in 1964. Brogie and Mossman (1983) did not report this species from the Niobrara Valley Preserve in 1982. There are no summer records west of Ainsworth.

Fall

summer→2, 4, 4 Oct

Departure is early, generally in late Aug and early Sep, although there are several later reports: 8 Oct 1995 Cherry Co (FN 50:76), 10 Oct 1981 Douglas-Sarpy Cos, 11 Oct 1986 Pierce Co, 15 Oct 1933 Webster Co (Ludlow 1935), 15 Oct 1976 Douglas Co, and 20 Oct 1993 Sarpy Co.

High counts include "hundreds" at Hastings 28 Jul 1971 (AB 26:83), a flock of 150 at O'Neill 28 Aug 1991 (Blake and Ducey 1991), and 59 at Arbor Day Farm 10 Aug 1998.

FINDING

Purple Martins are best located in summer in Missouri Valley towns and cities either by driving around until occupied martin houses are found or by checking ponds and lagoons for feeding birds.

Tree Swallow

Tachycineta bicolor

STATUS

Common regular spring and fall migrant statewide. Common regular breeder east, becoming uncommon west.

DOCUMENTATION

Specimen

UNSM ZM6452, 2 Jul 1902 Carns, Keya Paha Co.

DISTRIBUTION AND ECOLOGY

This species nests in old woodpecker holes, usually in willow or cottonwood stubs in riparian habitats (Rosche 1982; Johnsgard 1980). Breeding occurs most commonly in the Missouri Valley, although breeding there and elsewhere in the state is dependent on transitory nest-site requirements. In recent years westerly breeding has increased as a result of reservoir construction providing plentiful dead tree stubs (AB 43:1335) and also

by the provision of nest boxes, which this species readily utilizes.

Spring

15, 20, 20 Mar→summer

Tree Swallow is a regular migrant statewide but is much more common in the east. Rosche and Johnsgard (1984) listed this species only as a spring migrant in Keith Co, as did Lingle and Hay (1982) for their study area in Hall Co. Arrival is in late Mar, but departure is difficult to determine, although most migrants have departed by mid to late May.

High counts include "several hundred" at Harlan Co Res 9 Apr 1995, 134 in Sarpy Co 11 May 1996, and 75 at Wagon Train L 18 May 1996.

Summer

Tree Swallow nests throughout the Missouri Valley in riparian habitats, notably at Fontenelle Forest, but it has been reported regularly from most counties bordering the Missouri River. Breeding may occur south of the Platte River and east of Adams and Webster Cos, although there are regular summer reports only from Lancaster Co. There are summer reports from Clay Co in 1975, Gage Co in 1957 and 1964, and Saline Co in 1956.

In the northwest Rosche (1982) noted only 1 nesting site, in cottonwood and willow stubs at Whitney L, where 1 or 2 pairs nested each year. Subsequently, however, Rosche reported nesting in Sheridan Co in 1985 and Garden Co in 1987 (Ducey 1988). There are Jun and Jul reports from Sioux, Dawes, and Sheridan Cos as early as 1973, from Garden Co since 1978, and from Scotts Bluff Co since 1979. These dates compare with the dates for establishment of nesting birds in northeastern Colorado, where nesting was first confirmed in 1963, and the species now occurs regularly at 9 sites along the South Platte River (Andrews and Righter 1992). Rosche (1994a) indicated that breeding in the west and north has recently become more widespread.

Nesting in the north may not be as widespread

as in Bruner's time, but a few pairs currently breed along the Niobrara Valley. Mossman and Brogie (1983) found a few birds inspecting cavities at 4 locations in Cherry and Brown Cos in 1982, and Rosche reported nesting in Cherry Co in 1984 and Brown Co in 1980 (Ducey 1988). There are additional published summer reports for Cherry Co in 1980, Keya Paha Co in 1960, Brown Co in 1971, and Rock Co in 1985. Short (1961) reported birds near Valentine in Jun 1955 and suggested that it "may breed there." Away from the Niobrara Valley, Blake and Ducey (1991) found a nesting pair using a bluebird box in southern Holt Co. Short (1961) recorded a single bird near Halsey 13 Jun 1955; there is a single summer report for Logan Co from 1943; and Ducey (1988) cited a nesting record for Antelope Co around 1900. Apparently the first nesting record for Boone Co was 1 Jun 1997. Bray (1994), however, listed this species only as a migrant at Bessey Division, NNF.

Nesting in the Platte Valley has occurred west to Hall Co for many years (Short 1961; Johnsgard 1980), although it was not known to nest in Lincoln Co at least until 1947 (Tout 1947). A series of nest boxes along I-80 around the Lincoln-Dawson Co line had numerous occupants 21 May 1994 (WRS). There is a summer report from Adams Co for 1973 and summer reports from Phelps Co. and from Lincoln Co in 1991. It is currently a local summer resident in Keith Co (Rosche 1994a; Brown and others 1996) and was reported from Scotts Bluff Co without details 17 Aug 1993.

In the southwest there are recent summer records for Chase Co in 1989 and 1991 and for Frontier Co in 1992. There are few reports from the Republican Valley; up to 12 were at Harlan Co Res 6–21 Jun 1998, and adults were feeding young at Alma 25 Jun 1999.

Fall

summer→31, 31, 31 Oct

Recent fall counts are reflecting increasing numbers of migrants in locations where few if any were previously reported, such as 285 at Crescent L NWR 9 Aug 1995, where summering birds are not numerous. As far back as 1939 a flock of 200 was reported in Lincoln Co 25 Aug (Tout 1947), where breeding was not known at the time. It was not recorded in fall in Keith Co (Rosche and Johnsgard 1984) or Hall Co (Lingle and Hay 1982), although later studies (Lingle 1994; Rosche 1994a) indicated increased occurrence.

Identification of fall Tree Swallows in the west requires care, as breeders leave at about the same time as Violet-Green Swallows; the 2 species are similar in fall.

Data in Johnsgard (1980) indicate that migration peaks 26 Aug–7 Oct. Fall departure is mid to late Oct, although there are later reports from Douglas-Sarpy Cos 4 Nov 1990, 8 Nov 1983, and 14 Nov 1992. There are few reports of fall migrants away from the east; breeding birds leave after young are fledged (Rosche 1994a).

High counts include 200 at Schilling Refuge 23 Sep 1996 and 60 in Lancaster Co 20 Aug 1999.

FINDING

Tree Swallows are easy to find around ponds and lakes near the Missouri River in Apr and early May.

Violet-green Swallow

Tachycineta thalassina

STATUS

Rare regular spring and fall migrant west, rare casual central. Uncommon regular breeder west.

DOCUMENTATION

Specimen

UNSM ZM6449, 22 Jun 1901 Warbonnet Canyon, Sioux Co.

DISTRIBUTION AND ECOLOGY

This species is "characteristic of wooded canyons, where it nests in cracks and crevices on cliffs and canyon walls" (Rosche 1982). It also uses old woodpecker holes (Rosche 1982) and occasionally birdhouses (Johnsgard 1980). Breeding is restricted to the Pine Ridge in Sioux and Dawes Cos and the Wildcat Hills and other suitable habitat in Scotts Bluff Co (Rapp and others 1958).

Migrants are rarely seen away from the breeding areas, usually occurring with other swallows in flocks over water.

Spring

26, 27, 27 Apr→summer

Arrival is in late Apr, although there is an earlier report 17 Apr 1955 Scotts Bluff Co. Data in Johnsgard (1980) indicate that peak migration is 5–19 May.

Reports of migrants away from the breeding areas are few: 29 Apr 1956 Keith Co, 29 Apr 1989 Keith Co (Johnsgard 1990a), 2 May 1939 Lincoln Co, 11 May 1947 Keith Co, 12 May 1967 Garden Co, 14 May 1944 Lincoln Co, and 19 May 1941 Lincoln Co. There is an undocumented report 5 Jun 1974 Perkins Co.

The only documented report farther east is of one at Harlan Co Res 8 May 1997 (GH, WH). The few additional undocumented reports may be misidentified Tree Swallows. These are 11 May 1957 Adams Co, 15 May 1967 Brown Co, 1 Jun 1966 Brown Co, and 24–30 Jun 1964 Brown Co. Bruner and others (1904) noted that this species was "once taken during migration at West Point by Bruner," presumably the Cuming Co report cited by Johnsgard (1980). No specimen apparently exists, however, and this report remains unsubstantiated.

Summer

Rosche (1982) considered this species a "locally fairly common summer resident," most common in Sioux Co Pine Ridge canyons. The easternmost record of nesting may be that cited by Rosche (1982) of a pair using a woodpecker hole in Chadron SP in 1981.

Most published summer reports without breeding evidence are from Scotts Bluff Co; Rapp and others (1958) noted, without details, that it breeds in the Wildcat Hills in southern Scotts Bluff Co, as well as the "Scottsbluff area."

There are undocumented BBS reports of 7 birds on Route 21 in the period 1967–77; this route is in eastern Cheyenne and Morrill Cos, a little east of any suitable breeding habitat, and the birds may have been migrants. It was listed as summer resident without details in Sheridan Co in 1992. Another undocumented report is 10 Jul 1967 Brown Co.

Fall

summer→25, 26, 29 Aug

Departure is in mid-Aug, although there is a later report 5 Sep 1973 Sioux Co. South of Gering 5–12 birds considered migrants were noted 25–28 Jul 1994. An early undocumented report 8 Jul 1987 Keith Co may have been a fall transient (Rosche 1994a).

There are few reports away from breeding areas, all in the period 4–29 Aug: 9 Aug 1995 Garden Co, 11 Aug 1974 Sheridan Co (Rosche 1982), 26 Aug 1974 Perkins Co, 26 Aug 1979 Garden Co, 28 Aug 1993 Dawes Co, and 29 Aug 1986 Garden Co (Rosche 1994a). An easterly report 4 Aug 1990 Phelps Co is undocumented. It may have formerly been more common, as Tout (1947) considered it a migrant in Lincoln Co, where it appeared "in large flocks about the 1st of Aug," citing a flock of several hundred 5 Aug 1939.

High counts include 30 south of Gering 23 Aug 1998 and 15 there 2 Aug 1996.

Other Reports

19 Mar 1969 Scotts Bluff Co.

FINDING

Summering birds are most easily located flying around cliffs at Scotts Bluff NM and in Pine Ridge canyons such as Sowbelly Canyon, especially in Jun. Positive identification is difficult except by habitat, but occasionally close views can be had. Migrants should be looked for east of the breeding range in Aug, especially over reservoirs in the Panhandle or at the small lake at Chadron SP. In Aug most Tree Swallows are still on their breeding grounds.

Northern Rough-winged Swallow

Stelgidopteryx serripennis

STATUS

Common regular spring and fall migrant and breeder statewide.

DOCUMENTATION

Specimen

UNSM ZM6454, 7 Jun 1901 Gage Co.

DISTRIBUTION AND ECOLOGY

This species breeds statewide, usually as single
pairs or small groups rather than in colonies.
Preferred nesting habitat is dirt or sand
banks and road cuts near wooded rivers and
streams. Migrants occur over lakes and ponds
with other swallow species.

Spring

4, 4, 5 Apr→summer

Peak migration probably occurs 16 Apr–6 May,
according to data in Johnsgard (1980).

High counts include 317 in Dixon Co 11 May
1996; 250 at Hormel Park, Fremont, 5 May
1999; 188 in Sarpy Co 11 May 1996 (156 of
which were at Fontenelle Forest); and, in the
west, 90 at North Platte NWR 28 May 1995.

Summer

Breeding birds are distributed fairly evenly over
the state (Short 1961), except perhaps for parts
of the north and northwest. Rosche (1982)
considered it "uncommon" as a breeder in the
northwest, and Bray (1994) assigned it similar
status at Bessey Division, NNF. An unusual
concentration of nesting birds was a colony
of about 40 nests in Lancaster Co in 1982
(AB 36:994).

Fall

summer→14, 15, 18 Oct (east)

summer→3, 4, 4 Oct (elsewhere)

Peak migration probably is 23 Aug–15 Sep,
according to data in Johnsgard (1980). Away
from the east, departure is completed by early
Oct, but almost a month later in the east, in
late Oct. There are later reports 30 Oct 1980,
30 Oct 1988, 31 Oct 1986, and 8 Nov 1981
Douglas-Sarpy Cos and 21 Nov 1993 Sarpy Co.

In identifying late migrant Rough-winged
Swallows, care should be taken to eliminate
the possibility of immature Tree Swallow, a
late fall migrant.

High counts include "hundreds" near Nebraska
City 13 Aug 1996, 250 at Schilling Refuge 15
Sep 1996, and 155 in Otoe Co 15 Sep 1999.

FINDING

Rough-winged Swallows can most easily be
seen during fall migration, especially in early

Sep, with mixed flocks of swallow species.
Breeding birds can be located along wooded
river and streambanks, most conspicuously
in Jun.

Bank Swallow

Riparia riparia

STATUS

Common regular spring and fall migrant and
breeder statewide.

DOCUMENTATION

Specimen

UNSM ZM10606, 10 May 1913 Lincoln, Lancaster
Co.

DISTRIBUTION AND ECOLOGY

Breeding birds are fairly evenly distributed
statewide, nesting in colonies in sand and dirt
banks soft enough for nest holes. Excavated
sandpits are a favored site. Breeding colonies
are scarce in the Panhandle, probably due
to a scarcity of soft substrates for nest sites.
Colonies tend to move from year to year, as
breeding sites may be disturbed by water
levels. Migrants are found statewide, usually
with other swallows around ponds and lakes.

Spring

6, 8, 8 Apr→summer

Migrants are most common in the east, but
large numbers may appear anywhere. Peak
migration indicated by data in Johnsgard
(1980) is in the period 28 Apr–6 May. There is
an earlier report 1 Apr 1956 Scotts Bluff Co.

High counts include 100 in Sarpy Co 1 May 1995,
50 near Peru 25 May 1997, and 50 at Black
Island Area 15 May 1999.

Summer

Bank Swallow, while it breeds statewide, is
somewhat unpredictable and variable in its
occurrence; this is probably due to the often
temporary nature of its preferred breeding
sites. Colonies may be large, as indicated
by counts of about 1000 nests in cliffs at L
McConaughy (Brown and others 1996); 500
birds at Elbow Bend, along the Missouri River
in Sarpy Co, 15 Jun 1994; and "several hundred"
at Smith L, Sheridan Co, 18 May 1994.

Some authors (Ducey 1988; Johnsgard 1980) have indicated that this species is less common as a breeding bird westward, and Rosche (1982) considered it an uncommon summer resident in the northwest, although he noted that it is "decidedly more common some years than others." Tout (1947) did not list Bank Swallow for Lincoln Co. Published data indeed show few summer reports for the Panhandle, with Jun reports for only 7 years: Garden Co in 1967, 1978 and 1980; Dawes Co in 1973; and Scotts Bluff Co in 1959, 1961, and 1970.

Along the Niobrara River, Brogie and Mossman (1983) found no Bank Swallows in 1982 but noted that they were regular summer residents 30–50 years earlier, according to Youngworth (1955). At Bessey Division, NNF, Bray (1994) listed it as an "uncommon summer resident" and "possible nester."

Fall

summer→4, 4, 4 Oct

Peak migration probably occurs in mid-Aug, based on high counts (see below). Migrants depart by early Oct, although there are later reports 9 Oct 1968 Adams Co, 12 Oct 1965 Cass Co, 14 Oct 1963 Douglas-Sarpy Cos, and 15 Oct 1991 Douglas-Sarpy Cos.

High counts include 1820 at Crescent L NWR 15 Aug 1995, 800 at Funk Lagoon 24 Jul 1997, and 350 at Kiowa Springs 12 Aug 1994.

Other Reports

20 Mar 1955 Thayer Co, 20 Mar 1972 Hall Co, 29 Oct 1967 Brown Co.

FINDING

Bank Swallow is best located by checking fall flocks of migrating swallows around lakes and ponds in Aug and early Sep. Breeding birds are most likely to be found along rivers where sandpits have been excavated.

Cliff Swallow

Petrochelidon pyrrhonota

STATUS

Abundant regular spring and fall migrant and breeder statewide.

DOCUMENTATION

Specimen

UNSM ZM10610, 10 May 1913 Lancaster Co.

DISTRIBUTION AND ECOLOGY

Cliff Swallow breeds statewide but is most numerous in the west. It usually breeds in large colonies beneath cliff overhangs, concrete bridges and culverts, and overhangs on buildings, provided a suitable mud source is nearby. Migrants occur statewide, usually around lakes and ponds.

Spring

5, 6, 7 Apr→summer

Cliff Swallow is at times, especially in fall, an abundant migrant, as entire colonies tend to migrate together. Migration probably peaks 29 Apr–18 May, according to data in Johnsgard (1980). Arrival is in early Apr, although there are earlier reports 22 Mar 1976 Lincoln Co, 26 Mar 1994 Buffalo Co, and 28 Mar 1952 Adams Co.

High counts include 3500 between Scottsbluff and Kearney 17 May 1996 and 2000 at L Ogallala 9 May 1998.

Summer

BBS data indicate that this species is by far most numerous in the west (NBR 46:38). Breeding colonies are becoming more numerous in the east, however, as more large bridges are built; natural cliff sites are scarce away from the Panhandle. An exception is a cliff site at the mouth of the Niobrara River, which has been used for at least 150 years (Ducey 1988).

According to Brown and others (1996), eggs have been found 4 May–28 Jul in the Keith Co area.

High counts include colonies of up to 3700 nests in the Keith Co area (Brown and others 1996) and 1000 birds at the Pishelville Bridge on the Niobrara 31 May 1994, 740 at L McConaughy 19 May 1995, 702 near Broadwater 15 Jun 1995, and 500 on the Missouri River bridge at Nebraska City summer 1994.

Fall

summer→4, 4, 7 Oct

Data in Johnsgard (1980) indicate that peak migration is 20 Aug–15 Sep, although most colonies are abandoned by early Aug, when large flocks become conspicuous on telephone wires and over cornfields. Departure is in early Oct, although there are later reports 18 Oct 1982, 23 Oct 1981, 30 Oct 1988, and 31 Oct 1976 Douglas-Sarpy Cos.

High counts include 12 000 in the Funk Lagoon area 1 Aug 1999, 2000 there 21 Aug 1994, and 1250 there 18 Jul 1999.

COMMENTS

Brown and Brown have banded an incredible 87 000 Cliff Swallows in the Keith Co area since 1982; recoveries include southern Brazil, El Salvador, Colorado, Alberta, and Missouri (Brown and others 1996).

FINDING

Checking concrete bridges over streams and rivers with muddy edges typically will yield Cliff Swallows. The birds are usually conspicuous as they fly around their nest site.

Cave Swallow

Petrochelidon fulva

STATUS

Rare casual spring visitor Keith and Garden Cos.

DOCUMENTATION

Photograph

31 May 1991 Ash Hollow, Garden Co (Brown and Brown 1992; Brogie 1998).

Records

There are 3 records, all of juveniles netted in the Keith Co area by Charles Brown. One was netted 31 May 1991 at a concrete culvert about 6.45 km (4 mi) south of Highway 92 on Highway 26 in extreme southeast Garden Co. The bird was photographed and released later that day (Brown and Brown 1992) at Keystone L (Rosche 1994a). Brown and Brown (1992) suggested that this bird was hatched in central Texas in early spring 1991 and moved north with the migrating Cliff Swallows. They identified it as belonging to the race *pallida* (= *pelodoma*; Phillips 1986; Turner and Rose 1989), which breeds in central Texas and southeastern New Mexico. Another juvenile was found 25 Jun 1995 and captured 26 Jun in the same general area as the first record (FN 49:947, photograph; Brogie 1998); Brown and others (1996) indicated that this second bird was captured 10.5 km (6.5 mi) east of Cedar Point Biological Station along the Sutherland Canal. A third juvenile was netted at Keystone 1 Jul 1998; Brown hypothesized that it was hatched in Texas and came north with Cliff Swallows (Charles Brown, pers. comm. NK).

FINDING

Flocks of migrating and nesting Cliff Swallows should be scanned for Cave Swallows, adults of which can most easily be distinguished by their paler throats.

Barn Swallow

Hirundo rustica

STATUS

Common regular breeder and spring and fall migrant statewide.

DOCUMENTATION

Specimen

UNSM ZM6456, 27 Apr 1904 Falls City, Richardson Co.

DISTRIBUTION AND ECOLOGY

This species breeds in open areas throughout the state, placing its nests beneath overhangs such as eaves, more commonly on rural buildings, but sometimes also on urban dwellings. Occasionally nests are located beneath bridges. Brown and others (1996) stated that a nesting at a cliff site at L McConaughy in 1982, which may also have been active in 1981, was "one of few natural nesting locations on the Great Plains in modern times." Migrants are found statewide around ponds and lakes and over open agricultural areas.

Spring

1, 2, 3 Apr→summer

Barn Swallow is a common migrant statewide, at times abundant in fall. Data presented by Johnsgard (1980) indicate migration peaks 18–30 Apr.

Arrival is in early Apr, although there are earlier reports 20 Mar 1972 Hall Co, 22 Mar 1976 Lincoln Co, and 26 Mar 1971 Brown Co.

High counts include "hundreds" both in
southern Sioux Co and at Funk Lagoon 29
Apr 1995, 254 in Dixon Co 11 May 1996, and
190 at Crescent L NWR 17 May 1995.

Summer

Rapp and others (1958) suggested that this
species was least numerous in the west, and
BBS data confirm this (NBR 46:38).

Nesting may continue into early fall. A nest with
young was found in Valley Co 6 Sep 1989 (AB
42:99), and Brogie and Mossman (1983) found
fledglings as late as 6 Sep in the Niobrara
Valley Preserve.

Eggs were recorded by Brogie and Mossman in
the period 31 May–23 Jun, and nest card data
record eggs 18 May–13 Sep (N = 87).

Fall

summer→4, 5, 6 Nov

Data in Johnsgard (1980) indicate that migration
peaks 19 Sep–6 Oct. Fall departure is in early
Nov, although there are later reports 16 Nov
1980 Douglas-Sarpy Cos, 21 Nov 1993 Sarpy
Co, and 2 Dec 1983 Howard-Hall Cos.

High counts include 2700 at L Ogallala 3 Oct
1998, 2000 at Gadwall Basin 22 Sep 1996, and
1450 in Buffalo and Phelps Cos 19 Sep 1999.

Other Reports

8 Mar 1974 Adams Co.

FINDING

Barn Swallow is easy to find in farming areas,
where it nests on outbuildings, both occupied
and unoccupied. It is commonly seen foraging
over farm fields throughout the months of
May, Jun, and Jul.

Paridae (Chickadees, Titmice)

Carolina Chickadee

Poecile carolinensis

STATUS

Hypothetical.

COMMENTS

This species has been reported 3 times (Bray and
others 1986; NBR 63:49). A recording, which
cannot now be located, was made of a call
in Fontenelle Forest 5 Feb 1969, and 2 birds

banded 16 Jul 1974 in Fontenelle Forest were
identified as this species (Diggs and Diggs
1974). The latter report was rejected by the
NOURC due to "lack of specific measurements"
and failure to eliminate "immature, worn, or
molting Black-capped Chickadees" (Grenon
1991). A bird singing a 4-part song was
listened to carefully at Arbor Day Farm, Otoe
Co, 24 Mar 1995, but plumage characters could
not be discerned (NBR 63:49).

Carolina Chickadee is sedentary in Kansas
(Thompson and Ely 1992) and Missouri
(Robbins and Easterla 1992); the nearest
Carolina Chickadee range is about 325 km
(200 mi) from Nebraska in west-central
Missouri (Robbins and Easterla 1992).

Black-capped Chickadee

Poecile atricapilla

STATUS

Fairly common regular resident statewide.

DOCUMENTATION

Specimen

UNSM ZM6484, Nov 1893 Lancaster Co.

TAXONOMY

Two subspecies have been recorded: eastern
atricapilla breeds in southeastern Nebraska
(AOU 1957), although Bruner and others (1904)
suggested that most breeding chickadees
in eastern Nebraska are intergrades with
septentrionalis, the Rocky Mountain race,
which breeds over most of Nebraska (Bruner
and others 1904).

DISTRIBUTION AND ECOLOGY

Breeding birds are found statewide in mature
woodland that provides natural nesting
cavities; this includes riparian, upland, and
coniferous woodlands but also towns and
cities with mature trees. Wintering birds occur
statewide, most conspicuously at feeders in
towns and cities, but are most numerous in
the east.

Summer

Breeding birds are distributed throughout
the state but are least numerous in the
southwest and Panhandle. In the northwest it

is considered "uncommon to fairly common" in summer (Rosche 1982). BBS data (1967–77) show a marked decline in density in the south and west (NBR 46:38).

Winter

This species is common statewide in winter. It is generally assumed that birds from farther north winter in Nebraska (Rapp and others 1958; Johnsgard 1980), and indeed chickadees are more common in winter in the northwest (Rosche 1982) and become noticeable in locations where they are not conspicuous in summer, especially where feeders are provided. However, of 78 Nebraska banding records examined, for various times of the year, all recoveries were at the place of banding, as noted by Brown and others (1996) in Keith Co also. Nevertheless, significant fall movement of Black-capped Chickadees does occur (Robbins 1989), such as along the shore of L Superior in Minnesota (Janssen 1987); at Holiday Beach in Essex Co, Ontario, where from 1983 to 1996 most years had total counts 0–89 but in 3 years (1983, 1986, 1993) counts of 779–3894 were made, with peaks 25 Oct–5 Nov (Chartier 1999); and at Bessey Division, NNF (Zimmer 1913). CBC data show a far more even distribution of chickadees in winter than indicated in summer by BBS data, as well as a concentration in the east, suggesting that at least some movement of chickadees takes place in fall and winter. A count of 125 was made at Johnson Res 24 Oct 1999 (LR, RH).

High CBC totals are at Omaha, where 822 were counted 1987–88 and 657 in 1989–90.

FINDING

This species is usually the first to respond to squeaking in almost any woodland at any time of year. In general, chickadees are more conspicuous in the winter months, especially at feeders.

Mountain Chickadee

Poecile gambeli

STATUS

Uncommon casual winter visitor west, rare casual central.

DOCUMENTATION

Specimen

UNSM ZM17444, 23 Nov 1996 Wildcat Hills NC, Scotts Bluff Co (Brogie 1997).

Records

Despite several published reports prior to 1996–97, there was only 1 documented record, a single at Gering 5 Dec 1976 (Gubanyi 1996b).

Mountain Chickadee was reported in 6 of the winters in the period 1966–67 through 1976–77, including reports of birds at feeders and 3 CBC reports. Most reports were from Scotts Bluff Co in the winters of 1966–67, 1968–69, 1969–70, 1972–73, 1973–74, and 1976–77 in the period 6 Oct–23 May. Birds were reportedly photographed and banded at Scottsbluff Oct 1968 (AB 23:494), although these photographs have not been published. Other localities with reports during these winters were Sioux Co 5 Oct 1966, Lincoln Co 6 Nov 1968–6 Apr 1969 and 16 Nov 1976, and Dawson Co mid-Jan through mid-Mar 1969 (Shickley 1969b).

The only other report prior to 1996–97 was a single bird at Ogallala 24 Nov 1995 (RCR; FN 50:76).

In 1996 another irruption began when up to 7 were reported at Stateline Island 5–12 Oct 1996 (LM; Brogie 1997), and 3 were at a Scottsbluff feeder the same day. Several were subsequently noted at Scottsbluff and Gering feeders (Brogie 1997), and up to 5 were at feeders at the Wildcat Hills NC through 19 Apr 1997. Additional reports were of one at Riverside Park, Scottsbluff, 27 Oct 1996 (WRS; Brogie 1997) and a single as far east as North Platte 4–6 Dec 1996 (BP, LP). The final reports for the irruption were of singing birds 27 Apr 1997 at Scotts Bluff NM, 4 May 1997 at Winters Creek L, and another singing 3 May 1997 in Long Canyon (SJD).

One was reported as present at Harrison during winter 1998–99 (HKH), but no details were provided. The only other recent record was of a well-described bird at Oliver Res 1 Aug 1999 (MB).

COMMENTS

It appears that Mountain Chickadees enter Nebraska along the North Platte River Valley or possibly the South Platte River Valley. This species is rare on the eastern Colorado plains, but the few records are concentrated along major river valleys, including the South Platte (Andrews and Righter 1992). There are good numbers in the Laramie Mountains in Wyoming (Scott 1993), the source of several tributaries of the North Platte River; these tributaries apparently provide a habitat corridor into Nebraska.

An intriguing finding was described by Rosche (1974b), who found a population of chickadees in the Wildcat Hills of southern Scotts Bluff Co that resembled Black-capped Chickadees morphologically but sang Mountain Chickadee songs. Phillips (1986) was "aware of no authentic specimen" of hybrid Black-capped and Mountain Chickadees, although hybrids have been reported in central New Mexico.

FINDING

This species should be looked for in winter at feeders in Scottsbluff or other towns in the North Platte Valley, especially in years when numbers are found at lower elevations in the Rocky Mountain foothills.

Boreal Chickadee

Poecile hudsonica

STATUS

Hypothetical.

COMMENTS

There is 1 report, a bird at the Chet Ager NC feeder in Lincoln, from 16 Dec 1972 at least until 21 Jan 1973 (Bennett 1973a). A description was published, but it was not considered diagnostic, and no photograph or description of vocalizations has been published (Bray and others 1986).

Lending credibility to this report were 2 eastern South Dakota records during the same fall, one in Deuel Co 2–3 Nov 1972 and another that was banded and photographed in Brookings Co 12–18 Nov 1972 (SDOU 1991).

There are several records south of the normal range in Minnesota (Janssen 1987) and 4 for Iowa (Dinsmore and others 1984).

Tufted Titmouse

Baeolophus bicolor

STATUS

Fairly common regular resident southeast. Rare casual spring visitor elsewhere.

DOCUMENTATION

Specimen

UNSM ZM6547, 30 Nov 1900 Lancaster Co.

DISTRIBUTION AND ECOLOGY

Tufted Titmouse is restricted to mature oak-hickory woodland and forest in the southeast but may utilize immediately adjacent riparian woods. It utilizes natural cavities for nest sites. Where present it is fairly common year-round, although inconspicuous at times.

Spring

Reports away from the breeding range are few, primarily in spring, suggesting some spring movement westward in river valleys. There are westerly reports from the Republican Valley 29 Mar 1986 Franklin Co and 29 Mar 1986 Harlan Co. Birds moving up the Big Blue Valley may be the source of reports in the eastern and central Platte Valley: 11 Feb 1957 Hamilton Co, 12 Feb 1966 York Co, 11 Apr 1985 Adams Co, 13 Apr 1991 Hall Co, and 11 May 1991 Howard-Hall Cos and possibly 1 report farther west, 8 Apr 1976 Lincoln Co.

Farther north and west there are these reports: 5 Mar 1950 Antelope Co (see Summer for an old breeding report), 31 Mar 1967 Custer Co, and 23 May 1979 Dawes Co (Rosche 1982).

Summer

Currently the breeding range is restricted to counties bordering the Missouri River north to Washington Co and west in rapidly decreasing numbers to Lancaster, Saline (Short 1961), and Thayer Cos.

Curiously, there are no records for Platte Valley counties between Douglas/Washington and Hamilton, suggesting that the few spring reports from Hall and Adams Cos may be of birds moving up the Big Blue Valley.

The only summer reports north of Washington
Co in the Missouri Valley are cited by Bruner
and others (1904) for Burt and Dakota Cos.
Jorgensen (1988) noted that this species was
not observed in the Cauble Creek area of
Washington Co until 4 Apr 1988 but has
been present since. Bruner and others (1904)
mentioned breeding in Cuming and Antelope
Cos, and Bent (1946) noted "summer records"
in Webster Co, although Ludlow (1935)
listed only a single report, 1 May 1928, for
Webster Co.

Winter

This species is normally strongly resident, with
little movement away from breeding areas.
CBC data reflect the restricted breeding range;
there are reports west only to Gage and
Lancaster Cos and north to Washington Co.
The sharp decrease in numbers away from the
Missouri Valley is illustrated by the contrast
between the 816 birds counted at Omaha
1967–68 through 1991–92 and the 11 birds at
Lincoln during the same period.

There are 2 fall reports away from the breeding
range, neither documented: 7 Oct 1988
McPherson Co and 19 Oct 1968 Scotts
Bluff Co.

Other Reports

Details were provided for an individual
identified as a "Black-crested" Titmouse in
Otoe Co in fall 1994 (NBR 62:139). This form is
currently considered conspecific with Tufted
Titmouse and is resident in south Texas.
Although individuals have been seen some
160 km (100 mi) from the breeding range
in Texas, an occurrence in Nebraska would
be considered an "amazing record" (Mark
Lockwood, pers. comm. WRS).

The editor of NBR, R. G. Cortelyou, in a
discussion of species reported without
documentation, stated that he "was aware of
unpublished possible or probable records
in Nebraska of Black-crested Titmouse"
(NBR 41:80). No other information has been
published, and the editor did not publish the
report due to the lack of details.

FINDING

Tufted Titmouse is most easily located in forests
in the Missouri Valley, such as Indian Cave SP
and Fontenelle Forest, in April and May. The
distinctive, loud song is easily recognized, and
squeaking readily attracts these birds.

Sittidae (Nuthatches)

Red-breasted Nuthatch

Sitta canadensis

STATUS

Common regular resident north-central and
northwest, rare casual elsewhere. Uncommon
irregular winter visitor statewide.

DOCUMENTATION

Specimen

UNSM ZM6563, 20 Feb 1896 Harrison, Sioux Co.

DISTRIBUTION AND ECOLOGY

This species breeds most commonly in
ponderosa pine woodland in the central
Niobrara Valley (Brogie and Mossman 1983).
It is fairly common in similar habitat in the
Pine Ridge and Wildcat Hills. It also breeds
commonly in exotic coniferous woodland in
the Bessey and McKelvie Divisions of the NNF,
Thomas and Cherry Cos (Bray 1994).

Red-breasted Nuthatches are erratic, sometimes
common, winter visitors statewide, mostly in
coniferous habitats, including cemeteries and
plantings, but they also occur in deciduous
woodland and often visit feeders.

Summer

Records of breeding are relatively recent (Ducey
1988). The first indication of breeding was
in Jun 1955 when Short (1961) noted several
individuals exhibiting territorial behavior in
Dawes Co. Since then, however, confirmation
of breeding in the Pine Ridge has been elusive.
Rosche (1982) noted that in the Pine Ridge
this species was a "very rare summer visitor"
which presumably bred, but there was no
definite evidence. Dueker, however, (1982)
observed a bullsnake entering a nest hole
of this species near Chadron in May 1980.
Johnsgard (1980) noted that breeding is

probable in the Pine Ridge area and stated that "nest-building has been observed in Sowbelly Canyon, Sioux County." Summering birds are more frequent in the Pine Ridge after large fall or winter flights, but the species is "highly erratic" (Rosche 1982).

In recent years a population has become established in the Wildcat Hills, although the only evidence of breeding is of one carrying nesting material in spring 1995 (NBR 63:50). Summer reports from the Wildcat Hills are 22 Jun 1957, 30 Jun 1969, 5 Jul 1978, 9 Jul 1988, 20 Jul 1984, permanent resident in 1992, 8 birds seen 24 Jul 1994, and 14 birds seen 7 May 1995. The latter report was from Carter Canyon.

There are several recent confirmed records in the ponderosa pine woodland of the Niobrara Valley Preserve and nearby areas in Cherry, Keya Paha, and Brown Cos. In Cherry Co Manning (1981) found a nest in a "pine tree" and 2 weeks later, 2 Jun 1980, saw an adult feeding young in the nest hole and also removing a fecal sac. This nest was located at Steer Creek Campground in the McKelvie Division of the NNF, adjacent to the Niobrara River Valley. Short (1965) collected a paired male with enlarged gonads northwest of Ainsworth in Keya Paha Co in pine woodland 19 Jun 1964. Another pair was present. In 1982 Brogie and Mossman (1983) discovered adults feeding young in 1 Brown Co and 2 Keya Paha Co locations; Brogie and Mossman (1983) considered Red-breasted Nuthatch to be a "common breeding species in Ponderosa pine woods" in the Niobrara Valley Preserve.

Breeding occurs also in the cultivated pine forests of the Bessey Division, NNF, where the species was observed year-round by 1958 and now breeds commonly (Bray 1994).

There are only 5 other breeding season reports, all from areas essentially lacking pine woodland. These are 8 Jun 1978 Lancaster Co, 30 Jun 1992 Lancaster Co, 3 Jul 1986 Adams Co, 7 Jul 1988 Box Butte Co, 16 Jul 1978 Garden Co, and 30 Jul 1979 Garden Co. Five were heard in the Keith Co area early Jun 1994, and

a female netted 18 May had a brood patch; nesting is unknown in the area.

Winter

6, 11, 11 Aug→29, 30, 31 May

The above dates are from areas away from breeding locations.

This species is generally sedentary, although a variably significant portion of the population leaves the breeding areas during winter. The extent of this movement is dependent on the status of the cone crop in the breeding range (Bent 1946; Bock and Lepthien 1972).

In Nebraska winter visitors often occur outside the known breeding locations and forage in conifers and at feeders, sometimes remaining in an area for several weeks or months. According to CBC data, Red-breasted Nuthatch is essentially a low-density winter visitor, with small numbers present each year and erratic increases in certain years, such as 1980–81 and 1983–84. CBC data show little statewide variation in numbers of birds recorded per party-hour.

FINDING

It is easiest to see Red-breasted Nuthatch at feeders or in conifers in winter, but it is not difficult to find in coniferous woodland in the Pine Ridge or at the NNF at Halsey year-round.

White-breasted Nuthatch

Sitta carolinensis

STATUS

Common regular resident north and east, rare casual elsewhere. Uncommon regular winter visitor away from breeding range.

DOCUMENTATION

Specimen

UNSM ZM11608, 23 Mar 1895 Lancaster Co.

TAXONOMY

Two subspecies have been recorded: the eastern race *cookei*, which occurs in most of the eastern United States, including eastern Nebraska, and *nelsoni*, the Rocky Mountains race, which occurs in Nebraska where ponderosa pine woodland is present (AOU 1957; Rapp and others 1958). The races may meet in the central Niobrara Valley, but no likely contact zone in that area has been studied. Individuals reported recently in summer in deciduous woodland along the North Platte River in Scottsbluff have not yet been identified to subspecies, although vocalizations of the 2 races are noticeably different.

DISTRIBUTION AND ECOLOGY

The largest numbers are found in mature oak-hickory forest in the southeast. Lesser numbers occur westward in deciduous woodland in major river valleys and also in ponderosa pine woodland in the northwest. A key requirement in any habitat appears to be "old growth, large trees, with dead and dying" elements (Peterson 1990).

This species is a rare to uncommon winter visitor in areas where it does not occur in summer, noticeably in western towns and cities and river valleys wherever larger trees are present.

Summer

Published records of breeding (Ducey 1988) and reports of occurrence in summer months reflect this species' preference for mature forest and woodland in the eastern parts of Nebraska, the Niobrara Valley, and the Pine Ridge. There is a sharp decline in numbers away from these preferred habitats.

The species is common to uncommon along the entire Niobrara Valley, both in deciduous woodland in the east (Brogie and Mossman 1983) and ponderosa pine woodland in the west (Rosche 1982). At least 1 bird was noted at Smith I., Sheridan Co, 30 Aug 1998. Summer reports from the Wildcat Hills are lacking; reports there 10 Mar 1996 and 10 Apr 1996 may have been winter visitors. It breeds throughout the Loup drainage, west to Thomas Co, where it is an "uncommon permanent resident" (Bray 1994).

It appears to be rare in the Platte Valley west of Hall Co (Johnsgard 1979). However, it may be increasing, as Lingle and Hay (1982) considered it an uncommon permanent

resident but Lingle (1994) listed it as common. A few are present all year in the Gibbon area (LR, RH). Rosche (1994a) stated that the westernmost breeding site known to him was at Hershey. Rosche and Johnsgard (1984) listed it only as a winter resident in Keith Co, although it does breed rarely in the South Platte Valley in Colorado (Andrews and Righter 1992) and throughout Wyoming (Scott 1993). There are, however, recent summer records for Keith Co 4 Jun 1992, 18 Jun 1994, and 14 Aug 1994 (Brown and others 1996) and for Oshkosh 15 Jul 1987 (Rosche 1994a), as well as sightings at Stateline Island 29 Jun 1994, 9 Jul 1994, and 23 Aug 1996. Birds seen at L Minatare in Sep were "probably breeding" (SJD). These sightings suggest possible expansion into the North Platte Valley (NBR 64:120).

Rosche (1994b) stated that it breeds in the Republican Valley west to Dundy Co, apparently a somewhat recent phenomenon, as Ludlow (1935) considered it only a "winter visitor and rare spring migrant" with no dates after 6 May 1920. It does not occur, however, in the Colorado portion of the Republican River drainage (Andrews and Righter 1992).

Winter

While nuthatches are usually considered sedentary, those in northerly parts of the breeding range may move in fall to more hospitable wintering areas, where the species may be rare or absent as a breeder (Phillips 1986). In the Panhandle more than 90% of reports occur in the period 15 Sep–26 May, and as far east as Howard and Hall Cos more than 90% of records are in the period 7 Sep–20 May. Glandon and Glandon (1934b) described it as a "common winter resident" in Logan Co, and Tout (1947) listed it as an "irregular visitor" in Lincoln Co, occurring 16 Oct–13 Apr. Thus in large parts of southwestern and central Nebraska White-breasted Nuthatch is primarily a scarce winter visitor. There are no banding data to indicate the origin of these winter visitors, although Phillips (1986) suggested that *nelsoni* move "fairly regularly" to lowlands in winter. An early fall migrant appeared at Crescent L NWR 3 Sep 1995.

FINDING

This species is easily located year-round by its call in the preferred habitat of oak-hickory forest, such as at Fontenelle Forest and Indian Cave SP.

Pygmy Nuthatch

Sitta pygmaea

STATUS

Common regular resident northwest. Common regular winter visitor northwest, rare casual central.

DOCUMENTATION

Specimen

UNSM ZM11875, 16 Jun 1966 Monroe Canyon, Sioux Co.

DISTRIBUTION AND ECOLOGY

This species' distribution closely corresponds to that of ponderosa pine in the western United States. It is restricted to ponderosa pine in Nebraska, preferring open woodland in dry canyons and on ridgetops, nesting in dead snags (Rosche 1982).

Numbers vary in winter, at which time the range expands somewhat into areas immediately south and east of the breeding range, including deciduous habitats.

Summer

Until recently, confirmed breeding records were all from Sioux Co. The first was in 1957, when a juvenile male was collected 18 Jul from a family group 8 km (5 mi) north of Harrison (Ford 1959). A second breeding record was of food being carried to a nest hole in a dead ponderosa pine snag in Smiley Canyon 29 Jun 1970 (Rosche 1972). The only other reports involved one carrying food 12 Jun 1981, location not given but in either Sioux or Dawes Co (Wright 1982); a nest record for 1961 for the northern half of the Panhandle (Wensien 1962); and a confirmed nesting record in Sioux or Dawes Co during the period 1984–89 (Mollhoff 1997b).

Rosche (1982) considered this species to occur to about 8 km (5 mi) east of Chadron as a resident, while Johnsgard (1980) called it a "local but regular Permanent Resident" in the Pine Ridge, apparently nesting "east to Chadron State Park." One was reported 19.3 km (12 mi) south of Chadron 4 Aug 1994, and 18 were at Chadron SP 1 Aug 1996, but the range of confirmed breeding was extended to Sheridan Co when Mollhoff (1997b) found 2 active nests at Metcalf Area 31 May 1997; Mollhoff found an additional 15 pairs or trios and 6 nests in Monroe and West Ash Creek Canyons during the same month.

The only other breeding location is the Wildcat Hills. Following several suggestive observations, breeding was confirmed when a pair built a nest in an artificial snag outside the Wildcat Hills NC 16 May 1996 (NBR 64:57). Previous reports suggestive of breeding were a "possible nesting pair" in Scotts Bluff Co 23 Apr 1983 (AB 37:887) and a series of sightings in extreme northern Banner Co in the Wildcat Hills 21 May–30 Jun 1988: 2 birds were seen 21 Jun in pines 0.8 km (0.5 mi) south of where they were seen 21 May, and several were seen 0.8 km (0.5 mi) farther south 30 Jun, raising the "possibility that they nested in the Wildcat Hills" (NBR 56:73). There were subsequent reports in 1994 and 1995, including 8 seen there 28 Jul 1994 and 1 seen in Carter Canyon 7 May 1995.

High counts include 14 at Ponderosa Area 7 Oct 1995, 11 in Sioux Co 21 Aug 1994, and 11 in the Wildcat Hills 17 Nov 1995.

Winter

While this species is normally sedentary, flocks as large as 50–100 birds may form in fall and move short distances from breeding areas, especially to lower elevations (Bent 1946). At this time the birds may be found in deciduous woodland, especially along watercourses (Johnsgard 1980). Rosche (1982) stated that this species stages periodic flights eastward from the Laramie Mountains in Wyoming in fall, especially along the

Pine Ridge. A probable example of this phenomenon was the Crawford CBC count of 49 birds 16 Dec 1979. Pygmy Nuthatch was recorded on each of the CBCs held at Crawford (1973, 1976–79), ranging from 4 to 49 birds; it was recorded once on the Chadron CBC, 7 on 9 Jan 1972. A "family group" was reported in pines in Sheridan Co 15 Nov 1980 (AB 35:200). Away from the Pine Ridge, it has appeared fairly regularly in the Scottsbluff area in fall, with records 7 Sep–21 Dec. These may be Wildcat Hills birds, as there are no Scottsbluff CBC reports.

Pygmy Nuthatch is extremely rare away from the range of ponderosa pine. The only documented reports are of 3 at Smith L, Sheridan Co, 15 Nov 1980 (Rosche 1982) and as many as 4 that remained in pine and pin oak (*Quercus palustris*) woodland in Lincoln 31 Dec 1961–3 Feb 1962 (Harrington 1962). Undocumented reports are 30 Sep 1976 Lincoln Co, 28 Jan–7 Feb 1966 Lancaster Co, and a report without details from Kearney (Rapp and others 1958). An undated report by Aughey, apparently from the area around the mouth of the Niobrara River, was considered highly questionable (Sharpe 1993).

FINDING

Reliable areas are the top of Sowbelly Canyon and the Pants Butte Road adjacent to the east, both in Sioux Co. Late summer, Jul–Sep, is best, when family groups are wandering through the pines.

Certhiidae (Creepers)

Brown Creeper

Certhia americana

STATUS

Rare regular resident east and north. Uncommon regular spring and fall migrant and winter visitor east, becoming rare west.

DOCUMENTATION

Specimen

UNSM ZM6569, 25 Mar 1896 Lancaster Co.

TAXONOMY

Two subspecies have been recorded in Nebraska. The eastern subspecies *americana* occurs from the eastern edge of the Great Plains eastward (AOU 1957) and presumably is the race that breeds in the Missouri Valley. The Rocky Mountains race *montana* (AOU 1957) is a common breeding bird in the Black Hills of South Dakota (Peterson 1990) and presumably accounts for Pine Ridge summer records. It is unknown which race breeds in the central Niobrara Valley; the distribution of the 2 races in Nebraska resembles that of White-breasted Nuthatch subspecies.

DISTRIBUTION AND ECOLOGY

Brown Creeper breeds in very small numbers in the Missouri Valley, where it occupies swampy woodland with dead elms, a limited habitat in Nebraska. It also breeds occasionally in the central Niobrara Valley, where the single recorded nest was in a rotten ponderosa pine snag in mixed woodland (Brogie and Mossman 1983), and probably in the Pine Ridge, where it occurs in ponderosa pine woodland (Rosche 1982). Pine Ridge habitat may be marginal, as this species breeds in old-growth pine and spruce forest in the Black Hills of South Dakota (Peterson 1990). Absence of extensive nonfragmented habitat in Nebraska may limit breeding by this species, as it is considered a "forest-interior nesting species" (Dobkin 1994).

Migrants and wintering birds occur wherever mature woodland or stands of mature trees are found, mostly in the southeast.

Spring

winter→26, 29 Apr, 2 May

Departure of wintering birds and migrants takes place by the end of Apr; later reports may be of breeding birds.

Summer

Breeding has been confirmed in only 2 localities, Brown Co and Sarpy Co, and may also occur in the Pine Ridge, but regular breeding probably occurs only in Sarpy Co.

There are several Sarpy Co records, all from

Fontenelle Forest, beginning with the observation by Cortelyou (1975) that a female with a brood patch was banded 16 Jul 1974, and birds had been seen carrying food in 1974 and 1975 (AB 29:873). In 1975 birds were seen carrying sticks to a nest site. On 26 May 1975 2 birds were flushed from under loose bark on a dead tree, presumed to be an elm (Cortelyou 1975). The nest was 1.8 m (about 6 ft) from the ground and contained 2 nestlings. Breeding has been recorded in subsequent years also, and low numbers of summering birds are found most years.

There is some evidence for additional breeding in the Missouri Valley and Platte Valley. Ducey (1988) cited breeding prior to 1920 in Dakota Co, and there is a recent report from there for 14 May 1984, a late date for a migrant. One was singing at the Krimlofski Tract, Neale Woods, 24 May 1998 (BP, LP). There are reports for Cass Co 12 May 1966, 18 May 1995, 6 Jun 1997, and 30 Jun 1967 and from Schramm Park 18 May 1995.

The single Brown Co record was of an adult carrying food to a nest in a rotten ponderosa pine, with a male singing nearby (Brogie and Mossman 1983). The location was on the Niobrara River floodplain in mixed woodland also including American elm, cottonwood, Eastern red cedar (*Juniperus virginiana*), green ash, and ponderosa pine.

Rosche (1982) stated that Brown Creeper probably breeds in the Pine Ridge, where it is regular in summer in Sowbelly and King's Canyons, but breeding has not been confirmed there. It is listed as "rare" in summer at Chadron SP on the park's checklist of birds; a single was noted there 14 Jun 1998 (JS, WRS). Several were observed in late Aug 1994, too early to be fall migrants, in Sioux Co (FN 49:66). The species was listed as a permanent resident in Scotts Bluff Co in 1963 and Dawes Co in 1993, without details. It was also recorded in Scotts Bluff Co 9–15 May 1984 and in Lincoln Co 23 May 1983 and 26 Jun 1976. These late spring records may be of migrants, however.

Reports suggestive of breeding in the Platte Valley are from Adams, Polk, Hamilton, and Howard-Hall Cos. There are a few May reports from Adams Co and a 5 Jul 1973 report. There are also single May records 9 May 1966 Polk Co, 13 May 1972 Howard-Hall Cos, and 28 May 1950 Hamilton Co, as well as Jul reports from Howard-Hall Cos, 1 Jul 1989 and 5 Jul 1972. Taken together, these reports suggest a small breeding population in the central Platte Valley, where mature floodplain woodland exists.

May reports for Perkins and McPherson Cos must be of stragglers, as there is little suitable breeding habitat there. In Perkins Co there is a 29 May 1972 report, and in McPherson Co the reports are 2 May 1962 and 2 May 1966. Undocumented McPherson Co reports for 11 Aug 1980 and 26 Aug–1 Sep 1985 are questionable, as there are virtually no Aug reports other than of known breeding birds, and no breeding habitat exists in McPherson Co.

Fall

20, 22, 24 Sep→winter

Migrants and wintering birds arrive in late Sep, with peak movement probably in Oct and early Nov. It is uncommon in the Panhandle.

High counts include 7 in Omaha 13 Oct 1996, 6 in Bellevue 2 Nov 1996, and 6 at Oak Glen Area, Seward Co, 29 Oct 1999.

Winter

Brown Creepers generally winter within their breeding range, except for northern or high-altitude breeding birds. Nebraska breeding birds would be expected to remain near their breeding sites to winter, although young birds probably disperse to some extent. No banding data exist to confirm this, however.

This species is a low-density winter visitor, primarily in the southeast, although apparently as far west as Logan Co, where it was considered a "common winter resident" by Glandon and Glandon (1934b). It is scarce in the Panhandle in winter, especially away from the Pine Ridge, as indicated by CBC data, and also in the L McConaughy region (Rosche 1994a). It is an "uncommon winter resident" at Bessey Division, NNF (Bray 1994).

At least one example of social roosting in winter has been observed, that of about a dozen individuals clumped beneath an eave in "midwinter" during the 1960s near Wisner (W. Lueshen, pers. comm. RSS).

FINDING

This species is most easily found in late fall in Missouri Valley forests, often with mixed flocks of chickadees and other species.

Troglodytidae (Wrens)

Rock Wren

Salpinctes obsoletus

STATUS

Common regular spring and fall migrant west, becoming rare casual east. Common regular breeder west, rare central.

DOCUMENTATION

Specimen

UNSM ZM6601, 12 Jun 1901 Warbonnet Canyon, Sioux Co.

DISTRIBUTION AND ECOLOGY

Most numerous in xeric, rocky areas in the Panhandle and southwest, this species also occurs in diminishing numbers eastward in the Platte Valley to Lincoln Co and possibly Dawson Co. Exposed rocks appear to be a requirement, as nests are usually placed in crevices in rocks or in canyon or cliff walls, although it is known to nest in dirt banks or even trees.

It is an uncommon to casual migrant away from breeding areas east to Brown, Blaine, and Dawson Cos. Stragglers east of the regular range are usually found in such situations as riprap on dams or near concrete structures and also around dirt banks.

Spring

2, 6, 9 Apr→summer

In northern parts of the breeding range, including Nebraska, Rock Wrens are migratory. Arrival is in early Apr, and, away from breeding sites, reports are few after May.

There are few reports from the east: 17 Apr 1994 Seward Co, 28 Apr 1993 Sarpy Co (AB 47:428), 28–30 Apr 1996 Lancaster Co (also at the same location in 1995; NBR 64:57), 29 Apr–4 May 1965 Adams Co, 10 May 1996 Sarpy Co, 13 May 1972 Adams Co, and 20 May 1967 Adams Co.

Summer

Excluding an extralimital record of young fledged in Antelope Co in 1992 (Gubanyi 1996a; AB 46:1152), this species has been recorded breeding only in Sioux, Dawes, and Scotts Bluff Cos in the northwest and in Keith and Lincoln Cos in the western Platte Valley (Ducey 1988). In Keith Co it nests regularly in the L McConaughy area (Brown and others 1996). There are breeding season reports from Morrill, Cheyenne, Chase, Hitchcock, and Dawson Cos, suggesting that it probably breeds in small numbers east to Dawson Co.

The breeding range extended much farther to the north and east earlier in the 20th century, with records from Custer, Cherry, and Brown Cos (Youngworth 1955; Ducey 1989). The presence of a small breeding colony near Sioux City, Iowa, in the late 1890s and early 1900s (Dinsmore and others 1984) suggests the Niobrara Valley as a route for birds to travel eastward. A Rock Wren was seen near abandoned buildings in Brown Co in Apr and May 1982, but no evidence of nesting was observed (Brogie and Mossman 1983).

High counts include 15 in Sioux Co 1 Aug 1996.

Fall

summer→14, 15, 17 Oct

Most depart by mid-Oct, although there are later reports 29 Oct 1975 Scotts Bluff Co and 30 Oct 1980 Sioux Co; away from the breeding range migrants occur in Sep and Oct.

There are a few reports in the south and east away from breeding locations, suggesting some fall wandering: 6 Sept 1979 Sarpy Co (AB 34:177), 21 Sep 1912 Lancaster Co (specimen, UNSM ZM11649), 5 Oct 1992 Sarpy Co (Padelford

and Padelford 1993), 7 Oct 1986 Madison Co
(AB 41:111), and 15 Oct 1962 Adams Co. The
Sarpy Co records were on or near railroad
tracks that pass through Fontenelle Forest
(Padelford and Padelford 1993).

High counts include 8 in southwest Kimball Co
12 Sep 1998.

FINDING

This species is best looked for in rocky
escarpments in the Panhandle in May and
Jun, when it can be attracted by squeaking or
located by its characteristic call notes.

Canyon Wren

Catherpes mexicanus

STATUS

Rare casual fall, winter, and spring visitor
northwest and north.

DOCUMENTATION

Photograph

14 May–9 Jun 1992 Knox Co (Brogie 1992).

Records

There are 4 reports, 2 of which are documented:

20 Jan 1974 Smiley Canyon (RCR, DJR; AB 28:659;
Bray and others 1986)

14 May–9 Jun 1992 Lewis and Clark L, Knox Co
(Brogie 1992; Gubanyi 1996a; NBR 60:140)

A report without details has been published
for one seen at King's Canyon, Dawes Co, 11
Apr 1972 (Gates 1974), and Bruner and others
(1904) described, without details, a sighting 12
Aug 1903 in a canyon along the White River
between Glen and Andrews in Sioux Co.

COMMENTS

This species is widely distributed in the
western United States, including an isolated
population in the Black Hills of South Dakota
and scattered populations in Wyoming,
notably the Laramie Mountains (Scott
1993). Although the species is generally
nonmigratory, Wyoming birds depart in
winter (Scott 1993), and Johnsgard (1979)
stated that Canyon Wren is somewhat
migratory in South Dakota, with Black Hills
birds arriving on breeding territories in early
Apr. Some Black Hills birds are resident but

may move to lower elevations in winter to
avoid die-offs (Rosche 1982; Peterson 1990).
It has occurred in fall in extreme western
Kansas (Mark Robbins, pers. comm.).

Carolina Wren

Thryothorus ludovicianus

STATUS

Uncommon regular resident southeast. Rare
casual winter visitor elsewhere.

DOCUMENTATION

Specimen

UNSM ZM11643, 27 Dec 1911 Rulo, Lancaster Co.

DISTRIBUTION AND ECOLOGY

Currently the breeding range is restricted to
the southeast, where this species utilizes
edge habitats, especially brushy riverbottom
woodlands, although it may occur in
disturbed and urban habitats.

Summer

There were few reports prior to the winter of
1977–78. Breeding was recorded in Adams
Co in 1965, where young were seen 13 Jul
(NBR 34:41), and there were summer reports
there in 1966 and 1976 but none since 1977. A
specimen, HMM 2894, was collected at Inland,
Clay Co, 6 May 1923. There are breeding
records for Nuckolls Co in the early 1930s
(Ducey 1988), but otherwise there have been
no breeding records away from the Missouri
Valley. Summer reports from Thayer and
Hall Cos, along with the Adams, Clay, and
Nuckolls Cos reports (see above), suggest that
a breeding population existed in the Little
Blue drainage and north to the Platte Valley.
This population has apparently disappeared
since 1977. However, a small population has
recently established at Harlan Co Res, where 3
appeared in Dec 1998 and were still present 7
Nov 1999 (GH, WH, LR, RH).

Following a very cold winter in 1977–78, CBC
data show a 5-year period when no birds
were found. Only in recent years have CBC
numbers reached significant levels. Since 1977
breeding has been recorded only in Douglas
and Nemaha Cos, although there have been

summer or BBS records north to Dakota and Knox (AB 46:1152) Cos, and west to Gage, Lancaster (AB 30:974), Clay (NBR 63:50), and Harlan Cos. It was listed as a permanent resident in Lancaster Co in 1992 and was reported from Dakota Co 1 Jul–14 Aug 1992 and 21 May–30 Jun 1993.

Fledglings were banded at Bellevue 1 Jul 1994, and broods were being fed by adults in Sarpy and Cass Cos 11 May and 17 May 1995 and at Indian Cave SP and Wilderness Park, Lincoln, 15–23 May 1999.

Winter

This species is sedentary within its breeding range but expands its range north and west in years with mild winters. Data from states at the range periphery indicate that spring dispersal, coupled with ability to survive the subsequent winter, is the probable method by which the species expands its range. Nebraska reports fitting this pattern are few since the winter of 1977–78: 18 Apr 1978 Lincoln Co; 15 May 1994 Buffalo Co; 26 May 1994 Dodge Co.; fall 1995 at a Hall Co feeder; 12 Oct 1994 Buffalo Co; 29 Oct 1994 Cuming Co, a bird that remained through the winter at a feeder (Brogie 1997; NBR 62:140); 18 Dec 1998 (2) Harlan Co Res CBC; and 24 Dec 1993 Dakota Co, the "first local winter record" (BFH).

Prior to 1977 there were few such reports, and, while scattered through the year, they appear to fit the dispersal pattern suggested above. Two are from the Panhandle, both in winter: 7 Jan 1958 Scotts Bluff Co and 18 Jan 1956 Sheridan Co. The few reports from south-central Nebraska prior to 1977 are 31 Mar 1963, 26–30 April 1962, and 30 May 1960 Lincoln Co;, these suggest that a small breeding population existed there in the early 1960s. There is a report of one in the Tout yard in Lincoln Co 26 Aug–10 Sep 1938 (Tout 1947) and a Perkins Co report 9–10 Oct 1970.

FINDING

Carolina Wren can be readily attracted by squeaking. Good localities are Fontenelle Forest and Indian Cave SP during the breeding season, although it can be found easily year-round. It has a fondness for brushy ravines. It sings year-round.

Bewick's Wren

Thryomanes bewickii

STATUS

Rare casual spring visitor east and central, hypothetical west. Rare casual summer, fall, and winter visitor east and central.

DOCUMENTATION

Photograph

3 Apr 1993 Wild Rose Ranch, Hall Co (BJR; NBR 61:137).

TAXONOMY

Most, if not all, reports are presumed to be of *bewickii*, which breeds to the south and east of Nebraska (AOU 1957), although the race *pulichi* of central Oklahoma and Kansas (Phillips 1986; Pyle 1997) may occur also.

A specimen purportedly collected 21 Apr 1915 near Oshkosh by C. A. Black was examined by H. C. Oberholser (Swenk 1918b, 1933d) and determined to be of the southwestern United States race *eremophilus*. The specimen cannot now be located (Bray and others 1986). Possibly based on this specimen, Swenk (1933d) suggested that *eremophilus* occurred in western Nebraska as a summer resident, but there is no other evidence to support this. Indeed, data from northeastern Colorado (Andrews and Righter 1992) show that Bewick's Wren is but a rare spring and fall migrant, restricted to the foothills. It is unlikely that this specimen was collected in Nebraska.

DISTRIBUTION AND ECOLOGY

Where it is common in Missouri and Arkansas, Bewick's Wren occurs around abandoned farm buildings, junk piles, and brushy edges of woods (Robbins and Easterla 1992; James and Neal 1986). These authors believe that cleanup of such sites in recent years has caused the decline in Bewick's Wren populations in the eastern United States.

Spring

25, 26, 27 Mar→20, 20, 21 May (Ludlow 1935)

Bewick's Wren is generally sedentary, although birds at the northern edge of the breeding range are migratory.

About half of the Nebraska reports are from the 1950s and 1960s; there are only about 20 reports since 1969. Most of these are in spring, but there are 2 winter reports and a few others divided between summer and fall.

Since 1980 Bewick's Wren has occurred primarily as a spring visitor, with most reports in the period 3 Apr–20 May, several from rather far north in the state: 3–5 Apr 1993 Hall Co (AB 47:428); 15 Apr 1995 Clay Co; 1 May 1988 Knox Co; 3 May 1982 Brown Co; 4 May 1982 Cherry Co; 5 May 1984 Polk Co; about 15 May through at least 5 Jul 1999 Ashfall Fossil Beds SP, Antelope Co; 18–19 May 1985 Sarpy Co; and 20 May 1986 Douglas Co. There is an earlier report 15 Mar 1961 Nemaha Co and later reports 24 May 1938 Lincoln Co (undocumented), 27 May 1956 Dawes Co (the only Panhandle report, albeit undocumented), and 28 May 1966 Adams Co.

Summer

There is no indication that Bewick's Wren ever bred regularly in Nebraska. There are, however, 5 reports of nesting, beginning with Aughey's 1875 report of adults feeding young in Otoe Co, discussed by Swenk (1933d). The next reported nesting attempt was that of a pair in Kearney in spring 1931 (Swenk 1933d; Rapp and others 1958). One of these birds had wintered in the area (see Fall); the nesting was unsuccessful. The next 2 reports were from Nemaha Co in 1952 (NBR 21:23) and 1959 (NBR 38:15). The latter nesting attempt was successful, and the birds stayed until the first week of Jul. The most recent breeding report is from Gage Co, sometime prior to 1970 (Fiala 1970). No details were given. Johnsgard (1980) considered an old report of breeding in Lincoln Co "very questionable." No further details were given.

Aside from these breeding reports, there are a few additional summer reports, none documented: 20 Jun 1955 Webster Co, 29 Jun 1952 Webster Co, 10 Jun 1954 Saline Co, a BBS report from Loup Co that listed this species (NBR 46:52), 5 Jul 1965 Adams Co, 11 Aug (Johnsgard 1980), 12 Aug 1974 Lincoln Co, 13 Aug 1940 Lincoln Co (Tout 1947), and 13 Aug 1968 Lancaster Co.

Fall

There are few reports, all but 1 in Sep and Oct and generally in the east, although there are 2 from Lincoln Co. These include 5 Sep 1974 Lancaster Co, 6 Sep 1968 Lancaster Co, 14 Sep 1976 Lincoln Co, 18 Sep 1960 Cass Co, 23 Sep 1976 Lincoln Co, 29 Sep 1964 Cass Co, 1 Oct 1994 Otoe Co (LF, CF), 3 Oct 1971 Lancaster Co, 3 Oct 1972 Lancaster Co, 15 Oct 1916 Lancaster Co, 18 Oct 1998 Harlan Co Res (GH, WH), and 23 Oct 1965 Douglas-Sarpy Cos.

Winter

There are 3 reports that suggest attempted wintering: 21 Dec 1986 Douglas Co (description, NBR 48:74), one that appeared in Kearney 30 Dec 1930 and subsequently wintered (Swenk 1933d), and one that appeared in North Platte 21 Jan 1933 and remained until 30 Mar (Tout 1947).

FINDING

Perhaps the best chance for finding a Bewick's Wren is around junk piles or abandoned farmsteads, especially those with adjacent woodlots, in extreme southeast and south-central Nebraska in Apr, prior to the arrival of most House Wrens.

House Wren

Troglodytes aedon

STATUS

Common regular spring and fall migrant and breeder statewide.

DOCUMENTATION

Specimen

UNSM ZM6577, 24 May 1900 Monroe Canyon, Sioux Co.

DISTRIBUTION AND ECOLOGY

This species is ubiquitous in woodland in

the east and elsewhere in woodland and
residential areas wherever brushy cover
exists. It nests in a wide variety of nooks and
crannies, including nest boxes provided in
residential areas.

Spring

7, 7, 7 Apr→summer (east)

13, 15, 15 Apr→summer (elsewhere)

Arrival dates in the east are about 1 week earlier
than elsewhere, although there are a few
earlier reports away from the east: 1 Apr 1950
Logan Co, 7 Apr 1949 Keith Co, 8 Apr 1951
Logan Co, and 10 Apr 1945 Adams Co. There
are also a few extremely early undocumented
reports: 27 Feb 1988 Knox Co, 8 Mar 1980
Douglas-Sarpy Cos, 10 Mar (Johnsgard 1980),
20 Mar 1972 Howard-Hall Cos, 28 Mar 1992
Douglas-Sarpy Cos, and 31 Mar 1997 Crescent
L NWR, a westerly location. It is possible
that some of these reports are misidentified
Winter Wrens; overwintering of House Wren
has not been documented in Nebraska.

That the earliest birds are migrants and not
arriving summer residents is suggested
by South Dakota data (SDOU 1991), which
indicate that migrants arrive in mid-Apr and
earliest nesting is in mid-May. Even allowing
for females arriving some 9 days after males,
first nesting in mid-May implies arrival of
resident males no sooner than the first few
days in May, as nesting begins soon after
females arrive. This would indicate about
2 weeks between arrival of migrants and
resident birds, the same differential noted
in Nebraska data. Migration may continue
into May, however, as there are records of 2
birds banded in Nebraska in May 1985 being
recaptured in North Dakota and Alberta in Jul
1985.

High counts include 351 in Pierce Co 8 May 1999;
224 in Sarpy Co 11 May 1996, 167 of which
were at Fontenelle Forest; and 75 in Fontenelle
Forest 8 May 1999.

Summer

House Wren breeds statewide, although densities
are low in the southern Panhandle and in the

Sandhills. Two BBS routes in Cherry Co were
run 12 times with no House Wrens recorded
(NBR 46:38). BBS data (Robbins and others
1986) show highest densities (above 11 birds
per route) in the east and lowest (fewer than
4) in the west.

Summering birds show strong site fidelity. Of 14
birds banded in the period May–Aug, 12 were
recaptured 1–3 years later in the same place.

Fall

summer→22, 22, 25 Oct

Departure is mid to late Oct, although there
are a few later undocumented reports: 2 Nov
1996 Nebraska City, 5 Nov 1986 Lancaster
Co, 17 Nov 1981 Garden Co, 20 Nov 1982
Lancaster Co, 10 Dec 1993 Sarpy Co, 16 Dec
1989 Washington Co, 24 Dec 1963 Nemaha Co,
and 1 Jan 1957 Adams Co.

High counts include 15 at Arbor Day Farm 23 Sep
1998 and 11 at North Platte NWR 23 Aug 1996.

FINDING

Perhaps one of the easiest birds to find, House
Wrens make their presence obvious in May
by song and are easily attracted by squeaking
due to their curiosity and/or territoriality.
They are most numerous in Missouri Valley
woodlands.

Winter Wren

Troglodytes troglodytes

STATUS

Uncommon regular spring and fall migrant east,
rare casual central and west. Rare regular
winter visitor east, rare casual central and
west.

DOCUMENTATION

Specimen

UNSM ZM6573, 2 Dec 1901 Dunbar, Otoe Co.

TAXONOMY

According to the AOU (1957) and Rapp and
others (1958), the eastern subspecies *hiemalis*
occurs in Nebraska as a migrant and winter
resident.

Andrews and Righter (1992) considered the
race *pacificus* of the British Columbia coastal
region to be a rare winter visitor in the South

Platte Valley in northeastern Colorado based on 2 specimens; however, Phillips (1986) determined that specimens (possibly not the same specimens?) in the Denver Museum of Natural History were *hiemalis*.

DISTRIBUTION AND ECOLOGY

During migration this species is found in dense woodland thickets, usually near water; it is most numerous in the east and essentially limited to the Missouri Valley. Winter Wren is a marginal wintering species in Nebraska, possibly due to its preference for an insectivorous diet (Bent 1948).

Spring

winter→9, 10, 15 May

Numbers peak in Apr, and most depart by mid-May. There is a late report 30 May 1965 Adams Co.

High counts include 5 at Fontenelle Forest 25 Apr 1999.

Fall

12, 15, 15 Sep (AB 28:74)→winter

Migrants arrive in mid-Sep, with an apparent peak in Oct. There are 2 earlier undocumented reports 30 Aug 1964 and 4 Sep 1975, both from Lancaster Co.

High counts include 6 at Neale Woods 15 Oct 1999, 4 at L Ogallala 19 Nov 1999, and 3 at Krimlofski Tract, Washington Co, 6 Oct 1996.

Winter

Regular wintering is restricted to sheltered areas in the southeast, although there are scattered reports elsewhere, notably in the Pine Ridge and in Keith Co, of birds probably attempting to winter: 19 Nov 1999 (4 birds with 3 still present 2 Jan 2000) at L Ogallala (SJD), 23 Nov 1986 Keith Co (Rosche 1994a), 11 Dec 1988 Keith Co (AB 43:332; Rosche 1994a), 16 Dec 1973 Keith Co, 23 Dec 1978 Dawes Co, 5 Jan 1957 Cherry Co, 15 Jan 1959 Dawes Co, Jan 1971 Brown Co, 24–30 Jan 1993 Buffalo Co (AB 42:273), 1 Feb 1998 Willis (BFH), 5 Feb 1995 (2) Keith Co (SJD, WRS), 24 Feb 1974 Sioux Co, and "winter" 1986–87 Sioux Co (AB 41:299).

Rosche (1982) considered Winter Wren a "rare

winter visitor" in the northwest and suggested that few if any survive the winter, as there were no Mar records. The westernmost Mar reports are from Boone Co 27 Mar 1986 (AB 40:493) and Brown Co Mar 1971; these are the only Mar reports away from the Missouri Valley. Even in the southeast there are few Feb and Mar reports, indicating that the species does not overwinter (or survive in the attempt) every year. Winter Wrens were found 19 of 27 years (1968–96) on the Omaha CBC, with a high count of 6; the same number was found 19 Dec 1998.

FINDING

A regular location that is easily accessible is Stream Trail in Fontenelle Forest. Best times are mid-Apr and mid-Oct, at peak migration, when birds may be attracted into view by squeaking at dense, streamside tangles.

Sedge Wren

Cistothorus platensis

STATUS

Uncommon regular spring and fall migrant east, rare casual central. Rare regular breeder east. Uncommon regular summer visitor east, uncommon casual central.

DOCUMENTATION

Specimen

UNSM ZM6605, 28 Aug 1902 Lancaster Co.

DISTRIBUTION AND ECOLOGY

Most reports are from the east, usually in wet meadows. The few nesting records are from the eastern half of the state.

Sedge Wren has rather exacting breeding habitat requirements: damp grassy or sedge meadows; Bedell (1996) described a preferred habitat type as "sub-irrigated native meadow." Nests are well hidden and low in dense vegetation over damp ground or mud or at most a little standing water. Such habitat is extremely sensitive to weather conditions, and it appears that this species has the ability to search out suitable habitat and breed as late as Aug if necessary. In Nebraska territorial birds have also been found in lush midgrass and tallgrass prairie during summers with above-normal precipitation.

Spring

18, 20, 20 Apr→2, 3, 6 Jun

There is a well-defined movement in Apr and May, with apparently very small numbers remaining to breed, then a return migration beginning in mid-Jul and continuing into Oct. Early fall migrants are potential breeders (see Summer). Reports into early Jun may be late migrants or birds attempting to breed.

Reports away from the east are few: 20 April 1957 Thomas Co and 11 May 1957 Cherry Co. Although there are few reports from the western half of the state, it occurs regularly along the Middle Loup and Dismal Rivers in Thomas Co as an "uncommon migrant/summer resident" (Bray 1994).

Summer

Nesting dates from South Dakota (SDOU 1991) and Iowa (Dinsmore and others 1984) are in Jun and Jul, but dates for Kansas are later, in late Jul and Aug (Johnston 1965) and into Sep (Thompson and Ely 1992). A comment by Sherman (Dinsmore and others 1984) is of interest in this context: "migrants are seen [in Iowa] in mid-May, but then seem to disappear until Jul, when they nested." Thus migrants that pass through Nebraska in spring and are unsuccessful in locating suitable breeding habitat or indeed successfully raise a first brood but mowing or other intrusions cause failure of a second brood (Joe Roller, pers. comm. WRS) north of Nebraska apparently return and breed in Jul and Aug (and even Sep in Kansas) if suitable habitat can be found, a possibility noted by Lingle and Bedell (1989) and Bedell (1996). Bedell (1996) also suggested that late summer breeding in Nebraska may involve bachelor males moving south from northern breeding areas in hopes that if conditions are unsuitable to the north then returning females may arrive and allow breeding with these (now former) bachelor males. True fall migration (after nesting) in this species probably takes place Sep–Oct.

There are very few confirmed reports of breeding since 1960: a nest with 5 young was discovered in Thurston Co 24 Jun 1971 (Cink 1973b); a nest with eggs was found in Hall Co in 1988 at a location where 123 birds were present 26 Jul (Lingle and Bedell 1989; AB 43:126, 42:1309), and 7 nests with eggs were found in early Aug 1989 in Hall Co (AB 44:116). The 1971 nest record was the first for almost 70 years, since Wolcott (1902) noted nests at West Point and Lincoln prior to 1902. Aughey had found only 1 nest, that in a Dixon Co marsh in 1867 (Bruner and others 1904). Dates of nesting are unavailable for these old records.

Cink (1973b) listed instances of possible breeding in the early 1970s: several singing males were in a Wheeler Co marsh 9 Jun 1970; territorial birds and dummy nests were found north of Lincoln in Lancaster Co 8 Jul 1972; and territorial males and adults carrying food were found in a grass and sedge meadow near Clay Center 17 Jul 1971. All of these reports strongly suggest breeding, although no active nests were located. There is also a nest card from Lancaster Co for 1980, but, unfortunately, no date was noted (Bennett 1981).

There are about 30 mid-Jun through early Jul reports that may represent nesting birds; most are recent, the most westerly 1 Jul 1994 Valentine NWR, 27 Jun 1967 Brown Co, 12–21 Jun 1997 and all summer near Gibbon (LR, RH), 28 Jun 1984 Howard-Hall Cos, and 1 Jul 1964 Brown Co. Included are several recent reports (1996 and 1997) of territorial males in lush midgrass and tallgrass prairie habitat at Nine-Mile Prairie, Lancaster Co (JH). Reports from mid-Jul and Aug may either be migrants or breeding birds.

Fall

11, 12, 12 Jul→25, 26, 30 Oct

Migration apparently begins in mid-Jul where breeding did not occur, and departure is in mid to late Oct, although there is a later documented record, a specimen, WSC 196, 20 Nov 1976 Boone Co. Westerly reports include singles in Phelps Co 25 Aug 1996 and at Harvard Lagoon 13 Jul 1997.

High counts include 123 in Hall Co 26 Jul 1988 (Lingle and Bedell 1989), 60 in the eastern Rainwater Basin 7 Aug 1999, and 31 there 25 Jul 1999.

FINDING

The best chance of finding this species is in spring migration in grassy fields, especially in early May, in the eastern part of the state. It sings also in Jul and Aug and should be looked for at that time in breeding habitats such as sedgy grass meadows with 2.5 cm (about 1 in) of water. Such apparently territorial birds should be monitored for possible nesting attempts.

Marsh Wren

Cistothorus palustris

STATUS

Common regular breeder north, rare south. Rare regular winter visitor western Platte Valley.

DOCUMENTATION

Specimen

UNSM ZM6586, 2 Nov 1901 Nebraska City, Otoe Co.

TAXONOMY

There are 2 breeding subspecies. Evidence, both vocal and genetic, is accumulating that indicates these in fact are separate species (Kroodsma and Canady 1985; Kroodsma 1988; Sibley and Monroe 1990; AOU 1998). Ranges given by AOU (1957) suggest that *dissaeptus* occupies extreme eastern Nebraska, while *plesius* apparently occupies southwest Nebraska. Phillips (1986), however, uses *iliacus* for the birds occupying the western part of the range assigned to *dissaeptus* by AOU (the type of *dissaeptus* was collected in Massachusetts and that of *iliacus* in Indiana).

Kroodsma (1988) found 2 demonstrably different Marsh Wren song types are separated by a 160 km (100 mi) wide corridor with no suitable breeding habitat extending southeastward from the South Dakota line through O'Neill, effectively separating wrens

from Sandhills marshes to the west from birds of northeastern Nebraska marshes. Nesting birds at Twin Lakes, Rock Co, 24 May 1997 were western types (JGJ).

DISTRIBUTION AND ECOLOGY

Summer residents occur throughout most of the north, reaching greatest numbers in Sandhills cattail marshes. Migrants presumably occur statewide although are rarely seen outside breeding habitat. A few apparently survive the winter in marshes in southeast Garden Co, Keith Co, and possibly elsewhere.

Spring

10, 10 (Tout 1947), 11 Apr→14, 15, 18 May

Arrival is in mid-Apr, although there are earlier reports 5 Mar 1981 Garden Co, 13 Mar (Johnsgard 1980), 21 Mar 1999 (2) Funk Lagoon, 1 Apr 1956 Keith Co, 1 Apr 1995 Garden Co, and 4 Apr 1963 Douglas-Sarpy Cos. The Mar reports may have been overwintering birds (see Winter).

Data from the south, where Marsh Wrens do not breed, indicate that the last spring migrants pass through by mid-May. Reports after mid-May suggest breeding attempts.

Summer

While regular breeding is typically restricted to counties north of the Platte River, scattered breeding attempts may occur anywhere north of the southernmost 2 rows of counties (Ducey 1988; Kroodsma 1988; BBS data). The most southerly breeding record is a nest located by Cink in Lancaster Co 26 Aug 1970 (Cink and Fiala 1971a). This nest was in atypical habitat, a marsh overgrown by smartweed. There are breeding season reports (early fall migrants?) 30 Jun 1976 and 20 Aug 1974 Adams Co, late Jul 1983 (AB 37:1003) and 4 Aug 1998 Sarpy Co, 1 Jul and 6 Aug 1995 Arbor Lake, and 16 Jul 1995 Glenvil Basin.

High counts include 110 at Crescent L NWR 26 Jul 1995 and 45 there 8 Jun 1994.

Fall

summer→21, 22, 23 Nov

A later report was 30 Nov 1995 Garden Co.

It is difficult to determine the dates of early

fall departure, although there are no Jun reports south of the area of regular breeding; this is most noticeable in Lancaster and Douglas-Sarpy Cos, neither of which has a Jun report. Jul reports in these areas may therefore be early fall migrants; summering birds at Funk Lagoon were last seen 15 Aug in 1998.

Peak fall migration probably occurs in the second half of Sep and early Oct; Johnsgard (1980) found that half the fall reports were in the period 8 Sep–10 Oct. Tout (1947) stated that his earliest fall date was 19 Sep. Migration ends by early Nov; later reports may of birds attempting to winter.

Winter

Marsh Wrens apparently attempt each year to overwinter in the vicinity of L McConaughy in Keith and southeast Garden Cos (Rosche 1994a; FN 48:223), and there are also reports 6 Jan 1999 (6) Facus Springs (SJD), 20 Dec 1995 Winters Creek L, 21 Dec 1991 near Gering (Gubanyi 1996a), and 28 Jan 1997 Scotts Bluff Co. Midwinter reports are few, however: 6 Jan 1983 Garden Co (AB 37:314) and 9 Jan 1988 Keith Co (AB 42:285).

FINDING

Marsh Wrens can be found in Jun and Jul singing loudly at most Sandhills cattail marshes, notably at Crescent L NWR. Squeaking usually brings one into view.

Cinclidae (Dippers)

American Dipper

Cinclus mexicanus

STATUS

Rare casual fall, winter, and spring visitor. Accidental in summer.

DOCUMENTATION

Specimen

HMM 693 (pair), Jun 1903 Wauneta, Chase Co.

Records

There are 8 reports, of which 3 are documented:

2 Jun 1903 Frenchman Creek, Chase Co (cited above)

10 Oct 1967 Adams Co (Turner 1968; AB 22:59)

Dec 1969 Holt Co (Erwin 1970)

The 1903 record involved a pair collected along Frenchman Creek, near Wauneta in Chase Co, 2 Jun 1903 by James P. Allen. This pair was acquired by Brooking "after verifying the record" (Brooking, Notes), mounted as a display, and is currently in storage at the HMM (Bray and others 1986). Turner reported an American Dipper 10 Oct 1967 in Adams Co (Turner 1968; AB 22:59). His description mentioned "an odd-looking, slaty-gray bird, bearing some resemblance to a very plump thrush," and noted its calls. Another bird was seen by Erwin Dec 1969 along a cold-water stream running into the Niobrara River about 11.3 km (7 mi) downstream from the Highway 281 bridge in Holt Co (Erwin 1970). The description mentioned its "incessant dipping" and that it was a "brownish-gray, short-tailed bird about the size of a robin."

The other 5 reports do not contain enough information to judge the identification conclusively, but all may be correct. An unusual sighting was of one on a woodpile in downtown Stanton late Sep or early Oct 1982, seen by Jundt (Jundt 1984). The observer stated that the bird sat on the pile for 10 minutes at about 2 m (6–8 ft) distance and that he "had seen these birds on mountain streams and could hardly believe [his] eyes." One was seen on the Timber Reserve at Fort Robinson, in Sioux Co, 3 Oct 1960 by Suetsugu (Suetsugu 1961), but no description was given. One was seen on Otter Creek, Keith Co, by Paul Novak, apparently in Jan, but no other data were provided (NGP files). One was seen "sporting among the rocks" on the White River in Sioux Co in May 1891 by Bruner (Bruner and others 1904). Another report from the Niobrara River was one 13–16 km (8–10 mi) east of Valentine in Cherry Co 1 May 1977, seen by Maddux, which "exhibited the bobbing motion that is characteristic of the species" (Maddux 1977). The observer was familiar with dippers in Colorado.

COMMENTS

There are breeding populations as near as the northern Black Hills of South Dakota (SDOU 1991) and in the Laramie Mountains of Wyoming (Bent 1948; Dorn and Dorn 1990). Dippers are generally sedentary but on occasion in winter may move downstream to lower altitudes (Bent 1948). There are 2 records in northeastern Colorado: 31 Dec 1983 at Julesburg Res on the South Platte River and 1 Nov 1988 in Morgan Co (Andrews and Righter 1992), which, taken together with the Nebraska reports, suggest that these birds follow Nebraska rivers eastward, presumably during severe fall weather conditions. Some may linger into spring, such as the pair collected in Chase Co 2 Jun 1903.

FINDING

The most likely time to locate American Dipper appears to be the first week of Oct when early winter is severe in the eastern Rockies or especially in eastern Wyoming mountains such as the Laramies. Likely places would be the western reaches of rocky or fast-flowing rivers such as the White and Niobrara.

Regulidae (Kinglets)

Golden-crowned Kinglet

Regulus satrapa

STATUS

Fairly common regular spring and fall migrant statewide. Rare regular winter visitor southeast, becoming rare casual northwest.

DOCUMENTATION

Specimen

UNSM ZM6718, 13 Apr 1901 Lincoln, Lancaster Co.

DISTRIBUTION AND ECOLOGY

Golden-crowned Kinglet occurs statewide in all types of woodlands during migration, although most predictably in conifers, including cemeteries and similar plantings. Wintering birds are not numerous, but, while they may occur statewide, most are in the southeast and are usually found in coniferous habitats.

Spring

7, 7, 9 Mar→13, 14, 15 May (south)

20, 21, 21 Mar→13, 14, 15 May (elsewhere)

There is a discernible influx of migrants in early to mid-Mar, with earliest dates in the south, suggesting that many early spring reports there are of birds that wintered in the area or not far to the south. The earliest reports elsewhere, including the east, are about 2 weeks later.

Departure occurs mostly by the first week of May, with a few later reports: 21 May 1957 Keya Paha Co, 25 May 1994 Sarpy Co (FN 48:315), 28 May 1974 Adams Co, 4 Jun 1981 Adams Co, 15 Jun 1953 Thayer Co, 24 Jun 1988 (banded at place of recovery in Johnson Co 11 Nov 1987), and a report by Aughey in Dakota Co Jun 1865 (but see Sharpe 1993).

Fall

23, 24, 24 Sep→winter

Migrants arrive in late Sep, although there are earlier reports 13 Aug 1973 Adams Co, 21 Aug 1974 Lancaster Co, 10 Sep 1990 Sioux Co, 13 Sep 1911 Dawes Co (specimen, UNSM ZM10546), and 14 Sept 1962 Cass Co. It is difficult to determine fall departure, as numbers gradually decline into winter. CBC data indicate that by late Dec most birds are in the south and east, although a few still are present elsewhere.

High counts include 56 at L Ogallala 2 Jan 2000 and 19 there 28 Nov 1998.

Winter

Rosche (1982) observed that "very few of the individuals that attempt to winter are apparently successful, as there are only four early January records and two for February" in the northwest. More recently there are these additional reports from the Panhandle: 22 Jan 1993 Scotts Bluff Co and 5 Feb 1993 Sheridan Co.

Overwintering is rare elsewhere also but is most likely in the southeast, where a few birds may be found most years (Johnsgard 1980). A westerly report was of 1 at Funk Lagoon 18 Feb 1996; small numbers winter in the Kearney area (NBR 64:13), and 5 were at Johnson L 17 Jan 1999.

FINDING

This species is best located in Oct and Nov when squeaking will attract mixed flocks of chickadees and a few kinglets; later in fall most kinglets are Golden-crowned. It is most numerous in Missouri Valley woodland.

Ruby-crowned Kinglet

Regulus calendula

STATUS

Common regular spring and fall migrant east, becoming uncommon west. Rare regular winter visitor southeast, rare casual elsewhere.

DOCUMENTATION

Specimen

UNSM ZM6713, 13 Apr 1901 Lancaster Co.

DISTRIBUTION AND ECOLOGY

This species occurs in all types of woodland statewide and is generally common except possibly in the Panhandle, where it is rare to uncommon (Rosche 1982). Smaller brushy trees such as gray dogwood are preferred, but early in spring migration birds may be seen foraging on insects that have been attracted to sap released by buds in the crowns of large trees such as cottonwoods. This species is less likely to occur in conifers than is Golden-crowned Kinglet.

Spring

9, 12, 13 Mar→28, 28, 29 May

23, 30 Mar, 13 Apr→28, 28, 29 May (west)

Arrival is mid to late Mar; departure is in May. There are a few Jun reports, all from western parts of the state, as is the case with early fall reports: 4 Jun 1986 Grant Co, 1–18 Jun 1992 Dawes Co in a "patch of simulated boreal forest" (AB 46:1152), and 29 Jun 1954 Dawes Co. A report by Aughey, 5 Jun 1865 Dakota Co, was commented upon by Sharpe (1993).

High counts include 47 at Fontenelle Forest 11 Apr 1999, 35 in southeast Otoe Co 25 Apr 1999, and 12 in Saunders Co 22 Apr 1995.

Fall

17, 20, 22 Aug→7, 7, 8 Nov (north and west)

10, 11, 11 Sep→winter (south and east)

Migrants generally arrive in mid-Sep. Early dates in the north are somewhat earlier than those in the south, indicating a leisurely fall passage through the state.

Earlier reports from the north and west may be of migrants from the Black Hills of South Dakota, where the species is an uncommon breeder (SDOU 1991); these are 7–16 Aug 1977 McPherson Co and 25 July 1979 Lincoln Co. Early reports from the south and east are 29 Jul 1920 Lancaster Co (specimen, UNSM ZM6708), 20 Aug 1988 Lancaster Co, 21 Aug 1934 Webster Co (Ludlow 1935), 2 Sep 1998 Alda, and 6 Sep 1993 Lancaster Co.

Departure dates are obscured by birds lingering into Dec, especially in the southeast (see Winter). In the west migrants depart by early Nov, although 3–4 were at L Ogallala 28 Nov–5 Dec 1998 (SJD).

High counts include 51 in Douglas and Washington Cos 2 Oct 1999, 50+ at Branched Oak L 3 Oct 1999, and 35 in Douglas Co 13 Oct 1996.

Winter

Ruby-crowned Kinglets depart gradually in fall, with few remaining into Jan. Fewer than 100 have been recorded on CBCs since 1967–68, all from the southeast; few of these reports are documented, however. Virtually all late Dec–Feb reports are from the southeast, although there are these from the Panhandle: 29 Dec 1966 Scotts Bluff Co, 12 Jan 1979 Scotts Bluff Co, a "first winter record" in Dawes Co 16 Jan–17 Feb 1983 (AB 37:314), 21 Jan 1988 Dawes Co, and 22 Feb 1992 Scotts Bluff Co. Additional westerly reports are 27 Jan 1992 and 29 Jan 1961 Lincoln Co. There are no Jan reports from the north.

Other Reports

14 Feb 1972 Brown Co, 10 Dec 1976 Boone Co, 18 Dec 1962 McPherson Co, 27 Dec 1964 Brown Co.

FINDING

This species is easy to find in brushy woodland, especially in Apr and Oct, when squeaking attracts individuals so closely that binoculars are unnecessary.

Sylviidae (Gnatcatchers)

Blue-gray Gnatcatcher
Polioptila caerulea
STATUS
Uncommon regular spring and fall migrant east, becoming rare west. Uncommon regular breeder east, rare casual central and west. Uncommon casual summer visitor west.

DOCUMENTATION
Specimen
UNSM 11734, 1 May 1888 Peru, Nemaha Co.
DISTRIBUTION AND ECOLOGY
This species is generally restricted as a breeding bird to the interior of large tracts of oak-hickory forest, where it forages in the upper canopy. It is generally limited to the Missouri Valley in the extreme southeast, rarely north in recent years to Dakota Co, although numbers are increasing in the Panhandle. It also occurs rarely, primarily as a spring migrant, westward in the Republican Valley to the Orleans area and in the Platte Valley to about North Platte.
Spring
6, 7, 8 Apr→summer
Arrival is mid to late Apr, with an earlier report 30 Mar 1934 Jefferson Co.
Until recently this species was rarely encountered west of a line from Knox Co to Adams Co, with few reports: 22 Apr 1937 Lincoln Co, specimen, UNSM ZM6707 (Tout 1947); 28 April 1953 Thomas Co; 4 May 1996 Howard Co; and 12 May 1957 Logan Co. In 1998, however, 13 birds were found, suggesting local small breeding populations (see Summer): 5 birds at Oliver Res 8 May 1998 (SJD); 5 at Stage Hill Road, Wildcat Hills, 16 May 1998 (SJD); and 3 at Long Canyon 17 May 1998 (JGJ). A similar "explosion" in numbers was noted in the west in 1999, with at least 15 birds reported.
High counts include 12 in Sarpy Co 11 May 1996, 9 at Little Blue Area 20 Apr 1996, and 5 at Hummel Park, Douglas Co, 8 May 1995.

Summer
Except for recent reports from the Panhandle and from Dakota and Harlan Cos, summer reports have been confined to the area south and east of but including Douglas, Saunders, Saline, and Jefferson Cos.
There are few actual breeding records from this area. Ducey (1988) cited records from Douglas, Sarpy, Lancaster, and Gage Cos since 1960.
This species has recently expanded its range northward in eastern Colorado (Alan Versaw, pers. comm. WRS). Following a sighting 19 Jun 1993 of a single territorial bird in limber pines in western Kimball Co (WRS, incorrectly cited 16 Jun in AB 47:1123) and another, possibly the same bird, there 24 Jun 1995 (WRS), a family group was seen in the same area, including adults feeding dependent young, 22 Aug 1999 (JS, WRS).
In spring 1998 several birds were located in suitable breeding habitat in the Wildcat Hills and in Long Canyon (see Spring). At Stage Hill Road 3–6 were present 13 Jun and 20 Aug 1998, and as many as 11 were there 4 Sep 1999 (SJD). A single bird was in Box Butte Co 2 Jun 1996 (FN 50:966).
In Harlan Co a territorial pair was seen 8 km (5 mi) south and 1.6 km (1 mi) east of Orleans 6 Jun 1990 (WRS), and 3 birds were 3.2 km (2 mi) south of Orleans 23 Jun 1996 (WRS, JGJ), suggesting that a breeding population exists in this excellent riparian woodland.
Westward in the Platte Valley there is an 11 Jun 1985 report from Lincoln Co, and a female in breeding condition was netted in Keith Co 3 Jun 1993 (Rosche 1994a).
Elsewhere, Ducey (1989) suggested that this species was "extending its range into the lower Niobrara Valley" and would be "expected to breed there," although there is little recent evidence to support this.
Fall
summer→22, 25, 27 Sep
Departure is completed by mid-Sep. There is an undocumented report 11 Nov 1993 Sarpy Co.

West of Lancaster Co there are these reports: 29–30 Aug 1998 (1–2) Oliver Res, 30 Aug 1997 Morrill Co (DH), 30 Aug–2 Sep 1991 Sioux Co (not accepted by the NOURC; Gubanyi 1996a), 1 Sep and 6 Sep 1997 Stage Hill Road (SJD, JS), 7 Sep 1998 south of Crescent L NWR, 12 Sep 1997 Long Canyon (JGJ), 12 Sep 1998 southwest Kimball Co, 14 Sep 1962 Webster Co, 15 Sep 1965 Adams Co, 19 Sep 1981 Garden Co, and 22 Sep 1964 Adams Co.

Other Reports

2 March 1972 Dawes Co, undocumented and considered an "incredible date" by Rosche (1982).

FINDING

Gnatcatchers can be readily attracted by squeaking. Good locations are Fontenelle Forest and Indian Cave SP, and best time is probably Jun.

Turdidae (Thrushes)

Northern Wheatear

Oenanthe oenanthe

STATUS

Hypothetical.

COMMENTS

There are 2 published reports (Bray and others 1986), neither acceptably documented. Zimmer (1911a) reported a sighting 2 Dec 1910 near Squaw Mound in Dawes Co. He observed the bird closely for some time but was unable to collect it and stated that while he was "sure of the identity of the bird," it "cannot be included in our fauna positively" as he was unable to secure it as a specimen. The other report was of 5 birds with Horned Larks seen 4 Jan 1970 in Gage Co (Patton 1970). A brief description was given, but longspur species would seem more likely.

This species breeds in arctic North America but winters in Europe. A few individuals occur along the east and west coasts of North America in spring and fall, and there are stragglers to the interior, also in spring

and fall. There are about 7 Great Plains reports, 3 in the period 9–15 May and 4 in the period 12 Sep–22 Oct (some of these have not been accepted by local authorities). The Nebraska reports do not fit this pattern, although Northern Wheatear is a possibility in exposed rocky habitats, most likely in fall and early winter.

Eastern Bluebird

Sialia sialis

STATUS

Common regular spring and fall migrant and breeder east, becoming uncommon west. Uncommon regular winter visitor southeast, uncommon casual elsewhere.

DOCUMENTATION

Specimen

UNSM ZM6688, 23 Mar 1891 Lincoln, Lancaster Co.

TAXONOMY

Occasional hybridization with Mountain Bluebird has been recorded, including a Nebraska example, discussed below under Summer. A male Mountain Bluebird, far east of its normal range, mated with a female Eastern Bluebird in Riley Co, Kansas, in 1997 and fledged several nestlings (Cavitt and others 1998). Hybridization between these species occurred in less than 1% of more than 8000 pairs of the 2 species in southwestern Manitoba (Rounds and Munro 1982).

DISTRIBUTION AND ECOLOGY

This species is common in summer only in the east, particularly in oak savanna, and is rare, although increasing, in the Panhandle, where it is confined to areas with deciduous trees (Rosche 1982), usually along major river valleys. Nest boxes can have a significant impact on local breeding populations of this species, especially westward, where natural nest sites are scarce.

During migration it is somewhat more common westward. Rosche (1982) considered it rare in spring although sometimes fairly common in fall. Migrants occur in a wide variety of habitats, including the Sandhills (Rosche 1982).

By early winter most Eastern Bluebirds are confined to locales with fruiting trees and shrubs in the Platte and Missouri Valleys. Small concentrations may also be found along the more densely wooded floodplains of the Blue River drainage. Wintering numbers are variable, but this species is generally rare in midwinter, with numbers declining gradually after fall migration.

Spring

25, 25, 26 Mar→summer (west)

Due to a decline in numbers as winter progresses, a spring influx is detectable. Migrants appear first in the east in late Feb but not until late Mar in the Panhandle, where there are no Feb reports, although there are these early reports: 8 Mar 1970 Scotts Bluff Co, 11 Mar 1995 Sioux Co, and 18 Mar 1988 Dawes Co. In recent years spring migrants have become more numerous in the Panhandle.

Summer

The original range of oak savanna defines the current breeding range of Eastern Bluebird. This includes most of the southeast third of Nebraska east of a line from Knox Co to Harlan Co. West of this area the breeding range extends westward along the Niobrara Valley to the Niobrara Valley Preserve (Mossman and Brogie 1983), and there are a few summer reports from the Loup drainage, such as summering birds in Thomas Co 1 Jul–7 Sep 1993. Breeding birds occur throughout the Republican and South Platte Valleys (Johnsgard 1980), although it is rare west of Harlan Co and North Platte. It does, however, breed regularly on the North Platte floodplain west of Keystone in Keith Co (Rosche 1994a), and nesting has occurred in Scotts Bluff Co since 1986 (NBR 54:64), when a pair was present 24 May–2 Aug and used a box set out for Mountain Bluebirds.

Rosche (1982) stated that there are no definite nesting records for the Pine Ridge and included only 1 summer report from the

northwest, a pair at a hole in "cottonwood forest" at Box Butte Res summer 1981. More recently, Wilson and others (1985, 1986; AB 39:932) found an apparent hybrid male Eastern x Mountain Bluebird mated with a female Mountain Bluebird in Dawes Co 13 km (8 mi) south of Crawford in the summer of 1985. This pair fledged 2 broods from a nest box. Wilson and others (1986) suggested that hybridization in this case may have resulted from low numbers of summering Eastern Bluebirds in the area and the resulting difficulty in finding a mate. Four young were fledged at Box Butte Res 8 Jul 1994 (NBR 62:80, 112).

In 1993 this species was listed as a "Summer Resident" in Scotts Bluff and Dawes Cos. There were reports in Jun–Jul in Scotts Bluff Co in 1994, and 6 pairs and at least 1 chick were at the nest boxes at Wildcat Hills NC 13–14 Jun 1998 (AK, JS, WRS). Ducey (1988) cited nest record cards for Dawes Co, 1 in 1962 and 2 in 1974.

Taken together, the above records suggest that the provision of nest boxes for bluebirds in the Panhandle has attracted Eastern Bluebirds in recent years. Of about 20 Jun and Jul reports for the Panhandle through the 1990s, all but 3 occurred since the 1980s, and the earliest was as recent as 1974 in Sioux Co. There are, however, 3 old specimens for the Panhandle from Monroe and Warbonnet Canyons in northwest Sioux Co: 24 May 1900 (UNSM ZM6692 and ZM6693) and 21 Jun 1901, an immature (UNSM ZM6694).

Fall

Peak migration is in Oct; in the north and west most birds depart by late Oct. Departure in the south and east is difficult to detect as many birds linger into winter.

High counts include 200 between Center and Niobrara 3 Oct 1998, 52 in southeast Otoe Co 22 Oct 1997, 50 at Wildcat Hills NC 17 Oct 1998 (AK), 50 at Harlan Co Res 15 Nov 1998, and 50 near Fort Calhoun 8 Oct 1999.

Winter

CBC data indicate that by late Dec most Eastern Bluebirds are restricted to the south and east, generally at locations in the Platte and Missouri Valleys. A few birds overwinter there most years, but overwintering is unusual elsewhere. Small flocks of 10–15 were encountered in scattered floodplain vegetation in several locales along the Little Blue River in Thayer Co in early Jan 1997 (RSS). Although there are several midwinter reports from the north, there is only a single Jan report from the Panhandle, 20 Jan 1979 Scotts Bluff Co, and no Feb reports.

FINDING

Eastern Bluebirds are easiest to find in Missouri Valley oak savanna or oak-hickory woodland edge in May and Jun, as in and around Indian Cave SP. It is, of course, easy to locate at active nest-box trails.

Western Bluebird

Sialia mexicana

STATUS

Hypothetical.

COMMENTS

There are at least 6 published reports (Bray and others 1986), none acceptably documented. There are 2 spring reports, neither with details: 11 April 1964 Dawes Co (NBR 32:76) and 19–20 May 1972 Cherry Co (NBR 40:27). There are 4 fall reports. One was identified 26 Aug 1963 in Webster Co by "the brown patch across the back of one of a group when in flight" (NBR 32:61); this group was present for a few days (NBR 32:54). Aughey watched one in Aug feeding along the Niobrara River about 11.2 km (7 mi) from its mouth in Knox Co (Bruner and others 1904); the same observer reported it as "abundant" in Otoe Co, an observation considered extremely unlikely (Sharpe 1993). There are 2 other fall reports, neither with details: 5 Oct 1959 Webster Co and 19 Oct 1968 Scotts Bluff Co.

A report of "hundreds" along the Platte River near Benedict in fall 1980 (AB 35:200) almost certainly refers to Eastern or Mountain Bluebirds.

There are no South Dakota records and no recent records from Wyoming (Scott 1993). Andrews and Righter (1992) cited about 12 records on the northeastern Colorado plains in the periods Mar–Apr and Oct–Nov. Based on these records, 2 of the Nebraska reports seem plausible: 11 April 1964 Dawes Co and 19 Oct 1968 Scotts Bluff Co. This species should be looked for in the western Panhandle in early spring and late fall.

Mountain Bluebird

Sialia currucoides

STATUS

Common regular spring and fall migrant west and central, rare casual east. Fairly common regular breeder northwest. Uncommon regular winter visitor west and central, rare casual east.

DOCUMENTATION

Specimen

UNSM ZM6695, 24 May 1900 Monroe Canyon, Sioux Co.

TAXONOMY

See Eastern Bluebird.

DISTRIBUTION AND ECOLOGY

In summer this species is essentially restricted to the Panhandle, where it is fairly common in summer in open ponderosa pine woodland, particularly in the Pine Ridge and Wildcat Hills. It is often very common in migration (Rosche 1982), occurring rarely as far east as Adams Co (Rapp and others 1958). The breeding range is vacated in winter (Rosche 1982), but wintering birds occupy juniper habitats in the western half of Nebraska wherever food is available, although the Platte Valley is favored.

Spring

11, 12, 15 Feb→summer

Dates above are for the breeding range only.

Migrants arrive on the breeding range in late Feb. A pair was at a James Ranch, Sioux Co, nest box by 10 Mar 1995.

High counts of migrants include 100 south of Gering 10–12 Apr 1997, 65 at Crescent L NWR 16 Mar 1994, and 40 at Gering 20 Mar 1995.

Summer

Regular breeding has occurred only in Sioux, Dawes, and Scotts Bluff Cos, where the preferred breeding habitat of open ponderosa pine woodlands is found (Ducey 1988). There is a remarkable extralimital report of a pair feeding young 17 May 1992 at a picnic shelter at Red Willow Res, Frontier Co (WM; AB 46:444). There are summer reports from Sheridan Co, where suitable breeding habitat exists, 4 Jun 1986 and 20 Mar–12 Jun 1992, and at an unexpected location 7 Aug 1996 Kimball Co. The only other summer reports away from the breeding range are 28 Jun 1969 McPherson Co and 19 Jun 1955 Boone Co, neither documented.

Fall

summer→5, 7, 8 Nov

Dates above are for the breeding range only.

Migrants leave the breeding range by early Nov, although there is a later report 19 Nov 1966 Scotts Bluff Co.

High counts include 133 in southwest Kimball Co 13 Oct 1998, 31 in the western Panhandle 16 Oct 1999, and 22 in Dawes Co 8 Oct 1995.

Winter

2, 3, 3 Oct→12, 15, 18 Apr

Dates above are for the wintering range.

This is perhaps the only species that breeds and winters in different parts of Nebraska, possibly performing an "altitudinal migration" between summer and winter ranges. There are, however, no banding data to show that the same birds are involved. The summer range is completely abandoned in winter, with few reports: 1 Jan 1973 Dawes Co and 28 Jan 1956 Scotts Bluff Co.

Wintering birds are most commonly found along the Platte Valley, rarely east of Lincoln Co, although large numbers were wintering in Merrick Co near Hordville 1978–79 (Morris 1979). Lincoln Co appears to hold most wintering Mountain Bluebirds, especially

the extensive juniper area in the southeast part of the county. The North Platte CBC has recorded an average of 133 birds each year since 1987–88, with a high of 400 in 1988–89, and 100 were in cedar canyons in southeast Lincoln Co 6 Mar 1996.

Arrival in the winter range is in early Oct, although there are earlier reports 15 Sept 1963 Lincoln Co, 26 Sep 1992 Morrill Co, and 28 Sept 1987 Lincoln Co. Departure is in early to mid-Apr, although there are later reports 21 Apr 1941 Logan Co, 25 Apr Garden Co, 26 April 1946 Logan Co and undocumented reports 24 May 1987 Red Willow Co and 25 May 1960 Lincoln Co.

Winter reports away from the Platte Valley are few. Birds were found in Logan Co 1 Jan 1952, 1965, and 1967; there are 2 reports from Keya Paha Co 1 Jan 1957 and 6 Jan 1958, and it was listed as a "possible winter resident" at Bessey Division, NNF (Bray 1994), although the only specific report is of 2 there 16 Mar 1997.

Apparently this species does not normally winter in the Niobrara Valley.

There are 13 reports for the east in the period 6 Oct–17 Apr, including a specimen, HMM 28435, a male taken in Adams Co 26 Mar 1955.

FINDING

Mountain Bluebirds are fairly easy to find in open ponderosa pine woodland at the upper end of Pine Ridge canyons in May and Jun, such as Sowbelly Canyon. Wintering birds are present most years in juniper habitat south of North Platte.

Townsend's Solitaire

Myadestes townsendi

STATUS

Rare casual breeder northwest. Common regular winter resident and spring and fall migrant west, becoming rare casual east.

DOCUMENTATION

Specimen

UNSM ZM6699, 19 Feb 1896 Harrison, Sioux Co.

DISTRIBUTION AND ECOLOGY

During the winter solitaires migrate into

Nebraska from higher elevations in the mountain west and overwinter throughout much of western Nebraska, where dense stands of junipers are occupied. In the west this habitat occurs mostly on scarps overlooking major riparian systems where it may extend onto the floodplain. Such habitat occurs commonly along the Platte Valley east to Lincoln Co and to the Mariaville area on the Niobrara River. Appropriate habitat that supports good numbers of solitaires is also found at Bessey Division, NNF (Bray 1994). Stragglers are erratic in the east.

Spring

winter→28, 28 Mar (specimen, UNSM ZM6702), 2 Apr (east)

winter→28, 30 Apr, 1 May (west)

Departure from wintering grounds occurs by mid-Mar in the east and in the Panhandle by late Apr. There are later reports in the east 10–11 Apr 1997 Neale Woods, 7 May 1961 Nemaha Co, 19 May 1982 Lancaster Co, and 28 May 1994 Nemaha Co and in the Panhandle 9 May 1990 Scotts Bluff Co, 18 May 1980 Sioux Co, 20 May 1999 Scotts Bluff NM, 20 May 1999 Sowbelly Canyon, 22 May 1967 Dawes Co, 23 May 1970 Scotts Bluff Co, and 23 May 1992 Scotts Bluff Co. There are several late reports in central Nebraska, notably 13 May 1978 Hall Co, 13 May 1996 Thomas Co, 21 May 1970 McPherson Co, 22 May 1990 Burwell (RG), 24 May 1988 Lincoln Co, 25 May 1976 Hall Co, 25 May 1990 Hall Co, 26 May 1942 Adams Co, and 29 May 1937 Lincoln Co.

Summer

At the end of the 19th century this species occurred as a breeder in northern Sioux Co, occupying the various wooded riparian canyons that drain the Pine Ridge into the Hat Creek Basin (Bruner and others 1904). Although it is not clear that significant changes in habitat have occurred, today the species only occurs casually there during the breeding season.

Recent nesting attempts in Sioux and Dawes Cos have been unsuccessful. Two nests were found in Monroe Canyon in 1900, and a pair was seen the next summer (Crawford 1901). There were no reports until sightings of adults and flying young in 1979 and 1980, which Rosche suggested were not necessarily fledged in Nebraska but possibly nearby in the Black Hills of South Dakota: on 18–20 Jul 1979 an adult and a juvenile were in Monroe Canyon (Fickel 1979), and on 22 Jul 1980 a juvenile was seen flying strongly in Sowbelly Canyon (Rosche 1982). On 31 May 1986 Rosche found a nest with eggs (AB 40:493; Ducey 1988) in Hat Creek Canyon; this nest was unsuccessful (Rosche, pers. comm. WRS). A report for 10–30 Jun 1986 with no details probably refers to Rosche's sighting, and there are additional Sioux Co reports for 20 May 1991 when birds were "on territory" (AB 45:468), 11 Jun 1990, 27 Jun 1999, 1 Jul 1995, 11 Jul 1989, 2 Aug 1996, summer 1988 (AB 42:1309), and summers of 1992 and 1993. Two birds seen in Dawes Co 20 May 1991 "may have been breeding" (AB 45:468); other reports from Dawes Co are 25 May 1997, 13 Jun 1994, and 2 Jul 1987.

Rosche (1994a) cited a report of at least 1 bird in cedars east of Ogallala Jul 1948 and suggested that an occasional nesting attempt in the area "would not be too surprising."

Fall

25, 27, 29 Aug→winter (west)

18, 20, 21 Nov→winter (east)

Solitaires arrive in the Panhandle in late Aug, reaching the east in small numbers most years by late Nov. There are early reports in central Nebraska 23 Aug 1978 Lincoln Co and in the east 28 Sep 1995, 10 Nov 1966, and 11 Nov 1972 Douglas Co.

Winter

CBC data show that most solitaires winter in the Panhandle, with a few scattered fairly evenly throughout the rest of the state.

FINDING

This species is easy to find in winter in Platte Valley canyons with junipers from Lincoln Co westward. A good place is the south side of

Keystone L. It is easily attracted by squeaking, which also causes it to vocalize.

Veery

Catharus fuscescens

STATUS

Rare regular spring and fall migrant east and west, rare casual central.

DOCUMENTATION

Specimen

UNSM ZM11022 (*fuscescens*), 25 May 1909 Lincoln, Lancaster Co.

TAXONOMY

Two subspecies have been recorded (Rapp and others 1958), eastern *fuscescens* and western *salicicolus*. Rapp and others (1958) stated that *fuscescens* (specimen cited above) is a "rare migrant in the Missouri River Valley," while *salicicolus* "is more common in the western half of the state than in the east." The westernmost documented record of *fuscescens* is a specimen, #2043 in the Brooking collection, taken at Inland 3 May 1915 and identical to specimens at UNSM identified as this subspecies by Oberholser (Swenk, Notes before 1925).

Two birds with dark plumage were identified as Veeries by call at Fontenelle Forest 14 May 1995 (WRS); these were likely *salicicolus*. The AOU (1957) assigns breeding birds in the prairie provinces, Minnesota, Wisconsin, Illinois, and Iowa to *salicicolus*, and most Nebraska migrants are likely to be this race but are probably overlooked. Identification within this genus should be made with care.

DISTRIBUTION AND ECOLOGY

Spring migrants occur statewide, but most are reported in the Missouri Valley. There is also a sparse movement through the Panhandle, presumably including birds that breed in the Black Hills of South Dakota. Preferred habitat is dense undergrowth in damp forest, usually near flowing water, but elsewhere similar habitat in more open woodland is utilized. This species is rare in fall, when virtually all records are from the east.

Spring

26, 27, 28 Apr→3, 5, 5 Jun

Most arrive in early May, although early dates are in late Apr. Departure is at the end of May, with late dates in early Jun, although there are later reports away from the east 14 Jun 1972 Brown Co and 18 Jun 1981 Garden Co. Zeillemaker (RSS) netted 1–2 each spring in Garden Co while stationed there.

Mid-Jun and Jul reports in the east raise the possibility of breeding, although there is no direct evidence. These reports are 11 Jun 1978 Douglas-Sarpy Cos, 12 Jun 1983 Douglas-Sarpy Cos (AB 37:1003), 18 Jun 1992 Sarpy Co (AB 46:1152), 2–11 Jul 1977 Douglas-Sarpy Cos, 9 Jul–23 Aug 1981 Douglas-Sarpy Cos, and 8 Jul 1988 Howard-Hall Cos. This species is a rare breeder south and west to central Iowa (Dinsmore and others 1984) and breeds uncommonly in the Black Hills (Peterson 1990). There are no reports from western Nebraska in suitable breeding habitat however.

Seven birds were reported in Sarpy Co 11–31 May 1992, considered a good migration (AB 47:428).

Fall

2, 2, 3 Sep→21, 21, 22 Sep

Migration takes place in Sep, although there are earlier reports 18 Aug 1981 Garden Co, 20 Aug 1983 Howard-Hall Cos, 21 Aug 1986 Douglas Co (AB 41:111), 22 Aug 1987 McPherson Co, 23 Aug 1981 Garden Co, and 2 birds netted 28 Aug 1979 Garden Co (AB 34:178). Peterson (1990) indicated that Veeries leave breeding areas in the Black Hills by 1 Sep.

Migration ends in late Sep.

There are few reports away from the east. Rosche (1982) listed none in the northwest; the few Panhandle reports have been listed above, and there are these additional westerly reports: 4 Sep 1985, 10 Sept 1988, and 12 Sept 1974 Howard-Hall Cos and 20 Sept 1964 Lincoln Co. Zeillemaker (RSS) netted 1–2 each fall in Garden Co while stationed there.

Other Reports

Fourteen reports from Howard-Hall Cos outside
    expected migration periods, 14 Apr 1948 Keith
    Co, and a report of one seen in Fontenelle
    Forest 20 Dec 1980 on the Omaha CBC, which
    has not been examined by the NOURC.

FINDING

The best chance to see this rare migrant is
    mid-May in dense undergrowth near water
    in Missouri Valley forests such as Fontenelle
    Forest and Indian Cave SP.

Gray-cheeked Thrush

Catharus minimus

STATUS

Uncommon regular spring migrant east,
    becoming rare casual west. Rare casual fall
    migrant statewide.

DOCUMENTATION

Specimen

UNSM ZM11018, 11 May 1910 South Bend, Cass Co.

DISTRIBUTION AND ECOLOGY

This migrant thrush is most common in
    Missouri Valley deciduous forest, but it has
    been reported in spring in shaded deciduous
    woodland statewide (Johnsgard 1980). In
    fall it is virtually unrecorded away from the
    Missouri Valley.

Spring

25, 26, 27 Apr→3, 3, 6 Jun

Migrants occur late Apr–late May, although there
    are earlier undocumented reports 12 Apr 1999
    Dodge Park, Omaha; 15 Apr 1993 Boone Co;
    18 Apr 1999 Fontenelle Forest; 22 Apr 1958
    Adams Co; and 23 Apr 1961 Nemaha Co. There
    are about 22 Panhandle reports 3–30 May.

High counts include 55 in Sarpy Co 11 May 1996,
    including 20 at Fontenelle Forest and 15 at
    Cedar Island.

Fall

25 (Tout 1947), 29 Aug, 2 Sep→14, 15, 20 Oct

There are fewer than 30 reports. Migration
    occurs early Sep through mid-Oct. Away from
    the Missouri Valley there are few reports:
    25 Aug 1934 Lincoln Co (Tout 1947), 3 Sep
    1974 Dawes Co (Rosche 1982), 13–17 Sep 1982

Boone Co, 7 Oct 1973 Sioux Co, and 14 Oct
    1938 Lincoln Co (Tout 1947).

Other Reports

20 Mar 1972 Hall Co, 9 Apr 1949 Douglas
    Co, 22 Oct 1972 Adams Co, 28 Nov 1982
    Douglas-Sarpy Cos.

FINDING

This species is most likely in mid-May in
    Missouri Valley deciduous forests, such as
    Fontenelle Forest and Indian Cave SP. Look
    along hiking trails in shady areas near streams.

Swainson's Thrush

Catharus ustulatus

STATUS

Common regular spring and fall migrant
    statewide. Accidental breeder. Rare casual
    summer visitor west. Accidental in winter.

DOCUMENTATION

Specimen

UNSM ZM6673 (*swainsoni*), 23 May 1900 Monroe
    Canyon, Sioux Co.

TAXONOMY

The Pacific Coast race *ustulatus* has been
    collected as far east as southeast Iowa (AOU
    1957) and was reported from northwest
    Nebraska by Cary (1901). Virtually all
    Nebraska Swainson's Thrushes are *swainsoni*,
    however (AOU 1957), including the specimen
    cited above.

DISTRIBUTION AND ECOLOGY

Migrants occur in riparian woodland and forest.
    In the west it is also commonly found in
    city parks and farm shelterbelts. The single
    confirmed breeding record was in riparian
    growth in the Pine Ridge of Dawes Co
    (Rosche 1982).

Spring

21, 21, 22 Apr→6, 7, 7 Jun (east, south)

24, 26, 27 Apr →17, 18, 19 Jun (north, west)

Most pass through in May. The earliest Kansas
    specimen is 24 Apr (Thompson and Ely 1992).

Departure in the south and east is complete by
    early Jun but not until mid-Jun in the north
    and west. There are later reports in the south
    and east 10 Jun 1989 Lincoln Co, 13 Jun 1983

Adams Co, 15 and 17 Jun 1875 Dakota Co (but
see Sharpe 1993), 17 Jun 1981 Douglas Co,
and 19 Jun 1939 Lincoln Co (Tout 1947) and
undocumented reports 28 Jun 1969 Adams Co
and 30 Jun 1967 Adams Co. There are later
reports in the north and west 22 Jun 1995
Garden Co, 23 Jun 1969 McPherson Co, and 30
Jun 1967 Greeley Co. Later dates are discussed
under Summer.

High counts include 499 in Sarpy Co 11 May
1996, 272 of which were at Fontenelle Forest;
36 in Lancaster Co 10 May 1996; and 34 in
Sarpy Co 13 May 1995.

Summer

The only documented breeding record is of a
nest in Dawes Co 12 Aug 1973 (Rosche 1974a).
This nest was located in deciduous riparian
growth adjacent to ponderosa pines in West
Ash Canyon, but the 2 eggs, now specimen
UNSM ZM17073, were cold (AB 28:74). As
stated by Rosche (1974a), there had been
no prior evidence for breeding, although
it had been assumed based on Hudson's
collecting a female with enlarged ovaries
17 Jun 1938 in Squaw Canyon, Sioux Co
(Hudson 1939). Migrants are not uncommon,
however, as late as the second week of Jun
in the northwest (Rosche 1982); Jun reports
may also be the basis for purported nesting
in the central Niobrara Valley (Bates 1900;
Youngworth 1955).

There are Jun and Jul reports from Sioux Co
1976–79, although no evidence of breeding
was reported, as well as late Jun (but not Jul)
reports for the years 1967, 1972, 1974, and
1986, and for Jul (but not Jun) in 1975. Cary
(1901) considered breeding to be probable in
West Warbonnet Canyon based on summer
observations. That these summer reports
could be nonbreeding birds is indicated
by summer reports from unlikely breeding
locations 18 Jun 1981 and 27 Jul 1981 Garden
Co and 1–10 Jul 1989 Lincoln Co.

Fall

24 (Tout 1947), 25 (Ludlow 1935), 26 Aug→22, 23,
26 Oct

Arrival is in late Aug statewide, although there
are earlier reports 12 Aug 1971 and 12 Aug 1973
from Scotts Bluff Co. There is a later report
30 Oct 1977 Douglas Co. The latest specimen
dates for Kansas are in Oct (Thompson and
Ely 1992).

High counts include 4 at Wyuka Cem, Lincoln,
20 Sep 1998.

Winter

There is a single winter report, with "excellent
details," for 27 Jan–10 Feb 1980 in Washington
Co (AB 34:287; NBR 48:82).

COMMENTS

Care must be taken in identification of *Catharus*
thrushes during migration, especially in
the Panhandle, as various western races of
Catharus species are similar to each other.
A particular problem is the Pacific Coast
race of Swainson's Thrush, *ustulatus*, which
closely resembles the western race of Veery,
salicicolus. See the excellent discussion in
Zimmer (1985).

Other Reports

9 Mar 1981 Adams Co, 17 Mar 1955 Scotts Bluff
Co, 30 Mar 1949 Adams Co, 30 Mar 1949
Dawes Co, 1 Apr 1954 Boone Co, 3 Apr 1998
(2) Fontenelle Forest, 4 Apr 1939 Lincoln Co,
6 Apr 1943 Jefferson Co, 7 Apr 1984 Lancaster
Co, 11 Apr 1944 Jefferson Co, 11 Apr 1949
Douglas Co, 12 Apr 1937 Douglas Co, 12 Apr
1950 Douglas Co, 13 Apr 1956 Douglas Co, 15
Apr 1962 Douglas Co, 15 Apr 1973 Douglas Co,
3 Nov 1965 McPherson Co, 7 Nov 1973 Scotts
Bluff Co, 9 Nov 1983 Scotts Bluff Co, 10 Nov
1988 Scotts Bluff Co, 19 Nov 1983 Adams Co, 1
Dec 1962 Douglas Co.

FINDING

Swainson's Thrush is easy to find in shady
woodland and forest in the east during
migration peaks May and Sep. It may be
attracted by squeaking.

Hermit Thrush

Catharus guttatus

STATUS

Uncommon regular spring migrant east,

becoming rare west. Rare regular fall migrant east, becoming rare casual west. Rare casual winter visitor east, accidental central.

DOCUMENTATION

Specimen

UNSM ZM6670, 9 Apr 1900 Lancaster Co.

TAXONOMY

Two subspecies occur in Nebraska (Rapp and others 1958). Eastern birds presumably are *faxoni*, which breeds across most of the northern United States and Canada west to British Columbia. Migrants have been recorded as far west as Denver (AOU 1957); one was in Dawes Co 30 Sep 1994 (FN 49:66). The Rocky Mountain race, *auduboni*, a long-distance migrant that winters in central America, occurs rarely in western Nebraska and possibly farther east. The specimen cited above was identified by Oberholser as *auduboni*, but it does not appear to differ from examples of *faxoni* in the UNSM collection (JGJ, WRS). The few records of *auduboni* are of a single at Stateline Island 12 Oct 1996, which was described as "light gray, pale" (MB, NBR 64:121); 2 birds at Oliver Res 13 Oct 1998 (SJD); singles at Oliver Res and Gering 16 Oct 1999 (SJD); and another at Gering 1 Nov 1998 (SJD).

DISTRIBUTION AND ECOLOGY

In migration all woodland habitats are utilized, but a preference for dense, shaded forest allows for higher numbers in the east.

Spring

26, 30, 30 Mar→17, 18, 21 May (east)

26, 30, 30 Mar→29 May, 3, 4 Jun (elsewhere)

This species is an early migrant, arriving in early Apr and departing in mid-May in the east and somewhat later elsewhere. It migrates statewide but is rare in the west; Rosche (1982) observed only one in 8 years of fieldwork. Earlier reports that are likely migrants are 19 Mar 1971 Lancaster Co and 20 Mar 1985 Douglas Co.

High counts include 5 in Lancaster Co 1 May 1997 and 4 at Fontenelle Forest 4 Apr 1998.

Fall

22, 23, 23 Sep→29 Oct, 1, 4 Nov

Migrants arrive in late Sep and depart by late Oct. There are several Nov and Dec reports, probably individuals attempting to overwinter (see Winter).

There are a few Panhandle reports most, especially Oct reports, probably *auduboni* (see Taxonomy): 23 Sep 1978 Garden Co, 27 Sep 1975 Sioux Co, 3 Oct 1987 Sioux Co, 4 Oct 1986 Sioux Co, 6 Oct 1975 Sioux Co, 8 Oct 1989 Sioux Co, 12 Oct 1978 Garden Co, 13 Oct 1998 (2) Oliver Res (SJD), 16 Oct 1999 Oliver Res (SJD), 16 Oct 1999 Gering (SJD), 19 Oct 1979 Scotts Bluff Co, 20 Oct 1979 Garden Co, and 1 Nov 1998 Oliver Res (SJD).

High counts include 4 in Washington Co 13 Oct 1996.

Winter

There are several reports for Nov and Dec; most Dec reports are from CBCs. Midwinter reports are few and generally restricted to the Missouri Valley: one wintered in Lancaster Co 1974–75 (AB 29:710), and there are these additional reports: 2 Jan 2000 L Ogallala (SJD), 10 Jan 1976 Douglas Co, 10 Jan 1993 Otoe Co (WRS), 15 Jan 1985 Douglas Co (AB 39:184), 20 Jan 1994 Sarpy Co (FN 48:223), 1 Feb 1983 Douglas Co, 19 Feb 1992 Thomas Co (AB 46:285), and 22 Feb 1994 Sarpy Co (FN 48:223). Mar reports that are likely wintering birds are 1 Mar 1981 Douglas Co, 1 Mar 1994 Sarpy Co, and 9 Mar 1970 Lancaster Co. The only reports for the period Nov–Mar away from the east are the Thomas Co and L Ogallala reports above.

Other Reports

26 May 1968 Cass Co, 31 May 1959 Lancaster Co, 30 Jun 1976 Adams Co, 27 Jul 1980 Washington Co, 3 Sep 1989 Sioux Co, and 4 Sep 1977, 8 Sep 1973, 10 Sep 1981, and 11 Sep 1973 Douglas Co.

FINDING

This species can be most easily found in late Apr in dense forests in the east, such as Fontenelle Forest and Indian Cave SP. It is usually flushed from walking trails but often responds to squeaking by giving its low "chuck" note.

Wood Thrush
Hylocichla mustelina
STATUS
Uncommon spring and fall migrant east,
becoming rare casual west. Uncommon
breeder east and north. Rare casual summer
visitor away from breeding range.

DOCUMENTATION
Specimen
UNSM ZM11015, 10 May 1911 Lincoln, Lancaster
Co.

DISTRIBUTION AND ECOLOGY
Wood Thrush requires fairly extensive woodland
with moist conditions, a habitat that
disappears rapidly westward. Although most
commonly found breeding in oak-hickory
forest associated with the Missouri Valley,
it also occurs in extensive mature stands
of riparian forest, particularly in the lower
Platte, Elkhorn, and Niobrara Valleys. In the
Niobrara Valley it nests in small numbers in
deciduous and mixed woodlands at least as far
west as Brown Co (Brogie and Mossman 1983).

Since 1950 this species has experienced
significant range reduction by habitat loss
in eastern Nebraska, first by removal of
the dominant American elm by Dutch elm
disease in the 1960s and 1970s and more
recently by accelerated removal of large
timber stands along the Platte and Elkhorn
Rivers for agriculture, housing developments,
and sand-and-gravel operations.

Spring
25, 25, 26 Apr→24, 24, 27 May (specimen UNSM
ZM9459)
Departure dates above are from outside the
breeding range.
Migrants arrive in early May, although there is
an earlier undocumented report 23 Apr 1964
Douglas Co.
In areas of central Nebraska where it does not
breed it is a casual migrant (Rosche 1994a),
and there are few Panhandle reports: 10 May
1980 Garden Co, 12 May 1958 Sheridan Co, 17
May 1957 Sheridan Co, 17–19 May 1986 Sioux
Co, and 18 May 1980 Sioux Co.

High counts include 13 at Hormel Park, Fremont,
16 May 1999.
Summer
Currently the breeding range of this species
appears to be restricted to the Missouri Valley
and westward to the Niobrara Valley Preserve,
where Brogie and Mossman (1983) considered
it a "probable nester," noting 3 singing males
in Brown Co Jun–Jul 1982. Ducey (1983a)
cited Swenk's observation of a Wood Thrush
on a nest with 3 eggs in Long Pine Canyon,
Brown Co, in 1902, and Short (1961) found
Wood Thrush to be common in Holt Co
along the Niobrara River in 1955. Youngworth
(1955) found it a regular summer resident in
northeast Cherry Co. These records indicate
that Wood Thrush has probably nested in
this area for some time and possibly farther
west, as Short (1965) located a singing male
along the Niobrara River about 32 km (20
mi) west-southwest of Valentine during the
breeding season in 1965. There are additional
summer records for Brown Co for most of the
years 1964–72.

Rosche (1982) considered it an accidental
summer visitant in the northwest, citing 1
record only, a singing male in cottonwood
forest on the edge of Box Butte Res 19 Jun
1975. Ducey's (1988) citation of breeding in
Sioux Co is based on a statement in Bruner
and others (1904), which may refer only
to a nonbreeding bird. There are these few
additional summer reports for the Panhandle:
2 Jun 1980 Garden Co, 7–8 Jun 1985 Dawes Co,
and 8–9 Jun 1985 Sioux Co; all may be late
spring migrants.

There is a historical report of breeding in
Antelope Co (Ducey 1988) but no reports
since at least 1956 in summer for the upper
Elkhorn Valley. There is only 1 recent summer
report from the Loup drainage, 16–18 Jun
1983 Boone Co, although Short (1961) found
4 singing males on territory at Bessey
Division, NNF, Jun 1955, and Glandon (1956)
listed 1 record for Logan Co prior to 1955,
although no details were given. There is

1 summer report from McPherson Co 16
Jun 1984, and in Thomas Co its status was
listed as "uncertain, formerly a nesting
species" (Bray 1994), but no details were
provided.
For the Platte Valley, Ducey (1988) cited an article
by Tout (1935) implying breeding in that
area. Tout (1947) considered Wood Thrush
a "regular summer resident" in Lincoln Co,
present "every year in the big woods along
the Platte River east of [North Platte],"
presumably Stenger Grove, described by Tout
as located along the south side of the Platte
River about 16 km (10 mi) southeast of North
Platte. Short (1961) found several territorial
males west to Sutherland in 1956 and 1957, and
Brown and others (1996) stated that a female
with a brood patch was banded at L Ogallala
16 Jun. Short (1961) noted individuals in the
breeding season at various locations between
Adams and Lincoln Cos in 1955–57, but since
then the only reports from the central Platte
Valley are a few summer reports for Adams
Co 1964–76.
The only indication that it might occur in
summer in the Republican Valley is Ludlow's
(1935) statement that the species was a
"common summer resident" in Webster
Co; no dates later than 20 May were listed,
however. It is unusual as far west as Lancaster
Co, where it was reported at Wilderness
Park, Lincoln, 17 Jun 1995 and 9 May and 17
May 1996.
Fall
summer→6, 7, 10 Oct
Departure is complete by early Oct. Migrants are
casual away from the breeding range, and it is
difficult to establish dates for early arrivals. In
the central and west there are few reports: 24
Aug 1966 Custer Co; 2 Sep 1983 McPherson
Co; 14 Sep 1986 Sioux Co; 15 Sep 1982 Dawes
Co (AB 37:198); 15 Sep 1997 Bessey Division,
NNF (RG); 20–23 Sep 1974 Scotts Bluff Co;
21 Sep 1963 Adams Co; late Sep–early Oct
1947 Sioux Co (Rosche 1982); and 7 Oct 1983
McPherson Co.

Other Reports
29 Mar 1985 Howard-Hall Cos, 1 April 1953
Brown Co, 2 April 1974 Douglas Co, 4 April
1985 Douglas Co, 6 April 1943 Jefferson Co, 8
Apr 1948 Logan Co, 13 Apr 1964 Cass Co, 15 Apr
1947 Jefferson Co, 16 Apr 1990 Lancaster Co, 18
Apr 1952 Thayer Co, 18 Apr 1959 Lancaster Co.
FINDING
Wood Thrush can be found during May in the
southeast at Fontenelle Forest and Indian
Cave SP, where it is best located by song in
deeper woods.

American Robin
Turdus migratorius
STATUS
Abundant regular spring and fall migrant
statewide. Common regular breeder
statewide. Common but erratic regular winter
visitor statewide.
DOCUMENTATION
Specimen
UNSM ZM6652, May 1900 Sioux Co.
TAXONOMY
Two subspecies occur. The eastern race
migratorius breeds west to Gordon, North
Platte, and McCook, while *propinquis*, the
Rocky Mountain race, breeds east to Sioux
Co and Scottsbluff (AOU 1957). There is no
information on individuals from the area of
intergradation, although Bruner and others
(1904) stated that "most western Nebraska
birds were intermediate, with an occasional
propinqua."
Tout (1947) listed *propinquis* as an irregular
winter visitor to Lincoln Co, identified by the
lack of white spots at the tips of the outer tail
feathers.
DISTRIBUTION AND ECOLOGY
During migration this species occurs in all
terrestrial habitats but is least numerous in
forest interiors and open grasslands. Clearly a
savanna species, its original breeding habitat
was typical parkland, scattered trees with a
grassy understory, or perhaps woodland edge
adjacent to grasslands. It is now common

around human habitation, largely because urban and suburban yards and parks provide an adequate substitute for its original savanna habitat. It breeds statewide but is most numerous in the east. In winter American Robin can be abundant where a supply of native fruits such as Russian olive, hackberry, and juniper berries (Rosche 1982) is available, but it may also be absent from many areas of the state in some years.

Spring
The presence of overwintering flocks makes it difficult to discern arrival dates, although there is a noticeable increase of robins in late Feb and early Mar. Johnsgard (1980) noted that there is a concentration of early spring dates in the period 2 Feb–4 Mar; peak numbers were noted in Saunders Co 12 Mar–2 Apr 1994.

High counts include 577 in Pierce Co 8 May 1999, 500 in northern Saunders Co 16 Mar 1996, and 304 in Lancaster Co 27 Mar 1999.

Summer
This species breeds statewide. Ducey (1988) has summarized breeding records. BBS data indicate that breeding birds are most abundant in the east, with diminishing numbers westward, presumably as woodland and human habitation decrease (NBR 46:38). According to Bent (1949), the eastern subspecies *migratorius* spread westward as towns and parks were established. The tall- and mixed-grass prairie was not a favored habitat, a fact reflected by BBS data.

Mating was observed as early as 24 Feb in 1930 in Webster Co (Ludlow 1935).

Fall
Robins form flocks as early as Jul after breeding, and many of these flocks linger into winter, gradually departing as the supply of fruits is depleted. More than half of fall departure dates examined by Johnsgard (1980) fell in the period Oct 20–Dec 14.

High counts include "thousands" at Fontenelle Forest 6 Oct 1999, 1200 in Kearney Co 19 Sep 1999, and 350 in Thomas Co 16 Oct 1998.

Winter
CBC data demonstrate the erratic winter occurrence of this species. In CBCs from 1983–84 through 1992–93, the total count of robins has varied from a high of 10 410 in 1987–88 to a low of 213 in 1986–87. The highest individual count total was 9500 at Calamus-Loup in 1987–88.

An amazing count was an estimated 20 000 flying over in Brown Co 11 Jan 1988; 4800 were estimated in 6 minutes, and the flight lasted for 30 minutes (NGP). "Countless thousands" were at North Bend 5 Feb 1999.

FINDING
If you have not seen an American Robin, look on your lawn in late Mar, especially if you live in eastern Nebraska.

Varied Thrush
Ixoreus naevius
STATUS
Rare casual winter visitor statewide.
DOCUMENTATION
Specimen
UNSM ZM6661, 18 Dec 1935 North Platte, Lincoln Co.
DISTRIBUTION AND ECOLOGY
Most reports are from towns and cities, usually at feeders, and are distributed statewide. This species sometimes occurs with American Robin near a natural or provided food supply.
Winter
18, 20, 20 Oct→16, 30, 31 Mar
Of the 29 reports, 24 were in the period 1972–91. Prior to 1972 there were only 2 records, and there are only 3 since: 10 Jan–Mar 1993 Omaha, winter 1994–95 Omaha, and "prior to" 16 Oct–20 Oct 1999 Norfolk (MB). A flock of 3 appeared Dec 1935 at North Platte, remaining until Jan 1936, when a dead individual was found (NBR 4:9); one of these birds is the specimen cited above. This specimen was identified as the subspecies *meruloides*, breeding in the southern part of the species' range. Another Varied Thrush was reported at Omaha 10 Apr 1949.

Most reports are mid-Oct through late Apr, although there are earlier reports 20 Sep 1987 Sioux Co, 23 Sep 1980 Scotts Bluff Co, and 1 Oct 1976 Sioux Co and later reports 10 Apr 1949 Douglas Co, 1 May 1991 Douglas-Sarpy Cos, 6 May 1983 Douglas Co, 20 May 1991 Lincoln Co, and 7 Jun 1985 McPherson Co (Mollhoff 1987).

FINDING

This species is easy to observe when coming regularly to a feeder, where it tends to stay for extended periods. It is difficult to locate otherwise but could be looked for in cemeteries and other coniferous plantings.

Mimidae (Mockingbirds, Thrashers)

Gray Catbird

Dumetella carolinensis

STATUS

Common regular spring and fall migrant east, becoming uncommon west. Common regular breeder east, becoming uncommon northwest and rare southwest. Rare casual winter visitor east and central.

DOCUMENTATION

Specimen

UNSM ZM11666, 18 May 1907 Omaha, Douglas Co.

DISTRIBUTION AND ECOLOGY

Gray Catbird favors brushy habitats, often near water and associated with woodland edge. The stratum of low-growing gray dogwood that commonly develops in mature riparian forest is especially suitable catbird habitat. Overall, this habitat is best developed in the east, where numbers are greatest and the species is common. Numbers decline rapidly westward as brushy habitats diminish; in the northwest it is only a "locally rare" summer resident (Rosche 1982).

Spring

8, 9, 10 Apr→summer (east)

5, 7, 9 May→summer (west)

Migrants appear in mid-Apr; reports prior to about 20 Apr may be migrants or wintering

birds. There are a few reports in Mar that are probably wintering birds: 9 Mar 1998 Bellevue, 20 Mar 1972 Hall Co, 20 Mar 1989 Lincoln Co, 26 Mar 1955 Dawson Co, 28 Mar 1981 Scotts Bluff Co, and 1 Apr 1961 Gage Co. Reports earlier than these are discussed under Winter.

In the Panhandle, where the species is rare in summer, migrants are reported mostly in the second half of May, although there is an earlier report 6 Apr 1995 Garden Co.

High counts include 35 at Rowe Sanctuary 24 May 1998, 30 there 31 May 1997, and 26 at Wood Duck Area 12 May 1999.

Summer

There are breeding records (Ducey 1988) or summer records for most counties in the east but fewer records for central Nebraska. Breeding probably occurs in the central part of the state wherever habitat exists, as in the Niobrara Valley Preserve (Brogie and Mossman 1983) and the Bessey Division, NNF, where it is an "uncommon summer resident and nester" (Bray 1994). In the Panhandle confirmation of breeding is recent (Rosche 1994a), although there are several summer reports for Sioux, Dawes, Scotts Bluff, and Garden Cos. As recently as 1982, Rosche (1982) noted that while it "presumably breeds" in the northwest, there was no definite evidence. Nesting was not recorded in Keith Co until 1984 (Johnsgard 1990a); the species was only a migrant prior to then (Rosche and Johnsgard 1984). Rosche (1994a) indicated that regular breeding occurs westward in the Platte Valley to Sutherland and Paxton, and it now is a "common breeder" in the L Ogallala area (Brown and others 1996).

It is essentially absent from the arid southwest, south of a line from Banner Co to Furnas Co. None were found west of Harlan Co in the Republican Valley in Jun 1996 (WRS, JGJ).

Tout (1947) cited egg dates in Lincoln Co 28 May–30 Jul (N = 39).

Fall

summer→29, 29, 29 Aug (west)

summer→31 Oct, 1, 2 Nov (elsewhere)

Migration ends by late Oct in most of the state but before Sep in the Panhandle, where there are a few later reports: 14 Sep 1993 Scotts Bluff Co, 24 Sep 1993 Sheridan Co, 30 Sep 1994 Dawes Co, 1 Oct 1994 Scotts Bluff Co, and 3 Oct 1998 L Ogallala.

There are few Nov reports away from the east: 3 Nov 1961 Adams Co, 9 Nov 1990 Keith Co (AB 45:124), 10 Nov 1990 Holt Co, 13 Nov 1983 Adams Co, and 18 Nov 1983 Howard Co. Later reports are probably individuals attempting to overwinter (see Winter).

High counts include 12 in northern Saunders Co 6 Sep 1997.

Winter

Overwintering probably occurs casually, although there are no documented midwinter records. While there are several Nov reports in the east, Dec–Feb reports anywhere in the state are few: 2 Dec 1968 Douglas Co, 3 Dec 1962 Douglas Co, 8 Dec 1992 Douglas-Sarpy Cos, 12 Dec 1968 Nemaha Co, 20 Dec 1987 Hall Co, 5 Jan 1997 Buffalo Co, 7–8 Feb 1993 Douglas Co (AB 42:273), 17 Feb 1957 Gage Co, and 26 Feb 1957 Brown Co.

FINDING

Gray Catbird readily responds to squeaking and is easily located in dense thickets near streams in forest such as at Indian Cave SP and Fontenelle Forest, especially in May.

Northern Mockingbird

Mimus polyglottos

STATUS

Uncommon regular spring and fall migrant statewide. Uncommon regular breeder statewide. Rare casual winter visitor south.

DOCUMENTATION

Specimen

UNSM ZM11667, 31 May 1895 Lincoln, Lancaster Co.

TAXONOMY

Two subspecies occur. The eastern race, *polyglottos*, breeds north to eastern Nebraska (AOU 1957), while the western *leucopterus*

breeds north to southeast Wyoming and southwest South Dakota. *Leucopterus* "intergrades with *polyglottos* in the eastern plains area of Nebraska" (AOU 1957). Rapp and others (1958) suggested that *leucopterus* occurs in the Panhandle and east in the Platte Valley as far as Lincoln and Logan Cos. These subspecies were synonymized without comment by Phillips (1986).

Midwinter reports suggest that *leucopterus* may be more likely to winter than *polyglottos*, as there are more winter reports of Northern Mockingbird from the west.

DISTRIBUTION AND ECOLOGY

This species occupies open areas sparsely vegetated with scattered small to medium-size trees and brush in towns, cities, and farmsteads. While it is found statewide in these habitats, it is nowhere numerous. It avoids forest and treeless grassland.

Spring

This species is generally sedentary, but northern populations leave breeding ranges in winter. Migration is most noticeable in spring when birds return and establish territories. Arrival is in early Apr, although there are several Mar reports where wintering birds were not noted.

Summer

Available data (Ducey 1988) indicate that Northern Mockingbird breeds statewide, although there are more breeding records from the east (Johnsgard 1980). Rosche (1982) considered it a very rare summer resident in the northwest, "more irregular than formerly," possibly due to trampling of its riparian habitat. In recent years it appears to be increasing in numbers in the west.

The BBS routes with highest numbers of Northern Mockingbird are located in the extreme southeast (Johnson, Nemaha, Gage, and Pawnee Cos) and along or south of the Platte Valley (Gosper, Furnas, Keith, and Lincoln Cos). Ludlow (1935) considered this species a "common summer resident" in Webster Co.

A rather early nesting date was of a pair with

young near Gering 9 May 1994; 2 adults and an immature were in Lancaster Co 8 Jul 1995.

Fall

Departure times are difficult to determine, as there is a gradual movement southward in late fall (see Winter).

Winter

In fall and early winter numbers decline as birds gradually leave the state. The few winter reports are scattered statewide. There were 8 Nebraska CBC reports 1967–1993 and a few Feb reports: 7 Feb 1987 Lincoln Co, 9 Feb 1981 Douglas Co, 20 Feb 1990 Lincoln Co, and 28 Feb 1954 Boyd Co. The scarcity of Feb reports indicates that overwintering is unexpected. The several Mar reports may be wintering birds but are more likely early migrants (see Spring).

FINDING

This species can be located by driving back roads in the extreme southeast or southern Panhandle and listening for the song in May. Good locations are the shrub-dotted prairie areas in the vicinity of Burchard L and the limber pine area along the road south from exit 1 on I-80 in Kimball Co. Where present, mockingbirds are usually rather conspicuous.

Sage Thrasher

Oreoscoptes montanus

STATUS

Rare casual spring migrant west. Rare casual summer visitor west. Rare regular fall migrant west. Hypothetical in winter.

DOCUMENTATION

Specimen

UNSM ZM6640, 13 Jul 1901 Antelope Creek, Sioux Co.

DISTRIBUTION AND ECOLOGY

This species occurs in the western Panhandle in open grassy areas during migration, often where brush and tumbleweeds are present (Rosche 1982). The few summer reports are from areas in the extreme northwest where Western sagebrush is present (Rosche

1982), a habitat that is scarce in Nebraska (Ducey 1988).

Spring

Migration probably peaks in mid-Apr, based on the few reports: 23 Mar 1967 Sioux Co, 28 Mar–4 Apr 1999 (4) southern Kimball Co, 4 Apr 1999 west of Bushnell, 6 Apr 1991 Sioux Co, 12 Apr 1980 Sioux Co, 17 Apr 1935 Logan Co, 17 Apr 1998 (2) Kimball Co, 18 Apr 1987 Sioux Co (AB 41:457), 20 Apr 1997 Banner Co, 20 Apr 1997 Kimball Co, 16 May 1996 Sioux Co, and 18 May 1933 Sheridan Co. There is an undocumented report of 2 in Logan Co 17 Apr 1935 (Glandon and Glandon 1935b).

This species begins nesting in mid-Apr (Bent 1949); in 3 of the years listed above, there were also Jun reports (see Summer).

Summer

There is no evidence of successful nesting, although there are a few summer records from the limited areas of suitable breeding habitat. Rosche (1982) stated that most summer records in northwest Nebraska are from the Montrose–Orella–Sugarloaf Butte area of northern Sioux Co, where there are significant areas of Western sagebrush. Reports are 26 May 1981 Sioux Co; 1 Jun 1987 Sioux Co; 3 Jun 1980 Sioux Co; 26 Jun 1972 Sioux Co (Sugarloaf); 26 Jun 1972 Sioux Co (Montrose); 30 Jun 1967 Sioux Co; 4 Jul 1942 Lincoln Co; 11 Jul 1991 Kimball Co (considered a migrant by SJD; photograph, Gubanyi 1996a); 11 Jul 1992 Dawes Co (photograph, Gubanyi 1996a); 12 Jul 1893 Box Butte Co; 13 Jul 1901, specimen cited above, which is labeled a juvenile Northern Mockingbird but is a Sage Thrasher (JGJ, WRS); and 18 Jul 1996 (4) Kimball Co (BP, LP; Brogie 1997). The farthest east summer report was Lincoln Co 4 Jul 1942 (Tout 1947).

Fall

31 Jul, 1, 3 Aug→11, 11, 12 Oct

There are about 30 fall reports, half in Aug. Usually single birds are seen, but 13 were found 11 Aug 1996 at Oliver Res and at least 10 near exit 1 on I-80 10 Aug 1996, both locations in Kimball Co (SJD); 1 or more were present at

Oliver Res 10–31 Aug 1996, for a total of about 26 birds in fall 1996 (NBR 64:122). Ten were reported 1–29 Aug 1997. Four were found 26 Sep 1992 at Chimney Rock (RCR, DJR; NBR 61:85). In recent years more birds have been noted in fall; there was only 1 fall record prior to 1964.

Winter

There is 1 report of a wintering individual. A single bird was at Aurora Oct 1960–Feb 1961 (Swanson 1962). While no details of identification were presented, Sage Thrashers are known to winter on occasion east of the usual range. There are 2 such reports for central Iowa (Kent and Dinsmore 1996) and 1 for Missouri (Robbins and Easterla 1992).

FINDING

The best chance for finding this species is to check sage or native grassland areas in the extreme west in Aug. Fall migrants should be passing through at this time and may be located by driving back roads and checking fencerows in areas where weeds and tumbleweeds are present. A good area is the road south from exit 1 on I-80 in Kimball Co.

Brown Thrasher

Toxostoma rufum

STATUS

Common regular spring and fall migrant statewide. Common regular breeder statewide. Rare casual winter visitor southeast.

DOCUMENTATION

Specimen

UNSM ZM11670, 27 May 1907 Omaha, Douglas Co.

TAXONOMY

Two subspecies have been recorded. According to the AOU (1957), *rufum*, the eastern race, breeds west to western Minnesota, western Iowa, and western Missouri, while the western race, *longicauda*, breeds east to eastern Wyoming, southwestern Nebraska, and eastern Colorado. Thus the Brown Thrashers that breed throughout most of Nebraska are probably intergrades.

DISTRIBUTION AND ECOLOGY

This species is found in edge habitats where dense brush and small trees predominate; nests are located in areas of low light penetration (Johnsgard 1979). In the Panhandle this habitat is primarily riparian but also is found in towns and around farmsteads, while in the east woodland edge provides the best habitat. Numbers are highest in the east, where preferred habitat is most abundant.

Spring

2, 2, 3 Apr→summer (east)

19, 21, 24 Apr→summer (west)

Migrants in the southeast arrive in early to mid-Apr; arrival in the west is somewhat later. There are about 40 reports in the period 1 Jan–31 Mar that probably are of wintering birds or very early migrants (see Winter).

High counts include 60 at Kenesaw 26 Apr 1998, 33 in Dixon Co 13 May 1995, and 20 in Lancaster Co 10 May 1996.

Summer

Brown Thrasher breeds statewide, with lowest numbers in the Panhandle, especially the southwest corner; there is only 1 summer report from Kimball Co, 9–10 Jun 1985. Ducey (1988) cited a single record for Banner Co: R. E. Dawson reported it nesting in 1916. There have been no summer reports since then.

Tout (1947) cited egg dates for Lincoln Co (N = 14) 20 May–17 Jul.

Fall

Departure is gradual, with most birds leaving by mid-Oct, especially in the Panhandle, where later reports are unusual: 5 Nov 1978 Garden Co, 25 Nov 1963 Dawes Co, 12 Dec 1987 Scotts Bluff Co, 31 Dec 1974–1 Jan 1975 Scotts Bluff Co, and 31 Dec 1976 Sioux Co.

Elsewhere, Dec reports are numerous; 37 birds were reported on CBCs 1967–93, in addition to several other Dec reports. Jan reports are far fewer and are discussed under Winter.

Banding data indicate that Brown Thrashers that summer in Nebraska commonly winter in Texas and also that in fall some individuals wander north, at least to South Dakota.

High counts include 11 at Arbor Day Farm 22 Sep 1998.

Winter

There are several Jan reports, but overwintering is rare. Most late Jan–early Mar reports are probably overwintering birds; these reports are generally from the southeast, although there are these from farther west: 2 Jan 1978 Howard Co, 7 Jan 1957 Dawes Co, 26 Jan 1959 Antelope Co, and 19 Mar 1958 Dawes Co.

FINDING

Brown Thrashers are most conspicuous after arrival in spring, when they sing loudly from prominent perches. They are easy to locate at this time in edge habitats, especially in the east.

Curve-billed Thrasher

Toxostoma curvirostre

STATUS

Rare casual spring, summer, fall, and winter visitor.

DOCUMENTATION

Specimen

UNSM ZM6646, 2 May 1936 North Platte, Lincoln Co (Weakly 1936).

Records

There are 7 reports, 4 acceptably documented:

mid-Jan through 4 Apr 1969 McCook (Shickley 1969)

19 Apr–3 May 1936 North Platte (up to 5, including specimen cited above)

2 Jul 1996 near Gering (AK; Brogie 1997)

18 Nov 1962 40 km (25 mi) north of Mitchell, UNSM ZM13001 (Viehmeyer 1971)

Undocumented reports include breeding in the Lawrence Fork area of Morrill Co Jun 1965 which could not be confirmed by later searches of the area (Viehmeyer 1971). A dead individual was reportedly recovered and then disposed of when the 4H Camp building at Bessey Division, NNF, was opened in Mar 1970. It was assumed that this bird had been trapped in the building when it was closed for the winter in Oct and later died. Those present identified the bird as a Curve-billed Thrasher,

but no details were published (NBR 38:92). One was reported without details in Douglas Co 13 Aug 1998 (B).

COMMENTS

This species has exhibited a pattern of vagrancy, outlined by Newlon (1981), in which regular wintering occurs north of the breeding range in southeast Colorado and southwest Kansas, and vagrants appear farther north, notably in Nebraska, South Dakota, Iowa, and Minnesota. At least 1 of the Nebraska reports fits this pattern, although others are more indicative of summer vagrancy. Four of the 6 reports are for 1962–70, a time during which Curve-billed Thrasher was expanding its breeding range into Colorado (Andrews and Righter 1992). The indication of breeding in Morrill Co in 1965 is within this period; the habitat in Lawrence Fork is riparian with junipers on higher slopes (Viehmeyer 1971).

Ten of 13 records on the northeastern Colorado plains are in spring and summer (Andrews and Righter 1992), and there is a record from Norton Co, Kansas (immediately south of Furnas Co, Nebraska), 6 Jun 1981 (AB 35:956), presumably the same bird listed by Thompson and Ely (1992) 24 Jun in Norton Co, no year given.

FINDING

Curve-billed Thrasher may appear at feeders during late fall and winter in the southwest or southern Panhandle (Newlon 1981). Suitable habitat, especially in the southern Panhandle and in Dundy and Hitchcock Cos, should be searched in Jun for possible breeding birds.

Sturnidae (Starlings)

European Starling

Sturnus vulgaris

STATUS

Common regular resident east and central, becoming uncommon northwest. Abundant regular spring and fall migrant statewide. Fairly common regular winter visitor south and east, uncommon north and west.

DOCUMENTATION

Specimen

UNSM ZM6755, 7 Jun 1932 Saline Co.

DISTRIBUTION AND ECOLOGY

In summer starlings are found statewide, nesting in cracks and crevices in buildings both rural and urban, as well as unused or usurped woodpecker holes in dead trees. Numbers are higher where the human population is highest. Total numbers peak in fall when young birds flock with adults, and migrants enter the state from farther north. These flocks are often found around livestock operations, such as cattle feedlots. Winter numbers vary; many birds leave in colder weather, while those remaining are usually found near human habitation in towns or around farms.

Spring

In its native Europe this species migrates northeast in spring and southwest in fall, a pattern that has carried over to the central United States (Kessel 1953). Nebraska banding data fit this pattern, despite a strong resident component; 30 records of birds banded in Nebraska yielded 6 that were recovered elsewhere, 2 in Iowa and 1 each in Minnesota, Kansas, Oklahoma, and Nevada. The Nevada record was of a bird banded in Nebraska 20 Jan 1983 and recovered there 3 Dec 1990. There are 13 recoveries of birds banded outside Nebraska, all confirming the migratory pattern noted above; these were banded in North Dakota, South Dakota, Minnesota, Kansas, Colorado, and Utah.

In the north and west, where wintering birds are scarce, arrival dates are concentrated in the period mid-Apr to early May.

High counts include 1000 at Nebraska City 5 Mar 1999 and 1000 in Dixon Co 6 Mar 1999.

Summer

BBS data indicate that this species is most numerous as a breeding bird in the east and south, presumably related to human population distribution, less so in the north and west. It was considered "rare" at Bessey

Division, NNF, where none were seen during 1993 fieldwork (Bray 1994).

Banding data indicate that birds banded in Nebraska tend to be resident. Of 30 records of birds banded in Nebraska at all times of the year, all but 6 were recovered up to 8 years later at the same place although at various times of the year. One banded 6 Oct 1982 in Lancaster Co was recovered at the same place 2 Mar 1990.

Fall

Winter recoveries of birds that were banded in summer north of the recovery site show that fall migration was undertaken prior to Dec or Jan. Largest flocks are present in Dec; the largest CBC total was 101 384 at Lincoln in 1975–76, and 100 000 were at Clear Creek Marshes 14 Dec 1993. These large, early winter concentrations probably include significant numbers of migrants, as such flocks are rare in midwinter.

Flocks of juveniles appear in mid-Jul; a flock of 65 juveniles was in Lancaster Co 13 Jun 1999.

Winter

Winter distribution resembles that in summer but for different reasons. While summer distribution depends on availability of nest sites, winter distribution is limited by availability of food and shelter. This is best provided around human habitation, especially farm operations with livestock feedlots but also towns and cities.

COMMENTS

European Starling was first successfully introduced to North America in New York City in 1890–91 (Bent 1950) and reached Nebraska in 1930, when the first nesting occurred (NBR 1:15). The earliest specimen is 7 Jun 1932 Saline Co (UNSM ZM6755), described on its label as the "1st state record of an adult." Starlings were not recorded in northwest Nebraska until 1942 (Rosche 1982). Tout (1947) did not mention this species as having occurred in Lincoln Co by that time.

FINDING

This species is most obvious in late summer

when large flocks of young birds are commonly seen flying about in agricultural areas. Breeding birds are plentiful around cities in summer.

Motacillidae (Pipits)

American Pipit
Anthus rubescens
STATUS
Common regular spring and fall migrant statewide. Rare casual summer visitor west and central.
DOCUMENTATION
Specimen
UNSM ZM6724, 19 Oct 1899 Lancaster Co.
DISTRIBUTION AND ECOLOGY
Migrants occur statewide in open areas with sparse vegetation or bare dirt, probably most conspicuously on unvegetated damp margins of lakes and ponds. Like other open-area birds, American Pipits can appear common after heavy, late spring snowfalls. In the northwest it is more common in fall than in spring, at times becoming abundant (Rosche 1982).
Spring
13, 14, 15 Mar→21, 23, 23 May
Migration takes place mainly in Apr, although the first arrivals appear in Mar. There are earlier reports 5 Mar 1967 Lancaster Co and 10 Mar 1937 Adams Co.
High counts include 67 at Clear Creek Marshes 18 Apr 1998, 35 there 6 May 1995, and 30 near Elmwood 18 Apr 1997.
Summer
There are 3 reports: 30 Jun 1967 Garden Co, 22–23 July 1986 Sioux Co, and 6 Aug 1987 Lincoln Co. None are documented, but there is a 9 Jul record for Cloud Co, Kansas (Johnston 1965), and an early Aug record for the Colorado lowlands (Andrews and Righter 1992).
Fall
4 (Tout 1947), 6, 6 Sep→10, 12, 13 Nov
Migration is in late Sep and Oct, although there are earlier reports 14 Aug Keith Co (Rosche 1994a) and 24 Aug 1964 Lincoln Co.

There is no evidence for overwintering, but individuals have remained very late in fall. There are no documented records for Jan or Feb. American Pipits have been found on CBCS: 128 at Grand Island 16 Dec 1972 and 3 at L McConaughy 18–19 Dec 1992 (RCR; Gubanyi 1996a).
High counts include 75+ near Gibbon 23 Oct 1999, 68 at L McConaughy 3 Oct 1998, and 54 at Crescent L NWR 26 Oct 1994.
Other Reports
30 Dec 1967 Scottsbluff CBC, 1 Jan 1957 Adams Co, 30 Jan 1955 Adams Co.
FINDING
American Pipits are probably easiest to find around damp margins of reservoirs in fall, when water levels are often lower. The best time is probably early in migration, such as late Sep.

Sprague's Pipit
Anthus spragueii
STATUS
Uncommon regular spring and fall migrant central, rare east, rare casual west.
DOCUMENTATION
Specimen
UNSM ZM12164, 22 Apr 1909 Lancaster Co.
DISTRIBUTION AND ECOLOGY
This generally solitary species is found during migration in shortgrass or tallgrass prairie, usually dry, that has been grazed to 5–8 cm (2–3 in) (Thompson and Ely 1992), pastures (AOU 1998), and harvested cultivated fields such as alfalfa (Mark Robbins, pers. comm.) and wheat stubble (Andrews and Righter 1992).
Spring
27, 28, 28 Mar→18, 21, 23 May
Most reports are from central Nebraska, especially Adams and Webster Cos. Migration occurs mainly in Apr, although there are earlier reports 17 Mar 1963 Antelope Co and 17 Mar 1973 Sarpy Co (AB 27:635). Turner (1978) observed that this species "usually appears around Apr 20" in Kearney Co. Migration

ends in mid-May, although there is a later report from Bohemia Prairie 8–10 Jun 1985.

Sprague's Pipit is reported least often in the west: 23 Apr 1972, 10 May 1974, and 14 May 1961 Scotts Bluff Co.

Fall

14, 15, 17 Sep→25, 27, 30 Oct

Migration takes place in late Sep and Oct, although there are earlier reports 30 Aug 1985 Scotts Bluff Co and 7–9 Sep 1990 Thomas Co (Bray 1994) and a later report 8 Nov 1967 Adams Co.

A good count for this usually solitary species was 6–8 birds in Holt Co in late Sep 1988 (AB 43:126); up to 3 were there (south of Chambers) 1–17 Oct 1996 (Gubanyi 1996a).

There are few Panhandle reports: 30 Aug 1985 Scotts Bluff Co; 1 Oct 1920 Sioux Co, a specimen, UNSM ZM6731 (Dawson 1921; Rosche 1982); 2 Oct 1961 Scotts Bluff Co; and 7–12 Oct 1970 Scotts Bluff Co.

FINDING

This species should be looked for in short- to medium-grass fields during fall, especially early Oct, in central Nebraska. It has been suggested (AB 29:81) that it should be looked for "in shallow ruts made by vehicles crossing grassy fields."

Bombycillidae (Waxwings)

Bohemian Waxwing

Bombycilla garrulus

STATUS

Common but highly erratic regular winter visitor northwest, uncommon casual elsewhere.

DOCUMENTATION

Specimen

UNSM ZM12046, 7 Dec 1889 Lancaster Co.

DISTRIBUTION AND ECOLOGY

Greatest numbers are recorded in the northwest, where Rosche (1982) noted that "during the early part of the winter Russian Olive fruits are the primary food source; later towards the spring they move into areas where hackberry fruits and juniper berries are abundant."

Fruiting trees are common in towns, and waxwings of both species are often found there. In the east, where this species is only casual, individuals are usually found with Cedar Waxwings.

Spring

winter→8, 9, 13 Apr

Departure probably occurs by early Apr, although later reports are in May: 28 Apr 1990 Scotts Bluff Co, 1 May 1970 Scotts Bluff Co, 3 May 1962 Dawes Co, and 8 May 1958 Scotts Bluff Co. The latest record from South Dakota is 28 Apr (SDOU 1991), from Kansas 16 April (Thompson and Ely 1992), and from Colorado early May (Andrews and Righter 1992).

Fall

6, 7, 7 Nov (specimen UNSM ZM6738)→winter

The earliest arrivals are in mid-Nov, although there are earlier reports 14 Sep 1973 Adams Co, 25 Sep 1976 Scotts Bluff Co, Sep 1980 Dawes Co (AB 35:200), 10 Oct 1963 Lincoln Co, 26 Oct 1978 Scotts Bluff Co, and 31 Oct 1969 Adams Co, the latter undocumented.

Winter

This species appears in large numbers in some years and is absent in others; it is generally restricted to the northwest. Rosche (1982) listed incursion years and noted that in the 1970s and 1980s birds appeared in consecutive winters and were absent in the third, a pattern that repeated itself during that time period. It was considered a "common migratory visitor" in Logan Co (Glandon and Glandon 1934b).

CBCs in the northwest regularly reported Bohemian Waxwings, with a high count of 300 at Crawford 23 Dec 1978. The only CBC report since 1979 was 10 birds at L McConaughy 28 Dec 1996. One on the Omaha count 30 Dec 1989 was in the Iowa part of the count circle.

High counts include 419 at Monroe Canyon 21 Nov 1998, 362 at Harrison 29 Dec 1996, and 300 at Chadron Feb 1986 (AB 40:298).

Other Reports

15 May 1931 Webster Co (Ludlow 1935); 21 May 1939 Webster Co; 22 May 1973 Scotts Bluff

Co; 25 May 1969 McPherson Co; 25 May 1939 Logan Co; 26 May 1951 Adams Co; 27 May 1990 Douglas Co; 1 June 1957 Box Butte Co, considered "incredible" by Rosche (1982); and 18 Jun 1957 Brown Co.

FINDING

Western towns and cities in Dec and Jan would provide the best chance to see Bohemian Waxwings. Look for fruiting trees such as Russian olive, ash, and crab apples (*Malus* sp.). Later in the winter Bohemian Waxwings disperse into areas with junipers and may not be as obvious (Rosche 1982).

Cedar Waxwing

Bombycilla cedrorum

STATUS

Common regular spring and fall migrant statewide. Uncommon regular breeder east and west, rare central. Common regular winter visitor statewide.

DOCUMENTATION

Specimen

UNSM ZM12051, 8 Mar 1890 Lancaster Co.

DISTRIBUTION AND ECOLOGY

Cedar Waxwing flocks are most conspicuous during migration. In fall migrants utilize fruit plantings such as Russian olive and crab apples in towns and cities, and as these are consumed, wintering birds move to areas of junipers (Rosche 1982). Spring migrants are also found in junipers. Nesting birds, however, prefer open deciduous woodlands (Johnsgard 1979), even in the Pine Ridge, where Rosche (1982) indicated that pine forests are avoided by nesting birds.

Spring

Because Cedar Waxwings are observed year-round, timing of spring and fall migration is difficult to determine. Spring migration takes place mainly in Mar and Apr (Thompson and Ely 1992), although Robbins and Easterla (1992) suggest 2 peaks during spring migration, an early one in Mar (which may be difficult to discern due to the presence of wintering birds) and a later peak in late May

and early Jun. Robbins and Easterla (1992) suggest that the late peak is of birds wintering farther south than the early peak, perhaps as far south as Mexico and Central America.

High counts include 500 at Harlan Co Res 23 Apr 1999, 250–300 at Kearney 3 Apr 1999, and 85 in Otoe Co 3 Mar 1996.

Summer

Regular breeding probably occurs only in woodlands adjacent to the Missouri River and its southeast Nebraska tributaries and in the Panhandle. However, Cedar Waxwings are not common during the nesting season. Egg dates indicate that Cedar Waxwings breed in Jun and Jul (Johnstone 1965; SDOU 1991; Johnsgard 1979), although there is a Dundy Co record for 31 May (see below). Further confusion arises from the presence of small flocks in early Jun and late Jul, but these are presumed breeders still in migration (Robbins and Easterla 1992; Mark Robbins, pers. comm.).

Ducey (1988) cited recent breeding records only for Butler and Cass Cos in the east, but nesting occurred in Douglas Co in 1990 (AB 44:1153) and Platte Co in 1983 (AB 37:1003). There are also many breeding season reports in the east, especially in counties bordering the Missouri and Platte Rivers but notably also in Lancaster Co.

Johnsgard (1979) suggested that breeding occurred regularly in the Pine Ridge, and Rosche (1982) considered the species a rare to uncommon summer resident in the northwest. Ducey (1988) cited breeding records from Dawes and Scotts Bluff Cos.

Elsewhere, reports suggest breeding is possible wherever woodlands occur. Breeding season reports from the Platte Valley exist for Adams Co 1 Jul 1964 and 30 Jun 1965, Clay Co 26 Jun 1975, Dawson Co 4 Jun 1994, and Buffalo Co 4–7 present until 21 Jun 1998. Ducey (1988) cited a nesting record from Lincoln Co, and there are reports from there 2 Jul 1964, 30 Jun 1965, 14 Jul 1976, and summer 1992. Rosche (1994a) listed it as a "casual summer visitant" in the L McConaughy area with no evidence of nesting, although the species was listed as a permanent resident in 1992 and 1993, and Brown and others (1996) reported a nesting pair near Kingsley Dam May 1994.

In the central Niobrara Valley Brogie and Mossman (1983) did not detect nesting in the Niobrara Valley Preserve, although there are suggestive reports for Brown Co 30 Jun 1970, 1 Jul 1968, and 2–3 Jul 1994. Youngworth (1955) took specimens in northeast Cherry Co in 1932 and considered it a rare breeder there, and there is a recent summer report 1 Jul 1994. There are breeding season reports for Custer Co 30 Jun 1970, Logan Co 20 Jun 1966, and Valley Co 18 Jun 1994. In 1993 the species was considered a permanent resident in Thomas Co, although dates cited were 9 Jul–7 Sep, somewhat later than the expected time for breeding.

In the southwest 2 nests were located in Dundy Co 31 May 1985 (AB 39:321), a rather early date, and 5 birds were at Alma Jun–Jul 1998.

Fall

Migrants are observed as early as Aug, but most arrive in mid-Sep and peak in Oct (Robbins and Easterla 1992). Thompson and Ely (1992) noted that there is a "thinning" of numbers in Dec, at which time numbers vary depending on the wild fruit crop; in some years no Cedar Waxwings can be found.

High counts include 151 at Fremont 3 Nov 1998 and 115 in Thomas Co 10 Sep 1994.

Winter

Overwintering numbers are variable. CBC data indicate that most birds are located in southeast Nebraska in winter. The highest CBC total was 454 at Norfolk in 1988–89. Rosche (1982) noted that not since 1973–74 had Cedar and Bohemian waxwings been common at the same time.

FINDING

Flocks of Cedar Waxwings are most noticeable in Oct and Nov and again during periods of thaw in late Feb and Mar, when they can be located in parks, cemeteries, and towns with fruiting trees. Small groups may also be conspicuous

foraging in woody vegetation during late Apr when the vegetation is budding and flowering and insects are attracted to the sap and nectar.

Ptilogonatidae (Silky Flycatchers)

Phainopepla

Phainopepla nitens

STATUS

Accidental in winter.

DOCUMENTATION

Photograph

1 Jan 1983 Alliance, Box Butte Co (Thomas 1983).

Records

An adult male was photographed at Alliance 1 Jan 1983; it was first seen drinking at a birdbath and then was seen intermittently until 13 Feb (Thomas 1983).

COMMENTS

The regular summer range of this species extends north to southwest Utah, central New Mexico, and western Texas. It is migratory in the northern parts of its range and a rare postbreeding wanderer north of its range (NGS 1987). Andrews and Righter (1992) considered it a casual visitor on the eastern plains of Colorado, where there are 5 records, most of females or immatures, and most in urban areas. Four of the 5 records are in fall and winter (29 Aug–18 Dec), and the other is in May. There are 3 recent Kansas records in the period 3–30 Sep. Thus the Nebraska record extends this pattern to the north and suggests postbreeding wandering or reverse migration with subsequent attempted overwintering. There are no Wyoming or South Dakota records.

Parulidae (Wood-Warblers)

Blue-winged Warbler

Vermivora pinus

STATUS

Rare casual spring migrant east, becoming accidental west. Accidental in summer. Rare casual fall migrant east and central.

DOCUMENTATION

Specimen

UNSM ZM10813, 15 May 1888 Peru, Nemaha Co.

TAXONOMY

The hybrid form with Golden-winged Warbler, "Brewster's Warbler," was observed in Keya Paha Co 19 May 1982 (Brogie and Mossman 1983).

DISTRIBUTION AND ECOLOGY

Migrants use open areas with scattered brush and small trees. In recent years reports have been limited to the east.

Spring

25, 25, 26 Apr→28 May, 1, 3 Jun

Most reports are from the east during the period late Apr–early Jun, with later reports 11 Jun 1993 Sarpy Co (AB 47:1124), 11 Jun 1999 Fontenelle Forest (JG), and 18 Jun 1999 Keya Paha Co (PK).

There is only 1 report from the Panhandle, 20 May 1982 Sioux Co. There are few reports away from the east, most westerly 19 May 1993, a female netted in Keith Co (Brown and others 1996), 15 May 1982 Keya Paha Co (Brown and Mossman 1983), 22 May 1971 Brown Co, 17–18 May 1980 Thomas Co (Bray 1994), and 18 Jun 1999 Keya Paha Co (see above).

Considered "possibly the best [total] for one season," 6 birds were recorded 5–24 May 1995 in Sarpy Co.

Summer

The only summer report since 1970 is 6 Jul 1974 Douglas-Sarpy Cos.

The only evidence for breeding is a set of eggs ascribed to this species collected in Sarpy Co in 1901 (Ducey 1988). According to Bruner and others (1904), however, it was a "rather common summer resident and breeder in the wooded Missouri bottoms" and said to be "present about Omaha and Peru all summer." There are no breeding reports since that time, suggesting a significant range contraction in the Missouri Valley since the early 1900s. It was a regular breeder in the Kansas City area until the 1940s but now no longer breeds west of about 93 degrees latitude and north of the Missouri River in Missouri (Robbins and Easterla 1992), although 2 territorial males have been in Platte Co in northwest Missouri each year since about 1996 (Mark Robbins, pers. com.).

Occurrence as a summer resident in Webster Co (Rapp and others 1958) is likely an error (Johnsgard 1980).

Fall

28, 29, 29 Aug→9, 10, 16 Sep

There are only 11 fall reports, including an earlier report 18 Aug 1964 Cass Co. The only reports away from the Missouri Valley are 30 Aug 1970 Custer Co, 16 Sep 1961 Webster Co, 1 Sep 1994 Dakota Co, and 9–10 Sep 1995 Dixon Co.

FINDING

The best chance for finding this species would be to check woodland edge and brushy areas such as overgrown pastures in the extreme southeast, especially in and around Indian Cave SP in mid-May and along the railroad tracks at Fontenelle Forest.

Golden-winged Warbler

Vermivora chrysoptera

STATUS

Rare regular spring migrant east, rare casual central. Rare casual fall migrant east, accidental central and west.

DOCUMENTATION

Specimen

UNSM ZM16966, 15 May 1993 Lincoln, Lancaster Co (Brogie 1997).

DISTRIBUTION AND ECOLOGY

Virtually all reports are from the lower Missouri Valley during May, late Aug, and early Sep; fall reports are few, however. Migrants are usually found with other warblers in woodland edge habitats.

Spring

2, 4, 4 May→24, 25, 25 May

There is an earlier report 25 Apr 1982 Douglas-Sarpy Cos.

Most reports of this species are recent, reflecting a marked increase in numbers in recent years. Apart from a vague reference attributed to

Aughey (Bruner and others 1904) that it was "occasionally seen in eastern Nebraska," there were no reports until 1954, when one was found 17 May in Adams Co.

There are few reports away from the east, the single Panhandle record a singing male in Dawes Co 23 May 1997 (TLE). Other westerly reports are 11 May 1991 Keith Co (Rosche 1994a), 15 May 1998 Clear Creek Marshes (Betty Allen, pers. comm. SJD, JS), 17 May 1955 Keith Co, and 6 Jun 1992 Keith Co (Brown and others 1996).

In 1993 there was a major incursion, with 14 reports involving 15 birds, most at Fontenelle Forest (AB 47:428). Prior to 1993 there had only been about 30 reports in all. Another good year was 1995, when at least 12 birds appeared, including 3 at Fontenelle Forest 13 May.

Fall

22, 29 Aug, 1 Sep→11, 11, 11 Sep

There are only 16 reports in all. The only reports away from the Missouri Valley are the single Panhandle record, 14 Sep 1997 in Sowbelly Canyon (JGJ), and 7 Sep 1973 Thomas Co (AB 28:74). The northernmost report in the Missouri Valley is 11 Sep 1995 Washington Co (FN 50:76).

FINDING

As with other rare warblers, the best chance of locating a Golden-winged Warbler is with mixed flocks of migrants in early and mid-May in eastern Nebraska woodlands.

Tennessee Warbler

Vermivora peregrina

STATUS

Common regular spring migrant east, becoming rare west. Common regular fall migrant east, becoming rare casual west.

DOCUMENTATION

Specimen

UNSM ZM6808, 25 May 1901 Child's Point, Sarpy Co.

DISTRIBUTION AND ECOLOGY

This species is usually found fairly high above the ground in deciduous woodland. While it occurs statewide, it is most common in the east and declines in numbers westward.

Spring

15, 17, 18 Apr→9, 10, 10 Jun

Migration is from late Apr through early Jun.

In the Panhandle this species is regular but very rare in spring (Rosche 1994a). There are about 25 reports in the period 5 May–9 Jun. Brown and others (1996) cited 5 banding records 13–29 May in the Keith Co area.

High counts include 50+ at Dodge Park, Omaha, 11 May 1999; 45 in Dakota Co 17 May 1997; and 39 at Fontenelle Forest 15 May 1999.

Fall

12, 18, 18 Aug→24, 26, 27 Oct

Migrants occur from late Aug through mid-Oct, although there is a later report 1 Nov 1971 Lancaster Co (AB 26:83).

There are these Panhandle reports: 4 Sep 1992 Morrill Co, 5 Sep 1992 Dawes Co, 13 Sep 1997 Oliver Res (SJD), 17 Sep 1982 Dawes Co (AB 37:198), 24 Sep 1995 Sheridan Co (FN 50:76), 29 Sep–1 Oct 1994 Scotts Bluff Co, 1 Oct 1961 Scotts Bluff Co, and 2 Oct 1993 Scotts Bluff Co.

High counts include 10 in Sarpy Co 9 Sep 1995.

Other Reports

1 Apr 1957 Lancaster Co, 5 Apr 1972 Hall Co, 7 Apr 1976 Howard-Hall Cos, 29 Jun 1954 Boone Co, 30 Jun 1985 Howard-Hall Cos.

FINDING

Tennessee Warbler is easy to locate by song in mid-May in the Missouri Valley, when it sings loudly from deciduous trees even in populated areas.

Orange-crowned Warbler

Vermivora celata

STATUS

Common regular spring and fall migrant statewide.

DOCUMENTATION

Specimen

UNSM ZM6809, 5 May 1898 Beatrice, Gage Co.

TAXONOMY

Although the only subspecies documented

in Nebraska is *celata*, which breeds across northern Canada and migrates statewide through Nebraska, it is likely that the Rocky Mountain race *orestera* (Dunn and Garrett 1997) occurs in the Panhandle in fall. Many fall migrants there are brightly colored, especially yellowish underneath, and have gray crowns and napes that contrast with the greenish olive back, thus resembling Nashville Warblers without eyerings (SJD, WRS, JGJ).

DISTRIBUTION AND ECOLOGY

Orange-crowned Warblers are found in deciduous woodland and brushy habitats, often near the ground and usually in mixed-species flocks. It is among the few migrant warbler species that is to be expected throughout the state. In the northwest Rosche (1982) considered it an uncommon to fairly common migrant.

Spring

2 (UNSM ZM12004), 3, 3 Apr→31 May, 2, 3 Jun (Tout 1947)

Migrants arrive fairly early, in mid to late Apr, and most depart by the end of May, although there are later reports 7 Jun 1998 Funk Lagoon (LR, RH) and 14 Jun 1959 Boone Co.

High counts include 49 in Sarpy Co 13 May 1995, 42 in Scotts Bluff Co 1 May 1999, and 42 at Fontenelle Forest 2 May 1999.

Fall

10 (Tout 1947), 11, 12 Aug→31 Oct, 1, 1 Nov

Migrants arrive in mid-Aug and depart by late Oct, although there is an earlier report 19 Jul 1977 Keith Co, "probably . . . a very early autumn transient" (Rosche 1994a), and another reported near Funk Lagoon 28 Jul 1999 (GH, WH).

This species is one of the last warblers to leave in fall, and there are 4 late reports: 6 Nov 1975 Douglas-Sarpy Cos (AB 30:93), 12 Nov 1981 Douglas-Sarpy Cos, 18 Dec 1998 Harlan Co Res CBC (JGJ), and 21 Dec 1996 in Lincoln, Lancaster Co. It was described as "wintering as far north as Omaha" in 1979–80 (AB 34:287), but no details were given.

High counts include 42 in the eastern Rainwater Basin 27 Sep 1999, 40 in Otoe Co 18 Sep 1994, and 36 at Fontenelle Forest 11 Oct 1998.

FINDING

Orange-crowned Warblers are easiest to locate in late Apr or early May, when they often associate with Yellow-rumped Warblers in deciduous trees that are not yet leafed out. At this time there are few if any Tennessee Warblers to confuse identification.

Nashville Warbler

Vermivora ruficapilla

STATUS

Common regular spring and fall migrant east, becoming rare casual west.

DOCUMENTATION

Specimen

UNSM ZM6804, 4 May 1904 Falls City, Richardson Co.

TAXONOMY

Although only the eastern subspecies *ruficapilla* has been documented in Nebraska, observers in the Panhandle should be aware that the western subspecies *ridgwayi* may occur there. It differs in having a brighter yellow rump, whitish belly, and a habit of pumping its tail, a habit not shared by *ruficapilla*.

DISTRIBUTION AND ECOLOGY

This species is a common migrant in the east and casual in the Panhandle. It occupies deciduous woodland and brush, often foraging at low levels and in small brushy trees or even weedy fields.

Spring

9, 14, 18 Apr→30, 30 May, 2 Jun

Migration occurs between late Apr and late May. There are later reports, one at Wehrspann L 6 Jun 1998 (JWH) and a specimen, HMM 2601, taken at Inland 9 Jun 1917 (Swenk, Notes before 1925).

There are few Panhandle reports. It is possible that the western subspecies *ridgwayi* may account for some of these: 23–30 Apr 1972 Scotts Bluff Co, 11 May 1956 Sheridan Co, 18 May 1958 Sheridan Co, 22 May 1957 Sheridan Co, and 31 May 1959 Scotts Bluff Co.

High counts include 54 in Sarpy Co 13 May 1995 and 12 at Hormel Park, Fremont, 5 May 1999.

Fall

18, 21, 22 Aug→31 Oct, 3, 3 Nov

Migration is from late Aug through late Oct, rather late for a wood-warbler, although there is an earlier report 10 Aug 1967 Douglas-Sarpy Cos. An extraordinary late report involved up to 3 birds seen together in Omaha 18–20 Dec 1993 (FN 48:223).

There are 6 Panhandle reports, some possibly *ridgwayi*: 3 Sep 1961 Scotts Bluff Co, 26 Sep 1978 Scotts Bluff Co, 28 Sep 1974 Dawes Co (Rosche 1982), 2 Oct 1999 Oliver Res extensive white on underparts (SJD), 8 Oct 1994 Sheridan Co, and 12 Oct 1975 Sioux Co (Rosche 1982).

High counts include 20 at Wehrspann L 20 Sep 1999, 12 in Sarpy Co 9 Sep 1995, and 12 at Fontenelle Forest 12 Sep 1999.

COMMENTS

There are no documented breeding records. Reports of breeding in the east (Ducey 1988) are likely in error, as suggested by Johnsgard (1980). Aughey was said to have found a young bird newly fledged on 10 Jun 1865, questioned also by Sharpe (1993), and Carriker apparently shot a female with well-formed eggs in her ovary near Nebraska City 11 Jun 1900 (Bruner and others 1904). Bent (1953) stated that it was "reported to breed in northeastern Nebraska but no specific records," presumably a reference to Aughey. A nesting report from Webster Co for 1961 (Wensien 1962) probably referred to Yellow Warbler, which was unlisted.

FINDING

Nashville Warblers can be found easily in spring in deciduous woodland and brush. The best time is mid-May at locations such as Fontenelle Forest and Indian Cave SP.

Virginia's Warbler

Vermivora virginiae

STATUS

Rare casual spring migrant west.

DOCUMENTATION

Description

22 May 1995 Morrill, Scotts Bluff Co (ECT; Gubanyi 1996c).

Records

There are 4 reports, 2 documented:

17 May 1998 Bushnell Cem (JGJ)

22 May 1995 Morrill, Scotts Bluff Co (cited above)

Another report (NBR 32:67) apparently involved more than one bird first seen at the North Platte Fish Hatchery 26 April 1964 and again on 29 April by several observers, including at least one experienced observer familiar with the species. There is an old report without observation date or other details of one seen by Aughey along the Republican River in Hitchcock Co (Bruner and others 1904); Sharpe (1993) questioned whether Aughey had ever visited this site.

COMMENTS

This species breeds in the Rocky Mountains north to extreme southern Wyoming and is fairly common in Colorado (Andrews and Righter 1992). It has recently been discovered breeding in the Black Hills of South Dakota, and there are several records for Kansas, where it appears to be a rare migrant in the southwest (Thompson and Ely 1992). The 5 Kansas spring records are in the period 4–14 May, and the single fall record is for 20 Sep. Virginia's Warbler should be looked for in early May and Sep in the Panhandle and southwest.

Northern Parula

Parula americana

STATUS

Uncommon regular spring migrant east, becoming rare casual west. Uncommon regular breeder east. Rare regular fall migrant east, rare casual central and west.

DOCUMENTATION

Specimen

UNSM ZM6817, 20 Apr 1901 Havelock, Lancaster Co.

DISTRIBUTION AND ECOLOGY

This species prefers swampy riparian woodland as breeding habitat but in migration occurs in all types of woodland. It spends most of its time in the canopy. Most reports are from the Missouri Valley, where it occurs in summer in small numbers.

Spring

2, 2, 3 Apr→summer

Arrival is early, typical of the southern warbler species that breed in Nebraska. There appears to be a small movement up the Missouri Valley in spring; Northern Parula is a rare to uncommon migrant in eastern South Dakota (SDOU 1991). It is rare away from the east, especially in fall.

There are 14 reports for the Panhandle in the period 21 Apr–30 May.

High counts include 13 in Sarpy Co 13 May 1995 and 11 in Sarpy Co 11 May 1996.

Summer

Swampy riparian woodland is not an extensive habitat in Nebraska, and Northern Parula would not be expected to breed in any numbers. The few confirmed breeding records include young being fed in Sarpy Co in 1986 (Bennett 1987) and 5 Sep 1991 (AB 46:117). However, Northern Parula may be one of the few warbler species that are increasing in Nebraska. Since 1972, when it was first recorded in Sarpy Co in summer, it has occurred every year since 1975 except for 1977. There are also summer reports from Richardson and Nemaha Cos for 1982 and 1986, and it was "quite common," up to 6 present, at Indian Cave SP during Jun 1995. Missouri BBS data from 1967 to date show that Northern Parula has increased in numbers there also (Robbins and Easterla 1992).

There are scattered summer records elsewhere. One was at Ashford BSC, Thurston Co, 13 Jun 1998 (BFH). Another was singing at Valentine 4 Jul 1998 (NR). Short (1965) collected a singing male in Brown Co 21 June 1964, and it was reported there 12 Jun 1972 and 7 Jul 1968 but not in 1982 (Brogie and Mossman

1983). Rosche (1982) recorded a singing male at Chadron 5 Jun 1979. There are also reports from Adams Co for 1 Aug and 4 Sep 1973, suggestive of summering birds.

Fall

summer→24, 24 Sep, 2 Oct

Northern Parulas become inconspicuous in late summer and fall and depart by late Sep.

There are few fall reports away from the east, including 30 Jul–28 Aug 1977 Box Butte Res (RCR); 25 Aug 1967 Adams Co; 5 Sep 1979 Boone Co; 6 Sep 1998, male at Gering Cem (SJD); 7 Sep 1912, specimen, UNSM ZM11790, Thomas Co (Zimmer 1913); 12 Sep 1919, specimen, UNSM ZM6840, Sioux Co; 15 Sep 1975 Adams Co; 22 Sep 1997 Thomas Co (RG); and a very late immature female in Keith Co 4 Dec 1999 (photograph, SJD).

High counts include 4 in Washington Co 11 Sep 1995 and 4 in Washington and Douglas Cos 13 Sep 1998.

FINDING

The distinctive song facilitates location of this canopy species. It is most easily heard in swampy riparian woodland in the Missouri Valley in early May. Fontenelle Forest is probably the best site.

Yellow Warbler

Dendroica petechia

STATUS

Common regular breeder and spring and fall migrant statewide.

DOCUMENTATION

Specimen

UNSM ZM11952, 21 May 1895 Lancaster Co.

TAXONOMY

Two subspecies have been shown to occur in Nebraska. The eastern race *aestiva* breeds throughout the state, and during migration the northern race *amnicola* occurs (AOU 1957).

Rapp and others (1958) noted that *rubiginosa* also has occurred. A dark-colored individual seen in Logan Co 2 May 1934 (Glandon and Glandon 1934b) was identified as *rubiginosa*. A specimen, #2683 in the Brooking collection, taken 5 May 1920 at Inland was identified by Oberholser as *rubiginosa*, as were 3 additional records by Brooking in the same area (Swenk 1918b). This northwest coastal race seems unlikely, however, and these reports may refer either to the northern race *amnicola* or possibly the far northern breeder *parkesi*, both darker than *aestiva* (Dunn and Garrett 1997). Specimens identified as *rubiginosa* have been taken in Louisiana and Pennsylvania (Dunn and Garrett 1997). The northern *amnicola* and *parkesi* migrate later in spring and fall than does *aestiva*; both are darker than *aestiva* (Dunn and Garrett 1997).

The Rocky Mountain subspecies *morcomi* is likely as a migrant in the Panhandle, is greener above and paler yellow below, and has narrower and more blended chestnut streaks below than the northern races and *aestiva* (Dunn and Garrett 1997).

DISTRIBUTION AND ECOLOGY

This species breeds statewide in riparian brush and thickets, especially willows, and less commonly in upland thickets. In migration it occurs in brushy habitats throughout.

Spring

16, 17, 18 Apr→summer (except west)

30, 30 Apr, 1 May→summer (west)

Migrants arrive in mid to late Apr, somewhat later in the Panhandle, suggestive of the possibility that migrants there are largely northern birds of the races *amnicola* and *parkesi* or the Rocky Mountain race *morcomi* (see Taxonomy).

High counts include 127 in Sarpy Co 13 May 1995 and 85 at Crescent L NWR 28 May 1995.

Summer

BBS data indicate somewhat higher numbers in the north and east. They show stable numbers in Nebraska, and no evidence for a decline as has been noted in other parts of the United States (Tate 1986).

Tout (1947) cited egg dates (N = 34) for Lincoln Co 22 May–8 Jul.

Fall

summer→14, 18, 20 Oct

Most leave by late Sep, although late dates are in Oct. Birds in Sep and later may be migrants of the northern races *amnicola* and *parkesi*, as the breeding race *aestiva* leaves the breeding grounds early, starting in mid-Jul (Curson and others 1994); most Nebraska breeders may have departed by the end of Aug.

Rosche (1994a) noted that migrants in the L McConaughy area are most numerous in Aug, while information in Johnsgard (1980) suggests that peak fall migration is in late Aug and early Sep.

Other Reports

6 Nov 1966 Custer Co.

FINDING

Yellow Warblers occupy almost every willow patch in Nebraska but are most numerous where such habitat is extensive, such as along major rivers and streams. Squeaking will usually bring them into view after they are located by their persistent singing, especially in May and Jun.

Chestnut-sided Warbler

Dendroica pensylvanica

STATUS

Uncommon regular spring and fall migrant east, becoming rare casual west. Hypothetical breeder.

DOCUMENTATION

Specimen

UNSM ZM10828, 19 May 1910 Child's Point, Sarpy Co.

DISTRIBUTION AND ECOLOGY

Migrants occur in deciduous woodland edge and thickets, often at low levels, and are most numerous in the east.

Spring

1, 2, 3 May→4 (Glandon and Glandon 1935b), 4, 5 Jun

This species tends to be a later migrant than many other warbler species. Although Iowa banding data indicate that fall numbers are about 3 times higher than those in spring (Dinsmore and others 1984), the opposite appears to be the case in Nebraska.

Migration takes place in May and early Jun, although there is an earlier report 25 Apr 1936 Logan Co and later reports 12 Jun 1965 Douglas-Sarpy Cos, 15 Jun 1997 Buffalo Co (female) (LR, RH; Brogie 1998), 16 Jun 1992 Douglas-Sarpy Cos (AB 46:1152), 17 Jun 1973 Keith Co, and 19 Jun 1964 Douglas-Sarpy Cos.

Away from the east, most reports are of spring migrants; there are 10 reports from the Panhandle, all but 1 since 1973 and all in the period 13–29 May, except for a later report 17 Jun 1973 Scotts Bluff Co.

High counts include 7 in Dakota Co 17 May 1997; 6 at Hormel Park, Fremont, 16 May 1999; and 5 at Wilderness Park, Lincoln, 15 May 1999.

Summer

There are several mid-Jun reports (see Spring) that are probably late spring migrants.

There are 2 published reports of breeding, but neither is convincingly documented. Trostler was said to have collected a set of eggs near Omaha 23 Jun 1894, at which time this species was "frequently seen throughout summer" (Bruner and others 1904). These eggs apparently do not now exist, and there have been no reports of summering birds since that time, although "young" were observed in Scotts Bluff Co in 1975 (Bennett 1976). No further details were provided.

There are 2 Colorado breeding records (Andrews and Righter 1992) and 2 South Dakota summer records (SDOU 1991), and 2 nests were found in southern Missouri in 1978 (Robbins and Easterla 1992).

Fall

18, 20, 21 Aug→10, 14, 14 Oct

Migration is from late Aug through early Oct, with a later report 26 Oct 1964 Lincoln Co. The only reports west of Lancaster Co, including only 2 from the Panhandle, are 17 Aug 1964 McPherson Co, 13 Sep 1997 North Platte NWR (JGJ), 17 Sep 1982 Boone Co, 17 Sep 1997 Scotts Bluff Co (NK), 19 Sep 1974 Lincoln Co, 19 Sep 1998 Calamus Res (JGJ), 29 Sep 1975 Lincoln Co, 2 Oct 1977 Boone Co, 9–10

Oct 1977 Howard-Hall Cos, and 26 Oct 1964 Lincoln Co.

FINDING

Chestnut-sided Warblers are most likely to be found in Missouri Valley woodland around 15–20 May. The song is easily confused with those of some other species, and so all similar songs should be checked. These birds do not forage very high and are usually easy to see.

Magnolia Warbler

Dendroica magnolia

STATUS

Uncommon regular spring and fall migrant east, becoming rare casual west.

DOCUMENTATION

Specimen

UNSM ZM10780, 25 May 1909, Lincoln, Lancaster Co.

DISTRIBUTION AND ECOLOGY

Migrants are found in deciduous woodland, most commonly in the Missouri Valley but in small numbers farther west; it is casual in the Panhandle (Rosche 1982). Magnolia Warblers often occur in denser woodland than many other migrant warblers and forage at medium to high levels.

Spring

27, 28, 29 Apr→1 (Tout 1947), 4, 4 Jun

Migration is in the period late Apr–late May, with earlier reports 20 Apr 1955 Thayer Co, 21 Apr 1951 Lancaster Co, 21 Apr 1968 Custer Co, and 24 Apr 1962 Webster Co and later reports 10 Jun 1997 Bellevue (BP, LP), 12 Jun 1986 Douglas-Sarpy Cos, 13 Jun 1955 Thomas Co (Short 1961), and 16 Jun 1919 Buffalo Co, a specimen, HMM 2028A.

There are only 13 Panhandle reports in the period 4 May–1 Jun. Included are specimens taken 11 May 1919 in Scotts Bluff Co, UNSM ZM6838, and 28 May 1965 in Sheridan Co (Short 1966).

High counts include 5 at Dodge Park, Douglas Co, 15 May 1996; 4 in Dakota Co 17 May 1997; and 4 at Geneva Cem 15 May 1999.

Fall

22, 25, 28 Aug→22, 22, 24 Oct

Migrants occur from early Sep through mid-Oct and are among the later migrant warblers. There is an earlier report 12 Aug 1989 Douglas-Sarpy Cos.

There are these few Panhandle reports: 5 Sep 1981 Garden Co (AB 36:194), 7 Sep 1981 Garden Co (AB 36:194), 10 Sep 1977 Garden Co (AB 32:225), 12 Sep 1919 Monroe Canyon (Mickel and Dawson 1920), 16 Sep 1981 Garden Co (AB 36:194), and "fall" 1956 Box Butte Co (Rosche 1982). Westerly was one in Keith Co 7 Sep 1996 (FN 51:81).

High counts include 6 at Krimlofski Tract, Neale Woods, 13 Sep 1998 and 4 in Washington Co 11 Sep 1995.

FINDING

Most are found in mid-May in Missouri Valley woodland, often fairly high in taller trees in the interior of deciduous woodland or forest.

Cape May Warbler

Dendroica tigrina

STATUS

Rare casual spring migrant statewide. Rare casual fall migrant east.

DOCUMENTATION

Specimen

UNSM ZM6836, 27 May 1920 Lincoln, Lancaster Co.

DISTRIBUTION AND ECOLOGY

Migrants occur in deciduous woodland, although there is a tendency to frequent conifers. Most reports are from the Missouri Valley.

Spring

29 Apr, 1, 5 May→21, 22, 27 May

There are about 50 reports. The latest, 27 May 1920 Lancaster Co, is cited above. There are few reports since 1982, all in the narrow period 10–21 May: 10 May 1995 Sarpy Co, 11 May 1994 Douglas Co (FN 48:315), 12 May 1995 Sarpy Co, 12–13 May 1993 Lancaster Co, 13–14 May 1983 Douglas-Sarpy Cos (AB 37:888), 15 May 1983 Dawes Co (AB 37:888), 15 May

1984 Dakota Co, 15 May 1999 Geneva Cem, 16 May 1999 Bellevue, 16 May 1999 Douglas Co, 17–18 May 1999 Lincoln, and 19–21 May 1995 Saunders Co.

Away from the east, reports are few, especially in the Panhandle, where these are the only reports: 5 May 1975 Dawes Co (Rosche 1982), 15 May 1983 Dawes Co (AB 37:888), 20 May 1976 Dawes Co (Rosche 1982), and 20 May 1980 Scotts Bluff Co.

Fall

There are very few reports, all from Douglas-Sarpy Cos and all but one in the very narrow period 26 Sep–4 Oct: 26 Sep 1981, 27 Sep 1973 (AB 28:74), 28 Sep 1974, 29 Sep 1973, 29 Sep 1974, and 4 Oct 1975. The exception is an earlier report without details 3 Sep 1993 Douglas-Sarpy Cos.

FINDING

The best chance to find this rare migrant is 10–15 May in Missouri Valley woodland. This species may show a preference for oaks, where it feeds fairly high and often flutters after insects (Thompson and Ely 1992); it also has a fondness for conifers and might be looked for in cemeteries.

Black-throated Blue Warbler

Dendroica caerulescens

STATUS

Rare casual spring and fall migrant statewide.

DOCUMENTATION

Specimen

UNSM ZM6834, 19 Sep 1919 Monroe Canyon, Sioux Co (Mickel and Dawson 1920).

DISTRIBUTION AND ECOLOGY

Migrants prefer brush and woodland edge where they usually forage at low levels. Reports are statewide, especially in fall, although there are more reports from the east.

Spring

23, 30 Apr, 3 May→22, 24, 29 May

This species is recorded far more often in fall than spring; about two-thirds of the reports are in fall. Of 40 banded in Iowa, all but one were banded in fall (Dinsmore and others

1984), and Kansas data are similar (Thompson and Ely 1992).

There are about 21 reports for spring. These are in the period 23 Apr–29 May, with few since 1982: 3 May 1989 Cass Co, 5 May 1995 Keith Co (Brown and others 1996), 7 May 1984 Adams Co, 7 May 1999 Fontenelle Forest, 15 May 1994 Adams Co (NBR 62:113), mid-May 1996 Thomas Co, 17 May 1983 Douglas-Sarpy Cos, 19 May 1983 Douglas-Sarpy Cos, and 29 May 1995 Sarpy Co.

The only reports from the Panhandle are 7–9 May 1981 Sioux Co (Rosche 1982) and 15 May 1998 Crescent L NWR (SJD, JS).

Fall

31 Aug, 1, 4 Sep→13, 15, 20 Oct

There are about 50 fall reports from early Sep through mid-Oct.

Reports from the Panhandle are 19 Sep 1919 Sioux Co (specimen cited above), 21 Sep 1996 Sheridan Co (FN 51:81), an immature female 26 Sep 1999 Oliver Res (SJD), 29 Sep 1979 Garden Co (AB 34:178), 4 Oct 1979 Garden Co (AB 34:178), 6 Oct 1920 Sioux Co (Dawson 1921), 8 Oct 1920 Sioux Co (specimen, UNSM ZM6833), a male 10–13 Oct 1993 Sioux Co (Lemmon 1995), and 20 Oct 1979 Garden Co (AB 34:178).

Other Reports

5 Aug 1966 Cass Co.

FINDING

This species is most likely to be found in the Missouri Valley in brushy woodland edge in late Sep, although it occurs statewide in fall. It forages rather low and is most likely to be associated with other migrant warblers.

Yellow-rumped Warbler

Dendroica coronata

The subspecies *auduboni* and *coronata* were at one time considered separate species, Audubon's and Myrtle Warbler, respectively, but are now treated as conspecific due to interbreeding from southeastern Alaska to Alberta (AOU 1983). Curson and others (1994) suggest that a more appropriate treatment

might be to consider them as 2 of 4 allospecies within the superspecies *D. coronata*, the others *nigrifrons* of the southwest United States and *goldmani* of southern Mexico. Analysis of the hybrid zone supports this treatment (Barrowclough 1980; but see Rohwer and Wood 1998).

Of the 2 that occur in Nebraska, Audubon's Warbler is the more southerly breeder, ranging in summer from southeast Alaska south in mountain ranges to Mexico, while Myrtle Warbler breeds across Canada to Alaska, north of the range of Audubon's Warbler.

In this account Audubon's and Myrtle Warblers will be treated separately.

Yellow-rumped (Audubon's) Warbler

STATUS

Uncommon regular spring and fall migrant west, rare central, rare casual east. Uncommon regular breeder west.

DOCUMENTATION

Specimen

UNSM ZM6847, 21 May 1900 Sioux Co.

TAXONOMY

Intergrades between Audubon's and Myrtle Warbler occur on occasion; an example is a male collected 20 Apr 1920 in Lancaster Co, UNSM ZM6849, which has yellowish feathering on a white throat and limited whitish wingbars (JGJ, WRS).

DISTRIBUTION AND ECOLOGY

Breeding birds are restricted to ponderosa pine forest in the western Pine Ridge of Sioux Co (Rosche 1982), often with a preference for open habitats near deciduous trees (Johnsgard 1979). Migrants occur throughout the Panhandle in all types of woodland.

Spring

12, 12, 13 Apr→summer

The above dates are based on older data that specified Audubon's Warbler, as well as recent reports noting that *auduboni* was observed.

There are no reports of Audubon's Warbler prior to 12 Apr; an early report 1 Apr 1954 Dawes

Co without details may be a representative
of earlier-migrating Myrtle Warbler, possibly
even the race *hooveri* (see Myrtle Warbler
Taxonomy).

Data on Audubon's Warbler indicate that
it occurs eastward during migration in
declining numbers, becoming casual in the
east, with these few reports: 26 Apr 1999
Fontenelle Forest (DPA, JP), 5 May 1970
Lancaster Co, 6–7 May 1967 Lancaster Co, and
12 May 1959 Burt Co. In Webster Co it was
considered a "common migrant" by Ludlow
(1935), with reports 18 Apr–21 May, and at
Bessey Division, NNF, it occurs regularly
in spring and fall (Bray 1994). In Keith Co
banding data indicate Myrtle is 5 times more
numerous than Audubon's in spring (Brown
and others 1996).

High counts include 13 at Oliver Res 17 May 1999.

Summer

Audubon's Warbler breeds in the Pine Ridge,
notably in Sowbelly and Monroe Canyons,
where it is considered "locally uncommon"
(Rosche 1982). Although it is present every
year in Sioux Co in the breeding season, the
only breeding record since 1950 is a series
of specimens collected by Ford (1959) that
included a female with a brood patch and a
juvenile male taken 18 July 1957 8 km (5 mi)
west of Crawford. That it breeds east to Dawes
Co is suggested by an observation of young
in 1964, although no details were published
(NBR 33:12), and also by summer observations
there (Rosche 1972) and summer reports for
most of the years since 1980. Short (1961)
stated that there was "no doubt that the form
breeds throughout the Pine Ridge region
in suitable habitat, despite the presence of
breeding records from Sioux County only." It
was present in summer in Sheridan Co in 1992
and 1993.

There are summer reports from the Wildcat
Hills in Scotts Bluff Co 30 Jun 1974, 30 Jun
1994, and 4 Jul 1992.

There are a few summer reports away from
the breeding range, none with details on

subspecies: 30 Jun 1967 Greeley Co, 30 Jun
1967 Platte Co, 1–24 Jul 1976 Custer Co, 9 Jul
1989 Hall Co, 10 Jul–1 Aug 1989 Lincoln Co, 12
Jul 1981 Douglas-Sarpy Cos, 13 Jul 1981 Garden
Co, 13 July 1983 Lincoln Co, 16 Jul 1993 Lincoln
Co, and 11 Aug 1974 Lincoln Co.

Fall

12, 12, 13 Sep→9, 11, 16 Oct

Individuals appear in the Panhandle away from
breeding areas in mid-Sep; the early dates
above are away from the breeding range.
There are earlier reports of Audubon's 21
Aug 1999 Wildcat Hills NC (MB), 28 Aug 1980
Garden Co, 1 Sep 1997 Oliver Res (SJD), and 6
Sep 1998 Oliver Res (SJD).

The latest specimen date for Audubon's Warbler
is 27 Sep 1920 Sioux Co (UNSM ZM6844). Later
dates referable to *auduboni* are a Webster Co
report of 2 on 11 Oct 1934 (Ludlow 1935) and
Lincoln Co reports 24 Sep 1941 and 9 Oct
1940 (Tout 1947). Rosche (1994a) stated that
Audubon's is "much scarcer" than Myrtle in
the Keith Co area and departs earlier in fall.

FINDING

Audubon's Warbler can usually be found with a
little effort in Sowbelly or Monroe Canyons
in Jun when it is singing. It often responds
to squeaking. Migrants may be found in any
woodland in the Panhandle.

Yellow-rumped (Myrtle) Warbler

STATUS

Common regular spring and fall migrant
statewide. Hypothetical in summer. Rare
casual winter visitor statewide.

DOCUMENTATION

Specimen

UNSM ZM6857, 20 Apr 1901 Havelock, Lancaster
Co.

TAXONOMY

The westernmost populations of Myrtle Warbler
have been considered a separate subspecies
hooveri (Curson and others 1994; Dunn and
Garrett 1997), which has also been recorded
in Nebraska (Rapp and others 1958). This
subspecies breeds from Alaska south to

British Columbia and winters north and east to Colorado, Kansas, and Missouri (AOU 1957).

DISTRIBUTION AND ECOLOGY

Migrants occur statewide in all types of woodland and shrubby habitats. Wintering birds may occur in mature woodlands with flocks of wintering birds such as chickadees and kinglets and are generally restricted to the southeast.

Spring

9, 9, 13 Mar→1, 1, 2 Jun

Migrants appear first in the south and east and are not numerous until mid-Apr; the first birds may have overwintered in the area or nearby (see Winter). Departure occurs by late May, although there is a later report 15 Jun 1992 Lincoln Co.

High counts include 755 in Sarpy Co 11 May 1996, 260 at North Platte NWR 3 May 1995, and 255 in Lancaster Co 10 May 1996.

Summer

The few summer reports, none specified as to form, are listed under Audubon's Warbler.

Fall

25, 27, 28 Aug→6, 8 (Tout 1947), 13 Nov (except west)

25, 27, 28 Aug→29, 30, 31 Oct (west)

Most migrants pass through in late Sep and early Oct, although a few linger into Dec in the southeast and there are a few midwinter reports (see Winter). Late dates away from the southeast are few, generally in Dec: 2 Dec 1965 Lincoln Co, 6 Dec 1996 Lincoln Co, 16 Dec 1988 Boone Co (NBR 57:6), 18 Dec 1982 Howard-Hall Cos, 23 Dec 1994 L McConaughy CBC, 23 Dec 1995 L McConaughy CBC, and 30 Dec 1995 (8) Calamus-Loup CBC.

High counts include 129 in Washington-Douglas Cos 13 Oct 1996; 100 in Sarpy Co 7 Oct 1991 (AB 46:117); 100 in Lancaster Co 1 Oct 1995; 100 at Rowe Sanctuary, Buffalo Co, 13 Oct 1996; and 100 in southwest Dixon Co 27 Sep 1998.

Winter

In the southeast Myrtle Warbler is casual in Dec during the CBC period, and there are Jan reports as late as 15 Jan 1989 Washington Co, 17 Jan 1963 Douglas-Sarpy Cos, and 28 Jan 1986 Sarpy Co (AB 40:298). There are also these westerly reports: 1 Jan 1983 Scotts Bluff Co, 2 Jan 1971 Dawes Co (NBR 39:6), 2 Jan 1998 (6) below Kingsley Dam (SJD), 12 Jan 1964 Lincoln Co, and 17 Jan 1999 Johnson L (LR, RH), as well as a Feb report (see below). Some of these westerly reports may be of *hooveri*, which winters north and east to southeast Colorado, Kansas, and Missouri (AOU 1957).

There are these few Feb reports for the state: 3 Feb 1980 Washington Co, 8 Feb 1995 Seward Co, 13 Feb 1963 Lincoln Co, 14 Feb 1999 (4) Johnson L (LR, RH), 15 Feb 1995 Seward Co, 15 Feb 1998 (2) Indian Cave SP (JS), 17 Feb 1975 Douglas-Sarpy Cos, and 27 Feb 1999 Wolf L (TH).

While there are no definite records of overwintering, especially in the northwest (Rosche 1982), overwintering may occur in the southeast or possibly in the Platte Valley in Lincoln Co or Keith Co.

FINDING

This is perhaps the most common migrant warbler in Nebraska and is easily found by squeaking in woodlands statewide in Apr and early Oct.

Black-throated Gray Warbler

Dendroica nigrescens

STATUS

Accidental in spring.

Records

There are 8 reports, only 1 acceptably documented (SJD; Bray and others 1986): a male was at Oliver Res 1 May 1999 (SJD).

A bird seen 23–24 Aug 1967 at Tryon (Bassett 1968; AB 22:59) may have been correctly identified. There is 1 other fall report, 20 Sep 1979 Douglas-Sarpy Cos. There are 3 spring reports, only 1 at a likely date and location: 19 May 1974 Scotts Bluff Co. The others are 11 May 1971 Adams Co and 11 May 1975 Lancaster Co. Johnsgard (1980) listed additional reports for Cherry and Garden Cos, but no details were provided.

COMMENTS

This species breeds locally in southeast Colorado and north in the foothills to north-central Colorado (Andrews and Righter 1992). Spring migrants are rare but regular in southwest Kansas (Thompson and Ely 1992) and have been recorded east to Cedar Co, Iowa (Kent and Dinsmore 1996; Kent 1998), and Sioux Falls, South Dakota (SDOU 1991).

Thompson and Ely (1992) stated that it has occurred in Kansas in the period 3–22 September, and there are 2 South Dakota records for fall, 4 Aug 1973 in the Black Hills, where it does not breed, and 1 Sep 1986 in northeast South Dakota (SDOU 1991).

FINDING

Based on migration dates from Kansas, this species might be expected in Nebraska during the first half of May in the extreme southwest or the Panhandle. Habitat preferences include pines or junipers, although migrants occur in riparian brush also, where they forage low in the vegetation. Song might be helpful, although the bird should be seen since its song is very similar to that of Townsend's Warbler, a casual spring migrant in Nebraska.

Black-throated Green Warbler

Dendroica virens

STATUS

Uncommon regular spring migrant east, rare casual central, accidental west. Uncommon regular fall migrant east, becoming rare casual west.

DOCUMENTATION

Specimen

UNSM ZM12010, 14 Sep 1911 Crawford, Dawes Co (Zimmer 1912).

DISTRIBUTION AND ECOLOGY

This species is found in migration in deciduous woodland, especially riparian, where it forages at medium height. Most reports are from the east.

Spring

24, 28, 28 Apr→24, 24, 26 May

Migrants occur from late Apr through May, with later reports 14 Jun 1992 and 23 Jun 1975 Douglas-Sarpy Cos. Bruner and others (1904) cited reports by Aughey 5–6 Jun 1865 Dakota Co and 14 Jun 1875 Lancaster Co; these have been questioned by Sharpe (1993).

There is a single Panhandle report 22 May 1983 Sioux Co.

Fall

26, 30 Aug, 1 Sep→18, 21, 26 Oct

Migration begins in late Aug and ends in mid-Oct, with later reports 31 Oct 1969 McPherson Co, 5 Nov 1980 Douglas-Sarpy Cos (AB 35:200), and 18 Nov 1984 Wayne Co (AB 39:74).

There are few reports from the Panhandle: a specimen taken 14 Sep 1911 in Dawes Co (cited above), 29 Sep 1979 apparently in extreme western Dawes Co (Rosche 1982), 8 Oct 1920 Sioux Co (Dawson 1921), 13 Oct 1979 Garden Co (AB 34:178), and 19 Oct 1991 Sheridan Co (AB 46:117).

High counts include 5 at Krimlofski Tract, Neale Woods, 13 Sep 1998.

FINDING

This species is best looked for in deciduous woodland in the Missouri Valley in early May and in Sep. It is easier to identify in spring, when it can be located by song.

Townsend's Warbler

Dendroica townsendi

STATUS

Rare casual spring migrant west and central, accidental east. Rare regular fall migrant west, rare casual central.

DOCUMENTATION

Specimen

UNSM ZM12011, 19 Sep 1911 Crawford, Dawes Co.

DISTRIBUTION AND ECOLOGY

Migrants are found in oak, juniper, or pine woodlands (Johnsgard 1980) and riparian growth. Conifer plantings in cemeteries are also favored. Most reports are from the western two-thirds of the state in fall.

Spring

There are few reports. The only documented report is the lone report from the east, a male at Schramm Park 11 May 1996 (SG, SK; Brogie 1997). All but 2 of the remaining reports, surprisingly, are from the central part of the state: 27 Apr 1978 Buffalo Co, 5 May 1978 Buffalo Co, 8 May 1973 Buffalo Co, 11 May 1971 Adams Co, 15–17 May 1989 Sioux Co, 16 May 1964 Keith Co (Rosche 1994a), 16–17 May 1970 Thomas Co, 17 May 1998 L Minatare (BW, DW), and 17–18 May 1980 Thomas Co.

Fall

17, 19, 23 Aug→28, 29 Sep, 2 Oct

There is an earlier report 14 Jul 1965 Lincoln Co, a male observed for 45 minutes near North Platte (Nielsen 1965), and later reports 18 Oct 1991 Scotts Bluff Co (AB 46:117), 22 Oct 1997 Oliver Res (not accepted by the NOURC; Brogie 1998), and 11 Nov 1972 Perkins Co.

Townsend's Warbler is far more numerous in fall than in spring. In all, there are about 35 reports, most from the Panhandle. The easternmost reports are 14 Jul 1965 Lincoln Co (see above), 17 Aug 1966 Lincoln Co, 17 Sep 1972 Perkins Co, 28 Sep 1973 Perkins Co, and 11 Nov 1972 Perkins Co.

High counts include 3 netted in Garden Co 11 Sep 1977, 3 at Oliver Res 13 Sep 1997, and 3 there 30 Aug 1998. Five were netted in Garden Co 30 Aug–11 Sep 1979 (AB 34:178).

FINDING

The best time to find this species is probably the first half of Sep, and the reports indicate that Townsend's Warbler may occur anywhere in the western third of the state. Likely locations should be in the southern Panhandle and southwest. As this species has a preference for conifers, cemeteries should be checked.

Hermit Warbler

Dendroica occidentalis

STATUS

Hypothetical.

COMMENTS

The only published report involves a description of 2 birds seen in McPherson Co 21 Sep 1973 (Bassett 1974). While suggestive of this species, the description did not adequately eliminate Black-throated Green Warbler, which it resembles in fall (Bray and others 1986).

All records except 1 from nearby states are in spring. On the Colorado plains there are 7 records 27 Apr–17 May (Andrews and Righter 1992). The single Kansas record is 7 May (Thompson and Ely 1992), and the 2 Minnesota records are 3 May and 14 May (Janssen 1987). The only exception is the single Missouri record, an immature male collected in the extreme northwest part of the state on 20 Dec 1969 (Robbins and Easterla 1992).

The data suggest that this species might occur in Nebraska in early May, presumably in the western Panhandle in conifers.

Blackburnian Warbler

Dendroica fusca

STATUS

Uncommon regular spring and fall migrant east, rare casual central and west.

DOCUMENTATION

Specimen

UNSM ZM6893, 15 May 1920 Ashland, Cass Co.

DISTRIBUTION AND ECOLOGY

Blackburnian Warbler is found in deciduous woodland, with a tendency to forage high in the trees. It is most numerous in the Missouri Valley.

Spring

28, 30 Apr, 2 May→29, 30, 31 May

Migration takes place in May. There is a later report 5 Jun 1961 Douglas-Sarpy Cos.

Reports from the Panhandle are surprisingly numerous; there are about 10 in the period 9–24 May and additional reports 9 Jun 1981 and 29 Jun 1981, both from Garden Co (AB 35:956). There are 2 summer records for the Black Hills in South Dakota, including possibly territorial birds in 1967 (SDOU 1991).

High counts include 6 at Fontenelle Forest 17 May 1996 and 4 there 12 May 1995. A season total of 7 was the "most ever for a spring season" in Sarpy Co in 1995 (BP, LP).

Fall

13, 18, 20 Aug→5, 8, 8 Oct

Migration is from late Aug through early Oct, with later reports of an immature male 12 Oct 1997 Ash Hollow SHP (SJD; Brogie 1998), 14 Oct 1964 Cass Co, 19 Oct 1974 Douglas Co (AB 29:81), 26 Oct 1999 Schramm Park (RK), and 29 Oct 1983 Douglas-Sarpy Cos.

Away from the east this species is very rare in fall, with only 10 reports. Panhandle reports are 23 Aug 1996 Morrill Co, 29 Sep 1979 Dawes Co (Rosche 1982), 4 Oct 1970 Scotts Bluff Co, and 12 Oct 1997 Ash Hollow SHP (cited above).

High counts include 10 at Jewell Park, Bellevue, 27 Aug 1996.

Other Reports

5 Apr 1972 Howard-Hall Cos, 12 Apr 1972 Brown Co.

FINDING

The best opportunity to see this species is in mid-May in deciduous woodland along the Missouri Valley. It forages rather high and has a weak song; thus it may be difficult to locate. It is usually discovered with other species in mixed flocks.

Yellow-throated Warbler

Dendroica dominica

STATUS

Uncommon regular breeder and spring and fall migrant east. Hypothetical elsewhere.

DOCUMENTATION

Photograph

28 Nov–24 Dec 1980 Omaha, Douglas Co (BJR; NBR 49:33).

DISTRIBUTION AND ECOLOGY

Summering birds occupy sycamore stands in riparian deciduous woodland, foraging near the tops of the trees. The species is probably limited as a breeding bird, therefore, to areas in the lower Missouri Valley, but it has been reported in summer in riparian woodland in the Platte Valley.

Spring

11, 11, 12 Apr→summer

This species arrives in mid-Apr, with an earlier date 5 Apr 1992 Sarpy Co. A northerly report was 25 May 1997 Cuming Co. Documented reports are limited to the Missouri Valley, except for a specimen taken at Oshkosh 20 May 1917 and identified as the expected subspecies *albilora* by Oberholser; it was #2892 in the Brooking collection (Swenk, Notes before 1925), but it cannot now be located (WRS).

There are at least 36 reports from central Nebraska and 3 from the Panhandle; the only documented record is of one at Geneva Cem 14 May 1999 (JGJ, SJD), although Rosche (1982) noted that one reported at Antioch 6 May 1931 was "well described," and there are 17 spring and 4 fall records for Colorado (Andrews and Righter 1992).

Reports from the Panhandle should be carefully scrutinized, as it is possible that Grace's Warbler may occur there. There are 14 spring and 5 fall records for Grace's Warbler on the Colorado foothills and plains (Andrews and Righter 1992), about the same number of records as for Yellow-throated Warbler.

High counts include 5 at Krimlofski Tract, Neale Woods, 19 Apr 1998; 3 at Fontenelle Forest 5 Apr 1992 (AB 46:444); 3 there 13 May 1995; and 3 there 25 Apr and 4 May 1999.

Summer

Neither Johnsgard (1979, 1980) nor Ducey (1988) listed this species as breeding in Nebraska. Since 1981, however, it has been reported in summer from sycamore stands at Fontenelle Forest and since 1997 at Krimlofski Tract, Neale Woods. At Fontenelle Forest, singing territorial males are now regular. Confirmation of breeding at this location was obtained in 1989 when birds were observed gathering nesting material on 4 May (AB

43:501), establishing the northwest limit of the breeding range for the species. A family group of 4, with adults feeding begging young, was seen at Fontenelle Forest 6–9 Sep 1999 (BP, LP).

There are several recent undocumented breeding season reports from Lincoln and Adams Cos. Reports from a BBS route in Keith-Lincoln Cos, a BBS route in Wheeler Co, and 20 Jun 1953 Cherry Co are likely to be misreported Common Yellowthroats.

Fall

summer→14, 28 Sep, 1 Oct

Yellow-throated Warblers become inconspicuous by Aug, although there are later dates. There is an extraordinary record of one at a feeder in Omaha, Douglas Co, 28 Nov–24 Dec 1980 (BJR, cited above; AB 35:314).

There are few reports away from the east, none documented, and most, if not all, likely misreported Common Yellowthroats.

COMMENTS

Out-of-range reports may be misreported Common Yellowthroats. Reporters commonly make the error of confusing the 2 species' names.

FINDING

The best place to see this species is at Fontenelle Forest on North Stream Trail, where singing males are conspicuous in early and mid-May. Observation is difficult however, as these birds forage high in sycamore trees.

Pine Warbler

Dendroica pinus

STATUS

Hypothetical in spring. Rare casual fall migrant east, becoming accidental west.

DOCUMENTATION

Photograph

23 Aug 1986 Fontenelle Forest, Sarpy Co (Grenon 1990).

DISTRIBUTION AND ECOLOGY

This species prefers pines during migration but also occurs in deciduous woodland, especially in fall (Dinsmore and others 1984).

Spring

Pine Warblers apparently migrate in very low numbers through the Missouri Valley; it is a rare migrant in extreme eastern South Dakota (SDOU 1991) and a casual transient in western Missouri (Robbins and Easterla 1992). There are even 2 spring and 7 fall records for the Colorado plains (Andrews and Righter 1992). Nevertheless, its occurrence in Nebraska is poorly documented.

There are 17 spring reports, none with acceptable identification details. A report 9 May 1987 Sarpy Co was considered "unaccepted" by the NOURC (Mollhoff 1989a). The remaining reports are in the period 17 Apr–27 May; most likely to be correct are reports 2–8 May 1984 Polk Co (AB 38:931; BP, LP) and one banded in Lincoln 14 May 1976 (NBR 44:51).

Fall

There are 21 fall reports from the east in the period 22 Aug–2 Nov and a few later reports discussed below. Other than a record of a singing male that remained in a pine grove at Oliver Res 29 Aug–13 Oct 1998 (SJD, BP, LP; Brogie 1999), the westernmost report is 2 Nov 1986 Wayne Co (AB 41:112; NBR 57:71). Most are relatively recent, in contrast to the spring reports. Only 2, however, have been accepted by the NOURC: 23 Aug 1986 Douglas-Sarpy Cos (Grenon 1990) and the 1998 bird at Oliver Res.

Reports likely to be correct are 22 Aug 1987 Washington Co (Jorgensen 1987), 1 Sep 1985 Sarpy Co (AB 40:137), 8 Sep 1975 Lancaster Co (NBR 47:58; AB 30:94), one banded 14 Sep 1984 Lancaster Co (AB 39:74), 17–22 Sep 1974 Lancaster Co (NBR 43:34; AB 29:81), 2 Oct 1987 Douglas Co (AB 42:99), 10 Oct 1994 Sarpy Co (BP, LP; NBR 62:143), and 2 Nov 1986 Wayne Co (cited above).

Pine Warbler has a propensity to appear at feeders late in fall. One was at the feeding station at Fontenelle Forest 10–15 Dec 1984 (NBR 53:6, 14; AB 39:84), and there is an additional undocumented report 1 Jan 1950 Antelope Co.

Reports not accepted by the NOURC are 8 Aug 1989 Sarpy Co (Grenon 1990) and 12 Sep 1986 Douglas-Sarpy Cos (Mollhoff 1987).

FINDING

Probably the most likely time to find Pine Warbler is the first half of Sep. Suitable coniferous habitat is scarce, but cemeteries should be checked.

Prairie Warbler
Dendroica discolor
STATUS

Rare casual spring migrant statewide. Accidental in summer and fall.

DOCUMENTATION
Photograph

26 May–1 Jun 1982 Niobrara Valley Preserve, Brown Co (DP, MB; Brogie and Mossman 1983; Bray and others 1986).

DISTRIBUTION AND ECOLOGY

This species is usually associated with small to medium-size eastern red cedars (Robbins and Easterla 1992) and brushy overgrown fields. One observed in southeast South Dakota 16 Jun occupied a dogwood thicket surrounded by grassland and sumac thickets (Dean and others 1995). In Kansas it occupies the brushy edge habitat between oak-hickory woodland and adjacent old fields (Johnston 1965), preferably when vegetation is less than 4.6 m (15 ft) in height (Thompson and Ely 1992).

Spring

There are 14 reports, surprisingly scattered statewide, in the period 21 Apr–1 Jun. Documented records are:

7 May 1999 Wild Rose Ranch, Hall Co (JT)

12 May 1996 Wilderness Park, Lincoln (DS; Brogie 1997)

18 May 1996 Fontenelle Forest (RAB, BP, LP; Brogie 1997)

19 May 1974 Lincoln (NBR 42:73; AB 28:821)

26 May–1 Jun 1982 Niobrara Valley Preserve (cited above)

Of the other reports that are probably correct, the earliest was one repeatedly seen singing in Jewell Park, Bellevue, 21 Apr–10 May 1979

(NBR 47:43; AB 33:786). Westerly reports are of a singing male at Oshkosh 25 May 1978 (Rosche 1994a; AB 32:1027; NBR 52:33) and another in Box Butte Co 21 May 1991 (RCR, DJR; AB 45:469). Northerly records are of one singing in Brown Co 26 May–1 Jun 1982 (Brogie and Mossman 1983); this bird was photographed 31 May (cited above).

Additional undocumented reports are 1 May 1957 Buffalo Co, 3 May 1959 Lancaster Co, 12 May 1962 Adams Co, 20 May 1980 Douglas-Sarpy Cos, and 1 Jun (Johnsgard 1980).

Summer

The only record is a singing male at Bessey Division, NNF, Thomas Co, 5 Jul 1997 (JGJ; Brogie 1998).

Fall

There is a single documented record:

6 Sep 1998 Gering Cem (SJD)

Additional undocumented reports are 22 Sep 1983 Douglas-Sarpy Cos (AB 38:220) and 23 Sep 1972 McPherson Co.

There is 1 record for the northeastern Colorado plains (Andrews and Righter 1992).

COMMENTS

Bruner and others (1904) cited several reports from the 19th century by Aughey and others, including 2 of nesting. These have been questioned (Johnsgard 1979, 1980; Sharpe 1993), although the existence of these old breeding records for this species and also for Blue-winged Warbler, both of which occupy similar habitat that has since disappeared from Nebraska, is intriguing.

FINDING

A search of edge habitats near oak-hickory woodland, especially near old fields containing scattered cedars, would provide the best chance of locating this species. The distinctive song is the best way to find a Prairie Warbler.

Palm Warbler
Dendroica palmarum
STATUS

Uncommon regular spring migrant east,

becoming rare casual west. Rare casual fall migrant statewide.

DOCUMENTATION

Specimen

UNSM ZM12018, 8 May 1917 Lancaster Co.

TAXONOMY

There are 2 subspecies, both of which probably occur in Nebraska, although western *palmarum* and intergrades with eastern *hypochrysea* probably account for most Nebraska reports (Curson and others 1994; AOU 1957). Eastern *hypochrysea* is yellowish below; it probably is rare but regular in eastern Nebraska. Possible *hypochrysea* were one with obvious yellowish underparts 30 Oct 1988 in Washington Co (JGJ), a late date for the species; 2 identified as *hypochrysea* at Branched Oak L 27 Apr 1997 (JS); one reported without details as an "eastern" in Hall Co 10 May 1997; one in southeast Otoe Co 2 May 1999 (WRS); and a group of 7 in Dixon Co 2 May 1999 (JJ).

DISTRIBUTION AND ECOLOGY

Palm Warblers prefer brushy habitats typically not associated with woodlands and thus are not usually found with other migrating warblers. In migration individuals often are found in grassy areas with short brush, commonly with migrating White-throated Sparrows. Most reports are from the east.

Spring

16, 18, 22 Apr→24, 29 May, 2 Jun

Reports are from mid-Apr through May, although there are earlier reports 5 Apr 1972 Hall Co and 5 Apr 1992 Sarpy Co (AB 46:444).

Reports from the Panhandle are few: 3 May 1980 Sheridan Co (Rosche 1982), 5 May 1984 Sheridan Co (AB 38:931), 5 May 1994 Dawes Co, 5 May 1996 (3–4) Dawes Co (FN 50:299), 7 May 1993 Sheridan Co (AB 47:428), 8 May 1992 Sheridan Co (AB 46:444), 11 May 1996 (3) Sheridan Co (FN 50:299), and 24 May 1962 Dawes Co.

High counts include 26 in Sarpy Co 11 May 1996, including 10 at Wehrspann L.

Fall

30 Aug, 2, 9 Sep→11, 13, 17 Oct

There are later reports 30 Oct 1988 Washington Co and 10 Nov 1966 Douglas-Sarpy Cos.

There are only about 30 fall reports in all, including these few away from the east: 21 Sep 1980 McPherson Co, 23 and 29 Sep 1979 Dawes Co (Rosche 1982), 28 Sep 1996 Sheridan Co (FN 51:81), 1 Oct 1988 Sheridan Co (AB 43:126), 8 Oct 1994 Sheridan Co, and 13 Oct 1996 Rowe Sanctuary, Buffalo Co.

Other Reports

26 Mar 1950 Antelope Co.

FINDING

The best chance for finding Palm Warbler is in early or mid-May in the Missouri Valley. Migrants are likely to be associated with sparrows foraging in leaf piles around brushy thickets rather than with other warblers.

Bay-breasted Warbler

Dendroica castanea

STATUS

Rare spring migrant east, rare casual central and west. Uncommon regular fall migrant east, rare casual central, accidental west.

DOCUMENTATION

Specimen

UNSM ZM10831, 14 May 1913 Lincoln, Lancaster Co.

DISTRIBUTION AND ECOLOGY

Migrants occur at medium heights and above in deciduous and mixed woodland. Johnsgard (1980) stated that this species has a preference for conifers during migration. Most records are from the Missouri Valley.

Spring

28, 29 Apr, 3 May→28, 28, 29 May

Migration occurs in May, although there are 2 later reports, 19 Jun 1974 Perkins Co and 26 Jun 1957 Scotts Bluff Co (see Comments).

Bay-breasted Warbler is somewhat cyclical in its spring occurrence. In the period 1982–91 it was reported only in 1982 and 1991, but since 1993 it has been reported each year.

Apart from the Scotts Bluff Co report, there

are these additional westerly reports: 9 May 1953 Thomas Co, 15 May 1978 Garden Co, 24 May 1997 Wheeler Co, 26 May 1978 Garden Co (Rosche 1994a), and 20 May 1956 Logan Co.

Fall

14, 14, 16 Aug→12, 14, 14 Oct

Migration is from mid-Aug through early Oct.

In fall Bay-breasted Warbler is regular in the east, in contrast to its rarity in spring. This pattern has been noted also in Missouri (Robbins and Easterla 1992) and Iowa (Dinsmore and others 1984). Although the presence of hatching-year birds significantly increases fall numbers, Dinsmore and others (1984) suggested that the surprisingly 10-fold greater numbers banded in fall versus spring may be due to lower foraging in fall, a characteristic that would allow easier observation. Also, the song is rather nondescript and easy to overlook, adding to the difficulty of locating this species in spring.

Since 1964 all fall reports except these few are from the east: 8 Sep 1997 Garden Co, the only Panhandle record (WRS, JS; Brogie 1998); 20 Sep 1978 Boone Co; and 9 Oct and 10 Oct 1979 Howard-Hall Cos.

High counts include 6 in Washington Co 11 Sep 1995 and 3 at Krimlofski Tract, Neale Woods, 13 Sep 1998. Fourteen were recorded in Sarpy Co 14 Aug–9 Sep 1991 (AB 46:116).

COMMENTS

The Jun reports cited above for Perkins and Scotts Bluff Cos support the possibility of extralimital nesting such as has occurred in Colorado (Andrews and Righter 1992) and Manitoba (Sealy 1979), apparently in response to abundant food supplies of tent caterpillars or spruce budworms.

Other Reports

19 Apr 1969 Douglas-Sarpy Cos.

FINDING

The best chance of finding a Bay-breasted Warbler is in fall in the Missouri Valley between early and mid-Sep. Good places are Fontenelle Forest and Indian Cave SP. Mixed flocks of migrant warblers should be carefully scrutinized.

Blackpoll Warbler

Dendroica striata

STATUS

Fairly common regular spring migrant east, becoming uncommon west. Rare casual fall migrant statewide.

DOCUMENTATION

Specimen

UNSM ZM6873, 15 May 1901 Beatrice, Gage Co.

DISTRIBUTION AND ECOLOGY

Migrants are found at medium to high levels in deciduous woodland and brush. It is far less numerous in fall than in spring.

Spring

27, 28, 29 Apr→4, 4, 7 Jun

Migrants occur from late Apr through early Jun. Some earlier undocumented reports may have resulted from confusion with Black-and-white Warbler, normally a much earlier migrant than Blackpoll Warbler. Earlier reports are 18 Apr 1985 Lancaster Co, 19 Apr 1969 Douglas-Sarpy Cos, 21 Apr 1931 Sheridan Co (Rosche 1982), 22 Apr 1933 Adams Co, 23 Apr 1931 Sheridan Co (Rosche 1982), and 23 Apr 1954 Douglas Co. Later reports include 10 Jun 1998 Kingsley Dam (LP, BP), 11 Jun 1971 Lancaster Co, 19 Jun 1981 Garden Co, 20 Jun 1988 Chase Co, and 21 Jun 1998 Wehrspann L (JWH).

In contrast to Bay-breasted Warbler, which is rare in spring but uncommon in fall, Blackpoll Warbler is fairly common in spring and only casual in fall. Identification of fall Blackpoll Warblers should be made very carefully, as most migrate eastward to the Atlantic Coast. Indeed, banding data from Iowa (Dinsmore and others 1984) show that 84% of Blackpoll Warblers occurred in spring, which is remarkably similar to Nebraska sight reports, where 88% of the reports are for spring.

High counts include 31 at Geneva 17 May 1999, 28 in Sarpy Co 13 May 1995, and 14 there 11 May 1996.

Fall

21, 27, 28 Aug→26, 26, 29 Oct

Fall departure is rather late for a wood-warbler. There are about 45 fall reports, although many are probably misidentifications. There are several for the Panhandle: 2 Sep 1979 Garden Co, 3 Sep 1981 Garden Co, 7 Sep 1998 Crescent L NWR, 15 Sep 1974 Dawes Co, 24 Sep 1998 Oliver Res, 30 Sep 1979 Dawes Co, 1 Oct 1994 Sheridan Co, 9 Oct 1994 Dawes Co, and 13 Oct 1979 Garden Co (AB 34:178).

Other Reports

12 Mar 1980 Adams Co, 8 Apr 1942 Jefferson Co, 10 Apr 1953 Brown Co, 10 Apr 1972 Brown Co, 29 Oct 1983 (15) Sarpy Co (AB 38:220).

FINDING

Blackpoll Warblers are easiest to find in spring in Missouri Valley deciduous woodland, such as at Fontenelle Forest and Indian Cave SP. The song is distinctive, and foraging birds are often rather low in the trees. Peak spring migration is probably mid-May.

Cerulean Warbler

Dendroica cerulea

STATUS

Rare regular spring migrant east, rare casual central, accidental west. Rare regular breeder east. Rare regular fall migrant east, hypothetical central.

DOCUMENTATION

Specimen

UNSM ZM6860, 4 May 1901 Child's Point, Sarpy Co.

DISTRIBUTION AND ECOLOGY

This canopy species prefers mature deciduous forest, either riparian or upland; in Nebraska this habitat is generally restricted to the Missouri Valley, and breeding numbers are very low.

Spring

26, 30 Apr, 1 May→summer

Arrival is in early May.

There are fewer than 15 reports away from the Missouri Valley in the period 26 Apr–

6 Jun; all but 2 are from the southeast and the Platte Valley west to Lincoln Co. The exceptions are an undocumented report 2 May 1978 Sioux Co and the specimen collected by Short (see Summer) 6 Jun 1964 Sheridan Co.

High counts include 5 at Fontenelle Forest 4 May 1992 (AB 46:444).

Summer

In recent years all summer reports of Cerulean Warbler are from Nemaha-Richardson, Douglas, Sarpy, and Dakota Cos, although the only documentation of breeding since 1978 (Ducey 1988) was of a female gathering nesting material at Ashford BSC, Thurston Co, 17–18 May 1997 (BFH, WRS). Since 1981 this species has been reported from Ponca SP in 1984, 1985, and 1994, when 2 were there 20 May (AB 48:315), and from Douglas-Sarpy Cos in 1983, 1984, 1986, 1987, 1990, 1991, and each year since 1993. Two were on Trail 9 at Indian Cave SP 6 Jun 1995.

The only breeding season report away from these counties was of a male collected by Short (1965) along the Niobrara River near Rushville in Sheridan Co 6 Jun 1964; while this specimen had enlarged testes, it was considered to be a migrant.

Old nesting data of Aughey for Dakota Co in 1865 has been questioned (Sharpe 1993).

Fall

summer→28 Aug, 3, 4 Sep

There are few fall reports. The limited data indicate that fall departure takes place by early Sep. Away from the Missouri Valley there are few reports, none documented: 27 Aug 1978 Lincoln Co, 2 Sep 1973 Adams Co, and 3 Sep 1975 Adams Co.

FINDING

Cerulean Warblers are best located by song on the breeding range, the best time mid to late May. Observation is difficult, as this species occupies the canopy in mature forest. Presumed breeding sites are Fontenelle Forest and Ponca SP.

Black-and-white Warbler

Mniotilta varia

STATUS

Fairly common regular spring and fall migrant statewide. Uncommon regular breeder north.

DOCUMENTATION

Specimen

UNSM ZM6797, 6 May 1901 Child's Point, Sarpy Co.

DISTRIBUTION AND ECOLOGY

Migrants occur statewide in deciduous woodland, especially riparian. Breeding birds occupy rather open woodland (Johnsgard 1980), coniferous, deciduous, or mixed, and are numerous in the lower Niobrara Valley (Brogie and Mossman 1983) and uncommon in summer in the Pine Ridge (Rosche 1982).

Spring

24 (RCR, DJR; Rosche 1994a), 26, 30 Mar→29, 29 (Ludlow 1935), 30 May (except west)

26, 28, 28 Apr→summer (west)

Migration begins in Apr and ends by late May. Arrival is somewhat later in the Panhandle. A few early Jun records may represent migrants but are cited under Summer. Migrants occur statewide, although numbers are lower to the west; Rosche (1982) considered this species uncommon in the northwest.

High counts include 110 in Sarpy Co 11 May 1996, including 58 at Fontenelle Forest; 16 in Lancaster Co 10 May 1996; and 14 at Dodge Park, Omaha, 10 May 1995.

Summer

Short (1961) summarized the breeding status of Black-and-white Warbler as "probably most common . . . along the relatively well-wooded Niobrara River, in the Missouri Valley area and in parts of northwestern Nebraska."

It is still plentiful in the Niobrara Valley west to northeastern Cherry Co, as Brogie and Mossman (1983) found "numerous singing males and territorial birds" during summer 1982 along the Niobrara in the Niobrara Valley Preserve, where it apparently has bred regularly for many years (Bruner and others

1904). There are recent summer records for Cherry Co 9–16 Jun 1985 and 16 Jun 1977.

In the Pine Ridge of Sioux and Dawes Cos Rosche (1982) considered Black-and-white Warbler to be an "uncommon summer visitant" that "presumably breeds." A pair of adults feeding just-fledged young was seen at Coffee Park, Sowbelly Canyon, 12 Jun 1979 (RSS), apparently the only documentation of breeding in the Pine Ridge. During the nesting season individuals occur in deciduous vegetation in Pine Ridge canyons. There are recent summer reports for Sheridan Co 4 Jun 1986 and 19–29 Jun 1985.

Evidence for current breeding in the Missouri Valley is poor and is based only on these recent summer reports from Douglas-Sarpy Cos: 9 Jun 1998 Bellevue (BP, LP), 20 Jun–1 Jul 1988, 2 Jul 1987, 5 Jul 1982 (AB 36:944), 21 Jul 1995 when 2 were singing at Fontenelle Forest, 23 Jul 1969, and 8 Aug 1980. Elsewhere in the Missouri Valley there are these 2 reports: 6 Jun 1986 Indian Cave SP and 8 Jun 1985 Dakota Co.

Breeding has occurred in Thomas Co, where Ford (1959) stated that on "several occasions we observed adults feeding fledglings" in 1957 along the Middle Loup River. Short (1961) found it common there in 1955 also, but there are no recent summer reports from this area.

Finally, there are a few scattered Jun and Jul reports that might indicate sporadic nesting: 1 Jun 1969 Custer Co, 2 Jun 1961 Scotts Bluff Co, 8 Jun 1962 Harlan Co, 15 Jun Keith Co (Brown and others 1966), 1 Jul 1965 Adams Co, and 3 Jul 1981 Lincoln Co.

Fall

12, 14, 16 Aug→15, 17, 20 Oct

Arrival is in late Aug, and departure is completed by mid-Oct, although there are later reports 29 Oct 1983 Douglas-Sarpy Cos, 10 Nov 1972 Douglas Co (AB 27:81), and an extremely late report of one possibly attempting to overwinter 25 Jan 1981 Sarpy Co (AB 35:314).

High counts include 14 at Krimlofski Tract,

Neale Woods, 13 Sep 1998 and 10 at Ponca SP 3 Sep 1996.

FINDING

Black-and-white Warblers are rather conspicuous due to their nuthatchlike foraging habits during migration, especially in early and mid-May in the Missouri Valley. They should be looked for with mixed flocks of migrants.

American Redstart

Setophaga ruticilla

STATUS

Common regular spring and fall migrant east, becoming uncommon west. Common regular breeder northwest, north, and east.

DOCUMENTATION

Specimen

UNSM 6917, 18 Jun 1902 Springview Bridge, Keya Paha Co.

TAXONOMY

Breeding birds are of the eastern subspecies *ruticilla*, which presumably migrates throughout Nebraska, while western *tricolora* occurs as a migrant (Rapp and others 1958). These subspecies are poorly differentiated (Dunn and Garrett 1997).

DISTRIBUTION AND ECOLOGY

Summer habitat is deciduous woodland, notably cottonwoods with a well-developed deciduous understory that is utilized for nesting sites. Breeding is limited to the Missouri and Niobrara Valleys and the Pine Ridge. Migrants occur throughout the state in deciduous woodland but are most numerous in the east.

Spring

24, 25, 26 Apr→29, 30, 31 May

Migration takes place in late Apr and May. There are earlier reports 17 Apr 1938 Webster Co, 17 Apr 1982 Lincoln Co, 22 Apr 1974 Douglas-Sarpy Cos, and 22 Apr 1984 Adams Co and later reports from areas outside the breeding range 3 Jun 1992 Frontier Co and 7 Jun 1935 Lincoln Co (Tout 1947). Departure dates are difficult to determine; some early Jun reports from known breeding areas may be of migrants. Rosche (1994a) considered this species a migrant in the Keith-Garden Cos area, citing spring records as late as 15 Jun. High counts include 30 in Sarpy Co 13 May 1995, 18 in Dakota Co 17 May 1997, and 17 at Ponca SP 16 May 1999.

Summer

Since 1981 published data indicate that American Redstart breeds primarily in the Missouri Valley (Richardson-Nemaha, Douglas, Washington, Dakota, and Knox Cos), Niobrara Valley (Brown Co), and the Pine Ridge (Rosche 1982), a distribution similar to that outlined by Short (1961). It is difficult to evaluate Jun and early Aug reports relative to the possibility of nesting, as this species is "known to summer but not breed in many areas in the western United States" (Rosche 1994a).

It seems clear that since the 1970s there has been a decline in breeding numbers at least in the western parts of the southeastern oak-hickory region, the Loup drainage, and probably elsewhere. Around 1900 American Redstart was considered an "abundant breeder and summer resident over the state" (Bruner and others 1904), and as recently as 1958 Rapp and others (1958) listed it as a "common migrant and summer resident throughout the state." By 1980 (Johnsgard 1980) the range was restricted to the Missouri and Niobrara Valleys, the Pine Ridge, and "perhaps locally in the Sandhills, and sporadically in the Platte valley west to Adams or perhaps Phelps Cos." Since 1980 it appears that breeding numbers have declined further.

In the Pine Ridge of Sioux and Dawes Cos it was considered by Rosche (1982) to be a fairly common summer resident. A nest was found in Sowbelly Canyon 16 Jun 1986 (Huser 1986). There have been no summer reports away from Dawes Co or Sioux Co since the 1970s, when it was recorded in Garden, Scotts Bluff, and Box Butte Cos in Jun or Jul, until a pair was found at Scottsbluff 9 Jul 1999 (AK). Brogie and Mossman (1983) found several singing males in the Niobrara Valley Preserve in 1982, suggesting that it was a probable nester there, and it was reported from Valentine NWR 16 Jul 1994.

The only recent reports from the oak-hickory region of the southeast or the Platte Valley are Lancaster Co 7 Jun 1992, probably a late migrant, and one carrying food at Indian Cave SP 26 Jun 1995. Until the 1970s, however, there were summer reports west to Adams, Gage, Clay, Saline, and Lancaster Cos. Reports for Adams Co may be related to oak-hickory habitat rather than Platte Valley cottonwood woodland. There is a summer report 30 Jun 1971 Lincoln Co and another for "summer" 1994 Buffalo Co.

Jun and Aug reports from McPherson Co suggest breeding but may also be late spring and early fall migrants: 8 Jun 1980, 12 Jun 1982, 5 Aug 1985, 11 Aug 1982, and 16 Aug 1977. Rosche (1994a) considered records as early as 12 Aug in Keith-Garden Cos to be of migrants, although one was banded in the L Ogallala area 11 Jul 1992 (Brown and others 1996). There is no evidence for breeding in the Loup drainage since Ford (1959) found it to be a common breeding bird in Thomas Co and Short (1961) found it common at Halsey; Bray (1994) found no evidence of breeding there in 1993.

Fall

11, 12, 14 Aug→29, 30 Sep, 2 Oct

Migrants occur from mid-Aug through late Sep. There are additional later dates, but all are from Scotts Bluff Co, possibly birds from northwestern United States or Canadian breeding populations: 5 Oct 1974, 9 Oct 1965, 9 Oct 1966, 15 Oct 1983, and 19 Oct 1964. There is only a single Oct report for the east, at Arbor Day Farm 13 Oct 1999 (LF, CF); the only other Oct report is 2 Oct 1977 Boone Co.

High counts include 12+ at Krimlofski Tract, Neale Woods, 13 Sep 1998.

Other Reports

5 Apr 1942 Webster Co, 10 Apr 1974 Adams Co.

FINDING

This species is most common in spring, especially mid to late May, in the Missouri Valley. Its short, explosive whistled song locates it. It responds well to squeaking, and its active habits make it easy to spot.

Prothonotary Warbler

Protonotaria citrea

STATUS

Uncommon regular spring migrant east, rare casual central, accidental west. Uncommon regular breeder east. Rare casual fall migrant east and central, accidental west.

DOCUMENTATION

Specimen

UNSM ZM6799, 25 May 1901 Child's Point, Sarpy Co.

DISTRIBUTION AND ECOLOGY

This species nests in cavities, including dead snags found in woodlands adjacent to standing or slow-moving water. Oxbows surrounded by dense riparian woodland and flooded forest are the preferred habitats in eastern Nebraska. Almost all recent observations have been in flooded riparian habitat in Fontenelle Forest. Individuals are rarely seen outside breeding habitat. It is likely that this warbler was once more numerous when meanderings of the Missouri River created and modified oxbows. Bank-stabilization programs along the river during the latter half of the 20th century have certainly curtailed natural oxbow development; that, coupled with the systematic removal of associated riparian forest, has likely contributed to the rarity of this species in the state.

Spring

28, 28, 28 Apr→summer

Migrants arrive at the end of Apr, although there are earlier reports from the breeding range 12 Apr 1981 Douglas-Sarpy Cos, 18 Apr 1991 Douglas-Sarpy Cos, 20 Apr 1965 Douglas-Sarpy Cos, 22 Apr 1966 Cass Co, and 23 Apr 1989 Douglas-Sarpy Cos.

Virtually all reports are from the Missouri Valley, although this species has occurred twice in the Panhandle. A specimen was picked up in Chadron 6 May 1957 after a heavy rainstorm (NBR 25:51). It was supposedly placed in the collection at Chadron State College but could not be located there by Rosche (1982). Another was seen in Scotts Bluff Co 15 May 1998 (WRS). Brown and others (1996) considered this species a "rare and apparently often overlooked spring transient" in the L Ogallala area, citing 2 males and a female netted 17–18 May 1996. Other westerly reports are 30 Apr 1966 Brown Co, 1 May 1985 Boone Co, 17 May 1967 Custer Co, and 24 May 1968 Lincoln Co.

High counts include 6 at Fontenelle Forest 11 May 1994 and 6 in Sarpy Co 13 May 1995.

Summer

All Jun and Jul reports of this species are from Sarpy Co, with these exceptions: 12–19 Jun 1998 Krimlofski Tract, Neale Woods; 17 Jun 1966 Cass Co; and 22 Jun 1985 Knox Co. Sarpy Co is the northernmost extent of the breeding range for the species, but, curiously, there is no recent evidence that it breeds in the Missouri Valley south of Sarpy Co. Bruner and others (1904) indicated that it nested near Nebraska City in Otoe Co and near Omaha around 1900. Recent evidence of nesting includes an observation of an adult carrying nesting material to a hole in a dead tree in Sarpy Co 27 July 1973 (NBR 47:57), fledglings being fed in Fontenelle Forest 25 Jun 1986 (AB 40:1224), and a pair carrying nesting material there 1 Jun 1995.

Fall

summer→1, 1, 4 Sep

Departure probably takes place by late Aug, with late dates in early Sep, although there are later reports 10 Sep 1984 Douglas-Sarpy Cos, 12 Sep 1973 Adams Co, 24 Sep 1990 Douglas-Sarpy Cos, 3 Oct 1962 Cass Co, and 4 Oct 1964 Brown Co.

There is 1 Panhandle report, an adult male 13 Sep 1997 at L Minatare (MB, DH, JGJ; Brogie 1998).

FINDING

Prothonotary Warblers are rather conspicuous most years in late May in the open swampy woodland in Fontenelle Forest. The loud song locates the birds, and they are easy to see in the open habitat.

Worm-eating Warbler

Helmitheros vermivorus

STATUS

Rare casual spring migrant statewide. Accidental in summer. Rare casual fall migrant east, accidental west.

DOCUMENTATION

Specimen

BMNH 41749, 14–16 Oct 1957 Scotts Bluff Co.

DISTRIBUTION AND ECOLOGY

This species occurs in deciduous woods, with a preference for deep ravines near streams or rivers (Robbins and Easterla 1992). It forages on the ground in shady undergrowth (Johnsgard 1980). All recent reports are from the Missouri Valley.

Spring

19, 21, 28 Apr→24, 25, 30 May

There are about 39 reports, including these from the Panhandle: 16 May 1979 Garden Co (AB 33:786) and 24 May 1959 Alliance City Park (NBR 27:50). Also westerly was one banded at L Ogallala 14 May 1997. There is a late spring record, a singing male present at Schramm Park 2 May–13 Jun 1994 (AB 48:315; NBR 62:83, 113).

Totals of 4 birds for the spring period were reported for 1995, 1996, and 1999.

Summer

One was banded at Cedar Point Biological Station, Keith Co, 4 Jul 1992 (Brown and others 1996), the only modern summer record. Prior reports are those cited by Bruner and others (1904), including a sighting by Aughey in the southeast in Jun 1875, Jun and Jul occurrences around Omaha, and several reports from the Lincoln area, notably "all summer in deep woods near Roca in 1903," where "breeding" was "almost certain."

Fall

There are few fall reports: 27 Aug 1934 Webster Co (Ludlow 1935), 3–5 Sep 1977 Douglas-Sarpy Cos, 8 Oct 1963 Cass Co, and 14–16 Oct 1957 Scotts Bluff Co. The latter record was of "several" feeding in a chokeberry thicket "about October 14, 15, and 16, 1957," when a specimen (cited above) was found dead and presented to the BMNH; it was a hatching-year bird (NBR 27:73, Klicka, pers. comm. WRS).

FINDING

The best possibilities for locating this species might be around May 10 in densely wooded ravines or hollows in Fontenelle Forest and Indian Cave SP. The song will locate this species, although while very similar to that of Chipping Sparrow, the latter does not usually occupy typical Worm-eating Warbler habitat.

Swainson's Warbler

Limnothlypis swainsonii

STATUS

Rare casual spring migrant east.

DOCUMENTATION

Description

27 May 1927 Sarpy Co (LOI 1923).

Records

There are 5 published reports. Four are in the period 11–27 May and the other a specimen purportedly taken 9 Apr 1905 by C. A. Black at Kearney and mounted by him. The bird was in his yard, "hopping around on the ground under some cedar and maple trees"; the specimen was disposed of to Charles K. Worthen (*Auk* 23:227) but relocated by Brooking after Worthen's death about 1920. That it was the same specimen collected by Black was confirmed by Black based on the mounting used. It is now on display (specimen 2834) in the HMM (Brooking 1933b; Bray and others 1986).

We have serious concerns about the provenance of this specimen, particularly because of its very early date of occurrence and because of the presence in the Brooking collection of other unique specimens. Swainson's Warbler does not overshoot its breeding range as often

as other southern warblers, and when it does it is usually in late Apr–May (Dunn and Garrett 1997). The nearest breeding populations are in southern Missouri and eastern Oklahoma, and birds reaching Nebraska are spring overshoot migrants. There are no acceptable records for Iowa or South Dakota, but there are 4 for Colorado's eastern plains in the period 12 May–8 Jun (Andrews and Righter 1992) and 5 for eastern Kansas in the period 4–23 May (Thompson and Ely 1992). All records for Nebraska, Kansas, and Colorado are in the period 4 May–8 Jun, except for the 9 Apr Nebraska specimen.

Three other Nebraska reports are accompanied by descriptions that are suggestive of this species. One was seen by C. A. Mitchell near Camp Gifford, possibly within present-day Fontenelle Forest, 27 May 1927 (LOI 1923) in a marshy strip along the Missouri River in small willows partly inundated by floodwaters. Another was seen in a small bush in the Aurora city park by Mrs. Kermit Swanson 16 May 1956 (Swanson 1957). A fourth report is of one seen and heard singing for about an hour in Wilderness Park, Lincoln, 15 May 1977 by a Mr. Toews. It was heard singing again about 2 hours later (NBR 45:34).

Finally, one was reported without details 11 May 1952 at Stuart (NBR 20:83).

FINDING

This species is found in dense brushy thickets usually near or over water, and these habitats should be checked in mid to late May in the Missouri Valley. Backwaters of the Missouri River with dense growth over water would be best. The loud song usually indicates the presence of this species, but it can be very difficult to see.

Ovenbird

Seiurus aurocapillus

STATUS

Fairly common regular spring and fall migrant statewide. Fairly common regular breeder north and east.

DOCUMENTATION

Specimen

UNSM ZM6885, May 1900 Sioux Co.

TAXONOMY

The area where the breeding ranges of the 2 races, western *cinereus* and eastern *aurocapillus*, meet in Nebraska is uncertain. *Cinereus* breeds in the Panhandle, and as far east as "central [*sic*] Nebraska (Camp Sheridan)" (AOU 1957). This locality is in extreme northwest Sheridan Co, some 19 km (12 mi) south of the White River. Tordoff collected a singing male in Sioux Co 19 Jul 1957, a specimen determined to be *cinereus* (Ford 1959).

DISTRIBUTION AND ECOLOGY

This species breeds in dense oak-hickory woodlands associated with the Missouri and Niobrara River valleys and in deciduous woodlands in the canyons of the Pine Ridge. It is also found in appropriate oak habitat associated with the lower Elkhorn River and the Platte River in Douglas and Sarpy Cos. It is uncommon to absent in riparian forest except as a migrant. It is most numerous in the Missouri Valley and in the Pine Ridge and apparently absent from the Republican Valley and most of the Platte Valley.

Spring

22, 24, 24 Apr→26, 26 May, 2 Jun

Migration is from late Apr through late May, the latter in areas where breeding does not occur, including a later date 6 Jun 1935 Lincoln Co (Tout 1947).

High counts include 58 in Sarpy Co 11 May 1996, including 44 at Fontenelle Forest; 8 at Neale Woods, Douglas Co, 9 May 1995; and 5 in Dawes Co 14 May 1994 (AB 48:315).

Summer

In the west Rosche (1982) described Ovenbird as a "fairly common summer resident" breeding in Sioux and Dawes Cos. A good count was of 14 birds at 5 Panhandle locations 10–25 Jun 1973 (AB 27:889). There are several Jun and Jul reports for Sheridan, Scotts Bluff, and Garden Cos. Although it is usually accepted that

Ovenbird requires rather extensive woodland for breeding, this may not apply to the western race *cinereus*, which may be adapted to woodlands of sharply limited extent. Andrews and Righter (1992) noted that in Colorado it breeds in "foothill riparian thickets."

Elsewhere Ovenbirds are found in summer in the Missouri and Niobrara Valleys. Ovenbirds breed throughout the Missouri Valley in appropriate habitat, with many summer records. Brogie and Mossman (1983) found numerous singing males and territorial birds in the Niobrara Valley Preserve in 1982 and considered the species a "probable nester." There are several summer reports for these counties as well as Holt and Cherry Cos in the Niobrara Valley. Johnsgard (1980) suggested that it bred west along the Niobrara "at least" to Cherry Co. A high count of 200 was made in the Niobrara Valley Preserve in Brown Co 17 Jun 1995, and 40 were there 28 Jun 1997.

Perhaps surprisingly, Ovenbird does not breed in the Platte (Tout 1947; Short 1961), Loup, or Republican drainages. There are, however, 2 Jul reports (but none for Jun) in McPherson Co, 16 Jul 1988 and 20 Jul 1980, and one was in Thomas Co 30 Jun–2 Jul 1993 (Bray 1994). There are no Jun records and only 2 Jul records for the south: 10 Jul 1973 and 13 Jul 1975, both in Adams Co. Jul reports cited here may be early fall wanderers; Rosche (1994a) cited reports from Garden-Keith Cos as early as 15 Jul, but there are no Jun records from that area. Brown and others (1996) also noted that there was no evidence for nesting in the L Ogallala area, despite netting 12 birds there 2 Jul–7 Aug, all with brood patches.

Fall

13, 15, 16 Aug→3, 3, 5 Oct

Migration begins in mid-Aug, although several Jul reports from areas where breeding is not known may also represent fall migrants (see above). Fall migration ends in late Sep, although there are later reports 10 Oct 1985 Douglas-Sarpy Cos, 11 Oct 1974 Perkins Co, and 26 Oct 1991 McPherson Co (AB 46:117).

This species is most numerous and easily located by song in late May and early Jun in extensive, rather open oak-hickory woodland in the Missouri Valley. Indian Cave SP may have the best numbers of Ovenbirds, although the birds can be difficult to see.

Northern Waterthrush

Seiurus noveboracensis

STATUS

Fairly common regular spring migrant east, becoming rare west. Hypothetical in summer. Fairly common regular fall migrant east, becoming rare casual west.

DOCUMENTATION

Specimen

UNSM ZM10770, 8 May 1909 South Bend, Cass Co.

TAXONOMY

The western race *notabilis* presumably migrates throughout Nebraska (AOU 1957), but the status of the eastern race *noveboracensis* is unclear. Rapp and others (1971) suggested that if *noveboracensis* migrates through Nebraska it would be limited to the Missouri Valley. The 2 races are similar (Dunn and Garrett 1997).

DISTRIBUTION AND ECOLOGY

Migrants occur in wooded swamps and brushy riparian habitats and are most often seen along the edges of streams or ponds. While they occur statewide, most are recorded in the Missouri Valley.

Spring

20, 22, 23 Apr→27, 27, 28 May

Migration is from late Apr through May. Early undocumented reports may result from misidentification of Louisiana Waterthrush, a very early migrant. Later records include a specimen, HMM 28197, taken at Holstein 8 Jun 1953, and one at Rock Creek State Fish Hatchery, Dundy Co, 13 Jun 1997 (JS, WRS).

This species is a regular migrant statewide, although it is rare westward (Rosche 1982).

High counts include 28 in Sarpy Co 11 May 1996, including 21 at Fontenelle Forest; 21 at Dodge Park, Douglas Co, 10 May 1995; and 20 at Fontenelle Forest 9 and 10 May 1994.

Summer

There is a report of a female with an active brood patch banded in Fontenelle Forest 15 Jun 1983 (Green 1985b), and this species was reported at this location through 18 Jun 1983 and in Jul 1982 (AB 36:994), although no details were provided; another was noted in Douglas-Sarpy Cos 20 Jun 1993. A report for 1 Jul 1979 in Sioux Co was questioned by Rosche (1982). This species may appear in Jul as an early fall migrant, as in Colorado (Andrews and Righter 1992).

There is no evidence that this species ever bred in Nebraska. Ducey (1988) cited early references to breeding in Omaha and near Ponca in Dixon Co, but these probably refer to Louisiana Waterthrush or some other species. An often cited report is that Northern Waterthrush has bred in Sioux Co (AOU 1983; Bent 1953), apparently due to a statement made by Ridgway (1902) when he described the western race *notabilis* and mentioned Sioux Co as part of the southeastern extent of the breeding range. It is not known on what basis Ridgway made this statement.

Fall

7, 7, 8 Aug→24, 27, 30 Sep

Migrants arrive in mid-Aug and depart by late Sep, although there are later reports 12 Oct 1961 Platte Co, 20 Oct 1984 Douglas-Sarpy Cos, and 22 Oct 1962 Cass Co. An extremely late report was of one seen 8 Dec 1993 at Fontenelle Forest (AB 48:223), and possibly the same bird was noted there the next fall 13 Oct–21 Dec 1994 (Alt 1994; NBR 62:144, 150).

There are few reports away from the east; about 16 reports are all in the period 21 Aug–24 Sep, only 4 from the Panhandle: 15 Aug 1997 Carter Canyon (SJD), 25–27 Aug 1990 Sioux Co, 27 Aug 1999 Oliver Res (SJD), and 5 Sep 1992 Dawes Co.

High counts include 8 in Washington Co 11 Sep 1995.

Other Reports

30 Mar 1981 Douglas-Sarpy Cos, 3 Apr 1983 Douglas-Sarpy Cos, 5 Apr 1992 Sarpy Co (AB 46:444), and 10 Apr 1976, 10 Apr 1978, 11 Apr 1953, 12 Apr 1977, and 14 Apr 1991 Douglas-Sarpy Cos.

FINDING

This species is easiest to find in the Missouri Valley in mid-May in swampy areas with brush. It is often seen along small waterways and around ponds. The song is loud enough to draw attention but can be confused with that of other species.

Louisiana Waterthrush

Seiurus motacilla

STATUS

Rare spring migrant east, rare casual central, hypothetical west. Rare regular breeder east. Rare fall migrant east, hypothetical central.

DOCUMENTATION

Specimen

UNSM ZM6887, 18 May 1901 Child's Point, Sarpy Co.

DISTRIBUTION AND ECOLOGY

Louisiana Waterthrush breeds along small active streams in gullies in deciduous woodland and is restricted as a breeding species to the lower Missouri Valley. Breeding populations are very small, although territorial birds appear most years at favored locations such as Fontenelle Forest and Indian Cave SP.

Spring

30, 31 Mar, 1 Apr ➤summer

This species is a very early migrant, arriving in early Apr. Early dates cited include one (30 Mar) that was originally reported as Northern Waterthrush that we believe is indeed Louisiana Waterthrush.

Migrants are casual away from the breeding range. There are few reports since 1981: 20 Apr 1997 Rock Creek Station, Jefferson Co; 21 Apr 1987 Lancaster Co; 29 Apr 1995 Dakota Co; 4–14 May 1995 Cass Co; 6 May 1984 Polk Co; 9 May 1998 Wyuka Cem, Lincoln; 10 May 1997 Lancaster Co; 12–13 May 1973 Thomas Co (listed as "hypothetical" in Bray [1994]); 13 May 1991 Hall Co; and 17 May 1995 Sarpy Co.

There is 1 undocumented report for the Panhandle, 14 May 1967 Garden Co, and Tout (1947) listed spring reports for Lincoln Co 4 May 1920 and 7 May 1933.

Twelve were reported in 1996, all in the southeast.

Summer

Ducey (1988) listed this species as an occasional breeder along the Missouri Valley. Louisiana Waterthrush was at one time apparently more numerous there (Bruner and others 1904), although the extent of the breeding range is unchanged, with the likely exception of summering birds at Platte River SP. In recent years breeding has been confirmed north as far as Washington Co, where birds were present 7–31 Jul 1988 and seen carrying food 7 Jul (AB 42:1309). One was in extreme northern Thurston Co 18 Jun 1999. In Sarpy Co, the site of most recent summer reports, Bedell (1982) found a nest 5 Jun 1982 in Fontenelle Forest, and fledglings were being fed there 2 Jul 1983 (AB 37:1003). Young were being fed in Richardson Co 25 Jul 1988 (AB 42:1309), and one was carrying food at Indian Cave SP 26 Jun 1995.

Fall

summer→9, 9, 10 Sep

All reports are from Douglas-Sarpy Cos, with departure in early Sep, although there is a later report 24 Sep 1971. Tout (1947) listed an undocumented report for Lincoln Co 30 Aug 1932.

FINDING

Louisiana Waterthrush can be located most years in ravines with running streams in Fontenelle Forest or Indian Cave SP or along the stream at Platte River SP. Walking the streambeds and listening for the song or loud call note in late Apr or early May should yield this species. Care is needed in identification of silent birds, as migrant Northern Waterthrushes may occupy the same habitat.

Kentucky Warbler

Oporornis formosus

STATUS

Rare regular spring migrant east, rare casual central, accidental west. Rare regular breeder east. Rare regular fall migrant east, hypothetical central.

DOCUMENTATION

Specimen

UNSM ZM6888, 6 May 1901 Child's Point, Sarpy Co.

DISTRIBUTION AND ECOLOGY

Kentucky Warbler occupies dense upland forest, primarily oak-hickory, where a shrubby forest-floor understory has developed beneath a closed canopy. This habitat is present in larger tracts of upland forest that are associated with the Missouri Valley and is the locale of most reports.

Spring

28, 29, 30 Apr→summer

Migrants appear in early May and have occurred north to Dakota Co. This species is rare away from the Missouri Valley, with only a single Panhandle report 28 Apr 1981 Garden Co (AB 35:839). Other reports away from the Missouri Valley are 5 May 1984 Howard Co, 14 May 1911 Webster Co (Ludlow 1935), 15 May 1951 Adams Co, 16 May 1950 Adams Co, and 23 May 1967 Greeley Co.

High counts include 3 at Schramm Park 17 May 1995.

Summer

As stated by Ducey (1988), the "lack of recent breeding records makes current status uncertain," although the difficulty of locating nests of this hard-to-see species is a factor. Reports of nesting include a female flushed from a nest 30 May 1936 in Cass Co (Hudson and others 1937), "nesting birds" in 1972 at Fontenelle Forest (AB 26:874), and birds "apparently nesting" in 1995 at Indian Cave SP (JG).

Kentucky Warblers are present in suitable breeding habitat most years in the Missouri Valley, where the population appears to be increasing (NBR 62:83). There are reports north to Dakota Co 8 Jun 1985 and 30 Jun–1 Jul 1989. It was reported in Washington Co 21 Jun–2 Jul 1988 and is found most years in Fontenelle Forest and at other locations in Douglas and Sarpy Cos (AB 48:315). There are also summer reports since 1981 for Nemaha and Richardson Cos, presumably at Indian Cave SP. There are no summer reports away from the Missouri Valley.

It may have been more common formerly, as Bruner and others (1904) designated it a common summer resident and breeder in wooded ravines in the Missouri River bluffs, rarely as far west as Lincoln.

Fall

summer→19, 21, 23 Sep

Departure ends in mid to late Sep, with a later report 8 Oct 1969 Douglas-Sarpy Cos.

Other Reports

4 Oct 1964 Brown Co.

FINDING

Indian Cave SP, Fontenelle Forest, and Hummel Park, Omaha, are the most likely places to see this species, which is usually easy to detect from its loud, ringing song (which must be separated from that of Carolina Wren, however). Damp areas in upland forest should be checked in mid-May and Jun.

Connecticut Warbler

Oporornis agilis

STATUS

Rare regular spring migrant east, rare casual central. Rare casual fall migrant east, accidental central.

DOCUMENTATION

Photograph

25 May 1988 near Chambers, Holt Co (Grenon 1990).

DISTRIBUTION AND ECOLOGY

This species forages on the ground in dense undergrowth or leafy ground cover in deciduous woodland. Most reports are from the east.

Spring

8, 9, 10 May→31 May, 1, 2 Jun

There are about 32 reports, including an undocumented late report 6 Jun 1975 McPherson Co. Additional westerly reports include a specimen, now lost, collected 27 May 1933 in northeast Cherry Co about 17.7 km (11 mi) west of Valentine (NBR 1:62; Bray and others 1986); one banded in Keith Co May 1993 (Rosche 1994a); and another banded there 2 Jun 1990 (Brown and others 1996).

Fall

Although there are about 45 fall reports 26 Aug–26 Oct, only 3 are documented:

1 Sep 1995 Fontenelle Forest (BP, LP; NBR 63:110)

28 Sep 1990 adult female banded Bellevue (RG)

3 Oct 1987 Thomas Co (Mollhoff 1989a)

The number of Nebraska fall reports is questionable. It has been generally believed that most birds migrate eastward to the Atlantic Coast and are thus "extremely rare in the Midwest south of the Great Lakes" (Dunn and Garrett 1997). Four specimens in the Iowa State University collection identified as Connecticut Warblers were shown by Robert Mengel to be in fact Mourning Warblers, indicating the difficulty of identification in fall (James Dinsmore, Stephen Dinsmore, pers. comm.). Iowa banding data show that of 49 banded, only 13 were banded in fall (Dinsmore and others 1984). Kent and Dinsmore (1996) stated, "It is surprising that so many, presumably nonsinging birds are seen [in Iowa] in fall," but Robbins and Easterla (1992) stated, "This warbler is largely overlooked in fall [in Missouri] because it is not vocal and few observers work its habitat in late Aug or Sept, when most should pass through the state."

Other Reports

25 Apr 1962 Douglas-Sarpy Cos, 26 Apr 1970 Lincoln Co, 5 May 1963 Douglas-Sarpy Cos.

FINDING

This species is extremely difficult to see due to its habit of foraging underneath leafy ground cover and because it rarely flies. It is best located by its loud chipping song and should be looked for around 20 May in the Missouri Valley in deciduous woodland with abundant leafy ground cover.

Mourning Warbler

Oporornis philadelphia

STATUS

Fairly common regular spring migrant east, becoming rare casual west. Fairly common regular fall migrant east, hypothetical central, accidental west.

DOCUMENTATION

Specimen

UNSM ZM10773, 25 May 1909 Lincoln, Lancaster Co.

DISTRIBUTION AND ECOLOGY

Mourning Warbler is found during migration in brushy areas, usually with thick ground cover, and along the edges of dense woodlands as well as in interior thickets. Most reports are from the east.

Spring

29 Apr, 1, 2 May→11, 12, 15 Jun

This species is a late migrant, with Jun reports not uncommon. Migration takes place from early May through early Jun.

There are 2 documented Panhandle records: a specimen, USNM 481623, was collected 27 May 1965 by Short (1966) in Sheridan Co, and another at Morrill 27 May 1997 was described (ECT; Brogie 1998). Westerly banding records are of 3 birds with "no hint of an cycring" at L Ogallala 16 May 1996, 17 May 1993, and 25 May 1993 (Brown and others 1996).

There are also undocumented Panhandle reports 12 May 1974 Scotts Bluff Co, 19 May 1992 Dawes Co (AB 37:888), and 27 May 1984 Sioux Co.

High counts include 10 in Douglas Co 15 May 1996, 3 at Fontenelle Forest 14 May 1994, 3 at Neale Woods 18 May 1995, and 3 at Geneva Cem 16 May 1999.

Thirty-eight were reported 15–30 May 1996, and 25–30 were reported 9–29 May 1995.

Fall

21, 22, 22 Aug→21, 24, 26 Oct

Migration is from late Aug through late Oct, with a later report 2 Nov 1982 Lancaster Co. Westerly reports, undocumented, are 29 Sep–3 Oct 1975 Lincoln Co and 7 Oct 1966 McPherson Co.

The only report from the Panhandle is of an adult male in Kimball Co 29 Aug 1998 (LP, BP, SJD).

High counts include 3 in Sarpy Co 1 Sep 1995.

COMMENTS

Indications that Mourning Warbler has bred in Nebraska appear to derive from a statement by Bruner and others (1904) that Aughey saw adults feeding young in southeast Nebraska in the 19th century. This observation now seems to have been in error (Sharpe 1993). There is no other indication that breeding ever occurred in Nebraska.

Other Reports

20 Mar 1972 Hall Co.

FINDING

Mourning Warbler can usually be found in the Missouri Valley around 15–20 May by checking undergrowth in or along the edges of riparian thickets or woodland. An excellent location is Raccoon Hollow at Neale Woods, Douglas Co. It forages low, often in weeds, and may pop up in response to squeaking.

MacGillivray's Warbler

Oporornis tolmiei

STATUS

Uncommon regular spring migrant west, becoming rare casual east. Accidental in summer. Rare regular fall migrant west, rare casual central, hypothetical east.

DOCUMENTATION

Specimen

UNSM ZM9517, 29 May 1932 Cherry Co.

DISTRIBUTION AND ECOLOGY

MacGillivray's Warbler frequents dense brushy thickets and weed patches during migration, including riparian thickets along watercourses

and draws. Most reports are from the Panhandle.

Spring

5, 10, 10 May→3, 4, 8 Jun

Migration is from mid to late May, although there is an earlier report 25 Apr 1995 at Crescent L NWR. This species is a regular migrant in small numbers in the northwest (Rosche 1982) and straggles eastward. Its distribution in Nebraska may be thought of as a mirror image of that of Mourning Warbler.

In the L Ogallala area 9 birds were banded 16 May–4 Jun, all with "obviously incomplete, rather thick white eyerings" (Brown and others 1996).

There are 3 documented records from the east: a specimen, UNSM ZM17033, "documented itself" in Lancaster Co by flying into a window 10 May 1994 (TEL; AB 48:315; Brogie 1997); one was seen at Clay Center Cem 15 May 1999 (JGJ); and one was trapped, measured, and banded in Cedar Co 20 May 1986 (Brogie and Stage 1987). There is an additional undocumented report from the east 10 May 1964 Platte Co.

High counts include 4 at Oliver Res 17 May 1998.

Summer

MacGillivray's Warbler is an uncommon breeder in the Black Hills of South Dakota (SDOU 1991), where it occupies riparian deciduous habitats with shrubs (Peterson 1990). In the 19th century it might have bred in northern Sioux Co, where it was found in riparian thickets along streams draining northward into the Hat Creek drainage (Cary 1902). Rosche (1982) stated that Cary's assumption "must be discounted" since "the nearest known breeding habitat is in the higher reaches of the Black Hills." The only recent report suggestive of breeding was of 2 birds at Gilbert-Baker Campground, Monroe Canyon, 25 Jul 1999, one of which may have been a young bird (EB). We believe that habitat is indeed suitable in the western Pine Ridge, and breeding would not be unexpected.

Fall

15, 17, 21 Aug→4, 4, 5 Oct

Migration is from late Aug through early Oct, with later reports 10 Oct 1974 Perkins Co and 13 Oct 1979 Scotts Bluff Co.

The easternmost documented record is a specimen, wsc 794, collected at Albion 13 Sep 1982.

There are no accepted fall records farther east; one at Fontenelle Forest 14 Oct 1990 (wrs) was judged hypothetical (Class IV) by the NOURC (Grenon 1991), and a specimen, UNSM ZM10776, taken at Roca and labeled as this species is not now identifiable.

High counts include 5 at Oliver Res 29 Aug 1998.

FINDING

This species should be looked for in riparian thickets in the Panhandle, especially where willows occur. The best time is probably around 20 May.

Common Yellowthroat

Geothlypis trichas

STATUS

Common regular spring and fall migrant and breeder statewide.

DOCUMENTATION

Specimen

UNSM ZM6904, Jun 1898 Beatrice, Gage Co.

TAXONOMY

It is possible that 3 subspecies occur as breeding birds. The race *brachydactyla* (now treated as a northerly segment of the eastern race *trichas*) breeds throughout most of the state, west at least to Cherry and Logan Cos (AOU 1957), while the western *occidentalis* breeds east to northeastern Colorado, including Windsor, Jackson Res, and Wray (AOU 1957). Dunn and Garrett (1997) showed *occidentalis* reaching extreme southwest Nebraska in the Dundy Co area and the subspecies *campicola* occupying the Panhandle and northwest Nebraska east to about the Valentine area.

Because the ranges of 3 subspecies meet in Nebraska, most breeding birds in the state are probably intergrades, except possibly in the extreme east. Bruner and others (1904) stated that "the more western specimens [of *brachydactyla*] in Nebraska are so nearly intermediate between [*brachydactyla*] and [*occidentalis*] that it is difficult to satisfactorily place them."

Migrants of *occidentalis*, *campicola*, and *brachydactyla* probably occur statewide, although *occidentalis* is less common in the east (Bruner and others 1904). According to Dunn and Garrett (1997), males of *campicola* and *occidentalis* possess brighter yellow coloration on the underparts than *brachydactyla*, even tending to orange, and females of *occidentalis* and *campicola* have the least amount of yellow on the underparts, tending to creamy whitish throughout.

DISTRIBUTION AND ECOLOGY

This species breeds commonly statewide in wet meadows and brushy swamp or marsh edge habitats, particularly those associated with cattail marshes. There is a gradual decline in density of breeding birds westward. Migrants also occur in the same habitats as well as in drier brushy areas and may become locally abundant at migration peaks.

Spring

20, 21, 22 Apr→summer

Migrants arrive in late Apr, although there are earlier reports, possibly wintering birds, 1 Apr 1943 Jefferson Co, 4 Apr 1987 Adams Co, 5 Apr 1972 Hall Co, 5 Apr 1991 Douglas-Sarpy Cos, 8 Apr 1956 Gage Co, 11 Apr 1958 Nemaha Co, 15 Apr 1963 Gage Co, 15 Apr 1991 Hall Co, and 18 Apr 1937 Lincoln Co.

High counts include 119 in Sarpy Co 11 May 1996 and 73 in Sarpy Co 13 May 1995.

Summer

BBS data show that numbers decrease from east to west, probably reflective of available habitat. Rosche (1994a) noted that willow and other brush currently growing in the channel of the North Platte River due to low flows "have greatly benefited this species."

High counts include 34 at Crescent L NWR 7 Jun 1995.

Fall

summer→15, 16, 16 Oct

Departure is completed by early to mid-Oct. There are surprisingly few later reports: 24 Oct 1974 Lancaster Co, 26 Oct 1964 Lincoln Co, and 29 Oct 1972 Lancaster Co. This species may linger very late in fall on occasion; it is a casual winter resident on the eastern Colorado plains (Andrews and Righter 1992).

High counts include 9 at Arbor L 3 Oct 1998.

FINDING

This may be one of the easiest species to find. In May it is usually the first bird to respond to squeaking at any marsh edge. Its distinctive and often-repeated song indicates its presence.

Hooded Warbler

Wilsonia citrina

STATUS

Rare casual spring migrant statewide. Hypothetical in fall.

DOCUMENTATION

Specimen

CSC 17 Jun 1982 Chadron, Dawes Co (Probasco 1983).

DISTRIBUTION AND ECOLOGY

At the northwestern extreme of its breeding range in Nebraska, this species is found in extensive moist hardwood forests that have dense understory, often in ravines that support a brook. In migration it may occur in more open brushy situations. Most reports are from the east, although it has been recorded statewide.

Spring

21, 23, 24 Apr→25, 25, 27 May

Most reports are in the period 21 Apr–25 May; exceptions are the specimen cited above, collected 17 Jun 1982 at Chadron by a cat, and a singing male at Hummel Park, Omaha, 20 Jun 1999 (see Comments).

There are about 42 spring reports, including major influxes in 1991–95, when there were 18 reports, including 4 birds at Fontenelle

Forest on an unspecified single day in the period 1–10 May 1993 (NBR 61:138). Hooded Warbler may be expanding its range into Nebraska.

There is a surprising number of Panhandle reports for a species at the northwest edge of its summer range. These include a bird netted and photographed at Crescent L NWR 2 May 1981 (AB 35:839), 7–9 May 1995 Sioux Co, 15 May 1971 Scotts Bluff Co, 16 May 1994 L Ogallala (Brown and others 1996), 17 May 1961 Dawes Co, 18 May 1957 Sheridan Co, 20 May 1996 L Ogallala (Brown and others 1996), and 28 May 1994 Scotts Bluff Co (AB 48:315).

Other westerly reports are 6 May 1999 Phelps Co, 12 May 1956 Cherry Co, 16 May 1999 Geneva Cem (JGJ), 18 May 1955 Cherry Co, 18 May 1983 Keith Co (Rosche and Johnsgard 1984), and a pair 28–29 May 1930 Webster Co (Ludlow 1935).

Fall

There are very few reports, none documented: 7 Aug 1966 Adams Co, 21 Aug 1989 Sioux Co, 25 Aug 1967 Adams Co, Sep 1874 along the Nemaha River (Aughey 1878), 8 Sep 1973 Douglas-Sarpy Cos, 10 Sep 1975 Adams Co, and 12 Sep 1967 Adams Co. In addition, there is a report with few details of one in a Logan Co yard 10 Nov–early Dec 1965 (Glandon and Glandon 1966).

COMMENTS

Bruner and others (1904) suggested that "breeding is very probable" for this species in southeast Nebraska, based on an observation by Aughey along the Nemaha River Sep 1874. This supposition was repeated by Rapp and others (1958), but there is no evidence that this species ever nested in the state.

FINDING

Perhaps the best chance to find this rare species is in extensive areas of dense understory in deciduous forest in the Missouri Valley. Best time is mid-May, and likely locations are Indian Cave SP, Schramm Park, and Fontenelle Forest. The song is rather loud although not particularly distinctive.

Wilson's Warbler

Wilsonia pusilla

STATUS

Fairly common regular spring and fall migrant statewide. Hypothetical in summer.

DOCUMENTATION

Specimen

UNSM ZM10790, 4 Sep 1909 Roca, Lancaster Co.

TAXONOMY

Rapp and others (1958) suggested that western *pileolata* is an uncommon migrant in the west and eastern *pusilla* a common migrant in the east. Zimmer (1913) noted that both occurred in fall in Thomas Co. In recent years this species has been found to be a fairly common migrant in the west, most birds appearing to be brightly colored examples of western *pileolata* (SJD, WRS), which is indeed more common in fall on the western Great Plains than in spring (Dunn and Garrett 1997).

DISTRIBUTION AND ECOLOGY

Migrants are found statewide in deciduous brushy areas and woodland edge, often foraging at low levels and often showing a preference for willows.

Spring

24, 25, 26 Apr→29, 29, 30 May

Migration is from late Apr through late May. There are earlier reports 14 Apr 1976 Adams Co, 15 Apr 1979 Garden Co, 17 Apr 1974 Lancaster Co, 20 Apr 1980 Garden Co, and 21 Apr 1972 Adams Co. There are a few Jun reports, most from the west: 3 Jun 1970 Perkins Co, 3 Jun 1980 Garden Co, 4 Jun 1910 Webster Co (Ludlow 1935), 4 Jun 1969 Dawes Co, 4 Jun 1983 Scotts Bluff Co, 6 Jun 1998 Fontenelle Forest, 7 Jun 1998 Funk Lagoon (LR, RH), and 13 Jun 1999 Enders Res (MB). There are also a few late Jun and early Jul reports (see Summer).

This species is a statewide migrant, although less common in the west (Johnsgard 1980). Rosche (1982, 1994a) pointed out that Wilson's Warbler may not be seen in spring every year in the west, with most records after strong southeasterly storms; Tout (1947)

made a similar observation for Lincoln Co. This suggests that the western race *pileolata* is uncommon in spring and in some years may be supplemented by wind-driven eastern *pusilla*. In fall Wilson's Warbler is common statewide.

High counts include 14 in Sarpy Co 13 May 1995.

Summer

The western race *pileolata* breeds in the Colorado mountains but not in the Black Hills of South Dakota, although there is a single summer record for South Dakota, 10 July 1977 at Rapid City (SDOU 1991).

There are 5 summer reports for Nebraska, none documented: 23 Jun 1980 Douglas-Sarpy Cos, 30 Jun 1969 Dawes Co (reportedly present 4–30 Jun), 30 Jun 1970 McPherson Co, 1 Jul 1964 Dawes Co, and 1 Jul 1966 Douglas-Sarpy Cos.

Fall

12, 13, 14 Aug→20, 23, 24 Oct

Migration is from mid-Aug through early Oct, although there are earlier reports 3 Aug 1963 Webster Co, 8 Aug 1985 Lancaster Co, and 9 Aug 1995 Garden Co and an undocumented late report 20 Nov 1976 Boone Co.

High counts include 41 in the southern Panhandle 20 Sep 1999, 39 at Oliver Res 1 Sep 1997, and 36 there 13 Sep 1999.

FINDING

Wilson's Warbler responds well to squeaking and should not be hard to find in small to medium-height willow thickets in mid-May in the east or early Sep in the west.

Canada Warbler

Wilsonia canadensis

STATUS

Rare regular spring migrant east, rare casual central. Rare regular fall migrant east, rare casual central, accidental west.

DOCUMENTATION

Specimen

HMM 1793, 17 May 1916 Spring Ranch, Clay Co.

DISTRIBUTION AND ECOLOGY

Migrants occur in deciduous woodland, with

a preference for riparian undergrowth and edge. This species is virtually restricted to the east.

Spring

6, 7, 7 May→28, 29, 30 May

Migration is in the latter two-thirds of May, although there is an earlier report 28 Apr 1964 Nemaha Co and later reports 1 Jun 1950 Gage Co and 6 Jun 1970 Douglas-Sarpy Cos.

This species is less numerous in the interior United States in spring than in fall. Curson and others (1994) stated that Canada Warblers tend to follow the Atlantic Coast in spring rather than moving through the interior, the latter utilized more in fall. In Iowa, of 422 Canada Warblers banded, 302 were in fall (Dinsmore and others 1984). This species is one of the last spring migrants to arrive.

Away from the Missouri Valley there are only 12 reports, none from the Panhandle: May 1980 Thomas Co (Bray 1994); 6 May 1961 Garfield Co; 8 May 1978 Cherry Co; 14 May 1937 Lincoln Co; 14 May 1999 Geneva Cem (SJD); 15 May 1999 Wilderness Park, Lincoln (SJD); 17 May 1916 Clay Co (specimen HMM 1793); 19 May 1996 L Ogallala, banded male (Brown and others 1996); 20 May 1935 Adams Co; 23 May 1953 Lancaster Co; 25 May 1957 Hall Co; and 29 May 1975 McPherson Co.

A good spring total was 9 in 1996.

Fall

10, 18, 18 Aug→5, 9, 9 Oct

Migration is from mid-Aug through early Oct.

There are few reports away from the east, including a single Panhandle report 22 Aug 1987 Sheridan Co (AB 42:99) and these others: 3–4 Sep 1981 Boone Co (specimen WSC 679), 10 Sep 1964 Adams Co, 12 Sep 1982 Boone Co, 16 Sep 1964 Adams Co, 16 Sep 1982 Boone Co, and 18 Sep 1969 McPherson Co.

FINDING

The best chance to see this species is in the Missouri Valley 15–25 May. It should be looked for in brushy riparian situations such as at Fontenelle Forest. Its song is not particularly distinctive but is fairly loud and lengthy and should attract an observer's attention. This species may be just as easily found in fall in the same places; best time is early Sep.

Yellow-breasted Chat

Icteria virens

STATUS

Uncommon, locally common, regular breeder and spring and fall migrant west, becoming rare casual east.

DOCUMENTATION

Specimen

UNSM ZM6909, 3 Jun 1901 Monroe Canyon, Sioux Co.

TAXONOMY

Two subspecies occur in Nebraska. Eastern *virens* bred commonly in the east prior to about 1960, especially in the Missouri Valley but also in the valleys of the Little Blue and Elkhorn Rivers (Swenk 1940); it has declined in recent years and may be extirpated as a breeder. The western race *auricollis* is locally common in central, western, and northern Nebraska.

DISTRIBUTION AND ECOLOGY

Yellow-breasted Chat is found in dense thickets and brush, often riparian. Breeding populations in the east have declined markedly in recent years, and it has disappeared from the southeast.

Spring

25, 26, 28 Apr→summer

This species is now only casual in the east, where it is not reported every year. Spring migrants arrive in late Apr.

High counts include 9 in Sowbelly Canyon 21 May 1996.

Summer

Short (1961) considered this species common and abundant in central and western Nebraska and "scarce or absent to locally common" in the east; this remains an accurate description, although it probably is no longer "locally common" anywhere in the east. There are few reports during Jun and Jul in the east since 1981: 9 Jun–16 Jul 1990 Knox Co, 14 Jun

1997 Platte River sp (gw), 27 Jun 1986 Nemaha Co, 27 Jun 1986 Richardson Co, 29 Jun 1988 Lancaster Co, 5 Jul 1982 Douglas-Sarpy Cos, and 22–29 Jul 1989 Knox Co. This is in strong contrast with the Panhandle and north, where it has been reported every year (Brogie and Mossman 1983; Rosche 1982). It is a fairly common breeding bird in the L Ogallala area (Brown and others 1996) and in brush thickets along the Middle Loup River in Thomas Co (rss, tb).

This species was formerly considered to be most common in the eastern part of the state; Bruner and others (1904) listed it as "abundant" along the eastern edge. By 1958 Rapp and others (1958) listed the species as fairly common throughout but most numerous in the east, although by 1935 in Webster Co it was only an uncommon migrant (Ludlow 1935). Recently, 3 birds were found 12 Jun 1999 along the south side of Harlan Co Res (lr, rh). Beginning in the 1970s, breeding populations were scattered and only locally common. Those have declined to near extirpation. Although Ducey (1988) proposed that destruction of nesting areas along the Missouri River due to riverside development is a cause for this decline, many habitats in southeast Nebraska that formerly sustained active breeding populations, including specific sites in Douglas and Sarpy Cos, have remained relatively unchanged. An alternative explanation for the extirpation of the chat in eastern Nebraska is that this population was part of the eastern subspecies *virens*, and this race has experienced difficulties in its wintering range similar to those experienced by other eastern neotropical migrants (Terborgh 1989). Chat populations in eastern Kansas are also nearing extirpation (Thompson and Ely 1992), although it still occurs at several sites in the extreme east (Mark Robbins, pers. comm.) and in northwest Missouri (David Easterla, Mark Robbins, pers. comm.).

Fall

summer→4, 6, 9 Oct

Departure is protracted, with no clear conclusion of migration. Most birds depart by early Oct, but there are later reports 16 Oct 1961 Cass Co and 27 Oct 1960 Cass Co.

FINDING

Yellow-breasted Chat is most easily found in riparian brush in western canyons such as at Ash Hollow shp, along the Niobrara Valley, and in the Pine Ridge. Its distinctive song locates the birds, which can be difficult to see even though they respond well to squeaking. In late May and Jun males may also sing sporadically throughout the night.

Thraupidae (Tanagers)

Hepatic Tanager

Piranga flava

STATUS

Accidental in winter. Hypothetical in summer.

Records

A male at the feeders of Dr. and Mrs. David Mlnarik in West Point 18 Dec 1998–6 Jan 1999 was photographed 6 Jan (bp, lp, wrs). The only other record northeast of the normal range for this species is of one in western Illinois 23 Nov 1981 (Bohlen 1989).

A "red bird" in a large spruce in a yard in Scottsbluff 13 Jun 1983 was thought to be of this species due to apparent dark mandibles, although the light conditions were judged to be inadequate to identify the bird (nbr 51:78). No other details were published, and its identification remains questionable.

COMMENTS

As recently as 1978 this species first appeared in Las Animas Co in southeast Colorado and began breeding there in 1980. It may now be breeding in Fremont Co in central Colorado. There are 2 spring records for the northeast Colorado plains: 9 May 1977 Boulder Co and 15 May 1982 Bonny Reservoir, Yuma Co (Andrews and Righter 1992). The latter record is about 65 km (40 mi) from

Nebraska. Hepatic Tanager is found in ponderosa pine habitats in Colorado. There are several Wyoming records, all but one in May (Scott 1993).

Summer Tanager
Piranga rubra
STATUS
Locally uncommon regular breeder east. Rare regular spring migrant east, becoming rare casual west. Rare regular fall migrant east, accidental central.
DOCUMENTATION
Specimen
HMM 1858, 10 May 1917 Fairbury, Jefferson Co.
TAXONOMY
An adult male collected by L. Skow at Omaha 20 May 1892 (USNM 128032) is labeled a "possible hybrid" between this species and Scarlet Tanager; we have not had the opportunity to see the specimen.

There is a published report of the "western race" of Summer Tanager occurring at Bessey Division, NNF, but no details were provided (*Omaha World Herald* 11 Jan 1999).
DISTRIBUTION AND ECOLOGY
Summer Tanager occurs in mature oak-hickory forest, preferring stands with numerous tall oaks, such as white oak (*Quercus alba*). Although the species may be found foraging in the upper canopy, it usually is found in medium-size trees associated with dense understory. Breeding birds have been found in the Missouri Valley north to Sarpy Co and within the historic range of oak-hickory forest.
Spring
3, 4, 4 May→summer
Arrival is in early May, with earlier reports, undocumented, 25 Apr 1960 Logan Co and 26 Apr 1962 Webster Co. There are about 12 reports in all away from the east in the period 6–24 May.

The few Panhandle reports are 6 May 1961 Box Butte Co, 8 May 1992 Sheridan Co (AB 46:444; NBR 60:141), and 24 May 1975 Scotts Bluff Co.

Summer
In recent years breeding has been documented in Sarpy Co in 1980, when Padelford and Padelford (1980b) found adults feeding young in a nest in a burr oak at Schramm Park; in 1983 (Ducey 1988); and in Saunders Co in 1985 when adults carrying nesting material were observed (Bennett 1986). The only regular summer population appears to be that at Schramm Park, where it has been reported most years since 1987.

In recent years there have been scattered summer reports from the Missouri River valley north as far as Dakota Co without evidence of breeding.

Reports away from the Missouri Valley are as follows. A "breeding pair" was reported in cottonwoods in Thomas Co 16 Jun–13 Jul 1993, although no evidence for nesting was noted (Bright 1993; AB 47:1124; NBR 61:127, 137); Bray (1994) did not list this report but noted a singing male present in jackpines 13 Jun–21 Jul 1993. Another was reported in Thomas Co 19 Apr 1957 without details (Bray 1994). A pair was reported at North Platte in summer 1993 and again 6 Sep 1994 (NBR 62:144).

A tanager in Scotts Bluff Co 13 Jun 1983, originally identified as an Hepatic Tanager but later thought to have been a Summer Tanager (NBR 51:78), is probably best considered unidentified. A Summer Tanager was reported on a BBS route in the period 1967–77, possibly in the Long Pine Canyon area of Brown and Rock Cos, but no details have been published.
Fall
summer→19, 26, 28 Sep
There are few reports, most from the breeding location at Schramm Park; exceptions are the only central Nebraska report 19 Sep 1984 Lincoln Co (NBR 53:6); 15 Aug 1966 Richardson Co; 9 Sep 1996 Fontenelle Forest; 15 Sep 1998 Camp Brewster, Sarpy Co (BP, LP); and 19 Sep 1999 Indian Cave SP (JG).

Departure is completed by late Sep, although there are later reports 12 Oct 1987 Douglas-

Sarpy Cos, 30 Oct 1985 Sarpy Co (AB 40:137), and a female at a Bellevue feeder 20 Nov–27 Dec 1993 (AB 48:223).

FINDING

This species is most likely in extensive deciduous woodland with large oaks, the best time probably late May when males are singing and trying to establish territories. It is most consistent at Schramm Park. The distinctive call note is the best indicator of the presence of this species.

Scarlet Tanager

Piranga olivacea

STATUS

Uncommon regular spring migrant east and central, rare casual west. Uncommon regular breeder east and north. Uncommon regular fall migrant east, rare casual central, accidental west.

DOCUMENTATION

Specimen

UNSM ZM11098, 20 Jun 1907 Lancaster Co.

TAXONOMY

See Western Tanager. Summer reports of Scarlet Tanager in the Pine Ridge (see Summer) increase the possibility of hybridization with Western Tanager (Short 1961; Tordoff 1950; NBR 59:96), although there are no recorded instances in Nebraska.

DISTRIBUTION AND ECOLOGY

Scarlet Tanager breeds most commonly in deciduous hardwood forests in the Missouri Valley, in the Niobrara Valley west to Brown Co, and in small numbers in Platte Valley floodplain forest west to the Grand Island area. Although migrants occur statewide, they are least numerous in the Panhandle.

Spring

25, 26, 26 Apr→29 (Ludlow 1935), 29, 31 May

Late dates above are outside the breeding range. Arrival is in late Apr, although there are undocumented reports 18 Apr 1991 Douglas-Sarpy Cos and 20 Apr 1943 Jefferson Co. The earliest South Dakota date is 3 May (SDOU 1991).

Later dates from the Panhandle are discussed under Summer.

High counts include 6 in Sarpy Co 13 May 1995, 5 at Schramm Park 30 May 1996, and 4 there 15 May 1995.

Summer

In the east breeding occurs throughout the Missouri Valley, where there are breeding and summer reports for most counties. Breeding apparently occurs in the Platte Valley in the Hall Co area, as there are numerous Jun and Jul reports from there in recent years, but there are no data from counties to the east along the Platte River between Hall and Sarpy Cos. In the area south of the Platte River and west to Adams and Webster Cos there are few summer reports: 7 Jun 1963 Jefferson Co, 15 Jun 1954 Gage Co, 15 Jun 1970 Lancaster Co (Cink 1971), 19 Jun 1975 Clay Co, 1 Jul 1976 Lancaster Co, 2 Jul 1991 Richardson Co (MC), and 31 Jul 1976 Lancaster Co. Cink (1971) stated that "apparently there are no breeding records for Lancaster County." In the western Platte Valley a nest was found with adults nearby in Lincoln Co 12 Jun 1938 (Tout 1947), and an adult male was banded at L Ogallala 11 Jul 1992 (Brown and others 1996).

Breeding also occurs along the Niobrara River at least to Brown Co. Brogie and Mossman (1983) considered it uncommon in the Niobrara Valley Preserve in 1982 and found a nest there 22 June 1982; a recent Brown Co report was 26 Jun 1993 (AB 47:1124). There have been virtually no summer reports along the Niobrara Valley west of Brown Co since Ford (1959) and Short (1961) found birds west to the Valentine area, although one was east of Valentine in Cherry Co 19 Jul 1995.

Early Jun reports from Thomas and Logan Cos in the Loup drainage are probably late spring migrants, although it reportedly nested at Halsey until 1958 (Bray 1994).

A singing male in Sioux Co 7–9 Jun 1991 (NBR 59:96) and reports 9 Jun 1981 Garden Co and 28 Jun 1973 Sioux Co may have been spring migrants. Rosche (1982) listed this species as a "casual spring transient" in the northwest.

Fall

summer→6, 10, 10 Oct

Departure is in early Oct. There is a late report of 2 females at Johnson Res 10 Nov 1996 (LR, RH).

There are only these few fall reports away from known breeding areas: 12 Sep 1979 Boone Co, 15–16 Sep 1959 Lincoln Co, and 24 Sep 1994 Sheridan Co (RCR, DJR; NBR 62:144), the latter the only fall Panhandle report.

Other Reports

25 Mar 1953 Logan Co, 7–8 Apr 1976 Howard-Hall Cos, 8 Apr 1955 Boyd Co, 18 Apr 1944 Lincoln Co.

FINDING

Scarlet Tanager is not hard to find in the Missouri Valley, especially in May and Jun at Indian Cave SP and Fontenelle Forest. Its loud, burry song indicates its presence, although its tendency to forage rather high can make it hard to see.

Western Tanager

Piranga ludoviciana

STATUS

Rare regular spring and fall migrant west, rare casual central and east. Fairly common regular breeder west.

DOCUMENTATION

Specimen

UNSM ZM7051, 26 May 1900 Monroe Canyon, Sioux Co.

TAXONOMY

Short (1961) and Ford (1959) raised the possibility that Scarlet Tanager might occur far enough west in the Niobrara Valley to come into contact and hybridize with Western Tanager. Short (1961) suggested that the Niobrara Valley was the best possibility for such contact on the Great Plains, cited a hybrid specimen between the 2 species (Tordoff 1950), and described an individual seen near Chadron in late Jun 1955 that may have been a hybrid. However, neither Short

nor Ford could find either species between Chadron and Valentine, and Western Tanager has not been reported from that stretch of the Niobrara Valley in summer since.

DISTRIBUTION AND ECOLOGY

This species is restricted as a breeding bird to ponderosa pine forest in the Pine Ridge of Sioux and Dawes Cos, although it forages into adjacent deciduous woodland, a habitat also utilized by migrants (Rosche 1982). Migrants occur on occasion east of the breeding range.

Spring

3, 3, 6 May→summer

Spring arrival is in early May.

Away from the Panhandle this species is a casual migrant. Easternmost reports are 5 May 1995 at Neale Woods, Douglas Co (Brenneman 1995); 8 May 1994 at Indian Cave SP (NBR 62:84); 13 May 1995 Sarpy Co (NBR 63:64); 24 May 1942 Jefferson Co; and 29 May 1957 Lancaster Co. Reports from central Nebraska are in the period 8–29 May.

High counts include 5 at Harrison 27 May 1995.

Summer

According to Rosche (1982), Western Tanager is most common in western Pine Ridge canyons but occurs east at least to King's Canyon in Dawes Co. There is no recent evidence of breeding (Ducey 1988), but this species is a regular summer resident, with Jun or Jul reports for each year since 1984.

There are very few reports away from the Pine Ridge after early Jun: 11 Jun 1995 Kimball Co (JG); 18 Jun 1860 Snake River, Cherry Co (Coues 1874); and 27 Jun 1967 Brown Co. Early reports 2 Jun 1980 Scotts Bluff Co and 4 Jun 1986 Sheridan Co are probably late spring migrants, but 2 early fall reports 13 Jul 1972 and 7 Aug 1976 in Scotts Bluff Co are more difficult to explain. These may represent failed breeders, postbreeding wanderers, or very early fall migrants, although the possibility of breeding in ponderosa pine habitat in the Wildcat Hills or in Sheridan Co cannot be discounted.

Fall

summer→ 3, 6, 10 Oct

Departure is complete by early Oct.

Western Tanager is a casual migrant in the east, with few reports: 10 Sep 1963 Jefferson Co and 14 Oct 1981 Pierce Co (specimen WSC 750). Reports from central Nebraska are in the period 18 Aug–10 Oct, but there is a female specimen taken by G. Shickley at North Platte 20 Nov 1964 (USNM 488495).

FINDING

Checking ponderosa pines in upper Sowbelly Canyon and upper Monroe Canyon in Sioux Co should yield this species in late May and Jun. It may be difficult to locate, but its call and burry song, similar in pattern to that of Summer Tanager, will help. It also responds to squeaking.

Emberizidae (Towhees, New World Sparrows, Longspurs)

Green-tailed Towhee

Pipilo chlorurus

STATUS

Rare regular spring migrant west, rare casual central, accidental east. Rare casual fall migrant west and central, accidental east. Accidental in winter.

DOCUMENTATION

Specimen

UNSM 7242, 11 Sep 1919 Monroe Canyon, Sioux Co (Mickel and Dawson 1920).

DISTRIBUTION AND ECOLOGY

Green-tailed Towhee is found in migration in dense deciduous brush, usually riparian (Rosche 1982), but also in arid situations with sagebrush (Johnsgard 1980; Thompson and Ely 1992). Migration reports are from the extreme west, while winter reports are from the east.

Spring

25 (AB 48:315), 30 Apr, 1 May→2 (FN 50:967), 4, 6 Jun

Migration takes place in May. There are earlier reports 10 Mar 1985 Scotts Bluff Co, possibly a wintering bird, and 21 Apr 1961 Box Butte Co and a late report 30 Jun 1973 Scotts Bluff Co.

There are about 33 reports in all, most from the Panhandle, although a few are from farther east: 30 Apr 1974 Perkins Co, 3 May 1955 Keith Co, 15 May 1972 Perkins Co, 31 May 1963 Logan Co, 4 Jun 1985 Lincoln Co, and 6 Jun 1982 McPherson Co.

The only report from the east is one at a very early date, possibly a wintering bird, 17 Apr 1984 Dixon Co (NBR 52:68).

Fall

There are very few fall reports, including one as far east as Douglas-Sarpy Cos 4 Oct 1983 (AB 38:220), and these others: 6 Sep 1969 McPherson Co, 11 Sep 1919 Sioux Co (specimen cited above), 14–16 Sep 1971 Perkins Co, 19 Sep 1987 Scotts Bluff Co, 19 Sep 1998 Kimball Co (SJD), 20 Sep 1999 (2) southwest Kimball Co (SJD), 21–29 Sep 1976 Sioux Co, and 16 Oct 1985 Scotts Bluff Co.

Winter

This species has a tendency to winter at feeders east of its regular range; the winter records are all in this category. One was at a Lincoln feeder during the last 2 weeks of Feb 1978 (AB 32:372). The following winter 3 appeared, including 1 in Lincoln (AB 34:287), 1 in York, and 1 in Raymond (AB 34:287). The latter spent 5 Jan–17 Apr 1980 at a feeder 1.6 km (1 mi) east of Branched Oak L (NBR 48:85, 89); another, possibly the same, wintered there 1 Nov 1984–2 Mar 1985 (NBR 53:6, 58).

FINDING

The best chance to find this species is in the extreme west around mid-May. Likely locations might be the more southerly canyons such as Carter Canyon, the Wildcat Hills, behind the Gering Cem in Scotts Bluff Co, and in Long Canyon in Banner Co. It can be difficult to see but usually responds to squeaking at brush piles.

Spotted Towhee

Pipilo maculatus

STATUS

Common regular breeder statewide, except
   southeast. Rare regular winter visitor south.

DOCUMENTATION

Specimen

UNSM ZM7252, 24 May 1900 Monroe Canyon,
   Sioux Co.

TAXONOMY

According to the AOU (1957) 2 subspecies have
   occurred in Nebraska. Breeding birds and
   most wintering birds are of the western Great
   Plains race *arcticus*, but an individual of the
   western race *montanus* was taken 5 Oct 1915 at
   North Platte (Swenk 1918b; Tout 1947).

See Eastern Towhee for relationships and
   hybridization with that species in Nebraska.

DISTRIBUTION AND ECOLOGY

This species occupies riparian edge habitats
   in western Nebraska (Rosche 1982). Urban
   situations and smaller woodlands may also be
   occupied.

Spring

25, 28, 29 Mar→summer (central)

9, 10, 14 Apr→summer (west)

winter→10, 10, 13 May (southeast)

In central and western Nebraska, where
   wintering is rare, arrival is in late Mar
   and early Apr, with central birds arriving
   significantly earlier. Wintering birds in the
   east depart by mid-May.

High counts include 45 at Lilley 9 May 1998, 35 at
   Kenesaw 26 Apr 1998, 12 at North Platte NWR 5
   May 1996, and 12 in Harlan Co 8 May 1999.

Summer

BBS data show fairly even distribution within
   the virtually statewide range, with perhaps
   slightly higher numbers in northern Nebraska.

Tout (1947) cited egg dates (N = 6) for Lincoln
   Co 22 May–25 Jun.

Fall

24, 28, 30 Sep→winter (southeast)

Most departing birds leave the summer range
   by late Oct, although lingering birds make

determination of departure dates difficult.
   Significant movement into the southeast
   begins in late Sep; it was "common" in
   Omaha in early Oct 1975 (AB 30:94). The first
   "spotted" birds appear in Missouri in early
   Oct (Robbins and Easterla 1992).

High counts include 60 near Gibbon 10 Oct 1998,
   40 in northern Lancaster Co 17 Oct 1998, and
   10 near Gibbon 8 Oct 1999.

Winter

Although rather late to leave the summer range
   in fall, CBC data indicate that this species
   is still rather evenly distributed statewide
   in Dec, except for the Panhandle, which is
   usually vacated by then. Subsequently, there is
   a general southeastward movement.

There are 10 winter reports from the Panhandle
   and northern Nebraska, most in Scotts Bluff
   Co, with these others: 1 Jan 1957 Logan Co, 14
   Feb 1993 Antelope Co, 23 Feb 1957 Thomas
   Co, and 3 Mar 1958 Sheridan Co. Panhandle
   reports 13 Mar 1957 and 23 Mar 1967 Dawes
   Co may have been wintering birds or early
   migrants.

FINDING

These birds respond quickly to squeaking and
   are easily located in riparian and brushy
   areas in central and western Nebraska in Jun
   and Jul. Observers should take note of and
   report the presence of birds showing hybrid
   characteristics.

Eastern Towhee

Pipilo erythrophthalmus

STATUS

Fairly common regular breeder southeast, rare
   extreme south and southwest. Rare casual
   winter visitor southeast.

DOCUMENTATION

Specimen

UNSM ZM7238, 19 Jan 1900 Saltillo, Lancaster Co.

TAXONOMY

This species and Spotted Towhee have recently
   gained specific status after being considered
   subspecies of "Rufous-sided Towhee" for

some years. The "presence of at least 20% pure phenotypes throughout the contact zone indicates that these towhees do not interbreed freely and are better treated as separate species" (DeBenedictis 1996).

Both Eastern and Spotted Towhees breed in Nebraska, although the contact zone between them is complex. Areas where essentially pure birds of one or the other species occur without hybrids are limited to the southeast and Panhandle. Spotted Towhees occur, however, throughout most of the state from the mouth of the Niobrara (Bruner and others 1904) southward (Bray 1994) to the Buffalo Co area (NBR 63:110) and westward, while Eastern Towhee breeds primarily in the southeast.

The westward extent of the range of Eastern Towhee in the Republican Valley is unclear. Bruner and others (1904) stated that Eastern Towhee bred west along the Kansas border only to about Franklin Co, and Ludlow (1935) considered it an "occasional summer resident" in Webster Co. "Pure" Eastern Towhees were found (along with hybrids) west only to the Orleans area 23 Jun 1996 (WRS, JGJ). In northeast Colorado it is considered a "rare to uncommon resident" both in the extreme northeast portion of the South Platte Valley and extreme eastern Yuma Co, although most records are for spring and fall (Andrews and Righter 1992). A specimen in the Cornell University collection (CU26757) was collected along the South Platte River 8 km (5 mi) east-southeast of Crook, Colorado; it looks much like an Eastern Towhee, with a few small spots probably not visible in the field (Greg Butcher, pers. comm. WRS). There is, however, only a single Nebraska report from areas adjacent to these Colorado locations, that from Perkins Co in fall (see Fall).

Hybrids occur in the Blair area and northward to the Niobrara River mouth (Short 1961); both species breed at Ponca SP in Dixon Co (NBR 64:124). Sibley and West (1959) placed the clinal midpoint between the species along the Missouri River between Blair and the mouth of the Niobrara River. Birds in most of central Nebraska are intermediates (Short 1961; DeBenedictis 1976); Sibley and West (1959) found no difference in "hybrid index" between birds at Blair and in Lincoln Co. Brown and others (1996) noted that of 61 birds netted at L Ogallala, 32 resembled Easterns and 29 Spotteds, although many showed intermediate characters and were difficult to assign. Thus a large area of central Nebraska north of the Platte Valley contains hybrids and pure Spotted Towhees; hybrids appear to be far less common in the range of Eastern Towhee.

To summarize, current information indicates that Eastern Towhees breed westward to a line from Cedar Co through Platte, Hall, and Harlan Cos (NBR 64:124).

DISTRIBUTION AND ECOLOGY

This species is found in a variety of brushy habitats, but it prefers woodland edge and occasionally occurs in urban situations. Similar habitats are occupied by wintering birds, although denser thickets are utilized for protection. Wintering birds are restricted to the extreme southeast. Summer distribution is discussed under Taxonomy.

Spring

9, 10, 12 Apr→summer

Migrants generally arrive in mid-Apr, but data are few. Ludlow (1935) listed it as a common migrant in Webster Co, occurring in the period 5 Apr–9 May, and a specimen, HMM 2022, was taken at Inland 3 May 1915. Migrants are reported regularly west to Hall Co.

Summer

Eastern Towhee is fairly common but scattered in the extreme southeast and occurs throughout the Missouri Valley north to the Ponca SP area. See Taxonomy for a discussion on western range limits.

Numbers reported on BBS routes are low and limited to an area bounded by Cass, Otoe, and Gage Cos. "Many" were noted in the Gibbon area 28 May 1995, somewhat westerly for pure

Eastern Towhees; these may have been late migrants.

Fall

There are few data on departure dates. Departure is leisurely, with some remaining to winter. The last Eastern Towhee to leave Ponca SP in 1996 was noted 2 Sep. In northwest Missouri only wintering birds remain after Nov (Robbins and Easterla 1992).

Most migrants are reported from the east, westward to the Buffalo-Phelps Cos area, although an Eastern Towhee was in Perkins Co 23 Sep 1972 (NBR 41:41), and one in Knox Co 5 Oct 1997 was only the observer's second there (MB).

Winter

Eastern Towhee is rare in winter and is probably restricted to the extreme southeast when present; it is usually outnumbered there by wintering Spotted Towhees. The specimen cited above provides an unusual midwinter record; also unusual were a northerly report as late as Dec of a single on the Norfolk CBC 17 Dec 1994 and another in Washington Co 2 Jan 1999 (NR).

COMMENTS

Observers can contribute significant information by carefully noting plumage types and vocalizations of towhees summering in their localities. DeBenedictis (1976) stated that frequency of pairs of like towhees and unlike towhees is "the most useful data which birders could provide to ornithologists."

FINDING

Eastern Towhees are not hard to find in brushy woodland edge in the southern Missouri Valley in May and Jun. Indian Cave SP is a good location, and the birds are easily located by their distinctive song.

Canyon Towhee

Pipilo fuscus

STATUS

Hypothetical.

COMMENTS

There is 1 report, published as "Brown" Towhee.

A bird seen 18 Sep 1975 at Gering, Scotts Bluff Co, was with a flock of migrating sparrows in the observer's yard (Brashear 1976). The description given was considered adequate for identification as a Brown Towhee by Bray and others (1986), based on a "brown cap" and streaking in the throat area. The NOURC later deleted the taxon from the official list of the birds of Nebraska, as the description was considered inadequate to identify the bird to either of the newly erected taxa, Canyon Towhee or California Towhee (Grenon 1991). This decision was subsequently affirmed (Brogie 1998).

Canyon Towhee is a resident northeast as far as southeast Colorado, inhabiting pinyon-juniper woodlands and shrublands as well as cholla grasslands but has occurred north as far as Boulder Co, where there may have been a small disjunct breeding population in the 1960s through mid-1980s (Andrews and Righter 1992), about the time of the Nebraska report in 1975. Canyon Towhee also occurs in extreme southwest Kansas, where it is a casual winter visitor, and it may have nested there (Thompson and Ely 1992).

Cassin's Sparrow

Aimophila cassinii

STATUS

Rare but erratic regular breeder and spring and fall migrant west and southwest.

DOCUMENTATION

Specimen

UNSM ZM13090 (egg fragments), Jun 1974 Perkins Co (Bray and others 1986).

DISTRIBUTION AND ECOLOGY

Cassin's Sparrow is a grassland species, but it does not occur in the absence of shrubs (Hubbard 1977), preferring sandsage and rabbitbrush (*Gutierrezia sarothrae*) grasslands in Colorado (Andrews and Righter 1992). Most Nebraska records are from grasslands containing good stands of sandsage, primarily in the southwest and southern Panhandle (Rosche 1994b).

Spring

18, 21 (Labedz 1986), 27 May→summer

Although there are few reports, available data indicate that arrival is in late May, with the earliest report a singing bird near Bushnell 18 May 1995 (SJD).

Summer

There are fewer than 35 reports of Cassin's Sparrow, all in the last 35 years. Difficulty of identification and its rather remote habitat may have precluded its earlier discovery.

The first report, from Adams Co 7 May 1965 (Turner 1965), was far east of the expected range, and minimal details were provided. Subsequently, birds were reportedly seen in the Crescent L NWR area in Garden Co about 1970, but no details were published (NBR 42:56).

The year 1974 saw a remarkable influx of Cassin's Sparrows in the southwest, with 7 reports in Sioux, Morrill, Keith, Lincoln, Hayes, Perkins, and Dundy Cos. The Sejkora family found nesting Cassin's Sparrows on their ranch near Grant 2 Jun 1974, identifying the birds by comparing recordings to known songs of the species (NBR 42:56; AB 28:922). These birds were present until 16 Aug (NBR 43:37), and fragments of an unhatched egg were collected (UNSM ZM13090). The same observers found singing males in nearby Hayes and Lincoln Cos at Hayes Center and Wellfleet (NBR 42:56). Cassin's Sparrows were also found independently by other observers in Morrill Co (NBR 42:56), Dundy Co (NBR 42:56), Keith Co (NBR 43:37), and possibly Sioux Co (NBR 46:58).

Since 1974 it has become apparent that Cassin's Sparrow occurs most years in the southwest, with regular summering north to Box Butte Co (Rosche 1994b), where young were being fed by adults near Kilpatrick L 5 Jul 1991 (AB 45:468, 1134); it occurs occasionally east to Garden Co, where an apparent territorial male was seen near Lisco 21 May 1986 (Labedz 1986), Hitchcock Co (NBR 46:64; AB 43:1335), and Keith Co, where a small colony of 3

nests and 6 displaying males was discovered in Sandhills prairie along an ungrazed abandoned road right-of-way in 1993 (Bock and Scharf 1994). Most reports, however, have come from sandsage habitat in Dundy Co.

Additional documented summer records include specimens collected in sandy grasslands in Deuel Co 12 Jun 1976, although no evidence of breeding there was noted (Faanes and others 1979), and photographs of a bird netted in sandsage grassland in Dundy Co 22 Jun 1989 (Brogie 1989b).

Fall

summer→16, 18 Aug

Departure is in Aug, with few dates available; identification problems of birds away from breeding habitat may contribute to the lack of fall reports.

FINDING

Cassin's Sparrows should be looked for in sandsage prairie in Dundy Co or at Kilpatrick L, locations where the species occurs most years. They are easy to locate if present in Jun by their song and skylarking habits.

American Tree Sparrow

Spizella arborea

STATUS

Abundant regular spring and fall migrant and winter visitor statewide.

DOCUMENTATION

Specimen

UNSM ZM7404, 12 Dec 1895 Fort Robinson, Dawes Co.

TAXONOMY

Two subspecies occur as winter visitors, western *ochracea* and eastern *arborea* (AOU 1957). Bruner and others (1904) stated that the western race winters statewide, with intergrades in the east, where a few pure eastern race birds are found.

DISTRIBUTION AND ECOLOGY

Wintering flocks are found statewide in brushy woodland edge situations, in weedy fields, and in towns and cities wherever cover exists. Numbers are highest in the east and north.

Spring

winter→May 5, 5, 6

Late dates above are undocumented; there are numerous later undocumented reports for May that are questionable, although there are substantiated late dates for Iowa 14 and 18 May, both in 1991 (Kent and Dinsmore 1996). The possibility of misidentification of sparrows should be kept in mind when evaluating published early and late dates for virtually any species. Regarding American Tree Sparrow, Dinsmore and others (1984) stated, "Early and late birds should be carefully documented, as the dates occur at a time when the similar appearing Chipping Sparrow is more likely [than American Tree Sparrow]."

Johnsgard (1980) noted that half the reports are in the period 27 Mar–22 Apr, indicating peak movement.

Fall

22, 23, 23 Sep→winter

Johnsgard (1980) stated that half the fall reports are in the period 12 Oct–2 Nov; these dates probably approximate peak movement. There are 11 earlier undocumented reports for Sep that are questionable. Rosche (1994a) listed an early date 9 Oct in the Keith Co area.

High counts include 2000 in Buffalo Co 19 Nov 1995 and 1000 near Gibbon 28 Nov 1998.

Winter

This species is one of the more common Nebraska winter birds, occurring statewide. A winter site fidelity study by Labedz (1990b) showed 21% recaptured during the winter of banding, 11% the following winter, and 3% in the third winter.

CBC data indicate higher numbers in the north and east, probably a reflection of more extensive suitable habitat. An indication of potential winter numbers were counts of "several thousand daily" in Lancaster Co during winter 1982–83 (AB 37:314); 4290 were in Phelps and Kearney Cos 31 Jan 1999, 3000 were in northern Phelps Co 18 Jan 1998, and 2218 were reported on the Branched Oak–Seward CBC 20 Dec 1994.

COMMENTS

There are Jun and Jul reports from Dawes Co for the years 1985–90, as well as additional summer reports 15 Jul 1979 Lincoln Co, 16 Aug 1989 Sioux Co, 18 Aug 1979 Scotts Bluff Co, and 23 Aug 1989 Scotts Bluff Co. These summer reports are all from the west, but none are documented (NBR 59:9). Rosche (1982) presented no evidence for summering. The breeding range is far from Nebraska.

Bruner and others (1904) cited Aughey's observation in the 1860s that "a few breed here each summer," apparently an error in identification.

FINDING

American Tree Sparrows are very easy to find in winter; flocks can be located in most weedy fields adjacent to brush or woodland edge. Squeaking attracts them, and many perch in the open for easy observation.

Chipping Sparrow

Spizella passerina

STATUS

Abundant regular spring and fall migrant statewide. Common regular breeder north, east, and west, rare casual south-central and southwest. Accidental in winter.

DOCUMENTATION

Specimen

UNSM ZM7426, 27 Apr 1901 Lincoln, Lancaster Co.

TAXONOMY

Apparently 2 subspecies occur in Nebraska, although documentation of their status is lacking. Bent (1968) stated that the eastern race, *passerina*, breeds south from northeast Minnesota to eastern Kansas, while the western race, *arizonae*, breeds from northeast Manitoba south to northern Colorado, central Nebraska, and south (Rising 1995). Thus the western birds in Nebraska, especially those in the Pine Ridge, are probably *arizonae*, while Missouri Valley birds are probably *passerina*, which is the race breeding in eastern Kansas (Johnston 1965). Some authors (Byers and others 1995; AOU 1957) consider *arizonae* part of the northern race *boreophila* (Rising 1995).

DISTRIBUTION AND ECOLOGY

Breeding birds nest most commonly in conifers, utilizing trees planted in urban areas and windbreaks as well as native conifers (Rosche 1982), and are most numerous in the Pine Ridge (Rosche 1982), Niobrara Valley, Loup drainage, and urban areas where coniferous trees have been planted and matured. Nesting in deciduous trees is rare, although a nest was reported in a silver maple (*Acer saccharinum*) in Omaha in 1969 (LCH).

Migrants are found in open areas, including grasslands, although they are most often seen in open areas with scattered trees and shrubs.

Spring

23, 23, 23 Mar→24, 31, 31 May (southwest)

23, 23, 23 Mar→summer (elsewhere)

Chipping Sparrows typically arrive in early Apr, with earlier dates in Webster Co 6 Mar 1930, 11 Mar 1931, 12 Mar 1929, and 13 Mar 1927 (Ludlow 1935). In areas of the south where breeding does not generally occur, spring migrants depart by late May.

High counts include 508 in Sarpy Co 11 May 1996, 305 in Lancaster Co 10 May 1996, and 164 in Pierce Co 8 May 1999.

Summer

Chipping Sparrow has adapted to the arrival of Europeans by nesting in planted trees, usually conifers, around human settlement, both rural and urban (Thompson and Ely 1992). Bent (1968) suggested that the "population of today is many times that of pre-Columbian North America." This change has resulted in some increase in its breeding range in Nebraska from essentially east of the 98th meridian and in the Pine Ridge (presumably 2 subspecies, see Taxonomy) around 1900 (Bruner and others 1904) into most other parts of Nebraska. As late as 1947 it was only a migrant in Lincoln Co (Tout 1947). Short (1961) pointed out that most of this spread had been in towns and that the species was still scarce in the southwest (see below).

Chipping Sparrow is a common breeding bird in the northwest (Rosche 1982) and in Scotts Bluff Co. It is also common in the central Niobrara Valley (Brogie and Mossman 1983), at least west to central Cherry Co (Ducey 1983b), in the Loup drainage (Bray 1994), and in most of the east, especially the Missouri Valley. Breeding evidence is lacking for the Sandhills; it is described as a migrant in southern Holt Co (Blake and Ducey 1991).

There is no evidence for breeding in the Platte Valley west of Grand Island and in the area south of the Platte Valley (Johnsgard 1980). The western edge of the breeding range is at Alma, where adults were feeding begging young 8 Aug 1999 (GH, WH). It is listed only as a migrant in the Keith Co area (Rosche 1994a; Brown and others 1996). There are only these few summer reports for southern Nebraska west of Grand Island: 24 May 1987 Red Willow Co, 31 May 1982 Chase Co, 31 May 1991 Kearney Co, 19–26 Jun 1989 Chase Co, 23 Jun 1991 Buffalo Co, and 5 Aug 1979 Lincoln Co. While Ludlow (1935) listed this species as a "common summer resident," no summer dates or nesting dates were given. Reports for early and mid-Jul and later are probably migrating juveniles (Andrews and Righter 1992).

Nest card data (N = 7) indicate eggs present 23 May–19 Aug.

Fall

summer→10, 13, 13 Nov

20, 20, 22 Aug→10, 13, 13 Nov (south)

Migration begins in the south in late Aug, although several Jul dates may also refer to early migrants (Andrews and Righter 1992). Fall migration ends statewide by early Nov.

Reports after mid-Nov should be documented. There are about 35 reports for the period mid-Nov through mid-Mar, the only one documented an immature at a feeder in Chadron 28 Nov 1993 (NBR 62:4); the latest specimen date is 30 Oct 1909 (UNSM ZM10463). At this time of year descriptions should note gray rump and nape and black lores; American Tree Sparrow has a brown rump and pale lores.

High counts include 330 in Sheridan Co 24 Sep 1994, 100 in Thomas Co 13 Oct 1996, and 70 at Gering 24 Sep 1996.

Winter

There is a single wintering report: one was at an Omaha feeder 29 Dec 1982–31 Jan 1983 (AB 37:314). There are about 20 additional undocumented reports in the period 15 Dec–15 Feb. There is a single record for winter for South Dakota (SDOU 1991) but at least 8 for Missouri (Robbins and Easterla 1992). Typically, wintering Chipping Sparrows are single immatures at feeders.

FINDING

Chipping Sparrow is easily found in spring, especially in the east, and is easy to identify at that time while in breeding plumage. Its song is loud and often heard in urban areas around conifer plantings.

Clay-colored Sparrow

Spizella pallida

STATUS

Common regular spring and fall migrant statewide. Hypothetical breeder. Hypothetical in winter.

DOCUMENTATION

Specimen

UNSM ZM7429, 17 May 1890 Lincoln, Lancaster Co.

DISTRIBUTION AND ECOLOGY

Migrants are most common in grasslands, meadows, and brushy patches, often in disturbed areas that support weedy annuals such as dandelions that are blooming and seeding in abundance, but migrants may be found almost anywhere except deep woods. Occurrence is statewide, although larger numbers are seen in the western half of the state.

Spring

10, 10, 10 Apr→29, 30, 30 May

Migration is from mid-Apr through late May. There are earlier undocumented reports 30

Mar 1943 Jefferson Co, 2 Apr 1939 Jefferson Co, 2 Apr 1946 Jefferson Co, 2 Apr 1969 Lincoln Co, 3 Apr 1961 Nemaha Co, 4 Apr 1967 Lincoln Co, 6 Apr 1934 Jefferson Co, and 8 Apr 1949 Douglas Co. There is a later report 5 Jun Keith Co (Brown and others 1996).

High counts include "hundreds" in Buffalo Co 9 May 1996, "hundreds" in western Nebraska 11–14 May 1999, 155 at Arbor Day Farm 14 May 1997, 100 in Sarpy Co 11 May 1996, and 64 in Scotts Bluff Co 7 May 1994.

Summer

Although the AOU (1983) stated that this species breeds south to "southeastern Wyoming, eastern Colorado, western Kansas (casually), southern Nebraska, northern Iowa," the few reports are not conclusive for Nebraska. Ducey (1988) cited 3 breeding records: a statement by Bruner (1896) that it bred in Cherry Co, a report of breeding near Rushville around 1956 (NBR 24:11), and a report cited by Johnsgard (1980) from Hall Co, where "young" were observed in 1973 (Bennett 1974). None of these reports are satisfactory, however. In addition, Mossman and Brogie (1983) found 2 singing, territorial birds during the 1982 breeding season, both in Brown Co sandhills, and Youngworth (1955) considered this species a rare breeder in the same general area. Bent (1968) indicated breeding south to southern Nebraska, with Red Cloud and Belvidere as localities. There are no data to confirm this. There are Jun and Jul reports for Adams, Cass, Douglas-Sarpy, Howard-Hall, Lincoln, and Perkins Cos, none documented, although 1 bird netted 8 Jul at L Ogallala had a brood patch (Brown and others 1996).

Johnsgard (1980) suggested that Clay-colored Sparrow might breed locally in northern counties as it was regular in South Dakota, but it now breeds regularly only in northeast South Dakota (SDOU 1991). There are, however, summer reports from the northwest (Sioux, Dawes, Scotts Bluff, and Garden Cos) for 8 of the years since 1966, but no details have been published. The number of reports and the locations suggest that breeding may be occurring, however, particularly in Sioux Co grasslands. Dorn and Dorn (1990) indicated that Clay-colored Sparrow has been observed in summer in southeast Wyoming, although breeding has not been confirmed; Scott (1993) stated that there are no breeding records for Wyoming.

Fall

10, 12, 13 Aug→13, 13, 15 Nov

Migration is from mid-Aug through early Nov. Reports in mid to late Jul are probably early migrants: 21 Jul 1975 Lincoln Co, 23 Jul 1978 Lincoln Co, and a specimen, UNSM ZM7436, taken 4 Aug 1914 at Mitchell. There are later reports 24 Nov 1968 Adams Co and a documented report of 3 on the Scottsbluff CBC 18 Dec 1976 (NBR 45:12).

High counts include 100 at Crescent L NWR 30 Aug 1998, 65 in Kearney Co 11 Sep 1999, and 45 in Phelps Co 27 Sep 1995.

Winter

Winter reports are few and undocumented: 28 Dec 1961 Adams Co, 1 Jan 1962 Adams Co, and 1 Jan 1964 Scotts Bluff Co.

There are 2 records in late Dec for Missouri, including 1 in northwest Missouri (Robbins and Easterla 1992).

Other Reports

5 Feb 1968 Adams Co, 20 Feb 1955 Adams Co, 3–10 Mar 1978 Howard-Hall Cos.

FINDING

This species is common and conspicuous in western grasslands during fall migration, when flocks can be seen along fencerows in late Sep and early Oct. Care should be taken in identification, however, as Chipping and Brewer's Sparrows may also be present in these flocks.

Brewer's Sparrow

Spizella breweri

STATUS

Uncommon regular spring and fall migrant west, rare casual central. Fairly common regular breeder western Panhandle.

DOCUMENTATION

Specimen

UNSM ZM7444, 12 Jul 1901 Indian Creek, Sioux Co.

DISTRIBUTION AND ECOLOGY

Brewer's Sparrow breeds in shortgrass prairie with sage or other short shrubs in the western Panhandle, where it is fairly common although somewhat restricted due to its habitat preference. Migrants occur throughout the Panhandle but are rare eastward.

Spring

10, 15, 15 Apr→summer

Migrants occur primarily in the western Panhandle and occasionally eastward to Sheridan (Rosche 1982), Lincoln, and Keith Cos.

Arrival is in mid-Apr. The few reports east of the breeding range, all from Lincoln and Keith Cos, indicate a spring migration period 18 Apr–19 May.

Summer

The breeding range is limited to high-plains grasslands in the Panhandle, usually where Western sagebrush and occasionally greasewood (*Sarcobatus vermiculatus*) are found (Rosche 1982); the specimen cited above was taken in such habitat in extreme northern Sioux Co along Indian Creek 12 Jul 1901.

Johnsgard (1980) indictated that this species might breed south to Kimball Co along the western edge of the Panhandle; Colorado data (Andrews and Righter 1992) suggest that Brewer's Sparrow should indeed be looked for in all counties bordering Colorado, and several reports since 1976 confirm this. Singing males were present in sandy grassland south of Chappell in 1976 (Faanes and others 1979). There is a recent summer report 11 Jul 1992 Cheyenne Co; 2 singing males were found in southwest Kimball Co 16 Jun 1997 in atypical habitat, one in a young hedgerow in (CRP) bromegrass (*Bromus* sp.) and the other in roadside clumps of CRP bromegrass (JS, WRS); and a pair with 3 fledged young was in extreme southwest Kimball Co 12 Aug 1997 (BFH). There is a population in Banner Co just north of Long Canyon (WRS, JGJ) and also near Kilpatrick L (Rosche 1994a). In addition, there are reports 22 Jun 1973, 7 Aug 1992, and 17 Jul 1994 Dawes Co.

There are Jun–early Aug specimens indicating that breeding occurred in Scotts Bluff Co 1913–16 (UNSM ZM7439, ZM7441, ZM7443, ZM7440) and in Kimball Co 1919 (UNSM ZM7442).

There are several additional summer reports in the literature, none documented. Likely correct, however, was a BBS report in central Sioux Co in the period 1967–77 where 22 birds were found. Two birds were reported on a BBS route in Cheyenne and Morrill Cos and a single on a route in western Cherry Co during the same period. Breeding in Cherry Co (Bruner 1901) and Scotts Bluff Co (Ducey 1988) was reported without details. A breeding report for Howard Co (NBR 41:8) is at an extraordinary location and consists simply of "an observed nest"; this record should be disregarded.

Fall

summer→6, 9, 12 Oct

Departure is in late Sep. The only documented reports east of the breeding range are specimens, UNSM ZM7432, taken at Long Pine 20 Aug 1919, and HMM 28373, a female taken at North Platte 3 Oct 1954. There are undocumented reports from McPherson and Scotts Bluff Cos; Rosche (1994a) cited undocumented reports that it was an "uncommon migrant" Jul–Aug 1977 in Keith Co; Tout (1947) cited without details 2 fall reports for Lincoln Co 6 Sep 1937 and 6 Oct 1926; and one was reported without details at the NOU meeting in Blaine Co 9 Oct 1999.

Fall plumages are very similar to those of Clay-colored Sparrow, and identification of birds out of range requires care.

High counts include 45 in southwest Kimball Co 22 Aug 1999, 42 there 29 Aug 1998, and 38 in western Banner Co 6 Sep 1998.

Other Reports

3 May 1977, 4 May 1971, 10 May 1941, 10 May 1975, 21 May 1978, and 24 Sep 1976 Adams Co.

FINDING

Brewer's Sparrow is easy to find in early Jun south of Highway 20 in western Sioux Co along the Wyoming border. Singing males can be observed here perched on the upper branches of sagebrush bushes.

Field Sparrow

Spizella pusilla

STATUS

Common regular spring and fall migrant east, becoming rare west. Common regular breeder east, becoming rare casual west. Rare casual winter visitor southeast.

DOCUMENTATION

Specimen

UNSM ZM7452, Apr 1888 Peru, Nemaha Co.

TAXONOMY

Two subspecies are thought to breed in Nebraska, although their ranges are uncertain. Western *arenacea* apparently breeds in the north and west; 3 specimens at UNSM appear to be of this race, all from north-central Nebraska (JGJ). According to Bruner and others (1904), *arenacea* is the breeding bird in the Niobrara Valley, east to Niobrara, and in the Loup drainage, at least south to Broken Bow.

Although it has been suggested that eastern *pusilla* breeds in the lower Missouri Valley, with northern limits uncertain (Bruner and others 1904), there may be no "pure" *pusilla* breeding in Nebraska; examination of specimens at UNSM indicates that birds breeding in most of eastern and at least south-central Nebraska may be intergrades between *arenacea* and *pusilla* (JGJ).

Most migrants in the Panhandle are likely *arenacea*; one was in Cheyenne Co 8 May 1998 (SJD), and Colorado birds are presumed to be *arenacea* (Andrews and Righter 1992).

DISTRIBUTION AND ECOLOGY

This species breeds in edge habitats and overgrown fields with small to medium-sized shrubs and trees; it is most numerous in the north and east. In Keith Co, at the west edge of the summer range, it occupies "juniper-clad canyons" and floodplain woodlands containing junipers (Brown and others 1996). Migrants occur statewide in a wide range of brushy or weedy habitats and are common in the east, becoming rare westwardly. A few linger late in fall, mainly in the east, but overwintering is unusual.

Spring

20, 21, 28 Apr→29, 29 May, 4 Jun (west)

10, 10, 10 Apr→summer (elsewhere except southeast)

In the southeast migration begins in early Apr, possibly late Mar; about 45 reports from the southeast 24 Feb–7 Apr, if correctly identified, may be wintering birds or early migrants. Away from the southeast, migrants arrive in mid-Apr, although there are earlier undocumented reports 11 Mar 1961 Logan Co, 12 Mar 1934 Logan Co, 30 Mar 1986 Polk Co, 31 Mar 1966 Brown Co, 31 Mar 1986 Boone Co, 1 Apr 1953 Antelope Co, 2 Apr 1969 Lincoln Co, 3 Apr 1960 Keya Paha Co, 4 Apr 1955 Antelope Co, 4 Apr 1956 Antelope Co, 4 Apr 1963 Antelope Co, 5 Apr 1967 Greeley Co, and 6 Apr 1964 Brown Co. Departure is discernible in the west, where summer residents are only casual.

In the Panhandle, where the species is a rare migrant, there are about 15 reports.

High counts include 72 in Pierce Co 10 May 1997, 39 in Dixon Co 11 May 1996, and 22 in Sarpy Co 13 May 1995.

Summer

Field Sparrow is a common breeding bird in the east and north, although it is rare west of central Cherry Co (Ducey 1983b). Brogie and Mossman (1983) found it breeding commonly west to eastern Cherry Co in 1982, and it was common in brushy sandhill areas in Thomas Co in 1957 (Ford 1959) and still is today

(Bray 1994). The Smith L area in Sheridan Co is probably the farthest west summering location in the north; breeding there has not been documented, however (Rosche 1982). There are 2 reports from the Pine Ridge, 30 Jun 1966 Dawes Co and a male in summer 1992 in Sioux Co (AB 46:1152), and one from the southern Panhandle, a territorial bird in limber pines in western Kimball Co 24 Jun 1995 (WRS).

It probably breeds throughout the Loup drainage, although singing males in Grant Co in 1984 were in a "new Sandhills locality" (AB 38:1037).

In the Platte Valley it breeds west to Keith Co, where it is a local summer resident west to the Keystone area (Rosche 1994a). There are no summer records west of Keith Co in the Platte Valley, but it is common in the central Platte Valley (Lingle 1994).

Field Sparrows apparently occupy the entire area south of the Platte Valley; they are common eastward and occur west in some numbers to the Colorado boundary. High counts from this area include 250 in southeast Lincoln Co 5 Jul 1995 and 175 in Frontier Co 5 Jul 1995. A territorial pair was in western Hitchcock Co in summer 1990 (WRS, TB, DR), and there are recent summer reports from Chase Co 25 Jun 1988 and 24 Jun–3 Jul 1989.

A nest in Johnson Co contained 3 eggs 16 Jun 1982.

Fall

15, 20, 20 Aug→18, 22, 22 Nov

Arrival of migrants can be detected in the south, and migration is completed by mid-Nov. Dec and later reports are discussed under Winter.

There are only about 17 fall reports from the Panhandle; these are in the period 21 Aug–2 Oct.

High counts include 100 at Rowe Sanctuary 13 Oct 1996, 30 in Saunders Co 20 Aug 1994, and 30 in Thomas Co 11 Sep 1994.

Winter

Field Sparrows winter north to the southern borders of Nebraska and Iowa in small numbers, most often as single birds or very small groups with American Tree Sparrows. Details of identification should accompany such reports, as wintering Field Sparrows can be confused with immature White-crowned Sparrows or American Tree Sparrows. Since 1967–68 Field Sparrows have been reported on at least 15 CBCs, but counts of more than 1 or 2 birds are highly suspect. There are these additional Dec reports, including westerly locations: 2 were seen 14 Dec 1974 on the Scottsbluff CBC, a single was in Garden Co 18 Dec 1982 (Rosche 1994a), and 1 was at Grand Island 16 Dec 1995. Farthest north were 2 on the Norfolk CBC 21 Dec 1985.

There are about 20 Jan–early Feb reports, very few after 15 Jan: 20 Jan 1990 Polk Co, 27 Jan 1983 Douglas-Sarpy Cos, 29 Jan 1954 Lancaster Co, 5 Feb 1966 Adams Co, 11 Feb 1979 Boone Co, and 12 Feb 1969 Platte Co. There is a documented record 11 Jan 1986 Antelope Co (NBR 54:62), a northerly winter location.

Other Reports

Twenty-three on 16 Dec 1989 DeSoto Bend NWR CBC, 20 on 17 Dec 1977 Hastings CBC, 11 on 21 Dec 1985 Grand Island CBC, 5 on 19 Dec 1981 Grand Island CBC.

FINDING

Field Sparrow can be found in the Missouri Valley in May in grassy areas with small trees near woodland, such as at Indian Cave SP. Similar habitat in the Niobrara Valley also should be checked. The distinctive song helps locate this species.

Vesper Sparrow

Pooecetes gramineus

STATUS

Common regular spring and fall migrant statewide. Common regular breeder north and northwest, uncommon east, rare elsewhere. Accidental in winter.

DOCUMENTATION

Specimen

UNSM ZM7340, 12 Apr 1890 Lincoln, Lancaster Co.

TAXONOMY

There are 2 subspecies. The eastern subspecies *gramineus* breeds along the eastern edge of Nebraska; it was rare in the 1950s (Rapp and others 1958) but may be increasing due to its adaptation to nesting in fields prepared for planting of corn and soybeans. Its range in Nebraska may correspond to that of nonirrigated corn. Elsewhere, especially in the north and west, the breeding birds are probably western *confinis*.

DISTRIBUTION AND ECOLOGY

This is generally a breeding bird of grasslands, usually either disturbed Sandhills prairie with small shrubs and brush (Rosche 1982) or in the east near fencerows in fields planted to crops (Dinsmore and others 1984). Breeding birds are most common in the north and northwest, uncommon in the east, and rare to absent elsewhere.

Migrants are distributed fairly evenly throughout the state in grasslands or open cultivated areas.

Spring

4, 5, 9 Mar→5, 6 (Rosche 1994a), 9 May

Migration takes place essentially in Apr, with extreme dates in early Mar and earlier dates 22 Feb 1961 Lincoln Co and 24 Feb 1951 Gage Co; this species may linger late in fall and reappear very early in spring. It is difficult to detect spring departure in the east and north due to the presence of summering birds; late dates are determined away from breeding areas.

High counts include 40 in southeast Gosper Co 25 Apr 1999, 30 in Dixon Co 11 Apr 1995, and 21 in Sarpy Co 10 Apr 1997. After a snowstorm in western Nebraska 26 Apr 1994, a fallout of "thousands" was noted (RCR; AB 48:315).

Summer

Breeding Vesper Sparrows are most numerous in the northwest and the northern Missouri Valley; these areas are probably occupied by different subspecies (see Taxonomy).

Highest numbers on BBS routes are in western Cherry and eastern Sheridan Cos, and good numbers breed in extreme northern Sioux Co (Rosche 1982). Density declines somewhat east of these areas; Brogie and Mossman (1983) considered the species rare in summer, although a probable nester, in the Niobrara Valley Preserve.

Elsewhere in the north and west scattered breeding probably occurs as indicated by summer reports, but numbers are low. A count of 28 in Morrill Co 12 May 1995 would have been rather late as migrants and may have been summering birds. In Thomas Co at Bessey Division, NNF, Bray (1994) found it to be an "uncommon summer resident." Vesper Sparrow does not breed on the northeast Colorado plains (Andrews and Righter 1992), although it apparently does so in most of Wyoming (Scott 1993).

Vesper Sparrow also breeds uncommonly in the east in the counties bordering the Missouri River. It is most numerous northward but unreported from Nemaha and Richardson Cos, although it breeds in adjacent northwest Missouri (Mark Robbins, pers. comm.). The Missouri Valley birds are probably of the eastern race *gramineus*, which has apparently adapted to nesting in rowcrop fields (Dinsmore and others 1984). A nest in Dixon Co with eggs 5–14 Jun 1992 was on the ground in a cultivated corn/soybean field (Cornell Nest Record data).

Vesper Sparrow appears to be absent or very rare in summer in most of the Loup drainage and south of the Platte Valley; Ludlow (1935) noted that a pair nested in Webster Co 6 Jun 1926. There are few other summer reports from this area: 3 Jun 1989 Dundy Co, 12 Jun 1973 Howard Co, 30 Jun 1977 Lincoln Co, 30 Jun 1978 Lincoln Co, 30 Jun 1993 Thomas Co, and 16 Jul 1984 McPherson Co.

Fall

30, 30, 31 Aug→29 Nov (specimen, UNSM ZM10748), 3, 4 Dec (specimens, cited below)

Migration is mainly in Sep and Oct, with an early date 9 Aug Keith Co, a location where breeding does not occur (Rosche

1994a); reports 15 Aug 1987 Lincoln Co and 15 Aug 1990 Lincoln Co are probably migrants also.

A northerly late report is 24 Nov 1966 Custer Co, and there are several CBC reports, only one documented: 15 Dec 1984 (6) Scotts Bluff, 18 Dec 1982 (4) Scotts Bluff Co, 2 Jan 1971 (4) Kearney, and 3 Jan 1988 DeSoto Bend NWR (in Washington Co). The latest report is 7 Jan 1996 Dakota Co. Documented late records include the taking of 3 specimens (UNSM ZM10744, ZM10750, ZM10751) 4 Dec 1909 in Lancaster Co, and one was in Antelope Co 4 Jan 1998 DH. There is no evidence that this species has overwintered, however.

High counts include 542 in southwest Kimball Co 24 Sep 1998, 462 there 26 Sep 1999, and 72 at Crescent L NWR 26 Sep 1996.

FINDING

Vesper Sparrow is not hard to find during spring migration, especially in Apr, almost anywhere in grasslands with fencerows or in agricultural areas. It is also common in similar habitat in the west Sep–Oct.

Lark Sparrow

Chondestes grammacus

STATUS

Common regular spring and fall migrant statewide. Common regular breeder north and west, fairly common south and east.

DOCUMENTATION

Specimen

UNSM ZM7348, 16 Jun 1901 Badlands, Sioux Co.

TAXONOMY

Two subspecies occur. According to Bruner and others (1904), *grammacus* breeds in eastern Nebraska west to the 98th meridian (just east of Grand Island), while birds breeding elsewhere in the state are *strigatus*. Swenk (Notes before 1925) noted that specimens taken in the Kearney area were mostly *strigatus*. Thus *grammacus* appears to be the breeding race of the original tallgrass prairie area, while the rest of the state is occupied by *strigatus*.

DISTRIBUTION AND ECOLOGY

Breeding habitat consists essentially of
grasslands and open areas, often with small
shrubs and usually with patches of bare
ground, although breeding birds also occur in
open coniferous woodland (Rosche 1982; Ford
1959). Lark Sparrows are most common in
the Sandhills, particularly in pastures where
low shrubs such as yucca (*Yucca* sp.) are
found, but in pastures where yucca has been
selectively removed through grazing, Lark
Sparrow may be absent. It is least numerous
in south-central Nebraska and only locally
distributed in the east, occupying pastures
with widely scattered small trees and bushes.

Migrants occur almost anywhere but are usually
encountered in open areas, either grassland or
agricultural.

Spring

25, 25, 26 Mar→summer

Migrants appear in early Apr, although earliest
dates are in late Mar.

High counts include 78 in Pierce Co 8 May 1999
and 42 in Morrill Co 14 May 1995.

Summer

While Lark Sparrow breeds throughout the state,
it is by far most numerous in the north and
west, especially the Sandhills and shortgrass
prairie areas. Brogie and Mossman (1983)
considered it abundant in the Niobrara Valley
Preserve in 1982. Lingle (1994) considered it
common in the central Platte Valley, although
published breeding season reports are few:
27 Jun 1988 Hall Co, 30 Jun 1988 Adams Co,
30 Jun 1989 Hall Co, and 23 Jun 1991 Buffalo
Co. There are several summer reports in the
southwest, especially from sandsage areas,
and it is probably fairly common in the
Republican Valley east to about Webster Co
(Ludlow 1935).

High counts include 900 in Chase Co around 1
May 1999 and 118 at Crescent L NWR 26 Jul
1995.

Fall

summer→19, 22, 22 Oct

Summering birds departed James Ranch, Sioux

Co, 21 Aug 1995. Migration ends by mid-Oct,
with later dates 7 Nov 1994 Otoe Co, 13 Nov
1968 Douglas-Sarpy Cos, and 22 Nov 1980
Douglas-Sarpy Cos.

High counts include 89 in southwest Kimball Co
2 Sep 1999.

COMMENTS

There are few winter reports, none documented:
20 Dec 1986 Scotts Bluff Co CBC, 8 Feb 1953
Scotts Bluff Co, 17 Feb 1965 Douglas-Sarpy
Co, 29 Feb 1948 Dawes Co, and 8 Mar 1952
Holt Co.

There is a single record in late Dec for Missouri
(Robbins and Easterla 1992) and a few winter
records from southern and eastern Kansas
(Thompson and Ely 1992).

FINDING

Driving through the Sandhills in Jun will
yield this species, as it is very common
along roadsides. Its song is not particularly
noticeable, but the bird is easily identified
as it flushes and shows white corners on the
rounded tail. It is especially numerous at
Bessey Division, NNF, near Halsey, and can be
found along most of the secondary roads of
the area.

Black-throated Sparrow

Amphispiza bilineata

STATUS

Rare casual spring migrant west and east. Rare
casual winter visitor east.

DOCUMENTATION

Photograph

17 Dec 1973 Omaha, Douglas Co (Meier 1974; AB
28:659).

TAXONOMY

The bird that appeared at Omaha in winter
1997–98 was attributed to the southern Great
Plains subspecies *bilineata* based on the
extensive white spots on the outer rectrices
(Green 1997; Rising 1995).

Spring

There are 3 records, 2 documented:

31 Mar–9 Apr 1993 Lancaster Co (Gubanyi 1996a;
NBR 61:131)

26 Jun 1972 Toadstool Park, Sioux Co (RCR, DJR; Rosche 1972)

The third report is of one near Keystone in Keith Co 26 May 1984 (Rosche and Johnsgard 1984). The Lancaster Co bird may have wintered in the vicinity, as spring vagrants do not usually appear as early as 31 Mar (see Comments).

Winter

There are 3 records, all documented:

4 Dec 1973–6 Feb 1974 Omaha (cited above)

early Dec 1997–12 Apr 1998 Bellevue (Green 1997; Marilyn Jensen, pers. comm. WRS; Brogie 1998)

2 Jan–early Mar 1993 southwest Dixon Co (JJ; Gubanyi 1996a; NBR 61:92, 95; AB 42:273)

COMMENTS

Black-throated Sparrow breeds in cholla grasslands and shortgrass semidesert in southeastern Colorado and also occurs north to Wyoming in the eastern foothills, primarily as a spring migrant (Andrews and Righter 1992). In Wyoming it is a rare spring migrant and summer resident (Dorn and Dorn 1990; Scott 1993). Two Nebraska records, 26 May and 26 Jun, fit this pattern.

Elsewhere in the region this species has also occurred as a fall vagrant and winter visitor north and east of the normal range. Thompson and Ely (1992) suggested that it probably occurs in Kansas during late fall, citing a record in Finney Co 25 Nov 1952 and a sight record in Sedgewick Co 27 Oct 1979. There is 1 winter record for Missouri, 15–20 Jan 1993 in the northwest (FN 47:265); a single South Dakota record, at Vermillion the last 2 weeks of Dec 1971 (SDOU 1991); and 1 for Iowa, at Waterloo 16 Mar–9 Apr 1993 (Kent and Dinsmore 1996).

Hunn (1978) suggested that drought conditions in spring within the normal breeding range triggered vagrancy in this species.

FINDING

While the easternmost records for this species are of wintering birds, this dispersal pattern is unpredictable, and it might be worthwhile to search arid areas such as the sandsage prairie in the southwest and the western Panhandle for spring migrants, preferably during the first half of May in years when it is dry to the south and west of Nebraska.

Sage Sparrow

Amphispiza belli

STATUS

Accidental in fall.

DOCUMENTATION

Description

6 Aug 1989 Sowbelly Canyon, Sioux Co (Stage and Stage 1990).

Records

The single record was from about 1.6 km (1 mi) below Coffee Park in lower Sowbelly Canyon, Sioux Co, where a nonsinging bird was seen 6 Aug 1989 (cited above). This adult bird flew from an area of sagebrush beside the road and was observed at about 15 m (49 ft) distance on top of brush 3–4 m (10–13 ft) high. The NOURC accepted this record (Grenon 1990), although the possibility of an immature Black-throated Sparrow has been suggested (AB 44:117).

COMMENTS

Sage Sparrow occurs as a migrant, mostly in Apr, along the foothills on the eastern Colorado plains (Andrews and Righter 1992), but the only northeastern Colorado record close to Nebraska is in fall. Fall vagrants may occur farther east than spring vagrants. This is supported by records in southwest Kansas, where, although the species' status is unclear, it appears to be a fall vagrant or rare winter visitor (Thompson and Ely 1992). There are specimen records for southwest Kansas in Morton Co 1 Nov 1956 and Seward Co 11 Jan 1957. Thompson and Ely (1992) discounted sight records for May and Jul in Morton and Comanche Cos as possible molting Lark Buntings but noted recent sight records for Morton Co in the period 31 Dec–8 Jan.

Lark Bunting
Calamospiza melanocorys

STATUS

Abundant regular spring and fall migrant west, becoming rare casual east. Abundant regular breeder west, uncommon north. Hypothetical in winter.

DOCUMENTATION

Specimen

UNSM ZM12193, 30 May 1895 Lincoln, Lancaster Co.

DISTRIBUTION AND ECOLOGY

Lark Bunting is characteristic of shortgrass and midgrass native prairie, especially where small shrubs are present to provide nesting sites (Rosche 1982). River valleys and agricultural areas within native grasslands are generally avoided (Andrews and Righter 1992). Breeding occurs also in agricultural areas where weeds and brushy shrubs occur but much less commonly and mainly at the periphery of the breeding range.

Migrants occur anywhere in open grassland or agricultural areas. Population irruptions occur occasionally when small wandering flocks and territorial birds appear in somewhat appropriate habitat well east of the normal range. Examples were noted in Lancaster Co in 1965 and 1966 and in the early 1980s in Douglas and Sarpy Cos (see Summer).

Spring

4, 6, 7 Apr→summer

Migrants appear in late Apr. There are earlier undocumented reports 4 Mar 1948 Dawson Co, 17 Mar 1968 Logan Co, 20 Mar 1990 Dawes Co, 23 Mar 1957 Scotts Bluff Co, and 31 Mar 1966 Logan Co.

There are a few reports since 1981 from areas where spring migrants are probably at least casual, such as Cedar, Knox, Platte, and Polk Cos, but also these from farther to the southeast: 23 Apr 1988 Lancaster Co, 3 May 1987 Douglas-Sarpy Cos, 8 May 1992 Cass Co, 11 May 1997 Kearney Co, and 16 May 1991 Lancaster Co (AB 45:468).

High counts include 300 in Sioux Co 3 May 1995 and 250 in the western Panhandle 15 May 1998.

Summer

Greatest numbers are reached in the west and north, although breeding populations tend to be erratic and locally distributed. Even prior to the conversion of tallgrass prairie to agriculture, Lark Bunting was rare in the east (Bruner and others 1904), as is the case today. The current eastern limit of the breeding range corresponds approximately with the eastern edge of the midgrass prairie region, although it is irregular in these areas. There is no definite evidence of nesting at the Bessey Division, NNF, Thomas Co (Bray 1994). Brogie and Mossman (1983) found Lark Bunting to be only occasional in northeastern Cherry Co in 1982, with 1 breeding record. Since 1981 summer or breeding reports exist east to Holt, Boone, Howard, Hall, and Phelps Cos. A BBS route along the Rock-Brown Co line yielded "several" 16 Jun 1995.

East of these counties, breeding habitat is limited, but there are scattered summer reports since 1981: 4–20 Jun 1989 Douglas-Sarpy Cos, 8 Jun–24 Jul 1985 Polk Co, 10 Jun 1982 Sarpy Co (AB 36:994), 16 Jun 1989 Sarpy Co (AB 43:1335), 22 Jun 1989 Cedar Co, 30 Jun–4 Aug 1988 Lancaster Co, 30 Jun 1988 Adams Co, and 8 Jul 1995 Franklin Co.

High counts include 400 in Kimball Co 15 Jun 1997, 350 in Sioux Co 14 Jun 1997, and 220 in Box Butte Co 16 Jun 1995.

Fall

summer→10, 11, 14 Oct

Migration ends by mid-Sep, with later dates 2 Nov 1989 Dawes Co and 16 Nov 1987 Dawes Co.

Reports from the east are few: 7 Sep 1985 Pierce Co and 11 Oct 1975 Douglas-Sarpy Cos.

High counts include 600 in Box Butte Co 12 Aug 1994 and 200 at Crescent L NWR 30 Aug 1998.

Winter

There is an undocumented report 1 Jan 1949 Dawson Co.

It occurs irregularly in winter on the northeast Colorado plains (Andrews and Righter 1992). "A few flocks" remain into late Dec in Kansas (Thompson and Ely 1992), and a single basic- or immature-plumaged bird was in Davies Co, Missouri, 19 Jan 1997 (Jacobs 1997).

FINDING

Lark Buntings are easy to find in the west in May and Jun. Males are distinctive due to their plumage but also because of their skylarking flights on breeding territories. A drive through any native grasslands in the Panhandle should yield this species.

Savannah Sparrow

Passerculus sandwichensis

STATUS

Common regular spring and fall migrant statewide. Locally uncommon regular breeder west. Hypothetical in winter.

DOCUMENTATION

Specimen

UNSM ZM7285, 10 Oct 1885 Lincoln, Lancaster Co.

TAXONOMY

According to the AOU (1957), the northwestern United States subspecies *nevadensis* breeds in western Nebraska and apparently migrates through at least western and central Nebraska. Swenk (Notes before 1925) stated that most migrants in the Inland and Kearney area were *nevadensis*, but a 14 Apr 1914 specimen was of an "eastern" subspecies, probably *oblitus*, as currently recognized (Byers and others 1995). The subspecies *oblitus* breeds commonly in northern Iowa but as yet has not been shown to do so in northeastern Nebraska. It is probably a regular migrant through eastern Nebraska.

Rapp and others (1958) listed the Alaskan and western Canadian race *anthinus* for Nebraska, presumably occurring as a migrant in the west, without providing further information. There is no evidence that the northeastern

subspecies *savanna* has occurred in Nebraska, but it has been recorded rarely in winter in Kansas (AOU 1957; Johnston 1965).

DISTRIBUTION AND ECOLOGY

Migrants are found statewide and frequent open grasslands and agricultural areas and occasionally woodland edge. Summering habitat consists of damp swales in grasslands (Rosche 1982). Breeding is local and probably limited to the west.

Spring

10, 11, 14 Mar→24, 24, 25 May

Migration is from late Mar through late Apr. The earliest specimen date is 22 Mar 1902 Lancaster Co (UNSM ZM7291). Late May dates are indicative of potential summering birds.

High counts include 160 in Pierce Co 8 May 1999, 40 at Rowe Sanctuary 26 Apr 1997, and 39 in Dixon Co 11 May 1996.

Summer

The breeding range of this species barely reaches the northern and western edges of Nebraska (Johnsgard 1979). Breeding is documented only from a few locations in the Panhandle, although there are scattered summer reports from elsewhere in the state.

Rosche (1994a) considered a colony at Kilpatrick L to be the "largest colony known in Nebraska." Singing birds were at this location as early as 1990 (AB 44:1154). An adult was flushed from a nest with 4 eggs in southern Sioux Co 15 Jul 1994 (NBR 62:114, 146), and there is a breeding record for Mitchell (AOU 1957). Rosche (1982) presumed that breeding occurred in the sandhills of Sheridan Co but noted that there was no definite evidence. Singing birds were found most years since 1973 about 4.8 km (3 mi) north of Walgren L (RCR) and in 4 years out of 10, including 26 Jun 1981, just east of Lakeside (RCR).

Other Panhandle locations for which there are summer reports include Garden and Dawes Cos; in addition, a specimen, UNSM ZM7288, was collected at Mitchell 7 Jul 1916, and 3 individuals were singing in wet meadows at Facus Springs in 1987 (AB 41:1457) through

1990 (RCR). Farther east, in the Sandhills, there are summer reports from McPherson, Thomas, and Greeley Cos, and 3 birds were singing in southern Holt Co 25 May 1991 (LEB).

In the north there are old late May reports for Logan and Brown Cos and recent summer reports for Cedar Co, although in 1982 Brogie and Mossman (1983) found no evidence of breeding in the Niobrara Valley Preserve. There are several late spring and summer reports from Lincoln and Keith (Rosche 1994a) Cos, and a few from the east where breeding seems unlikely: 29 May 1992 Lancaster Co, 31 May 1975 Douglas-Sarpy Cos, 30 Jun 1965 Gage Co, 14 Jul 1985 Polk Co, 28 Jul 1974 Lancaster Co, 3 Aug 1987 Lancaster Co, and "summer" 1993 Sarpy Co.

Fall

18, 20, 21 Aug→20, 21, 22 Nov

Migration is from late Aug through early Nov, with earlier dates 9 Aug Keith-Garden Cos (Rosche 1994a) and 10 Aug 1994 Garden Co. The latest specimen date is 2 Nov 1907 Lancaster Co (UNSM ZM10444).

High counts include 500 in Hamilton Co 5 Oct 1996, 370 at Crescent L NWR 26 Sep 1996, 245 in Phelps and Kearney Cos 14 Oct 1995, and 175 at Crescent L NWR 25 Sep 1994.

Winter

Savannah Sparrow winters casually in Kansas (Thompson and Ely 1992) and northern Missouri (Robbins and Easterla 1992), but there is no documented evidence for wintering in Nebraska. There are reports from CBCS 17 Dec 1989 Lincoln, 18 Dec 1998 Harlan Co Res (SJD), 22 Dec 1977 (6) DeSoto Bend NWR, and 29 Dec 1964 (2) Scottsbluff and these undocumented Jan reports: 14 Jan 1952 Thayer Co, 20 Jan 1999 Otoe Co, 22 Jan 1955 Gage Co, and 25 Jan 1953 Douglas-Sarpy Cos.

FINDING

Savannah Sparrows are common fall migrants in relatively short or disturbed grasslands and can be easily flushed by walking in such habitat virtually anywhere during Oct. Identification often needs care, as immatures of other species provide opportunity for confusion. Breeding birds should be looked for in damp swales in the western Sandhills from Crescent L NWR northward and westward.

Grasshopper Sparrow

Ammodramus savannarum

STATUS

Fairly common regular spring and fall migrant statewide. Common regular breeder west, fairly common elsewhere.

DOCUMENTATION

Specimen

UNSM ZM7306, 12 May 1891 Lincoln, Lancaster Co.

TAXONOMY

Most belong to the western subspecies *perpallidus* (AOU 1957), although intergrades with eastern *pratensis* occur along the eastern edge of the state (Bruner and others 1904). Bruner and others (1904) considered the occurrence of pure *pratensis* questionable.

DISTRIBUTION AND ECOLOGY

Although Johnsgard (1980) suggests that numbers are limited in tallgrass prairie, it is the most abundant prairie passerine in many eastern Nebraska prairies, including Allwine Prairie in Douglas Co, a restored tallgrass habitat (RSS). Densities as high as 2.6 per ha (1.05 per ac) have been recorded in similar habitat in Wisconsin (Wiens 1969) and also in lush Sandhills meadow at Valentine NWR (Cole 1976), but Anderson (1992) found densities of only 0.2–0.4 per ha (0.08–1.62 per ac) in dry upland Sandhills prairie in Nebraska. In areas where agriculture is extensive, breeding occurs in alfalfa and hay fields (Lingle and Hay 1982). It is most numerous in the north and west, where native grasslands are extensive; it may be sensitive to habitat fragmentation (Dobkin 1994). Migrants occur throughout the state, usually in grasslands and pastures.

Spring

31, 31 Mar, 1 Ap →summer (east, central)

27 Apr, 1, 2 May→summer (west)

Ludlow (1935) noted that "this species arrives at Red Cloud earlier than in most Nebraska localities," but the Webster Co reports are undocumented except for the earliest Nebraska specimen, 31 Mar 1925 (UNSM ZM7305), which was taken at Red Cloud.

Arrival in the Panhandle is much later, in early May, with an earlier undocumented report 10 Apr 1950 Dawes Co.

Summer

This species breeds throughout the state, but it is most numerous in native shortgrass and mixed or Sandhills prairie and least numerous in areas of the east and south where cropping has severely reduced habitat. In these areas, however, Grasshopper Sparrow has adapted to alfalfa fields, where it sometimes breeds abundantly (Lingle and Hay 1982).

It is still common in parts of Nebraska where suitable habitat exists. Brogie and Mossman (1983) considered it to be abundant in the central Niobrara Valley in 1982, and it is a "common summer resident" at Bessey Division, NNF (Bray 1994). At Crescent L NWR 185 were counted 13 Jul 1994, 60% of which were young of the year (NBR 62:115). A high count of "at least" 500 was made in the Calamus Res area 2 Jul 1995 (NBR 63:81).

This species may attempt late nestings; nests with eggs were found 11 Aug 1989 in Hall Co (AB 44:116) and 1 Aug 1973 in Sheridan Co (AB 28:76), and an adult with 2 juveniles was seen 10 Sep in Garden Co (NBR 62:146).

Fall

summer→15, 16, 16 Oct

Migration ends in mid-Oct, although there are later dates 25 Oct 1962 Webster Co, 26 Oct 1964 Douglas-Sarpy Cos, 3 Nov 1996 Saunders Co, and 6 Nov 1965 Adams Co.

High counts include 200 in northern Sioux Co 30 Aug 1997 and 44 at Crescent L NWR 27 Sep 1995.

Other Reports

3 Mar 1991 Douglas-Sarpy Cos, 4 Mar 1967 Custer Co, a series of Mar reports with the earliest 8 Mar in Webster Co (Ludlow 1935), 19 Mar 1966 Adams Co, 24 Mar 1935 Adams Co.

FINDING

Grasshopper Sparrows can usually be located by song or by driving roads in grasslands such as in the Sandhills and watching for the birds on fence lines. Best time is early Jun when males are actively singing.

Baird's Sparrow

Ammodramus bairdii

STATUS

Rare casual spring and fall migrant west and central, accidental east.

DOCUMENTATION

Specimen

UNSM ZM7299, 17 Aug 1936 Stapleton, Logan Co.

DISTRIBUTION AND ECOLOGY

Preferred habitat during migration is probably native prairie, especially dense shortgrass prairie (Pulich 1988) or sparse grassland with clumps of grass in well-drained areas like hilltops (Mark Robbins, pers. comm.). These habitats are most common from Sheridan and Garden Cos westward. This species is probably regular in occurrence in such habitat but is difficult to find.

Spring

Published data show spring reports in the period 24 Apr–23 May. The very few documented records are in the short period 24 Apr–16 May. These reports suggest that migrants may appear statewide and not always in expected habitat.

There are 7 documented records. One Baird's Sparrow was photographed by 2 observers at different times and remained in atypical habitat, a nonnative grassy roadside verge near a prairie wetland basin, in York Co 26 Apr–10 May 1998 (JGJ, JS). One was photographed on the roadside at Kissinger Basin 24 Apr 1999 (JGJ). A singing male was seen 4 May and 6 May 1990 in Cass Co

(Grenon 1991; AB 44:456). A specimen "very critically studied" by Swenk was collected near Kearney 6 May 1914 and mounted for the Olson collection (Swenk, Notes before 1925). A single bird was at Clear Creek Marshes 9 May 1998 (SJD). A singing male was in Banner Co 11 May 1996 (JGJ; Brogie 1997; NBR 64:65). A specimen taken at Overton 16 May 1911 was #2835 in the Brooking collection (Swenk, Notes after 1925); Swenk had listed this specimen earlier as taken 16 May 1901, probably in error (Swenk, Notes before 1925). There is a mounted specimen, HMM 14237, collected at Juniata, but it is undated. Brooking (Notes) listed one in his collection taken by Black at Kearney May 1917, without details; this specimen was apparently not recorded by Swenk.

Fall

Published reports are mainly in the period 25 Sep–25 Oct, although there are scattered dates outside this period, including the specimen cited above, 17 Aug 1936 Logan Co (Glandon and Glandon 1937), and a recent report 9 Sep 1994 Custer Co (NBR 62:123). The only documented records are the Logan Co record, 2 birds at Clear Creek Marshes 26 Sep 1999 (SJD), and 1 banded at Crescent L NWR 15 Oct 1980, for a first fall refuge record (FZ; NBR 49:19; AB 35:200).

COMMENTS

A statement that "the Rosches may have run down the only known location for breeding Baird's Sparrows in the Region, with 3 territorial males in Sioux [Co], NE, June 27" (FN 50:965), is unsupported by any evidence.

Specimen data from Colorado (Andrews and Righter 1992) and Kansas (Thompson and Ely 1992) show spring dates 25 Apr–14 May and fall dates 22 Aug–14 Oct. Zimmer (1985) stated that Baird's Sparrow is a rather late spring migrant, arriving in late May on the breeding grounds and still migrating through southern Arizona in mid-Apr. In extreme eastern Wyoming Baird's Sparrow occurs in May and early Jun (Scott 1993).

South Dakota (SDOU 1991) and Wyoming (Scott 1993) data indicate that fall migrants occur during Sep and Oct, with late dates for South Dakota 15, 20, and 25 Oct. The Nebraska specimen cited above was taken 17 Aug, a rather early date.

Spring migration probably peaks during the first week of May, and fall migration probably takes place from mid-Aug to mid-Oct.

Other Reports

Three reports were not accepted by the NOURC: 25 Mar 1988 Lancaster Co (Grenon 1990), 12 Apr 1997 Sarpy Co (NBR 65:95; Brogie 1998), 27 Sep 1996 Scotts Bluff Co (Brogie 1997).

FINDING

The best time and location to look for this species is probably around 5 May and 15 Sep in ungrazed shortgrass prairie or western Sandhills prairie.

Henslow's Sparrow

Ammodramus henslowii

STATUS

Rare casual spring and fall migrant east and central. Rare regular summer visitor southeast. Hypothetical breeder.

DOCUMENTATION

Specimen

UNSM ZM7332, 22 Apr 1899 Lincoln, Lancaster Co.

DISTRIBUTION AND ECOLOGY

Henslow's Sparrow occupies moist grasslands throughout most of its range, but in Kansas the largest breeding population is in dry upland prairie with considerable thatch build-up in Riley Co (Thompson and Ely 1992; Zimmerman 1988). Rising (1995) noted that burning renders grassland unsuitable for this species for 1 or 2 years, due to a preference for standing dead vegetation and matted dead grass, but ultimately enhances suitability by reducing woody vegetation. Similarly, the largest breeding concentration known for Missouri was some 200 pairs on an unburned part of Taberville Prairie where dewberry was common (Easterla 1967).

Nebraska summer records are generally in dry, unburned prairie also.

Spring

22, 22 (specimen, UNSM ZM7332), 25 Apr→summer

Migration essentially occurs in May, with early dates 6–15 Apr 1960 Webster Co and 11 Apr 1997 Washington Co. The westernmost reports are 22 Apr 1955 Keith Co, 4 May 1947 Logan Co, 5 May 1967 Lincoln Co, and 29 May 1963 Lincoln Co.

Summer

There are about 75 reports of this species, but few are documented, and there are no documented breeding records.

Regarding documented records, in addition to the specimen cited above, there are additional specimens, UNSM ZM7333 and ZM7334, taken at Lincoln 6 Apr 1919 and 18 May 1920, respectively. A singing bird was at Nine-Mile Prairie, Lancaster Co, 8 Jul 1951 (Baumgarten 1953); it was assumed to be nesting, although no others were seen and no evidence for nesting was observed. A single singing bird was at Burchard L, 6 May 1985 (Wright 1985; Brogie 1997; AB 39:321); this bird was photographed (NBR 53:82) and later seen by others through 31 May (NBR 54:50). It was singing from small, scattered shrubs in an upland prairie gully. At this site in 1996 as many as 6 were singing on 2 Jun and courtship behavior was observed (WRS), and at a similar site elsewhere at Burchard L 2 singing males were noted 1–4 Jun 1996 (Brogie 1997). In 1995 up to 6 males were singing in spring-burned prairie at Burchard L through 26 Jun; more forbs were present than on the 1985 site (FN 49:948). The presence of singing birds near Bennet 25 May 1994 was documented (AB 48:315, NBR 62:85), and the same year singing birds near Denton 11 Jun were seen by many observers (NBR 62:115). A singing male was recorded at Meadowlark L, northeast Seward Co, 14 Jul 1997 (JG; Brogie 1998); at least 2 singing

birds were there 30–31 May 1998 (JG, WRS, JS); and another was there 10 May 1999 (JG). Two were found just north of Spring Creek Prairie, Lancaster Co, 1 and 16 May 1999, and one was singing at Nine-Mile Prairie, Lincoln, 2 Jun 1999 (JS). Farthest west summer record was of one singing at Harvard Marsh 25 Jul 1999 (JGJ).

Trostler was said to have collected a set of eggs and a female in Douglas Co prior to 1900 (Bruner and others 1904), and there is a report without details from Keya Paha Co 23 Jun 1962. The latter is of interest in that there are 4 summer reports for southeast South Dakota, including 2 for 1965 (SDOU 1991). Johnsgard (1979, 1980) indicated that breeding may have occurred in Washington Co, but no details were provided.

Fall

summer→18, 20, 23 Oct

Migration occurs in Sep and Oct. There are about 25 reports, the westernmost 6 from Adams and Hall Cos 25 Aug–15 Oct.

FINDING

Singing males should be looked for in May and Jun at Burchard L. Preferred habitat is unburned prairie with dead stalks from the previous year's growth. Other similar prairie areas in the east should be checked also.

Le Conte's Sparrow

Ammodramus leconteii

STATUS

Fairly common regular spring and fall migrant east, rare central, accidental west. Hypothetical in summer.

DOCUMENTATION

Specimen

UNSM ZM7313, 20 Apr 1901 Havelock, Lancaster Co.

DISTRIBUTION AND ECOLOGY

Migrating LeConte's Sparrows occur in dense grasslands, often associated with damp areas but also drier upland habitat. Most reports are from the east, where it can be common at times, especially in fall. Away from the east, there are fewer than 40 spring reports and about 20 fall reports.

Spring

30, 31 Mar (Swenk, Notes before 1925), 2 Apr (specimen, UNSM ZM12194)→20, 21, 22 May

Migration is from mid-Apr through mid-May, although there is a very early record, a specimen, UNSM ZM10448, taken in Lancaster Co 17 Mar 1909.

There is 1 report from the Panhandle, 12 May 1967 Garden Co. Other westerly reports are 28 Apr 1949 Lincoln Co, 8 May 1946 Logan Co, 8 May 1948 Keith Co, 8–14 May 1933 Lincoln Co (Tout 1947), and 19 May 1964 Lincoln Co.

Summer

Nesting occurs no closer than northeast South Dakota, so summer reports are unexpected. There are several, but none are documented: 7 Jun 1966 Cass Co, 1 Jul–16 Aug 1964 Brown Co, 12 Jul 1992 Lincoln Co, 25–27 Jul 1978 McPherson Co, 5 Aug 1973 Howard Co, 17 Aug 1964 Cass Co, 26 Aug 1968 McPherson Co, 28 Aug 1965 Cass Co, and 5 Sep 1974 Douglas-Sarpy Cos.

Fall

14, 15, 20 Sep→28, 28, 30 Oct

Migration is from mid-Sep through late Oct, with a late report 9 Nov 1967 Adams Co. It was cited as occurring on the Lincoln CBC Dec 1976 (AB 31:348), although it was not reported that year in the NBR (NBR 45:7).

Most migrants pass through the Missouri Valley, but in recent years reports indicate that it is regular in the Rainwater Basin also.

High counts include 40 at Jack Sinn Marsh 27 Oct 1998, 40 at Kirkpatrick Basin 25 Sep 1999, and 24 north of Ames 3 Oct 1998.

FINDING

LeConte's Sparrow can usually be located in dense grassy areas in the Missouri Valley in fall, especially during the first 10 days of Oct. Birds usually respond to squeaking and will often perch on a shrub where they can be readily observed.

Nelson's Sharp-tailed Sparrow

Ammodramus nelsoni

STATUS

Rare regular spring migrant east, accidental central. Uncommon regular fall migrant east, rare casual central, accidental west.

DOCUMENTATION

Specimen

UNSM ZM7335, 8 Oct 1904 Lincoln, Lancaster Co (*Auk* 22:210).

DISTRIBUTION AND ECOLOGY

This species inhabits marshes and is usually associated with or near cattails, although it occurs also in grassy swales near water. Most of the few reports are from the east.

Spring

The few reports are in the period 25 Apr–30 May; it is likely to have been overlooked due to its habitat and late migration timing. The only documented spring records are of one 12 May 1995 at Jack Sinn Marsh (JG; Gubanyi 1996c); one banded 15 May 1980 in Lancaster Co (AB 34:793); a specimen, HMM 2587, collected 25 May 1919 at Inland (Hudson 1934b); and 2 specimens, UNSM ZM10734 and ZM10735, collected 30 May 1910 in Lancaster Co (Proc NOU 5:36).

Fall

7 (Rosche 1982), 9 (Hudson 1934b), 15 Sep→18, 21, 23 Oct

This species has been reported more often in fall than in spring in recent years. There are about 25 reports in the period 7 Sep–23 Oct, as well as later reports 29 Oct 1995 Lancaster Co and 7 Nov 1999 (16) Sandpiper Basin (JGJ).

Fall specimens are one taken 9 Sep 1934 in Cass Co (Hudson 1934b) and one taken 8 Oct 1904 at Lincoln, cited above, considered the first state record (*Auk* 22:210).

Westernmost reports are 7 Sep 1973 Sheridan Co (Rosche 1982; AB 28:76), 5 Oct 1985 Blaine Co (AB 40:137), 6 Oct 1991 Knox Co (AB 46:117), and 14 Oct 1994 Gosper Co (NBR 62:146). The Sheridan Co report is the only one for the Panhandle; this species is unrecorded in Colorado (Andrews and Righter 1992).

High counts include 16 at Sandpiper Basin 7 Nov 1999 and 6 in Washington Co 16 Oct 1994. It may become "common" at times, such as in fall 1975 and 1976 in Lancaster Co (AB 30:94, 31:196).

COMMENTS

Brown and others (1996) suggested that this species is a "hypothetical rare summer vagrant" in the Keith Co area but noted that "several sparrows thought to be this species were netted in summer 1994 in marshes on the north side of Keystone Lake. Photographs were identified by others as Savannah Sparrow."

Other Reports

29 Mar 1968 Cass Co.

FINDING

Fall migration yields more reports, suggesting a search of cattail marshes, such as at Jack Sinn Marsh or eastern Rainwater Basin marshes, during Oct. This species usually responds to squeaking.

Fox Sparrow

Passerella iliaca

STATUS

Uncommon regular spring and fall migrant east, becoming rare casual west. Rare casual winter visitor south and southeast.

DOCUMENTATION

Specimen

UNSM ZM7498, 8 Nov 1900 Lincoln, Lancaster Co.

TAXONOMY

Until recently, birds occurring in Nebraska were considered subspecies of Fox Sparrow, *P. iliaca*, but there is growing evidence that in fact there are as many as 4 species involved in the *P. iliaca* complex (Rising 1995). The Nebraska taxa belong to 2 of these incipient species: Red Fox Sparrow, *P. iliaca*, breeds across the arctic tundra and includes the subspecies *iliaca*, the reddish eastern form, and *zaboria*, the grayer western form. These are generally not separable in the field, although grayer birds are probably *zaboria* (Rising 1995). Another incipient species, Slate-colored Fox Sparrow, *P. schistacea*, of the Rocky Mountains, is rather darker with less rusty coloration than *P. iliaca*.

Essentially all Nebraska observations are of Red Fox Sparrow; both races, *iliaca* and *zaboria*, presumably migrate through Nebraska (Bent 1968). Definite records of *zaboria* are lacking, although a bird seen 12 Oct 1990 in Keith Co was described as "belonging to one of the grayish western races" (Rosche 1994a). Red Fox Sparrows migrate primarily through the eastern United States but straggle as far west as the Colorado plains, where they are rare migrants (Andrews and Righter 1992).

The Rocky Mountain form, Slate-colored Fox Sparrow, *P. schistacea*, is "accidental" on the eastern Colorado plains, with 1 spring, 1 fall, and 2 winter records (Andrews and Righter 1992). The type specimen of *schistacea* was collected about 40 km (25 mi) east of the northeast corner of Colorado, apparently in Keith Co, Nebraska (AOU 1957). There are no other reports of *schistacea* from Nebraska.

DISTRIBUTION AND ECOLOGY

Migrants occur in woodland edge and other brushy habitats, often riparian in nature.

Spring

3, 3, 4 Mar→5, 6, 7 May

Migration is from early Mar through early May, with an earlier report 26 Feb 1998 at Fontenelle Forest. Peak movement is during the first 10 days of Apr.

There are several later reports, none documented. The earliest and northernmost of these may be correct, although latest records for Iowa are 23 Apr and 6 May (Kent and Dinsmore 1996) and for South Dakota 12, 14, and 18 May (SDOU 1991). Reports that may be correct are 10 May 1962 McPherson Co and 12 May 1962 Scotts Bluff Co.

Fox Sparrow is a regular migrant only in the east. In the northwest Rosche (1982) considered it accidental, and on the northeast Colorado plains it is listed as a rare migrant by Andrews and Righter (1992). The few Panhandle reports are undocumented: 30 Mar 1957 Dawes Co, 12 May 1962 Scotts Bluff Co, and 21 May 1977 Scotts Bluff Co.

High counts include 22 at Wood Duck Area 1 Apr 1998.

Fall

17, 19, 20 Sep→winter

Migration is from late Sep through late Nov, although fall departure is protracted, especially in the southeast. There is an early undocumented report 2–19 Aug 1967 Brown Co and additional reports 5 Sep 1962 McPherson Co and 8–9 Sep 1972 McPherson Co.

There are several reports after Nov, including several CBC reports, with a high count of 7 at Omaha 18 Dec 1982, and a few Jan reports as late as 21 Jan 1979 Adams Co, 26 Jan 1993 Douglas-Sarpy Cos, and 31 Jan 1973 Adams Co (see Winter).

This species is rare in the Panhandle, with reports 20 Sep–4 Oct 1986 Sioux Co, 8 Oct 1977 Scotts Bluff Co, 13 Oct 1979 Scotts Bluff Co, 14 Oct 1977 Sioux Co, and 23 Oct and 2 Nov 1979 Garden Co (AB 34:178). It was observed during the CBC period at Scottsbluff in Dec 1971.

High counts include 32 at Boyer Chute NWR 18 Oct 1998, 18 at Wood Duck Area 28 Oct 1997, and 10 at Boyer Chute NWR 20 Oct 1996.

Winter

There are Jan reports that may be of wintering birds (see above) and a few reports of birds remaining throughout the winter: one was reported from Scotts Bluff Co 29 Dec 1964–1 Feb 1965, another was at Bellevue 29 Dec 1984–28 Feb 1985 (AB 39:184), and 1–2 wintered at a Lincoln feeder 1997–98. Fox Sparrow winters uncommonly in eastern and southern Kansas (Thompson and Ely 1992), rarely to northern Missouri (Robbins and Easterla (1992), and rarely on the eastern plains of Colorado (Andrews and Righter 1992).

Other Reports

9 May 1954 Gage Co, 10 May 1980 Adams Co, 12 May 1965 Cass Co, 15 May 1955 Gage Co, 15 May 1958 Gage Co, 19 May 1990 Polk Co, 21 May 1977 Scotts Bluff Co, 22 May 1971 Brown Co, 3 Jun 1988 Lincoln Co, 7 Jun 1975 Clay Co, 29 Jun 1967 Brown Co.

FINDING

This species can be most easily found during peak migration in brushy edge habitat in the Missouri Valley, usually with other sparrows. It responds to squeaking, and its size and coloration help identify it. Best times are early Apr and the last half of Oct.

Song Sparrow

Melospiza melodia

STATUS

Common regular spring and fall migrant east, becoming uncommon west. Common regular breeder north and east, locally uncommon elsewhere. Uncommon regular winter visitor south and east, rare elsewhere.

DOCUMENTATION

Specimen

UNSM ZM7540 (labeled *fallax*), 23 Oct 1890 Lincoln, Lancaster Co.

TAXONOMY

Two subspecies were listed by the AOU (1957) as breeding in Nebraska: eastern *euphonia* in the Missouri Valley and northern *juddi* along the northern edge of the state. Information presented by Cink (1975b) suggested that breeding Song Sparrows in Lancaster Co are *juddi*, indicating that the recent increase of sporadic breeding records throughout much of the state may be due to a southward spread of *juddi* rather than westward spread of *euphonia*, as well as the possibility of a few breeding pairs of *montana* in the extreme west and southwest. There are, however, no published data to support the presence of *montana*, although Bailey and Neidrach (1967) listed *montana* as the breeding race in Colorado, where Song Sparrows occur throughout the northeastern plains (Andrews and Righter 1992). Three birds that were "dark chestnut with gray napes" were at Cochran L, Scotts Bluff Co, 27 Oct 1996; they may have been examples of *montana* (WRS), and Rosche (1982) stated that "very dark reddish-brown birds" occur occasionally during fall migration in the northwest, presumably of "northwestern" races. According to Bent (1968), *montana* occurs casually in migration east to Crawford.

Most wintering Song Sparrows outside the southeast are probably *juddi*, which is migratory (Byers and others 1995) and winters from southern Canada south to Texas. *Euphonia* probably winters only in the east, within its breeding range, while a few *montana* may occur in the extreme west (Bent 1968).

The specimen cited above was identified as an example of *fallax*, a subspecies currently understood to breed in southwest Utah and southeast Nevada (Rising 1995; Byers and others 1995); we cannot confirm this identification.

DISTRIBUTION AND ECOLOGY

Territorial Song Sparrows are most commonly found in brushy or weedy riparian habitats, along streams, and around ponds or lakes (Cink 1975b) but may frequent wet ditches that support stands of willow or other shrubby vegetation. Willow islands of the lower Platte River support good numbers. The species is most numerous along the eastern and northern edges of the state. Similar habitats are occupied in winter, although open water is usually required, such as small streams and spring-fed marshes. Wintering birds are most common in southern and especially eastern Nebraska. Migrants are common statewide in brushy riparian and edge habitats.

Spring

Song Sparrow occurs statewide during migration, although it becomes uncommon westward (Rosche 1994a). Arrival and departure dates are difficult to determine due to the presence of wintering and summering birds. Data suggest, however, that spring migration is in progress by mid-Mar and is essentially concluded around the end of Apr. Reports from the Keith Co area, where the species is a winter visitor (Rosche 1994a), include a late spring date 26 Apr, while the last migrant at Box Butte Res was noted 5 May 1994.

High counts include 250 at Funk Lagoon 18 Apr 1999, 50 at Gibbon 19 Apr 1997, 50 at Wood Duck Area 15 Apr 1997, and 50 there 2 Apr 1998. It was "abundant" in northern Saunders Co 19 Mar–1 Apr 1994.

Summer

Although Johnsgard (1979) indicated that Song Sparrow was breeding regularly in the Missouri Valley and adjacent counties, available evidence suggests that breeding is a relatively recent phenomenon in Nebraska. Bruner and others (1904) speculated that "a few breed, especially in northern Nebraska," although no evidence was provided. Rapp and others (1958) made no mention of breeding. The first documentation of breeding was that of Wood (1965), who found a nest near Plattsmouth. From that time the breeding population increased along the Missouri Valley and began expanding westward, first along the Platte Valley and then into adjacent drainages, although summering Song Sparrows were found in the Platte Valley as far west as Polk Co in 1956 and 1957 by Short (1961). Breeding was first documented in Lancaster Co by Cink (1975b), who suggested that Song Sparrow was increasing its range as a breeding bird at least partially as a result of the establishment of reservoirs and irrigation infrastructure. Song Sparrows reported on BBS routes 1967–77 were virtually restricted to the east and north, with 86% of the birds in those areas.

Song Sparrow remains a sporadic breeder away from the Missouri and lower Platte Valleys. Most such reports are from the east and the Loup drainage, where there are reports from Custer, Wheeler, Greeley, Loup, Howard, and Boone Cos. It was not considered a breeding bird at Bessey Division, NNF (Bray 1994). It breeds rarely westward in the Platte Valley to Hall Co (Lingle 1994; Lingle and Hay 1982), and there are Platte Valley summer reports west as far as Lincoln Co. Summering birds have been noted at Funk Lagoon since 1995; 15 were counted 22 Jun 1996 (FN 50:967). This species has also been reported in eastern Cherry, Brown, and Holt Cos along the Niobrara River, although Brogie and Mossman (1983) found no summering birds in the Niobrara Valley Preserve in 1982.

Rosche (1994a) stated that this species is "not at all a common bird anywhere in western Nebraska at any time of the year" and noted that the "only known habitat in western Nebraska with summering Song Sparrows" is at Oliver Res. Young were reported from Sioux Co in 1966 (Sharpe 1967), and there are several summer reports for Scotts Bluff Co, as well as a report 27 Aug 1993 Dawes Co.

It apparently does not breed in the Republican Valley; a breeding record for Red Cloud (AOU 1957) could not be verified by Cink (1975b), and it was listed only as a migrant in Webster Co by Ludlow (1935). It breeds in northeast Kansas, south to the Kansas River (Thompson and Ely 1992; Mark Robbins, pers. comm.).

Cornell Nest Record data show a nest in Boone Co had eggs 19 May 1983, with young fledged 10 Jun.

Fall

Migration begins in late Aug and concludes in late Nov. Recent reports from the Panhandle are presumably migrants, as breeding is not known at the locations reported: 30 Aug 1998 Smith L, Sheridan Co; 30 Aug 1998 (10) Crescent L NWR; 3 Sep 1997 Scotts Bluff Co; 7 Sep 1996 Crescent L NWR; 10 Sep 1994 Hay Springs; 13 Sep 1997 north of Harrison; 22 Sep 1994 Scotts Bluff Co; and 3 Oct 1998 south of Gering. In the Keith Co area Rosche (1994a) listed an early fall date of 8 Aug.

High counts include 55 at Harvard Lagoon 25 Sep 1996 and 40 in Buffalo Co 25 Oct 1998.

Winter

CBC data show that by early winter most Song Sparrows are found in the east and south, although overwintering birds may occur anywhere where open water in the form of springs, seeps, or small streams exists. The highest CBC total is 135 at Omaha in 1967, compared to high counts at other points in the state of 20 at Hastings in 1969, 14 at Norfolk in 1985, and 13 at Crawford in 1977.

FINDING

Song Sparrow is easiest to find during spring migration, especially in early May in the Missouri Valley, when it sings loudly. Good locations include Fontenelle Forest and Indian Cave SP. It can generally be found year-round in similar habitat, except during severe winters.

Lincoln's Sparrow

Melospiza lincolnii

STATUS

Common regular spring and fall migrant east, becoming uncommon west. Rare casual winter visitor southeast.

DOCUMENTATION

Specimen

UNSM ZM7513, 22 Apr 1890 Lincoln, Lancaster Co.

TAXONOMY

Rapp and others (1958) listed 2 subspecies, generally indistinguishable in the field: *lincolnii*, which breeds across Canada and occurs statewide in Nebraska during migration, and the Rocky Mountain race *alticola*, which is probably a rare migrant in western Nebraska. There are no known Nebraska specimens of *alticola*; none are labeled as such in the UNSM collection (TEL). *Alticola* is, however, a rare migrant in western Kansas (Johnston 1965).

DISTRIBUTION AND ECOLOGY

During migration Lincoln's Sparrow frequents dense brush and thickets, usually close to water. It occurs statewide but is more numerous in the east.

Spring

29, 29, 30 Mar→30, 31, 31 May

Migration is from early Apr through mid-May. There are several earlier undocumented reports, which may be either birds that wintered close to Nebraska or simply misidentified Song Sparrows. The earliest dates for Lincoln's Sparrow in Iowa are 26 Mar and 7 and 8 Apr (Kent and Dinsmore 1996) and for South Dakota 6, 8, and 12 Apr (SDOU 1991). Migrants arrive in eastern Colorado around 1 Apr (Andrews and Righter 1992). Early reports from the southeast, if correctly identified, are likely to be birds that wintered in Kansas, as there are no Feb records for Nebraska (see Winter). Reports in this category are: 5 Mar 1967 Lancaster Co, 6 Mar 1974 Douglas-Sarpy Cos, 7 Mar 1998 Buffalo Co, 10 Mar 1999 Harlan Co, 11 Mar 1989 Lancaster Co, 14 Mar 1971 Douglas-Sarpy Cos, 14 Mar 1991 Douglas-Sarpy Cos, 16 Mar 1942 Jefferson Co, 19 Mar 1937 Lincoln Co, 20 Mar 1949 Jefferson Co, 22 Mar 1959 Thayer Co, 23 Mar 1984 Polk Co, 24 Mar 1981 Douglas-Sarpy Cos, 24 Mar 1991 Lancaster Co, 25 Mar 1975 Douglas-Sarpy Cos, and 28 Mar 1954 Thayer Co.

Late reports are 10 Jun 1991 Douglas-Sarpy Cos, 11 Jun 1974 McPherson Co, 15 Jun 1974 Scotts Bluff Co, and 20 Jun 1986 Lancaster Co.

High counts include 44 in Sarpy Co 11 May 1996, 30 in Scotts Bluff Co 27 Apr 1994, and 30 at Kenesaw 26 Apr 1998.

Fall

22, 27, 30 Aug (Rosche 1982)→28, 30, 30 Nov

Migration is from early Sep through late Oct, with an earlier report 12 Aug 1977 Sioux Co and many reports in Nov.

In the south and east Lincoln's Sparrow occasionally lingers into Dec and early Jan. There are about 25 Dec reports, many from CBCs, with high counts 6 at Omaha 16 Dec 1972 and 5 there 19 Dec 1998. Most of these are from Douglas-Sarpy and Lancaster Cos but

also north to Norfolk and Dakota Co and west to Grand Island and Kearney (see Winter).

High counts include 100 at Rowe Sanctuary 13 Oct 1996, 75 in southwest Dixon Co 27 Sep 1998, and 50 in Dixon Co 3 Oct 1994.

Winter

There are a few Jan reports but none for Feb, suggesting that successful overwintering probably does not occur. The latest Jan dates are 10 Jan 1988 Lancaster Co (AB 42:285), 11 Jan 1967 Douglas-Sarpy Cos, and 23 Jan 1988 Douglas Co (AB 42:285). One was rather far west for the date at L Ogallala 2 Jan 2000 (SJD).

Other Reports

2 Jul 1973 Douglas-Sarpy Cos, 14 Jul 1985 Polk Co.

FINDING

Lincoln's Sparrow can be most easily found in fall migration in the Missouri Valley, usually with other sparrows in weedy fields and brush near water. Best time is early to mid-Oct, when it is often rather conspicuous.

Swamp Sparrow

Melospiza georgiana

STATUS

Common regular spring and fall migrant east, becoming uncommon west. Uncommon regular breeder central. Rare casual winter visitor southeast and Platte Valley.

DOCUMENTATION

Specimen

UNSM ZM7526, 15 Dec 1889, Lincoln, Lancaster Co.

DISTRIBUTION AND ECOLOGY

Breeding habitat is marshes, usually cattail marshes with shrubs or small trees present for singing perches (Rosche 1982), although territorial birds have been found in brushy riparian habitat in central Nebraska (AB 43:1335). Migrants occur in dense brushy habitats, usually but not necessarily riparian, and marshes and are most numerous in the east. Wintering birds are most often associated with cattail marshes containing open or running water, such as springs or seeps, and may be found wherever such habitat exists, especially in the southeast and in the central Platte Valley.

Spring

28, 29, 30 Mar→25, 29, 31 May

Earlier dates may be wintering birds (see Winter). Late dates above are at nonbreeding locations.

Away from the east, migrants are rare and not often detected away from breeding sites. In the east they are more numerous and are often found in flocks with other sparrow species. Migration is from late Mar through mid-May, peaking around Apr 20 (Johnsgard 1980).

High counts include 7 at Fontenelle Forest 28 Apr 1998.

Summer

The distribution of breeding and summer records closely parallels the occurrence of cattail marshes in the Sandhills and Loup drainage, although breeding colonies tend to be locally distributed and birds are often absent from seemingly suitable habitat.

Summer reports exist in the northeast for Knox Co, where birds seen 22 May 1994 were "maybe breeders" (AB 48:315); Antelope Co; Boone Co; Holt Co (Blake and Ducey 1991); Rock Co, where breeding was reported in 1985 (AB 39:932); and Brown Co, where it may have been summering in 1994 (NBR 62:147). Farther west there are reports for Cherry Co (Brogie and Mossman 1983), where breeding occurred at Cody Lake in 1988 (AB 42:1309); Garden Co (Rosche and Johnsgard 1984), where it is a regular summer resident at Clear Creek Marshes (Rosche 1994a) and was recorded at Crescent L NWR 10 Aug 1994 (NBR 62:147); Sheridan Co (Rosche 1982), where nesting occurred in 1982 (AB 36:994); Blaine Co, where nesting occurred in 1991 (AB 45:1134); and Logan Co, where there are also specimens, UNSM ZM7533 and ZM7532, taken 11 Aug 1936 and 2 Sep 1936, respectively. Johnsgard (1980) listed Howard Co as a breeding site, presumably based on the discovery in 1969 of a small breeding colony in a marsh along the Middle Loup River (Sharpe 1970). This record

was incorrectly listed for Hall Co by Ducey (1988).

Elsewhere, breeding localities are few. There appears to be a small breeding colony established in the kiosk area at Funk Lagoon, where there have been reports since 16 Jun 1991 (LR, RH), the best count 20 on 5 Jul 1999 (LR, RH). Breeding probably occurs in Lincoln Co also, where there are reports 19 Jun 1948, 23–29 Jun 1980, 24–26 Jun 1991 (TB), 6 Aug 1995 at a cattail marsh along the Platte River east of North Platte, and 14 Aug 1992. It also breeds at Jack Sinn Marsh, Lancaster Co (TEL).

There are few summer reports elsewhere: 4 Jun 1992 Keith Co, 6 Jul 1988 Stanton Co, 1 Aug 1976 Lancaster Co, and 10 Aug 1991 Keith Co. Territorial males have been detected singing in late May along the wooded north edge of Great Marsh in Fontenelle Forest in several different years (RSS). In all cases, singing persisted for 10 days or less, and no further evidence of breeding activity was observed.

Fall

13, 16, 17 Sep→winter

Migration is from mid-Sep through early Nov, although there are a few earlier reports away from known breeding areas: 23 Aug 1986 Pierce Co, 27 Aug 1982 Douglas-Sarpy Cos, and 6 Sep 1987 Polk Co. There are numerous Nov reports and several in Dec and Jan (see Winter).

CBC data since 1967 record Swamp Sparrows most years; reports are generally confined to the area south of Washington Co and east of Lancaster Co, with a few exceptions: 26 Dec 1987 Dakota Co (Sioux City, Iowa, CBC), 27 Dec 1985 Boone Co (Beaver Valley CBC), 19 Dec 1991 Keith Co (Lake McConaughy CBC), and 18–19 Dec 1992 Keith Co (Rosche 1994a). Highest CBC total was 13 birds at Lincoln 17 Dec 1978.

High counts include 30 at Neale Woods 15 Oct 1999, 24 in Lancaster Co 15 Oct 1996, and 22 in Knox Co 5 Oct 1997.

Winter

Swamp Sparrows linger late in fall if conditions allow, although reports suggesting wintering are mostly from Douglas-Sarpy Cos, where there are several Jan and Feb reports: 17 Jan 1988, 18 Jan 1975, 23 Jan 1989, 1 Feb 1986, 5 Feb 1992, and 17 Feb 1964. A report 2 Mar 1981 Douglas-Sarpy Cos was probably a wintering bird, as may have been others: 5 Mar 1999 Neale Woods, 10 Mar 1996 Otoe Co, 24 Mar 1984 Douglas-Sarpy Cos, and 24 Mar 1991 Lancaster Co. Wintering birds were reported from Douglas and Sarpy Cos 1987–88 (AB 42:285).

Dec reports elsewhere may indicate attempts to overwinter, but there are few reports of survival after 1 Jan; the latest dates are 27 Dec 1985–1 Jan 1986 Boone Co, 29 Dec 1964 Scotts Bluff Co, and 1 Jan 1957 Lincoln Co. The only westerly midwinter reports are of one wintering in Lincoln Co 1976–77 (AB 31:348) and singles at L Alice 25 Jan 1996 and Stateline Island 7 Feb 1997.

FINDING

Swamp Sparrows are easiest to find during migration in the Missouri Valley, when the birds are not confined to marshes and are more easily approached. Peak migration times are Apr and Oct, when Swamp Sparrows can be found in riparian brush, usually with other sparrows. Breeding birds should be looked for in cattail marshes in the Sandhills and Loup drainage; the song is helpful in locating this species.

White-throated Sparrow

Zonotrichia albicollis

STATUS

Common regular spring and fall migrant east, becoming rare west. Hypothetical in summer. Rare regular winter visitor southeast.

DOCUMENTATION

Specimen

HMM 2015, 2 Jun 1915 Inland, Clay Co.

TAXONOMY

This species occurs as a tan-striped morph and the more familiar white-striped morph, as well as intermediates (Byers and others 1995; Rising 1995). These morphs are not population specific and occur throughout the range (Byers and others 1995). First winter birds all have tan crown stripes, but some retain these into adulthood (Bent 1968). A tan morph bird wintered in a Buffalo Co yard 1995–96 (NBR 64:15), and another was in the same yard 11 Oct 1998 (LR, RH). All of the 13 reported in the Panhandle 20 Sep–16 Oct 1999 were tan morph birds (SJD).

DISTRIBUTION AND ECOLOGY

This ground forager can be found in brushy woodland edge situations, both during migration and (rarely) in winter. Successful overwintering may depend on feeders. Migrants occur statewide but are more numerous in the east.

Spring

4, 5, 6 Mar→31, 31 May, 2 Jun (specimen cited above)

White-throated Sparrow is uncommon as a migrant away from the east and is rare in the Panhandle, especially in spring (Rosche 1982).

Migration is from mid-Mar through late May, with an earlier report 29 Feb 1968 Adams Co and later dates 10 Jun 1983 Adams Co, 10 Jun 1991 Douglas-Sarpy Cos, 15 Jun 1975 Clay Co, and 22 Jun 1974 Adams Co. Earlier reports may be of wintering birds (see Winter).

High counts include 197 in Sarpy Co 11 May 1996, 61 in Lancaster Co 4 May 1996, and 50 at Hummel Park, Douglas Co, 1 May 1997.

Summer

There are undocumented reports 2 Aug 1981 Lincoln Co and 3 Aug 1994 Otoe Co, possibly early fall migrants. There are no comparable records for South Dakota (SDOU 1991) or Kansas (Thompson and Ely 1992), but there is a 10 Jul record for Missouri (Robbins and Easterla 1992) and Iowa records 18 Jul 1987, 20 Jul 1984, and 5 Aug 1988 (Kent and Dinsmore 1996).

Fall

14, 15, 16 Sep→winter

Migrants arrive in mid to late Sep, with an earlier report 8 Sep 1967 Adams Co. Departure is gradual, with most birds gone by late Nov, although Dec reports are fairly common, especially in the east, where late dates are difficult to determine. During the CBC period, most are found in the southeast, where high counts are 42 on the Lincoln CBC 17 Dec 1989 and 34 at the same location 16 Dec 1990. During the CBC period northwesternmost reports are from Sioux City, Iowa (4 on 15 Dec 1984), Loup Co (1 on 29 Dec 1989 on Calamus-Loup CBC), and Kearney (1 on 29 Dec 1979 and 2 on 31 Dec 1982).

In the north there are as many as 9 reports for 1 Jan (coincident with the beginning of the "Spring" seasonal occurrence reporting period for NBR), and one was at L Ogallala 2 Jan 2000 (SJD). In the west, where it is rare in fall, late dates include 10 Nov 1973 Sioux Co, 13 Nov 1994 Dawes Co, and 21–23 Nov 1978 Garden Co.

High counts include 38 in Saunders Co 8 Oct 1994, 32 at Fontenelle Forest 11 Oct 1998, and 30 there 16 Oct 1997.

Winter

As indicated, White-throated Sparrow probably overwinters occasionally, essentially from Omaha and Lincoln southward; most wintering birds are reported from feeders. Many remain through Dec into the first part of Jan, but Feb reports are rare. Nevertheless, there are several midwinter reports indicative of wintering; all but the following are from Antelope, Boone, Howard, and Buffalo Cos eastward: 6 Feb 1986 Scotts Bluff Co and an "unprecedented" record of 2 wintering at a Chadron feeder 1993–94 (AB 48:223).

FINDING

This species can be located during migration in brushy habitats and woodland edge, usually with other sparrows, in the Missouri Valley. Best times are Apr and Oct; the whistled song is a good clue in spring.

White-crowned Sparrow

Zonotrichia leucophrys

STATUS

Common regular spring and fall migrant statewide. Rare casual summer visitor statewide. Uncommon regular winter visitor statewide.

DOCUMENTATION

Specimen

UNSM ZM7489, 19 Apr 1890 Crete, Saline Co.

TAXONOMY

Two subspecies occur. The eastern race *leucophrys* (black lores, pinkish bill; Dunn and others 1995) is generally limited to the eastern third of the state (Bruner and others 1904), although one was reported by Dawson (1921) in Monroe Canyon 30 Sep 1920 (but see the subspecies *oriantha* below). It occurs as a migrant and rarely in winter and is "not nearly as common" as *gambelii* (Swenk, Notes before 1925). The northwestern race *gambelii* (pale lores, orange bill; Dunn and others 1995) is found statewide in migration, wintering primarily in the Panhandle. In South Dakota the 2 races occur in equal numbers in the extreme east, while *gambelii* is more frequent elsewhere (SDOU 1991). Ludlow (1935) considered both races common spring migrants in Webster Co, *leucophrys* in the period 19 Apr–21 May, with an early date 13 Mar 1930, and *gambelii* 19 Apr–9 May.

Some authors consider Rocky Mountain birds *oriantha* distinct from *leucophrys* (Rising 1995); both have black lores, but their breeding ranges do not overlap, and they probably cannot be separated in the field. A dark-lored bird in Monroe Canyon 20 Sep 1920 (Dawson 1921) may have been *oriantha*, and there is a specimen, UNSM ZM7494, a male collected by Mickel and Dawson 23 Jun 1916 at Mitchell, that is likely to be *oriantha* also. It has a very dark reddish brown bill, is as pale on the flanks as any other specimen in the collection, and has extensive black coloration in the loral area, which extends to the gape (WRS).

DISTRIBUTION AND ECOLOGY

White-crowned Sparrow occurs statewide in brushy habitats, especially woodland edge. It is most numerous during migration and during winter in the west, although small numbers winter in the south and east.

Spring

winter→31 May, 2, 3 Jun

Because of wintering birds, early spring dates are difficult to determine.

Migration is from mid to late Mar through May, although there are later reports: 8 Jun 1986 Grant Co, 8 Jun 1990 Cedar Co, 9 Jun 1919 Kimball Co (specimen, UNSM ZM7483), 14 Jun 1988 Hall Co, and 15 Jun 1975 Clay Co.

High counts include 275 at Lake McConaughy 29 Apr 1994, 91 (*gambelii*) at Enders Res 2 May 1998, and 62 in Sarpy Co as late as 11 May 1996.

Summer

There are a few reports from the summer months, most from the northwest, but there is no evidence that this species has bred in Nebraska. Ducey (1988) cited a statement by Aughey that he had found young of this species in Dixon Co "at the edge of the woods, and breeding here must be rare." This was probably an error in identification. It does not breed in the Black Hills of South Dakota, and there is only one record for Jun–Aug for South Dakota (SDOU 1991). Rosche (1982) listed White-crowned Sparrow only as a migrant in the northwest.

Summer reports include 23 Jun 1916 Scotts Bluff Co (specimen, UNSM ZM7494; Mickel and Dawson 1920), 24 Jun–1 Jul 1988 Dawes Co, 29 Jun 1974 Lincoln Co, 30 Jun 1967 Greeley Co, 30 Jun 1967 Garden Co, 30 Jun 1967 Sioux Co, 4 Jul 1960 Dawes Co, 6 Jul 1989 Dawes Co, 7 Jul 1973 Adams Co, 8 Jul 1990 Dawes Co, and 17 Jul 1991 Scotts Bluff Co.

Fall

25, 28 Aug, 2 Sep→winter

Migration generally takes place from late Aug through mid-Nov. Departure is difficult to determine, as many birds linger into winter (see Winter).

High counts include 160 at Crescent L NWR 28 Sep 1994, 85 in Scotts Bluff Co 30 Sep 1995, and 75 at Crescent L NWR 2 Oct 1996.

Winter

Most White-crowned Sparrows winter south of Nebraska, and overwintering within the state is uncommon; Johnsgard (1980) considered it a "locally common" winter visitor statewide. Wintering occurs regularly in the Scottsbluff area, where CBC totals are the highest for the state, including 33 in 1988–89 and 31 in 1986–87. In the east, where this species is far less numerous in winter than in the west, there is only 1 midwinter report north of the Platte Valley, an immature at Creighton 12–13 Jan 1999 (MB). Farther west there are these midwinter reports: 1 Jan 1991 Sioux Co, 5 Jan 1967 Custer Co, 15 Jan 1953 Dawes Co, 3 Feb 1960 Antelope Co, 11 Feb 1983 Boone Co, 9 Mar 1957 Brown Co, and 10 Mar 1985 Sioux Co.

Rosche (1994a) noted that "there is always a marked decrease after long, severe cold spells after late January each year."

CBC data since 1967 indicate that in Dec about 86% of birds reported were in the Panhandle, with the rest in the southeast.

FINDING

White-crowned Sparrow is easiest to find during migration, especially in fall, almost anywhere. Best times are Apr and Oct, and woodland edge in the Panhandle probably has the greatest numbers.

Harris's Sparrow

Zonotrichia querula

STATUS

Common regular spring and fall migrant east, becoming rare west. Rare casual summer visitor south and east. Common regular winter visitor southeast, becoming rare west.

DOCUMENTATION

Specimen

UNSM ZM7459, 25 Apr 1889 Lincoln, Lancaster Co.

DISTRIBUTION AND ECOLOGY

Harris's Sparrow occupies weed patches and

brushy areas that provide an abundant seed source. These habitats may be found along woodland edges, along fencerows, or in fallow fields. Best concentrations of ideal habitat are found in the southeastern part of the state, where wintering is common. Numbers of overwintering birds and their location may vary considerably during a season and from year to year as snow cover changes. Small numbers may be sustained locally at feeders. In urban areas individuals may appear at feeders after late Feb–Mar snowstorms. Although migrants occur statewide, they are most numerous in the east.

Spring

winter→29, 30, 31 May

Arrival of migrants from south of Nebraska is difficult to detect in the southeast, although there is an influx beginning in late Mar where wintering birds are not numerous.

Migrants depart by late May, although there are later reports 3 Jun 1991 Cherry Co (AB 45:1134), 8 Jun 1991 Knox Co, 9 Jun 1978 Douglas-Sarpy Cos, 10 Jun 1976 Adams Co, and 15 Jun 1975 Clay Co (see Summer).

High counts include 369 in Pierce Co 8 May 1999, 235 in Lancaster Co 10 May 1996, and 235 in Sarpy Co 11 May 1996.

Summer

There are summer records from South Dakota (SDOU 1991), Kansas (Thompson and Ely 1992), Iowa (Kent and Dinsmore 1996), and a few from Nebraska, presumably wintering birds that did not migrate. These are 29 Jun 1970 Adams Co; 9 Jul 1976 Lancaster Co; 20 Jul 1919, a specimen, UNSM ZM7476, Lancaster Co (Mickel and Dawson 1920); 3 Aug 1987 Lincoln Co; and 13 Aug 1975 Adams Co.

Ducey (1988) cited a statement by Aughey that young were seen "frequently" in northeast Nebraska, probably in the 1870s. This is almost certainly an error on Aughey's part (Sharpe 1993).

Fall

10 (Ludlow 1935), 12, 12 Sep→winter

Harris's Sparrow is a common migrant in the east but uncommon to rare in the Panhandle. It was considered rare in spring in the northwest by Rosche (1982), and Lingle and Hay (1982) indicated that it was uncommon as a spring migrant but common in fall at Mormon Island Crane Meadows in Hall Co. As of about 1900 there were no records of Harris's Sparrow west of Cherry Co (Bruner and others 1904).

Although migrants normally arrive in late Sep, there are earlier reports 1 Sep 1990 Buffalo Co, 1 Sep 1990 Holt Co, 4 Sep 1966 Lancaster Co, 6 Sep 1984 Douglas-Sarpy Cos, and 9 Sep 1965 Scotts Bluff Co. Most probably depart in early Nov, but many birds linger through Dec, and wintering is common in the southeast.

High counts include 250 at Harlan Co Res 11 Nov 1995, 200 at Wood Duck Area 28 Oct 1997, and "hundreds" in Dakota Co 23 Oct 1999.

Winter

Fall migrants leave in Nov, but good numbers remain in the south and east in Dec, as indicated by CBC data. Wintering is regular in the southeast, generally from the Platte Valley south, although a few apparently overwinter in the Platte Valley as far west as Scotts Bluff Co; Rosche and Johnsgard (1984) listed Harris's Sparrow as a winter visitor in Keith Co. North of the Platte Valley overwintering is local and generally rare, especially in the northwest, where Rosche (1982) noted that there are "unusually few observations after late January."

FINDING

Harris's Sparrow is easy to find in the Missouri Valley and virtually statewide in fall. Best time is Oct, when squeaking at almost any brushy spot will cause some to fly up into view.

Golden-crowned Sparrow

Zonotrichia atricapilla

STATUS

Rare casual spring and fall migrant west and central.

DOCUMENTATION

Photograph

25–26 Nov 1984 Alliance, Box Butte Co (Thomas 1984; Mollhoff 1989a).

DISTRIBUTION AND ECOLOGY

This species occurs east of its regular winter range primarily in spring and fall with migrating flocks of White-crowned Sparrows and occupies the same brushy habitats as that species.

Spring

There are 3 reports in the period 16 Apr–14 May, none documented but all likely correct (see Comments). One trapped and banded at Halsey 7 May 1950 was with Harris's Sparrows (Smith 1950), but no other details were provided. Another was reported attending a feeder near Elsmere for 2 weeks prior to 30 April 1962 (Held 1962). One reported seen on a spring count near Scottsbluff 14 May 1966 was described as having "yellow on the crown of the head" (Banghart 1966).

Fall

Of 4 reports, 3 are documented:

25–26 Nov 1984 immature, Alliance (cited above)

3 Dec 1999 near Brownlee (JED)

18 Dec 1998 immature Harlan Co Res (JGJ, SJD)

The other report was of one in a yard in McPherson Co 7 Oct 1966; the observer had seen the species before in San Francisco (Bassett 1967).

Johnsgard (1980) mentioned that one had been reported on the Scottsbluff CBC in 1979, but this apparently was an error due to a garbled data transmission (NBR 52:77).

COMMENTS

Despite the poor documentation of 4 of the reports cited above, all were at expected times, fitting the pattern of eastward vagrancy seen in neighboring states, and all seem likely to be correct. Golden-crowned Sparrow is an easterly vagrant in fall and spring, with a few wintering birds recorded (Dinsmore and others 1984). Most records are at feeders, but it also occurs with White-crowned Sparrows (Thompson and Ely 1992).

There are 5 records from the northeastern Colorado plains (Andrews and Righter 1992) but none for Wyoming (Scott 1993) or South Dakota (SDOU 1991). There are 7 sight reports for Kansas, all in the period 1 Jan–13 May (Thompson and Ely 1992), and 2 for Iowa, both in May (Kent and Dinsmore 1996).

FINDING

Golden-crowned Sparrows should be looked for in migrating flocks of *gambelii* (pale-lored) White-crowned Sparrows, especially in the Panhandle. Most vagrants are immatures, and these can be difficult to differentiate from immature White-crowned Sparrows. Immature Golden-crowned Sparrows have dark bills and lack the dark eye-line of White-crowned Sparrows, and they possess varying amounts of yellow on the forecrown.

Dark-eyed Junco

Junco hyemalis

Seven forms of Dark-eyed Junco, now treated as subspecies but until recently considered to involve 4 distinct species (AOU 1957; Miller 1941), have been recorded in Nebraska (Moser 1946; Rapp and others 1958). The taxa are classified by the AOU (1957) into 4 species: White-winged Junco, *J. aikeni*; Slate-Colored Junco, *J. hyemalis hyemalis* and *J. h. cismontanus*; Oregon Junco, *J. oreganus montanus*, *J. o. mearnsi* (Pink-sided Junco), and *J. o. shufeldti*; and Gray-headed Junco, *J. caniceps caniceps*. These 4 species will be discussed separately herein.

Dark-eyed ("White-winged") Junco

STATUS

Fairly common regular resident northwest. Uncommon regular winter visitor west, rare casual elsewhere. Rare regular spring and fall migrant west, rare casual elsewhere.

DOCUMENTATION
Specimen
UNSM ZM7366, 12 Dec 1895 Fort Robinson, Dawes Co.

TAXONOMY

There is no evidence to support breeding of any race other than *aikeni* in Nebraska, although Rapp and others (1958) listed "Slate-colored Junco" (which includes both *hyemalis* and *cismontanus*) as an uncommon summer resident in the Pine Ridge, a statement repeated by Short (1961); the reported presence of 6 "Slate-colored" Juncos in Monroe Canyon 28 Jul 1995 was considered to be a "curious surprise" (FN 49:948).

Bruner (1896) mentioned the collection of a specimen of *aikeni* with "decidedly pinkish sides," presumably in Sioux Co. This bird may have been a hybrid with *mearnsi*, which breeds in the Big Horn Mountains of northern Wyoming and whose range approaches within about 32 km (20 mi) of that of *aikeni* in Montana, at a location where extensive interbreeding occurs (Miller 1941).

DISTRIBUTION AND ECOLOGY

Breeding habitat is ponderosa pine in the Pine Ridge in Sioux and Dawes Cos. Migrants and wintering birds occur in brushy woodland edge habitats and at feeders in Panhandle towns and cities.

Spring

Because many "White-winged" Juncos are resident, it is difficult to determine dates of migration. However, birds wintering in Colorado depart by mid-Apr (Andrews and Righter 1992).

Summer

"White-winged Juncos" are fairly common in summer in the Pine Ridge (Ducey 1988), especially in the westerly canyons such as Monroe (Rosche 1982) and Warbonnet (Bruner and others 1904).

Ducey (1988) listed a report of breeding in Holt Co in 1901, but no details are available. This report was not cited by the AOU (1957).

Fall

"White-winged" Juncos wintering south of the breeding range into Colorado arrive around mid-Oct (Andrews and Righter 1992).

Winter

Many of the summering birds are resident (AOU 1957), but in some winters there is considerable movement south of the summer range, most notably to Colorado (AOU 1957), where it occurs mid-Oct through mid-Apr (Andrews and Righter 1992). The eastern limit of regular wintering in Nebraska appears to be Keith Co (Rosche 1994a; Brown and others 1996).

Occasionally there is significant eastward movement, for example, a flock of about 40 in Lincoln Co 6 Jun 1911, 1 of which was examined in hand (Tout 1947). Although Thompson and Ely (1992) show winter records statewide in Kansas, Nebraska reports away from the Panhandle are few and undocumented. Some, if not most, easterly reports may in fact be variants of *hyemalis* with exaggerated wingbars (Robbins and Easterla 1992); *aikeni*, however, is a larger and paler bird overall than *hyemalis*. Easterly reports are 1 Jan 1955 Boone Co; 5 Jan 1947 Douglas Co; 18 Jan 1945 Douglas Co; 10 Feb 1972 Thayer Co; 12 Feb 1917, a specimen, #2782A in the Brooking collection taken at Inland (Brooking, Notes); 12 Feb 1957 Logan Co; 19 Feb 1974 Sarpy Co (AB 28:659); 22 Feb 1935 Adams Co; 20–24 Mar 1940 Lincoln Co (Tout 1947); 4 Apr 1955 Keya Paha Co; and winter 1972–73 Douglas Co (AB 27:636).

FINDING

"White-winged Junco" should be looked for in Pine Ridge canyons, where it occurs in pines and usually can be attracted by squeaking at any time of year, although in winter it is most easily found at feeders, such as those at the Wildcat Hills NC near Scottsbluff.

Dark-eyed ("Slate-colored") Junco

STATUS

Abundant regular winter visitor statewide.

Abundant regular spring and fall migrant statewide.

DOCUMENTATION

Specimen

UNSM ZM7380, 16 Oct 1907 Lancaster Co.

TAXONOMY

While most documented records are of *J. h. hyemalis*, *J. h. cismontanus* has reportedly been collected at Long Pine (listed as *J. h. shufeldti* by Bruner and others 1904) and twice at Lincoln (Moser 1946; AOU 1957), but its status in Nebraska is not well known. It resembles *hyemalis*, but the head is darker than the back, females have sides washed with brown or pink (Bent 1968), and the breast demarcation is concave (Rising 1995). The Long Pine specimen, UNSM ZM7401, fits this description (WRS, JGJ). A similar bird was at the Wildcat Hills NC feeders 27 Oct 1996 (WRS). *Cismontanus* may in fact be a "stable population of apparent hybrids between Slate-colored and Oregon Juncos" (Rising 1995).

Spring

winter→2, 4, 5 Jun

Most depart by mid-May, although peak migration is probably in Apr (Johnsgard 1997). It was "abundant" in Saunders Co 12 Mar–18 Apr 1994. Later reports are 11 Jun 1973 McPherson Co, 11 Jun 1981 Garden Co, 14 Jun 1960 Scotts Bluff Co, and 16 Jun 1974 Lancaster Co.

High counts include 140 in Otoe Co 25 Mar 1998 and 92 at Branched Oak L 7 Mar 1999.

Fall

1, 4, 4 Sep→winter

Data used here include those juncos reported unspecified as to form; "Slate-colored" Junco is by far the most numerous form statewide. Migrants arrive statewide in mid-late Sep, and peak movement is in Oct (Johnsgard 1997).

High counts include 210 in Lancaster Co 14 Nov 1997, 200 near Heartwell 30 Oct 1999, and 130 at Wolf L 30 Oct 1999.

Winter

Numbers of wintering juncos vary considerably from year to year as pointed out by Rosche (1982) for the northwest and as shown by CBC data. The high CBC total in the period 1982–1991 was 7189 in 1989–90 and the low 2047 in 1984–85. During the CBC period juncos are much more numerous in the east than elsewhere. A high count was "thousands" in Howard Co 8 Jan 1995.

FINDING

"Slate-colored" Juncos are ubiquitous in winter statewide and easily observed in brushy areas as well as at feeders.

Dark-eyed ("Oregon") Junco

STATUS

Abundant regular spring and fall migrant west, becoming rare east. Abundant regular winter visitor west, becoming rare east.

DOCUMENTATION

Specimen

UNSM ZM7383 (labeled *shufeldti*), 6 Oct 1920 Monroe Canyon, Sioux Co.

TAXONOMY

Most "Oregon" Juncos reported in Nebraska are of the form *J. h. montanus*, breeding in southwest Canada southeast to western Montana (AOU 1957), but "Pink-sided" Junco, *J. h. mearnsi*, breeding in Alberta, Saskatchewan, Montana, and Wyoming, is also common in the west. The form *shufeldti* is generally treated as part of *montanus*, or, formerly, a subspecies of Oregon Junco (AOU 1957). The only documented record of *shufeldti* is the specimen cited above, but its taxonomic status has been somewhat variable over the years and, as it is currently constituted (Rising 1995), it probably is not to be expected in Nebraska.

DISTRIBUTION AND ECOLOGY

Wintering birds and migrants are generally found in brushy areas and woodland edge with other junco forms. "Oregon" Juncos are much more common westward.

Spring

winter→19, 19, 22 Apr (Tout 1947)

There are later reports 28 Apr 1935 Lincoln Co (Tout 1947) and 30 Apr 1995 Sioux Co.

Fall

30 Sep, 6 (specimen UNSM ZM7386), 6 (Ludlow 1935) Oct→winter

There are earlier reports 19 Sep 1919 Sioux Co (specimen UNSM ZM7387, labeled *mearnsi*, Mickel and Dawson 1920) and 19 Sep 1934 Lincoln Co (Tout 1947).

The limited data available indicate that "Oregon" Junco arrives later in fall and leaves earlier in spring than "Slate-colored" Junco, especially in the east.

Winter

Rosche (1982) stated that "Pink-sided" Juncos (pink-sided birds with gray hoods, dark lores, and concave breast demarcation) are the most common form in the northwest in winter, followed by "Oregon" juncos (black-hooded birds with concave breast demarcation) and "Slate-colored" Juncos. A similar mix was reported by Rosche (1994a) in the Keith Co area, although Tout (1947) stated that "Pink-sided" Juncos were "not common" in Lincoln Co, occurring in the period 10 Oct–22 Apr.

"Oregon" Juncos are far more common in the Panhandle than easterly, as indicated by comparison of data for 10 of the CBC years 1967–91 when the forms were listed separately; on the Scottsbluff CBC 444 of 714 juncos were "Oregons," while at Omaha only 29 of 8229 were "Oregons."

FINDING

Both "Oregon" and "Pink-sided" Juncos are easy to find in junco flocks in the west, especially at feeders such as those at the Wildcat Hills NC.

Dark-eyed ("Gray-headed") Junco

STATUS

Rare casual winter visitor statewide.

DOCUMENTATION

Specimen

UNSM ZM10385, 19 Apr 1911 Red Cloud, Webster Co.

TAXONOMY

This taxon breeds southward in the Rocky Mountains from southern Wyoming, including the Laramie Mountains where it hybridizes with *mearnsi* (Miller 1941). There have been at least 45 reports of "Gray-headed" Junco in Nebraska, but it is likely that most are in fact examples of hybrids between various forms or brightly marked examples of *mearnsi*, especially in light of information provided by Sutton (1975), who collected several birds resembling this form in Oklahoma but found none to be "Gray-headed" Juncos. "Gray-headed" Juncos have no pinkish wash on the flanks.

Winter

There are only 5 documented records, despite about 45 reports and the statement by Tout (1947) that *caniceps* was an "irregular winter resident in Lincoln Co, rather common in some years, rare during others." Andrews and Righter (1992) considered it a rare winter visitor on the eastern Colorado plains.

Documented Nebraska records are as follows. One was at Scotts Bluff NM 12 Sep 1998 (SJD). Another in a Bellevue yard unaccompanied by other juncos 2 Oct 1977 was said to be "like the slightly larger juncos of Colorado with the large rufous patch on the back" (NBR 46:34). A specimen, cited above, was taken 19 Apr 1911 at Red Cloud. One hit a window in Lincoln 27 Apr 1982 and was rehabilitated, banded, and released 18 May (NBR 50:60). A report from Lancaster Co 13 Apr 1982 (AB 36:870) probably referred to the same bird. One was in Grant Co 4 Jun 1986 (NBR 54:47, 81; Mollhoff 1987). A further report is Jun 1981 Garden Co (AB 35:896).

The Nebraska reports as a whole are in the period 12 Sep–25 May, except for 2 Jun reports cited above and an undocumented report 1 Jul 1973 Dawes Co, which was not listed by Rosche (1982).

FINDING

This form should be looked for with other

juncos in winter; any sightings should be carefully documented.

McCown's Longspur

Calcarius mccownii

STATUS

Uncommon regular spring and fall migrant west, rare casual elsewhere. Fairly common regular breeder western Panhandle. Hypothetical in winter.

DOCUMENTATION

Specimen

UNSM ZM7565, 20 Jun 1901 Indian Creek, Sioux Co.

DISTRIBUTION AND ECOLOGY

Breeding birds are fairly common in the western Panhandle wherever the preferred habitat of grazed shortgrass prairie with significant bare patches is found. Most such sites are at a higher altitude than surrounding grasslands. Rosche (1982) observed that the nesting range in Sioux Co extends farther east in dry years but retreats westward when conditions are wet. Migrants occur regularly only as far east as Sheridan Co (Rosche 1982).

Spring

21, 23, 25 Mar→summer

Migrants arrive in early Apr, with earlier reports 5 Mar 1960 Dawes Co, 10 Mar 1956 Cherry Co, 12 Mar 1960 Dawes Co, and 18 Mar 1979 Sioux Co. It was "numerous" in Sioux Co by 18 Apr 1995, and 50 were at Agate 20 Apr 1995.

There are few reports east of the Panhandle: 23 Mar 1958 Douglas Co, 26 Mar 1936 Adams Co, 30 Mar 1965 Lincoln Co, a male with other longspur species during a snowstorm 10 Apr 1997 Knox Co (MB; Brogie 1998), a specimen collected from a flock of Chestnut-collared Longspurs 11 Apr 1919 at Lincoln (Mickel and Dawson 1920), and 21 Apr 1992 Douglas-Sarpy Cos also during a snowstorm.

High counts include 1893 in southwest Kimball Co 17 Apr 1999, 1732 in southern Kimball Co 19 Apr 1998, and 50 at Agate 20 Apr 1995.

Summer

Grazed shortgrass prairie is most extensive in central and western Sioux Co, south of the Pine Ridge escarpment, and breeding numbers are greatest in these areas (Rosche 1982; Johnsgard 1980). High counts include "hundreds" in Sioux Co 13 Jul 1991 (AB 45:1134) and 380 there 19 Jul 1990 (AB 44:1154). Breeding also occurs in western Scotts Bluff, Banner, and Kimball Cos where suitable habitat exists (Rosche 1994b); Huser (1986) noted that nesting McCown's Longspurs were found in Kimball Co "a few miles" north and west of Bushnell, and young were reported in Kimball Co 16 Jun 1991 (AB 45:1134). A "sizeable breeding population" was noted 1993–95 21 km (13 mi) south and 13 km (8 mi) west of Kimball (WRS). Although breeding evidence has not been published for Scotts Bluff Co or Banner Co, there is a series of summer records for Scotts Bluff Co for 1965–69.

Away from the Panhandle, this species is casual, with most reports of migrants. Rosche (1994a) considered the L McConaughy area to be "well to the east of the normal Nebraska range."

Cornell Nest Record data show that nests 13 km (8 mi) west of Harrison and 9.7 km (6 mi) west and 4 km (2.5 mi) north of Agate had eggs 9 Jun 1985 and 27 Jun 1977, respectively.

Fall

summer→25, 27, 27 Oct

Departure is in late Oct, with later reports 8 Nov 1980 Sioux Co, 22 Nov 1986 Sioux Co, and 26 Nov (Rosche 1982). There are few reports east of the Panhandle: 30 Sep 1973 McPherson Co, 3 Oct 1976 near Lewellen (Rosche 1994a), and 19 Oct 1964 Adams Co.

High counts include 1230 in southwest Kimball Co 13 Oct 1998.

Winter

There are 13 published reports in the period 15 Nov–5 Feb, all east of the Panhandle, none documented, and all prior to 1974. Wintering is not expected in Nebraska. In Kansas McCown's Longspur is sporadic in winter in the west (Thompson and Ely 1992). There are no South Dakota (SDOU 1991) or Iowa (Kent and Dinsmore 1996) records for the winter period. There are, however, 2 Missouri records for midwinter (Robbins and Easterla 1992).

FINDING

McCown's Longspur is conspicuous along the Wyoming border south of Highway 20 in Sioux Co in Jun. Migrants and wintering birds should be looked for in the Panhandle and southwest in Oct and Apr; they usually occur with Lapland Longspurs (Thompson and Ely 1992). In the east flocks of Lapland Longspurs alongside roads after Mar and early Apr snowstorms should be checked for McCown's Longspurs.

Lapland Longspur

Calcarius lapponicus

STATUS

Abundant regular spring and fall migrant and winter visitor statewide.

DOCUMENTATION

Specimen

UNSM ZM7581, 25 Oct 1890, Lincoln, Lancaster Co.

TAXONOMY

Two subspecies occur in Nebraska; the wintering ranges of both western *alascensis* and eastern *lapponicus* include the entire state (AOU 1957; Glandon and Glandon 1937).

DISTRIBUTION AND ECOLOGY

Wintering birds are common to abundant statewide, occurring in open habitats with exposed bare ground, both in relatively short grasslands and cultivated fields. Flocks are often seen along plowed roadsides after snowstorms.

Spring

winter→15, 15, 15 Apr

Migrants depart by mid-Apr. Peak migration is probably in late Mar and early Apr, when 10 000 were in Adams Co 22 Mar 1993 (AB 47:428), "at least" 10 000 were along 9.7 km (6 mi) of road in Knox Co 10 Apr 1997, and 2000 were in Dixon Co 11 Apr 1995. There are late reports 24 Apr 1995 Adams Co and 28 Apr 1951 Gage Co.

Fall

9 (Rosche 1994a), 9, 9 Oct→winter

Migrants appear in mid-Oct, with an earlier report 25 Sep 1960 McPherson Co. Peak migration is probably in late Nov and Dec; large numbers seen during the CBC period often decline markedly during midwinter (see Winter).

High counts include "thousands" in the eastern Rainwater Basin 7 Nov 1999, more than 1000 in Platte Co 12 Nov 1986 (AB 41:112), and 1000 near Wakefield 11 Nov 1998.

Winter

Lapland Longspur occurs statewide in winter, often with Horned Larks but in pure flocks also, especially when large numbers are present. Both Rosche (1982) and Thompson and Ely (1992) refer to the propensity for large numbers to congregate over towns during snowfalls. The largest flocks occur in the west; Rosche (1982) stated that peak numbers are found in wheat stubble and sugar beet fields near Alliance and Hemingford.

This species moves around during the winter in response to variations in extent of snow cover, resulting in the large variance in numbers from year to year. CBC data reflect this; in the period 1967–93 the high count was 23 452 at Lincoln 17 Dec 1978. These birds were in mixed flocks with Horned Larks and were "gone by January" (NBR 47:37), suggesting that migration or at least significant movement is still in progress in mid-Dec. By contrast, none were reported on CBCs in 1990–91.

High counts include 50 000 in Dodge Co 16 Jan 1999, 10 000 in Dixon Co 20 Feb 1997, and 5000 in York Co 28 Jan 1995.

Other Reports

10 May 1975 Adams Co, 3 Jun 1980 Scotts Bluff Co.

FINDING

Lapland Longspurs are best looked for along roadsides after snowstorms during the winter months, when they are usually mixed with Horned Larks. Large pure flocks are more often seen in the west and may be located by their sheer size when present.

Smith's Longspur

Calcarius pictus

STATUS

Uncommon casual spring migrant southeast. Uncommon casual fall migrant statewide. Accidental in winter.

DOCUMENTATION

Specimen

UNSM ZM7609, 25 Oct 1890 Lincoln, Lancaster Co.

DISTRIBUTION AND ECOLOGY

Smith's Longspur is limited in spring to the southeast corner of the state; at this season and in this region it occurs in areas of short, thick grass such as airports, golf courses, or pastures that have been grazed or mowed, but it also can be found in harvested cornfields (David Easterla, pers. comm. WRS). In general, it occurs in the same types of fields preferred by Sprague's Pipit (Mark Robbins, pers. comm. WRS). In fall and winter grasslands of all types are utilized. Fall migrants have been reported statewide.

Spring

There are about 10 reports in the period 21 Mar–3 May, all in the southeast: 21 Mar–2 Apr 1963 Webster Co; 31 Mar 1974 Lancaster Co (NBR 42:77); 4 after a snowstorm north of Blair in Washington Co 10 Apr 1997 (JGJ); 1 after a snowstorm in southeast Washington Co 10 Apr 1997 (JGJ); 12 Apr 1999, a molting male in Lancaster Co (JS); 14 Apr 1952 Webster Co; a single bird seen during a blizzard 15 Apr 1986 Douglas Co (Mollhoff 1987); specimens UNSM ZM7606–08 taken 20 Apr 1901 in Lancaster Co (Bray and others 1986); 28 Apr 1957 Adams Co; and 3 May 1933 Jefferson Co.

Although it seems likely that migrants regularly pass through Nebraska, essentially east of a line from Washington Co to Webster Co, since 1980 this species has been reported very few times. In all there are fewer than 30 published reports for the state; these reports indicate that Smith's Longspur occurs during migration both in spring and fall and possibly in winter (see Winter). There are only a few reports for Colorado (Andrews and Righter 1992) and only 1 for Wyoming (Scott 1993). This information, taken together with data from Missouri, Kansas, Iowa, and South Dakota, indicates that spring migrants occur most often in the original tallgrass prairie region from eastern Nebraska eastward, but, as such habitat is now extremely limited in Nebraska, birds are found in shortgrass locations (see Distribution and Ecology).

Fall

There are fewer than 15 reports; these are in the period 18 Sep–22 Nov and are distributed statewide. Recent reports suggest that males still in alternate plumage may appear early in Sep: one was at Minot, North Dakota, 6 Sep 1997; a male was at Denver, Colorado, the same day; and another was reported in northwest Iowa 6 Sep 1996. The latter report was not accepted by the Iowa Ornithologists' Union Records Committee, however (Kent 1997).

Nebraska reports are 18 Sep 1975 Scotts Bluff Co, 27 Sep 1981 Sheridan Co (Rosche 1982; AB 36:194), 5 Oct–22 Nov 1965 Adams Co, 9 Oct 1962 Webster Co, 23 Oct 1964 Adams Co, 24 Oct 1985 Cass Co (AB 40:137), specimens UNSM ZM7609 and ZM7610 taken 25 Oct 1890 Lancaster Co (Bray and others 1986), 29 Oct 1966 Adams Co, 12 Nov 1966 Custer Co, 13–17 Nov 1963 Webster Co, 3 on the Lincoln CBC 17 Dec 1978, and 2 on the Lincoln CBC 20 Dec 1992 (NBR 61:13).

A report of a flock of 15 in Morrill Co 2 Sep 1984 was not accepted by the NOURC (Mollhoff 1987; NBR 53:81).

Winter

There are only 2 reports for the winter period, 1 documented: 4 were seen at Pawnee L 12 Feb 1988 (Mollhoff 1989a). The other report is 5 Feb 1959 Nemaha Co.

Records from surrounding states suggest that Smith's Longspur winters as close as east-central Kansas, where it is regular in the Flint Hills (Thompson and Ely 1992).

Other Reports

22 May 1966 Sioux Co, "prior to 23 May" 1961 Sheridan Co.

FINDING

Nebraska data as well as those from neighboring states indicate that migration peaks in Oct and from late Mar through Apr. During these periods sites with short, thick grass should be checked, such as airports, golf courses, or thick pastures that have been mowed or grazed. In Mar and Apr roadsides should be checked after snowstorms; reports indicate that spring migrants occur east of a line from Webster Co to Washington Co.

Chestnut-collared Longspur

Calcarius ornatus

STATUS

Common regular spring and fall migrant west, becoming rare east. Common regular breeder west, locally uncommon north-central. Hypothetical in winter.

DOCUMENTATION

Specimen

UNSM ZM7614, 20 Mar 1900 Mullen, Hooker Co.

DISTRIBUTION AND ECOLOGY

This species breeds in native grasslands somewhat taller and more moist than those used by McCown's Longspur (Rosche 1982). It is most numerous in the Panhandle, but it also occurs in small numbers eastward near the South Dakota border possibly as far as the Missouri Valley (Johnsgard 1979).

Migrants are occasionally found in the east but occur most often in the Panhandle, where they frequent a variety of grassland and agricultural habitats but are most common in wheat stubble (Rosche 1982).

Spring

12, 14, 15 Mar→25, 28, 30 Apr

Migration occurs mid-Mar through mid-Apr, with an earlier report of 25 on 6 Mar 1994 in Dixon Co and later reports 7 May 1980 near Valentine (NBR 48:73) and 8 May 1951 Antelope

Co, both locations where breeding is possible. Rosche (1994a) considered a specimen, UNSM ZM7611, collected 5 May 1919 near Oshkosh to be a migrant.

Migrants are regular in the Panhandle but are casual elsewhere. Easterly reports are few, although the species may be conspicuous after spring blizzards such as those on 14–15 April 1986 (NBR 54:68) and 10 Apr 1997. Reports are of 42 birds north of Blair in Washington Co 10 Apr 1997 (JGJ), 14 Apr 1986 Dakota Co, 14–15 Apr 1986 Douglas-Sarpy Cos (AB 40:493), 15 Apr 1985 Dakota Co, 15 Apr 1986 Pierce Co, and 21 Apr 1992 during a snowstorm Douglas-Sarpy Cos.

High counts include 1045 in southwest Kimball Co 16 Apr 1997 and 500+ near Crescent L 3 May 1978 (AB 32:1028).

Summer

Chestnut-collared Longspur breeds throughout the Panhandle but is most numerous in "certain pastures on the edge of the sandhills region" (Rosche 1982). It breeds south to Kimball Co, where breeding birds were found a few miles northwest of Bushnell in 1986 (Huser 1986), and it was reported in Kimball Co 26 May–10 Jun 1985 and on a BBS route in central Kimball Co in 1993 (WRS). It has been reported there each year since and was considered common in Kimball Co 24–25 Jun 1995. There are summer reports for Scotts Bluff Co 1 Jun 1969, 1–7 Jul 1968, and 17 Jul 1966 and for Morrill Co 29 May 1992, and there is a specimen, HMM 21796, taken at Oshkosh 7 Jun 1915.

The breeding range as outlined by Johnsgard (1980) included southern Sioux and Box Butte Cos northeastward to Sheridan Co and "perhaps north to Cherry County." Youngworth (1955) reported breeding birds in northeast Cherry Co in 1932, and Ducey (1988) cited breeding season reports by Rosche eastward to pastures on the edge of the sandhills in Cherry Co and certain meadows in Keya Paha Co. Two males were seen in a "historical nesting locality" in Keya Paha

Co 29 May 1989 (AB 43:501), and Brogie and Mossman (1983) had reports 11–14 May 1982 in the same general area. Summering birds were reported a few kilometers northeast of O'Neill in Holt Co 21 May–17 Aug 1991 (Blake and Ducey 1991) and again in 1992, when 20 were seen 5 May (AB 45:468). In Holt Co 7 were noted 22 May 1994 (NBR 62:86), and "about 35" were noted in Holt Co 14–31 Jul 1996 (FN 50:967). One reported 17 Sep 1989 in Holt Co may have been summering.

Fall

summer→21, 21, 23 Oct

Departure ends by late Oct; there are later reports 29 Oct 1996 at Crescent L NWR and 8 Nov 1980 Sioux Co. There is insufficient information to determine early dates for fall migration.

Easterly reports are fewer than in spring: 24 Sep 1964 Adams Co, 25 Sep–10 Oct 1965 Adams Co, 17–19 Oct 1963 Webster Co, 19 Oct 1964 Adams Co, 19–22 Oct 1962 Webster Co, and 23 Oct 1967 Adams Co.

High counts include 753 in southwest Kimball Co 13 Oct 1998 and 150 at Clear Creek Marshes 26 Sep 1999.

Winter

There are a few reports for Dec–Feb that raise the possibility that this species occasionally winters; most of these reports, if correct, may be late fall or early spring migrants: 20 Dec 1990 Cass Co, 1 Jan 1964 Gage Co, 4 Jan 1955 Hamilton Co, 4 Jan 1957 Webster Co, 8 Jan 1958 Hamilton Co, 9 Jan 1956 Keya Paha Co, 25 Jan 1956 Cherry Co, 15 Feb 1956 Gage Co, 17 Feb 1970 Lincoln Co, and 20 Feb 1957 Cherry Co. It was listed as "commonly seen in winter" in Logan Co (Glandon and Glandon 1934b).

None of these reports are documented, especially those from northern parts of the state. A report of 50 on the Whitney CBC 31 Dec 1972 appears to be an error. Information from Missouri (Robbins and Easterla 1992) and Kansas (Thompson and Ely 1992) indicates that wintering in Nebraska would

be extremely unlikely. The latter authors suggested that many or most Kansas winter reports were doubtful due to confusion with winter-plumaged Lapland Longspur.

Other Reports

10 May 1964 Gage Co.

FINDING

Breeding birds can be found in the northwest in summer in the following locations, although variation in moisture levels causes populations to fluctuate and often relocate: pastures to the south and east of Walgren L; 4.8 km (3 mi) north and 3.2 km (2 mi) east of the junction of Highways 2 and 385 in Alliance; and along the Wyoming border several kilometers south of Highway 20, past the main breeding areas of McCown's Longspur and where the prairie grassland is less heavily grazed. There also is a good population most years at the historical marker 16.1 km (10 mi) south of Harrison. In the east, as with other longspurs, migrants should be looked for along plowed roadsides in Apr after heavy snowstorms.

Snow Bunting

Plectrophenax nivalis

STATUS

Uncommon regular spring and fall migrant and winter visitor statewide.

DOCUMENTATION

Specimen

UNSM ZM7618, 17 Nov 1897 Lincoln, Lancaster Co.

DISTRIBUTION AND ECOLOGY

This species is often found with Horned Larks and Lapland Longspurs in exposed, often windswept open areas and along roadsides; it also frequents exposed sandy areas near water, such as reservoir beaches and parking lots.

Spring

winter→10, 11, 12 Mar

Departure is completed by early to mid-Mar, with later reports 17 Mar 1955 Adams Co, 23 Mar 1971 Dawes Co, and 23 Mar 1975 Lincoln Co.

Fall

26, 27, 27 Oct (AB 38:220)→winter

Arrival is in early Nov; there is an early report 19 Oct 1976 Lancaster Co.

Winter

Snow Bunting is near the southern edge of its winter range in Nebraska, although in recent years it has become more common, especially around reservoirs, a phenomenon discussed by Thompson and Ely (1992). Of about 130 reports considered here, only 23 were prior to 1970. Rapp and others (1958) considered Snow Bunting a "very irregular" winter visitor that was absent from the state for long intervals. This is not now the case, as it occurs every year, although not usually in large numbers. Highest CBC total was 23 birds in 1981–82; Snow Buntings are reported in about a third of the CBC years.

High counts include 3000 in Sheridan Co 24 Feb 1978 (AB 32:372), 3000 in Sheridan Co 7 Feb 1997 (FN 51:767), 1000 in Wayne Co 7 Jan 1997, and "hundreds" in the Alliance area 24 Jan–6 Feb 1982 (AB 36:308).

FINDING

Snow Bunting should be looked for with flocks of Horned Larks and longspurs along roadsides in snowy weather, especially in the northern part of the state Dec–Feb. Sandy or stony beaches at reservoirs are also likely places.

Cardinalidae (Cardinals, Grosbeaks, New World Buntings)

Northern Cardinal

Cardinalis cardinalis

STATUS

Common regular resident south and east, uncommon north, locally rare west. Rare regular spring, fall, and winter visitor outside breeding range.

DOCUMENTATION

Specimen

UNSM ZM7063, Dec 1884 Unadilla, Otoe Co.

DISTRIBUTION AND ECOLOGY

This species is found in brushy habitats both in and at the edge of woodland and in urban areas, often riparian-associated. It is common in the southeastern two-thirds of the state but is at best local in the Panhandle.

Spring

Normally a resident species, at the edge of its range Northern Cardinal exhibits some migratory tendencies. According to Bent (1958), "records show a decided trend of movement northeastward and northward in the fall and late summer, which may account for . . . the eventual northward spread of the species." There are winter records for most areas in Nebraska beyond the breeding range; such birds may establish breeding populations if winter conditions allow survival. Data from Scotts Bluff Co show that until the recent establishment of a small breeding population it had been recorded most often between early Apr and late May.

Johnsgard and Rosche (1984) considered it mainly a spring and fall migrant in Keith Co. Rosche (1994a), however, noted that fluctuations in numbers in the same area were apparently related to the severity of winter conditions.

Summer

Since the early 1900s Northern Cardinal has gradually expanded its range from the southeast (Bruner and others 1904) across the state westward, primarily along major river valleys (Short 1961) as they develop mature stands of woody vegetation. As of 1958 it was breeding commonly in the eastern half of Nebraska and was uncommon to rare westward, with records as far west as the Panhandle (Rapp and others 1958).

Ducey (1988) and Rosche (1994a) have suggested that expansion westward in the Platte Valley is related to an increase in riparian woody habitat, as controlled low river flows have allowed encroachment of vegetation onto previously bare areas in the Platte River channels. This is, however, but the most recent contributing factor to the westward expansion of riparian woodland. The reduced frequency of fire after European settlement allowed the natural succession to woodlands along the Platte River from about Grand Island westward (see Introduction). Damming of the North Platte River, for example, did not occur until about 1940 with the building of Kingsley Dam, yet the range expansion of the cardinal was already well under way (Tout 1947). It was first reported in Lincoln Co at Sutherland 2 Nov 1933 (Tout 1947) but not until 6 Apr 1941 at North Platte and 13 Apr 1954 at Scottsbluff. Its occurrence is sporadic in the North Platte Valley west of Oshkosh (Rosche 1994a).

Several were present in Ash Hollow in extreme southeast Garden Co Apr–May 1997 (SJD). It was found at 6 Morrill Co locations in 1985 where previously only 1 location was occupied (AB 39:322). After first being recorded in Scotts Bluff Co in 1954, there were several Dec and Jan reports through 1976, followed by reports only in spring and summer through 1985. There were no additional reports until 1992, when it was listed as a permanent resident, although breeding was not noted until a male was seen feeding young near the North Platte River east of Scottsbluff 31 Jul 1994 (NBR 62:114). A small population has been present ever since at the same location, with the maximum count 5 in fall 1995 (AK). There is a recent report for Cheyenne Co 10 Jun 1990.

Short (1961) found Northern Cardinal in the South Platte Valley as far as Colorado in Jun 1956; it currently occurs there mainly as a migrant and winter resident (Andrews and Righter 1992). It is fairly common at Ogallala.

As long ago as the early 1900s it occurred west in the Republican Valley to Harlan Co (Bruner and others 1904); Ludlow (1935) noted that it first appeared in Webster Co in 1915. There are recent reports 1 May 1992, 3 Jun 1989, and 12 Jun 1999 from Dundy Co and 22 Jun 1994 from the North Fork of the Republican River just west of Dundy Co (Andrews and Righter 1992). It was apparently common along Frenchman Creek in eastern Chase Co

as early as 1945 (Maddux 1989) and has been reported from the area in recent years; 4–5 were at Enders Res 7–12 Jun 1999.

The first record for the upper Elkhorn Valley is a midwinter specimen, UNSM ZM10621, collected at Neligh 11 Jan 1909. Ducey (1988) cited Bruner's (1896) statement that it bred at West Point, establishing its presence there prior to 1900. It was breeding in Wayne Co by 1953 (Gates 1953) and in Boone Co by 1984 (Bennett 1984).

Northern Cardinal has been recorded in the Loup drainage for some time. It was recorded at Stapleton as early as spring 1933 and subsequently 14 Oct 1934 (Glandon and Glandon 1935a), 1 Jan 1944, 13 Jan 1946, and 1 Jan 1948 and apparently regularly from that time until 30 Jun 1969; there have been no reports from Logan Co since then. The earliest report for Cherry Co is from Elsmere in the Loup drainage 25 May 1952. It was first reported at Halsey as early as the early 1940s, when 4–5 nesting pairs were present; Zimmer (1913) had earlier unverified reports in winter, however, and it is now a "common permanent resident" there (Bray 1994).

In the Niobrara Valley this species occurs west at least to Anderson Bridge, some 32.2 km (20 mi) west of Valentine, where Ducey (1988, 1989) noted an increase in numbers during the period 1984–88. Brogie and Mossman (1983) found several singing males and territorial birds at the Niobrara Valley Preserve in 1982, listing the species as an uncommon "probable nester." Short (1965) found no Northern Cardinals at his camp southwest of Merriman in 1964 but found a pair and collected the male (USNM 480458) south of Rushville in Sheridan Co in summer 1964. This pair was territorial, and the male had enlarged testes.

Data presented by Rosche (1982) for the Pine Ridge region suggest a tenuous hold at best; cattle tend to severely damage the preferred habitat of riparian thickets. Northern Cardinals were apparently seen

along Bordeaux Creek, a tributary of the
White River east of Chadron in Dawes Co,
in 1935 and 1936 by MacKinlay (1936). There
were no reports until 8 Apr 1954 and 2 May
1956 Dawes Co, followed by reports 22 May
1957 Sheridan Co and 2 Apr 1959, 7 Sep 1959,
and 20 May 1960 Box Butte Co. Apart from
the Sheridan Co record by Short described
above, there were no more reports in the Pine
Ridge region until 15 Nov 1987 Dawes Co, and
the only report since is of one seen 26 Jul 1997
about 19 km (12 mi) east of Chadron.

Fall
See Spring.
Winter
See Spring.
FINDING
Northern Cardinal is easy to find in the east. It
occurs in urban areas with brushy borders
and in woodland edge and is easily attracted
by squeaking almost any time of year. It is a
common feeder bird in winter.

Rose-breasted Grosbeak

Pheucticus ludovicianus
STATUS
Common regular spring migrant east,
uncommon central, rare west. Common
regular breeder east. Common regular fall
migrant east, rare central, accidental west.
DOCUMENTATION
Specimen
UNSM ZM12080, 31 May 1895 Lancaster Co.
TAXONOMY
This species hybridizes with Black-headed
Grosbeak where their ranges meet (Short
1961; Swenk 1936). Both pure birds and
hybrids are found in the zone of overlap
(Short 1961), an indication of some degree of
assortative mating, the primary reason why
these species have retained specific status
(Mayr and Short 1970; AOU 1983; Sibley and
Monroe 1990).
Individuals showing characters indicative of
hybridization or introgression are widespread
in central Nebraska. Short (1961) found

phenotypically pure Rose-breasted Grosbeaks
west to Burwell and Elm Creek but evidence
of hybridization in the Platte Valley from
Colfax to Dawson Cos, and in Saline, Adams,
Howard, and Holt Cos.
Hybrid specimens UNSM ZM7076 and UNSM
ZM7077 were collected at Grand Island 28 Jun
1930 and Inland 18 May 1920, respectively,
and another hybrid specimen, HMM 2897,
was taken at Inland 24 May 1923. Farther
to the west in the North Platte Valley at L
Ogallala, 2 of 5 banded birds that appeared
to be Rose-breasted Grosbeaks showed some
Black-headed Grosbeak characteristics, while
the reverse was true of 5 of 35 Black-headed
Grosbeak types (Brown and others 1996).
Hybrids, along with pure Black-headed
Grosbeaks, were common in the Republican
Valley between Orleans and Oxford 23 Jun
1996; pure Rose-breasted Grosbeaks were
noted west only to Orleans (WRS, JGJ). A
female Black-headed Grosbeak was with a
male Rose-breasted Grosbeak at Harlan Co
Res 13 May 1995 (NBR 63:56).
A hybrid male at Box Butte Res 8 Aug 1992 was
described by Rosche and Rosche (1993).
DISTRIBUTION AND ECOLOGY
This species occurs in fairly open deciduous
woodland with a well-developed understory
of gray dogwood, ashleaf maple, and
mulberry, most commonly in the east
but westward in diminishing numbers in
floodplain woodland to central Nebraska.
Spring
23, 24, 24 Apr→26, 29, 31 May
Late dates above are for areas away from the
breeding range.
Migrants generally appear in early May, although
early dates are in late Apr. There are earlier
undocumented reports 15 Apr 1959 Gage Co,
15 Apr 1960 Gage Co, 17 Apr 1934 Jefferson
Co, 18 Apr 1980 Douglas-Sarpy Cos, 20
Apr 1954 Thayer Co, 20 Apr 1991 Hall Co,
and 21 Apr 1956 Lancaster Co. These early
dates are questionable, as spring arrival in
Missouri (Robbins and Easterla 1992), Kansas

(Thompson and Ely 1992), and Iowa (Kent and Dinsmore 1996) is around 25 Apr, and the earliest date for Kansas is 13 Apr.

In areas west of the breeding range where this species is essentially a migrant departure is in late May. As far west as the Panhandle it is casual, with reports in the period 28 Apr–29 May. As far west as L Ogallala it can be fairly numerous in spring; 15 were counted 2 May 1997 (Brown and others 1996).

High counts include 61 in Sarpy Co 11 May 1996, 59 there 13 May 1995, 30 in Dakota Co 17 May 1997, and, unusual as far west as Sioux Co, 3 at James Ranch 12 May 1996.

Summer

Bruner and others (1904) stated that Rose-breasted Grosbeak bred in the eastern third of Nebraska, west to about Grand Island, whereas Rapp and others (1958), while noting that the species bred in the eastern half of Nebraska, stated that "there is evidence that this bird is moving westward through the major river valleys." Recent information does not support any further significant westward spread, however; BBS data (1967–77) show high numbers only in the east. Johnsgard (1979) considered the western extent of the breeding range difficult to discern due to hybridization with Black-headed Grosbeak but indicated that Rose-breasted Grosbeak bred west to Holt, Garfield, and Phelps Cos, similar to the western extent of the range of phenotypically pure Rose-breasted Grosbeaks noted by Short (1961).

Since 1960 there are breeding records west in the Platte Valley to Adams and Howard Cos (Ducey 1988), along with numerous summer reports from that area and Hall Co. It may summer regularly in the Gibbon area, where 3 females and a male summered in 1997; 4, including a juvenile 14 Jul, were there in 1998 (LR, RH), and there were summer reports 3 Jun 1992, 28 Jun 1991, and 1 Aug 1999. Additional summer reports farther west in the Platte Valley are 12 Jun 1951 Keith Co, 14 Jun 1950 Keith Co, and 10 Jul 1982 Lincoln Co along with a record from Perkins Co 14 Jun 1972, but these records may be of unmated late spring migrants. Rosche (1994a) considered this species a "possible accidental summer resident" in the Keith Co area, based on a single unsubstantiated summer report 18 Jul–19 Aug 1977. There are additional westerly reports 11 Jun 1981 and 12 Jun 1980 Garden Co and 6 Jun 1972, 6 Jun 1984, and 7 Jun 1985 Scotts Bluff Co.

There are few summer reports from the northwest, none with evidence of breeding: a singing male was in Dawes Co 19 Jul 1979 (Rosche 1982), one was at a feeder northeast of Chadron 20 Jun 1998, and it was reported in Dawes Co 10 May 1991 (AB 45:468) and 8 Aug 1992.

Breeding in the Niobrara Valley currently seems limited to the lower portion, particularly Knox Co, where there are several recent summer reports. The meeting point of the breeding ranges of Rose-breasted and Black-headed Grosbeaks in the Niobrara Valley has apparently fluctuated over the years. In 1902 Swenk (Ducey 1983a) found no Rose-breasted Grosbeaks in the region where Keya Paha, Brown, and Rock Cos meet, but in the 1930s he found the species breeding in Brown Co (Swenk 1936). Youngworth (1955) found Rose-breasted Grosbeak breeding in northeast Cherry Co at about the same time. Brogie and Mossman (1983) found Rose-breasted Grosbeak common in the Niobrara Valley Preserve, but it appeared only as a spring migrant in May, and "evidently none stayed to nest." There are these few recent summer reports from this area: 12 Jun 1988 Keya Paha Co, 16 Jun 1977 Cherry Co, 1 Jul 1969 Brown Co, 4 Jul 1989 Holt Co, and 9 Aug 1964 Brown Co.

Recent information is lacking also for the Loup drainage, except for a recent report 28 Jun 1993 Thomas Co; it currently is considered an "uncommon migrant" of uncertain status there (Bray 1994). The presence of a male, female, and immature in Logan Co as long ago as 2 Jul 1936 was considered evidence of breeding nearby (Glandon and Glandon 1937).

There is little recent data from the Republican Valley, where Swenk (1936) found it west to Inavale; it was found west to Orleans 23 Jun 1996 (WRS, JGJ).

Fall

summer→7, 10, 11 Oct

Departure is complete by early Oct, with later reports 19 Oct 1966 Douglas-Sarpy Cos, 21 Oct 1966 McPherson Co, 25 Oct 1975 Lincoln Co, 29 Oct 1984 Lancaster Co, and 29 Oct 1989 Douglas-Sarpy Cos and an earlier report 15 Sep 1983 Garden Co (AB 38:220), the westernmost and only Panhandle report.

Rose-breasted Grosbeaks appear occasionally at feeders in late fall and early winter; these birds apparently attempted to winter: 11 Nov 1974 Douglas-Sarpy Cos, 22 Nov 1973 through mid-Jan 1974 at feeders in Douglas-Sarpy Cos (AB 28:76, 659), 31 Dec 1980 Boone Co, and 31 Dec 1980 Lancaster Co.

High counts include 40 in Sarpy Co 9 Sep 1995.

Other Reports

25 Mar 1972 Hall Co, 25 Mar 1984 Howard Co, 6 Apr 1943 Jefferson Co, 10 Apr 1962 Antelope Co, 10 Apr 1974 Howard Co, 12 Apr 1963 Gage Co, 12 Apr 1976 Howard Co, 12 Apr 1985 Lincoln Co.

FINDING

Rose-breasted Grosbeak is easy to find in May and Jun in the Missouri Valley; it is easily attracted by squeaking.

Black-headed Grosbeak

Pheucticus melanocephalus

STATUS

Common regular spring and fall migrant and breeder west and central, rare casual east. Rare casual winter visitor east.

DOCUMENTATION

Specimen

UNSM ZM7085, 13 Jul 1901 Sioux Co.

TAXONOMY

This species hybridizes with Rose-breasted Grosbeak where the ranges of the 2 species

meet in major river valleys (Short 1961; Swenk 1936); see Rose-breasted Grosbeak.

DISTRIBUTION AND ECOLOGY

Black-headed Grosbeak occupies riparian deciduous woodland dominated by trees of a shrubby aspect such as box elder, green ash (*Fraxinus pennsylvanica*), and Siberian elm. The breeding range includes most of the western half of Nebraska; it occurs on occasion as a migrant and winter visitor east of the breeding range.

Spring

23, 24, 24 Apr→summer

Spring arrival is typically in early May, with earliest reports in late Apr.

Migrants occur regularly east to Antelope, Boone, Howard, Hall, and Adams Cos, and there are a few reports farther east, only one since 1981: 25 May 1985 Saunders Co (AB 39:321). Other easterly reports are in the period 23 Apr–28 May, including a specimen, UNSM ZM10410, taken at Lincoln 20 May 1911.

Summer

According to Johnsgard (1980), this species breeds commonly in the west and eastward in the Niobrara Valley to Rock Co, in the Loup drainage to Garfield Co, and along the Platte Valley to Hall Co, the same easterly limits given by Short (1961). It is common in the Niobrara Valley Preserve, where Brogie and Mossman (1983) found several nests in 1982, and probably in the Loup drainage, where it was noted in Loup Co 18 Jun 1994 and was listed as an "uncommon summer resident and nester" in Thomas Co (Bray 1994), although "rather more common" than Rose-breasted Grosbeak (RG).

Its range in the Platte Valley has retracted to some extent in recent years. Ducey (1988) indicated that it was more widespread formerly and cited breeding records east to Lancaster and Jefferson Cos, apparently in the 1920s. Ludlow (1935) noted that it was more common in summer than Rose-breasted Grosbeak in Webster Co. Rosche (1994a) described it as only a casual summer resident in the Keith Co area, stating that it was "probably well east of its more normal range" there. There are, however, recent reports from Clear Creek Marshes 16 Jul 1994 and Lincoln Co 10 Jun 1993 and 13–14 Aug 1992, and since 1991 it has been reported regularly in summer from Buffalo and Phelps Cos.

It probably breeds regularly east along the Republican Valley to the Harlan Co Res area, where it was reported 29 May 1994 and, several were noted (along with hybrids) 23 Jun 1996 (WRS, JGJ). An adult male was at Limestone Bluffs, Franklin Co, 3 Jul 1998 (LR, RH).

There are few summer reports from the east since the 1920s: 11 Jun 1937 Lancaster Co, 28 Jun–1 Jul 1971 Lancaster Co, 10 Jul 1973 Lancaster Co, and 30 Jun 1964 Hamilton Co. Tout (1947) cited egg dates (N = 8) for Lincoln Co 17 May–17 Jun.

Fall

summer→25, 25, 27 Sep (west, central)

summer→29, 30 Sep, 1 Oct (east)

Departure is in mid to late Sep, with late dates in most of the state in late Sep. Farther east, where the species is casual in fall, there are later reports and even a few winter records.

There is a very late report, undocumented, 16 Nov 1982 Douglas-Sarpy Cos, although hybrids appeared at Lancaster Co feeders 23 Aug–19 Sep 1980, 9 Aug–22 Sep 1981, and 23 Aug–27 Sep 1982, and there are occasional winter reports also in the east (see Winter).

Winter

This species has a tendency to occur in winter at feeders east of its normal range. There are these Jan reports: 11 Jan–May 1970 Douglas-Sarpy Cos, Jan 1981 Cass Co (AB 35:314), and 5 Jan 1986 Lincoln Co. The 1970 bird was at an Omaha feeder and stayed until the owners departed on vacation in May (NBR 38:85).

FINDING

This species can be found fairly easily in riparian habitats in the west in late May and Jun. It is best located by song, which is very similar to that of Rose-breasted Grosbeak, and it responds to squeaking.

Blue Grosbeak

Guiraca caerulea

STATUS

Uncommon regular spring and fall migrant and breeder statewide.

DOCUMENTATION

Specimen

UNSM ZM7091, 8 Aug 1900 Beatrice, Gage Co.

TAXONOMY

Two subspecies have been reported to occur, but available information regarding the eastern subspecies is conflicting. The AOU (1957) stated that the western race *interfusa* occurred east to east-central Nebraska and did not list the eastern race *caerulea* as breeding in the state. Furthermore, Dinsmore and others (1984) thought that the Blue Grosbeaks breeding in western Iowa were of the western race *interfusa*, which coincides with the opinion of Bruner and others (1904) that western *lazula* (= *interfusa*) bred statewide. However, eastern *caerulea* was listed as occurring in "the eastern part of the state" by Rapp and others (1958) and has been stated to breed "north sparingly to . . . Nebraska" (Dwight and Griscom 1927). Evidence for occurrence of this subspecies in Nebraska is lacking, however.

DISTRIBUTION AND ECOLOGY

Blue Grosbeak is an open-country species that utilizes isolated islands of brushy habitat in both riparian and upland locations. Low-growing woody vegetation along fencerows in agricultural habitat often attracts Blue Grosbeak, as do thickets of wild plum in grassland habitats, including Sandhills grasslands. It tends to be local in distribution, with no particular center of abundance.

Spring

30, 30 Apr, 1 May→summer

Migrants arrive in early May.

Summer

BBS data indicate relatively even distribution statewide, but slightly higher numbers were recorded in the north and west. Blue Grosbeak is not numerous anywhere in Nebraska but occurs throughout the state in suitable habitat.

It is probably least numerous in the southeast; Johnsgard (1980) indicated that breeding is "highly local" in eastern counties. In the northwest Rosche (1982) considered it rare as a breeder, occurring mainly in the White River drainage, and unrecorded in Sioux Co. Since then, however, there are several summer reports in Sioux Co. Brogie and Mossman (1983) found it uncommon in the Niobrara Valley Preserve in 1982, and Bray (1994) considered it a "common summer resident" in Thomas Co.

Fall

summer→1, 3, 3 Oct

Departure is completed by late Sep, with later reports 10 Oct 1968 Brown Co, 10 Oct 1970 Scotts Bluff Co, and 13 Oct 1968 Dawes Co.

Other Reports

4 Jan 1948 Lincoln Co, 18 Mar 1955 Boyd Co, 9 Apr 1943 Jefferson Co, 12 Apr 1927 Webster Co (Ludlow 1935), 18 Apr 1958 Lincoln Co, 20 Apr 1942 Jefferson Co.

FINDING

This species can be located in dry brush almost anywhere in Jun, but the best places are probably in central and southwest Nebraska. Look for them in isolated plum thickets in the east-central and southern Sandhills. Males sing from prominent perches, and individuals usually respond to squeaking.

Lazuli Bunting

Passerina amoena

STATUS

Uncommon regular spring and fall migrant west, becoming rare casual east. Uncommon regular breeder west.

DOCUMENTATION

Specimen

UNSM ZM10346, 11 Jul 1910 Glen, Sioux Co.

TAXONOMY

This species and Indigo Bunting hybridize where their summer ranges meet in the northern Great Plains (Sibley and Short 1959).

However, because of sympatric occurrence without interbreeding in the southwest United States and nonrandom hybridization in the northern Great Plains, they are generally regarded as separate species (Mayr and Short 1970; AOU 1983; Sibley and Monroe 1990).

It is possible that some reports of Lazuli Bunting east of the rather restricted Nebraska breeding range may be of hybrids that resemble Lazuli Bunting more than Indigo Bunting. Short (1961) found hybrids in an 800 km (500 mi) wide zone from Blair west to Greeley, Colorado, and stated that hybrids should be expected statewide although more commonly in the west. Nine of 18 birds collected by Short (1961) in eastern Nebraska showed backcross characteristics, although this would have been difficult to discern in the field. A bird resembling an Indigo Bunting with a white abdomen, a common hybrid type, was as far east as Schramm Park 26 May 1997 (JS).

Johnsgard (1979) noted that most Platte Valley birds were hybrids; Brown and others (1996) stated that of 27 birds netted that appeared to be Lazuli Buntings, 6 showed some hybrid characteristics; and Rosche (1994a) indicated that all territorial males at Clear Creek Marshes appeared to be hybrids, based on their resemblance to Indigo Buntings but with white abdomens. The presence of similar individuals in the Pine Ridge area led Rosche (1982) to note that hybrids were far more prevalent there than pure Indigo Buntings.

DISTRIBUTION AND ECOLOGY

Lazuli Bunting occupies woodland edge and open woodland habitats associated with shrubs and brush, usually riparian in nature, in the west. It is most common in the northern Panhandle and along the Niobrara River east to the Niobrara Valley Preserve.

Spring

27, 28, 29 Apr→28, 29, 29 May

Migrants arrive in early May, with early dates in late Apr and an earlier report 22 Apr 1990 Dundy Co (AB 44:456). Migrants have been reported statewide, although reports from the east are few: 29 Apr 1992 Lancaster Co, 5 May 1993 Sarpy Co (AB 47:428), 8 May 1999 Conestoga L, 9 May 1999 Lincoln, 11 May 1986 Lancaster Co, 11 May 1997 Bellevue, 12–16 May 1994 Adams Co, 16 May 1997 L North, 17 May 1991 Lancaster Co, and 21 May 1990 Washington Co.

High counts include 9 near Gering 20 May 1996, 8 there 8 May 1995, and 8 there May 1997.

Summer

Currently, breeding is probably restricted to Sioux and Dawes Cos, where it is uncommon (Rosche 1982); the Scotts Bluff Co area, where there are numerous recent summer records; and along the Niobrara River. The eastern limit in the Niobrara Valley appears to be the Niobrara Valley Preserve; Johnsgard (1980) stated that Lazuli Bunting bred east to eastern Cherry Co, and Brogie and Mossman (1983) found several singing males, albeit outnumbered 40:1 by Indigo Buntings, and considered it to have "certainly nested" in the preserve (Mossman and Brogie 1983).

There are a few scattered summer reports east of the range outlined above, none accompanied by evidence of nesting. One was found on a BBS route in Knox and Antelope Cos between 1967 and 1977. There are additional reports from Garden Co 2 Jun 1978 and 2 Jun 1981, possibly late migrants, and 1 Jul 1977, as well as 30 Jun 1965 Lincoln Co, 30 Jun 1966 Lincoln Co, 4 Jul 1989 Chase Co, territorial singing males 5 Jul 1987 and 25 Jul 1984 Keith County (Rosche 1994a), and Jul–Aug 1977 Keith Co (Rosche and Johnsgard 1984) and easterly reports 5 Jun 1962 Platte Co, 11 Jun 1976 Douglas-Sarpy Cos, and 19 Jun 1983 Otoe Co.

Fall

summer→9, 10, 12 Sep (specimen, UNSM ZM7103)

Departure is by early Sep, although there are later reports 21 Sep 1980 Garden Co, 21 Sep 1997 Oliver Res (SJD), 3 Oct 1998 immature male L Ogallala (SJD), 9 Oct 1972 Perkins Co, 10 Oct 1975 Sarpy Co with "excellent details"

(AB 30:94), 12 Oct 1961 Scotts Bluff Co, and 13 Oct 1916, a male specimen, HMM 2829, taken at Inland.

There are very few reports in fall away from the Panhandle: 17 Aug 1936 Logan Co (specimen, UNSM ZM7102), 18 Aug 1960 Webster Co, 27 Aug 1973 Perkins Co, 29 Aug 1970 Perkins Co, 31 Aug 1985 McPherson Co, and 2 Oct records listed above.

Other Reports

6 Apr 1953 Logan Co, 9 Apr 1979 Scotts Bluff Co.

FINDING

The best chance to locate this species is in riparian deciduous habitats in Pine Ridge canyons in May and Jun. Indigo Buntings and hybrids occur there also (Rosche 1982), so song alone is not reliable. All singers of Lazuli-Indigo song types should be tracked down until one turns out to be a Lazuli.

Indigo Bunting

Passerina cyanea

STATUS

Common regular spring and fall migrant and breeder east, becoming rare west.

DOCUMENTATION

Specimen

UNSM ZM7099, 15 Aug 1899 Lancaster Co.

TAXONOMY

This species hybridizes with Lazuli Bunting; Short (1961) indicated that *Passerina* buntings throughout Nebraska show evidence of hybridization. See Lazuli Bunting.

DISTRIBUTION AND ECOLOGY

This species and Lazuli Bunting occupy similar habitat; brushy woodland edge and disturbed habitats are preferred. Indigo Bunting breeds statewide, but numbers decline significantly away from the Missouri and lower Niobrara Valleys.

Spring

24, 26, 27 Apr→summer

Migrants arrive in late Apr.

High counts include 27 in Sarpy Co 13 May 1995; 15 at Wilderness Park, Lincoln, 20 May 1995; and 12 in Dakota Co 17 May 1997.

Summer

Around 1900 this species apparently bred only in the east, westward to the Grand Island area (Bruner and others 1904). As recently as 1958 this was still the case (Rapp and others 1958), although a specimen, UNSM ZM10413, was taken at Glen, on the White River in Sioux Co, as early as 18 Jul 1910. This specimen appears to be a phenotypically pure male (JGJ, WRS).

Indigo Bunting has spread westward in the 20th century along major river valleys, most notably those of the Niobrara and Platte; these are the 2 areas where contact with Lazuli Bunting has resulted in the most obvious hybridization.

The greatest numbers breed along the Missouri River; the 4 BBS routes with highest numbers are in counties adjacent to the Missouri River, with fewest birds in the south and west.

According to Johnsgard (1980), Indigo Buntings now breed throughout the Platte Valley to Colorado, although breeding west of Keith Co along the North Platte is limited; most males in the Keith Co area are hybrids with white abdomens (Rosche 1994a; Johnsgard 1990), although Brown and others (1996) recorded only 4 birds with hybrid characteristics among 28 netted at L Ogallala. Rosche (1994a) noted that Indigo Bunting has increased as a breeding species in the Keith-Garden Cos area during the 1980s. It has been reported in Scotts Bluff Co several years since 1981, and there is a report for Garden Co 10 Jun 1990.

Indigo Bunting presumably breeds throughout the Republican Valley, as there are recent summer records for Furnas, Red Willow, and Chase Cos, and it breeds along the north fork of the Republican River in Yuma Co, Colorado (Andrews and Righter 1992). By 1935 it was a common summer resident in Webster Co (Ludlow 1935).

Breeding occurs in Thomas Co (Short 1961), although as a species it is not common there in summer (Bray 1994). Breeding presumably occurs throughout the Loup drainage, but the species is rare or absent in the western Sandhills (Johnsgard 1980). Rosche (1982) has "a few records" for the Niobrara Valley and Smith L in Sheridan Co. There is a single summer report for Logan Co, 30 Jun 1965.

Away from the Missouri Valley, Indigo Bunting is most numerous in the Niobrara Valley. It outnumbers Lazuli Bunting 40:1 as far west as eastern Cherry Co (Mossman and Brogie 1983) and was present there in 1900 (Bruner and others 1904). It apparently occurs in small numbers in the Niobrara Valley farther west (Rosche 1982), although as recently as the late 1950s it was not found west of Valentine in Cherry Co (Short 1961).

In the northwest Indigo Bunting is apparently more numerous than it is in the central Niobrara Valley, as Rosche (1982) noted that it occurs in about the same numbers as Lazuli Bunting in the Pine Ridge, being most common in the Chadron section of the White River drainage. Rosche (1982) "strongly suspects," however, that there are "few, if any, pure Indigo Buntings present during the nesting season," and most territorial males show some white on the abdomen.

Fall

summer→12, 13, 13 Oct (Missouri Valley)

summer→15, 16, 18 Sep (elsewhere)

A later report from the Missouri Valley is 22 Oct 1984 Douglas-Sarpy Cos. Away from the Missouri Valley there are these later reports: 23 Sep 1909 Lancaster Co (specimen, UNSM ZM10655), 25 Sep (Rosche 1982), 2 Oct 1992 Dawes Co, 28 Oct 1976 Howard Co, and an immature that remained at a feeder in southwest Dixon Co 19 Dec 1998–2 Jan 1999 (JJ).

The Missouri Valley may be a major migration corridor for Indigo Buntings from farther north, as late dates there are in early Oct, whereas the breeding range away from the Missouri Valley is vacated by mid-Sep.

Other Reports

19 Mar 1959 Burt Co, 24 Mar 1966 Scotts Bluff Co, 14 Apr 1941 Lincoln Co, 15 Apr 1943 Jefferson Co.

FINDING

In May and Jun Indigo Buntings are common in Missouri Valley brushy edge habitats. Males are easily attracted by squeaking, while females are more secretive.

Dickcissel

Spiza americana

STATUS

Common regular spring and fall migrant and breeder east, becoming rare west. Rare casual winter visitor east.

DOCUMENTATION

Specimen

UNSM ZM12132, 22 May 1895 Lancaster Co.

DISTRIBUTION AND ECOLOGY

Dickcissel is found in midgrass and tallgrass prairies (Andrews and Righter 1992); cultivated fields including alfalfa, clover, and timothy (Johnsgard 1980; Rosche 1982; Ducey 1989); old fields; and prairie that has been invaded by mixed shrubs (Zimmerman 1982). Dickcissels are abundant in the southeast and become less numerous westward; they are absent through much of the Sandhills, except along roadsides with well-developed fence-line vegetation, and are rare in the Panhandle. None were noted in open prairie during studies in 1990–91 at Valentine NWR (Anderson 1992).

Optimum habitat in pre–European settlement times may have been prairie-woodland borderlands that would have been in a constant state of flux because of periodic fires and climatic fluctuations. These borderlands would be equivalent to contemporary old fields that Zimmerman (1982) has demonstrated to be optimum present-day breeding habitat for Dickcissels in Kansas. He found that even though nesting success is the same for nests located in prairie and for those in old fields, densities are considerably higher in old fields, partly because males tend to be polygynous there but monogamous or mateless in open grasslands. Unequal sex ratios in favor of males have also been

documented in Dickcissels (Fretwell 1977) and may account for many mateless territorial males in grassland habitat.

Spring

15, 15, 18 Apr→summer (south, east)

30, 30 Apr, 2 May→summer (north)

23, 25, 25 May→summer (west)

Migrants arrive in late Apr. A few very early undocumented reports from the southeast may be of wintering birds, if correctly identified: 5 Mar 1977 Otoe Co, 29 Mar 1980 Douglas-Sarpy Cos, 4 Apr 1980 Lancaster Co, and 7 Apr 1943 Jefferson Co.

Migrants are rare in the Panhandle, generally arriving in late May and Jun. Early Panhandle arrivals are 15 Apr 1950 Dawes Co, 18 Apr 1960 Scotts Bluff Co, and 20 Apr 1964 Scotts Bluff Co.

High counts include 76 in Sarpy Co 13 May 1995 and 35 in Otoe Co 31 May 1997.

Summer

BBS data show abundance in the east and rarity in the Panhandle. There are few reports of breeding in the northwest, none apparently conclusive. Rosche (1982) indicated that he had "never observed a female nor any courtship or possible nesting activity" and that singing males "often did not arrive until mid-Jun or even mid-Jul, sang for a week, and departed." Ducey (1988) listed nesting records from Dawes Co in 1965 and 1967, and it apparently bred in Sioux Co around 1900 (Bruner and others 1904). Nesting was apparently observed in Scotts Bluff Co in 1964 (NBR 33:9).

There are, however, numerous summer reports for the Panhandle, most from Scotts Bluff, Garden, and Dawes Cos. The only report from south of the Platte Valley in the Panhandle is 21 Jun 1988 Kimball Co. No Dickcissels were found at Crescent L NWR 7–11 Jun 1965 (Sharpe and Payne 1966), although 14 were there 8 Jun 1994.

Brogie and Mossman (1983) found no Dickcissels until 11 Jun in 1982 in the Niobrara Valley Preserve area, but singing males were seen in

Jun and Jul and it was considered a "probable nester." Prior to 1934 it was considered an "abundant summer resident" in Logan Co (Glandon and Glandon 1934b).

The absence of mates for males occupying grassland territories in Kansas (Zimmerman 1983) has led to speculation that much of the central and western Nebraska population in grasslands is unproductive. Although territorial males are present nearly every breeding season in central Nebraska, it is quite possible they are unpaired. This hypothesis may also explain why individuals arrive almost a month later in central and western Nebraska than in the east and, when present, often no earlier than late Jun in the Panhandle (Rosche 1994a, 1982; Ducey 1988).

Another factor contributing to suppression of Dickcissel populations is the high incidence of cowbird (*Molothrus ater*) nest parasitism. Hergenrader (1962) found that about 53% of the roadside nests that he found in eastern and south-central Nebraska contained 1–3 cowbird eggs. Zimmerman (1983) found that in Kansas parasitism was significantly higher in grassland than in old fields and parasitism contributed significantly to lowered Dickcissel productivity.

Nest card data (N = 33), all from the east, indicate presence of eggs 27 May–9 Aug. This species may nest late on occasion, as evidenced by young being fed in a nest in Douglas Co 27 Sep 1979 (AB 34:178) and nests with eggs or young in Hall Co 8 Aug 1989 (AB 44:116).

Fall

summer→27, 28, 30 Oct

Most birds leave by late Sep, but Oct dates are not unusual in the southeast.

Winter

Overwintering birds generally appear as singles at feeders in the east. There are reports as far north as Wayne Co in 1985–86 (AB 40:299). One was at a Bellevue feeder 15 Jan–29 Feb 1988 (AB 42:285), and another was at a Douglas Co feeder 10–17 Jan 1985 (AB 39:184). Very late, possibly attempting to winter, were singles on the Grand Island CBC 16 Dec 1995 and at a Lancaster Co feeder through 31 Dec 1984 (AB 38:184).

Other Reports

A Sioux Co report in 1964 (NBR 33:9) was described as "rather confusing" by Rosche (1982).

FINDING

This species can be found with ease in summer in the east, where birds sing loudly and continuously from prominent perches on fencerows. Females are secretive, however, and not often seen.

Icteridae (Blackbirds, Orioles)

Bobolink

Dolichonyx oryzivorus

STATUS

Common regular spring and fall migrant statewide. Common regular breeder north, uncommon elsewhere.

DOCUMENTATION

Specimen

UNSM ZM6932, 29 Jul 1902 Carns, Keya Paha Co.

DISTRIBUTION AND ECOLOGY

Preferred breeding habitat for this species is native wetland meadows. It was once quite common in lowland, riparian prairie that existed along most major river systems and in moist lowlands throughout the Sandhills. Drainage of many of these meadows and conversion of much of this habitat to cropland have contributed to a major population reduction in the 20th century, particularly in the east and south-central areas of the state, where it now occurs only locally. Rosche (1994a) and Ducey (1988) noted that Bobolink has begun using alfalfa fields in some locations, but there is no information on breeding success in this habitat.

Migrants occur in similar habitats as breeding birds, although in fall large flocks frequent cattail marshes.

Spring

28, 30 Apr, 1 May→summer

Arrival is in early May, with undocumented earlier reports 21 Apr 1957 Douglas-Sarpy Cos and 24 Apr 1960 Adams Co. Migrants occur statewide but are most common in central Nebraska (Rapp and others 1958; Johnsgard 1980).

High counts include 100 at Rowe Sanctuary 24 May 1998, 45 in Sarpy Co 11 May 1996, and 26 in Dixon Co 11 May 1996.

Summer

BBS data indicate that the species is most common in the Sandhills. Breeding occurs as far west as Agate along the Niobrara Valley in wet years (Rosche 1982). Breeding probably occurs throughout the Platte Valley, although while there are several summer reports from Scotts Bluff Co, there is no apparent evidence of breeding there (Ducey 1988). In Keith Co and southern Garden Co it is a local summer resident (Rosche 1994a). Farther east it breeds commonly in the central Platte Valley (Lingle 1994; Bruner and others 1904; Lingle and Hay 1982). In the east breeding is regular in scattered locations south to Lancaster and Sarpy (AB 36:994) Cos, with a few summer reports farther south, including 12 Jun 1997 Cass Co, 5 Jul 1965 Cass Co, 2 locations in Cass Co in 1999, 3 birds near Lorton in Otoe Co 9 Jun and 4 Jul 1994, 1–2 in Otoe Co 11 Jun 1995, and a single in Otoe Co in 1999. One at Mallard Haven, Fillmore Co, 6 Jul 1999 (BP, LP) may have been an early migrant. The only summer report from the southern 2 rows of counties prior to Jul, when migrants can occur, is 3–25 Jun 1989 Dundy Co.

Fall

summer→20, 24, 27 Sep

Reports are fewer than for spring, possibly as the males molt into drab winter plumage in Jul and are far less conspicuous, but also because migration is primarily nocturnal and during the daytime the birds frequent marshes and are easily overlooked. A male at Rowe Sanctuary was still in breeding plumage as late as 2 Aug 1997. Most migration apparently occurs in late Jul and Aug, ending by mid-Sep;

Rosche (1982) noted that large roosts occur in Sandhills marshes in Aug, and flocks were noted in Phelps Co 29 Jul 1995.

High counts include 89 in the eastern Rainwater Basin 29 Aug 1999.

Other Reports

1 Jan 1949 Saline Co, 13 Mar 1950 Logan Co, 20 Mar 1957 Sheridan Co, 20 Mar 1972 Hall Co, 2 Apr 1952 Harlan Co, 5 Apr 1976 Lincoln Co.

FINDING

Bobolinks can be best located in late May and Jun when the males are very active and perch on tall grass stems in wet meadows and pastures. Largest numbers occur around the periphery of Sandhills marshes and in irrigated alfalfa pastures in major river valleys.

Red-winged Blackbird

Agelaius phoeniceus

STATUS

Abundant regular spring and fall migrant and breeder statewide. Uncommon regular winter visitor south and east, rare elsewhere.

DOCUMENTATION

Specimen

UNSM ZM6980, 10 Oct 1889 Lincoln, Lancaster Co.

TAXONOMY

There is little information to indicate the subspecies of Red-winged Blackbird breeding in most of the state. According to the AOU (1957), the Plains subspecies *fortis* breeds in western Nebraska, northern *arctolegus* breeds south to southern South Dakota and possibly extreme northern Nebraska, and southeastern *phoeniceus* breeds in southeast Nebraska and most of Kansas. Swenk (Notes before 1925) indicated that birds breeding in the Inland area were at the maximum wing length for *predatorius* (= *pheonicius*), with short bills, shorter than the minimum for *predatorius*, but not thickened as in *fortis*. Most Nebraska Red-winged Blackbirds are probably intergrades (see Spring).

DISTRIBUTION AND ECOLOGY

Red-winged Blackbird breeds in a variety of

habitats, from marshes to upland roadsides, although mainly in the vicinity of water or damp situations.

Migrants and wintering birds depend on marsh vegetation for roosting but forage almost anywhere, from livestock feedlots to urban areas.

Spring

Most summering birds do not overwinter in the state; there are no banding or other data to indicate otherwise. Ohio data (Bent 1968) show that breeding birds (*phoeniceus*) were replaced in winter by northern birds (*arctolegus*). Nebraska banding data indicate that breeding birds from North and South Dakota (presumably *arctolegus*) pass through eastern Nebraska during migration, while birds breeding in or migrating through the Panhandle (presumably *fortis*) winter a short distance to the south in Colorado or on occasion in Nebraska. Banding data also indicate that some Nebraska migrants come from states to the northwest, including Wyoming and Montana. These birds are probably *fortis* also. The type specimen of *fortis* was collected at Omaha (Bruner and others 1904); this race is a regular migrant as far east as western Iowa (DuMont 1934).

Males appear first in spring, arriving in late Feb and early Mar. Early males were on territory in the Sandhills 8 Feb 1998 (wrs, js). The first arrivals are usually individual males that may have wintered close to or in Nebraska, followed by migrant males and then resident males (Bent 1968). Migrant females arrive soon after and then resident females. There may be 2 weeks between arrival of territorial males and females. In Holt Co in 1990 the first birds seen were 26 Feb, with flocks noted 11 Mar (Blake and Ducey 1991). In north-central Nebraska males were on territory by 4 Apr in 1982, but females were not seen until 14 Apr (Brogie and Mossman 1983). A flock of about 1000 females was in the eastern Rainwater Basin as late as 4 May 1997 (wrs).

High counts include 38 000 in Scotts Bluff Co 6 Mar 1994, 35 000 at Kiowa Springs 8 Mar 1998, 5000 in Dakota Co 11 Apr 1995, and 5000 in Dixon Co 13 Mar 1997.

Summer

This species is ubiquitous, perhaps with lowest densities in the Panhandle, as indicated by bbs data. Largest numbers nest at the margins of permanent marshes statewide, usually over shallower water than Yellow-headed Blackbird when the two occur together.

Tout (1947) cited egg dates involving "hundreds" of nestings in Lincoln Co in the period 21 May–23 Jun.

Fall

Flocks composed of young birds and females form in Jul and are joined later by males. "Several thousand" were in Buffalo Co 19 Jul 1997 and 4000 at Funk Lagoon 8 Aug 1998. The birds are inconspicuous during Aug, when molting takes place in marsh vegetation. Flocks reappear in Sep, and migration takes place in Oct and Nov (Bent 1968). A flock of 100 000 was at Clear Creek Marshes 27 Oct 1996. By late Dec most have left the state, although large roosts may still be present during the cbc period in the south and east: 500 000 were at North Platte in 1977, 255 806 were at Omaha in 1984, and 187 065 were there in 1987. Flocks of this size have not been recorded in Jan.

Winter

Relatively few overwinter, although flocks remain at favored locations if food and shelter are available (Rosche 1982, 1994a). A flock of 12 500 was in a sheltered creek bottom in Pawnee Co 8 Feb 1997 (jgj, wrs), and 1500 were near Lewellen 9 Jan 1986 (Rosche 1994a). Late Jan and early Feb records of large flocks are unusual.

FINDING

Males are conspicuous in damp spots along roadsides from late Mar into Jun virtually statewide. Large flocks containing males and females are common around marshes in Oct and Nov when the birds return to roost in the late afternoon.

Eastern Meadowlark

Sturnella magna

STATUS

Common regular spring and fall migrant east, becoming uncommon west. Common regular breeder east, becoming locally common west. Rare casual winter visitor southeast.

DOCUMENTATION

Specimen

UNSM ZM6952, 2 Mar 1901 Lincoln, Lancaster Co.

DISTRIBUTION AND ECOLOGY

Breeding birds are most numerous south of the Platte River in the southeast, where they occupy pastures, prairie remnants, and other grasslands. There is also a significant population breeding throughout the Sandhills, where wet meadows and other low-lying grasslands are utilized, often in association with Bobolinks (Rosche 1982). Elsewhere, populations are generally local, considered rare in the Panhandle (Rosche 1982), and are restricted to low-lying pastures and meadows associated with river and stream valleys, as pointed out by Lanyon (1956) and Rohwer (1972).

Wintering birds are also associated with grasslands and are limited to the extreme southeast section of the state when present.

Spring

10, 14, 16 Mar→summer (north, west)

Movement begins in the south and east in late Feb, probably involving birds that wintered in or near the state. True migrants arrive in early Mar, although an earlier report was of 11 vocalizing in Kearney Co 16 Feb 1996. Arrival in the north and west is somewhat later, around mid-Mar, although there are some earlier dates (see Winter).

High counts include 40 in central Nebraska 1 Apr 1996.

Summer

Eastern Meadowlark breeds commonly in the southeast in an area approximately bounded by Thurston, Colfax, Polk, Hall, and Thayer Cos (BBS data; Johnsgard 1980; Lingle and Hay 1982). The species is also locally common throughout the Sandhills, including the western sections in Sheridan Co, where it is "fairly common" in wet meadows (Rosche 1982), and in Garden Co, where there are several summer reports, including counts of 173 in 1996 at Crescent L NWR. At the western edges of the range, more Eastern Meadowlarks are present in wetter years (Rosche 1994a).

Elsewhere in the state it is local as a breeding bird. It is common in the Platte Valley at least as far west as Mormon Island Crane Meadows in Hall Co, where it was listed as abundant in the summers of 1980 and 1981 (Lingle and Hay 1982), and it occurs farther west in relatively small numbers to Lincoln and Keith Cos (Rosche and Johnsgard 1984; Rosche 1994a). North of the Sandhills it was not recorded in 1982 by Brogie and Mossman (1983) in the Niobrara Valley Preserve, although it is a fairly common summer resident in a small area of "valley meadows" in south-central South Dakota around LaCreek NWR (SDOU 1991). It is "occasional" in summer in Dawes Co (Rosche 1982), where there are 3 published summer reports, 1 Jun 1954, 12 Jun 1973, and 30 Jun 1975, and there is a recent summer report for Box Butte Co, 29 May 1992. Perhaps the westernmost regular breeding site is Snake Creek Meadows near Kilpatrick L (Rosche 1994b).

South of the Platte River and westward there are few reports. There is a summer report from Dundy Co 3 Jun 1989, and one was singing at Funk Lagoon 12 Jun 1999 (LR, RH).

The presence of Eastern Meadowlark in the Sandhills may be relatively recent, although the evidence is inconclusive. Eastern Meadowlark was collected near present-day Columbus 1856–57 (Coues 1874), but Bruner and others (1904) stated that this species is "rare, breeding doubtful" in Nebraska, with possible nesting in the Omaha and Lincoln areas. By 1934 it was a "summer resident," nesting in Logan Co (Glandon and Glandon 1934b), and by 1950 there were at least isolated colonies statewide (Lanyon 1956). By 1958 it had reached the entire state, as Rapp and others (1958) considered it a "common breeder" throughout, although "uncommon to the west."

Fall

summer→25, 26, 29 Oct

Departure is difficult to detect, as singing has ceased and birds are largely silent and hard to identify except by calls. By Dec, however, those remaining are probably confined to the southeast (see Winter), although one was singing in northern Lancaster Co 28 Nov 1998. The few fall dates available away from the east suggest fall migration takes place mainly in Sep and early Oct.

Later reports away from the winter range are few, including 10 Nov 1986 Pierce Co; a specimen, UNSM ZM11032, 13 Nov 1909 Lincoln; a specimen, UNSM ZM6957, 18 Nov 1931 Lincoln; 16 Dec 1972 Hall Co; and 17 Dec 1988 Lincoln Co.

Winter

Although documentation of wintering is lacking, information from Missouri (Robbins and Easterla 1992) and Kansas (Thompson and Ely 1992) suggests that wintering is possible and probably does occur in the extreme southeast. There are, however, no specimens at UNSM later than 18 Nov (see Fall).

Published data indicate that wintering takes place north to the vicinity of the Platte River and west at least to Hall, Adams, and Franklin Cos, although documented reports are lacking. There are numerous Jan and Feb reports from this area, and Eastern Meadowlark was listed as common in winter in Hall Co by Lingle and Hay (1982). Reports for 31 Dec–early Feb away from this area should be considered doubtful until documentation can be obtained, especially in light of the virtual absence of Nov and Dec reports from the same areas.

Gross stated (Bent 1958) that early birds arriving in winter prior to the regular Mar migrants are essentially winter vagrants, often young birds.

Reports apparently of this type are 11 Feb 1968 Lincoln Co, 13 Feb 1999 (7 vocalizing) Standing Bear L (NR), 16 Feb 1956 Keya Paha Co, 18 Feb 1956 Sheridan Co, 20 Feb 1947 Keith Co, and 1 Mar 1950 Logan Co.

FINDING

Eastern Meadowlark can be found (and identified) in grassy areas in the southeast in May and Jun when they are actively singing and calling. As the call notes are diagnostic, they should be heard to confirm identification. Pastures in Gage, Johnson, Nemaha, Pawnee, and Richardson Cos hold the best numbers; Burchard L is excellent. Winter identification is difficult (Wilson 1983; Zimmer 1985), but observers should attempt to document such sightings; winter road-killed specimens should be examined as to species and then preserved.

Western Meadowlark

Sturnella neglecta

STATUS

Abundant regular spring and fall migrant and breeder statewide. Uncommon regular winter visitor southeast, becoming rare northwest.

DOCUMENTATION

Specimen

UNSM ZM11049, 30 Oct 1890 Lincoln, Lancaster Co.

DISTRIBUTION AND ECOLOGY

Western Meadowlark breeds commonly statewide in all types of grasslands and agricultural areas. It is least common in the extreme southeast, where Eastern Meadowlark breeds also, and in that area the 2 species segregate into habitats according to moisture, with Western Meadowlark occupying drier habitat than Eastern Meadowlark (Lanyon 1956).

Overwintering is uncommon, most birds occurring in the south and east, with the same habitats occupied as in summer.

Spring

Migration timing is difficult to determine, but movement of small flocks is noticeable in Mar, independent of weather conditions (Robbins and Easterla 1992). One calling as far north as Knox Co 14 Feb 1999 (MB) may have been an early migrant or a wintering bird.

High counts include 551 in Pierce Co 8 May 1999.

Summer

This species is a common to abundant breeder throughout the state, except for the extreme southeast, as indicated by BBS data. As many as 2000 were at Crescent L NWR 30 Aug 1998. While Ducey (1988) cited no breeding records from the 4 counties in the southeast corner of Nebraska since 1920, there are numerous summer reports since 1981, and the species undoubtedly breeds there in good numbers.

Fall

Movement is discernable in Oct independent of weather (Robbins and Easterla 1992), although there is a southward drift later in the year in response to adverse weather conditions. Counts indicative of migratory movement are 300 at Gibbon 25 Oct 1998 and 200 at Nelson Basin 27 Oct 1996.

Departure from the state varies from year to year. This is illustrated by a banding recovery in Arkansas 30 Nov 1940 of an individual banded in Lincoln Co, Nebraska, 21 Jan 1940. This bird apparently wintered in Nebraska 1939–40 but had already left Nebraska and arrived in Arkansas, some 800 km (500 mi) to the southeast, by Nov of the subsequent winter.

Winter

Western Meadowlarks winter statewide, although numbers vary considerably from year to year, and they may be absent from large parts of the state in some years. CBC data indicate that by late Dec Western Meadowlarks are still fairly evenly distributed over most of the state, with lowest numbers in the north.

The highest single-location CBC total was 578 at Lincoln in 1972. The highest total number of Western Meadowlarks on all counts, 1401, was also in 1972; this contrasts with the low of 58 in 1984.

COMMENTS

Bruner and others (1904) suggested that Western Meadowlarks from "the lake region of Cherry County" differ in some respects from both Eastern Meadowlarks and Western Meadowlarks from elsewhere in Nebraska, although no details were given on these differences.

FINDING

This species is easily located in summer virtually anywhere in grassland, usually by song, but identification is clinched by the call notes. Conclusive identification in winter is unreliable based on plumage characters alone (Wilson 1983).

Yellow-headed Blackbird

Xanthocephalus xanthocephalus

STATUS

Common regular spring and fall migrant and breeder statewide. Rare casual winter visitor statewide.

DOCUMENTATION

Specimen

UNSM ZM6964, 18 May 1901 Nebraska City, Otoe Co.

DISTRIBUTION AND ECOLOGY

As a breeding bird this species is restricted to deeper, generally permanent marshes (Johnsgard 1980) with extensive growths of cattails and similar vegetation. It appears to select those areas of the marsh with permanent standing water and actively excludes Red-winged Blackbirds from those sites. Breeding sites may include marshes that develop in backwaters to rivers such as the Loup, upper Niobrara, and upper Elkhorn. Migrants occur in flocks, often with other blackbirds, usually in agricultural areas.

Spring

8, 8, 8 Mar→summer

Migrants arrive in mid to late Mar, although most pass through in Apr. Late flocks are sometimes noted; 1000 were in the eastern Rainwater Basin 4 May 1997, and 200 females were still in a flock near Funk Lagoon 25

May 1997 (WRS). An early report was of one at Kiowa Springs 23 Feb 1996; it may have wintered in the area, although there are several late Feb reports (see Winter).

High counts include 10 000 at Funk Lagoon 19 Apr 1998, 2000 at Clear Creek Marshes 22 Apr 1995, and 1000+ at North Lakes Basin 23 Apr 1999.

Summer

The Sandhills area is the center of abundance of this species, although it is also very common in the Panhandle (Rosche 1982; Rapp and others 1958); 3000 were at Crescent L NWR 30 Aug 1998. Smaller breeding populations occur throughout Nebraska wherever there are suitable deep marshes, although from year to year these populations vary in size and may be absent depending on water conditions (Blake and Ducey 1991).

BBS data indicate increasing numbers of this species in Nebraska. It is possible that flood and fire control has allowed cattail growth to increase, providing more suitable habitat for this species on the Great Plains.

Fall

Large flocks of molting birds appear at favored localities during Jul (Brown and others 1996). Most fall migrants leave the state by late Oct, with a cluster of departure dates in early Nov. Rosche (1994a) noted that a major fall staging area is the marshes at the west end of L McConaughy, where 2000–3000 were counted 10 Sep 1988. There are several later dates, including CBC data, that indicate that a few fall migrants may linger as late as early Jan. As many as 6 were reported on the DeSoto Bend NWR CBC 22 Dec 1979, and 5 were there 20 Dec 1981. Jan reports are from as far north as Scotts Bluff Co 25 Jan 1956, Logan Co 3 Jan 1948, McPherson Co 1 Jan in 1967–70 and 1973, and Boone Co 1 Jan 1986.

Winter

The only documented records of overwintering include a single at a feeder at Crescent L NWR 9 Dec–21 Feb 1980–81 (AB 35:314) and one at Ogallala 1993–94 (Rosche 1994a).

There are few additional midwinter reports: 8 Feb 1984 Sarpy Co (AB 39:184), 9 Feb 1976 Douglas-Sarpy Cos (AB 30:737), and 9 Feb 1999 Lincoln (JS). There are also a few late Feb reports that are probably early spring migrants: 18 Feb 1978 Douglas-Sarpy Cos (AB 32:371), 22 Feb 1998 Funk Lagoon, 23 Feb 1985 Brown Co, 24 Feb 1946 Adams Co, 25 Feb 1952 Dawson Co, 25 Feb 1984 Dakota Co, 26 Feb 1980 Lincoln Co, and 1 Mar 1980 Boone Co.

FINDING

In May Yellow-headed Blackbirds are conspicuous in cattail marshes in the Sandhills. This species is aggressively territorial and vocal at this time and is difficult to overlook.

Rusty Blackbird

Euphagus carolinus

STATUS

Uncommon regular spring migrant east, becoming uncommon casual west. Common regular fall migrant east, becoming uncommon casual west. Rare regular winter visitor east.

DOCUMENTATION

Specimen

UNSM ZM9568, 31 Oct 1919 Niobrara Valley Preserve, Cherry Co.

DISTRIBUTION AND ECOLOGY

This species occurs during migration and winter in deciduous woodland near water, usually in river or stream bottoms. It is rare and irregular in the Panhandle and uncommon in the east during migration. Overwintering is rare and mainly in the eastern half of the state.

Spring

winter→1, 1, 2 May

Rusty Blackbird is an uncommon migrant in the east, becoming less common westward; it is considered "accidental" in spring in the northwest (Rosche 1982) and is not common in the Hall Co area (Lingle 1994).

Migration begins in Feb in the east, although peak migration probably occurs in early Apr. Migrants depart by late Apr; the many undocumented May and later reports probably result from confusion with Common Grackle or Brewer's Blackbird. The latest specimen date for spring in Kansas is 17 Apr, and there are no Kansas sight records later than 5 May (Thompson and Ely 1992). The latest spring date for Missouri is 23 Apr (Robbins and Easterla 1992), for Iowa 28 Apr (Kent and Dinsmore 1996), and for South Dakota 16 May (SDOU 1991).

In the Panhandle Rusty Blackbird is rare, with few reports: 2 Mar 1956 Sheridan Co (possibly a wintering bird), 27–28 Mar 1999 (1–2) Clear Creek Marshes (SJD), 10 Apr 1950 Dawes Co, 12 Apr 1961 Scotts Bluff Co, 14 Apr 1958 Scotts Bluff Co, 20 Apr 1960 Dawes Co, 24 Apr 1949 Dawes Co, and 30 Apr 1961 Dawes Co.

High counts include 500 in Washington Co 24 Feb 1983 (AB 37:314), 100 at Sinninger Lagoon 16 Apr 1995, and 40 at Fontenelle Forest 5 Apr 1998.

Fall

30 Sep, 1, 3 Oct→winter

Migrants are more widespread in fall than in spring. Arrival is in early Oct, with earlier reports 15 Sep 1933 Lincoln Co (Tout 1947) and 19–21 Sep 1988 Cedar Co. Peak fall migration probably occurs in late Oct and Nov. Rusty Blackbirds are still fairly common in the southeast in early Dec; CBC data indicate that they are widespread away from the Panhandle in mid to late Dec, as exemplified by counts of 99 and 25 as far north as the Beaver Valley in 1986 and 1985, respectively, and 156 as far west as Grand Island in 1990. High CBC totals are 626 at Omaha in 1984 and 357 there in 1971.

There are a few Jan reports from the north, where wintering is not expected (see Winter): 4 Jan 1951 and 6 Jan 1955 Antelope Co; 7 Jan 1988, 10 Jan 1981, and 11 Jan 1986 Boone Co; and 16 Jan 1952 Antelope Co.

This species is rare in the Panhandle, with fewer than 15 reports, most from Scotts Bluff Co and all in the period 12 Sep–5 Oct, except for a specimen, UNSM ZM7023, taken at Chadron 3 Nov 1919.

High counts include 50 at DeSoto Bend NWR 30 Nov 1994.

Winter

This species winters regularly in the southeast and occasionally elsewhere. A high count was of 200 wintering in Sarpy Co 1984–85 (AB 39:184). It winters some years in Holt Co (Blake and Ducey 1991), and there are early Feb reports in Pierce and Dakota Cos. In Dodge Co 20 were present 31 Dec 1998, but only 2 remained 11 Feb (DFP, JP).

Rosche (1982) considered Rusty Blackbird a casual winter visitor in the northwest; there are, however, more reports for the winter period than for fall, suggesting that some winter reports involve misidentifications. A specimen, UNSM ZM10595, collected at Crawford 12 Sep 1911 and labeled as this species is in fact a fall male variant Brewer's Blackbird, as illustrated in the NGS *Field Guide* (1987; JGJ). There are 11 Panhandle reports in the period 21 Nov–2 Mar, the latest 20 Jan 1979 Garden Co (Rosche and Johnsgard 1984), 21 Jan 1972 Dawes Co (Rosche 1982), and 2 Feb 1958 Dawes Co, but few survive or remain after 20 Jan in the Keith Co area (Rosche 1994a).

COMMENTS

There are numerous undocumented reports for the period May–Aug, including one putative record of breeding in Hall Co in 1972 (Ducey 1988) that should be disregarded (Johnsgard 1979). There are 5 recoveries of birds identified as Rusty Blackbirds that were banded in Arkansas and recovered in Nebraska in May and Jun: 9 May 1934 Antelope Co, 29 May 1932 Madison Co, Jun 1940 Antelope Co, 10 Jun 1931 Saunders Co, and 12 Jun 1931 Fillmore Co. We consider these records highly questionable.

Other Reports

11 May 1957 Scotts Bluff Co, 14 May 1960 Scotts Bluff Co, 21 May 1977 Lancaster Co, 21 May 1978 Lancaster Co, 23 May 1957 Sheridan Co, 23 May 1973 Lancaster Co, 24 May 1947 Lincoln Co, 26 May 1980 Lancaster Co, 31 May 1987 Douglas-Sarpy Cos, 5 Sep 1974 Lancaster Co.

FINDING

Rusty Blackbirds can be best located in Nov in the Missouri Valley. Small flocks are conspicuous in riparian deciduous woodland, where they are usually separate from flocks of other blackbird species. Identification is easiest in fall also, when rusty feather edgings are conspicuous.

Brewer's Blackbird

Euphagus cyanocephalus

STATUS

Common regular spring and fall migrant west, becoming rare east. Common regular breeder northwest. Rare winter visitor south.

DOCUMENTATION

Specimen

UNSM ZM7025, 20 Jun 1901 Indian Creek, Sioux Co.

DISTRIBUTION AND ECOLOGY

Breeding birds occupy grassland habitats, usually somewhat disturbed (Johnsgard 1980), with scattered scrubby vegetation (Rosche 1982) or scattered ponderosa pines, primarily in the northern Panhandle.

Migrants and wintering birds are found in open habitats, sometimes with other blackbird species, but show a preference for farmyards and feedlots. Migrants are most numerous westward, while wintering birds are most often reported in the south.

Spring

20, 21, 23 Feb→1, 1, 2 Jun (south, east, north)
22, 28 (Rosche 1994a), 31 Mar→summer (west)

As with other blackbird species, migration begins in late Feb, and peak migration probably occurs in mid to late Apr (Rosche 1994a). "Several hundred" were in Scotts Bluff Co 27 Apr 1994. Not as numerous in the east, high counts there include 100 in Clay Co 19 Apr 1997 and 100 in York Co 27 Apr 1999. Away from the breeding range, migration ends in late May, although there are later reports (see Summer).

Migrants arrive much later in the Panhandle, in early Apr, although there is an earlier report 15 Mar 1965 Dawes Co.

Summer

Brewer's Blackbird breeds primarily in Sioux Co, particularly on the high grasslands in the Harrison area (Rosche 1982). Documented breeding records away from Sioux Co are few, although the species was listed as a summer resident in Sioux, Dawes, Sheridan, Box Butte, and Morrill Cos in 1993 (NBR 61:80, 134). In Morrill Co young were seen by Rosche in 1987 (Ducey 1988), and a female was feeding an immature south of Redington 25 Jun 1995 (WRS). At the same location in 1998, 15+ pairs were present, and an active nest and 3 females carrying nest material were noted 15–16 May 1998 (WM). Nest building was observed in Dawes Co 12 May 1993 (AB 47:1124).

Although there are several summer reports from Dawes Co, a small number from Sheridan and Scotts Bluff Cos, and a single Kimball Co record near Bushnell 6 Jul 1995, the breeding range in Nebraska may be decreasing as it has in recent years on the plains of eastern Colorado, where Brewer's Blackbird has been largely replaced by Common Grackle (Andrews and Righter 1992). Early in the 20th century Brewer's Blackbird expanded its range eastward, particularly in Minnesota and Wisconsin (Bent 1968). There is no evidence that a similar expansion occurred in Nebraska, although prior to 1934 Glandon and Glandon (1934b) considered Brewer's Blackbird a "very common summer resident" in Logan Co.

There are no documented records of breeding elsewhere in Nebraska. Reported nestings in Hall, Johnson, and Lancaster Cos (Bennett 1969, 1972, 1973b, 1975) probably result from misidentifications of Common Grackle; the Johnson Co report was of nesting in a conifer in a suburban yard, typical of Common Grackle. The 1974 Hall Co report (Bennett 1975) was based on birds seen "carrying food"; Johnsgard (1979) suggested that this report may have been of a Rusty Blackbird, although there is no evidence to suggest any species other than Common Grackle.

There are about 50 undocumented Jun and Jul reports of Brewer's Blackbird away from the breeding range outlined above; most, including BBS data, are from the east and south and are questionable, except for those few from westerly locations where wandering birds may occur in summer, as is the case in western Kansas (Thompson and Ely 1992). There are no extralimital breeding records for South Dakota (SDOU 1991). Even in the Panhandle there are very few Jun and Jul reports away from the breeding areas: 4 Jun 1986 and 5–12 Jun 1992 Sheridan Co and 30 Jun 1965, 30 Jun 1982, and 23 Jul 1983 Scotts Bluff Co.

Fall

26, 27, 28 Aug→winter (north, south, east)

Migrants appear statewide in late Aug and early Sep, although there is an early report 14 Aug 1994 at Axtell. A flock of 100 was near Crawford 1 Sep 1997. Peak movement probably occurs in early Oct, and most migrants leave the state by the end of Nov.

There are several reports after Nov and as late as Jan, but overwintering is rare (see Winter). CBC data indicate that by late Dec individuals and small flocks are still distributed statewide, although most are in the south and east. Highest CBC totals were 283 at Lincoln in 1986, 206 at DeSoto Bend NWR in 1989, and 180 at Hastings in 1976.

High counts include 2000 at Oshkosh 2 Oct 1976 (Rosche 1994a) and 600 near Minatare 26 Sep 1999.

Winter

Brewer's Blackbird probably overwinters occasionally, although the evidence is based mainly on a number of Jan and Feb reports from the south and east, most undocumented. There are no records of individual birds wintering, for example, at a feeder. Northerly midwinter reports are few: 10 Jan 1981 Boone Co, one remained to 10 Jan 1998 from a flock of 33 at Wolf L (TH), 11 Jan

1946 Logan Co, 11 Jan 1986 Boone Co, 15 Jan 1998 (2) Gering (SJD), 17 Feb 1935 Logan Co, and 23 Feb 1985 Boone Co, the latter two possibly early migrants. Blake and Ducey (1991) stated that "a few overwinter some years" in southeast Holt Co. There are also very few midwinter reports for the Panhandle: 7 Jan 1991 and 6 Feb 1956 Scotts Bluff Co.

FINDING

The best areas to locate Brewer's Blackbird are in the breeding range in grasslands around Harrison, where pairs can be seen on fence lines in Jun and Jul in pastures with small shrubs. Identification of this species in fall and winter can be problematic and should be done with care.

Common Grackle

Quiscalus quiscula

STATUS

Abundant regular spring and fall migrant and breeder statewide. Uncommon regular winter visitor southeast.

DOCUMENTATION

Specimen

UNSM ZM7029, 25 Mar 1890 Lincoln, Lancaster Co.

DISTRIBUTION AND ECOLOGY

Common Grackle breeds in woodland edge habitats and is partial to areas with planted trees, especially conifers, such as windbreaks, residential areas, and parks. Breeding birds occur statewide but are most numerous in southern and eastern Nebraska where human population is greatest.

Migrants usually occur in agricultural areas with other blackbirds, although wintering birds are most commonly found as single birds or small groups at feeders.

Spring

28 Feb, 3, 5 Mar→summer

In the southeast, where a few birds overwinter, spring movement begins in mid-Feb, although 125 were noted at Nebraska City 6 Feb 1998. Peak movement probably occurs in late Mar

(Johnsgard 1980). In the north and west, where overwintering is rare, movement is discernible by early Mar.

High counts include 2000 at Arbor Day Farm 9 Apr 1998, 1000 in the eastern Rainwater Basin 16 Apr 1997, and 780 in central Nebraska 31 Mar 1996.

Summer

While this species breeds statewide, BBS data (1967–77) indicate that greatest density is reached in the south and east. Brogie and Mossman (1983) found it abundant in the Niobrara Valley Preserve, with 74 nests located in their study area in 1982. In the northwest Rosche (1982) considered it a common summer resident.

Fall

Juveniles begin flocking in early Jul (Brown and others 1996), and migration probably peaks during Oct (Johnsgard 1980). By late Dec CBC data show that over 85% of Common Grackles are in the east.

High counts include "tens of thousands" in Custer Co 6 Oct 1995, 9000 at Funk Lagoon 18 Oct 1996, and 6500 at Kiowa Springs 16 Oct 1999.

Winter

There are many midwinter reports in the south and east, but such reports are unusual elsewhere. Rosche (1982) noted that in the northwest "a few attempt to winter each year, but few are successful as numbers decline markedly after early February." There are very few Feb reports for the north and west: 5 Feb 1999 (100) Funk Lagoon, 6 Feb 1953 Boyd Co, 9 Feb 1952 Boyd Co, 9 Feb 1966 Sioux Co, 10 Feb 1962 Logan Co, 16 Feb 1985 Morrill Co (AB 39:184), and 23 Feb 1980 Keith Co (Rosche and Johnsgard 1984).

FINDING

Common Grackle is easy to find almost anywhere in spring when the species is obvious as males actively pursue females in towns and around farmsteads. Large flocks appear in fall.

Great-tailed Grackle
Quiscalus mexicanus

STATUS

Common regular spring and fall migrant southeast, uncommon casual elsewhere. Common regular breeder south-central, uncommon casual elsewhere. Uncommon casual winter visitor east.

DOCUMENTATION

Specimen

UNSM ZM17601, May 1998 Ceresco, Saunders Co.

TAXONOMY

The measurements (Tom Labedz, pers. comm. WRS) of the specimen cited above are somewhat intermediate between *monsoni* and *prosopidicola*, although the tail is long for *prosopidicola*. All specimens taken to date in northwest Missouri, where the first nesting was recorded in 1979, have been of the eastern race *prosopidicola* (Robbins and Easterla 1992).

DISTRIBUTION AND ECOLOGY

Breeding in Nebraska is usually associated with wetlands, especially cattail marshes, and it has been suggested that "wet habitat types may be important for pioneering birds" (Faanes and Norling 1981). However, an early successful site involved nesting in dense spruces in a parklike setting (NBR 45:18, 35, 46:22), a habitat that is most often utilized within the species' core range (Pruitt and McGrowan 1975). Most Nebraska reports are from the south-central and southeast part of the state, where breeding occurs in widely scattered colonies, often involving new sites.

Spring

6, 7, 7 Mar→summer

Migrants arrive in mid-Mar, with earlier dates 1 Mar 1988 Lancaster Co and 3 Mar 1992 Polk Co. Most reports are east of a line connecting Phelps and Knox Cos, with westernmost reports 30 Apr 1993 Keith Co, 6 May 1996 Grant Co, 11–12 May 1999 (1–2) Facus Springs (SJD), 17 May 1996 (7) Morrill Co, and 18 May 1997 L Ogallala.

High counts include 185 in Clay Co "spring" 1992

(AB 46:444), 130 at Funk Lagoon 10 Mar 1999, and 112 at Eckhardt Lagoon 16 Mar 1996.

Summer

This species was first recorded in Nebraska in 1976 in Phelps Co, when several birds were seen in spring and summer around the headquarters of the Sacramento-Wilcox Game Management Area; nesting was suspected but not confirmed (Longfellow 1979). There are extensive conifer plantings in this area.

The second and third records involved confirmed nesting. At the Hastings Ammunition Depot in Adams Co, several pairs were seen in spruces near the officers' club from 11 May 1977 into summer (AB 31:1019; NBR 45:18, 35). Some 10–15 pairs were later counted, with at least 6 young present. It was believed by the observers that birds may have been at this site for 2 or 3 years, and an impressive count of 100+ birds was made here 21 May 1978 (NBR 46:22). At the same time a small colony with 2 nests was found 14 May 1977 in a clump of willows in a small lake at Boystown, Douglas Co (BJR; AB 31:1019; NBR 45:18, 37); at least 4 males and 2 females were present through early Jul.

Since 1977 nesting has been regularly reported from an area of southeast Nebraska bounded by Phelps, Buffalo, Hall, and Lancaster Cos, with most reports from the Rainwater Basin. Reports of nesting elsewhere in the state are few, but expansion into the northeast and westward in the Platte Valley is taking place. The first Boone Co nesting record was of fledged young 1 Jun 1997 (WM), and a colony was established at Wood Duck Area in 1997 (DH). A report of nesting in Platte Co in 1981 was later withdrawn (NBR 49:28, 53). Expansion westward in the Platte Valley from the Rainwater Basin population is also occurring. Birds were displaying at North Platte in 1980 (AB 34:909), and breeding was recorded there in 1991 (AB 45:468); single males were seen at Clear Creek Marshes 26 Apr and 17 May 1986 (Rosche 1994a),

and 2 pairs nested successfully at Ogallala in 1993 (Rosche 1994a; AB 47:1124). By 1989 individuals had been seen west to Morrill Co in the Platte Valley (AB 43:501), and a pair was seen at Meadows Ranch in Cheyenne Co 28 Apr 1994. The only breeding records from the Panhandle are of a female feeding what appeared to be a juvenile male at the Box Butte Res campground 12 Aug 1988 (WRS, SJD) and nest building in progress at Kiowa Springs 17 May 1998 (SJD). Up to 7 were at Kiowa Springs by 19 Apr 1999, and 26 were there 2 Oct 1999 (SJD). According to Rosche, this species was "holding its own in the west, but not markedly increasing" in 1994 (AB 32:87). A reference to breeding in Scotts Bluff Co (Bennett 1984) was subsequently withdrawn (NBR 53:24). Great-tailed Grackle is a local summer resident on the eastern Colorado plains (Andrews and Righter 1992).

Fall

Flocks form by early Aug; examples are 115 in Otoe Co 21 Aug 1998 and 65 at Ong 20 Aug 1994. At this time individuals may appear at nonbreeding localities. Migration ends by mid-Oct, although there are several later reports. Westerly reports include 1 in Scotts Bluff Co 20 Aug 1997 and a male at Kiowa Springs 21 Nov 1998. Northerly reports include a large group of 49–52 at the marshes near Niobrara 13–16 Dec 1998.

As is the case with other blackbird species in Nebraska, there are a few reports as late as Jan, but overwintering is rare. Curiously, most late fall reports are from Lancaster Co southeastward; these sightings suggest movement toward northwest Missouri, where large numbers have wintered in recent years (Robbins and Easterla 1992). One reported on the Omaha CBC 22 Dec 1984 was in Iowa (WRS), but it has been recorded from Lincoln during CBC week twice, in 1986 and 1990.

High counts include 200 in Sarpy Co 27 Nov 1997 and "about 150" in Lancaster Co 19–20 Oct 1990 (AB 45:124).

Winter

To date there are few reports of overwintering; these are from Lincoln in 1985–86 (AB 40:299) and in 1995–96 when up to 32 wintered at Oak Park L and 9 Feb 1999 when 14 were there. The only other midwinter report was of 10 in the northeast in Knox Co 25 Jan 1998 (MB, JS). Wintering may become more regular in the southeast, based on the numerous late fall reports.

COMMENTS

A grackle not identified to species but fitting the description of a female Boat-tailed Grackle, with dark eyes, was reported in Lancaster Co 26 Mar 1978 (Ott 1978). According to Pruitt (1975), "first-year" Great-tailed Grackles have dark eyes also.

Faanes and Norling (1981) gave measurements of a male specimen purportedly collected in Buffalo Co 13 Jun 1979; this specimen has never been available to other researchers and appears to be lost (Bray and others 1986).

FINDING

Great-tailed Grackles can be found at favored breeding sites in May and Jun, especially at Funk Lagoon. Colonies may move from year to year, however, and checking wetland areas in the Rainwater Basin is a worthwhile strategy.

Brown-headed Cowbird

Molothrus ater

STATUS

Abundant regular spring and fall migrant statewide. Common regular breeder statewide. Rare regular winter visitor southeast.

DOCUMENTATION

Specimen

UNSM ZM11867, 10 Oct 1889, Lincoln, Lancaster Co.

TAXONOMY

The eastern subspecies *ater* breeds in the east and the western *artemisiae* breeds in the west (AOU 1957), but the limits of the respective breeding ranges are unknown.

DISTRIBUTION AND ECOLOGY

Breeding birds, which parasitize a wide range of species, necessarily have adapted to a wide range of habitat types from grassland to woodland including, in the northwest, forested canyons (Rosche 1982). Although preferred habitat is edge and thickets (Johnsgard 1980), probably no forest in Nebraska is extensive enough that this species does not penetrate its interior; indeed, forest fragmentation in recent years has allowed this species access to interior forest species (Dobkin 1994).

Migrants and wintering birds are usually found in open agricultural areas with other blackbirds, although overwintering is rare and mostly restricted to the southeast.

Spring

27 Feb, 2, 3 Mar→summer (except west)

9, 10, 13 Apr→summer (west)

Most migration is in Apr (Johnsgard 1980), although early arrival is somewhat obscured in the southeast due to the occasional presence of small numbers of wintering birds or early arrival of birds that wintered close to Nebraska. Elsewhere, arrival is in early Mar, somewhat later in the Panhandle, where birds arrive early to mid-Apr, with earlier dates 7 Mar 1968 Dawes Co, 8 Mar 1953 Scotts Bluff Co, 21 Mar 1984 Sioux Co, 23 Mar 1956 Scotts Bluff Co, and 1 Apr 1954 Scotts Bluff Co.

High counts include 600 in Pierce Co 10 May 1997, 200 in the eastern Rainwater Basin 14 Apr 1997, and 191 in Dixon Co 11 May 1996.

Summer

This species is a host parasite and probably associates with the same host species as in Kansas, where 121 have been reported (Lowther 1984, 1988). Hosts are generally from the following groups: flycatchers, vireos, warblers, and finches (Bent 1968).

Breeding densities are greatest in the east, according to BBS data.

Eggs were recorded 10 May–6 Jun in the L Ogallala area (Brown and others 1996).

Fall

summer→13, 13, 15 Oct (north, west)

Flocking of young birds to roosts begins as early as 18 Jun (Brown and others 1996), and sizeable flocks appear in Aug and Sep statewide. In the west many adults disappear after the breeding season ends in Jul, while young birds linger into Sep; virtually all birds depart the west during Sep (Rosche 1994a).

Final departure in the north and west occurs by mid-Oct, with a few later reports 25 Oct 1986 Sioux Co, 26 Oct 1979 Sioux Co, 8 Nov 1976 Sioux Co, 30 Nov 1986 Boone Co, and 3 Dec 1976 Scotts Bluff Co (see Winter). Except for the west and north, CBC data indicate that a few birds linger through Dec, especially in the southeast.

High counts include 12 000 at Funk Lagoon 18 Oct 1996, "several thousand" between Odessa and Funk Lagoon 5 Sep 1999, and 2000 near Johnson Lagoon 19 Sep 1999.

Winter

Overwintering is rare. Most winter reports are of small numbers with wintering flocks of Red-winged Blackbirds and Common Grackles around livestock lots. Away from the east and south midwinter records are scarce. There are no Jan or Feb records from the Panhandle and very few from the north, all from Boone Co: 1 Jan 1986, 23 Jan 1981, 3 Feb 1985, and 11 Feb 1983. Most overwintering occurs south of the Platte Valley in the east; midwinter reports are few outside this area and generally define the extent of the winter range: 1 Jan 1981 Washington Co, 9–13 Jan 1999 South Sioux City, 21 Jan 1953 Webster Co, 23 Jan 1960 Lincoln Co, 27 Jan 1980 Washington Co, 4 Feb 1962 Lincoln Co, and 25 Feb 1984 Dakota Co.

FINDING

This species is most conspicuous in May and Jun when displaying males are commonly seen on fence lines statewide. Sizeable flocks occur in Aug and Sep.

Orchard Oriole

Icterus spurius

STATUS

Common regular spring and fall migrant and breeder statewide.

DOCUMENTATION

Specimen

UNSM ZM8770, 14 Jun 1898 Lincoln, Lancaster Co.

DISTRIBUTION AND ECOLOGY

Orchard Oriole breeds statewide in open deciduous woodland and edge habitats, especially brushy thickets in riparian situations but also around farmsteads. Migrants occur in similar habitats.

Spring

21, 24, 24 Apr→summer

Migrants generally arrive in early May, with earlier dates 12 Apr 1950 Antelope Co, 15 Apr 1995 Sioux Co, 17 Apr 1956 Nemaha Co, and 17 Apr 1969 Brown Co.

High counts include 55 in the Keith Co area 18 May 1986 (Rosche 1994a).

Summer

This species has expanded its breeding range westward in the 20th century (Short 1961). Bruner and others (1904) stated that it was a "common breeder west to the 100th meridian [line from Keya Paha to Furnas Cos] or a little farther." Ducey (1988) cited no breeding records for the Panhandle prior to 1920, while Rapp and others (1958) considered it a common summer resident in the eastern two-thirds and especially the eastern quarter of Nebraska and added that its status in the west was "poorly known."

By 1980 Johnsgard (1980) noted that it was breeding "virtually statewide, less commonly in the extreme west." Currently it is as common a breeding bird in the west as it is in the east; 19 were counted in Dundy Co 22 Jun 1994. It has become more common in the central Platte Valley; in 1982 Lingle and Hay (1982) considered it rare in their Hall Co study area, but by 1994 it had become common (Lingle 1994).

A pair with a brood was at Oliver Res 11 Aug 1997.

Fall

summer→23, 24, 26 Sep

Orchard Orioles leave the breeding grounds in Aug, with the last birds leaving the southeast by late Sep. There are later dates 9 Oct 1965 Adams Co, 9 Oct 1979 Douglas Co, and 29 Oct 1989 when a male was seen in Sarpy Co (AG; AB 44:116).

High counts include 30 in the Keith Co area on both 9 and 10 Aug 1986 (Rosche 1994a) and 19 at Box Butte Res 12 Aug 1994.

Other Reports

10 Mar 1961 Logan Co, 21 Mar 1948 Webster Co, 2 Apr 1959 Gage Co, 4 Apr 1956 Adams Co, 12 Nov 1961 Douglas Co.

FINDING

Males are conspicuous in Jun, especially in the Panhandle where riparian vegetation tends to be less dense. Orchard Orioles can usually be located anywhere in Jun in deciduous thickets near water.

Hooded Oriole

Icterus cucullatus

STATUS

Hypothetical.

COMMENTS

One was reported seen near Columbus on 2 days, including 12 May 1965 (Armstrong 1965). Identification was based on an orange crown and "fiery orange" hood, rump, and underparts. While neither Baltimore nor Bullock's Orioles nor their hybrids possess "orange crowns" (although first fall birds have greenish orange crowns), no further details were provided for this report.

Hooded Oriole has occurred as a winter vagrant north to Oregon on the West Coast (Roberson 1980) but east of the Rocky Mountains has occurred no farther north than east Texas and Louisiana (White 1990). Thus there is no indication of vagrancy on the Great Plains. A banded bird recovered in Kansas identified as a Hooded Oriole was considered to have been incorrectly identified (Thompson and Ely 1992).

Baltimore Oriole

Icterus galbula

STATUS

Common regular spring migrant east and central, rare west. Common regular breeder statewide except west, where rare casual. Common regular fall migrant east and central, rare casual west.

DOCUMENTATION

Specimen

UNSM ZM7017, 30 Nov 1900 Lancaster Co.

TAXONOMY

Hybridization occurs between this species and Bullock's Oriole (Rising 1970, 1996), causing their taxonomic status to be controversial (AOU 1998), including their treatment as a single species, Northern Oriole, by the AOU (1983). Recently, however, the AOU (1995) returned both to full species status based on absence of free interbreeding at some locations where there is contact and variation in the extent of hybridization geographically; furthermore, "the two species differ in plumage, vocalizations, physiologic traits, molt, nesting ecology, and allozyme frequency to a degree greater than in any other birds generally treated as single species" (DeBenedictis 1996), and the extent of the hybrid zone in Kansas appears to be stable (Rising 1996). Recent work with cytochrome-b phylogeny of the orioles (AOU 1998) suggests that the closest relative of Baltimore Oriole is actually Black-backed (Abeille's) Oriole and that Bullock's was a member of the Streak-backed Oriole complex.

Short (1961) considered the hybrid zone rather wide in Nebraska, with its center a line from Big Springs to just west of Valentine. Short found pure Baltimore Orioles west to Hastings, St. Paul, and Spencer and pure Bullock's Orioles east only to Chadron and Big Springs. However, hybrids outnumbered pure

Baltimores east as far as Blair, and "obvious hybrids" occur in the northwest in summer (Rosche 1982). Youngworth (1955) and Mossman and Brogie (1983) found Baltimore Orioles predominant in the Niobrara Valley Preserve, although hybrids were seen during the breeding season. Both Baltimore and Bullock's Orioles have been reported at Bessey Division, NNF, in Thomas Co (Bray 1994). In Keith Co Brown and others (1996) found that 116 of 176 birds examined (not all were nesting birds) were phenotypically Baltimore Orioles, and 41 showed evidence of hybridization, the remainder apparent Bullock's Orioles.

DISTRIBUTION AND ECOLOGY

Baltimore Oriole breeds east of the Panhandle in tall deciduous trees, usually cottonwoods, although also in elms (Rosche 1982). Preferred nesting habitat is riparian, but this species also occurs commonly in both rural and urban residential situations.

A study by Rising (1969) indicated that Baltimore Orioles have a "higher metabolism at high temperatures than do Bullock's Orioles and that they are less tolerant of high temperatures" and that "the contact zone between the two generally coincides with the climatic boundary separating hot, arid western summer climates from cooler, more-humid eastern summer climates."

Migrants occur statewide, but they are rare in the Panhandle in spring and casual there in fall (Rosche 1982). Migrants occur most often in wooded edge habitats but can appear almost anywhere, including open grassland.

Spring

12, 15, 15 Apr→summer

This species arrives in late Apr, with an earlier report 3 Apr 1981 Adams Co. Such early spring arrivals may be birds that wintered somewhere in the United States, an increasingly common recent phenomenon.

In the Panhandle, where migrants are rare, reports are in the period 12–29 May, except for a few summer reports (see Summer).

High counts include 124 in Sarpy Co 13 May

1995, 119 there 11 May 1996, and 50 at Gibbon 6 Jun 1998.

Summer

This species breeds commonly throughout Nebraska, except for the Panhandle, where "obvious hybrids" are frequent in the breeding season (Rosche 1982). There are a few Panhandle reports of birds identified as Baltimore Orioles: 15 Jun 1963 Scotts Bluff Co, 24 Jun 1995 Oliver Res, 30 Jun 1964 Scotts Bluff Co, 30 Jun 1967 Garden Co, 30 Jun 1972 Scotts Bluff Co, and 30 Jul 1971 Scotts Bluff Co. There is a specimen, UNSM ZM7012, taken at Scottsbluff 28 Jun 1916.

It is apparent that this species has spread westward since 1900, possibly at the expense of Bullock's Oriole; around 1900 it occurred west to central Nebraska (Bruner and others 1904) but now is found throughout Nebraska except for the extreme west (Rosche 1982).

BBS data show that this oriole has increased in Nebraska. Near Crook, Colorado, between the 1950s and 1970s, the percentage of Baltimore Orioles increased while the percentage of intergrades and Bullock's Orioles decreased (Corbin and others 1979).

Fall

summer→1, 2, 4 Oct

Departure is essentially completed by late Sep, but individuals may linger on occasion, with late reports 14 Oct 1986 Lancaster Co, 26 Oct 1971 Lancaster Co, 1 Nov 1974 Sarpy Co (AB 29:710), and 9 Nov 1999 near Milford (PK).

There are a few extremely late reports from the east: one was at an Omaha feeder 8–30 Nov 1981 and later (AB 36:194); one was in Omaha from 21 Nov 1973 (AB 28:76); one was photographed feeding on berries at DeSoto Bend NWR 24 Nov 1982 (NBR 51:14); a male was in a Bellevue yard 8 Nov–12 Dec 1981 feeding on fallen apples (NBR 50:4); a specimen, UNSM ZM7017 (cited above), was collected at Emerald 30 Nov 1900; one was at a Sarpy Co feeder 30 Nov 1979–December (AB 34:178, 287); and one that had been present in

Cass Co since 1 Dec 1986 was caught by a cat 14 Jan 1987 (AB 41:299).

There is a single fall report from the Panhandle, 3 Sep 1966 Scotts Bluff Co.

FINDING

Baltimore Orioles are most conspicuous in the Missouri Valley in May and Jun when males are actively defending territories. The loud song locates the birds, although they can be hard to see in tall trees.

Bullock's Oriole

Icterus bullockii

STATUS

Fairly common regular spring and fall migrant west, accidental elsewhere. Fairly common regular breeder west.

DOCUMENTATION

Specimen

UNSM ZM7010, 19 Jun 1901 Indian Creek, Sioux Co.

TAXONOMY

See Baltimore Oriole.

DISTRIBUTION AND ECOLOGY

This species breeds in riparian woodlands in the Panhandle, preferring cottonwoods and elms (Rosche 1982), occurring as a migrant in similar habitats. It also is found in cottonwoods and elms in towns and cities (Rosche 1982). It occurs in more arid areas than Baltimore Oriole, as discussed under that species.

Spring

29 Apr, 1, 2 May→summer

Arrival is in early May, with earlier dates 13 Apr 1972 Scotts Bluff Co and 15 Apr 1990 Scotts Bluff Co.

There are several reports east of the Panhandle, all in the period 30 Apr–25 May; none are documented. Easternmost are 30 Apr 1952 Platte Co, 3 May 1953 Platte Co, 8 May 1955 Thayer Co, 20 May 1959 Thayer Co, and 25 May 1955 Gage Co. Swenk (Notes before 1925) recorded a specimen in the Olson collection taken at Kearney Jun 1915 but provided no details.

It seems likely that Bullock's Oriole occurs in migration east in small numbers to a line connecting Valentine and McCook (Bruner and others 1904), although this is difficult to determine due to intergrades.

Summer

Short (1961) found pure Bullock's Oriole eastward only as far as Dawes and Deuel Cos, and in recent years this range may have contracted somewhat as Baltimore Oriole has extended its range westward into northeast Colorado (Andrews and Righter 1992; Corbin and Sibley 1977). Recent Nebraska data are scarce, however, due to inclusion of Bullock's Oriole in reporting as "Northern Oriole."

Reports of this species east of the Panhandle are few, and most recent occurrences probably represent hybrids exhibiting Bullock's Oriole characters. Brogie and Mossman (1983) found only hybrids and Baltimores in the Niobrara Valley Preserve, as did Rosche (1994a) and Brown and others (1996) in the Keith Co area, where Bullock's Oriole was thought to have decreased as a breeding bird in recent years (Brown and others 1996). Bray (1994) made no mention of recent reports of Bullock's Oriole in the Thomas Co area, although it was said to have occurred there some 50 years ago. Tout (1947) listed this species as a "rare summer resident" in Lincoln Co but provided no evidence.

Easterly summer reports, none documented, include 7 Jun 1953 Brown Co, 9 Jun 1957 Gage Co, 22 Jun 1954 Gage Co, 30 Jun 1969 Boone Co, 30 Jun 1969 Logan Co, and 30 Jun 1972 Perkins Co.

Fall

summer→11 (Rosche 1994a), 12, 12 Sep (Rosche 1982)

Departure is in early Sep, with a late report 3 Oct 1969 Scotts Bluff Co. There is only 1 report of Bullock's Oriole east of the Panhandle in fall, although young birds, the most apt to wander eastward, are difficult to distinguish from same-age Baltimore Orioles, and intergrades also present an identification problem. There was a group of 5 birds, including an adult Bullock's and 4 immature or female birds, at Alma 11 Aug 1999 (GH, WH).

COMMENTS

The current status of this species is poorly known as a result of reporting as Northern Oriole for the past several years. Observers should carefully study putative Bullock's Orioles east of the Panhandle and record location of the sighting and presence or absence of hybrid characters.

FINDING

In Jun Bullock's Orioles are not hard to find in riparian stands of cottonwoods, such as those at Box Butte Res and Oliver Res.

Scott's Oriole

Icterus parisorum

STATUS

Rare casual summer visitor central.

DOCUMENTATION

Description

20–24 Jun 1975 Hall Co (Stoppkotte 1975a; Brogie 1998).

DISTRIBUTION AND ECOLOGY

This species has been recorded in central Nebraska only, generally in riparian habitats. Preferred habitat within its normal range at its nearest point in Colorado is isolated pinyon-juniper and juniper groves in semidesert shrublands, although vagrants are recorded in riparian woodlands (Andrews and Righter 1992).

Summer

There are 5 reports, 1, a singing adult male, acceptably documented:

20–24 Jun 1975 Hall Co (cited above)

The 4 additional reports are 31 May 1951 Adams Co, 17 and 25 Jun 1983 Chase Co (Pennington 1983; Brogie 1998), 18 Jun 1978 McPherson Co, and 2 Jul 1997 Hall Co (RS; Brogie 1998), all of which fit the pattern of casual summer occurrence and are most likely correct. The 1975–83 reports were in a period following a significant population expansion in Colorado (Andrews and Righter

1992) and southwestern Wyoming (Scott 1993).

The reports are somewhat later than the 7 records of Scott's Oriole from eastern Colorado, Kansas, and Minnesota. The single Kansas record was of a female collected in Morton Co 16 Apr 1977 in open cottonwoods (Thompson and Ely 1992), the Minnesota record was of an immature male in Duluth from 23 May through mid-Jun 1974 (Janssen 1987), while the 5 Colorado records from the eastern slope and plains nearby are all for the first half of May (Andrews and Righter 1992).

FINDING

This species should be looked for in riparian woodlands in the west in May and Jun. A song resembling that of Western Meadowlark emanating from riparian woodland frequented by other orioles should be investigated. Spring overshoot migrants are likely to appear in May and become territorial in Jun.

Fringillidae (Finches)

Brambling

Fringilla montifringilla

STATUS

Accidental in spring.

DOCUMENTATION

Photograph

14–19 Apr 1999 Scottsbluff, Scotts Bluff Co (SJD).

Records

A female appeared at the feeder of Dean and Phyllis Drawbaugh in Scottsbluff 14 Apr 1999 and remained until 19 Apr.

COMMENTS

There are at least 50 records of this species in North America, all in the period Oct–Apr and most from northern states and Canada (Mlodinow and O'Brien 1996). Most of the records are from northwestern North America (AOU 1998). The Nebraska record was part of an irruption into British Columbia and the northwestern United States that involved about 17 birds in the period 28 Oct 1998–24 Apr 1999.

Gray-crowned Rosy-Finch
Leucosticte tephrocotis
STATUS
Rare regular winter visitor west.
DOCUMENTATION
Specimen
UNSM ZM7133, 15 Feb 1896 Harrison, Sioux Co.
DISTRIBUTION AND ECOLOGY
This species appears erratically during winter in the western Panhandle, usually along roadsides in open areas, sometimes at feeders, but most predictably at certain rocky outcrops used for roost sites.
Winter
25 (Green 1985a), 26, 29 Oct→20, 27 Mar, 2 Apr
Fall arrival is in late Oct and departure in Mar. There is an early report 1 Oct 1977 Sioux Co and a late report 16 Apr 1989 Sioux Co.
This species forms flocks in winter that wander around the northwestern Great Plains, notably in eastern Wyoming (Scott 1993) and northeastern Colorado where most are found on the Pawnee Grasslands (Andrews and Righter 1992). However, there are no Kansas records (Thompson and Ely 1992). Occasionally, large flocks occur in western Nebraska, such as 300 near Crawford 10 December 1910 (Zimmer 1911a) and up to 2000 in northern Sioux Co 25–27 October 1984 (Green 1985a), but many Nebraska sightings are of individuals with flocks of Pine Siskins and American Goldfinches (Rosche 1982; Hughson 1990).
Almost all reports are from Sioux and Scotts Bluff Cos, with few elsewhere: 26–28 Oct 1972 Perkins Co; 6 Dec 1983 Lincoln Co; 7 Dec 1910, a specimen, UNSM ZM10696, Dawes Co; 29 Dec 1996, flock of about 100 near Whitney; 1 Jan 1984 Cheyenne Co (Wittrock 1984); 7 Feb 1922, a pair of specimens, HMM 22176, taken at Oshkosh (Swenk, Notes after 1925); and 11 Feb 1922, a specimen, UNSM ZM7131, collected at Oshkosh (Rosche 1994a).

Up to 75 wintered at a Chadron feeder 1996–97.
Other Reports
2 Nov 1944 Gage Co (Patton 1945), 3 Feb 1967 Brown Co, 3–7 Mar 1966 Brown Co. A report 7 Nov 1993 Hall Co was not accepted by the NOURC (Gubanyi 1996b).
FINDING
This species can be looked for by driving open country roads where rocky outcrops occur in western parts of Scotts Bluff and Sioux Cos in late Oct and early Nov, when flocks of seed-eating finches should be checked and flocks containing only this species are likely to occur. Flocks tend to roost in rocky outcroppings by midafternoon and are difficult to see later in the day. Later in the winter individuals may be located at feeders. Scotts Bluff NM has yielded several reports.

Pine Grosbeak
Pinicola enucleator
STATUS
Uncommon casual winter visitor statewide.
DOCUMENTATION
Specimen
UNSM ZM10416, 30 Nov 1910 Crawford, Dawes Co.
TAXONOMY
Two subspecies have been recorded. Eastern *leucurus*, which breeds in central Canada, has been recorded west to Grand Island and Long Pine, and the Rocky Mountains subspecies *montana* has been recorded in the Panhandle (Rapp and others 1958).
DISTRIBUTION AND ECOLOGY
Wintering birds are associated with fruiting trees, either coniferous or deciduous, although most are seen in conifers. Reports are distributed throughout the state.
Winter
5, 6, 7 Nov→11, 11, 15 Mar
There are about 80 reports of Pine Grosbeak. The reports are in the period 11 Oct–28 May, but fall arrival dates cluster in early Nov and

spring departure dates in early to mid-Mar. There are earlier fall reports 11 Oct 1966 Lincoln Co and 30 Oct 1932 Adams Co and later spring reports 28 Mar 1936 Adams Co, 6 Apr 1970 Lincoln Co, 13 Apr 1970 Scotts Bluff Co, and 20 Apr 1974 Lincoln Co.

Most reports are from "invasion" winters, including 1910–11, 1932–33 (many reports), 1962–63, 1963–64, 1966–67, 1969–70 (many reports), and 1970–71. The species was described as "numerous" in Dodge Co in winter 1946–47 through May.

Since 1980 there are few reports: 5 Nov 1987 Adams Co, 6 Nov 1984 Douglas-Sarpy Cos, 12 Nov 1981 Douglas Co, 16 Dec 1984 Lancaster Co, 18 Dec 1983 Lincoln Co, and 22 May 1988 Scotts Bluff Co (NBR 56:73; see below).

Pine Grosbeak is a casual visitor in summer in the Black Hills of South Dakota (SDOU 1991), suggesting that May reports in Nebraska, most of which are from the north and west, may be of similar stragglers. Reports in this category are 27 Apr 1950 Webster Co, 6 May 1972 Brown Co, 7 May 1936 Lincoln Co, 7 May 1936 Logan Co, 16 May 1938 Lincoln Co, 22 May 1988 (a well-described female) Scotts Bluff Co (NBR 56:73), 22 May 1967 Brown Co, and 28 May 1943 Logan Co. There is a 15 May sight record for Christian Co in southern Missouri (Robbins and Easterla 1992).

Other Reports

5 Sep 1949 Webster Co, 25 Sep 1960 Loup Co.

FINDING

This species should be looked for Nov–Mar in cemeteries with fruiting conifers or in areas with native conifers such as the central Niobrara Valley. It is very rare and generally unreported in "noninvasion" years.

Purple Finch

Carpodacus purpureus

STATUS

Uncommon regular spring and fall migrant and winter visitor east, becoming rare west. Hypothetical in summer.

DOCUMENTATION

Specimen

UNSM ZM7120, 1891 Lincoln, Lancaster Co.

TAXONOMY

Nebraska birds, including those wintering in the Panhandle, are presumably of the eastern race *purpureus*, as are listed Colorado specimens (Bailey and Niedrach 1967).

DISTRIBUTION AND ECOLOGY

Wintering birds, as well as migrants, are associated with riparian woodland and are most common in southern and eastern Nebraska. Rosche (1982) noted that Purple Finches favor Russian olive, juniper, and ashleaf maple and also commonly attend feeders. It has been suggested that the recent explosive House Finch expansion has caused a decline in Nebraska Purple Finch numbers (Rosche 1994a; FN 51:82).

Spring

winter→11, 12, 13 May

Departure is in early May, with later reports from the north and west 15–17 May 1994 Thomas Co and 19 May 1955 Scotts Bluff Co. There are very late undocumented reports (prior to the House Finch influx) 25 May 1987 Polk Co, 5 Jun 1977 Lincoln Co, and 17 Jun 1975 Clay Co.

Summer

A report suggesting that an immature banded 20 Apr in Sarpy Co had hatched in Nebraska has been questioned (Mollhoff 1989b).

Fall

29, 30, 30 Sep→winter

Purple Finches occur statewide during migration but are most common in the east. Arrival is in early Oct, although there are earlier reports 12 Sep 1986 Lincoln Co, 16 Sep 1990 Phelps Co, and 23 Sep 1979 Garden Co. Migration probably peaks in late Oct (Johnsgard 1980).

High counts include 40 at Wolf L 27 Nov 1999 and 25 in Saunders Co 25 Nov 1995.

Winter

Early authors considered Purple Finch to occur only in eastern Nebraska (Rapp and others 1958; Bruner and others 1904). More recently, Johnsgard (1980) suggested that this species is uncommon in winter in the Pine Ridge, and Rosche (1982) described it as "uncommon to fairly common" in the northwest, although "erratic and irregular." Data from the Crawford CBCs held in 1973 and 1976–79 showed 7 birds found in 166 party-hours, comparable to the density recorded on CBCs in southern Nebraska.

CBC data indicate that during the second half of Dec most Purple Finches are in southern and eastern Nebraska.

FINDING

Purple Finches are most readily located during Nov and Dec in Missouri Valley riparian woodland edge. This species is readily attracted by squeaking and commonly appears at feeders.

Cassin's Finch

Carpodacus cassinii

STATUS

Accidental breeder northwest. Uncommon casual summer visitor west. Rare regular winter visitor northern Panhandle, rare casual elsewhere in west and central.

DOCUMENTATION

Specimen

UNSM ZM10486, 7 Dec 1910 Crawford, Dawes Co (Mickel and Dawson 1920). This specimen and UNSM ZM10487 are inadvertently labeled Purple Finch.

DISTRIBUTION AND ECOLOGY

Cassin's Finch occurs in ponderosa pine habitat in the northern parts of the Panhandle (Rosche 1982). It also regularly attends feeders. Numbers vary greatly from year to year.

Summer

Although Cassin's Finch breeds regularly in the Black Hills of South Dakota (SDOU 1991), there is only 1 breeding record for Nebraska, perhaps because of lack of its preferred breeding habitat of high-elevation coniferous forests (Andrews and Righter 1992). Rosche (1982) found a female feeding 2 juveniles in Chadron SP 25 Jul 1980.

There are 6 reports for late summer, apparently wandering postfledging young birds presumably from the Black Hills or even Nebraska locations. One was netted in Garden Co 24–25 Jul 1979; there are Sioux Co reports 1 Aug 1997, 5–6 Aug 1989 (AB 44:116), and 15 Aug 1981; Rosche (1982) listed an early observation date of 28 Aug; and one was in Scotts Bluff Co 30 Aug 1985. A male was at a Chadron feeder 25 Jun 1987 (AB 41:1457).

Winter

Most reports are in the period 1 Jan–28 May in or near the Panhandle. Fall reports are erratic, with no cluster of arrival dates. There are only these reports prior to 1 Jan: 18 Sep 1919, specimen UNSM ZM7118 (Mickel and Dawson 1920); up to 14 at Wildcat Hills NC 6 Oct 1996–20 Apr 1997 (LM; Brogie 1997); 27 Oct 1997 Wildcat Hills NC (SJD); 2 Nov 1996, specimen UNSM 17484, Wildcat Hills NC (Brogie 1998); 22 Nov 1936 Lincoln Co (Tout 1947); 7 Dec 1910 Dawes Co (specimen cited above); 7 Dec 1910, specimen UNSM 10487, Dawes Co; 17 Dec 1993 Keith Co; 18 Dec 1963 Dawes Co; and 19 Dec 1919, specimen UNSM ZM7117, Dawes Co.

In years when they occur Cassin's Finches typically arrive in small numbers, increasing to as many as the 52 seen 10 May 1979 at Crawford and then disappearing in May (Rosche 1982). An exception may be the influx of 1996–97, when several appeared in early Oct and wintered in Scotts Bluff Co.

There are several Nebraska reports east of the Panhandle, but the only documented record is of one at an Ogallala feeder 17 Dec 1993 (DJR). Cassin's Finch has not been reported as a vagrant far east of its range (White 1990) except for 2 South Dakota records, one banded at Pierre 8 Apr 1973 and another wintering in Shannon Co 19 Jan–7 May 1980 (SDOU 1991).

Other Reports

A series of undocumented reports from Logan Co in the period 15 Jan–30 May agrees with Panhandle data, but the same observers list an additional report for Oct 1933 (Glandon and Glandon 1934b), an unlikely date.

A report from Lancaster Co 7–9 Dec 1982 had "The Cassin's Finches . . . in with Purple Finches" (NBR 51:3).

Additional reports include 1 Jan 1956 Keith Co, 3 Jan 1953 Thomas Co, 31 Mar 1953 Boone Co, 18 Apr 1953 Lincoln Co, and 24 Apr–1 May 1961 Lincoln Co.

FINDING

The best chance of finding Cassin's Finch is in Apr and early May in open ponderosa pine habitat in the Pine Ridge such as at the heads of Monroe and Sowbelly Canyons. Squeaking readily attracts these birds. In recent years the feeders at Wildcat Hills NC have been good locations to find this species.

House Finch

Carpodacus mexicanus

STATUS

Common, locally uncommon, regular resident statewide.

DOCUMENTATION

Specimen

HMM 2425, 29 Jan 1916 Haigler, Dundy Co.

DISTRIBUTION AND ECOLOGY

House Finch is restricted in Nebraska to urban habitats, where it has been established for many years in towns and cities in the Panhandle and western Platte Valley and has spread slowly eastward to Buffalo Co; the eastward spread in the North Platte Valley has been attributed to the presence of "dense planted evergreens in residential areas . . . and feeding stations" (Rosche 1994a). In the 1990s the species became established in towns and cities in eastern Nebraska, partly because of the continued eastward movement of the western population and also because of an explosive westward range expansion of the introduced eastern United States population. Wherever it occurs it soon becomes a common and conspicuous resident.

Summer

As of 1900 House Finch was unrecorded in Nebraska (Bruner and others 1904), although it must have appeared soon after, as one was collected by G. A. Williams in Haigler 29 Jan 1916 (specimen cited above) and another in Bushnell 2 Jun 1919 by C. E. Mickel (specimen UNSM ZM7121; Mickel and Dawson 1920; Bray and others 1986). These birds presumably were derived from the western House Finch population.

The spread of House Finch is apparently accomplished by dispersal in fall of birds that visit feeders in towns and cities and then breed the next summer or within very few years thereafter (Rosche 1994a). First records at most locations are indeed in fall or winter and in some cases predate the establishment of breeding populations by several years. The first report for Adams Co was 28 Dec 1968, and breeding began there about 1984; for Hall Co corresponding dates are 4–15 Dec 1973 (NBR 42:27) and 1983; for Douglas Co 12 Dec 1977 (NBR 46:23) and 1988; for Lancaster Co early September 1980 (NBR 49:6, 18) and 1988. House Finch was first recorded in Dawes Co in 1973 and seen every year since at feeders, although the first summer records were in 1982 (AB 36:994).

Eastward spread of the species has been primarily in the North Platte Valley. It became established in Scottsbluff apparently well before 1958 (Olson 1956; LOI 18:2; Ducey 1988; Rapp and others 1958), and an isolated population was established in North Platte in the 1920s (NBR 21:38; Gates 1955; Johnsgard 1979). Between these locations, establishment has been recent, at Oshkosh in 1977 (Rosche 1994a; Brown and others 1996) and the Kingsley Dam area in 1985 (Rosche 1994a). A population became established in Kearney prior to 1980, although there is no specific information published. Breeding began in Hall Co in 1983 (Coons 1983; Bennett 1984) and at about the same time in Adams Co, where the first summer reports were in 1984. Reports from neighboring Kearney and Phelps Cos are recent, despite the presence of the long-established colony in Kearney, although individuals were reported at a Minden feeder in late summer 1983–85 (NBR 51:91, 53:41, 54:19).

Prior to 1980 colonies had become established in McCook as well as in Kimball and Haigler. House Finch first appeared in Ogallala in 1977 (Rosche 1994a), although there was an earlier report for Keith Co 20 Mar 1964. Summering birds had been reported in Chase and Perkins Cos by 1982, and by 1990 the southern Panhandle and southwest were occupied, east to Kearney Co, including Republican Valley counties. Establishment in Pine Ridge towns is recent; nesting was first detected in Chadron in 1982 (AB 36:994), and the species had become common there 2 years later (AB 38:1036). Colonization of the Sandhills is recent; nesting birds were noted at Hyannis in 1993 (Rosche 1994a).

At the eastern edge of the state breeding first occurred in Sarpy and Lancaster Cos in 1988 (Green 1988; Bennett 1989; NBR 58:9). Nesting may have occurred in Douglas Co in 1987, when individuals appeared at a feeder in Omaha 26–30 Jun (NBR 55:56); summering birds have been reported in Omaha since. Knox Co was considered a "new locality" in 1989 (AB 43:501). Since 1990 House Finch has rapidly established itself throughout the Missouri Valley. The source of birds in Polk Co in 1993 is conjectural.

The westward spread of House Finches descended from a group released in New York in the early 1940s has been well documented (Woods, in Bent 1968). Summer birds were first recorded in Iowa in 1982 (Dinsmore and others 1984), with nesting in 1986; the "invasion of Iowa" was said to be complete by 1992 (Dinsmore 1992). Thus it appears that eastern House Finches reached the Missouri Valley of Iowa in 1989 and 1990, coinciding with establishment of the species in Nebraska counties along the Missouri Valley. The derivation of birds breeding in Sarpy and Lancaster Cos in 1988, as well as summer birds

in Douglas Co at the same time, is conjectural. There are 3 possible sources, all reaching the Omaha area at about the same time: eastern birds from adjacent Iowa counties, eastern birds spreading northward along the Missouri Valley from Missouri, and western birds spreading east along the Platte Valley. It is of interest that the first breeding in northwest Missouri was in 1988 at both St. Joseph and Maryville (Robbins and Easterla 1992).

Currently, summering House Finches are fewest in the Sandhills, central Niobrara Valley, upper Loup drainage, and lower Little Blue drainage, even though the species has been reported for several years from some of these areas. There are undocumented reports as early as 31 Jan 1942 in Webster Co and 1 Mar 1966 in Brown Co. House Finch was reported at Halsey 4 Oct 1981 (NBR 49:65) and at Elsmere 4–5 Oct 1986 (NBR 54:79). There are recent Holt Co reports for 31 Oct 1991 and 1 Jan–30 Jun 1992, suggesting that breeding may be occurring there. The species was not listed for Holt Co in 1990 by Blake and Ducey (1991). There is a single Custer Co report 1 Mar 1992. Five birds were counted on the Norfolk CBC in Madison Co 16 Dec 1989, the only report for the Elkhorn drainage west of Cuming Co.

CBC data show highest counts in cities where the species has been long established. Highest single count is 235 from Scottsbluff in 1989–90, with other high counts 134 in North Platte in 1983–84 and 110 in Kearney in 1989–90.

Winter

House Finches often disappear from their summer locales, forming small flocks that may winter in the area or possibly wander some distance away; this is the method by which the species has expanded its range.

CBC data in recent years in states where House Finch has increased rapidly suggest that House Sparrow populations may be declining (Silcock 1994); whether there is a direct relationship between these trends has not been elucidated.

FINDING

House Finches are conspicuous in towns and cities, especially at feeders in winter, and their warbling song is ubiquitous once learned.

Red Crossbill

Loxia curvirostra

STATUS

Common regular resident northwest. Common regular spring and fall migrant north and northwest, rare elsewhere. Common regular winter visitor north and northwest, rare elsewhere.

DOCUMENTATION

Specimen

UNSM ZM7187, winter 1886 Fairbury, Jefferson Co.

TAXONOMY

Several subspecies have been reported in Nebraska, but the taxonomy of Red Crossbill is notoriously complex, and there may in fact be more than 1 species involved (Groth 1993; AOU 1998). Studies by Groth (1993) have established 7 types of Red Crossbill based on calls and morphology. Type 2 (Groth 1993) is large billed and is associated with ponderosa pine. Its range includes the Black Hills of South Dakota. The relationship between the call types and traditional subspecies is not yet clear.

Breeding birds in Nebraska are presumably the eastern Rocky Mountain subspecies *benti*, which breeds in the "pine hills of southeast Montana, eastern Wyoming, western North and South Dakota, and Colorado" as outlined by Griscom (Bent 1968). An adult male of this race was collected 19 Jul 1957 (Ford 1959) in the Pine Ridge west of Crawford. Bruner and others (1904) suggested that *minor* was "common in summer in Sioux County"; based on Griscom (1937), these birds are now included in *benti*.

During nonbreeding seasons specimens of *bendirei* (western interior mountains), *minor* (northeastern United States and Canada), and *sitkensis* (Alaska) have been taken in

Nebraska (AOU 1957; Rapp and others 1958; Bruner and others 1904; Mickel and Dawson 1920; Swenk, Notes after 1925). Bruner and others (1904) cited a specimen considered "definite *stricklandi*" taken in Neligh 9 Dec 1898; this race breeds in the southwest United States. Two collected in Webster Co 29 Oct 1913 were sent to UNSM and identified as eastern *pusilla* (Ludlow 1935).

DISTRIBUTION AND ECOLOGY

Red Crossbills in Nebraska occur in summer almost exclusively in ponderosa pine woodland; distribution and abundance are directly related to the status of the pine seed crop. Regular breeding is restricted to Sioux and Dawes Cos, where the species is usually common, but breeding may be absent in years when the pine seed crop is poor.

Wintering birds occur statewide, usually in coniferous habitats like cemeteries and pine plantings, but their appearances are notoriously unpredictable.

Spring

winter→1, 2, 2 Jun

Departure of wintering birds is complete by late May, even in the extreme southeast, where late reports include 23 May 1961 Lancaster Co, 25 May 1997 Cuming Co, and 31 May 1997 (5) Nebraska City. Johnsgard (1980) indicated that half the published spring reports are in the period 19 Mar–19 May, suggesting peak movement at that time.

Summer

Red Crossbill is a regular nester in the Pine Ridge, although numbers vary markedly; nesting occurs only in years with good seed crops (Rosche 1982). Prior to the 1970s breeding was apparently only sporadic, with 1 report in 1961 near Chadron (Wensien 1962). Neither Bruner and others (1904) nor Rapp and others (1958) listed the species as a breeder. Since 1974, when Rosche found a nest with eggs near Crawford on 24 Mar (AB 28:659; NBR 47:58), breeding has become regular (Rosche 1982). Ducey (1988) indicated that breeding occurred farther eastward in the

Niobrara Valley, although while he reported birds in Jun at Anderson Bridge in Cherry Co and in the Niobrara Valley Preserve in Brown Co in the mid-1980s, no evidence of nesting was obtained. A pair of adults with 4-5 juveniles was seen in the Keya Paha Co portion of the Niobrara Valley Preserve 18 Jun 1995, although Brogie and Mossman (1983) found only 1 bird during their study in 1982, that on 7 Jun in extreme eastern Cherry Co. Juveniles were present in early Aug at Bessey Division, NNF (Bray 1994).

Characteristic of this species is its propensity to nest in unexpected locations, usually after incursion years. There are 4 such Nebraska records cited by Ducey (1988) for Custer, Douglas, Lancaster, and Adams Cos. The Custer Co report was of a nest being brooded by a female at Broken Bow 20 Mar 1920; this nesting was unsuccessful, as the female was found dead a few days later (Swenk 1921). The Douglas Co report was of a female on a nest, accompanied by a male, in a box elder in Elmwood Park, Omaha, 28 Mar 1920; the nest was destroyed in a storm (Swenk 1921). The Adams Co report involved 3 birds, "one of which was obviously a young of the year" at Hastings 27 Jul 1966 (AFN 20:581). No further details were provided; it should be noted that by Jul young of the year are quite mobile. A pair raised young in Lincoln Mar–Apr 1932; the male fed the young by regurgitation (LOI 66:11).

South Dakota data suggest that nesting occurs in the period Dec–Aug, and late summer records, from Jul on, may in many cases be early fall migrants or dispersing birds (SDOU 1991) rather than evidence of breeding. Nesting was in progress as early as late Feb 1989 in Dawes Co (AB 43:332). Reports from Holt Co in Jul 1990 (Blake and Ducey 1991) probably are migrants, as are other Jul and Aug reports away from the breeding range (see Migration). On the other hand, reports in Jun from locations away from the known breeding range in Sioux and Dawes Cos but

which possess ponderosa pine habitat might indicate breeding. Most are from Scotts Bluff Co: 12 Jun–1 Jul 1969, 30 Jun 1974, 21 Jun 1981, 30 Jun–1 Jul 1984, 1 Jul 1987, 11 Jun 1989, and 30 Jun and 24 Jul 1994. Suggestive of breeding in the area was the presence of a flock of as many as 50 birds, including many juveniles, at the Wildcat Hills NC Mar–Jun 1998. Similar reports exist for Sheridan Co 4 Jun 1986 and 5–12 Jun 1992, Brown Co 12 Jun 1988, Thomas Co 30 Jun and 3 Jul–7 Sep 1993, and Cherry Co 19 Jul 1993. Early summer reports from Garden Co 21 Jun 1981, Adams Co 14 Jun 1951, Douglas Co throughout summer in 1990 (AB 44:1154), Lancaster Co 6 Jun 1991 and 8 Jun 1982, and Alma (a male) 1 Jul 1998 are not as easily explainable.

Fall

11, 16, 20 Jul→winter

Red Crossbill probably has no set migration pattern, but if the pine seed crop is poor, small flocks leave the breeding grounds in late summer; birds may appear throughout Nebraska as early as late Jul and Aug but may be derived from breeding populations far removed from the state. Published data, including those cited by Johnsgard (1980), suggest increased movement in Oct, although the very few reports from northern Nebraska after mid-Oct suggest passage through that region without wintering: 31 Oct 1981 Thomas Co, 10 Dec 1967 Brown Co, and 24–31 Dec 1986 Boone Co.

CBC data indicate erratic but broad distribution across the state in late Dec, although there are no CBC reports from the north. High counts are 62 at Lincoln 1969-70, 44 at Crawford 1973-74, and 34 at Lincoln 1972-73.

Winter

Red Crossbill is perhaps less common in midwinter than in fall and spring, but numbers are highly variable (Rosche 1982).

FINDING

This species can usually be easily found in Sowbelly and Monroe Canyons year-round. It is readily attracted by squeaking, when small flocks noisily surround the squeaker, observing from the tops of the pines.

White-winged Crossbill

Loxia leucoptera

STATUS

Accidental in summer. Uncommon casual winter visitor statewide.

DOCUMENTATION

Specimen

UNSM ZM7225, winter 1886 Fairbury, Jefferson Co.

DISTRIBUTION AND ECOLOGY

This species is essentially restricted to coniferous habitats while in Nebraska, including plantings and cemeteries but with a marked preference for hemlocks (*Tsuga* sp.). It may occur anywhere in the state, but most reports are from the east, possibly a reflection of observer concentration.

Summer

The only documented record is of an immature at Harrison 8 Aug 1996 (MB; NBR 64:128). There are 3 undocumented summer reports: 14 Jun (Johnsgard 1980), a report cited by Rosche (1982) of 3 seen 11.3 km (7 mi) south of Crawford 29 Jun–4 Jul 1974, and "several" present in Thomas Co 17–18 Aug 1981 (AB 36:194).

Summer reports from the Pine Ridge are of interest in that White-winged Crossbill may rarely breed in the Black Hills of South Dakota (SDOU 1991); Nebraska birds may be postbreeding wanderers.

Winter

4, 7, 10 Nov→20, 25 Apr, 1 May

Of about 70 reports for the state, most are in the period 4 Nov–1 May, about half of these in Nov. There are earlier reports 16 Oct 1970 Douglas Co and 23 Oct 1919 Jefferson Co (specimen cited above) and additional May reports 5–19 May 1974 Scotts Bluff Co and a single at a feeder the month of May 1989 in Madison Co (AB 53:501). Most reports are from the east, although since 1981 this species has occurred infrequently, with most reports from the winters of 1984–85, 1989–90, 1995–96, and 1997–98. The only report from winters other than these since 1981 is 7–8 Nov 1994 Scotts Bluff Co.

Highest counts were in late 1989: 30 on the Norfolk CBC 16 Dec 1989 and 20 in Douglas Co 6 Nov 1989 (AB 44:116).

FINDING

This species should be looked for in Nov in cemeteries with good plantings of hemlocks and spruce, especially in years when Red Crossbills are present; the 2 species often flock together.

Common Redpoll

Carduelis flammea

STATUS

Uncommon regular winter visitor north, rare south.

DOCUMENTATION

Specimen

UNSM ZM7139, 11 Dec 1895 Crawford, Dawes Co.

DISTRIBUTION AND ECOLOGY

Common Redpoll occurs in flocks in open areas where seeds such as sunflower, ragweed, and thistle can be found. It is most common in northern and western Nebraska but occurs statewide in some years. It reaches Nebraska during the coldest parts of the winter and in some years may be absent. It visits feeders on occasion.

Winter

31, 31 Oct, 1 Nov→3, 3, 5 Apr

Arrival is in early Nov, although there are earlier reports 7 Oct 1987 (20) Knox Co (AB 42:99), 19 Oct 1976 Lincoln Co, and 24 Oct 1965 Adams Co.

Most depart by late Mar, with later reports 9 Apr 1954 Dawes Co, 10 Apr 1952 Cass Co, 15 Apr 1982 Lancaster Co, 23 Apr 1976 Lancaster Co, and 30 Apr 1978 Douglas Co.

CBC data show that most Common Redpolls are in northern and western Nebraska in late Dec. High CBC totals are 303 at Crawford 1977–78, 230 at Greeley 1969–70, and 103 at Norfolk 1985–86. However, none were reported on CBCs in about a third of the years since 1966.

High counts include "several hundred" in Boone Co 3–6 Feb 1982 (AB 36:308), 800 in Dawes Co 18 Mar 1978 (AB 32:1027), and 400 in Holt Co 1 Mar 1973 (AB 26:626).

Other Reports

10 May 1949 Dawes Co, 10 May 1972 Brown Co, 14 May 1962 Platte Co, 15 May 1971 Adams Co, 20 May 1967 Brown Co, 22 May 1969 Brown Co, 22 May 1971 Brown Co, 25 May 1969 Scotts Bluff Co, 30 May 1970 Adams Co, 8 Aug 1970 Adams Co, 4–13 Sep 1963 Cass Co.

FINDING

This species should be looked for in midwinter in open areas of the north and west among flocks of finches feeding in patches of favored seed plants. It also appears at feeders.

Hoary Redpoll

Carduelis hornemanni

STATUS

Rare casual winter visitor statewide.

DOCUMENTATION

Photograph

27 Feb–15 Mar 1972 Offutt Base L, Sarpy Co (AB 26:26; NBR 40:85, 44:35).

DISTRIBUTION AND ECOLOGY

In Nebraska this species occurs with Common Redpolls in years when the latter are numerous.

Winter

Hoary Redpoll has been reported 8 times, 7 in the period Jan–29 Feb, the other on 20 May. The reports are probably all correct, although the May report was an extremely late date; the bird was described only as having an "unstreaked white rump" and was not with any other redpolls (Young 1968). The winters of 1977–78 and 1995–96 produced multiple reports. Records are:

Jan 1970 Custer Co (NBR 38:74)

17 Jan 1996 Dawes Co (RCR; AB 50:188)

29 Jan–29 Feb 1996 Dixon Co (JJ)

5 Feb 1982 Ames (Manning 1982)

13–17 Feb 1978 Douglas Co (Bray and others 1986; AB 32:372)

25–26 Feb 1978 Lancaster Co (AB 32:372; NBR 46:83)

27 Feb–15 Mar 1972 Offutt Base L (photograph cited above)

Nebraska is at the southern limit of occurrence of Hoary Redpoll. There are no records for Kansas (Thompson and Ely 1992) or Colorado (Andrews and Righter 1992), and it is rare in South Dakota, where it usually occurs only after January (SDOU 1991), and south of Sheridan in Wyoming (Scott 1993). Identification should be made with care, noting the immaculate rump, short bill, and limited or absent streaking on the flanks and undertail coverts (Zimmer 1985; Knox 1988).

FINDING

Hoary Redpolls occur with flocks of Common Redpolls, generally in "invasion" years of the latter. Flocks should be checked for very pale birds, which then should be carefully identified. Birds at feeders are far more easily studied.

Pine Siskin

Carduelis pinus

STATUS

Common regular spring and fall migrant statewide. Common regular breeder northwest, rare casual elsewhere. Common regular winter visitor statewide.

DOCUMENTATION

Specimen

UNSM ZM7161, 6 Feb 1897 Lincoln, Lancaster Co.

DISTRIBUTION AND ECOLOGY

Regular breeding occurs in the Pine Ridge, where Pine Siskin is a characteristic breeding species of ponderosa pine habitat. It may also utilize deciduous habitats (Rosche 1982; Johnsgard 1979). Breeding has occurred on several occasions outside the Pine Ridge; in those instances the species typically selects planted conifers such as pines and spruces.

Migrants and wintering birds occur almost anywhere that the preferred food of weed seeds can be obtained; this species often flocks with redpolls and goldfinches and readily attends feeders.

Spring

winter→15, 18, 18 Jun

Departure of most birds that have wintered outside of the Pine Ridge breeding range is generally complete by late May. There are, however, a number of late Jun reports that may refer to birds that nested; these birds generally depart the nest location by the end of Jun (see Summer).

Summer

It appears that Pine Siskin became a common regular breeder in the Pine Ridge only during the 20th century, as Bruner and others (1904) listed it as a migrant and winter visitor only. According to Swenk (1929), the first record of nesting was in Sarpy Co in 1904. Although Rapp and others (1958) listed it as an uncommon to rare resident in the Pine Ridge, it is now generally common although irregular (Rosche 1982). In 1993 Pine Siskin was listed as a permanent resident in Sioux, Dawes, and Sheridan Cos. There are no records of breeding in the Pine Ridge prior to Ford's (1959) collecting 2 adult females with active brood patches west of Crawford in Sioux Co in Jul 1957.

Away from the Pine Ridge, regular breeding probably occurs in Scotts Bluff Co, Thomas Co, and Kearney. There are Jun and Jul reports for most years since 1963 in Scotts Bluff Co, although no documented breeding has been reported. At Bessey Division, NNF, it was listed as a permanent resident in 1993, with sightings 2 Jul–7 Sep, and was considered a "common permanent resident" by Bray (1994). As many as 451 were banded there 17 Apr–31 May 1997 (RG); many had brood patches. There has apparently been a small nesting population on the campus of UNK for several years.

Swenk (1929) and Ducey (1988) cited several breeding records away from the Panhandle; these were from all areas of the state, and some predate the start of regular breeding in the Pine Ridge. These sporadic statewide breeding records generally occur after large influxes of Pine Siskins in the previous winter and typically occur rather early; the birds involved usually depart early, phenomena noted also in Kansas (Thompson and Ely 1992) and South Dakota (SDOU 1991). Birds that nested in Lancaster Co in 1978 and 1985 were last seen 10 and 15 Jun, respectively, and young had fledged from nests in Omaha in 1978 and 1987 by 20 May (Cornell Nest Record data).

Since 1967 most reported nestings have occurred after major influxes following the winters of 1975–76, 1977–78, and 1984–85 and were reported from Adams (Bennett 1977), Wayne (Schock 1983), Douglas (Bennett 1980a; Wright 1985a), Lancaster (Bennett 1980a), Sarpy (Bennett 1980a; AB 39:322), and Chase (Bennett 1986) Cos. Exceptions were in Douglas Co in 1984 (AB 38:931) and 1987 (Cornell Nest Record data) and in Keith Co; the latter involved the netting of several individuals with active brood patches and 2 juveniles during the period 9 Jun–21 Jul 1993 at Keystone L (Scharf and others 1993) and the netting of juveniles in the area 17–21 Jun 1994, as well as 17 females with brood patches 9 Jun–4 Jul (Brown and others 1996). Scharf and others (1993) noted that their study "strongly suggests that this species may have bred."

Fall

6, 8, 9 Aug→winter

Migration begins in Sep, although there are earlier reports in Jul–Aug, which are likely dispersers from breeding locations near or in Nebraska; individuals known to have bred locally generally depart by late Jun (see Summer). Such early reports include 8 Jul 1970 Adams Co, 15 Jul 1979 Garden Co, 22 Jul 1961 Douglas Co, 25 Jul 1972 Lancaster Co, and 1 Aug 1998 (2) Cass Co. The peak of true migration probably occurs in late Oct. Johnsgard (1980) noted that half of the fall records analyzed were in the period 1 Oct–18 Nov, and migration in South Dakota is in the period mid-Oct through mid-Nov.

High counts include 143 at Oliver Res 30 Nov 1998 and 85 in the Wildcat Hills 19 Nov 1995.

Winter

Pine Siskins winter statewide in varying numbers from year to year. CBC data since 1967 show the highest total of 1133 birds in 1987–88 and the lowest only 17 birds in 1991–92; the data indicate that on average in late Dec most birds are in the Panhandle and fewest in northern Nebraska. High non-CBC counts include 300 in Keith Co 8 Jan 1982 (Rosche 1994a).

FINDING

Pine Siskin is most predictable in May and Jun in ponderosa pines in the Pine Ridge, such as at Sowbelly Canyon. These birds are easily attracted by squeaking. In some years very few are present, however. In winters when good numbers enter the state, siskins are easily seen at feeders.

Lesser Goldfinch

Carduelis psaltria

STATUS

Rare casual summer visitor northwest.

DOCUMENTATION

Photograph

20 May 1984 Monroe Canyon, Sioux Co (Swanson 1984; AB 38:929).

TAXONOMY

Two of 5 documented Nebraska records are of black-backed males and 2 are of green-backed males. Traditionally, 2 subspecies have been recognized, the black-backed Rocky Mountain *psaltria* and the green-backed western *hesperophilus* (AOU 1957), but recent opinion holds that green back color may not be grounds for subspecific delineation but merely an example of polymorphism (Pyle 1997).

Summer

There are 8 reports, 5 documented:

20 May 1984 Monroe Canyon (cited above)

26 May 1999 Cedar Canyon, Scotts Bluff Co (JH, RB)

1 Jun–7 Jul 1986 Crawford (NBR 54:56, 64; Mollhoff 1987)

9 Jul 1988 Carter Canyon (Kenitz 1988)

21–27 Aug 1999 Oliver Res (MB, SJD, BP, LP)

The only other report likely to be correct is of one in early Sep 1995 near Gering.

The records are all rather recent, as is the case in South Dakota, where all but 1 of about 5 records date from 1982. The South Dakota records are from the southwest part of the state in the period 5 May–10 Sep (SDOU 1991), 3 within 50 km (about 30 mi) of Sioux Co, Nebraska. In Wyoming Lesser Goldfinch is an uncommon summer resident from Apr to Oct in the Cheyenne-Laramie area and is presumed to nest (Scott 1993). Thus it seems that this species may appear as a summer resident in southeast Wyoming, extreme western Nebraska, and extreme southwest South Dakota.

Other Reports

There are 2 reports that are rather late for this species. "Several" birds were reported in Hall Co 7 Nov 1993, but the details were considered unacceptable by the NOURC (Gubanyi 1996a). A bird with a black cap and greenish back, "probably" this species, was near Gibbon 16 Nov 1996 (LR, RH; NBR 64:128).

FINDING

Possibly the best chance to find this species is to check flocks of goldfinches in the western Panhandle in Jun and Jul. While the records to date are of males, females have probably been overlooked due to their similarity to American Goldfinch females.

American Goldfinch

Carduelis tristis

STATUS

Common regular spring and fall migrant and breeder statewide. Abundant regular winter visitor statewide.

DOCUMENTATION

Specimen

UNSM ZM7163, 5 Nov 1889 Lincoln, Lancaster Co.

TAXONOMY

Two subspecies occur. Most birds are eastern *tristis*, which breeds west to "central Nebraska, . . . Minnesota" and winters north to "central Nebraska, southeastern South Dakota," while western *pallidus* breeds east to "northwest Nebraska (Springview)" and probably winters in extreme western Nebraska (AOU 1957).

DISTRIBUTION AND ECOLOGY

American Goldfinch can be found foraging in almost any habitat, although it is least common in coniferous forest (Rosche 1982); it is also uncommon in interior deciduous forest except for the early budding season, when they may forage high in the crowns of trees upon insects attracted to the buds. During the breeding season goldfinches are most often associated with open areas and woodland edge. For nest sites they select stands of shrubby vegetation, including large weed patches such as ragweed or dense stands of young willow. Breeding populations are smallest in the Panhandle.

Wintering birds are ubiquitous and commonly utilize feeders around human habitation.

Spring

It is difficult to determine arrival and departure dates, as some birds are always present. In South Dakota migration periods are mid-Apr through mid-May and mid-Oct through mid-Nov (SDOU 1991). Rosche (1982) noted that in the northwest American Goldfinch is most common in fall.

Summer

BBS data show that breeding populations are highest in the east. Nevertheless, Rosche (1982) considered it a "fairly common" summer resident in the northwest.

American Goldfinch is a late nester, waiting until thistledown is available as a nesting material, thus avoiding Brown-headed Cowbird parasitism (Brown and others 1996). Most

nesting takes place in the period Jun–Sep (SDOU 1991), and winter flocks do not break up until well into Jun.

Cornell Nest Record data show young fledged from a nest in Custer Co 28 Sep 1982 and from a nest in Lancaster Co 4 Sep 1989.

Fall

See Spring.

Winter

Winter populations are augmented by northern birds (Thompson and Ely 1992). While there are variations in numbers from year to year, CBC data do not show the marked swings characteristic of other "winter finch" species. CBC data since 1967 show a high count of 1948 in 1987–88 and a low of 214 in 1972–73. Winter distribution is fairly even statewide.

FINDING

This species is most easily found at feeders in winter, although many remain at feeders as late as Jun in spring, at which time males have acquired the bright nuptial plumage. At any time goldfinches can be readily attracted by squeaking.

Evening Grosbeak

Coccothraustes vespertinus

STATUS

Hypothetical breeder northwest. Rare casual summer visitor northwest. Common regular winter visitor northwest, fairly common casual elsewhere.

DOCUMENTATION

Specimen

UNSM ZM7114, 29 Oct 1900 Long Pine, Brown Co.

TAXONOMY

Two subspecies have been recorded (Rapp and others 1958). Eastern *vespertinus*, which breeds from Alberta east, mainly in Canada, winters "south to southwest South Dakota, Kansas, central Missouri" (AOU 1957); birds wintering in eastern Nebraska probably are of this race. There is a single banding recovery in Lancaster Co 18 Apr 1986 of a bird banded in New York 16 Mar 1979. The Rocky Mountain subspecies *brooksi* breeds from British Columbia south and east to "western Montana, western Wyoming, central Colorado" and winters east to southwest South Dakota and Oklahoma (AOU 1957) and apparently also rarely in the Black Hills and possibly northwest Nebraska. Rosche (1982) banded Evening Grosbeaks in Crawford, and the 2 recoveries confirm that northwest Nebraska winter birds are mostly *brooksi*. One banded 3 Apr 1974 was recovered 21 Jul 1974 in northwest Montana, and another, banded 5 Mar 1980, was recovered twice in the Rapid City, South Dakota, area, first on 24 Apr 1983 and then on 5 Jun 1986.

DISTRIBUTION AND ECOLOGY

Winter birds are found in riparian habitats where fruiting trees such as Russian olive and ashleaf maple occur (Rosche 1982). Feeders are commonly utilized also, where it prefers sunflower seeds. This species is found most commonly in the Panhandle but occasionally occurs elsewhere in the state.

Spring

winter→27, 28 Apr, 2 May (east)

winter→24, 26, 29 May (west)

In the east spring departure is about a month earlier than elsewhere, possibly because of differing subspecies (see above), although there are later dates 8 May 1999 Fort Niobrara NWR, 14 May 1962 Hamilton Co, 20 May 1986 Douglas Co (AB 40:493), and 6 Jun 1965 (about 12 birds) Thomas Co (RSS)

Summer

Breeding occurs rarely in the northern Black Hills of South Dakota (SDOU 1991; Peterson 1990) and may also occur in the Pine Ridge, especially in recent years in Dawes Co. Rosche (1982) cited a statement without details by Blinco (1946) that nesting had occurred at Chadron, and in recent years there have been several summer occurrences suggestive of breeding. There is a report 13 Sep 1963 Dawes Co, possibly an early migrant. Rosche (1982) described a family group that attended his Crawford feeders 4 Aug–early Oct 1976; the young were capable of flight and therefore

may not have fledged in Nebraska, but the record is highly suggestive. There is a report 22 Aug 1986 Dawes Co, and a pair with young appeared at the Rosche feeder in 1989 (AB 43:1335). At least 1 bird summered in Dawes Co in 1990, but no young were seen (AB 44:1154). However, a juvenile appeared at the Rosche feeder 13 Jul 1991 (AB 45:1134), up to 3 birds were there during Jun and Jul 1993 (AB 47:1124), and a female was there 28 Jul 1994 (NBR 62:115). The species was listed as a permanent resident in Dawes Co in 1993 (NBR 61:134).

Fall

25, 28, 28 Sep→winter (west, central)

28, 28 Oct, 6 Nov→winter (east)

In the western half of the state arrival is in late Sep; the species occurs every year only in the Panhandle. There are earlier reports 19 Aug 1978 Lancaster Co (AB 33:193), 24 Aug 1982 Lincoln Co, 3–5 Sep 1965 Douglas Co, 12 Sep 1975 Adams Co, and 13 Sep 1986 Lincoln Co.

In the east, where the species is only casual in occurrence, arrival is about a month later than in the western and central part of the state, with early dates in late Oct. This may be due to differing timing of movements of eastern and western subspecies.

High counts include 30 in Washington Co 29 Nov 1985 (AB 40:137) and 20 in Sowbelly Canyon 28 Nov 1998.

Winter

Evening Grosbeak is regular only in the Panhandle. Of the 426 birds counted on CBCs 1968–92, 79% were at Crawford or Scottsbluff. High count was 126 at Crawford in 1977.

Elsewhere in the state occurrence is irregular, especially in eastern Nebraska where, despite the large number of observers, since 1982 the species has been seen less often than elsewhere in the state. High CBC total from the east is only 9 at Omaha in 1977.

Banding recoveries (see above) suggest that at least some of the (relatively few) birds summering in the Black Hills of South Dakota move into the Nebraska Pine Ridge in winter.

Other Reports

9 Jun 1952 Dawson Co, 10 Jun 1974 Adams Co, 27 Jun 1967 Brown Co, 29 Jul 1987 Lincoln Co.

FINDING

When an incursion of Evening Grosbeaks occurs, they are easily seen at winter feeders, but the best chance of locating this species on a regular basis is to check riparian areas in the Pine Ridge in winter. Evening Grosbeaks can be rather conspicuous in the leafless trees and occur regularly at feeders in Pine Ridge towns, notably Crawford and Harrison.

Passeridae (Old World Sparrow)

House Sparrow

Passer domesticus

STATUS

Abundant regular resident statewide.

DOCUMENTATION

Specimen

UNSM ZM10430, 31 Dec 1908 Lincoln, Lancaster Co.

DISTRIBUTION AND ECOLOGY

This species is found in and around both urban and rural human habitation. It was likely much more common in the earlier part of the 20th century when horses were commonly used as work animals and for carrying people into towns and cities. The undigested pieces of grain in the horse droppings were an important food source for House Sparrows, particularly during the colder months. It may be currently less abundant in towns and cities than formerly, possibly due to the absence of horses and a recent influx of House Finches, and appears to be most numerous in rural areas, especially around farmsteads. It is unusual to find House Sparrows very far from human activity.

This is not a native species to North America. Ducey (1988) cited unpublished research by Swenk that showed that the first House Sparrows were released at Nebraska City in the 1870s and subsequently spread rapidly, primarily along railroads, until the Wyoming border was reached in northwest Nebraska in 1899 (Rosche 1982).

Summer

Nesting is usually in the form of a loose colony, with nests placed in virtually any cavity or in trees. BBS data (1967–77) confirm that distribution of these birds parallels that of the human population in Nebraska, with highest densities in the south and east.

Winter

Winter flocks are particularly conspicuous around livestock feeding operations and grain elevators. CBC data indicate a rather even distribution statewide, with no evidence of any significant seasonal shift of populations when data are compared to BBS data; highest CBC totals are from eastern population centers, such as 5500 birds at Omaha in 1967, 4901 at Norfolk in 1985, and 4638 at Lincoln in 1974.

The recent westward expansion of North American House Finches may be affecting House Sparrow numbers; CBC data in Iowa are suggestive of a decline in House Sparrow numbers as House Finch numbers have increased (Silcock 1994). In the winter of 1995–96 observers in Dakota and Dawes Cos noted a similar phenomenon (NBR 64:17).

FINDING

This species can be seen easily year-round in towns and cities, usually foraging on the ground. Flocks are conspicuous in winter.

APPENDIX 1
Abbreviations

NGS	National Geographic Society
NM	National Monument
NMNH	National Museum of Natural History, New York
NNF	Nebraska National Forest
NOU	Nebraska Ornithologists' Union
NOURC	Nebraska Ornithologists' Union Records Committee
NWR	National Wildlife Refuge
Proc NOU	Proceedings of the Nebraska Ornithologists' Union
PSM	Puget Sound Museum
Res	Reservoir
SDBN	*South Dakota Bird Notes* (South Dakota Ornithologists' Union)
SDOU	South Dakota Ornithologists' Union
SHP	State Historical Park
SL	Sewage Lagoon(s)
SP	State Park
SRA	State Recreation Area
UNK	University of Nebraska–Kearney
UNSM	University of Nebraska State Museum
USDI	U.S. Department of the Interior
USNM	U.S. National Museum, Washington DC
WMA	Wildlife Management Area (State of Nebraska)
WPA	Waterfowl Production Area (U.S. Department of the Interior)
WSC	Wayne State College, Wayne NE

OBSERVERS

AG	Alan Grenon
AK	Alice Kenitz
B	Nebraska Birdline
BFH	Bill F. Huser
BJR	B. J. Rose
BP	Barbara Padelford
BR	Bob Russell
BS	Bonnie Simons
BW	Bruce Walgren
CD	Chuck Dummer
CF	Carol Falk
CNK	Clem N. Klaphake
DB	Duane Bright
DCE	David C. Ely
DFP	Don F. Paseka
DG	Daryl Giblin
DH	David Heidt
DJR	Dorothy J. Rosche
DLS	David L. Swanson
DP	Deb Paulson
DR	Douglas Rose
DS	Don Shower
DTW	Daniel T. Williams
DW	Donna Walgren
EB	Elliot Beddows
ECT	Edna Clare Thomas
EV	Eric Volden
FZ	Fred Zeillemaker
GH	Glen Hoge
GL	Gary Lingle
GW	Gertrude Wood
HKH	Helen Hughson
JD	Joan Dummer
JED	James E. Ducey
JG	Joseph Gubanyi
JGJ	Joel G. Jorgensen
JH	Jeff Hall
JJ	Jan Johnson
JJH	Jeff J. Huebschmann
JK	Jim Kovanda
JM	Jim Mountjoy
JP	Jan Paseka
JS	John Sullivan
JT	Jerry Toll
JW	Jeff Wallace

JWH	John W. Hall
KE	Kim Eckert
KM	Klare Meltvedt
KN	Kenny Nichols
LB	Laurel Badura
LCH	Larry C. Holcomb
LE	Larry Einemann
LEB	Loren E. Blake
LF	Laurence Falk
LM	Larry Malone
LN	LaDonna Nichols
LP	Loren Padelford
LR	Lanny Randolph
LS	Lona Shafer
MB	Mark Brogie
MBH	Mary B. Hunt
MC	Mary Clausen
MG	Margaret Giblin
m.ob.	multiple observers
NK	Nick Komar
NR	Neil Ratzlaff
PK	Paul Kaufman
PL	Paul Lehman
PW	Patrice Wallace
RAB	Roland A. Barth
RB	Russ Benedict
RCR	Richard C. Rosche
RG	Ruth Green
RH	Robin Harding
RK	Ray Korpi
RS	Ron Storey
RSS	Roger S. Sharpe
SG	Sue Gentes
SJD	Stephen J. Dinsmore
SK	Sandy Kovanda
SY	Shaw Young
TA	Tom Aversi
TB	Tanya Bray
TEL	Thomas E. Labedz
TH	Thomas Hoffman
TLE	Ted Lee Eubanks
WH	Wanda Hoge
WM	Wayne Mollhoff
WRS	W. Ross Silcock

APPENDIX 2
Gazetteer

High Plains The shortgrass prairie area essentially confined to the Panhandle, west of the Sandhills.

Pine Ridge An area of escarpments and ponderosa pine woodland in northern Sioux, Dawes, and northwestern Sheridan Cos. The Pine Ridge is an outlier of the Black Hills of South Dakota.

Rainwater Basin An area containing numerous scattered playa wetlands in the south-central region of the state. The region is often divided into the eastern portion (Clay, Fillmore, Hamilton, York and parts of Seward, Thayer, Polk, and Adams Cos) and the western portion (Kearney, Phelps, and parts of Harlan and Gosper Cos).

Sandhills An extensive area of sand dunes covered with mixed-prairie grassland in the north-central region of the state. Most of the Sandhills is included in Cherry, southern Sheridan, northern Garden, southern Brown, southern Rock, Grant, Hooker, Thomas, Blaine, Loup, Garfield, Wheeler, Arthur, McPherson, Logan, northwestern Custer, northern Keith, and northern Lincoln Cos.

Wildcat Hills An area of minor escarpments and ponderosa pine woodland south of the North Platte River in Scotts Bluff, Banner, and Morrill Cos.

LOCATIONS MENTIONED
IN SPECIES ACCOUNTS
*Not included in this list are towns mentioned
in the species accounts that can be found on
standard maps and atlases.*

Ackley Valley Marsh, Keith Co: private owner-
ship about 6.45 km (4 mi) north of L
McConaughy

Alma SL, Harlan Co: at south edge of Harlan Co
Res, west of Highway 183

Ames, Dodge Co: unincorporated

Anderson Bridge, Cherry Co: 55 ha (137 ac) WMA
on Niobrara River

Andrews, Sioux Co: defunct town southeast of
Harrison

Antelope Canyon, Sioux Co: extreme west end
Pine Ridge

Antioch, Sheridan Co: unincorporated

Arbor Day Farm, Otoe Co: National Arbor Day
Foundation property adjacent to Arbor Lodge
SHP

Arbor L, Lancaster Co: 25 ha (63 ac) WMA, saline
wetland

Arbor Lodge SHP, Otoe Co

Arikaree River: joins Republican River in
southwest Dundy Co

Ash Hollow SHP, Garden Co: rocky bluffs with
extensive junipers

Ayr L, Adams Co: 81 ha (200 ac) private basin

Ballard's Marsh, Cherry Co: 632 ha (1561 ac)
WMA, 29 km (18 mi) south of Valentine

Beaver Creek: joins Loup River near Genoa,
Nance Co

Beaver L, Sarpy Co: 81 ha (200 ac) reservoir

Belmont, Dawes Co: defunct town south of
Crawford

Bingham, Sheridan Co: unincorporated

Blake Ranch, Holt Co: southeast of Chambers

Blue Creek, Garden Co: drains Crescent L and
joins North Platte River

Bluestem L, Lancaster Co: 132 ha (326 ac)
reservoir

Bluewing Marsh, Clay Co: 65 ha (160 ac) WPA,
part of 121 ha (300 ac) basin

Bohemia Prairie, Knox Co: 280 ha (680 ac) WMA,
upland prairie and woodland

Bordeaux Creek, Sioux Co: east of Chadron

Bordeaux Pass, Scotts Bluff Co

Box Butte Res, Dawes Co: 656 ha (1600 ac)
reservoir, 248 ha (612 ac) riparian woodland
SRA

Boyer Chute NWR, Washington Co: Missouri
River floodplain restoration

Branched Oak L, Lancaster Co: 730 ha (1800 ac)
reservoir, 130 ha (326 ac) SRA

Brickton, Adams Co: unincorporated

Bronco L, Box Butte Co: drained former lake 3.2
km (2 miles) east of Alliance

Bufflehead Pond, Buffalo Co: 4.85 ha (12 ac)
WMA basin

Bull Canyon, Banner Co

Burchard L, Pawnee Co: 61 ha (150 ac) reservoir
and WMA, tallgrass prairie

Calamus Res, Loup and Garfield Co: 2075 ha
(5124 ac) reservoir, 2100 ha (5188 ac) SRA
Sandhills prairie

Camp Opal Springs, Lincoln Co: in Medicine
Creek Valley

Capitol Beach, Lancaster Co: on Salt L, Lincoln

Carns, Keya Paha Co: defunct town southeast of
Springview

Carson L, Garfield Co: large lake in northwest
part of county

Carter Canyon, Scotts Bluff Co

Carter L, Douglas Co: 120 ha (300 ac) public lake
formerly Cut-Off L

Cauble Creek, Washington Co: in Blair

Cedar Island, Sarpy Co: Omaha utility well-field
area closed to public

Cedar Point Biological Station, Keith Co:
University of Nebraska property on
Keystone L

Chadron SP, Dawes Co: 328 ha (801 ac)
ponderosa pine woodlands

Chet Ager NC, Lancaster Co: in city of Lincoln

Child's Point, Sarpy Co: locality now within
Fontenelle Forest

Chimney Rock, Morrill Co: National Historic
Site

Clear Creek Marshes, Keith and Garden Co: 2520 ha (6224 ac) WMA on North Platte River

Clear L, Cherry Co: 1050 ha (425 ac) lake within Valentine NWR

Cochran L, Scotts Bluff Co: 18 ha (45 ac) lake southwest of Melbeta

Cody L, Cherry Co: 140 ha (350 ac) lake north of Cody

Conestoga L, Lancaster Co: 93 ha (230 ac) reservoir

County Line Marsh, York and Fillmore Co: 165 ha (408 ac) WPA basin

Crescent L, Garden Co: large private lake at south edge of Crescent L NWR

Crescent L NWR, Garden Co: 16 560 ha (40 900 ac) Sandhills prairie with many lakes

Crystal Cove L, Dakota Co: 12 ha (30 ac) Missouri River oxbow in South Sioux City

Cunningham L, Douglas Co: 160 ha (390 ac) reservoir

Deadhorse Burn, Dawes Co: ponderosa pine burn at west edge of Chadron SP

Deep Well Basin, Hamilton Co: 31.5 ha (78 ac) WMA within 81 ha (200 ac) basin

DeSoto Bend NWR, Washington Co: 3167 ha (7823 ac) floodplain cropland and oxbow lake

DeWitt L, Box Butte Co: Sandhills lake 16 km (10 mi) northeast of Alliance

Dodge Park, Douglas Co: Omaha park in northeast part of city

Dorey L, Cherry Co (exact location not known by us)

Dunlap, Dawes Co: defunct town on Niobrara River east of Box Butte Res

Eagle Canyon, Keith Co: on south side of L McConaughy

East Hat Canyon, Sioux Co: northeast of Harrison

East Odessa Area, Buffalo Co: 32 ha (78 ac) WMA along Platte River

Eckhardt Basin, Clay Co: 70 ha (174 ac) WPA basin

Elmwood Park, Douglas Co: city park

Elsmere, Cherry Co: unincorporated

Elwood Res, Gosper Co: 490 ha (1200 ac) reservoir

Enders Res, Chase Co: 690 ha (1700 ac) reservoir, 1140 ha (2818 ac) SRA, shortgrass prairie

Facus Springs, Morrill Co: WMA, saline wetland 16 km (10 mi) west of Bridgeport

Fahrenholtz Ponds, Dixon Co: 6.45 km (4 mi) east of Dixon

Father Hupp Marsh, Thayer Co: 65 ha (160 ac) WMA, formerly Prairie Marsh

Fontenelle Forest, Sarpy Co: private forest reserve, admission charged

Fort Kearny SHP, Kearney Co: south side of Platte River

Fort Niobrara NWR, Cherry Co: 7837 ha (19 122 ac) Sandhills prairie and riparian woodland

Fort Robinson SP, Dawes and Sioux Co: 4715 ha (11 500 ac) Pine Ridge

Freeman Lakes, York Co: 77 ha (188 ac) WPA within 145 ha (350 ac) basin

Frenchman Creek, Chase and Hitchcock Co: joins Republican River

Funk Lagoon, Phelps Co: 770 ha (1900 ac) WPA basin

Gadwall Basin, Hamilton Co: 36 ha (90 ac) WMA basin

Gavin's Point Dam, Cedar-Knox Co: dam creates Lewis and Clark L

Gering SL, Scotts Bluff Co: east edge of Gering

Gifford Farm, Sarpy Co: adjacent to Fontenelle Forest

Gifford Peninsula, Sarpy Co: location of Gifford Farm

Gilbert-Baker Area, Sioux Co: 995 ha (2457 ac) ponderosa pine woodland in Pine Ridge

Gleason Lagoon, Kearney Co: 230 ha (569 ac) WPA basin

Glen, Sioux Co: defunct town southwest of Crawford

Glenvil Basin, Clay Co: 49 ha (120 ac) WPA basin

Goose L, Garden Co: lake within Crescent L NWR

Gordon SL, Sheridan Co: at town of Gordon

Grandpa's Steakhouse Pond, Buffalo Co: sandpit south of Kearney adjacent to Platte River

Griess Basin, Fillmore Co: 6 ha (15 ac) WPA basin

Hackberry L, Cherry Co: lake within Valentine NWR

Halsey Forest, Thomas Co: see Nebraska National Forest, Bessey Division

Hansen Marsh, Clay Co: 130 ha (320 ac) WPA basin

Harlan Co Res, Harlan Co: 5466 ha (13 338 ac) reservoir, 1595 ha (3940 ac) riparian woodlands

Harold W. Anderson Area, Howard Co: 110 ha (272 ac) WMA along Loup River

Harvard Lagoon, Clay Co: 600 ha (1484 ac) WPA basin; formerly Inland Lagoon

Hastings Ammunition Depot, Adams Co: east edge of Hastings

Hat Creek, Sioux Co: north of Pine Ridge

Holmes L, Lancaster Co: city lake in Lincoln

Hoover L, Sheridan Co: 65 ha (160 ac) lake

Hormel Park, Dodge Co: city park in Fremont

Hughson Ranch, Sioux Co: private ranch (= Wind Springs Ranch) in southern part of county

Hultine Basin, Clay Co: 372 ha (920 ac) WPA basin

Hummel Park, Douglas Co: city park in north Omaha

Inavale, Webster Co: unincorporated

Indian Cave SP, Richardson and Nemaha Co: 1135 ha (2800 ac) upland forest

Indian Creek, Sioux Co: in extreme northwestern part of county

Inland, Clay Co: unincorporated

Inland Lagoon, Clay Co: now known as Harvard Lagoon

Jack Sinn Marsh, Lancaster and Saunders Co: 406 ha (1002 ac) saline wetland

James Ranch, Sioux Co: western part of Fort Robinson SP

Jewel Park, Sarpy Co: city park in Bellevue

Johnson Lagoon, Phelps Co: 234 ha (578 ac) WPA basin

Johnson Res, Dawson and Gosper Co: 1135 ha (2800 ac) reservoir

Keystone, Keith Co: unincorporated

Keystone Diversion Dam, Keith Co: creates Keystone L and L Ogallala

Keystone L, Keith Co: below Kingsley Dam, adjoins L Ogallala

Kilpatrick L, Box Butte Co: 65 ha (160 ac) private reservoir

Kingsley Dam, Keith Co: dam creates L McConaughy

Kiowa Springs, Scotts Bluff Co: 205 ha (506 ac) WMA saline wetland

Kissinger Basin, Clay Co: 138 ha (341 ac) WMA basin

L Alice, Scotts Bluff Co: 567 ha (1400 ac) reservoir

L Babcock, Platte Co: 243 ha (600 ac) reservoir

L George, Rock Co: 32 ha (80 ac) lake south of Bassett

L Maloney, Lincoln Co: 701 ha (1732 ac) reservoir

L McConaughy, Keith Co: 14 450 ha (35 700 ac) reservoir

L Minatare, Scotts Bluff Co: 930 ha (2300 ac) reservoir

L North, Platte Co: 81 ha (200 ac) reservoir, adjacent to L Babcock

L Ogallala, Keith Co: 133 ha (329 ac) borrow lake below Kingsley Dam

Lakeside, Sheridan Co: unincorporated

L Yankton, Cedar Co: 101 ha (250 ac) borrow lake below Gavin's Point Dam

Lawrence Fork, Morrill and Banner Co: creek flowing into North Platte River

Lewis and Clark L, Knox Co: 12 960 ha (32 000 ac) reservoir, 497 ha (1227 ac) wetland

Little Blue Area, Thayer Co: 123 ha (303 ac) riparian woodland WMA along Little Blue River

Little L Alice, Scotts Bluff Co: small reservoir within North Platte NWR

Long Canyon, Banner Co: west of Harrisburg

Long Pine Canyon, Brown Co: contains Long Pine Creek

Mallard Haven, Fillmore Co: 375 ha (927 ac) WPA basin

Maloney Canal, Lincoln Co: inlet to L Maloney

Marsh L, Cherry Co: within Valentine NWR

McClanahan Canyon, Scotts Bluff Co: in Wildcat Hills

McKissick Island, Nemaha Co: old Missouri River oxbow now on Missouri side

Meadow, Sarpy Co: defunct town on Platte River across from Louisville

Meat Animal Research Center, Clay Co: large area of grassland west of Clay Center

Merritt Res, Cherry Co: 1177 ha (2906 ac) reservoir

Milburn, Custer Co: defunct town in northern part of county

Monroe Canyon, Sioux Co: Pine Ridge canyon north of Harrison

Montrose, Sioux Co: defunct town north of Pine Ridge

Mormon Island Crane Meadows, Hall Co: wildlife area south of Alda along Platte River

Neale Woods, Douglas and Washington Co: private upland and floodplain forest with public access

Nebraska National Forest, Bessey Division, Thomas Co: near Halsey, mix of sandhills and planted conifers (also known as Halsey Forest)

Nebraska National Forest, McKelvie Division, Cherry Co: west of Valentine

Nelson Basin, Hamilton Co: 65 ha (160 ac) WPA basin

Nine-Mile Prairie, Lancaster Co: restored tallgrass prairie in northwest Lincoln

Niobrara Valley Preserve, Cherry, Keya Paha, and Rock Co: 22 670 ha (56 000 ac) Nature Conservancy area along Niobrara River

North Harvard Basin, Clay Co: Private basin 0.8 km (0.5 mi) west and 4 km (2.5 mi) north of Harvard

North Hultine Basin, Clay Co: 178 ha (440 ac) WPA basin, formerly Sandpiper WPA

North L Basin, Seward Co: 140 ha (346 ac) WMA basin

North Platte NWR, Scotts Bluff Co: 2043 ha (5047 ac) area including reservoirs L Alice, Little L Alice, and Winters Creek L and upland grasslands

North Platte SL, Lincoln Co: large lagoons at east edge of North Platte

North Platte State Fish Hatchery, Lincoln Co: 4.8 km (3 mi) south of North Platte

Oak Glen, Seward Co: 256 ha (633 ac) WMA upland forest

Oak Park, Lancaster Co: city park in Lincoln

Offutt Base L, Sarpy Co: small military-owned lake just south of Bellevue

Olive Creek L, Lancaster Co: 71 ha (175 ac) reservoir

Oliver Res, Kimball Co: 109 ha (270 ac) reservoir

Omadi Bend, Dakota Co: 15 ha (38 ac) riparian area along Missouri River oxbow

Orella, Sioux Co: defunct town 24 km (15 mi) northwest of Crawford

Oshkosh SL, Garden Co: at southeast edge of Oshkosh

Overton L, Holt Co: private lake in southwest part of county

Palmer L, Sheridan Co: private lake 3.2 km (2 mi) north of Antioch

Parkview Cem, Adams Co: in city of Hastings

Pauline, Adams Co: unincorporated

Pawnee L, Lancaster Co: 300 ha (740 ac) reservoir

Pearson Basin, Fillmore Co: 49 ha (120 ac) private basin in extreme southwest

Pelican L, Cherry Co: lake within Valentine NWR

Pintail Marsh, Hamilton Co: 194 ha (478 ac) WMA basin

Pishelville, Knox Co: defunct town 9.7 km (6 mi) south of Verdel

Platte River SP, Cass Co: woodland along Platte River

Plum Creek Canyon, Brown Co: tributary of Niobrara River

Ponca SP, Dixon Co: 336 ha (830 ac) upland forest along Missouri River

Ponderosa Area, Dawes Co: 1481 ha (3659 ac) WMA in Pine Ridge

Pony L, Rock Co: 142 ha (350 ac) private lake

Powell, Jefferson Co: unincorporated

Prairie Dog Creek, Harlan Co: joins south side of Harlan Co Res

Pumpkin Creek, Banner-Morrill Co: south of Wildcat Hills

Real Basin, Fillmore Co: 65 ha (160 ac) WPA basin

Redington, Morrill Co: unincorporated, southwest of Bridgeport

Riverside Park, Scotts Bluff Co: adjacent to Riverside Zoo in Scottsbluff

Robideaux Pass, Scotts Bluff Co: southwest of Scottsbluff

Rock Creek L, Dundy Co: 67 ha (165 ac) SRA riparian area with 20 ha (50 ac) lake

Rolland Basin, Fillmore Co: 52 ha (129 ac) WPA basin

Rowe Sanctuary, Kearney Co: (= Lillian Annette Rowe Sanctuary), south edge of Platte River

Sacramento-Wilcox Basin, Phelps Co: 1224 ha (3023 ac) WPA basin complex and upland grassland

Salt L, Lancaster Co: 81 ha (200 ac) residentially developed lake

Schilling Refuge, Cass Co: 3236 ha (1310 ac) WMA adjacent to Platte and Missouri Rivers at northeast edge of Plattsmouth

Schramm Park, Sarpy Co: SRA, riparian and oak-hickory forest reserve along Platte River

Scotts Bluff NM, Scotts Bluff Co: large rock outcropping at southwest edge of Scottsbluff

Sherman Res, Sherman Co: 1050 ha (2600 ac) reservoir

Sinninger Lagoon, York Co: 24 ha (60 ac) WPA connecting 2 private basins

Smiley Canyon, Sioux Co: at west edge of Fort Robinson SP

Smith L, Garden Co: lake within Crescent L NWR

Smith L Area, Sheridan Co: 117 ha (290 ac) WMA lake and upland area south of Rushville

Snake Creek Meadows, Box Butte Co: wetlands near Kilpatrick L

Sowbelly Canyon, Sioux Co: northeast of Harrison

Squaw Canyon, Sioux Co: northwest of Harrison

Stagecoach L, Lancaster Co: 80 ha (195 ac) reservoir in 246 ha (607 ac) SRA

Standing Bear L, Douglas Co: 55 ha (135 ac) reservoir in city of Omaha

Stateline Island, Scotts Bluff Co: 53 ha (130 ac) WMA south of Henry

Sugarloaf Butte (Hill), Garden Co: northwest of Oshkosh

Summit Res, Burt Co: 77 ha (190 ac) reservoir

Sutherland Res, Lincoln Co: 1223 ha (3020 ac) reservoir

Swan L, Holt Co: 95 ha (235 ac) lake

Tamora Basin, Seward Co: private basin just south of Tamora

Taylor Ranch, Howard Co

Theesen Lagoon, Clay-Adams Co: 65 ha (160 ac) WPA basin and adjacent private cattle feedlot runoff pond in Adams Co

Timber (Wood) Reserve, Sioux Co: west edge of Fort Robinson SP

Twin Lakes, Seward Co: 2 reservoirs totaling 85 ha (210 ac)

Two Rivers Park, Douglas Co: 263 ha (650 ac) SRA/ WMA along Platte River

University L, Sioux Co: 12 ha (30 ac) reservoir in south part of county

University of Nebraska Agricultural Experiment Station, Lincoln Co: south of North Platte

Valentine NWR, Cherry Co: 28 954 ha (71 516 ac) prairie and lakes area

Victoria Springs, Custer Co: 24 ha (60 ac) WMA along Victoria Springs Creek

Victor L, Gosper Co: 96 ha (238 ac) WPA basin

Waco Basin, York Co: 64 ha (159 ac) WPA basin

Wagon Train L, Lancaster Co: 128 ha (315 ac) reservoir

Walgren L, Sheridan Co: 40 ha (100 ac) impoundment south of Hay Springs

Warbonnet Canyon, Sioux Co: northwest of Harrison

Wehrspann L, Sarpy Co: 99 ha (245 ac) city of Omaha reservoir

Weis Lagoon, Fillmore Co: 40 ha (100 ac) WPA basin

Wellfleet L, Lincoln Co: 20 ha (50 ac) WMA reservoir and 26 ha (63 ac) upland area

West Ash Canyon, Dawes Co: south of Whitney

West Hat Creek Canyon, Sioux Co: northeast of Harrison

West Warbonnet Canyon, Sioux Co: northwest of Harrison

Whitman, Grant Co: unincorporated

Whitney L, Dawes Co: 364 ha (900 ac) reservoir northwest of Whitney

Wildcat Hills NC, Scotts Bluff Co: in Wildcat Hills south of Scottsbluff

Wilderness Park, Lancaster Co: city park in southwest Lincoln

Wild Rose Ranch, Hall Co: along south side of Platte River

Winters Creek L, Scotts Bluff Co: 95 ha (235 ac) lake within North Platte NWR

Wolf L, Saunders Co: sandpit area east of Morse Bluff

Wood Duck Area, Stanton Co: 270 ha (668 ac) WMA along Elkhorn River

Youngson Basin, Kearney Co: WPA

Zorinsky L, Douglas Co: 103 ha (255 ac) reservoir in western part of Omaha

APPENDIX 3
Appended Species

Crested Tinamou
Eudromus elegans
COMMENTS
An introduction of 89 birds made on the True Howard Ranch near Benkelman 11 Sep 1970 (NBR 39:39) was unsuccessful (Johnsgard 1980).

White-tailed Tropicbird
Phaethon lepturus
COMMENTS
One reported flying along the South Platte River in Lincoln Co 13–14 May 1973 (NBR 41:46) was later suggested to be a tern trailing a fishing line (NBR 41:79). The report was considered inconclusive (Bray and others 1986). There are no records of this species any closer to Nebraska than Pennsylvania, western New York, and Arizona (AOU 1983).

Swan Goose
Anser cygnoides
COMMENTS
Five were taken by hunters in Keith Co 8 Dec 1984 (NBR 53:4), and another was at Funk Lagoon 4 Apr 1999 (LR, RH, GH). These are certainly escaped birds, as there is no pattern of vagrancy of wild birds into North America (Wright 1985a).

Bar-headed Goose
Anser indicus
COMMENTS
A pair was seen with Canada Geese in Keith Co Feb 1985, presumed escapees (NBR 53:3). There is no pattern of vagrancy of this species into North America (AOU 1983).

Ruddy Shelduck

Tadorna ferriginea

COMMENTS

One was present at Kearney L, Buffalo Co, 24 Feb 1997 (B). It was considered an escapee based on its behavior, including "aggressive begging, following people around, and approaching cars and people fearlessly" (David Rintoul, pers. comm. WRS).

Chukar

Alectoris chukar

COMMENTS

Several attempts were made to establish this species in Nebraska with the release between 1938 and 1943 of about 6500 birds, but these were unsuccessful, possibly due to predation (Mathisen and Mathisen 1960b). Ducey (1988) cited nesting reports from Logan, Madison, and Sheridan Cos in the 1940s. A notable release was of 700 birds in southern Scotts Bluff Co in 1969 (Johnsgard 1997); there were consistent reports from Scotts Bluff Co from this time until 1977, as well as a more recent report 19 Mar 1984, which was apparently derived from private releases made in the Robideaux Pass area 1981–84 (NBR 52:58). A population apparently persisted for some time in the Sheridan Co sandhills (Johnsgard 1979), although none were seen after 1977 (Rosche 1982), and possibly in the Pine Ridge in the Fort Robinson area in the 1970s (Rosche 1982). One seen in southern Sioux Co 4 Apr 1998 was "probably part of a flock set free by Rick Kaan some years ago" (HKH).

Reports from Lincoln Co 8 Apr 1974, 14 Apr 1977, and 18 May 1986 were probably locally released birds, as are reports from Douglas and Sarpy Cos 1963–65.

A report from Boone Co fall 1980 in fact referred to Gray Partridge (Mollhoff 1986a).

Common Quail

Coturnix coturnix

COMMENTS

According to Mathisen and Mathisen (1960b),

some 76 600 were released in Nebraska beginning in 1957. Many migrated southward out of Nebraska, including one recovered in Chihuahua, Mexico, but nesting occurred to some extent. Birds released in Frontier and Dawson Cos may have wandered into Lincoln Co, where 2 were shot in fall 1962 (Shickley 1968). There have been no reports since, however.

California Quail

Callipepla californica

COMMENTS

The only release of these birds was of about 70 in Butler, Madison, Knox, and Douglas Cos in 1939; although a nesting was reported in Madison Co in 1939 (Halbert 1940), the populations apparently died out soon after (Mathisen and Mathisen 1960b).

Scaled Quail

Callipepla squamata

COMMENTS

About 33 were released by a private individual in Loup Co in 1957; these birds initially did well, with as many as 400 birds known in subsequent years in Loup, Garfield, Holt, Rock, and Blaine Cos (Mathisen and Mathisen 1960b). Soon, however, they died out, as none were reported after 1962 (Johnsgard 1979). There were nesting reports from northwestern Nebraska and Lincoln Co (Wensien 1962; Ducey 1988), but these populations apparently died out also (Rosche 1982).

American Oystercatcher

Haematopus palliatus

COMMENTS

This species was listed by Bruner (1896) based on a specimen reportedly taken in northeastern Nebraska in spring 1889 by Charles Vaughn. However, Bruner and others (1904) stated that this species "should be omitted from our list" as the evidence was not sufficient for its inclusion. American Oystercatcher has been reported in the interior of North

America only once, a sight record for Idaho (AOU 1998).

Skua sp.
Catharacta sp.
COMMENTS
A dying bird though to be a Skua was picked up alongside a road in Sheridan Co 23 Aug 1968 (Gates 1968). Subsequent examination of the specimen, now UNSM ZM 12309, revealed that it was a Parasitic Jaeger (Gates 1969).

Ringed Turtle-Dove
Streptopelia risoria
COMMENTS
There are 5 reports of free-flying birds, all probable escapees or released birds. There is no evidence of nesting. A single banded bird was noted at a feeder in a Plattsmouth yard 3–16 Mar 1968 (Schneider 1968). Another rather tame bird which also had a "private band" was noted in Bellevue 7 Aug 1974–early Jan 1975 (NBR 42:62, 43:13, 38). One was noted in Crete for about 4 months from 19 Aug 1978 (Gross 1978). A free-flying bird was reported in Douglas-Sarpy Cos late Apr 1979 (NBR 47:42). A single bird was reported without details at Valley View tennis courts, Omaha, 22 Feb 1998 (B).

Recently the ABA Checklist Committee removed this species from the ABA Checklist, as there no longer appear to be any established populations in North America (DeBenedictis 1994). Indeed, there is some question as to whether it is a legitimate taxon, the current consensus being that it is a domestic strain of African Collared-Dove *S. roseogrisea* (DeBenedictis 1994).

Budgerigar
Melopsittacus undulatus
COMMENTS
A free-flying yellow morph bird was seen at a feeder in Kearney 8 Aug 1991 (NBR 60:32). It was considered "probably an escape[e]."

Another was seen flying near Wakefield 18 Aug 1999, also "surely an escape[e]" (JJ).
This species is considered established in Florida (AOU 1983); there are no established populations anywhere near Nebraska.

Black-chinned Hummingbird
Archilochus alexandri
COMMENTS
There are no documented records of this species for Nebraska, although a female *Archilochus* hummingbird was observed at Kimball 4 Aug 1996 (FN 51:81). There are no records from the northeast Colorado plains (Andrews and Righter 1992) or eastern Wyoming (Scott 1993).
Bray and others (1986) discussed the provenance of a mounted specimen identified as a Black-chinned Hummingbird collected at Kearney Aug 1903. While the specimen is indeed a female Black-chinned Hummingbird, Swenk (1918b, 1920) noted that it appeared to have been collected in Arizona and the accompanying collection data had been switched with that of a Broad-tailed Hummingbird collected at Kearney Aug 1903 (in itself a significant record). Swenk (1920) removed Black-chinned Hummingbird from the list of species recorded in Nebraska.

Ivory-billed Woodpecker
Campephilus principalis
COMMENTS
Bruner (1896) cited a report by G. A. Coleman that this species was rare in the vicinity of Peru. Bruner and others (1904), however, stated that "the Peru record of the 'Ivory-billed Woodpecker' refers to [Pileated Woodpecker]."

Green Jay
Cyanocorax yncas
COMMENTS
The only report is a mention by Ducey (1989), who considered the report questionable, of its occurrence in the Panhandle along the

Niobrara River in the period 1921–60; no other details were given.

This species is not known to wander from its south Texas range; the report cited above is almost certainly an error.

Bushtit
Psaltriparus minimus
COMMENTS

The editor of NBR, in a discussion of species reported without documentation, stated that he "was aware of unpublished possible or probable records in Nebraska of . . . Wrentit" (NBR 41:80). The editor, R. G. Cortelyou, noted later (pers. comm. WRS) that the observer of the "Wrentit" had made an error and had intended to report Bushtit.

It is not unlikely that this species could occur in Nebraska. It regularly appears in the eastern Colorado foothills north to the Wyoming border, and there is a record from Weld Co about 73 km (45 mi) from the southwestern corner of Nebraska (Andrews and Righter 1992). There is also a record of 4 netted in Ellis Co, Kansas (Thompson and Ely 1992), some 130 km (80 mi) south of Harlan Co, Nebraska.

It should be looked for in winter months in towns and juniper areas in the southern Panhandle and in southwestern Nebraska.

Wrentit
Chamaea fasciata
COMMENTS

A report of this species was corrected to one of Bushtit (NBR 41:46; Bray and others 1986). No details were published.

Red-crested Cardinal
Paroaria coronata
COMMENTS

One reported at a feeder in Lincoln Sep 1972–30 Apr 1973 was determined to be an escapee from the Lincoln Zoo (Bennett 1973c).

This South American species is established in Hawaii (AOU 1983).

Abert's Towhee
Pipilo aberti
COMMENTS

A description of one reported 29 Sep 1980 at Tryon (Bassett 1980; NBR 49:14) was not considered diagnostic (Bray and others 1986), nor has it been included in the Official List of the Birds of Nebraska (NBR 56:86). This species is sedentary in the southwest United States and has no history of vagrancy.

Audubon's Oriole
Icterus graduacauda
COMMENTS

A description furnished by the observer to the NOURC of one in McPherson Co 7 Jun 1985 was not sufficient to confirm the identification as Audubon's Oriole; this is a sedentary south Texas species that has no history of vagrancy (Mollhoff 1987). The bird was likely a Black-headed Grosbeak (NBR 53:59).

Brown-capped Rosy-Finch
Leucosticte australis
COMMENTS

A bird found on the Tristate CBC 15 Dec 1984 near Nebraska City was tentatively identified as a female of this species (NBR 53:23). It could not be relocated, and it was concluded that it may have been an aberrantly plumaged female House Sparrow. There are no records of this species east of the Rocky Mountains.

European Goldfinch
Carduelis carduelis
COMMENTS

One was reported without details at a feeder in Scribner 13 and 17 Feb 1996 (Carson 1996). This species is listed by the AOU (1983) but does not meet the ABA requirements for an established introduced species (ABA 1996). The AOU (1983) indicated that reports of this species, widely scattered, are probably of escaped cage birds. The only established population, on Long Island, New York, is probably now extirpated (AOU 1983).

APPENDIX 4
Additional Species

The following 5 species include one inadvertently omitted from the original text (Painted Bunting), 2 which were included in the original text as Hypothetical but are now confirmed (Arctic Tern and Dusky Flycatcher), and 2 species new to the state of Nebraska since 1 Jan 2000 (Reddish Egret and Black Rosy-Finch).

Reddish Egret
Egretta rufescens
STATUS
Accidental in fall.
DOCUMENTATION
Photograph
27 Sep-15 Oct 2000 L McConaughy (SJD).
Record
An immature white morph individual was at the west end of L McConaughy 27 Sep–15 Oct 2000 (SJD, WRS, MB, JF). The bird was photographed (SJD) and the pictures examined by Rich Paul, who confirmed the identification and, although ageing is difficult in this species, thought the bird to be an immature.
COMMENTS
The white morph is now rare, with only 1% of birds along the Texas Gulf of Mexico coast being white morph birds (Palmer 1962).
This species occurs rarely north of Texas. It has been recorded in southeastern Colorado and southern Illinois and is accidental further north (AOU 1996). There is a record for Iowa of a juvenile 5–8 Sep 1993 (Kent and Dinsmore 1996), but no records for other states adjacent to Nebraska.

Arctic Tern

Sterna paradisaea

STATUS

Accidental in fall.

DOCUMENTATION

Photograph

20 Sep 2000 molting adult L Minatare (SJD).

Records

There is a single documented record, cited above.

There are two additional reports. A first basic bird was reported with minimal details at Gavin's Point Dam 6 Dec 1998 (DLS), a very unusual date for this species (see below). Aughey mentions that "a few were seen in Dixon Co in May 1866" (Bruner 1896), a report discounted by Bruner et al (1904), who noted that no specimen was taken and that "the chances for a mis-identification are too great to warrant the continual inclusion of *paradisaea* in our list."

This species is quite likely to be recorded in Nebraska, however, as there is a record for the eastern Colorado plains 12 Jun 1991 (Andrews and Righter 1992), 6 records for the Duluth area in Minnesota in the period 18 May-17 Jun (Janssen 1987), and an adult and a juvenile were found in central Iowa 18 and 19 Aug 1999 (SJD).

There are also two fall records for the eastern Colorado plains: 11–12 Sep 1979 and 16 Sep 1912 (Andrews and Righter 1992). Any "non-Forster's" terns seen in Nebraska in late May and early Jun should be carefully scrutinized.

Dusky Flycatcher

Empidonax oberholseri

STATUS

Rare regular fall migrant west and central. Hypothetical in spring.

DOCUMENTATION

Photographs and measurements

31 Aug 2000 Oliver Res (SJD, WRS).

DISTRIBUTION AND ECOLOGY

Spring

One was reported 17 May 1992 in Kimball Co (WRS); the documentation provided the NOURC was judged Class IV (Hypothetical) due to the difficulty of identifying silent *Empidonax* flycatchers.

Fall

Mist-netting during the fall of 2000 showed that this species is probably a regular migrant in the Panhandle. Singles were captured, measured, and photographed at Oliver Res 31 Aug (SJD, WRS), 7 Sep (SJD, BP, LP), and 20 Sep (SJD). During the same fall 4 additional sight records were made, all probably correct. One was carefully studied at Wellfleet L 21 Jul (WRS), and additional singles were at Oliver Res ("probable") 2–3 Sep (SJD, JG), Mud Springs, Morrill Co (SJD), and Wind Springs Ranch, Sioux Co 20 Sep (SJD, HKH).

COMMENTS

This species breeds in the Black Hills of South Dakota (Peterson 1990, Johnsgard 1979) and would be expected to migrate through the Nebraska Panhandle. The earliest South Dakota records are 11, 20, and 25 May (SDOU 1991). In Kansas it is a "low-density transient" in the west, with specimens collected 29 Apr–17 May; it was considered probably the "most regular of the western *Empidonaces*" in Kansas (Thompson and Ely 1992). There are, however, no specimens from the eastern plains of Colorado (Andrews and Righter 1992), although Leukering (pers. comm. WRS) found that as many as 55% of fall *Empidonax* flycatchers banded at his station at Barr Lake in eastern Colorado after 20 Aug were Dusky Flycatchers.

Black Rosy-Finch

Leucosticte atrata

STATUS

Accidental in winter.

DOCUMENTATION

Description

12 Nov 2000 southwest Sioux Co (SJD).

TAXONOMY

Until recently the rosy-finches were considered conspecific, based on limited hybridization between Gray-crowned and Black Rosy-Finches from west-central Idaho to central Montana, but they are now split (AOU 1998).

Record

A single male was found with a flock of 216 Gray-crowned Rosy-Finches in southwest Sioux Co 12 Nov 2000 (SJD).

COMMENTS

This species occurs in winter on the eastern Wyoming plains but is rare as far east as the Nebraska border. The only record farther east (other than the record cited above) is one in Ohio (AOU 1998).

Painted Bunting

Passerina ciris

STATUS

Rare casual spring visitor. Accidental in fall.

DOCUMENTATION

Specimen

UNK, 26 Apr 1960 Kearney, Buffalo Co (Brown 1960; Wetmore 1961).

TAXONOMY

The only information available for Nebraska involves the specimen cited above; it was considered an intermediate between eastern *ciris* and western *pallidior* (Wetmore 1961). Birds reaching Nebraska would be expected to be *pallidior*, which breeds north into Kansas, while *ciris* breeds no closer than southern Missouri (AOU 1957).

DISTRIBUTION AND ECOLOGY

All but 1 of the 10 spring reports are from Adams Co westward and in or near the Platte and North Platte Valleys. This species occurs in open brushlands and scrubby habitats, but also appears at feeders.

Spring

The 10 reports are in the period 26 Apr–4 Jun; of these, 6 are acceptably documented, although all may be correct.

26 Apr 1960 Kearney specimen UNK (cited above)

12 May 1996 male and two females Morrill Co (Brogie 1997)

14 May 1966 Scotts Bluff Co (SY; NBR 35:21)

16 May 1967 male Fontenelle Forest (NBR 25:78)

30–31 May 1996 female Hall Co (EV; Brogie 1997; NBR 64:64)

4 Jun 1973 male Lincoln Co (Janssen 1973)

The four undocumented reports are: a "sight record" in Scotts Bluff Co in May 1927 (Johnsgard 1980), one reportedly seen but not described in Parkview Cemetery, Hastings, 19 May 1962 (NBR 30:51), one reportedly seen but undescribed 19 May 1982 near Kingsley Dam in Keith Co (Rosche 1994a; Rosche and Johnsgard 1984) and later withdrawn by the original observer (Brown et al 1996), and a report of a female seen 21 May 1988 in Scotts Bluff Co (Korpi 1990) which was not accepted by the NOURC (Grenon 1990).

Fall

There is one report, documented: 12–14 Nov 1983 male and possible female Grand Island (NBR 52:24).

COMMENTS

The occurrence of this species in Nebraska fits the pattern established in Colorado, where it is a casual spring migrant on the eastern plains, with seven records in May and early Jun (Andrews and Righter 1992). In Wyoming there are three records, 17 May 1991, 23–26 May 1956, and 4 Jun 1975 (Scott 1993; Dorn and Dorn 1990). There are no records for South Dakota (SDOU 1991) and it is hypothetical in Iowa, with a sight report from the northwest 30 May 1956 and an eastern feeder sighting in Jan considered to be an escaped cage bird (Kent and Dinsmore 1996).

FINDING

Patterns of occurrence suggest that this species is most likely to occur in the central or western Platte Valley in May. It should be looked for in brushy riparian habitats.

Bibliography

Abbott, D. F. 1998. Re: Western Sandpiper. *ID-Frontiers listserv. Birdwg01@listserv.arizona.edu.* (31 May 1998).

Ager, J. H. 1946. A sight record of the Purple Gallinule for Adams County. NBR 14:19–20.

Ahlbrandt, T. S., J. B. Swinehart, and D. G. Maroney. 1983. The dynamic Holocene dune fields of the Great Plains and Rocky Mountain Basins, U.S.A. In M. E. Brookfield and T. S. Ahlbrandt, eds., Eolian sediments and processes. Amsterdam: Elsevier.

Albertson, F. W., and J. E. Weaver. 1945. Injury and death or recovery of trees in prairie climate. Ecological Monographs 15:393–433.

Aldrich, J. W., and H. Friedmann. 1943. A revision of the Ruffed Grouse. Condor 45:85–103.

Allen, R. P. 1952. The Whooping Crane. National Audubon Society Research Report No. 3. New York: National Audubon Society.

Alt, J. 1994. Northern Waterthrush. NBR 62:150.

[ABA] American Birding Association. 1996. ABA checklist: birds of the continental United States and Canada. 5th ed. Colorado Springs, CO: ABA.

[AOU] American Ornithologists' Union. 1957. The AOU check-list of North American birds. 5th ed. Baltimore: Port City Press.

———. 1983. The AOU check-list of North American birds. 6th ed. Lawrence, KS: Allen Press.

———. 1995. Fortieth supplement to the American Ornithologists' Union check-list of North American birds. Auk 112:819–30.

———. 1997. Forty-first supplement to the American Ornithologists' Union check-list of North American birds. Auk 114:542–52.

————. 1998. The AOU check-list of North American birds. 7th ed. Lawrence, KS: Allen Press.

Ammann, G. A. 1957. The Prairie Grouse of Michigan. Michigan Department of Conservation Technical Bulletin. Lansing: Michigan Department of Conservation.

Anderson, D. 1957. Excerpts from letters to the editors. NBR 25:45–46.

Anderson, J. 1992. Upland territorial birds and cattle grazing regimes in Nebraska Sandhills prairie. Master's thesis, University of Nebraska–Omaha.

Andrews, R., and R. Righter. 1992. Colorado birds. Denver: Denver Museum of Natural History.

Antholz, S. 1975. Prairie chickens, Upland Sandpipers. NBR 43:59.

Armstrong, J. R. 1960. Letters to the editor. NBR 38:56.

————. 1965. Hooded Oriole. NBR 33:65.

Audubon, M. R., and E. Coues. 1960. Audubon and his journals. Vols. 1 and 2. New York: Dover Publications Reprint.

Audubon Society Master Guide to Birding. 1983. Vol. 3, Warblers to sparrows. New York: Alfred A. Knopf.

Aughey, S. 1878. Notes on the nature of the food of the birds of Nebraska. 1878 Report of the United States Entomological Commission. Washington DC: Government Printing Office.

————. 1880. Sketches of the physical geography and geology of Nebraska. Omaha, NE: Daily Republican Book and Job Office.

Bailey A. M., and R. J. Neidrach. 1967. Pictorial checklist of Colorado birds. Denver: Denver Museum of Natural History.

Baird, S. F., J. Cassin, and G. N. Lawrence. 1858. Reports of explorations and surveys to ascertain the most practible [sic] and economic route for a railroad from the Mississippi River to the Pacific Ocean [etc]. Vol. 9. Washington DC: A. O. P. Nicholson.

Baker, M. F. 1953. Prairie chickens in Kansas.

Wichita: Publications of the University of Kansas Museum of Natural History and State Biological Survey.

Banghart, M. A. 1966. Golden-Crowned Sparrow. NBR 34:76.

Banks, R. C. 1977. The decline and fall of the Eskimo Curlew, or why did the curlew go extaille? American Birds 31:127–34.

————. 1983. A review of the nomenclature of populations of the Greater White-fronted Goose, Anser albifrons. Unpublished U.S. Fish and Wildlife Service Report. Available from author.

————. 1986. Subspecies of the Glaucous Gull, Larus hyperboreus (Aves: Charadriiformes). Proceedings of the Biological Society of Washington 99:149–59.

Barrowclough, G. F. 1980. Genetic and phenotypic differentiation in a Wood Warbler (genus Dendroica) hybrid zone. Auk 97:655–68.

Bassett, O. 1967. Golden-Crowned Sparrow. NBR 35:24.

————. 1968. Black-throated Gray Warbler. NBR 36:15–16.

————. 1974. Hermit Warbler reported in Nebraska. NBR 42:42.

————. 1980. Abert's Towhee reported. NBR 48:89.

Bates, J. M. 1900. Additional notes and observations on the birds of northern Nebraska. Proc NOU 1:15–18.

————. 1901. Additional observations on the birds of northwestern Nebraska. Proc NOU 2:73–75.

Baumgarten, H. E. 1953. Henslow's Sparrow at Lincoln. NBR 21:25.

Baumgarten, H. E., and W. F. Rapp Jr. 1953. Two new birds added to the Nebraska list. NBR 21:2–3.

Baumgartner, F. M., and A. M. Baumgartner. 1992. Oklahoma bird life. Norman: University of Oklahoma Press.

Baxter, W. L., and C. W. Wolfe. 1973. Life history and ecology of the Ring-necked Pheasant in

Nebraska. Lincoln: Nebraska Game and Parks Commission.

Bedell, P. A. 1982. Louisiana Waterthrush nesting in Fontenelle Forest. NBR 50:88–89.

———. 1987. Early fall migration of Sedge Wrens. NBR 55:86–88.

———. 1996. Evidence of dual breeding ranges for the Sedge Wren in the central Great Plains. Wilson Bulletin 108:115–22.

Bellinghiere, S. 1980. Saw-whet Owl. NBR 48:24.

Bellrose, F. C. 1976. Ducks, geese and swans of North America. Harrisburg, PA: Stackpole Books.

Benckeser, H. R. 1950. Keith County winter notes. NBR 18:66–67.

———. 1955. Keith County notes. NBR 23:22.

———. 1956. Notes from Keith County. NBR 24:26.

Bennett, E. V. 1968. 1967 Nebraska nesting survey. NBR 36:35–42.

———. 1969. 1968 Nebraska nesting survey. NBR 37:39–46.

———. 1970. 1969 Nebraska nesting survey. NBR 38:3–9.

———. 1971. 1970 Nebraska nesting survey. NBR 39:10–15.

———. 1972. 1971 Nebraska nesting survey. NBR 40:9–15.

———. 1973a. Boreal Chickadee. NBR 41:43.

———. 1973b. 1972 Nebraska nesting survey. NBR 41:3–9.

———. 1973c. Red-Crested Cardinal. NBR 41:43–44.

———. 1974. 1973 Nebraska nesting survey. NBR 42:3–10.

———. 1975. 1974 Nebraska nesting survey. NBR 43:13–19.

———. 1976. 1975 Nebraska nesting survey. NBR 44:8–11.

———. 1977. 1976 Nebraska nesting survey. NBR 45:3–5.

———. 1978. 1977 Nebraska nesting survey. NBR 46:13–16.

———. 1980a. 1978 Nebraska nesting survey. NBR 48:39–43.

———. 1980b. 1979 Nebraska nesting survey. NBR 48:67–68.

———. 1981. 1980 Nebraska nesting survey. NBR 49:8–11.

———. 1982. 1981 Nebraska nesting survey. NBR 50:38–42.

———. 1983. 1982 Nebraska nesting survey. NBR 51:26–32.

———. 1984. 1983 Nebraska nesting survey. NBR 52:47–50.

———. 1985. 1984 Nebraska nesting survey. NBR 53:46–49.

———. 1986. 1985 Nebraska nesting survey. NBR 54:31–35.

———. 1987. 1986 Nebraska nesting survey. NBR 55:31–35.

———. 1988. 1987 Nebraska nesting survey. NBR 56:35–40.

———. 1989. 1988 Nebraska nesting report. NBR 57:34–41.

———. 1990. 1989 Nebraska nesting report. NBR 58:38–45.

Bent, A. C. 1923. Life histories of North American wild fowl. Bulletin of the United States National Museum 126. New York: Dover Publications Reprint 1962.

———. 1926. Life histories of North American marsh birds. Bulletin of the United States National Museum 135. New York: Dover Publications Reprint 1963.

———. 1927. Life histories of North American shore birds. Parts One and Two. Bulletin of the United States National Museum 142 (Part One) and 146 (Part Two). New York: Dover Publications Reprint 1962.

———. 1932. Life histories of North American Gallinaceous birds. Bulletin of the United States National Museum 162. New York: Dover Publications Reprint 1963.

———. 1937. Life histories of North American birds of prey. Part One. Bulletin of the United States National Museum 167. New York: Dover Publications Reprint 1961.

———. 1938. Life histories of North American birds of prey. Part Two. Bulletin of the United

States National Museum 170. New York: Dover Publications Reprint 1961.

———. 1939. Life histories of North American woodpeckers. Bulletin of the United States National Museum 174. New York: Dover Publications Reprint 1964.

———. 1940. Life histories of North American cuckoos, goatsuckers, hummingbirds, and their allies. Bulletin of the United States National Museum 176. Two parts. New York: Dover Publications Reprint 1964.

———. 1942. Life histories of North American flycatchers, larks, swallows, and their allies. Bulletin of the United States National Museum 179. New York: Dover Publications Reprint 1963.

———. 1946. Life histories of North American jays, crows, and titmice. Bulletin of the United States National Museum 191. Two parts. New York: Dover Publications Reprint 1964.

———. 1948. Life histories of North American nuthatches, wrens, thrashers, and their allies. Bulletin of the United States National Museum 195. New York: Dover Publications Reprint 1964.

———. 1949. Life histories of North American thrushes, kinglets, and their allies. Bulletin of the United States National Museum 196. New York: Dover Publications Reprint 1964.

———. 1950. Life histories of North American wagtails, shrikes, vireos, and their allies. Bulletin of the United States National Museum 197. New York: Dover Publications Reprint 1965.

———. 1953. Life histories of North American wood warblers. Bulletin of the United States National Museum 203. Two parts. New York: Dover Publications Reprint 1963.

———. 1958. Life histories of North American blackbirds, orioles, tanagers, and their allies. Bulletin of the United States National Museum 211. New York: Dover Publications Reprint 1965.

———. 1968. Life Histories of North American cardinals, grosbeaks, buntings, towhees, finches, sparrows, and their allies. Bulletin of the United States National Museum 237. Three parts. New York: Dover Publications Reprint 1968.

Bergmann, D. L., and J. M. Bergmann. 1995. American Woodcock use of a nest box. Prairie Naturalist 27:172–73.

Bettis, E. A., III, J. C. Prior, G. R. Hallberg, and R. L. Handy. 1986. Geology of the Loess Hills region. Proceedings of the Iowa Academy of Science 43:78–85.

Bevill, W. V., Jr. 1970. Effects of supplemental stocking and habitat development on abundance of Mexican Ducks. M.S. thesis, New Mexico State University, Las Cruces.

Bicak, T. K. 1977. Some etho-ecological aspects of a breeding population of Long-billed Curlews (*Numenius americana*) in Nebraska. Master's thesis, University of Nebraska–Omaha.

Black, C. A. 1922. Some bird notes from central and western Nebraska. Wilson Bulletin 34:43.

———. 1933. An addition to the Nebraska bird list: the Northern Louisiana Heron. NBR 1:31.

———. 1941. The Northern Louisiana Heron in Nebraska: a correction. NBR 9:16.

Blake, L., and J. E. Ducey. 1991. Birds of the eastern Sandhills in Holt County, Nebraska. NBR 59:103–32.

Bledsoe, A. H., and D. Sibley. 1985. Patterns of vagrancy of Ross' Gull. American Birds 39:219–27.

Bliese, J. C. W. 1975a. A Monk Parakeet in the Kearney area. NBR 43:42.

———. 1975b. Snowy Plovers. NBR 43:43.

Blinco, G. 1946. Notes on Grosbeaks and Bohemian Waxwings at Chadron, Dawes County. NBR 14:45.

Blom, R. 1996. Re: barrovianus Glaucous Gulls. *ID-Frontiers listserv. birdwg01@listserv.arizona.com.* (28 Dec 1996).

Blus, L. J., and J. A. Walker. 1966. Progress report on the Prairie Grouse nesting study in the Nebraska Sandhills. NBR 34:23–30.

Bock, C. E., and L. W. Lepthien. 1972. Winter eruptions of Red-breasted Nuthatches in North America 1950–1970. American Birds 26:558–61.

Bock, C. E., and W. C. Scharf. 1994. A nesting

population of Cassin's Sparrows in the Sandhills of Nebraska. Journal of Field Ornithology 65:472–75.

Bohlen, H. D. 1989. The birds of Illinois. Bloomington and Indianapolis: Indiana University Press.

Brashear, J. 1976. Brown Towhee reported. NBR 44:30.

Bray, T. E. 1994. Habitat utilization by birds in a man-made forest in the Nebraska Sandhills. M.S. thesis, University of Nebraska–Omaha.

Bray, T. E., B. K. Padelford, and W. R. Silcock. 1986. The birds of Nebraska: a critically evaluated list. Bellevue, NE: Bray, Padelford, and Silcock.

Brenneman, J. 1995. Western Tanager. NBR 63:61.

Bright, D. 1993. White-eyed Vireo and Summer Tanager. NBR 61:137.

———. 1994. Correction in the Nebraska Bird Review 61:137. NBR 62:63.

Brogie, M. A. 1986. California Gull in Keith County, Nebraska. NBR 54:77.

———. 1989a. Notes on Black-legged Kittiwakes in Cedar County, Nebraska. NBR 57:53–54.

———. 1989b. Cassin's Sparrow in Dundy and Chase Counties, Nebraska. NBR 57:67–71.

———. 1990. Laughing Gull in Cedar County, Nebraska. NBR 58:99–100.

———. 1992. Sighting of a Canyon Wren in Knox County, Nebraska. NBR 60:163–65.

———. 1997. 1996 (Eighth) Report of the NOU Records Committee. NBR 65:115–26.

———. 1998. 1997 (Ninth) Report of the NOU Records Committee. NBR 66:147–59.

———. 1999. 1998 (Tenth) Report of the NOU Records Committee. NBR 67:141–52.

Brogie, M. A., and E. M. Brogie. 1985. A Red Phalarope in Pierce County, Nebraska. NBR 53:71–73.

———. 1987a. Barrow's Goldeneye in Keith County, Nebraska. NBR 55:44.

———. 1987b. Black Rail in Knox County, Nebraska. NBR 55:40–41.

———. 1987c. A Red-naped Sapsucker in Sioux County, Nebraska. NBR 55:40.

———. 1988. Red Phalarope. NBR 56:22–24.

———. 1989. An Iceland Gull in Cedar County, Nebraska. NBR 57:52–53.

Brogie, M. A., and M. J. Mossman. 1983. Spring and summer birds of the Niobrara Valley Preserve, Nebraska: an annotated checklist. NBR 51:44–51.

Brogie, M. A., and D. A. Stage. 1987. MacGillivray's Warbler in Cedar County, Nebraska. NBR 55:41–42.

Brooking, A. M. Notes. Bird specimen records. Manuscript in NOU Archives.

———. 1933a. The Laughing Gull and the Yellow-crowned Night-Heron at Inland, Clay County. NBR 1:130.

———. 1933b. The loss and recovery of the first Nebraska specimen of the Swainson's Warbler. NBR 1:132.

———. 1934. A recent Nebraska record of the American Brant. NBR 2:60.

———. 1942a. The vanishing birdlife of Nebraska. NBR 10:43–47.

———. 1942b. A record of the Roseate Spoonbill for Nebraska. NBR 10:52.

———. 1944. Nesting of the White-necked Raven in Kearney County. NBR 12:40.

Brown, C. R., and M. B. Brown. 1992. First record of the Cave Swallow in Nebraska. NBR 60:36–40.

Brown, C. R., M. B. Brown, P. A. Johnsgard, J. Kren, and W. C. Scharf. 1996. Birds of the Cedar Point Biological Station area, Keith and Garden Counties, Nebraska: seasonal occurrence and breeding data. Transactions of the Nebraska Academy of Sciences 23:91–108.

Brown, E. 1947. Nests of the White-necked Raven in Kearney County. NBR 15:49.

Brown, G. W. 1960. Painted Bunting at Kearney. NBR 28:53.

———. 1975. Pileated Woodpecker. NBR 43:20.

———. 1979. Marsh birds. NBR 47:67.

Browning, M. R. 1974. Comments on the winter distribution of Swainson's Hawk (*Buteo swainsoni*) in North America. American Birds 28:865–67.

———. 1995. Do Downy Woodpeckers migrate? Journal of Field Ornithology 66:12–24.

Bruner, L. 1896. A list of Nebraska birds, together

with notes on their abundance, migrations, breeding, food-habits, etc. Nebraska State Horticultural Society 27th Annual Report, 57–163.

———. 1901. Birds that nest in Nebraska. Proc NOU 2:48–61.

———. 1902. A comparison of the bird-life found in a Sandhills region of Holt County in 1883–84 and in 1901. Proc NOU 3:58–63.

Bruner, L., R. H. Wolcott, and M. H. Swenk. 1904. A preliminary review of the birds of Nebraska, with synopses. Omaha: Klopp and Bartlett.

Busch, D. E. 1979. Successful Golden Eagle nest in southwest Nebraska. NBR 47:57.

Butler, R. W., F. S. Delgado, H. De La Cueva, V. Pulido, and B. K. Sandercock. 1996. Migration routes of the Western Sandpiper. Wilson Bulletin 108:662–72.

Byers, C., J. Curson, and U. Olsson. 1995. Sparrows and buntings. A guide to the sparrows and buntings of North America and the world. Boston and New York: Houghton Mifflin.

Carriker, M. A., Jr. 1900. Some notes on the nesting of the raptors of Otoe County. Proc NOU 1:29–34.

Carriker, M. A., Jr. 1902. Notes on the nesting of some Sioux County birds. Proc NOU 3:75–89.

Carson, E. 1996. Merlin, European Goldfinch, and other species. NBR 64:35.

Cary, M. 1900. Some bird notes from the Upper Elkhorn. Proc NOU 1:21–29.

———. 1901. Birds of the Black Hills. Auk 18:231–38.

———. 1902. Some general remarks on the distribution of life in northwest Nebraska. Proc NOU 3:63–75.

Cavitt, J. F., A. T. Pearse, and D. A. Rintoul. 1998. Hybridization of Mountain Bluebird and Eastern Bluebird in northeastern Kansas. Kansas Ornithological Society Bulletin 49:21–25.

Cecil, R. 1997. Field reports—winter 1996–97. Iowa Bird Life 67:57–63.

Chartier, A. 1999. Re: Chickadee irruptions. *Birdchatlistserv. birdchat@listserv.arizona.edu.* (16 Apr 1999).

Chu, P. C. 1998. A phylogeny of the gulls (*Aves: Larinae*) inferred from osteological and integumentary characters. Cladistics 14:1–43.

Cink, C. L. 1971. Some interesting summer bird records for Lancaster County in 1970. NBR 39:58–59.

———. 1973a. Louisiana Heron in Clay County. NBR 41:14–15.

———. 1973b. Summer records of the Short-billed Marsh Wren in Nebraska. NBR 41:17–19.

———. 1973c. The Yellow Rail in Nebraska. NBR 41:24–27.

———. 1975a. Some waterfowl breeding records for Lancaster County. NBR 43:40–41.

———. 1975b. The Song Sparrow as a Nebraska breeding bird. NBR 43:3–8.

———. 1977. White-necked Raven. NBR 45:16.

Cink, C. L., and K. L. Fiala. 1971a. Breeding record of the Long-billed Marsh Wren from Lancaster County. NBR 39:74–75.

———. 1971b. First specimen of Poor-Will from Lancaster County; a summary of Poor-Will migration. NBR 39:70–71.

Clark, W. S. 1987. A field guide to hawks, North America. Boston: Houghton Mifflin.

Clausen, M. K. 1990. Mountain Plover sighted in Kimball County. NBR 58:98–99.

Cole, T. 1976. A comparative study of two grassland bird communities. Master's thesis, University of Nebraska–Omaha.

Collister, A. 1947. Long-Crested Jay in Lincoln County. NBR 15:18.

———. 1948. Nesting of the Double-crested Cormorant in Nebraska. NBR 16:22–26.

Conover, H. B. 1934. Records of the Red Phalarope and American Scoter from Cherry County, Nebraska. NBR 2:38.

Cooch, G. 1961. Ecological aspects of the Blue-Snow Goose complex. Auk 78:72–89.

Cook, H. J. 1939. The American Goldeneye in Sioux County in mid-November. NBR 7:13.

Cooke, F., and G. Cooch. 1968. The genetics of polymorphism in the goose *Anser caerulescens*. Evolution 22:289–300.

Coons, V. 1976. White-necked Raven reported. NBR 44:38.

———. 1983. House Finch nest at Grand Island. NBR 51:96.

Corbin, K. W., and C. G. Sibley. 1977. Rapid evolution in orioles of the genus *Icterus*. Condor 79:335–42.

Corbin, K. W., C. G. Sibley, and A. Ferguson. 1979. Genic changes associated with the establishment of sympatry in orioles of the genus *Icterus*. Evolution 33:624–33.

Cortelyou, R. G. 1960. Letters to the editor. NBR 28:55.

———. 1964. Apparent local differences in the calls of the Acadian Flycatchers. NBR 32:82–83.

———. 1967. Wild Turkeys in the Fontenelle Forest area. NBR 45:6–7.

———. 1975. A Brown Creeper's nest in Nebraska. NBR 43:80–83.

———. 1979. A possible Black Rail. NBR 47:67.

Coues, E. 1874. Birds of the Northwest: a handbook of the ornithology of the region drained by the Missouri River and its tributaries. Washington DC: Government Printing Office.

Cox, M. K., and W. L. Franklin. 1989. Terrestrial vertebrates of Scotts Bluff National Monument, Nebraska. Great Basin Naturalist 49:597–613.

Cox, T. A. 1983. Mountain Plover. NBR 51:94.

Cramp, S., and K. E. L. Simmons, eds. 1980. Handbook of the birds of Europe, the Middle East, and North Africa; the birds of the western Palearctic. Vol. 2, Hawks to bustards. Oxford and London, England, and New York: Oxford University Press.

Crawford, J. C., Jr. 1901. Results of a collecting trip to Sioux County. Proc NOU 2:76–79.

Curson, J., D. Quinn, and D. Beadle. 1994. Warblers of the Americas. Boston and New York: Houghton Mifflin.

Czaplewski, M. M. 1989. Least Terns at Lake McConaughy. NBR 57:95–96.

Dawson, R. W. 1921. Fall migration in northwestern Nebraska in 1920. Wilson Bulletin 33:35–37.

Day, M. 1947. Long-Crested Jay at Superior, Nuckolls County. NBR 15:18.

Dean, K. L., D. L. Swanson, and J. S. Martin. 1995. Warbler hunting in southeastern South Dakota. South Dakota Bird Notes 47:52–61.

DeBenedictis, P. A. 1976. Gleanings from the technical literature. Birding 8:241–45.

———. 1977. Gleanings from the technical literature. Birding 9:238–41.

———. 1982. Gleanings from the technical literature. Birding 14:51–55.

———. 1990. Gleanings from the technical literature—Thayer's Gull. Birding 22:197–200.

———. 1994. ABA checklist report, 1992. Birding 26:93–102.

———. 1995. Red Crossbills, one through eight. Birding 27:494–501.

———. 1996. Fortieth supplement to the AOU check-list. Birding 28:228–31.

Delacour, J., and S. D. Ripley. 1975. Description of a new subspecies of the White-fronted Goose *Anser albifrons*. American Museum Novitates 2565.

Desmond, M., and J. Savidge. 1990. Solitary vs. gregarious nesting in Burrowing Owls. NBR 58:77.

Diggs, H., and F. Diggs. 1974. Carolina Chickadees taken in Fontenelle Forest. NBR 42:57.

Dinan, J. J. 1996. 1996 survey of Bald and Golden Eagles in Nebraska. NBR 64:26–29.

Dinsmore, J. J. 1992. Field reports—summer 1992. Iowa Bird Life 62:104–12.

Dinsmore, J. J., T. H. Kent, D. Koenig, P. C. Petersen, and D. M. Roosa. 1984. Iowa birds. Ames: Iowa State University Press.

Dinsmore, S. J. 1996a. A late Baird's Sandpiper in Keith County. NBR 64:79–80.

———. 1996b. Red-necked Grebe. NBR 64:131.

———. 1996c. Snowy Plovers in the Rainwater Basin. NBR 64:71.

———. 1996d. Whimbrel. NBR 64:131–32.

———. 1997a. Mountain Plover nest in Kimball County. NBR 65:131.

———. 1997b. Red-necked Grebe at Sutherland Reservoir. NBR 65:99.

————. 1999a. Clark's Nutcracker at Lake McConaughy. NBR 67:79–80.

————. 1999b. Neotropic Cormorant at Sutherland Reservoir. NBR 67:72–73.

————. 1999c. Red-throated Loon at Lake McConaughy. NBR 67:81.

————. 1999d. Roseate Spoonbill in Otoe County. NBR 67:80–81.

————. 1999e. Two winter records of Pacific Loons at Lake McConaughy. NBR 67:16–17.

Dinsmore, S. J., and W. R. Silcock. 1993a. First record of a Ross' Gull for Nebraska. NBR 61:88–89.

————. 1993b. Lesser Black-backed Gull at Sutherland Reservoir. NBR 61:87.

————. 1995a. Cattle Egrets nesting in Keith County. NBR 63:89–90.

————. 1995b. First nesting records for Mississippi Kite in Nebraska. NBR 63:88–89.

————. 1995c. Laughing Gulls in western Nebraska. NBR 63:66–67.

————. 1998. Parasitic Jaeger at Lake McConaughy. NBR 66:114–15.

Di Silvestro, R. 1975. Monk Parakeets. NBR 43:60.

Dobkin, D. S. 1994. Conservation and management of neotropical migrant landbirds in the northern Rockies and Great Plains. Moscow: University of Idaho Press.

Dorn, J. L., and R. D. Dorn. 1990. Wyoming birds. Cheyenne, WY: Mountain West Publishing.

Douglas, A. V. 1985. Rufous Hummingbird. NBR 53:80.

Ducey, J. E. 1980a. Black Rail in Lancaster County. NBR 48:88.

————. 1980b. Immature Yellow-crowned Night Herons. NBR 48:87–88.

————. 1981. Breeding season occurrence of Sharp-shinned Hawks in southeast Nebraska. NBR 49:44–45.

————. 1982. The 1982 Least Tern and Piping Plover breeding season on the lower Platte River, Nebraska. NBR 50:68–72.

————. 1983a. Notes on the birds of the lower Niobrara River in 1902 as recorded by Myron H. Swenk. NBR 51:37–44.

————. 1983b. Some birds of the Anderson Bridge Wildlife Management Area in Cherry County, Nebraska. NBR 51:62–63.

————. 1984a. Some 1983 summer birds of the Missouri National Recreation River, Nebraska. NBR 52:36–40.

————. 1984b. Cattle Egrets and White-faced Ibises nesting at Valentine Refuge. NBR 52:76.

————. 1988. Nebraska birds, breeding status and distribution. Omaha, NE: Simmons-Boardman Books.

————. 1989. Birds of the Niobrara River valley, Nebraska. Transactions of the Nebraska Academy of Sciences 27:37–60.

————. 1999. History and status of the Trumpeter Swan in the Nebraska Sand Hills. North American Swans 28:31–39.

Ducey, J. E., and J. Schoenenberger. 1991. Some birds of the Pony Lake area of the eastern Sandhills, Nebraska. NBR 59:55–58.

Dueker, J. 1982. Bullsnakes. NBR 50:48.

DuMont, P. A. 1933. A definite Nebraska record for the Black Gyrfalcon. NBR 1:12.

————. 1934. A revised list of the birds of Iowa. University of Iowa Studies. Studies in Natural History 15(5).

Dunn, J. L., and K. L. Garrett. 1997. A field guide to warblers of North America. Boston and New York: Houghton Mifflin.

Dunn, J. L., K. L. Garrett, and J. K. Alderfer. 1995. White-Crowned Sparrow subspecies: identification and distribution. Birding 27:182–200.

Dwight, J., and L. Griscom. 1927. A revision of the geographical races of the Blue Grosbeak (*Guiraca caerulea*). American Museum Novitates 257.

Dwyer, M. 1988. Additional reports from Thomas County, spring 1988. NBR 56:99.

Easterla, D. A. 1967. A breeding population of Henslow's Sparrows in southwestern Missouri. Bluebird 34(1):18–19.

Eckert, K. R. 1993. Identification of Western and Clark's Grebes, the Minnesota experience. Birding 25:304–10.

————. 1996. Birding by hindsight: a second look at Western (and Eastern) Sandpipers. Loon 68:121–24.

Edgell, M. C. R. 1984. Trans-hemispheric movements of Holarctic *Anatidae*: the Eurasian Wigeon (*Anas penelope*) in North America. Journal of Biogeography 11:27–39.

Egger, M. 1978. Flammulated Owl? NBR 46:70.

Eiche, A. 1901. Breeding of the Snowy Heron and Swallow-tailed Kite. Proc NOU 2:96.

Einemann, L. L. 1977. A Groove-billed Ani and other birds in Cuming and Dodge Counties. NBR 45:13–14.

———. 1980. Gray (Hungarian) Partridge nest. NBR 48:88.

Ely, C. 1970. Migration of Least and Traill's Flycatchers in west-central Kansas. Bird-Banding 41:198–204.

Ely, C., and A. X. Dzubin. 1994. Greater White-fronted Goose, *Anser albifrons*. The birds of North America, No. 131. Washington DC: Academy of Natural Sciences.

Epperson, C. 1976. Food remains from a Barn Owl nest in Nebraska. NBR 44:54–57.

Erwin, W. J. 1970. Dipper observation in Holt County. NBR 38:17–18.

Evans, R. D., and C. W. Wolfe. 1965. Scissor-tailed Flycatcher nesting in Clay County. NBR 33:14.

———. 1967. Nest parasitism between duck and pheasants. NBR 35:47.

Faanes, C. A. 1990. A recent record of Eskimo Curlew in Nebraska. Prairie Naturalist 22:137–38.

Faanes, C. A., B. A. Hanson, and H. A. Kantrud. 1979. Cassin's Sparrow—first record for Wyoming and recent range extensions. Western Birds 10:163–64.

Faanes, C. A., and W. Norling. 1981. Nesting of the Great-tailed Grackle in Nebraska. American Birds 35:148–49.

Faeh, J. 1975. Anhinga seen in Nebraska again. NBR 43:41–42.

Fall, B. A. 1988. Brant at Funk Lagoon. NBR 56:77–78.

Fiala, K. L. 1970. The birds of Gage County, Nebraska. NBR 38:41–72.

———. 1971. Pileated Woodpecker in Otoe County. NBR 39:3.

Fichter, E. 1946. Possible sight record of Eastern Glossy Ibis in Nebraska. NBR 14:44.

———. 1947. Scattered notes. NBR 15:49–50.

Fickel, T. 1979. Townsend's Solitaires in Sioux County, Nebraska. NBR 47:68–69.

Flowerday, C. 1986. Sand Hills represent evidence of droughts in ancient climate. Resource Notes 1:10–12.

Ford, N. L. 1959. Notes on summer birds of western Nebraska. NBR 27:6–12.

Fox, A. 1952. Missouri River bird notes. NBR 20:21–22.

Frémont, J. C. 1988. The exploring expedition to the Rocky Mountains, 1845. Washington DC: Smithsonian Institution Press.

Fretwell, S. D. 1977. Is the Dickcissel a threatened species? American Birds 31:923–32.

Fritz, D. 1986. Prairie-chickens. NBR 54:67–68.

Fuller, H. C. 1961. Excerpts from letters. NBR 29:57.

Gabig, P. J. 1986. Canada Goose: managing the comeback. Lincoln: Nebraska Game and Parks Commission.

Galloway, S/Sgt., and Mrs. Galloway. 1956. Excerpts from letters to the editors. NBR 24:16.

Garabrandt, G. 1978. A history of land use in the oak-hickory woodland of Fontenelle Forest. Master's thesis, University of Nebraska–Omaha.

Garrett, F. D. 1944. A survey of nesting birds in the Fontenelle Forest. NBR 12:25–31.

Garthright, W. C. 1984. Nesting Common Moorhen in Lancaster County. NBR 52:74–75.

———. 1985a. Fillmore County. NBR 53:76–77.

———. 1985b. Red-necked Grebe. NBR 53:77.

Gates, D. 1950. Glossy Ibis at North Platte. NBR 18:68.

———. 1951. Birds of Chadron State Park. NBR 19:19–22.

———. 1953. Birds at Wayne (Nebraska). NBR 21:4–5.

———. 1955. Common House Finch nesting in Lincoln County. NBR 23:58.

———. 1965. Excerpts from letters. NBR 33:67.

———. 1966. Comparative arrival dates of selected migratory birds in selected counties in Nebraska. NBR 34:66–69.

———. 1968. Skua. NBR 36:76.

———. 1969. Parasitic Jaeger, rather than Skua.

NBR 37:31.

———. 1970. Clark's Nutcracker. NBR 38:17.

———. 1974. Canyon Wren in Nebraska. NBR 42:16.

———. 1981. Lewis' Woodpecker in western Nebraska. NBR 49:33–34.

Gersib, R. A. 1991. Nebraska wetlands priority plan. Lincoln: Nebraska Game and Parks Commission.

Gersib, R. A., K. F. Dinan, J. D. Kauffeld, M. O. Onnen, P. J. Gabig, J. E. Cornely, G. E. Jasmer, J. M. Hyland, and K. M. Strom. 1992. Looking to the future: an implementation plan for the Rainwater Basin Joint Venture. Lincoln: Nebraska Game and Parks Commission.

Gifford, H. 1925. Occurrence of the Wood Ibis near Omaha, Nebraska. Wilson Bulletin 37:219.

Glandon, E. W. 1936. The Rufous Hummingbird at Stapleton, Logan County. NBR 4:83.

———. 1948. Birds of the Stapleton area of Nebraska. NBR 16:1–21.

———. 1956. Birds that nest in the Stapleton area. NBR 24:30.

———. 1960. Red-bellied Woodpecker in Logan County. NBR 28:29–30.

———. 1964. Excerpts from letters. NBR 32:16.

Glandon, E. W., and R. Glandon. 1934a. Additions to the list of Logan County birds. NBR 2:61.

———. 1934b. Notes on some Logan County birds. NBR 2:31–36.

———. 1935a. Further additions to the list of Logan County birds. NBR 3:29–31.

———. 1935b. A third list of additions to the list of Logan County birds. NBR 3:143–44.

———. 1937. More bird identifications for Logan County. NBR 5:29–31.

———. 1964. New records for Logan County. NBR 32:83.

———. 1966. Hooded Warbler in Logan County. NBR 34:44.

Graul, W. D. 1973. Adaptive aspects of the Mountain Plover social system. Living Bird 12:69–94.

Green, J. C., and R. B. Janssen. 1975. Minnesota birds: where, when, and how many.

Minneapolis: University of Minnesota Press.

Green, R. C. 1974. Notes from nature. A Bird's Eye View: October 1974. Omaha, NE: Omaha Audubon Society.

———. 1977. Notes from nature. A Bird's Eye View: January 1977. Omaha, NE: Omaha Audubon Society.

———. 1978. Yellow bellied Flycatcher. NBR 46:86.

———. 1982. White-tailed Kite in Garden County. NBR 50:10.

———. 1984a. Gyrfalcon. NBR 52:23.

———. 1984b. Whip-poor-wills. NBR 52:24.

———. 1985a. Gray-Crowned Rosy-Finches. NBR 53:44.

———. 1985b. Northern Waterthrush. NBR 53:43.

———. 1988. House Finches nesting in Sarpy County. NBR 56:82.

———. 1992. Brown Pelican in Nebraska. NBR 60:69.

———. 1994. Rufous Hummingbird. NBR 62:123.

———. 1997. Black-throated Sparrow banded in Omaha, Nebraska. NBR 65:179.

Greer, J. G. 1992. Tricolored Heron in Pottawattamie County. Iowa Bird Life 62:88–89.

Grenon, A. G. 1990. 1990 (Third) Report of the NOU Records Committee. NBR 58:90–97.

———. 1991. 1991 (Fourth) Report of the NOU Records Committee. NBR 59:150–55.

Griscom, L. 1937. A monographic study of the Red Crossbill. Proceedings of the Boston Society of Natural History 41:77–210.

Gross, M. 1978. Ringed Turtle Dove. NBR 46:86.

Groth, J. G. 1993. Evolutionary differentiation in morphology, vocalizations, and allozymes among nomadic sibling species in the North American Red Crossbill (*Loxia curvirostra*) complex. University of California Publications in Zoology 127:1–143.

Gubanyi, J. G. 1996a. 1992, 1993 (Fifth) Report of the NOU Records Committee. NBR 64:30–35.

———. 1996b. 1994 (Sixth) Report of the NOU Records Committee. NBR 64:38–42.

———. 1996c. 1995 (Seventh) Report of the NOU Records Committee. NBR 64:132–38.

Haecker, F. W. 1937. American Ravens and American Magpies in eastern Nebraska. NBR 5:33.

Haecker, F. W., and R. A. Moser. 1941. Present-day bird life along the Missouri River compared with Say's and Audubon's findings. NBR 9:31–35.

Haecker, F. W., R. A. Moser, and J. B. Swenk. 1945. Checklist of the birds of Nebraska. NBR 13:1–40.

Halbert, D. 1940. Notes on Nebraska-reared Chukar Partridges and Valley California Quail. NBR 8:87–88.

Hancock, B., J. E. Ducey, and R. Lock. 1981. Captive breeding of Barn Owls in Nebraska—initial efforts. NBR 49:31–33.

Hanson, M. L. 1952. Some oological records of Nebraska. NBR 20:31–35.

Harding, R. G. 1986. Waterfowl nesting preferences and productivity in the Rainwater Basin, Nebraska. M.S. thesis, Kearney State College, Kearney, NE.

Harrington, R. 1962. Pygmy Nuthatches at Lincoln. NBR 30:29.

Harrison, P. 1983. Seabirds, an identification guide. Boston: Houghton Mifflin.

Hartlaub, G. 1852. Descriptions de quelques nouvelles especies d'oiseaux. Rev. et Mag. de Zool., 2nd Series, 4:1–7.

Hatch, D. E., and P. A. Garrels. 1971. Turkey Vulture nesting records for Nebraska—a new record for Sheridan County. NBR 39:18–20.

Hayden, F. V. 1862. On the geology and natural history of the upper Missouri. Transactions of the American Philosophical Society 12:151–76.

Hayman, P., J. Marchant, and T. Prater. 1986. Shorebirds, an identification guide to the waders of the world. Boston: Houghton Mifflin.

Heindel, M. T. 1996. Field identification of the Solitary Vireo complex. Birding 28:458–71.

Heineman, L. 1951. Winter notes from Plattsmouth. NBR 19:49–50.

Heineman, P. T. 1967. King Rail. NBR 35:23.

Held, D. H. 1958. Cherry County. NBR 26:28–29.

Held, L. 1962. Golden-Crowned Sparrow at Elsmere. NBR 30:50.

Helsinger, M. J. 1985a. Black-necked Stilts nesting in Nebraska. NBR 53:72.

———. 1985b. Cinnamon Teal nest at Crescent Lake NWR. NBR 53:80.

Hergenrader, G. L. 1962. The incidence of nest parasitism by the Brown-Headed Cowbird (*Molothrus ater*) on the roadside nesting birds in Nebraska. Auk 79:85–88; reprinted NBR 30:20–23.

Hiatt, J. T., and J. C. W. Bliese. 1971. Activities of the Yellow-bellied Sapsucker in the Kearney, Nebraska, area. NBR 39:60–62.

Higgins, K. F., L. M. Kirsch, M. R. Ryan, and R. B. Renken. 1979. Some ecological aspects of Marbled Godwits and Willets in North Dakota. Prairie Naturalist 11:114–18.

Hilsabeck, J., and R. Bell. 1999. Marsh and water bird nesting at Squaw Creek National Wildlife Refuge 1992–1997. Bluebird 66:24–32.

Hoffman, T. A. 1973. Snowy Plover. NBR 41:19–20.

———. 1974. Yellow Rail. NBR 42:20.

Hoge, G. L. 1964. Poorwills near Powell. NBR 32:14.

Hoge, G. L., and W. Hoge. 1984. Southeast Nebraska. NBR 52:22.

Holcomb, L. C. 1967. Long-eared Owl nest in Nebraska. NBR 35:56–58.

———. 1970. Gray Jay in Douglas County. NBR 38:17.

Holmes, R. T. 1966. Breeding ecology and annual cycle adaptations of the Red-backed Sandpiper (*Calidris alpina*) in northern Alaska. Condor 68:3–46.

Holt, H. R. 1997. A birder's guide to Colorado. Colorado Springs, CO: American Birding Association.

Hubbard, J. P. 1977. The status of Cassin's Sparrow in New Mexico and adjacent states. American Birds 31:933–41.

Huber, R. R. 1987. Crescent Lake nesting records. NBR 55:75.

Hudson, G. E. 1933a. The Snowy Plover occurs again in Nebraska. NBR 1:31.

————. 1933b. A Nebraska record of the California Gull. NBR 1:61.

————. 1934a. The first record specimen for Nebraska of the Caspian Tern. NBR 2:37.

————. 1934b. The fourth taking of the Nelson Sharp-Tailed Sparrow in Nebraska. NBR 2:120.

————. 1937a. The American Raven and American Magpie at Peru, Nemaha County. NBR 5:13.

————. 1937b. A sight record of the Atlantic Kittiwake for Lincoln, Lancaster County. NBR 5:57.

————. 1939. Some ornithological results of a six-weeks' collecting trip along the boundaries of Nebraska. NBR 7:4–7.

————. 1956. Error. NBR 24:54.

Hudson, G. E., P. T. Gilbert, and O. Wade. 1937. The Summer Tanager and other summering birds at Rock Bluff, Cass County. NBR 5:8–9.

Hughson, H. K. 1990. Gray-Crowned Rosy-Finches. NBR 58:56.

Hunn, E. S. 1978. Black-Throated Sparrow vagrants in the Pacific Northwest. Western Birds 9:85–89.

Huntley, C. W. 1953. Cassin Kingbird at Brule. NBR 21:37.

————. 1958. Keith County. NBR 26:11–12.

————. 1960. Summer birds of Camp Opal Springs, 1958–1959. NBR 28:5–7.

————. 1971. Ross' Geese. NBR 39:37.

Huser, B. F. 1986. June notes for western Nebraska. NBR 54:82–83.

Hyland, J. M. 1968. Black Duck. NBR 36:81.

Jacobs, B. 1997. Seasonal survey: winter report. Bluebird 64:8–12.

Jacoby, K. 1976. The breeding birds of Nebraska—a temporal study. Master's thesis, University of Nebraska–Omaha.

James, D. A., and J. C. Neal. 1986. Arkansas birds: their status and distribution. Fayetteville: University of Arkansas Press.

James, E. 1972. Account of an expedition from Pittsburgh to the Rocky Mountains under the command of Major Stephen H. Long, from the notes of Major Long, Mr. T. Say, and other gentlemen of the exploring party. Barre, MA: Imprint Society.

Janssen, R. B. 1973. Painted Bunting. NBR 41:67–68.

————. 1987. Birds in Minnesota. Minneapolis: University of Minnesota Press.

Jaramillo, A. 1997. Re: California Gulls in Alberta. *ID-Frontiers listserv. birdwg01@listserv.arizona.edu*. (10 July 1997).

Jehl, J. R., Jr. 1963. An investigation of fall migrating dowitchers in New Jersey. Wilson Bulletin 85:115–47.

————. 1987. Geographic variation and evolution of the California Gull (*Larus californicus*). Auk 104:421–28.

Johnsgard, P. A. 1975. Waterfowl of North America. Don Mills, Ontario: Fitzhenry and Whiteside.

————. 1979. Birds of the Great Plains: breeding species and their distribution. Lincoln and London: University of Nebraska Press.

————. 1980. A preliminary list of the birds of Nebraska and adjacent Great Plains states. Lincoln, NE: Johnsgard.

————. 1990a. Additional observations of the birds of the Lake McConaughy region. NBR 58:52–54.

————. 1990b. First Nebraska kittiwake specimen. NBR 58:75.

————. 1997. The birds of Nebraska and adjacent plains States. Occasional Papers 6. Nebraska Ornithologists' Union. Lincoln, NE: Johnsgard.

Johnsgard, P. A., and R. Di Silvestro. 1976. Seventy-five years of changes in the Mallard–Black Duck ratios in eastern North America. American Birds 30:905–8.

Johnsgard, P. A., and R. Redfield. 1977. Sixty-five years of Whooping Crane records in Nebraska. NBR 45:54–56.

Johnsgard, P. A., and R. Wood. 1968. Distributional changes and interaction between prairie chickens and Sharp-Tailed Grouse in the Midwest. Wilson Bulletin 80:173–88.

Johnson, N. K. 1995. Speciation in Vireos. I. Macrogeographic patterns of allozymic variation in the *Vireo solitarius* complex in the contiguous United States. Condor 97:903-19.

Johnson, N. K., and J. A. Marten. 1988. Evolutionary genetics of Flycatchers. II. Differentiation in the *Empidonax difficilis* group. Auk 105:177-91.

Johnson, W. C. 1994. Woodland expansion in the Platte River, Nebraska: patterns and causes. Ecological Monographs 64:45-84.

Johnson-Mueller, N., and R. Morris. 1992. 1990 Nebraska nesting reports. NBR 60:64-69.

Johnston, H. C. 1933. The Clarke Nutcracker at Superior, Nuckolls County. NBR 1:130.

Johnston, R. F. 1965. A directory to the birds of Kansas. Miscellaneous Publication No. 41. Lawrence: University of Kansas Museum of Natural History.

Jones, A. H. 1939. The Williamson's Sapsucker at Hastings, Adams County. NBR 7:27-28.

Jones, R. 1963. Identification and analysis of Lesser and Greater Prairie-Chicken habitat. Journal of Wildlife Management 27:757-78.

Jorgensen, J. G. 1987. Blair backyard birder. NBR 55:75.

———. 1988. The birds of Cauble Creek and the wetlands. Blair, NE: Jorgensen.

———. 1996. A review of the status of *Limnodromus griseus*, the Short-billed Dowitcher, in Nebraska. NBR 64:74-78.

———. 1997. Shorebird migration in the eastern Rainwater Basin—spring 1997. NBR 65:133-35.

Jundt, B. P. 1984. Dipper. NBR 52:42.

Karl Bodmer's America. 1984. Omaha and Lincoln: Joslyn Art Museum and University of Nebraska Press.

Kaufman, K. 1990. A field guide to advanced birding. Boston: Houghton-Mifflin.

Kaul, R., and S. Rolfsmeier. 1993. Native vegetation of Nebraska. (Map). Lincoln: Conservation and Survey Division, University of Nebraska-Lincoln.

Kenitz, A. 1988. Lesser Goldfinch in Scotts Bluff County. NBR 56:82-83.

Kent, T. H. 1987. Eiders in Iowa. Iowa Bird Life 57:88-93.

———. 1997. Report of the Records Committee for 1996. Iowa Bird Life 67:81-83.

———. 1998. Report of the Records Committee for 1997. Iowa Bird Life 68:85-88.

Kent, T. H., and J. J. Dinsmore. 1996. Birds in Iowa. Iowa City and Ames, IA: Kent and Dinsmore.

Kessel, B. 1953. Distribution and migration of the European Starling in North America. Condor 55:49-67.

Kieborz, E. M. 1971. Black-backed Three-toed Woodpecker. NBR 39:39.

———. 1978. Brown Pelican. NBR 46:16.

Kimball, B. 1984. Chuck-will's-widow. NBR 52:24.

Kinch, C. 1963. Notes from Lexington. NBR 31:12-13.

———. 1964. Some Dawson County records. NBR 32:37-40.

———. 1968. Swans. NBR 36:16-19.

King, J. R. 1998. Re: Albertaensis California Gull. *ID-Frontiers listserv. birdwg01@listserv.arizona.edu.* (11 Jul 1988).

Kiser, M. 1985. Groove-billed Ani. NBR 53:78-79.

Knopf, F. L. 1986. Changing landscapes and the cosmopolitanism of the eastern Colorado avifauna. Wildlife Society Bulletin 14:132-42.

Knopf, F. L., and J. R. Rupert. 1996. Reproduction and movements of Mountain Plovers breeding in Colorado. Wilson Bulletin 108:28-35.

Knox, A. G. 1988. The taxonomy of redpolls. Ardea 76:1-26.

Korpi, R. T. 1990. Painted Bunting. NBR 58:54-56.

Kovanda, S., and J. Kovanda. 1986. Sharp-tailed Sandpiper. NBR 54:70.

Kroodsma, D. E. 1988. Two species of Marsh Wren (*Cistothorus palustris*) in Nebraska? NBR 56:40-42.

Kroodsma, D. E., and R. A. Canady. 1985. Differences in repertoire size, singing behavior, and associated neuroanatomy among Marsh Wren populations have a genetic basis. Auk 102:439-46.

Labedz, T. E. 1986. Cassin's Sparrow in Garden County. NBR 54:80-81.

———. 1987. A Nebraska specimen of Clark's Grebe *Aechmophorus clarkii*. NBR 55:36–37.

———. 1990a. A Black-Bellied Whistling-Duck specimen from Nebraska; a first state record. NBR 58:49–52.

———. 1990b. Results of a preliminary study showing evidence of winter site fidelity in migrating sparrows in Nebraska. NBR 58:81–82.

LaGrange, T. 1997. Guide to Nebraska's wetlands and their conservation needs. Lincoln: Nebraska Game and Parks Commission.

Lanyon, W. E. 1956. Ecological aspects of the sympatric distribution of meadowlarks in the north-central states. Ecology 37:98–108.

Larson, G. K. 1956. Notes. NBR 24:10.

Lawson, M. 1976. The climate of the Great American Desert. Lincoln: University of Nebraska Press.

Lemburg, W. W. 1976. White-necked Raven reported. NBR 44:38–39.

———. 1979. Woodcock nesting. NBR 47:59.

———. 1980. Mottled Duck. NBR 48:88.

Lemmon, S. 1995. Black-throated Blue Warbler. NBR 63:27.

Lesick, L., Jr. 1987. Yellow Rails. NBR 55:88.

Leukering, T. 1997. Re: Gull mantle colors. *ID-Frontiers listserv. birdwg01@listserv.arizona.edu.* (26 Mar 1997).

Linder, K. A., and S. H. Anderson. 1998. Nesting habitat of Lewis's Woodpeckers in southeastern Wyoming. Journal of Field Ornithology 69:109–16.

Lingle, G. R. 1981. Status of the American Woodcock in Nebraska with notes on a breeding record. Prairie Naturalist 13:47–51.

———. 1983. A new nesting species for Nebraska. NBR 51:86–87.

———. 1989. Winter raptor use of the Platte and North Platte River valleys in south central Nebraska. Prairie Naturalist 21:1–16.

———. 1994. Birding Crane River: Nebraska's Platte. Grand Island, NE: Harrier Publishing.

———. 1996. Another Common Crane in Nebraska with a summary of North American records. NBR 64:80–82.

Lingle, G. R., and P. A. Bedell. 1989. Nesting ecology of Sedge Wrens in Hall County, Nebraska. NBR 57:47–49.

Lingle, G. R., and M. A. Hay. 1982. A checklist of the birds of Mormon Island Crane Meadows. NBR 50:27–36.

Lingle, G. R., and T. E. Labedz. 1984. An exceptional "fall" migration of shorebirds along the Big Bend Reach of the Platte River. NBR 52:70–71.

Lingle, G. R., and K. L. Lingle. 1983. A second Black-shouldered Kite in Nebraska, with notes on food habits. NBR 51:58.

Lingle, G. R., K. J. Strom, and J. W. Ziewitz. 1986. Whooping Crane roost site characteristics on the Platte River, Buffalo County, Nebraska. NBR 54:36–39.

Lionberger, E. L. 1944. Nesting of the Western Turkey Vulture in Nuckolls County. NBR 12:41.

Lock, R. A. 1975. Young Mountain Plover seen in Kimball County. NBR 43:54–56.

———. 1978. A second Great Grey Owl record from Nebraska, and other recent owl records. NBR 46:16.

———. 1979. A second nesting record for Merlins in Nebraska. NBR 47:39.

———. 1980. Barn Owl management plan. Lincoln: Nebraska Game and Parks Commission.

———. 1987. Short-eared Owls in Lancaster County. NBR 55:46.

Lock, R. A., and R. Craig. 1975. Merlin nest in Nebraska. NBR 43:78–79.

Lock, R. A., and J. Schuckman. 1973. A Bald Eagle nest in Nebraska. NBR 41:76–77.

Long, J. L. 1981. Introduced birds of the world. New York: Universe Books.

Longfellow, S. 1979. A 1976 Great-tailed Grackle record, and some 1979 records. NBR 46:16.

Lowther, P. E. 1984. Catalog of Brown-headed Cowbird hosts from Kansas. Kansas Ornithological Society Bulletin 35:25–33.

———. 1988. Kansas Cowbird hosts, a catalogue update. Kansas Ornithological Society Bulletin 39:36–37.

Ludden, C. E. 1956. Birds through the years. NBR 24:34–37.

Ludlow, C. S. 1935. A quarter-century of bird migration records at Red Cloud, Nebraska. NBR 3:3–25.

Lueshen, W. 1963. Excerpts from letters. NBR 31:51.

MacKinlay, E. 1936. The Eastern Cardinal near Chadron, Dawes County. NBR 4:84.

Maddux, E. H. 1977. Dipper. NBR 45:52.

———. 1989. Have the Northern Cardinal and Red-bellied Woodpecker expanded their ranges in Nebraska recently, 1968–1987? NBR 57:87–92.

Manning, R. 1981. Red-breasted Nuthatch. NBR 49:11.

———. 1982. A Hoary Redpoll in Dodge County. NBR 50:37.

Marsh, E., and B. Marsh. 1986. Ivory Gull. NBR 54:70.

Marshall, J. 1967. Parallel variation in North and Middle American Screech-Owls. Western Foundation for Vertebrate Zoology Monograph 1.

Martin, P. R., and B. M. Di Labio. 1994. Identification of Common x Barrow's Goldeneye hybrids in the field. Birding 26:104–5.

Mathisen, J. 1961. Excerpts from letters. NBR 29:23–24.

———. 1962. The status of Wild Turkey in Nebraska. NBR 30:34–35.

Mathisen, J., and A. Mathisen. 1958. A study of bird habitats in Alliance City Park. NBR 26:22–25.

———. 1959. Sharp-tailed Grouse and Prairie Chickens. NBR 27:28.

———. 1960a. Nesting study of Long-eared Owls in Box Butte County, Nebraska. NBR 28:10–11.

———. 1960b. History and status of introduced game birds in Nebraska. NBR 28:19–22.

Maunder, V. 1966. Roseate Spoonbill. NBR 34:77.

Mayr, E., and L. L. Short. 1970. Species taxa of North American birds, a contribution to avian systematics. Publications of the Nuttall Ornithological Club No. 9.

Cambridge, MA: Nuttall Ornithological Club.

McClure, H. E. 1966. Some observations of vertebrate fauna of the Nebraska Sandhills, 1941 through 1943. NBR 34:2–15.

McDermott, J. F., ed. 1951. Up the Missouri with Audubon; the journal of Edward Harris. Norman: University of Oklahoma Press.

McKinley, D. 1965. The Carolina Parakeet in the upper Missouri and Mississippi River valleys. Auk 82:215–26.

McKinley, D. 1978. The Carolina Parakeet in the West and additional references. NBR 46:3–7.

McKinney, B. 1993. Bald Eagles nest at North Platte National Wildlife Refuge. NBR 61:138–39.

McLaren, I. 1999. Re: Summer Tanager puzzle. ID-Frontiers listserv. birdwg01@listserve.arizona.edu. (6 Feb 1999).

McMullen, J. 1989. Eddie, the fearless prairie-chicken. NBR 57:94.

Meier, M. 1974. Another Black-Throated Sparrow in Nebraska. NBR 42:18–19.

Mendell, D. P. 1983. Harlan's Hawk (*Buteo jamaicensis harlani*): a valid subspecies. Auk 100:161–69.

Menzel, K. E. 1964. White Ibis in Rock County. NBR 32:12–14.

———. 1970. Mexican Duck. NBR 38:89–90.

———. 1974. Hawk concentration at Valentine. NBR 42:19–20.

Mickel, C. E., and R. W. Dawson. 1920. Some interesting records of Nebraska birds for the year 1919. Wilson Bulletin 32:73–79.

Middleton, A. D. 1947. The Long-crested Jay in Lincoln County. NBR 15:41.

Miller, A. H. 1941. Speciation in the avian genus *Junco*. University of California Publications in Zoology 44:173–434.

Mlodinow, S. G., and M. O'Brien. 1996. America's 100 most wanted birds. Helena and Billings, MT: Falcon Press.

Mohler, L. L. 1944. Distribution of upland game birds in Nebraska. NBR 12:1–6.

———. 1946. Notes on the breeding and nesting of the Long-billed Curlew. NBR 14:31–39.

———. 1951. Lewis' Woodpecker in Sioux County. NBR 19:31.

Mollhoff, W. J. 1979. Great Grey Owl distribution, winter 1977–78. NBR 47:62–65.

———. 1983. Tower kills. NBR 51:92.

———. 1985. Dundy County. NBR 53:79.

———. 1986a. Correction to the 1980 migration reports. NBR 54:40.

———. 1986b. Late report of Common Raven. NBR 54:43.

———. 1987. First report of the NOU Records Committee. NBR 55:79–85.

———. 1988. The official list of the birds of Nebraska. NBR 56:86–96.

———. 1989a. Second report of the NOU Records Committee. NBR 57:42–47.

———. 1989b. "Young Purple Finch"; some questions. NBR 57:49–51.

———. 1997a. A probable nesting of Clark's Nutcracker (*Nucifraga columbiana*) in Nebraska. NBR 65:147–50.

———. 1997b. Notes on the nesting biology of Pygmy Nuthatches in Nebraska. NBR 65:150–59.

Morris, L. 1965. Glossy Ibis in York County. NBR 50:46.

———. 1979. Mountain Bluebirds (and others) near Hordsville. NBR 47:18.

———. 1984. York Co. NBR 52:23.

———. 1993. Gyrfalcon. NBR 61:136–37.

———. 1995. Notes on bird sightings in Nebraska. NBR 63:60.

Morris, R. 1992. Peregrine Falcon nesting success in Omaha, Nebraska. NBR 60:71.

Moser, R. A. 1946. The genus *Junco* in Nebraska. NBR 14:1–6.

Mossman, M. J., and M. A. Brogie. 1983. Breeding status of selected bird species on the Niobrara Valley Preserve, Nebraska. NBR 51:52–62.

Murray, B. W., W. B. McGillivray, J. C. Barlow, R. N. Beech, and C. Strobeck. 1994. The use of Cytochrome b sequence variation in estimation of phylogeny in the *Vireonidae*. Condor 96:1037–54.

National Geographic Society. 1987. Field guide to the birds of North America. 2nd ed. Washington DC: National Geographic Society.

Nebraska Blue Book. 1975. Lincoln: State of Nebraska.

Nebraska Ornithologists' Union Records Committee. 1986. The official list of the birds of Nebraska. NBR 54:86–96.

Neilson, W. H. 1965. Excerpts from letters. NBR 33:66–67.

———. 1968. Young ospreys. NBR 36:10.

Newfield, N. L. 1983. Records of Allen's Hummingbird in Louisiana and possible Rufous x Allen's Hummingbird hybrids. Condor 85:253–54.

Newlon, M. C. 1981. Curve-billed Thrashers in Iowa. Iowa Bird Life 51:21–24.

Nuechterlein, G. L. 1981. Courtship behavior and reproductive isolation between Western Grebe color morphs. Auk 98:335–49.

Nuechterlein, G. L., and R. W. Storer. 1982. The pair-formation displays of the Western Grebe. Condor 94:351–69.

Ohlander, B. G. 1979. Gyrfalcon; Prairie Falcon. NBR 47:58–59.

Olson, M. 1956. Reports of nests, nestlings, and fledglings. NBR 24:14.

Orendurff, C. F. 1941. The first wildlife inventory of Nebraska shelterbelts. NBR 9:7–8.

Orr, W., and J. Porter, eds. 1983. A journey through the Nebraska region in 1833 and 1834: from the diaries of Prince Maximilian of Wied. Nebraska History 64:325–453.

Ott, M. B. 1978. An uncommon grackle. NBR 46:84–85.

Padelford, L., and B. Padelford. 1973. Monk Parakeets in Omaha. NBR 41:15–17.

———. 1980a. A Ground Dove at DeSoto Bend National Wildlife Refuge. NBR 48:22.

———. 1980b. Summer Tanager nest. NBR 48:90.

———. 1993. Rock Wren. NBR 61:137.

Paine, E. 1988. An Inca Dove in Nebraska, in winter! NBR 56:3.

Palmer, R. S., ed. 1962. Handbook of North American birds. Vol. 1. Loons through flamingos. New Haven, CT, and London: Yale University Press.

————. 1976. Handbook of North American birds. Vol. 2. Waterfowl (Part 1). New Haven, CT, and London: Yale University Press.

————. 1988. Handbook of North American birds. Vol. 4. Diurnal raptors (Part 1). New Haven, CT, and London: Yale University Press.

Parker, J. 1982. Niche competition among members of a bark foraging guild in eastern Nebraska. Master's thesis, University of Nebraska–Omaha.

Parkes, K. C. 1988. The Ontario specimen of Carolina Chickadee. Ontario Birds 6:111–13.

Patton, F. 1945. Notes from Blue Springs, Gage Co. NBR 13:44.

————. 1970. Wheatears? NBR 38:18.

Paulson, D. R. 1993. Shorebirds of the Pacific Northwest. Seattle and London: University of Washington Press.

Peabody, N. 1974. Nebraska's trumpeters. Proceedings and Papers of the Third Trumpeter Swan Society Conference 47–49.

Pennington, L. I. 1983. Scott's Oriole. NBR 51:84.

Peters, K. L., and R. H. Schmidt. 1981. Barn Owls. NBR 49:25.

Peterson, R. A. 1990. A birdwatcher's guide to the Black Hills. Vermillion, SD: PC Publishing.

————. 1995. The South Dakota breeding bird atlas. South Dakota Ornithologists' Union. Aberdeen, SD: Northern State University Press

Pettingill, O. S., Jr., and N. R. Whitney. 1965. Birds of the Black Hills. Cornell Library of Ornithology Special Publication 1. Ithaca, NY: Cornell University Press.

Peyton, M. M. 1991. Snowy Plover sighting. NBR 59:98.

Peyton, M. M., and R. Knaggs. 1995. Bald Eagle counts from two viewing facilities, 1994–95 season. NBR 63:31–33.

Phillips, A. R. 1986. The known birds of North and Middle America. Part I. Denver: Phillips.

Phipps, K. 2000. A summer survey of the birds at two eastern Nebraska wetlands. NBR 68:2–7.

Pittaway, R. 1997. Re: California Gull— subspecies identification. ID-Frontiers listserv. birdwg01@listserv.arizona.edu. (29 March 1997).

Plissner, J. H., and S. M. Haig. 1997. 1996 International Piping Plover census. Report to U.S. Biological Survey, Biological Resources Division, Forest and Rangeland Ecosystem Science Center, Corvallis, OR.

Poague, K., and J. Dinan. 1997. International shorebird survey report for southeastern Nebraska—spring 1997. NBR 65:127–31.

Podany, M. 1996. Nest placement of the Ferruginous Hawk, Buteo regalis, in northwestern Nebraska. Master's thesis, University of Nebraska–Omaha.

Pound, R., and F. E. Clements. 1900. The phytogeography of Nebraska. 2nd ed. Lincoln, NE: Seminar.

Prather, B. 1997. The report of the Colorado Bird Records Committee (CBRC) for 1994. Colorado Field Ornithologists' Union Journal 31:23–31.

Prevett, J. P., and C. D. MacInnes. 1972. The number of Ross' Geese in central North America. Condor 74:431–38.

Pritchard, M. L., and C. G. Pritchard. 1958. 1957 nesting report. NBR 26:20–22.

————. 1959. 1958 nesting report. NBR 27:36–37.

————. 1960. 1959 nesting report. NBR 28:27–28.

————. 1961. 1956–1960 nesting reports. NBR 29:31–34.

Probasco, G. E. 1983. Hooded Warbler. NBR 51:95.

Pruitt, J., and N. McGowan. 1975. The return of the Great-tailed Grackle. American Birds 29:985–92.

Pulich, W. M. 1988. The birds of north central Texas. College Station: Texas A & M University Press.

Pyle, P. 1997. Identification guide to North American birds. Bolinas, CA: Slate Creek Press.

Raines, R. R., M. C. Gilbert, R. A. Gersib, W. S. Rosier, and K. F. Dinan. 1990. Regulatory planning for Nebraska Rainwater Basin wetlands. Prepared for the Rainwater Basin Advanced Identification Study. U.S. Environmental Protection Agency, Region 7, Kansas City, KS and U.S. Army Engineer District, Omaha.

Rakestraw, J. 1995. A closer look: Lesser Prairie-Chicken. Birding 27:209–12.

Rapp, W. F., Jr. 1953. The distribution of the Red-bellied Woodpecker in Nebraska. NBR 21:22.

Rapp, W. F., Jr., and J. L. C. Rapp. 1951. The Clapper Rail in Nebraska. NBR 19:38–39.

Rapp, W. F., Jr., J. L. C. Rapp, H. E. Baumgarten, and R. A. Moser. 1958. Revised checklist of Nebraska birds. Occasional Papers 5. Nebraska Ornithologists' Union, Crete, NE.

Rasmussen, R. L. 1973. Clark's Nutcracker. NBR 41:43.

Ratti, J. T. 1979. Reproductive separation and isolating mechanisms between sympatric dark- and light-phase Western Grebes. Auk 96:573–86.

———. 1981. Identification and distribution of Clark's Grebe. Western Birds 12:41–46.

Richardson, L. 1965. Excerpts from letters. NBR 33:66.

Ridgway, R. 1902. The birds of North and Middle America. Bulletin of the United States National Museum 50, Part 2.

Rising, J. D. 1969. A comparison of metabolism and evaporative water loss of Baltimore and Bullock's Orioles. Comparative Biochemistry and Physiology 31:915–25.

———. 1970. Morphological variation and evolution in some North American orioles. Systematic Zoology 19:315–51.

———. 1995. A guide to the identification and natural history of the sparrows of the United States and Canada. San Diego and London: Academic Press.

———. 1996. The stability of the oriole hybrid zone in western Kansas. Condor 98:658–62.

Rising, J. D., and F. W. Schueler. 1980. Identification and status of wood-pewees (Contopus) from the Great Plains: what are sibling species? Condor 82:301–8.

Ritchey, O. W. 1966. Gray (or Hungarian) Partridge. NBR 34:77.

Robbins, C. S., D. Bystrack, and P. H. Geissler. 1986. The Breeding Bird Survey: its first fifteen years, 1965–1979. U.S. Fish and Wildlife Service Publication 157.

Robbins, M. B. 1989. What's your name, my little chickadee? Birding 21:205–7.

Robbins, M. B., and D. A. Easterla. 1992. Birds of Missouri, their distribution and abundance. Columbia and London: University of Missouri Press.

Roberson, D. 1980. Rare birds of the West Coast of North America. Pacific Grove, CA: Woodcock Publications.

Roberson, D., and R. Carratello. 1996. First record of Glaucous-winged Gull for Nebraska. NBR 64:3–4.

Robertson, K. 1977. Jaeger. NBR 45:15.

Rohwer, S. A. 1972. Distribution of meadowlarks in the central and southern Great Plains and the desert grasslands of eastern New Mexico and west Texas. Transactions of the Kansas Academy of Science 75:1–19.

Rohwer, S., and C. Wood. 1998. Three hybrid zones between Hermit and Townsend's Warblers in Washington and Oregon. Auk 115:284–310.

Root, T. 1988. Atlas of wintering North American birds, an analysis of Christmas Bird Count data. Chicago: University of Chicago Press.

Rosche, R. C. 1972. Notes on the distribution of some summer birds in Nebraska. NBR 40:70–72.

———. 1974a. A Nebraska Swainson's Thrush nest. NBR 42:17.

———. 1974b. Mixed-up chickadees. NBR 42:80.

———. 1982. Birds of northwestern Nebraska and southwestern South Dakota, an annotated checklist. Crawford, NE: Cottonwood Press.

———. 1993. Another description of Ross' Gull. NBR 61:89–90.

———. 1994a. Birds of the Lake McConaughy area and the North Platte River valley, Nebraska. Chadron, NE: Richard C. Rosche.

———. 1994b. Birding in Western Nebraska. Birding 26:178–89, 416–23.

Rosche, R. C., and P. A. Johnsgard. 1984. Birds of Lake McConaughy and the North Platte River valley, Oshkosh to Keystone. NBR 52:26–35.

Rosche, R. C., and D. J. Rosche. 1993. Rose-Breasted x Black-Headed Grosbeak. NBR 61:91.

Rose, B. J. 1985. (photograph). NBR 53:3.

Rounds, R. C., and H. L. Munro. 1982. A review of hybridization between *Sialia sialis* and *S. currucoides*. Wilson Bulletin 94:219–23.

Rundquist, B. 1990. Wetlands in the Rainwater Basin: benefits are in the eye of the beholder. Resource Notes 5:11–16.

Ryan, R. R., and R. B. Renken. 1987. Habitat use by breeding Willets in the northern Great Plains. Wilson Bulletin 99:175–89.

Ryan, R. R., R. B. Renken, and J. J. Dinsmore. 1984. Marbled Godwit habitat selection in the northern prairie region. Journal of Wildlife Management 48:1206–18.

Scharf, W. C., J. Berigan, and J. Kren. 1993. Pine Siskins in breeding condition along the North Platte River, Keith County, Nebraska. NBR 61:144–45.

Schneider, J. 1968. Ringed Turtle Dove. NBR 36:76–77.

Schock, A. J. 1983. Pine Siskin nests at Wayne State College. NBR 51:89.

Schrad, D. 1981. (photograph) NBR 49:42.

Schreier, L. Barred Owl in Knox County. NBR 54:24–25.

Scott, O. K. 1993. A birder's guide to Wyoming. Colorado Springs, CO: American Birding Association.

Sealy, S. G. 1979. Extralimital nesting of Bay-breasted Warblers: response to Forest Tent Caterpillars? Auk 96:600–603.

Sejkora, K. 1978. Barrow's Goldeneye. NBR 46:17.

Senner, S. E., and E. F. Martinez. 1982. A review of Western Sandpiper migration in interior North America. Southwestern Naturalist 27:149–59.

Sharpe, R. S. 1964. Yellow-crowned Night Heron nesting in Nebraska. NBR 32:9–11.

———. 1966. Nesting report, 1965. NBR 34:41–43.

———. 1967. The 1966 nesting season. NBR 35:29–38.

———. 1968. The evolutionary relationships and comparative behavior of prairie chickens. Ph.d. diss., University of Nebraska–Lincoln.

———. 1970. Swamp Sparrow breeding on the Middle Loup River. NBR 38:18–19.

———. 1978. The origins of spring migratory staging by Sandhill Cranes and White-fronted Geese. Transactions of the Nebraska Academy of Sciences 6:141–44.

———. 1993. Samuel Aughey's list of Nebraska birds (1878): a critical evaluation. NBR 61:3–10.

Sharpe, R. S., and R. R. Payne. 1966. Nesting birds of the Crescent Lake National Wildlife Refuge. NBR 34:31–34.

Shickley, G. M. 1964. Knots in Lincoln County. NBR 32:62–63.

———. 1965. Calliope Hummingbird in Nebraska. Auk 82:650.

———. 1966. Excerpts from letters. NBR 34:40.

———. 1968. Additions to Lincoln County checklist. NBR 36:54–57.

———. 1969a. Curve-billed Thrashers. NBR 37:47–48.

———. 1969b. Mountain Chickadees. NBR 37:64.

Short, L. L., Jr. 1961. Notes on bird distribution in the central Plains. NBR 29:2–22.

Short, L. L., Jr. 1965. Bird records from northern Nebraska during the breeding season. NBR 33:2–5.

———. 1966. Some spring migrant and breeding records from northern Nebraska. NBR 34:18.

Shupe, S. 1985. Observations of the Barred Owl in southeastern Nebraska. NBR 53:37–38.

Shupe, S., and R. Collins. 1983. A winter roadside survey of hawks in eastern Nemaha County, Nebraska. NBR 51:19–22.

Sibley, C. G., and B. L. Monroe Jr. 1990. Distribution and taxonomy of birds of the world. New Haven, CT, and London: Yale University Press.

Sibley, C. G., and L. L. Short Jr. 1959. Hybridization in the buntings (*Passerina*) of the Great Plains. Auk 76:443–63.

Sibley, C. G., and D. A. West. 1959. Hybridization in the Rufous-sided Towhees of the Great Plains. Auk 76:326–28.

Sibley, D. 1997. Re: Gull mantle colors. *ID-Frontiers listserv. birdwg01@listserv. arizona.com.* (20 Mar 1997).

Sidle, J. G., J. J. Dinan, and B. K. Good. 1991. The 1991 census of Least Terns and Piping Plovers in Nebraska. NBR 51:19–22.

Silcock, W. R. 1979. Some comments on the "Breeding Birds of Nebraska." NBR 48:38.

———. 1989a. Barrow's Goldeneye at DeSoto Bend National Wildlife Refuge. Iowa Bird Life 59:89–90.

———. 1989b. Brant at DeSoto Bend National Wildlife Refuge. Iowa Bird Life 59:88.

———. 1992. An additional Nebraska record of Common Eider? NBR 60:149–50.

———. 1994. Christmas Bird Count 1993–94. Iowa Bird Life 64:33–44.

Silcock, W. R., T. E. Bray, and B. K. Padelford. 1986a. Mexican Duck in Nebraska. NBR 54:41–42.

———. 1986b. Mottled Duck in Nebraska. NBR 54:39–40.

———. 1986c. Records needed. NBR 54:40–41.

Silcock, W. R., and S. J. Dinsmore. 1999. Common Crane in Central Platte Valley, Nebraska, March 1999, and a discussion of prior North American records. NBR 67:28–30.

Skagen, S. K., and F. L. Knopf. 1994. Migrating shorebirds and habitat dynamics at a prairie wetland complex. Wilson Bulletin 106:91–105.

Smith, B. J., and K. F. Higgins. 1990. Avian cholera and temporal changes in wetland numbers and densities in Nebraska's Rainwater Basin area. Wetlands 10:1–5.

Smith, B. J., K. F. Higgins, and C. F. Gritzner. 1989. Land use relationships to avian cholera outbreaks in the Nebraska Rainwater Basin. Prairie Naturalist 21:125–36.

Smith, C. E. 1950. The Golden-crowned Sparrow in Nebraska. NBR 18:68.

———. 1957. Thomas County. NBR 25:27.

Smith, N. G. 1966. Evolution of some Arctic Gulls (*Larus*): an experimental study of isolating mechanisms. American Ornithologists' Union Monograph No. 4.

South Dakota Ornithologists' Union. 1991. The birds of South Dakota. Aberdeen, SD: Northern State University Press.

Squires, W. A. 1928. The eastward advance of the Arkansas Kingbird. Bird-Lore 30:330.

Stage, D., and L. Stage. 1990. Sage Sparrow in Sioux County. NBR 58:27–28.

Stephens, T. C. 1957. The birds of Dakota County, Nebraska. Revised and annotated by William Youngworth. Occasional Papers 3. Nebraska Ornithologists' Union, Crete, NE.

Stiles, F. G. 1972. Age and sex determination in Rufous and Allen Hummingbirds. Condor 74:25–32.

Stoppkotte, G. W. 1975a. Scott's Oriole reported. NBR 43:64–66.

———. 1975b. A Groove-billed Ani seen again in Nebraska. NBR 44:79–80.

Storer, R. W. 1965. The color phases of the Western Grebe. Living Bird 4:59–63.

Storer, R. W., and G. L. Nuechterlein. 1985. An analysis of plumage and morphological characters of the two color forms of the Western Grebe (*Aechmophorus*). Auk 102:102–19.

Sturmer, M. 1959. Nesting of a Scissor-tailed Flycatcher in Gage County. NBR 27:19–20.

Suetsugu, H. Y. 1961. Excerpts from letters. NBR 29:23.

Sutton, G. M. 1967. Oklahoma birds. Norman: University of Oklahoma Press.

———. 1975. A junco is a junco is a junco. NBR 43:66–67.

Svoboda, W. J. 1963. Excerpts from letters. NBR 31:15.

Swanson, C. 1961. Letters to the editor. NBR 29:40.

Swanson, K. S. 1953. Reported sight record of Great Black-backed Gull. NBR 21:9.

———. 1957. Hamilton County. NBR 25:27.

———. 1962. Birding areas near Aurora. NBR 30:18–20.

Swanson, P. 1984. Lesser Goldfinch reported in Nebraska. NBR 52:42.

Swarth, H. S., and H. C. Bryant. 1917. A study of the races of the White-fronted Goose

(*Anser albifrons*) occurring in California. University of California Publications in Zoology 17:209–22.

Sweet, J. T., and K. Robertson. 1966. Ross' Geese in Nebraska. NBR 34:70–71.

Swenk, M. H. Notes before 1925. Bird notes from A. M. Brooking of Hastings, C. A. Black of Kearney, and B. J. Olson of Kearney, based chiefly on their collections, up to January 1, 1925. Typed manuscript in NOU Archives.

———. Notes after 1925. Critical notes on specimens in Brooking, Black, and Olson collections made subsequent to January 1, 1925. Handwritten manuscript in NOU Archives.

———. 1902. Additional notes on Gage County birds. Proc NOU 3:107.

———. 1907. Two interesting Nebraska records. Auk 24:223.

———. 1915a. The birds and mammals of Nebraska; in Nebraska Blue Book. Lincoln: State of Nebraska.

———. 1915b. The Eskimo Curlew and its disappearance. Annual Report of the Smithsonian Institution for 1915: 325–40.

———. 1918a. The birds and mammals of Nebraska; in Nebraska Blue Book. Lincoln: State of Nebraska.

———. 1918b. Revisory notes on the birds of Nebraska. Wilson Bulletin 30:112–17.

———. 1920. The birds and mammals of Nebraska; in Nebraska Blue Book. Lincoln: State of Nebraska.

———. 1921. Nesting of the Red Crossbill in Nebraska. Wilson Bulletin 33:38–39.

———. 1929. The Pine Siskin in Nebraska: its seasonal abundance and nesting. Wilson Bulletin 41:77–92.

———. 1933a. The Ancient Murrelet wanders to Nebraska. NBR 1:14–15.

———. 1933b. A brief synopsis of the birds of Nebraska. I. Loons. NBR 1:87–90.

———. 1933c. A brief synopsis of the birds of Nebraska. II. Grebes. NBR 1:142–51.

———. 1933d. Minutes of the thirty-fourth annual meeting of the Nebraska Ornithologists' Union. NBR 1:90–103.

———. 1934a. A brief synopsis of the birds of Nebraska. III. Totipalmate Swimmers-Pelicans. NBR 2:128–36.

———. 1934b. The interior Carolina Paroquet as a Nebraska bird. NBR 2:55–59.

———. 1936. A study of the distribution, migration, and hybridism of the Rose-breasted Grosbeak and Rocky Mountain Black-headed Grosbeak in the Missouri Valley region. NBR 4:27–40.

———. 1937a. A history of Nebraska Ornithology. III. Period of the explorations of the early nineteenth century (1804–1854). NBR 5:51–56.

———. 1937b. A study of the distribution and migration of the Great Horned Owls in the Missouri Valley region. NBR 5:79–105.

———. 1938. Some additional observations on the races of the Great Horned Owl. NBR 6:7–8.

———. 1940. Distribution and migration of the chat in Nebraska and other Missouri Valley states. NBR 8:33–44.

Swenk, M. H., and R. W. Dawson. 1921. Notes on the distribution and migration of Nebraska birds. I. Tyrant Flycatchers (*Tyrannidae*). Wilson Bulletin 33:132–41.

Swenk, M. H., and E. Fichter. 1942. Distribution and migration of the Solitary Sandpiper in Nebraska. NBR 10:15–22.

Tate, D. J. 1969. Common Eider record for Nebraska. NBR 37:38.

Tate, J., Jr. 1966. An early record of the Ross' Goose in Nebraska. NBR 34:46–47.

———. 1969. Pileated Woodpecker and other birds at Indian Cave State Park. NBR 37:57–60.

———. 1986. The Blue List for 1986. American Birds 40:227–36.

Taylor, D. M., and C. D. Littlefield 1986. Willow Flycatcher and Yellow Warbler response to cattle grazing. American Birds 40:1169–73.

Terborgh, J. 1989. Where have all the birds gone?: essays on the biology and conservation of birds that migrate to the American tropics. Princeton, NJ: Princeton University Press.

Terres, J. K. 1980. The Audubon Society encyclopedia of North American birds. New York: Alfred A. Knopf.

Thomas, D. G. 1983. Phainopepla. NBR 51:19.

———. 1984. Golden-crowned Sparrow in Box Butte County. NBR 52:77.

Thompson, M. C., and C. Ely. 1989. Birds in Kansas. Vol. 1. Lawrence: University of Kansas Museum of Natural History.

———. 1992. Birds in Kansas. Vol. 2. Lawrence: University of Kansas Museum of Natural History.

Thorson, T. B. 1960. The distribution of field observations on Nebraska birds during the last fifty years. NBR 18:47–49.

Thwaites, R. G. 1904. Early western travels, 1748–1846. Vol. 6. Long expedition. Cleveland: Arthur H. Clark.

———. 1905. Early western Travels, 1748–1846. Vol. 15. Long expedition. Cleveland: Arthur H. Clark.

———. 1969. Original journals of the Lewis and Clark expedition. New York: Arno Press.

Tomback, D. F. 1995. A possible breeding record for Clark's Nutcracker in northwestern Nebraska in 1987. NBR 63:114–15.

Toochin, M. 1998. Possible anywhere: Tufted Duck. Birding 30:370–83.

Tordoff, H. B. 1950. A hybrid tanager from Minnesota. Wilson Bulletin 62:3–4.

Tout, W. 1902. Ten years without a gun. Proc NOU 3:42–45.

———. 1935. The spread of the Wood Thrush to North Platte, Lincoln County. NBR 3:28–29.

———. 1938. Birds of the Crescent Lake Migratory Bird Refuge. NBR 6:1–4.

———. 1947. Lincoln County birds. North Platte, NE: Tout.

Tremaine, M. 1962. Excerpts from letters. NBR 30:14.

———. 1970. Sandhill Cranes. NBR 38:23–25.

———. 1972. Black-necked Stilt. NBR 40:46.

———. 1974. Yellow Rail. NBR 42:77–78.

Tremaine, M., C. Booney, J. Redall, G. Viehmeyer, and R. G. Cortelyou. 1972. Sightings of the Common Crane (European) in Nebraska spring 1972. NBR 40:3–7.

Turner, A., and C. Rose. 1989. Swallows and Martins, an identification guide and handbook. Boston: Houghton Mifflin.

Turner, H. 1959. A monthly record of birds occurring at Bladen, Webster County, 1957 and 1958. NBR 27:21–24.

———. 1960. Rare birds at Bladen. NBR 28:53–54.

———. 1964. Boreal Owl in Webster County. NBR 32:12–14.

———. 1965. Two unusual birds reported. NBR 33:65.

———. 1968. Dipper. NBR 36:14–15.

———. 1975. Falcons. NBR 43:59.

———. 1978. Minden notes. NBR 46:19.

———. 1987. Minden notes. NBR 55:43.

Velich, R. 1958. Prairie chickens in eastern Nebraska. NBR 26:28.

———. 1970. Common Raven. NBR 38:92.

Viehmeyer, G. 1961a. Excerpts from letters. NBR 29:23.

———. 1961b. Calliope Hummingbird at North Platte. NBR 29:39–40.

———. 1971. The Dalton, Harrisburg, Kimball triangle. NBR 39:72–74.

Vierling, K. T. 1997. Habitat selection of Lewis' Woodpeckers in southeastern Colorado. Wilson Bulletin 109:121–30.

Wagner, H. J. 1966. Dakota County, May 30 through June 4, 1965. NBR 24:17.

Wampole, J., and E. Fichter. 1946. Western Grebes nesting at George Lake, Grant County. NBR 14:45.

Weakly, H. E. 1936. The Palmer Curve-billed Thrasher at North Platte, Lincoln County. NBR 4:54.

———. 1940. Some birds noted in western Nebraska during the fall and early winter of 1939. NBR 8:24–25.

Weaver, J. E. 1943. Replacement of true prairie with mixed prairie in Nebraska and Kansas. Ecology 24:421–34.

Welch, B. 1966. Cattle Egret. NBR 34:76.

Wensien, R. 1962. Nesting report, 1961. NBR 30:24–25.

———. 1964. Nesting report, 1963. NBR 32:45.

Wetmore. A. 1961. Excerpts from letters. NBR 29:58.

White, A. 1990. List of North American birds showing the states, provinces and territories in which each species has been found. Bethesda, MD: White.

Whitney, N. R., Jr., B. E. Harrell, B. K. Harris, N. Holden, J. W. Johnson, B. J. Rose, and P. F. Springer. 1978. The birds of South Dakota, an annotated checklist. Vermillion: South Dakota Ornithologists' Union.

Wiens, J. 1969. Ornithological Monographs No. 8. Lawrence, KS: American Ornithologists' Union, Allen Press.

Williams, G. P. 1978. Historical perspective of the Platte Rivers in Nebraska and Colorado. In W. D. Graul and S. J. Bissell, eds. Lowland river and stream habitat in Colorado, a symposium. Greeley: Colorado Chapter of the Wildlife Society and Colorado Audubon Council.

Williams, R. D. 1987. Prairie Grouse. NBR 55:73–74.

Wilson, B. L. 1983. Identifying meadowlarks in Iowa. Iowa Bird Life 53:83–87.

———. 1989. Williamson's Sapsucker in Omaha. NBR 57:30.

Wilson, B. L., J. Minyard, and H. Minyard. 1986. Hybrid bluebird update. NBR 54:26–27.

Wilson, B. L., J. Minyard, H. Minyard, and T. E. Bray. 1985. Hybrid bluebirds in the Pine Ridge. NBR 53:67.

Wittrock, G. 1984. Rosy Finch. NBR 52:68.

Wolcott, R. H. 1902. Short-billed Marsh Wren and Carolina Wren at Lincoln. Proc NOU 3:108.

Wolff, D. 1987. Wintering Killdeer. NBR 55:75.

Wood, D. 1965. Song Sparrows nesting at Plattsmouth. NBR 33:14.

Wood, R. 1970. Common Snipe. NBR 38:17.

Wright, R. 1982. Untitled. NBR 50:42.

———. 1983. Olivaceous Cormorant. NBR 51:18.

———. 1985a. Early Pine Siskin nest. NBR 53:81.

———. 1985b. Henslow's Sparrow. NBR 53:43–44.

———. 1985c. Re Swan Geese in Keith County. NBR 53:67–68.

———. 1991. Pileated Woodpecker responds to owl tape. NBR 59:98–99.

Wycoff, R. S. 1953. Snowy Plover. NBR 21:40.

Wyman, W. 1983. Gray Partridge. NBR 51:93.

Young, S. R. 1968. Hoary Redpoll. NBR 36:77–78.

Youngworth, W. 1955. Some birds of the Quicourt Valley. NBR 23:29–34.

———. 1957. The birds of Dakota County, Nebraska. Occasional Papers 3. Nebraska Ornithologists' Union, Crete, NE.

———. 1958. Dakota County spring notes. NBR 26:68.

Zimmer, J. T. 1911a. Some notes on the winter birds of Dawes County. Proc NOU 5:19–30.

———. 1911b. Some results of four years observation and collecting chiefly in the vicinity of Lincoln. Proc NOU 5:34–37.

———. 1912. Several interesting warbler records from Dawes County. Proc NOU 5:50.

———. 1913. Birds of the Thomas County Forest Preserve. Proc NOU 5:51–104.

———. 1945. Hammond's Flycatcher in Nebraska. NBR 13:41.

Zimmer, K. J. 1985. The Western birdwatcher: an introduction to birding in the American West. Englewood Cliffs, NJ: Prentice-Hall.

———. 1991. Plumage variation in "Kumlien's" Iceland Gull. Birding 23:254–69.

Zimmerman, J. L. 1982. Nesting success of Dickcissels (Spiza americana) in preferred and less preferred habitats. Auk 99:292–98.

———. 1988. Breeding season habitat selection by the Henslow's Sparrow (Ammodramus henslowii) in Kansas. Wilson Bulletin 100:17–24.

Index